T0213401

Lecture Notes in Computer Science 9801

Commenced Publication in 1973
Founding and Former Series Editors:
Gerhard Goos, Juris Hartmanis, and Jan van Leeuwen

More information about this series at http://www.springer.com/series/7407

Yifan Hu · Martin Nöllenburg (Eds.)

Graph Drawing and Network Visualization

24th International Symposium, GD 2016
Athens, Greece, September 19–21, 2016
Revised Selected Papers

Editors
Yifan Hu
Yahoo Research
New York, NY
USA

Martin Nöllenburg
TU Wien
Vienna
Austria

ISSN 0302-9743 ISSN 1611-3349 (electronic)
Lecture Notes in Computer Science
ISBN 978-3-319-50105-5 ISBN 978-3-319-50106-2 (eBook)
DOI 10.1007/978-3-319-50106-2

Library of Congress Control Number: 2016958998

LNCS Sublibrary: SL1 – Theoretical Computer Science and General Issues

Printed on acid-free paper

This Springer imprint is published by Springer Nature
The registered company is Springer International Publishing AG
The registered company address is: Gewerbestrasse 11, 6330 Cham, Switzerland

Preface

This volume contains the papers presented at the 24th International Symposium on Graph Drawing and Network Visualization (GD 2016), which took place September 19–21, 2016, in Athens, Greece. Graph drawing is concerned with the theory of and algorithms for the geometric representation of graphs, while network visualization has wide applications in understanding and analyzing relational datasets. Information about the conference series and past symposia is maintained at http://www.graphdrawing.org. This year, the conference was hosted by the Institute of Communications and Computer Systems, an affiliate of the National Technical University of Athens, with Antonios Symvonis as the chair of the Organizing Committee. A total of 99 participants from 16 countries attended the conference, whose venue was located in the beautiful historic center of Athens.

Paper submissions were divided into two main tracks and an additional poster track: Track 1 for combinatorial and algorithmic aspects of graph drawing and Track 2 for experimental, applied, and network visualization aspects. In each of the two tracks authors could submit full papers or short papers. All tracks were handled by a single Program Committee. In response to the call for papers, the Program Committee received a total of 112 submissions consisting of 99 papers (66 in Track 1 and 33 in Track 2) and 13 posters. More than 350 expert reviews were provided, of which more than 130 were contributed by external subreviewers. After often extensive electronic discussions, the Program Committee selected 44 papers and all 13 posters for inclusion in the scientific program of GD 2016. This resulted in an overall paper acceptance rate of 44% (48% in Track 1 and 36% in Track 2). Two new policies were introduced in the submission and publication process this year. Firstly, references no longer count toward the page limits of 12 pages for regular papers and six pages for short papers. Secondly, electronic versions of all accepted papers were made available through a conference index on the ArXiv repository before the conference.

There were two keynote talks at GD 2016. Roger Wattenhofer (ETH Zurich, Switzerland) talked about "Distributed Computing: Graph Drawing Unplugged" and Daniel Keim (University of Konstanz, Germany) talked about "The Role of Visual Analytics in Exploring Graph Data." Abstracts of both talks are included in the proceedings.

Springer sponsored awards for the best paper in each of Track 1 and Track 2, plus a best presentation award and a best poster award. The Program Committee voted to give the best paper award in Track 1 to the paper "Block Crossings in Storyline Visualizations" by Thomas C. van Dijk, Martin Fink, Norbert Fischer, Fabian Lipp, Peter Markfelder, Alex Ravsky, Subhash Suri, and Alexander Wolff, and in Track 2 to the paper "A Sparse Stress Model" by Mark Ortmann, Mirza Klimenta, and Ulrik Brandes. The participants of the conference voted to give the best presentation award to Martin Gronemann for his presentation of the paper "Bitonic st-orderings for Upward Planar Graphs" and the best poster award to Jonathan Klawitter and Tamara Mchedlidze for their poster entitled "Heuristic Picker for Book Drawings." Congratulations to all award winners for their excellent contributions!

Following the tradition, the 23rd Annual Graph Drawing Contest was held during the conference. The contest had two parts, each with two categories: the Creative Topics (Panama papers and Greek mythology family tree) and the Live Challenge on crossing minimization in book embeddings (Automatic Category and Manual Category). Awards were given in each of the four categories. We thank the Contest Committee for preparing interesting and challenging contest problems. A report about the contest is included in the proceedings.

Directly after GD 2016, a two-day PhD school on "Visualization Software" took place. We thank Antonios Symvonis for organizing this satellite event. A short report about the school is also included in the proceedings.

Many people and organizations contributed to the success of GD 2016. We thank the Program Committee members and the external reviewers for carefully reviewing and discussing the submitted papers and posters, and for putting together a strong and interesting program. Thanks to all the authors for choosing GD 2016 as the publication venue for their research. Further, we thank the local organizers, Kostas Karpouzis, Chrysanthi Raftopoulou, Antonios Symvonis, and Ioannis Tollis, and all the volunteers who put a lot of time and effort into the organization of GD 2016.

GD 2016 thanks its sponsors, "gold" sponsor Tom Sawyer Software, "silver" sponsors yWorks and Microsoft, and "bronze" sponsor Springer. Their generous support helps to ensure the continued success of this conference. We further thank the team behind EasyChair for providing an incredibly useful conference management system.

The 25th International Symposium on Graph Drawing and Network Visualization (GD 2017) will take place in September, 2017, in Boston, MA, USA. Fabrizio Frati and Kwan-Liu Ma will co-chair the Program Committee, Cody Dunne and Alan Keahey will chair the Organizing Committee.

October 2016

Yifan Hu
Martin Nöllenburg

Organization

Steering Committee

Therese Biedl	University of Waterloo, Canada
Giuseppe Di Battista	Roma Tre University, Italy
Tim Dwyer	Monash University, Melbourne, Australia
Fabrizio Frati	Roma Tre University, Italy
Michael T. Goodrich	University of California at Irvine, USA
Yifan Hu	Yahoo Research, USA
Giuseppe Liotta (Chair)	University of Perugia, Italy
Kwan-Liu Ma	University of California at Davis, USA
Martin Nöllenburg	TU Wien, Vienna, Austria
Roberto Tamassia	Brown University, USA
Ioannis G. Tollis	ICS-FORTH and University of Crete, Greece
Alexander Wolff	University of Würzburg, Germany

Program Committee

Patrizio Angelini	University of Tübingen, Germany
Therese Biedl	University of Waterloo, Canada
Walter Didimo	University of Perugia, Italy
Cody Dunne	Northeastern University, USA
David Eppstein	UC Irvine, USA
Jean-Daniel Fekete	Inria, France
Stefan Felsner	TU Berlin, Germany
Radoslav Fulek	IST Austria
Emden Gansner	Google, USA
Yifan Hu (Co-chair)	Yahoo Research, USA
Karsten Klein	University of Konstanz, Germany
Stephen Kobourov	University of Arizona, USA
Marc van Kreveld	Utrecht University, The Netherlands
Jan Kynčl	Charles University Prague, Czech Republic
Kwan-Liu Ma	UC Davis, USA
Tamara Mchedlidze	Karlsruhe Institute of Technology, Germany
Stephen North	Infovisible, Oldwick, USA
Martin Nöllenburg (Co-chair)	TU Wien, Vienna, Austria
Maurizio Patrignani	Roma Tre University, Italy
Helen Purchase	University of Glasgow, UK
Huamin Qu	HKUST, Hong Kong, SAR China

Günter Rote   FU Berlin, Germany
André Schulz   University of Hagen, Germany
Lei Shi   Chinese Academy of Sciences, China
Alex Telea   University of Groningen, The Netherlands

Organizing Committee

Kostas Karpouzis   National Technical University of Athens, Greece
Chrysanthi Raftopoulou   National Technical University of Athens, Greece
Antonios Symvonis (Chair)   National Technical University of Athens, Greece
Ioannis G. Tollis   ICS-FORTH and University of Crete, Greece

Contest Committee

Philipp Kindermann   University of Hagen, Germany
Maarten Löffler (Chair)   Utrecht University, The Netherlands
Lev Nachmanson   Microsoft Research, USA
Ignaz Rutter   Karlsruhe Institute of Technology, Germany

Additional Reviewers

Alam, Md. Jawaherul   Di Donato, Valentino   Klemz, Boris
Alt, Helmut   Di Giacomo, Emilio   Knauer, Kolja
Asinowski, Andrei   Dijk, Thomas C. van   Kriegel, Klaus
Balko, Martin   Dwyer, Tim   Kwon, Oh-Hyun
Barth, Lukas   Elias, Marek   Leaf, Nick
Beck, Fabian   Frati, Fabrizio   Li, Jianping
Bekos, Michael   Fu, Siwei   Li, Yifei
Binucci, Carla   Goethem, Arthur van   Lidicky, Bernard
Bläsius, Thomas   Grilli, Luca   Liu, Dongyu
Brandenburg, Franz   Grunert, Romain   Lubiw, Anna
Bryan, Chris   Guerra Gómez,   Löffler, Maarten
Buchin, Kevin   John Alexis   Mcguffin, Michael
Buchin, Maike   Haberkorn, Tasja   Meulemans, Wouter
Burch, Michael   Hoffmann, Frank   Mondal, Debajyoti
Chimani, Markus   Hossain, Md. Iqbal   Montecchiani, Fabrizio
Chou, Jia-Kai   Huang, Weidong   Mulzer, Wolfgang
Cibulka, Josef   Igamberdiev, Alexander   Mustaă, Irina-Mihaela
Cornelsen, Sabine   Kaufmann, Michael   Orden, David
Crnovrsanin, Tarik   Kerren, Andreas   Panse, Christian
Da Lozzo, Giordano   Keszegh, Balázs   Pizzonia, Maurizio
De Mesmay, Arnaud   Kim, Heuna   Pupyrev, Sergey
Derka, Martin   Kindermann, Philipp   Radermacher, Marcel
Di Bartolomeo, Marco   Kleist, Linda   Renssen, André van

Roselli, Vincenzo
Rutter, Ignaz
Scharf, Nadja
Schlipf, Lena
Schrezenmaier, Hendrik
Seiferth, Paul
Simonetto, Paolo

Sommer, Björn
Stasi, Despina
Strash, Darren
Tao, Jun
Toth, Csaba
Ueckerdt, Torsten
Verbeek, Kevin

Wang, Yang
Wang, Yong
Wiechert, Veit
Wismath, Steve
Wu, Yanhong

Sponsors

Gold Sponsor

Silver Sponsors

Bronze Sponsor

Keynote Presentations

Distributed Computing:
Graph Drawing Unplugged

Roger Wattenhofer

Distributed Computing Group, ETH Zurich, 8092 Zurich, Switzerland
wattenhofer@ethz.ch

Abstract. Computer networks and distributed systems are typically represented as graphs, and sooner or later everybody working in distributed computing is facing a graph drawing problem. In my talk I will discuss a few artifacts in distributed computing that are related to graph theory and graph drawing. The focus of my talk will be wireless communication networks. While a vertex in a wireless network is simply some kind of communication device, vertices are not necessarily connected by edges, but rather "unplugged." We discuss the following "family of open problems": How well can we draw a wireless network modeled by a UDG, QUDG, BIG, or UBG, using connectivity, interference, distance, angle, or multipath information, to understand which node is which? Does such a drawing help to design better routing or media access protocols?

The Role of Visual Analytics in Exploring Graph Data

Daniel A. Keim

Database and Visualization Group, Computer and Information Science,
University of Konstanz, Konstanz, Germany

Abstract. Sophisticated algorithms are the central part of most graph analysis and graph drawing methods. Many clearly specified problems can be solved using algorithmic methods, but in some cases fully automatic methods are not enough to understand the complex graph data and draw valid conclusions. Humans with their abilities - their background knowledge, their creativity, and their judgment? need to be an integral part of the analysis process. This is where the research field of visual analysis comes into play: It tries to integrate automatic data analysis methods with interactive visualization techniques to support the human in gaining new insights. In this presentation, we will discuss the role of the human in the process of exploring and analyzing large graphs, and will illustrate the exiting potential of current Visual Analytics techniques as well as their limitations with several application examples.

Contents

Layered and Tree Drawings

Visibility Representations

Beyond Planarity

Crossing Minimization and Crossing Numbers

Topological Graph Theory

Special Graph Embeddings

Dynamic Graphs

Large Graphs and Clutter Avoidance

A Distributed Multilevel Force-Directed Algorithm

Alessio Arleo$^{(\boxtimes)}$, Walter Didimo$^{(\boxtimes)}$, Giuseppe Liotta,
and Fabrizio Montecchiani

Università Degli Studi di Perugia, Perugia, Italy
alessio.arleo@studenti.unipg.it,
{walter.didimo,giuseppe.liotta,fabrizio.montecchiani}@unipg.it

Abstract. The wide availability of powerful and inexpensive cloud computing services naturally motivates the study of distributed graph layout algorithms, able to scale to very large graphs. Nowadays, to process Big Data, companies are increasingly relying on PaaS infrastructures rather than buying and maintaining complex and expensive hardware. So far, only a few examples of *basic* force-directed algorithms that work in a distributed environment have been described. Instead, the design of a distributed *multilevel* force-directed algorithm is a much more challenging task, not yet addressed. We present the first multilevel force-directed algorithm based on a distributed vertex-centric paradigm, and its implementation on Giraph, a popular platform for distributed graph algorithms. Experiments show the effectiveness and the scalability of the approach. Using an inexpensive cloud computing service of Amazon, we draw graphs with ten million edges in about 60 min.

1 Introduction

Force-directed algorithms are very popular techniques to automatically compute graph layouts. They model the graph as a physical system, where attractive and repulsive forces act on each vertex. Computing a drawing corresponds to finding an equilibrium state (i.e., a state of minimum energy) of the force system through a simple iterative approach. Different kinds of force and energy models give rise to different graph drawing algorithms. Refer to the work of Kobourov for a survey on the many force-directed algorithms described in the literature [25]. Although basic force-directed algorithms usually compute nice drawings of small or medium graphs, using them to draw large graphs has two main obstacles: (*i*) There could be several local minima in their physical models: if the algorithm falls in one of them, it may produce bad drawings. The probability of this event and its negative effect increase with the size of the graph. (*ii*) Their approach is computationally expensive, thus it gives rise to scalability problems even for graphs with a few thousands of vertices.

Research supported in part by the MIUR project AMANDA "Algorithmics for MAssive and Networked DAta", prot. 2012C4E3KT_001.

© Springer International Publishing AG 2016
Y. Hu and M. Nöllenburg (Eds.): GD 2016, LNCS 9801, pp. 3–17, 2016.
DOI: 10.1007/978-3-319-50106-2_1

To overcome the above obstacles, *multilevel* force-directed algorithms have been conceived. A limited list of works on this subject includes [13, 15, 18, 20, 21, 23, 33] (see [25] for more references). These algorithms generate from the input graph G a series (hierarchy) of progressively simpler structures, called *coarse graphs*, and then incrementally compute a drawing of each of them in reverse order, from the simplest to the most complex (corresponding to G). On common machines, multilevel force-directed algorithms perform quickly on graphs with several thousand vertices and usually produce qualitatively better drawings than basic algorithms [8, 19, 25]. Implementations based on GPUs have been also experimented [16, 24, 29, 34]. They scale to graphs with a few million edges, but their development requires a low-level implementation and the necessary infrastructure could be expensive in terms of hardware and maintenance.

The wide availability of powerful and inexpensive cloud computing services and the growing interest towards PaaS infrastructures observed in the last few years, naturally motivate the study of distributed graph layout algorithms, able to scale to very large graphs. So far, the design of distributed graph visualization algorithms has been only partially addressed. Mueller *et al.* [27] and Chae *et al.* [9] proposed force-directed algorithms that use multiple large displays. Vertices are evenly distributed on the different displays, each associated with a different processor, which is responsible for computing the positions of its vertices; scalability experiments are limited to graphs with some thousand vertices. Tikhonova and Ma [30] presented a parallel force-directed algorithm that can run on graphs with few hundred thousand edges. It takes about 40 minutes for a graph of 260, 385 edges, on 32 processors of the PSC's BigBen Cray XT3 cluster. More recently, the use of emerging frameworks for distributed graph algorithms has been investigated. Hinge and Auber [22] described a distributed basic force-directed algorithm implemented in the Spark framework, using the GraphX library. Their algorithm is mostly based on a MapReduce paradigm and shows margins for improvement: it takes 5 hours on a graph with 8, 000 vertices and 35, 000 edges, on a cluster of 16 machines, each equipped with 24 cores and 48 GB of RAM. A distributed basic force-directed algorithm running on the Apache Giraph framework has been presented in [7] (see also [5] for an extended version of this work). Giraph is a popular platform for distributed graph algorithms, based on a vertex-centric paradigm, also called the TLAV ("Think Like a Vertex") paradigm [11]. Giraph is used by Facebook to analyze the huge network of its users and their connections [12]. The algorithm in [7] can draw graphs with a million edges in a few minutes, running on an inexpensive cloud computing infrastructure. However, the design of a distributed multilevel force-directed algorithm is a much more challenging task, due to the difficulty of efficiently computing the hierarchy required by a multilevel approach in a distributed manner (see, also [5, 22]).

Our Contribution. This paper presents MULTI-GILA (Multilevel Giraph Layout Algorithm), the first distributed multilevel force-directed algorithm based on the TLAV paradigm and running on Giraph. The model for generating the coarse graph hierarchy is inspired by FM3 (Fast Multipole Multilevel Method), one of the most effective multilevel techniques described in the literature [8,18,19]. The basic force-directed algorithm used by MULTI-GILA to refine the drawing of each coarse graph is the distributed algorithm in [5] (Sect. 3). We show the effectiveness and the efficiency of our approach by means of an extensive experimental analysis: MULTI-GILA can draw graphs with ten million edges in about 60 min (see Sect. 4), using an inexpensive PaaS of Amazon, and exhibits high scalability. To allow replicability of the experiments, our source code and graph benchmarks are made publicly available [1]. It is worth observing that in order to get an overview of the structure of a very large graph and subsequently explore it in more details, one can combine the use of MULTI-GILA with systems like LAGO [35], which provides an interactive level-of-detail rendering, conceived for the exploration of large graphs (see Sect. 4). Section 2 contains the necessary background on multilevel algorithms and on Giraph. Conclusions and future research are in Sect. 5. Additional figures can be found in [6].

2   Background

Multilevel Force-Directed Algorithms. Multilevel force-directed algorithms work in three main phases: *coarsening*, *placement*, and *single-level layout*. Given an input graph G, the coarsening phase computes a sequence of graphs $\{G = G_0, G_1, \ldots, G_k\}$, such that the size of G_{i+1} is smaller than the size of G_i, for $i = 0, \ldots, k-1$. To compute G_{i+1}, subsets of vertices of G_i are merged into single vertices. The criterion for deciding which vertices should be merged is chosen as a trade-off between two conflicting goals. On one hand, the overall graph structure should be preserved throughout the sequence of graphs, as it influences the way the graph is unfolded. On the other hand, both the number of graphs in the sequence and the size of the coarsest graph may have a significant influence on the overall running time of the algorithm. Therefore, it is fundamental to design a coarsening phase that produces a sequence of graphs whose sizes quickly decrease, and, at the same time, whose structures smoothly change. The sequence of graphs produced by the coarsening phase is then traversed from G_k to $G_0 = G$, and a final layout of G is obtained by progressively computing a layout for each graph in the sequence. In the placement phase, the vertices of G_i are placed by exploiting the information of the (already computed) drawing Γ_{i+1} of G_{i+1}. Starting from this initial placement, in the single-level (basic) layout phase, a drawing Γ_i of G_i is computed by applying a single-level force-directed algorithm. Thanks to the good initial placement, such an algorithm will reach an equilibrium after a limited number of iterations. For G_k an initial placement is not possible, thus the layout phase is directly applied starting from a random placement.

Since our distributed multilevel force-directed algorithm is partially based on the FM3 algorithm, we briefly recall how the coarsening and placement phases are implemented by FM3 (see [17,18] for details). Let $G = G_0$ be a connected graph (distinct connected components can be processed independently), the coarsening phase is implemented through the SOLAR MERGER algorithm. The vertices of G are partitioned into vertex-disjoint subgraphs called *solar systems*. The diameter of each solar system is at most four. Within each solar system S, there is a vertex s classified as a *sun*. Each vertex v of S at distance one (resp., two) from s is classified as a *planet* (resp., a *moon*) of S. There is an *inter-system link* between two solar systems S_1 and S_2, if there is at least an edge of G between a vertex of S_1 and a vertex of S_2. The coarser graph G_1 is obtained by collapsing each solar system into the corresponding sun, and the inter-system links are transformed into edges connecting the corresponding pairs of suns. Also, all vertices of $G = G_0$ are associated with a *mass* equal to one. The mass of a sun is the sum of the masses of all vertices in its solar system. The coarsening procedure halts when a coarse graph has a number of vertices below a predefined threshold. The placement phase of FM3 is called SOLAR PLACER and uses information from the coarsening phase. The vertices of G_{i+1} correspond to the suns of G_i, whose initial position is defined in the drawing Γ_{i+1}. The position of each vertex v in $G_i \setminus G_{i+1}$ is computed by taking into account all inter-system links to which v belongs. The rough idea is to position v in a barycentric position with respect to the positions of all suns connected by an inter-system link that passes through v.

The TLAV Paradigm and the Giraph Framework. The TLAV paradigm requires to implement distributed algorithms from the perspective of a vertex rather than of the whole graph. Each vertex can store a limited amount of data and can exchange messages only with its neighbors. The TLAV framework Giraph [11] is built on the Apache Hadoop infrastructure and originated as the open source counterpart of Google's *Pregel* [26] (based on the *BSP model* [31]). In Giraph, the computation is split into *supersteps* executed iteratively and synchronously. A superstep consists of two phases: (i) Each vertex executes a user-defined vertex function based on both local vertex data and on data coming from its adjacent vertices; (ii) Each vertex sends the results of its local computation to its neighbors, along its incident edges. The whole computation ends after a fixed number of supersteps or when certain user-defined conditions are met (e.g., no message has been sent or an equilibrium state is reached).

Design Challenges and the GILA Algorithm. Force-directed algorithms (both single-level and multilevel) are conceived as sequential, shared-memory graph algorithms, and thus are inherently centralized. On the other hand, the following three properties must be guaranteed in the design of a TLAV-based algorithm: P1. Each vertex exchange messages only with its neighbors; P2. Each vertex locally stores a small amount of data; P3. The communication load in each supertsep (number and length of messages sent in the superstep) is small: for example, linear in the number of edges of the graph. Property P1 corresponds to an architectural constraint of Giraph. Violating P2 causes out-of-memory errors during the computation of large instances, which translates in the impossibility of storing large routing tables in each vertex to cope with the absence of global information. Violating P3 quickly leads to inefficient computations, especially on graphs that are locally dense or that have high-degree vertices. Hence, sending heavy messages containing the information related to a large part of the graph is not an option.

In the design of a multilevel force-directed algorithm, the above three constraints P1–P3 do not allow for simple strategies to make a vertex aware of the topology of a large part of the graph, which is required in the coarsening phase. In Sect. 3 we describe a sophisticated distributed protocol used to cope with this issue. For the same reason, a vertex is not aware of the positions of all other vertices in the graph, which is required to compute the repulsive forces acting on the vertex in the single-level layout phase. The algorithm described in [5], called GiLA, addresses this last issue by adopting a locality principle, based on the experimental evidence that in a drawing computed by a force-directed algorithm (see, e.g., [25]) the graph theoretic distance between two vertices is a good approximation of their geometric distance, and that the repulsive forces between two vertices u and v tend to be less influential as their geometric distance increases. Following these observations, in the GiLA algorithm, the resulting force acting on each vertex v only depends on its k-neighborhood $N_v(k)$, i.e., the set of vertices whose graph theoretic distance from v is at most k, for a predefined small constant k. Vertex v acquires the positions of all vertices in $N_v(k)$ by means of a controlled flooding technique. According to an experimental analysis in [5], $k = 3$ is a good trade-off between drawing quality and running time. The attractive and repulsive forces acting on a vertex are defined using Fruchterman-Reingold model [14].

3   The Multi-GiLA Algorithm

In this section we describe our multilevel algorithm MULTI-GiLA. It is designed having in mind the challenges and constraints discussed in Sect. 2. The key ingredients of MULTI-GiLA are a distributed version of both the SOLAR MERGER and of the SOLAR PLACER used by FM3, together with a suitable dynamic tuning of GiLA.

3.1   Algorithm Overview

The algorithm is based on the pipeline described below. The pruning, partitioning, and reinsertion phases are the same as for the GiLA algorithm, and hence they are only briefly recalled (see [5] for details).

Pruning: In order to lighten the algorithm execution, all vertices of degree one are temporarily removed from the graph; they will be reinserted at the end of the computation by means of an ad-hoc technique.

Partitioning: The vertex set is then partitioned into subsets, each assigned to a computing unit, also called *worker* in Giraph (each computer may have more than one worker). The default partitioning algorithm provided by Giraph may create partitions with a very high number of edges that connect vertices of different partition sets; this would negatively affect the communication load between different computing units. To cope with this problem, we use a partitioning algorithm by Vaquero *et al.* [32], called SPINNER, which creates balanced partition sets by exploiting the graph topology.

Layout: This phase executes the pipeline of the multilevel approach. The coarsening phase (Sect. 3.2) is implemented by means of a distributed protocol, which attempts to behave as the SOLAR MERGER of FM3. The placement (Sect. 3.3) and single-level layout (Sect. 3.4) phases are iterated until a drawing of the graph is computed.

Reinsertion: For each vertex v, its neighbors of degree one (if any) are suitably reinserted in a region close to v, avoiding to introduce additional edge crossings.

This pipeline is applied independently to each connected component of the graph, and the resulting layouts are then arranged in a matrix to avoid overlaps.

3.2 Coarsening Phase: DISTRIBUTED SOLAR MERGER

Our DISTRIBUTED SOLAR MERGER algorithm yields results (in terms of number of levels) comparable to those obtained with the SOLAR MERGER of FM3 (see also Sect. 4). The algorithm works into four steps described below; each of them involve several Giraph supersteps. For every iteration i of these four steps, a new coarser graph G_i is generated, until its number of vertices is below a predefined threshold. We use the same terminology as in Sect. 2, and equip each vertex with four properties called *ID*, *level*, *mass*, and *state*. The ID is the unique identifier of the vertex. The level represents the iteration in which the vertex has been generated. That is, a vertex has level i if it belongs to graph G_i. The vertices of the input graph have level zero. The second property represents the mass of the vertex and it is initialized to one plus the number of its previously pruned neighbors of degree one for the vertices of the input graph. The state of a vertex can receive one of the following values: *sun*, *planet*, *moon*, or *unassigned*. We shall call sun, planet, moon, or unassigned, a vertex with the corresponding value for its state. All vertices of the input graph are initially unassigned.

Sun Generation. In the first superstep, each vertex turns its state to sun with probability p, for a predefined value of p. The next three supersteps aim at avoiding pairs of suns with graph theoretic distance less than 3. First, each sun broadcasts a message containing its ID. In the next superstep, if a sun t receives a message from an adjacent sun s, then also s receives a message from t, and the sun between s and t with lower ID changes its state to unassigned. In the same superstep, all vertices (of any state) broadcast to their neighbors only the messages received from those vertices still having state sun. In the third superstep, if a sun t receives a message generated from a sun s (with graph theoretic distance 2 from t), again also t receives a message from s and the sun with lower ID changes its state to unassigned. This procedure ensures that all pairs of suns have graph theoretic distance at least three.

Solar System Generation. In the first superstep, each sun broadcasts an *offer message*. At the next superstep, if an unassigned vertex v receives an offer message m from a sun s, then v turns its state to planet and stores the ID of s in a property called *system-sun*. Also, v sends a *confirmation message* to s. Finally, v forwards the message m to all its neighbors. At the next superstep, every sun vertex processes the received confirmation messages. If a sun s received a confirmation message, s stores the ID of the sender in a property called *planet-list*. This property is used by each sun to keep track of the planets in its solar system. If a planet v receives an offer message, then such a message comes from the same sun stored in the system-sun property of v, and thus it can be ignored (recall that the theoretic distance between two suns is greater than two). If an unassigned vertex u receives one or more offer messages originated by the same sun s, then u turns its state to moon and stores the ID of s in its system-sun property. Furthermore, u stores the ID of all planets that forwarded the above offer messages in a property called *system-planets*. This property is used by each moon u to keep track of the planets adjacent to u and in the same solar system as u. Finally, u sends a confirmation message to its sun s through a two-hop message (that requires two further supersteps to be delivered), which will be sent to one of the planets stored in the system-planets property. If u receives offer messages from distinct vertices, then the above procedure is applied only for those messages originated by the sun s with greatest ID. For every offer message originated by a sun t with ID lower than the one of s, u informs both s and t of the conflict through ad-hoc two-hop messages. These messages will be used by s and t to maintain a suitable data structure containing the information of each path between s and t. At the end of this phase, all the galaxies of the generated sun vertices have been created and have diameter at most four. Also, some of the inter-system links have already been discovered, and this information will be useful in the following. The two steps described above are repeated until there are no more unassigned vertices. An example is illustrated in Fig. 1.

Inter-system Link Generation. In the first superstep, every planet and every moon broadcasts an *inter-link discovery message* containing the ID stored in the system-sun property of the vertex. In the next superstep, each vertex v processes the received messages. All messages originated by vertices in the same solar system are ignored. Similarly as in the previous step, for each inter-link discovery message originated from a sun t different from the sun s of v, vertex v informs both s and t of the conflict through two-hop messages that will be used

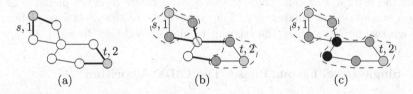

Fig. 1. Illustration for the coarsening phase. (a) Two suns s (ID 1) and t (ID 2) broadcast an offer message. (b) The dark gray vertices receive the offer messages, become planets, and forward the received offer messages. The striped vertex will then receive offer messages from both s and t, and (c) will accept the offer message of t due to the greatest ID of t. In (c) the final galaxies are enclosed by dashed curves, suns (planets, moons) are light gray (dark gray, black).

by s and t to maintain a suitable data structure containing the information of each path between s and t. Once all messages have been delivered, each sun s is aware of all links between its solar system and other systems. Also, for each link, s knows what planet and moon (if any) are involved.

Next Level Generation. In the first superstep, every sun s creates a vertex v_s whose level equals the level of s plus one, and whose mass equals the sum of the masses of all the vertices in the solar system of s. Also, an *inter-level edge* between s and v_s is created and will be used in the placement phase. In the next superstep, every sun s adds an edge between v_s and v_t, if t is a sun of a solar system for which there are $k > 0$ inter-system links. The edge (v_s, v_t) is equipped with a *weight* equal to the maximum number of vertices involved in any of the k links. Finally, all vertices (except the newly created ones) deactivate themselves.

3.3 Placement Phase: DISTRIBUTED SOLAR PLACER

We now describe a DISTRIBUTED SOLAR PLACER algorithm, which behaves similarly to the SOLAR PLACER of FM3. After the coarsening phase, the only active vertices are those of the coarsest graph G_k. For this graph, the placement phase is not executed, and the computation goes directly to the single-level layout phase (described in the next subsection). The output of the single-level layout phase is an assignment of coordinates to all vertices of G_k. Then, the placement phase starts and its execution is as follows.

In the first superstep, every vertex broadcasts its coordinates. In the second superstep, all vertices whose level is one less than the level of the currently active vertices activate themselves, and hence will start receiving messages from the next superstep. In the same superstep, every vertex v forwards the received messages to the corresponding vertex v^* of lower level through its inter-level edge. Then v deletes itself. At the next superstep, if a vertex s receives a message, then s is the sun of a solar system. Thanks to the received messages, s becomes aware of the position of all suns of its neighboring solar systems. Hence, s exploits this information (and the data structure containing information on the inter-system links), to compute the coordinates of all planets and moons in its solar system, as for the SOLAR PLACER. Once this is done, s sends to every planet u of its solar system the coordinates of u. The coordinates of the moons are delivered through two-hop messages (that is, sent to planets and then forwarded).

3.4 Single-Level Layout Phase: The GILA Algorithm

This phase is based on the GILA algorithm, the distributed single-level force-directed algorithm described in Sect. 2. Recall that the execution of GILA is based on a set of parameters, whose tuning affects the trade-off between quality of the drawing and speed of the computation. The most important parameter is the maximum graph theoretic distance k between pairs of vertices for which the pairwise repulsive forces are computed. Also, there are further parameters that affect the maximum displacement of a vertex, at a given iteration of the algorithm. The idea is to tune these parameters in order to achieve better quality for the coarser graphs, and shorter running times for the graphs whose size is closer to the original graph. Here we only describe how the parameter k has been experimentally tuned, since it is the parameter that mostly affect the trade-off between quality and running time. The other parameters have been set similarly. For the drawing of every graph G_i, the value of k is 6 if the number of edges m_i of G_i is below 10^3, it is 5 if $10^3 \leq m_i < 5 \cdot 10^3$, it is 4 if $5 \cdot 10^3 \leq m_i < 10^4$, it is 3 if $10^4 \leq m_i < 10^5$, it is 2 if $10^5 \leq m_i < 10^6$, and it is 1 if $m_i \geq 10^6$.

4   Experimental Analysis

We executed an experimental analysis whose objective is to evaluate the performance of MULTI-GILA. We aim to investigate both the quality of the produced drawings and the running time of the algorithm, also in terms of scalability when we increase the number of machines. We expect that MULTI-GILA computes drawings whose quality is comparable to that achieved by centralized multilevel force-directed algorithms. This is because the locality-based approximation scheme adopted by GILA (used in the single-level layout phase) should be mitigated by the use of a graph hierarchy. Also, we expect MULTI-GILA to be able to handle graphs with several million edges in tens of minutes on an inexpensive PaaS infrastructure. Clearly, the use of a scalable vertex-centric distributed framework adds some unavoidable overhead, which may make MULTI-GILA not suited for graphs whose size is limited to a few hundred thousand of edges. Our experimental analysis is based on three benchmarks called REGULARGRAPHS, REALGRAPHS, and BIGGRAPHS, described in the following.

The REGULARGRAPHS benchmark is the same used by Bartel et al. [8] in an experimental evaluation of various implementations of the three main phases of a multilevel force-directed algorithm (coarsening, placement, and single-level layout). It contains 43 graphs with a number of edges between 78 and 48,232, and it includes both real-world and generated instances [2]. See also Table 1 for more details. We used this benchmark to evaluate MULTI-GILA in terms of quality of the computed drawings. Since the coarsening phase plays an important role in the computation of a good drawing, we first evaluated the performance of our DISTRIBUTED SOLAR MERGER in terms of number of produced levels compared to the number of levels produced by the SOLAR MERGER of FM3. It may be worth remarking that, in the experimental evaluation conducted by Bartel et al. [8], the SOLAR MERGER algorithm showed the best performance in terms of drawing quality when used for the coarsening phase. Our experiments show that the number of levels produced by the two algorithms is comparable and follows a similar trend throughout the series of graphs. The DISTRIBUTED SOLAR MERGER produces one or two levels less than the SOLAR MERGER in most of the cases, and this is probably due to some slight difference in the tuning of the two algorithms. To capture the quality of the computed drawings, we compared FM3 (the implementation available in the OGDF library [10]) and MULTI-GILA in terms of average number of crossings per edge (CRE), and normalized edge

Table 1. REGULARGRAPHS: number of vertices (n), number of edges (m), average number of crossings per edge (CRE), normalized edge length std deviation (NELD).

NAME	n	m	FM3 CRE	FM3 NELD	MULTI-GILA CRE	MULTI-GILA NELD	NAME	n	m	FM3 CRE	FM3 NELD	MULTI-GILA CRE	MULTI-GILA NELD
karateclub	34	78	1.10	0.25	1.09	0.33	Grid_40_40_df	1,597	3,120	0.19	0.23	0.20	0.33
snowflake_A	98	97	0.00	0.25	0.11	0.21	Grid_40_40_sf	1,599	3,120	0.39	0.18	0.38	0.31
spider_A	100	160	3.06	0.24	2.86	0.27	ug_380	1,104	3,231	25.68	0.64	13.47	0.96
cylinder_010	97	178	0.35	0.16	0.72	0.08	esslingen	2,075	5,530	19.89	0.41	34.18	0.53
sierpinski_04	123	243	0.00	0.25	0.00	0.22	uk	4,824	6,837	0.07	0.36	0.06	0.65
tree_06_03	259	258	0.40	0.29	1.54	0.17	4970	4,970	7,400	0.01	0.23	0.01	0.46
rna	363	468	0.04	0.24	0.06	0.50	add20	2,395	7,462	60.38	0.50	100.44	0.50
protein_part	417	511	1.20	0.33	1.73	0.50	dg_1087	7,602	7,601	0.06	0.34	0.00	1.04
516	516	729	0.09	0.13	0.18	0.44	tree_06_05	9,331	9,330	8.63	0.47	19.65	0.93
Grid_20_20	400	760	0.00	0.13	0.00	0.23	add32	4,960	9,462	1.31	0.88	0.97	1.66
Grid_20_20_df	397	760	0.24	0.23	0.20	0.34	snowflake_C	9,701	9,700	0.00	0.64	0.00	0.40
Grid_20_20_sf	397	760	0.41	0.17	0.41	0.26	flower_005	930	13,521	48.76	0.61	45.24	0.61
dg_617_part	341	797	10.57	0.30	16.61	0.36	3elt	4,720	13,722	0.40	0.35	0.27	0.60
snowflake_B	971	970	0.00	0.42	0.00	0.39	data	2,851	15,093	2.15	0.39	2.52	0.64
tree_06_04	1,555	1,554	8.53	0.35	7.04	0.19	grid400_20	8,000	15,580	0.02	0.22	0.24	0.89
spider_B	1,000	1,600	7.03	0.24	8.26	0.73	spider_C	10,000	16,000	171.31	0.32	262.09	0.93
grid_rnd_032	985	1,834	0.00	0.15	0.00	0.30	grid_rnd_100	9,499	17,849	0.00	0.16	0.00	0.34
cylinder_032	985	1,866	0.46	0.19	0.44	0.39	sierpinski_08	9,843	19,683	0.09	0.44	0.03	0.70
cylinder_100	985	1,866	4.60	0.18	4.48	0.45	crack	10,240	30,380	0.00	0.26	0.00	0.42
sierpinski_06	1,095	2,187	0.06	0.34	0.03	0.63	4elt	15,607	45,878	0.52	0.39	0.30	0.62
flower_001	210	3,057	47.37	0.67	45.97	0.47	cti	16,840	48,232	10.19	0.39	10.26	0.71
Grid_40_40	1,600	3,120	0.00	0.15	0.00	0.32							

length standard deviation (NELD). The values of NELD are obtained by dividing the edge length standard deviation by the average edge length of each drawing. We chose FM3 for this comparison for two main reasons: (i) MULTI-GILA is partially based on distributed implementations of the SOLAR MERGER and of the SOLAR PLACER algorithms; (ii) FM3 showed the best trade-off between running time and quality of the produced drawings in the experiments of Hachul and Jünger [19]. The results of our experiments are reported in Table 1. The performance of MULTI-GILA is very close to that of FM3 in terms of CRE. In several cases MULTI-GILA produces drawings with a smaller value of CRE than FM3 (see, e.g., ug_380). Concerning the NELD, MULTI-GILA most of the times generates drawings with larger values than FM3. This may depend on how the length of the edges is set by the DISTRIBUTED SOLAR PLACER algorithm. However, also in this case the values of NELD follow a similar trend throughout the series of graphs. Figure 2 shows a visual comparison for some of the graphs. Similarly to FM3, MULTI-GILA is able to unfold graphs with a very regular structure and large diameter.

The REALGRAPHS and BIGGRAPHS sets contain much bigger graphs than REGULARGRAPHS, and are used to evaluate the running time of MULTI-GILA, especially in terms of strong scalability (i.e., how the running time varies on a given instance when we increase the number of machines). The REALGRAPHS set is composed of the 5 largest real-world graphs (mainly scale-free graphs) used in the experimental study of GILA [5]. These graphs are taken from the Stanford Large Networks Dataset Collection [3] and from the Network Data

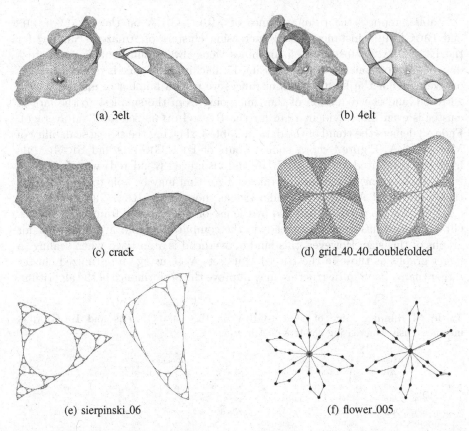

(a) 3elt

(b) 4elt

(c) crack

(d) grid_40_40_doublefolded

(e) sierpinski_06

(f) flower_005

Fig. 2. Layouts of some REGULARGRAPHS instances. For each graph, the drawing computed by FM3 (MULTI-GILA) is on the left (right).

Repository [4], and their number of edges is between $121,523$ and $1,541,514$. The BIGGRAPHS set consists of 3 very large graphs with up to 12 million edges, taken from the collection of graphs described in [28][1]. Details about the REALGRAPHS and BIGGRAPHS sets are in Table 2.

Table 2. Left: Details for REALGRAPHS. Right: Details for BIGGRAPHS benchmark. Isolated vertices, self-loops, and parallel edges have been removed from the original graphs. The graphs are ordered by increasing number of edges.

NAME	n	m	DESCRIPTION	NAME	n	m	DESCRIPTION
asic-320	121,523	515,300	circuit sim. problem	hugetric-10	6,600,000	10,000,000	Mesh
amazon0302	262,111	899,792	co-purchasing network	hugetric-20	7,100,000	10,700,000	Mesh
com-amazon	334,863	925,872	co-purchasing network	delaunay_n22	4,100,000	12,200,000	Triangulation
com-DBLP	317,080	1,049,866	collaboration network				
roadNet-PA	1,087,562	1,541,514	road network				

[1] See also http://www.networkrepository.com/.

Table 3 reports the running times of MULTI-GILA on the REALGRAPHS and BIGGRAPHS instances, using increasing clusters of Amazon. Namely, for the REALGRAPHS instances, 5 machines were always sufficient to compute a drawing in a reasonable time, and using 15 machines the time is reduced by 35% on average. For the BIGGRAPHS instances we used a number of machines from 20 to 30, and the reduction of the time going from the smallest to the largest cluster is even more evident than for the REALGRAPHS set (50% on average). Figure 3 depicts the trend of the data in Table 3, showing the strong scalability of MULTI-GILA. Figure 4 shows some layouts of REALGRAPHS and BIGGRAPHS instances computed by MULTI-GILA and visualized (rendered) with LAGO. It is worth observing that some centralized algorithm may be able to draw quicker than MULTI-GILA graphs of similar size as those in the REALGRAPHS set (see e.g. [16]). This is partially justified by the use of a distributed framework such as Giraph, which introduces overheads in the computation that are significant for graphs of this size. However, this kind of overhead is amortized when scaling to larger graphs as those in the BIGGRAPHS set. Also, using an optimized cluster rather than a PaaS infrastructure may improve the performance of the algorithm.

Table 3. Running time of MULTI-GILA on the REALGRAPHS and BIGGRAPHS instances, using increasing clusters of Amazon.

	Running time (s)				Running time (s)		
NAME	5 machines	10 machines	15 machines	NAME	20 machines	25 machines	30 machines
asic-320	1,626	1,102	1,281	hugetric-10	7,923	4,828	3,679
amazon0302	2,518	2,696	1,577	hugetric-20	9,891	8,243	4,445
com-amazon	3,400	3,395	2,242	delaunay_n22	8,160	3,301	3,932
com-DBLP	4,000	3,612	2,366				
roadNet-PA	3,813	2,369	2,241				

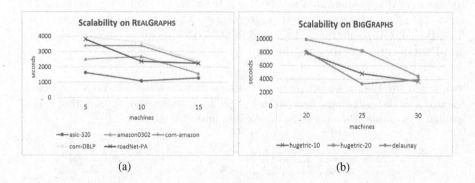

(a)   (b)

Fig. 3. Scalability of MULTI-GILA on REALGRAPHS instances.

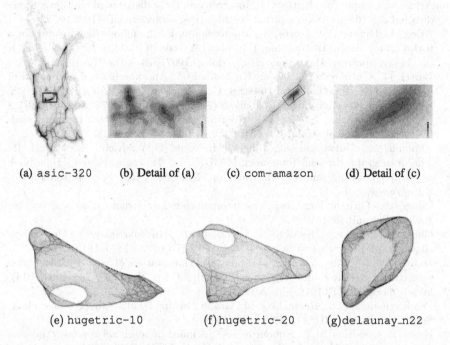

(a) `asic-320`   (b) Detail of (a)   (c) `com-amazon`   (d) Detail of (c)

(e) `hugetric-10`   (f) `hugetric-20`   (g) `delaunay_n22`

Fig. 4. Layouts of (a–d) REALGRAPHS instances and (e–f) BIGGRAPHS instances computed by MULTI-GILA and visualized (rendered) with LAGO.

5   Conclusions and Future Research

As far as we know, MULTI-GILA is the first multilevel force-directed technique working in a distributed vertex-centric framework. Its communication protocol allows for an effective computation of a coarse graph hierarchy. Experiments indicate that the quality of the computed layouts compares with that of drawings computed by popular centralized multilevel algorithms and that it exhibits high scalability to very large graphs. Our source code is made available to promote research on the subject and to allow replicability of the experiments. In the near future we will investigate more coarsening techniques and single-level layout methods for a vertex-centric distributed environment.

References

1. http://www.geeksykings.eu/multigila/
2. http://ls11-www.cs.tu-dortmund.de/staff/klein/gdmult10
3. http://snap.stanford.edu/data/index.html
4. http://www.networkrepository.com/
5. Arleo, A., Didimo, W., Liotta, G., Montecchiani, F.: A distributed force-directed algorithm on Giraph: design and experiments. ArXiv e-prints (2016). http://arxiv.org/abs/1606.02162

6. Arleo, A., Didimo, W., Liotta, G., Montecchiani, F.: A distributed multilevel force-directed algorithm. ArXiv e-prints (2016). http://arxiv.org/abs/1608.08522

7. Arleo, A., Didimo, W., Liotta, G., Montecchiani, F.: A million edge drawing for a fistful of dollars. In: Di Giacomo, E., Lubiw, A. (eds.) GD 2015. LNCS, vol. 9411, pp. 44–51. Springer, Heidelberg (2015). doi:10.1007/978-3-319-27261-0_4

8. Bartel, G., Gutwenger, C., Klein, K., Mutzel, P.: An experimental evaluation of multilevel layout methods. In: Brandes, U., Cornelsen, S. (eds.) GD 2010. LNCS, vol. 6502, pp. 80–91. Springer, Heidelberg (2011). doi:10.1007/978-3-642-18469-7_8

9. Chae, S., Majumder, A., Gopi, M.: Hd-graphviz: Highly distributed graph visualization on tiled displays. In: ICVGIP 2012, pp. 43: 1–43: 8. ACM (2012)

10. Chimani, M., Gutwenger, C., Jünger, M., Klau, G.W., Klein, K., Mutzel, P.: The open graph drawing framework (OGDF). In: Tamassia, R. (ed.) Handbook on Graph Drawing and Visualization, pp. 543–569. CRC, Boca Raton (2013). http://www.ogdf.net/

11. Ching, A.: Giraph: large-scale graph processing infrastructure on hadoop. In: Hadoop Summit (2011)

12. Ching, A., Edunov, S., Kabiljo, M., Logothetis, D., Muthukrishnan, S.: One trillion edges: graph processing at facebook-scale. PVLDB 8(12), 1804–1815 (2015)

13. Didimo, W., Montecchiani, F.: Fast layout computation of clustered networks: algorithmic advances and experimental analysis. Inf. Sci. 260, 185–199 (2014). http://dx.doi.org/10.1016/j.ins.2013.09.048

14. Fruchterman, T.M.J., Reingold, E.M.: Graph drawing by force-directed placement. Softw. Pract. Exp. 21(11), 1129–1164 (1991)

15. Gajer, P., Goodrich, M.T., Kobourov, S.G.: A multi-dimensional approach to force-directed layouts of large graphs. Comput. Geom. 29(1), 3–18 (2004)

16. Godiyal, A., Hoberock, J., Garland, M., Hart, J.C.: Rapid multipole graph drawing on the GPU. In: Tollis, I.G., Patrignani, M. (eds.) GD 2008. LNCS, vol. 5417, pp. 90–101. Springer, Heidelberg (2009). doi:10.1007/978-3-642-00219-9_10

17. Hachul, S.: A potential field based multilevel algorithm for drawing large graphs. Ph.D. thesis, University of Cologne (2005). http://kups.ub.uni-koeln.de/volltexte/2005/1409/index.html

18. Hachul, S., Jünger, M.: Drawing large graphs with a potential-field-based multilevel algorithm. In: Pach, J. (ed.) GD 2004. LNCS, vol. 3383, pp. 285–295. Springer, Heidelberg (2005). doi:10.1007/978-3-540-31843-9_29

19. Hachul, S., Jünger, M.: Large-graph layout algorithms at work: an experimental study. J. Graph Algorithms Appl. 11(2), 345–369 (2007)

20. Hadany, R., Harel, D.: A multi-scale algorithm for drawing graphs nicely. Discrete Appl. Math. 113(1), 3–21 (2001)

21. Harel, D., Koren, Y.: A fast multi-scale method for drawing large graphs. J. Graph Algorithms Appl. 6(3), 179–202 (2002)

22. Hinge, A., Auber, D.: Distributed graph layout with Spark. In: IV 2015, pp. 271–276. IEEE (2015)

23. Hu, Y.: Efficient, high-quality force-directed graph drawing. Mathematica J. 10(1), 37–71 (2005)

24. Ingram, S., Munzner, T., Olano, M.: Glimmer: Multilevel MDS on the GPU. IEEE Trans. Vis. Comput. Graph. 15(2), 249–261 (2009)

25. Kobourov, S.G.: Force-directed drawing algorithms. In: Tamassia, R. (ed.) Handbook of Graph Drawing and Visualization. CRC Press, Boca Raton (2013)

26. Malewicz, G., Austern, M.H., Bik, A.J., Dehnert, J.C., Horn, I., Leiser, N., Czajkowski, G.: Pregel: A system for large-scale graph processing. In: SIGMOD 2010, pp. 135–146. ACM (2010)

27. Mueller, C., Gregor, D., Lumsdaine, A.: Distributed force-directed graph layout and visualization. In: EGPGV 2006, pp. 83–90. Eurographics (2006)
28. Rossi, R.A., Ahmed, N.K.: An interactive data repository with visual analytics. SIGKDD Explor. **17**(2), 37–41 (2016). http://networkrepository.com
29. Sharma, P., Khurana, U., Shneiderman, B., Scharrenbroich, M., Locke, J.: Speeding up network layout and centrality measures for social computing goals. In: Salerno, J., Yang, S.J., Nau, D., Chai, S.-K. (eds.) SBP 2011. LNCS, vol. 6589, pp. 244–251. Springer, Heidelberg (2011). doi:10.1007/978-3-642-19656-0_35
30. Tikhonova, A., Ma, K.: A scalable parallel force-directed graph layout algorithm. In: EGPGV 2008, pp. 25–32. Eurographics (2008)
31. Valiant, L.G.: A bridging model for parallel computation. Commun. ACM **33**(8), 103–111 (1990)
32. Vaquero, L.M., Cuadrado, F., Logothetis, D., Martella, C.: Adaptive partitioning for large-scale dynamic graphs. In: ICDCS 2014, pp. 144–153. IEEE (2014)
33. Walshaw, C.: A multilevel algorithm for force-directed graph-drawing. J. Graph Algorithms Appl. **7**(3), 253–285 (2003)
34. Yunis, E., Yokota, R., Ahmadia, A.: Scalable force directed graph layout algorithms using fast multipole methods. In: ISPDC 2012, pp. 180–187. IEEE (2012)
35. Zinsmaier, M., Brandes, U., Deussen, O., Strobelt, H.: Interactive level-of-detail rendering of large graphs. IEEE Trans. Vis. Comput. Graph. **18**(12), 2486–2495 (2012)

A Sparse Stress Model

Mark Ortmann[✉], Mirza Klimenta, and Ulrik Brandes

Computer and Information Science, University of Konstanz, Konstanz, Germany
Mark.Ortmann@uni-konstanz.de

Abstract. Force-directed layout methods constitute the most common approach to draw general graphs. Among them, stress minimization produces layouts of comparatively high quality but also imposes comparatively high computational demands. We propose a speed-up method based on the aggregation of terms in the objective function. It is akin to aggregate repulsion from far-away nodes during spring embedding but transfers the idea from the layout space into a preprocessing phase. An initial experimental study informs a method to select representatives, and subsequent more extensive experiments indicate that our method yields better approximations of minimum-stress layouts in less time than related methods.

1 Introduction

There are two main variants of force-directed layout methods, expressed either in terms of forces to balance or an energy function to minimize [3,25]. For convenience, we refer to the former as spring embedders and to the latter as multi-dimensional scaling (MDS) methods.

Force-directed layout methods are in wide-spread use and of high practical significance, but their scalability is a recurring issue. Besides investigations into adaptation, robustness, and flexibility, much research has therefore been devoted to speed-up methods [20]. These efforts address, e.g., the speed of convergence [10,11] or the time per iteration [1,17]. Generally speaking, the most scalable methods are based on multi-level techniques [13,18,21,35].

Experiments [5] suggest that minimization of the stress function [27]

$$s(x) = \sum_{i<j} w_{ij}(\|x_i - x_j\| - d_{ij})^2 \tag{1}$$

is the primary candidate for high-quality force-directed layouts $x \in (\mathbb{R}^2)^V$ of a simple undirected graph $G = (V, E)$ with $V = \{1, \ldots, n\}$ and $m = |E|$. The target distances d_{ij} are usually chosen to be the graph-theoretic distances, the weights set to $w_{ij} = 1/d_{ij}^2$, and the dominant method for minimization is majorization [16]. Several variant methods reduce the cost of evaluating the stress

We gratefully acknowledge financial support from Deutsche Forschungsgemeinschaft under grant Br 2158/11-1.

Y. Hu and M. Nöllenburg (Eds.): GD 2016, LNCS 9801, pp. 18–32, 2016.
DOI: 10.1007/978-3-319-50106-2_2

function by involving only a subset of node pairs over the course of the algorithm [6,7,13]. If long distances are represented well already, for instance because of initialization with a fast companion algorithm, it has been suggested that one restrict further attention to short-range influences from k-neighborhoods only [5].

We here propose to stabilize the sparse stress function restricted to 1-neighborhoods [5] with aggregated long-range influences inspired by the use of Barnes & Hut approximation [1] in spring embedders [33]. Extensive experiments suggest how to determine representatives for individually weak influences, and that the resulting method represents a favorable compromise between efficiency and quality.

Related work is discussed in more detail in the next section. Our approach is derived in Sect. 3, and evaluated in Sect. 4. We conclude in Sect. 5.

2   Related Work

While we are interested in approximating the full stress model of Eq. (1), there are other approaches capable of dealing with given target distances such as the strain model [4,24,32] or the Laplacian [19,26].

An early attempt to make the full stress model scale to large graphs is GRIP [13]. Via a greedy maximal independent node set filtration, this multilevel approach constructs a hierarchy of more and more coarse graphs. While a sparse stress model calculates the layout of the coarsened levels, the finest level is drawn by a localized spring-embedder [11]. Given the coarsening hierarchy for graphs of bounded degree, GRIP requires $\mathcal{O}(nk^2)$ time and $\mathcal{O}(nk)$ space with $k = \log \max\{d_{ij} : i, j \in V\}$.

Another notable attempt has been made by Gansner et al. [15]. Like the spring embedder the maxent-model is split into two terms:

$$\sum_{\{i,j\}\in E} w_{ij}(\|x_i - x_j\| - d_{ij})^2 - \alpha \sum_{\{i,j\}\notin E} \log \|x_i - x_j\|$$

The first part is the 1-stress model [4,13], while the second term tries to maximize the entropy. Applying Barnes & Hut approximation technique [1], the running time of the maxent-model can be reduced from $\mathcal{O}(n^2)$ per iteration to $\mathcal{O}(m + n \log n)$, e.g., using quad-trees [30,34]. In order to make the maxent-model even more scalable Meyerhenke et al. [28] embed it into a multi-level framework, where the coarsening hierarchy is constructed using an adapted size-constrained label propagation algorithm.

Gansner et al. [14], inspired by the idea of decomposing the stress model into two parts, proposed COAST. The main difference between COAST and maxent is that it adds a square to the two terms in the 1-stress part and that the second term is quadratic instead of logarithmic. Transforming the energy system of COAST allows one to apply fast-convex optimization techniques making its running time comparable to the maxent model.

While all these approaches somewhat steer away from the stress model, MARS [23] tries to approximate the solution of the full stress model. Building on a result of Drineas et al. [9], MARS requires only $k \ll n$ instead of n single-source shortest path computations. Reconstructing the distance matrix from two smaller matrices and by setting $w_{ij} = 1/d_{ij}$, MARS runs in $\mathcal{O}(kn + n \log n + m)$ per iteration with a preprocessing time in $\mathcal{O}(k^3 + k(m + n \log n + k^2 n))$, and a space requirement in $\mathcal{O}(nk)$.

3   Sparse Stress Model

The full stress model, Eq. (1), is in our opinion the best choice to draw general graphs, not least because of its very natural definition. However, its $\mathcal{O}(n^2)$ running time per iteration and space requirement, and expensive processing time of $\mathcal{O}(n(m + n \log n))$, hamper its way into practice.

The reason sparse stress models are still in early stages of development is that the adaption to large graphs requires not just a reduction in the running time per iteration, but also the preprocessing time and its associated space requirement. Where these problems originate from is best explained by rewriting Eq. (1) to the following form:

$$s(x) = \sum_{\{i,j\} \in E} w_{ij}(\|x_i - x_j\| - d_{ij})^2 + \sum_{\{i,j\} \in \binom{V}{2} \setminus E} w_{ij}(\|x_i - x_j\| - d_{ij})^2 \quad (2)$$

As minimizing the first term only requires $\mathcal{O}(m)$ computations and all d_{ij} are part of the input, solving this part of the stress model can be done efficiently. Yet, the second term requires an all-pairs shortest path computation (APSP), $\mathcal{O}(n^2)$ time per iteration, and in order to stay within this bound $\mathcal{O}(n^2)$ additional space. We note that the 1-stress approaches presented in Sect. 2 of Gajer et al. [13] and Brandes and Pich [4] ignore the second term, while Gansner et al. [14,15] replace it. Discounting the problems arising from the APSP computation, we can see that the spring embedder suffered from exactly the same problem, namely the computation of the second term – there called repulsive forces. Barnes & Hut introduced a simple, yet ingenious and efficient solution, namely to approximate the second term by using only a subset of its addends.

To approximate the repulsive forces operating on node i Barnes & Hut partition the graph. Associated with each of these $\mathcal{O}(\log n)$ partitions is an artificial representative, a so called super-node, used to approximate the repulsive forces of the nodes in its partition affecting i. However, as these super-nodes have only positions in the euclidean space, but no graph-theoretic distance to any node in the graph they cannot be processed in the stress model. Furthermore, deriving a distance for a super-node as a function of the graph-theoretic distance of the nodes it represents would require an APSP computation, which is too costly, and since the partitioning is computed in the layout space, probably not a good approximation. Choosing a node from the partition as a super-node would not solve the problems, not least because the partitioning changes over time.

Therefore, adapting this approach cannot be done in a straightforward manner. However, the model we are proposing sticks to its main ideas. In order to reduce the complexity of the second term in Eq. (2), we restrict the stress computation of each $i \in V$ to a subset $\mathcal{P} \subseteq V$ of $k = |\mathcal{P}|$ representatives, from now on called pivots. The resulting sparse stress model, where $N(i)$ are the neighbors of i and w'_{ip} are adapted weights, has the following form:

$$s'(x) = \sum_{\{i,j\} \in E} w_{ij}(\|x_i - x_j\| - d_{ij})^2 + \sum_{i \in V} \sum_{p \in \mathcal{P} \setminus N(i)} w'_{ip}(\|x_i - x_p\| - d_{ip})^2 \quad (3)$$

Note that GLINT [22] uses a similar function, yet the pivots change in each iteration, no weights are involved, and it is assumed that d_{ip} is accessible in constant time.

Just like Barnes & Hut, we associate with each pivot $p \in \mathcal{P}$ a set of nodes $\mathcal{R}(p) \subseteq V$, where $p \in \mathcal{R}(p), \bigcup_{p \in \mathcal{P}} \mathcal{R}(p) = V$, and $\mathcal{R}(p) \cap \mathcal{R}(p') = \emptyset$ for $p, p' \in \mathcal{P}$. However, we propose to use only one global partitioning of the graph that does not change over time. Still, just like the super-nodes, we want that the pivots are representative for their associated region. In terms of the localized stress minimization algorithm [16] this means we want that for each $i \in V$ and $p \in \mathcal{P}$

$$\frac{\sum_{j \in \mathcal{R}(p)} w_{ij}(x_j^\alpha + d_{ij}(x_i^\alpha - x_j^\alpha)/\|x_i - x_j\|)}{\sum_{j \in \mathcal{R}(p)} w_{ij}} \approx x_p^\alpha + \frac{d_{ip}(x_i^\alpha - x_p^\alpha)}{\|x_i - x_p\|},$$

where α is the dimension. As the left part is the weighted average of all positional votes of $j \in \mathcal{R}(p)$ for the new position of i, we require p to fulfill the following requirements in order to be a good representative:

– The graph-theoretic distances to i from all $j \in \mathcal{R}(p)$ should be similar to d_{ip}
– The positions of $j \in \mathcal{R}(p)$ in x should be well distributed in close proximity around p.

We propose to construct the partitioning induced by $\mathcal{R}$ only based on the graph structure, not on the layout space, and associate each node $v \in V$ with $\mathcal{R}(p)$ of the closest pivot subject to their graph-theoretic distance. As our algorithm incrementally constructs $\mathcal{R}$, ties are broken by favoring the currently smallest partition. Given the case that $\mathcal{P}$ has been chosen properly and since all nodes in $\mathcal{R}(p)$ are at least as close to p as to any other pivot, and consequently in the stress drawing, it is appropriate to assume that both conditions are met.

Even if the positional vote of each pivot is optimal w.r.t. $\mathcal{R}(p)$, it is still not enough to approximate the full stress model. In the full stress model the iterative algorithm to minimize the stress moves one node at a time while fixing the rest. By setting node i's position in dimension α to

$$x_i^\alpha = \frac{\sum_{j \neq i} w_{ij}(x_j^\alpha + d_{ij}(x_i^\alpha - x_j^\alpha)/\|x_i - x_j\|)}{\sum_{j \neq i} w_{ij}},$$

it can be shown that the stress monotonically decreases [16]. However, in our model we move node i according to

$$x_i^\alpha = \frac{\sum_{j \in N(i)} w_{ij}\left(x_j^\alpha + \frac{d_{ij}(x_i^\alpha - x_j^\alpha)}{\|x_i - x_j\|}\right) + \sum_{p \in \mathcal{P} \setminus N(i)} w'_{ip}\left(x_p^\alpha + \frac{d_{ip}(x_i^\alpha - x_p^\alpha)}{\|x_i - x_p\|}\right)}{\sum_{j \in N(i)} w_{ij} + \sum_{p \in \mathcal{P} \setminus N(i)} w'_{ij}}. \quad (4)$$

Algorithm 1. Sparse Stress

Input: Graph $G = (V, E)$ with $w : E \to \mathbb{R}_{>0}$, and k number of pivots.
Output: α–dimensional layout $x \in (\mathbb{R}^\alpha)^V$

1 sample $\mathcal{P}$ with $|\mathcal{P}| = k$
2 calculate $\mathcal{R}$, all adapted weights w'_{ip}, and all d_{ip} via weighted MSSP
3 $x \leftarrow \texttt{PivotMDS(G)}$ [4]
4 rescale x such that $\sum_{\{i,j\} \in E} \|x_i - x_j\| = \sum_{\{i,j\} \in E} w_{ij}$
5 **while** *relative positional change* $> 10^{-4}$ **do**
6   **foreach** $i \in V$ **do**
7   **foreach** *dimension* α **do**
8          $t^\alpha \leftarrow \dfrac{\sum_{j \in N(i)} w_{ij} \left(x_j^\alpha + \frac{d_{ij}(x_i^\alpha - x_j^\alpha)}{\|x_i - x_j\|} \right) + \sum_{p \in \mathcal{P} \setminus N(i)} w'_{ip} \left(x_p^\alpha + \frac{d_{ip}(x_i^\alpha - x_p^\alpha)}{\|x_i - x_p\|} \right)}{\sum_{j \in N(i)} w_{ij} + \sum_{p \in \mathcal{P} \setminus N(i)} w'_{ij}}$
9          $x_i \leftarrow t$

This implies that in order to find the globally optimal position of i we furthermore have to find weights w'_{ip}, such that $\dfrac{w'_{ip}}{\sum_{j \in N(i)} w_{ij} + \sum_{p \in \mathcal{P} \setminus N(i)} w'_{ip}} \approx \dfrac{\sum_{j \in \mathcal{R}(p)} w_{ij}}{\sum_{i \neq j} w_{ij}}$. Since our goal is only to reconstruct the proportions, and our model only knows the shortest-path distance between all nodes $i \in V$ and $p \in \mathcal{P}$, we set $w'_{ip} = s/d_{ip}^2$ where $s \geq 1$. At the first glance setting $s = |\mathcal{R}(p)|$ seems appropriate, since p represents $|\mathcal{R}(p)|$ addends of the stress model. Nevertheless, this strongly overestimates the weight of close partitions. Therefore, we propose to set $s = |\{j \in \mathcal{R}(p) : d_{jp} \leq d_{ip}/2\}|$. This follows the idea that p is only a good representative for the nodes in $\mathcal{R}(p)$ that are at least as close to p as to i. Since the graph-theoretic distance between i and $j \in \mathcal{R}(p)$ is unknown, our best guess is that j lies on the shortest path from p to i. Consequently, if $d_{jp} \leq d_{ip}/2$ node j must be at least as close to p as to i. Note that $w'_{pp'}$ does not necessarily equal $w'_{p'p}$ for $p, p' \in \mathcal{P}$, and if $k = n$ our model reduces to the full stress model.

Asymptotic Running Time: To minimize Eq. (3) in each iteration we displace all nodes $i \in V$ according to Eq. (4). Since this requires $|N(i)| + k$ constant time operations, given that all graph-theoretic distances are known, the total time per iteration is in $\mathcal{O}(kn + m)$. Furthermore, only the distances between all $i \in V$ and $p \in \mathcal{P}$ have to be known, which can be done in $\mathcal{O}(k(m + n \log n))$ time and requires $\mathcal{O}(kn)$ additional space. If the graph-theoretic distances for all $p \in \mathcal{P}$ are computed with a multi-source shortest path algorithm (MSSP), it is possible to construct $\mathcal{R}$ as well as calculate all w'_{ip} during its execution without increasing its asymptotic running time. The full algorithm to minimize our sparse stress model is presented in Algorithm 1.

Table 1. Dataset: n, m, $\delta(G)$, $\Delta(G)$, and $D(G)$ denote the number of nodes, edges, the min. and max. degree, and the diameter, respectively. Column $\{deg(i)\}$ shows the degree and $\{d_{ij}\}$ the distance distribution. Bipartite graphs are marked with $*$ and weighted graphs with $**$

graph	n	m	$\delta(G)$	$\Delta(G)$	$D(G)$	$\{deg(i)\}$	$\{d_{ij}\}$	graph	n	m	$\delta(G)$	$\Delta(G)$	$D(G)$	$\{deg(i)\}$	$\{d_{ij}\}$
dwt1005	1005	3808	3	26	34			pesa	11738	33914	2	9	208		
1138bus	1138	1458	1	17	31			bodyy5	18589	55346	2	8	132		
plat1919	1919	15240	2	18	43			finance256	20657	71866	1	54	55		
3elt	4740	13722	3	9	65			btree (binary tree)	1023*	1022	1	3	18		
USpowerGrid	4941	6594	1	19	46			qh882	1764*	3354	1	14	32		
commanche	7920	11880**	3	3	438.00			lpship04l	2526*	6380	1	84	13		
LeHavre	11730	15133**	1	7	33800.67										

4   Experimental Evaluation

We report on two sets of experiments. The first is concerned with the evaluation of the impact of different pivot sampling strategies. The second set is designed to assess how well the different sparse stress models approximate the full stress model, in both absolute terms and in relation to the speed-up achieved.

For the experiments we implemented the sparse stress model, Algorithm 1, as well as different sampling techniques in Java using Oracle SDK 1.8 and the yFiles 2.9 graph library (www.yworks.com). The tests were carried out on a single 64-bit machine with a 3.60 GHz quad-core Intel Core i7-4790 CPU, 32 GB RAM, running Ubuntu 14.10. Times were measured using the System.currentTimeMillis() command. The reported running times were averaged over 25 iterations. We note here that all drawing algorithms, except stated otherwise, were initialized with a 200 PivotMDS layout [4]. Furthermore, the maximum number of iterations for the full stress algorithm was set to 500. As stress is not resilient against scaling, see Eq. (1), we optimally rescaled each drawing such that it creates the lowest possible stress value [2].

Data: We conducted our experiments on a series of different graphs, see Tab. 1, most of them taken from the sparse matrix collection [8]. We selected these graphs as they differ in their structure and size, and are large enough to compare the results of different techniques. Two of the graphs, *LeHavre* and *commanche*, have predefined edge lengths that were derived from the node coordinates. We did not modify the graphs in any way, except for those that were disconnected. In this case we only kept the largest component.

4.1   Sampling Evaluation

In Sect. 3 we discussed how vital the proper selection of the pivots is for our model. In the optimal case we would sample pivots that are well distributed over the graph, creating regions of equal complexity, and are central in the drawing of their regions. In order to evaluate the impact of different sampling strategies on the quality of our sparse stress model and recommend a proper sampling scheme, we compared a set of different strategies:

- random: nodes are selected uniformly at random
- MIS filtration: nodes are sampled according to the maximal independent set filtration algorithm by Gajer et al. [13]. Once $n \leq k$ the coarsening stops. If $n < k$, unsampled nodes from the previous level are randomly added
- max/min euclidean: starting with a uniformly randomly chosen node, $\mathcal{P}$ is extended by adding $\arg\max_{i \in V \setminus \mathcal{P}} \min_{p \in \mathcal{P}} \|x_i - x_p\|$
- max/min sp: similar to max/min euclidean except that $\mathcal{P}$ is extended according $\arg\max_{i \in V \setminus \mathcal{P}} \min_{p \in \mathcal{P}} d_{ip}$ [4]

Pretests showed that the max/min sp strategy initially favors sampling leaves, but nevertheless produces good results for large k. Thus, we also evaluated strategies building on this idea, yet try to overcome the problem of leaf node sampling.

- max/min random sp: similar to max/min sp, yet each node i is sampled with a probability proportional to $\min_{p \in \mathcal{P}} d_{ip}$
- k-means layout: the nodes are selected via a k-means algorithm, running at most 50 iterations, on the initial layout
- k-means sp: initially k nodes with max/min sp are sampled succeeded by k-means sampling using the shortest path entries of these pivots
- k-means + max/min sp: $\mathcal{P}$ is initialized with $k/2$ pivots via k-means layout and the remaining nodes are sampled via max/min sp

To quantify how well suited each of the sampling techniques is for our model, we ran each combination on each graph with $k \in \{50, 51, \ldots, 200\}$ pivots. For all tests the sparse stress algorithm terminated after 200 iterations. Since all techniques at some point rely on a random decision, we repeated each execution 20 times in order to ensure we do not rest our results upon outliers. To distinguish the applicability of the different techniques to our model, we used two measures. The first measure is the normalized stress, which is the stress value divided by $\binom{n}{2}$. While the normalized stress measures the quality of our drawing, we also calculated the Procrustes statistic, which measures how well the layout matches the full stress drawing [31]. The range of the Procrustes statistic is $[0, 1]$, where 0 is the optimal match.

The results of these experiments for some of the instances are presented in Figs. 1 and 2 (see the Appendix in [29] for the full set of data). In these plots each dot represents the median and each line starts at the 25%, 75% percentile and ends at the 5%, 95% percentile, respectively. For the sake of readability we binned each 25 consecutive sample sizes. Furthermore, the strategies were ordered according to their overall ranking w.r.t. the evaluated measure. For most of the graphs using k-means sp sampling yields the layouts with the lowest normalized stress value. There are only two graphs where this strategy performs worse than other tested strategies. The one graph where k-means sp is outclassed, yet only for large k by max/min sp, is *pesa*. The reason for this result is that k-means sp mainly samples pivots in the center of the left arm, see Table 4, creating twists. Max/min sp for small k in contrast mostly samples nodes on the contour of the arm, yet once k reaches a certain threshold the resulting distribution of the pivots prevents twists, yielding a lower normalized stress value.

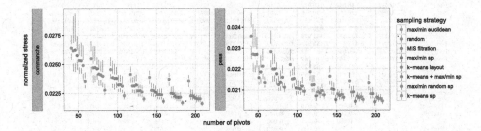

Fig. 1. Comparison of different sampling strategies and number of pivots w.r.t. the resulting normalized stress value

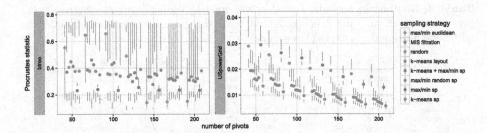

Fig. 2. Comparison of different sampling strategies and number of pivots w.r.t. the resulting Procrustes statistic

The explanation of the poor behavior for *lpship04l* is strongly related to its structure. The low diameter of 13 causes, after a few iterations, the max/min sp strategy to repeatedly sample nodes that are part of the same cluster, see Table 4, and consequently are structurally very similar. As k-means sp builds on max/min sp, it can only slightly improve the pivot distribution. The argument that the problem is related to the structure is reinforced by the outcome of the random strategy. Still, except for these two graphs k-means sp generates the best outcomes, and since this strategy is also strongly favorable over the others subject to the Procrustes statistics, see Fig. 2, our following evaluation always relies on this sampling strategy. However, we note that the Procrustes statistic for *btree* and *lpship04l* are by magnitudes larger than for any other tested graph. While for *lpship04l* this is mostly caused by the quality of the drawings, this is only partly true for *btree*. The other factor contributing to the high Procrustes statistic for *btree* is caused by the restricted set of operations provided by the Procrustes analysis. As dilation, translation, and rotation are used to find the best match between two layouts, the Procrustes analysis cannot

Table 2. Stress and Procrustes statistics: sparse model values are highlighted when no larger than minimum over previous methods

graph	full stress	sparse 200	sparse 100	sparse 50	maxent	MARS 200	MARS 100	GRIP	1-stress	PivotMDS
					stress					
dwt1005	10 729	10 940	11 081	11 329	21 623	17 660	20 134	52 517	12 495	14 459
1138bus	39 974	40 797	41 471	42 686	44 650	64 363	63 614	54 986	73 512	75 427
plat1919	18 572	18 840	19 072	19 719	23 850	53 246	64 166	113 765	75 973	82 865
3elt	422 940	426 564	430 200	437 051	585 967	503 600	754 134	934 206	555 934	634 401
USpowerGrid	702 055	720 642	731 187	749 464	1 021 457	766 535	783 888	1 495 373	1 111 216	1 123 698
commanche	654 694	677 220	699 890	749 609	1 507 654	2 761 605	3 145 489	1 539 767	2 085 818	2 157 943
LeHavre	439 188	433 030	441 986	454 785	1 231 283	12 012 307	12 570 692	8 658 371	1 255 474	1 305 577
pesa	1 373 514	1 417 449	1 452 975	1 495 512	10 423 779	3 563 772	8 281 116	2 957 738	3 486 176	3 325 889
bodyy5	3 547 659	3 566 636	3 585 087	3 630 380	5 248 755	6 385 559	4 072 905	10 389 846	4 245 006	4 715 728
finance256	6 175 210	6 415 761	6 474 787	6 582 890	8 151 335	7 267 598	8 643 239	19 817 355	12 257 268	11 380 089
btree	60 206	61 839	63 325	66 122	67 871	103 436	100 767	96 235	157 988	164 329
qh882	84 524	86 345	87 695	89 556	103 601	117 195	161 113	127 914	146 935	143 142
lpship04l	250 599	297 547	316 674	343 694	329 255	558 923	542 667	771 284	775 813	793 238
					Procrustes statistic					
dwt1005		0.001	0.005	0.003	0.027	0.008	0.018	0.263	0.004	0.008
1138bus		0.009	0.016	0.025	0.022	0.148	0.145	0.071	0.097	0.102
plat1919		0.000	0.000	0.001	0.015	0.026	0.031	0.236	0.045	0.051
3elt		0.001	0.001	0.002	0.026	0.009	0.029	0.199	0.017	0.023
USpowerGrid		0.006	0.008	0.012	0.068	0.014	0.018	0.256	0.051	0.051
commanche		0.001	0.002	0.005	0.039	0.026	0.167	0.092	0.066	0.066
LeHavre		0.001	0.001	0.001	0.012	0.163	0.173	0.256	0.010	0.010
pesa		0.009	0.010	0.010	0.095	0.025	0.070	0.017	0.021	0.021
bodyy5		0.000	0.000	0.000	0.012	0.011	0.003	0.100	0.004	0.007
finance256		0.009	0.006	0.005	0.013	0.007	0.018	0.206	0.042	0.041
btree		0.748	0.165	0.241	0.233	0.360	0.367	0.386	0.361	0.364
qh882		0.015	0.015	0.021	0.046	0.061	0.114	0.075	0.086	0.079
lpship04l		0.176	0.112	0.148	0.160	0.246	0.587	0.463	0.393	0.401

Table 3. Runtime in seconds: fastest sparse model yielding lower stress than best previous method, c.f. Table 2, is highlighted. Marked implementations written in C/C++ with time measured via `clock()` command

graph	full stress	sparse 200	sparse 100	sparse 50	maxent*	MARS 200*	MARS 100*	GRIP*	1-stress	PivotMDS
dwt1005	1.26	0.22	0.14	0.10	0.47	1.02	2.36	0.06	0.08	0.06
1138bus	2.20	0.25	0.14	0.09	0.91	3.16	1.96	0.20	0.06	0.04
plat1919	9.70	0.74	0.51	0.42	1.15	6.80	4.79	0.19	0.31	0.20
3elt	31.82	0.94	0.59	0.46	2.26	16.31	8.43	0.71	0.37	0.23
USpowerGrid	36.48	0.81	0.48	0.35	2.53	13.54	7.62	1.67	0.28	0.21
commanche	340.10	4.89	2.56	1.47	3.60	22.72	12.43	2.29	0.47	0.35
LeHavre	475.05	6.53	3.48	2.37	6.31	27.57	19.50	10.18	0.81	0.54
pesa	373.23	4.25	2.47	1.53	5.96	50.10	42.68	3.56	0.95	0.60
bodyy5	463.47	3.84	2.63	1.78	9.97	46.63	9.27	10.43	1.64	1.04
finance256	1016.92	7.32	4.41	3.09	14.76	32.16	24.66	12.12	2.51	1.60
btree	7.79	0.38	0.15	0.10	0.63	2.70	1.48	0.06	0.06	0.03
qh882	6.61	0.45	0.30	0.23	0.97	8.45	5.79	0.15	0.17	0.14
lpship04l	18.30	0.55	0.30	0.20	0.99	7.06	7.63	0.16	0.15	0.10

resolve reflections. Therefore, if in the one layout of *btree*, the subtree T_1 of v is drawn to the right of subtree T_2 of v and vice versa in the second drawing, although the two layouts are identical, the statistic will be high. This symmetry problem mainly explains the low performance w.r.t. *btree*.

4.2   Full Stress Layout Approximation

The next set of experiments is designed to assess how well our sparse stress model using k-means sp sampling, as well as related sparse stress techniques, resembles the full stress model. For this we compared the median stress layout over 25 repetitions on the same graph of our sparse stress model with $k \in \{50, 100, 200\}$, with MARS,[1] maxent,[2] PivotMDS, 1-stress, and the weighted version of GRIP.[3]

---

[1] https://github.com/marckhoury/mars.
[2] We are grateful to Yifan Hu for providing us with the code.
[3] http://www.cs.arizona.edu/~kobourov/GRIP/.

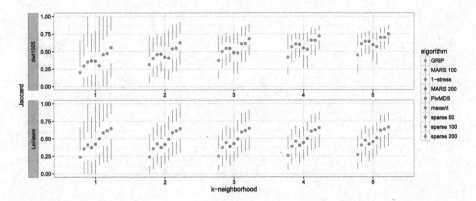

Fig. 3. The similarity of the Gabriel Graph of the full stress layout and the Gabriel Graph of the layout algorithms under consideration as a function of k. For each node of the graph the k-neighborhood in the Gabriel Graph of the full stress layout and the layout algorithm are compared by calculating the Jaccard coefficient. A higher value indicates that the nodes share a high percentage of common neighbors in the different Gabriel Graphs.

The number of iterations of our model as well as for MARS and 1-stress have been limited to 200. Furthermore, we tested MARS with 100 and 200 pivots and report the layout with the smallest stress from the drawings obtained by running `mars` with argument `-p` $\in \{1, 2\}$ combined with a PivotMDS or randomly initialized layout.

Besides comparing the resulting stress values and Procrustes statistics, we compared the distribution of pairwise euclidean distances subject to their graph-theoretic distances. Since the Procrustes statistic has problems with symmetries, as we pointed out in the previous subsection, we propose to evaluate the similarity of the sparse stress layouts with the full stress layout via Gabriel graphs [12]. The Gabriel graph of a given layout x contains an edge between a pair of points if and only if the disc associated with the diameter of the endpoints does not contain any other point. Since the treatment of identical positions is not defined for Gabriel Graphs, we resolve this by adding edges between each pair of identical positions. We assess the similarity between the Gabriel Graph of the full stress layout and the sparse stress layouts by comparing the k-neighborhoods of a node in the graphs using the Jaccard coefficient.

A further measure we introduce evaluates the visual error. More precisely we measure for a given node v the percentage of nodes that lie in the drawing area of the k-neighborhood, but are not part of it. We calculate this value by computing the convex hull induced by the k-neighborhood and then test for each other node if it belongs to the hull or not. This number is then divided by $n - |\{w \in V | d_{vw} \leq k\}|$. Therefore, a low value implies that there are only a few nodes lying in the region, while high values indicate we cannot distinguish non k-neighborhood and k-neighborhood nodes in the drawing. This measure is to a certain extend similar to the precision of neighborhood preservation [15].

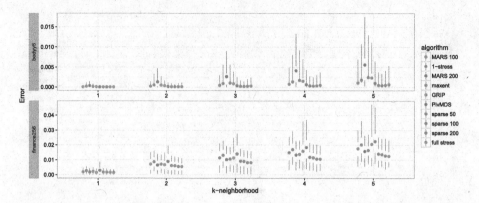

Fig. 4. Error charts as a function of k. For each node of the graph the convex hull w.r.t. the coordinates of the nodes in the k-neighborhood is computed. For each of the convex hulls the error is calculated by counting the number of non k-neighborhood nodes that lie inside or on the contour of this hull divided by their total number.

The results of all these experiments, see Tables 2 and 4, Figs. 3 and 4, and the Appendix in [29], reveal that our model is more adequate in resembling the full stress drawing than any other of the tested algorithm, while showing comparable running times that scale nicely with k, cf. Table 3. The error plots in Table 4 expose the strength of our approximation scheme. We can see that, while all approaches work very well in representing short distances, our approach is more precise in approximating middle and especially long distances, explaining our good results. As the evaluation clearly shows that our approach yields better approximations of the full stress model, we rather want to discuss the low performance of our model for *lpship04l* and thereby expose one weakness of our approach.

Looking at the sparse 50 drawing of *lpship04l* in Table 4, we can see that a large portion of nodes share a similar or even the same position. This is because *lpship04l* has a lot of nodes that share very similar graph-theoretic distance vectors, exhibit highly overlapping neighborhoods, and are drawn in close proximity in the initial PivotMDS layout. While our model would rely on small variations of the graph-theoretic distances to create a good drawing we diminish these differences even further by restricting our model to $\mathcal{P}$. Consequently, the positional vote for two similar non-pivot nodes i and j that lie in the same partition will only slightly differ, mainly caused by their distinct neighbors. However, as these neighbors are also in close proximity in the initial drawing of *lpship04l* the distance between i and j will not increase. Therefore, if the graph has a lot of structurally very similar nodes and the initial layout has poor quality, our approach will inevitably create drawings where nodes are placed very close to one another.

Table 4. Layouts and error charts of the algorithms. Each chart shows the zero y coordinate (black horizontal line), the median (red line), the 25 and 75 percentiles (black/gray ribbon) and the min/max error (outer black dashed line). The error (y-axis) is the difference between the euclidean distance and the graph-theoretic distance (x-axis). 1000 bins have been used for weighted graphs

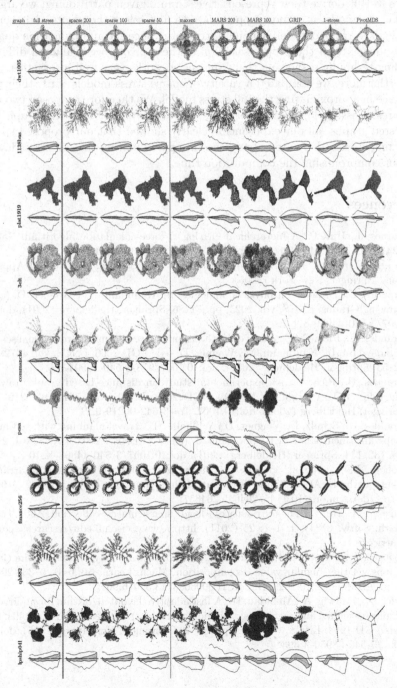

5   Conclusion

In this paper we proposed a sparse stress model that requires $\mathcal{O}(kn + m)$ space and time per iteration, and a preprocessing time of $\mathcal{O}(k(m + n \log n))$. While Barnes & Hut derive their representatives from a given partitioning, we argued that for our model it is more appropriate to first select the pivots and then to partition the graph only relying on its structure. Since the approximation quality heavily depends on the proper selection of these pivots, we evaluated different sampling techniques, showing that k-means sp works very well in practice.

Furthermore, we compared a variety of sparse stress models w.r.t. their performance in approximating the full stress model. We therefore proposed two new measures to assemble the similarity between two layouts of the same graph. For the tested graphs, all our experiments clearly showed that our proposed sparse stress model exceeds related approaches in approximating the full stress layout without compromising the computation time.

References

1. Barnes, J., Hut, P.: A hierarchical O($n \log n$) force-calculation algorithm. Nature **324**(6096), 446–449 (1986). http://dx.doi.org/10.1038/324446a0
2. Borg, I., Groenen, P.J.: Modern Multidimensional Scaling: Theory and Applications. Springer, New York (2005)
3. Brandes, U.: Drawing on physical analogies. In: Kaufmann, M., Wagner, D. (eds.) Drawing Graphs. LNCS, vol. 2025, pp. 71–86. Springer, Heidelberg (2001). doi:10. 1007/3-540-44969-8_4
4. Brandes, U., Pich, C.: Eigensolver methods for progressive multidimensional scaling of large data. In: Kaufmann, M., Wagner, D. (eds.) GD 2006. LNCS, vol. 4372, pp. 42–53. Springer, Heidelberg (2007). doi:10.1007/978-3-540-70904-6_6
5. Brandes, U., Pich, C.: An experimental study on distance-based graph drawing. In: Tollis, I.G., Patrignani, M. (eds.) GD 2008. LNCS, vol. 5417, pp. 218–229. Springer, Heidelberg (2009). doi:10.1007/978-3-642-00219-9_21
6. Brandes, U., Schulz, F., Wagner, D., Willhalm, T.: Travel planning with self-made maps. In: Buchsbaum, A.L., Snoeyink, J. (eds.) ALENEX 2001. LNCS, vol. 2153, pp. 132–144. Springer, Heidelberg (2001). doi:10.1007/3-540-44808-X_10
7. Cohen, J.D.: Drawing graphs to convey proximity: an incremental arrangement method. ACM Trans. Comput. Hum. Interact. **4**(3), 197–229 (1997). http://doi.acm.org/10.1145/264645.264657
8. Davis, T.A., Hu, Y.: The university of Florida sparse matrix collection. ACM Trans. Math. Softw. **38**(1), 1: 1–1: 25 (2011). http://www.cise.ufl.edu/research/sparse/matrices
9. Drineas, P., Frieze, A.M., Kannan, R., Vempala, S., Vinay, V.: Clustering large graphs via the singular value decomposition. Mach. Learn. **56**(1–3), 9–33 (2004). http://dx.doi.org/10.1023/B:MACH.0000033113.59016.96
10. Frick, A., Ludwig, A., Mehldau, H.: A fast adaptive layout algorithm for undirected graphs (extended abstract and system demonstration). In: Tamassia, R., Tollis, I.G. (eds.) GD 1994. LNCS, vol. 894, pp. 388–403. Springer, Heidelberg (1995). doi:10. 1007/3-540-58950-3_393

11. Fruchterman, T.M.J., Reingold, E.M.: Graph drawing by force-directed placement. Softw. Pract. Exp. **21**(11), 1129–1164 (1991). http://dx.doi.org/10.1002/spe. 4380211102

12. Gabriel, R.K., Sokal, R.R.: A new statistical approach to geographic variation analysis. Syst. Zool. **18**(3), 259–278 (1969)

13. Gajer, P., Goodrich, M.T., Kobourov, S.G.: A multi-dimensional approach to force-directed layouts of large graphs. In: Marks, J. (ed.) GD 2000. LNCS, vol. 1984, pp. 211–221. Springer, Heidelberg (2001). doi:10.1007/3-540-44541-2_20

14. Gansner, E.R., Hu, Y., Krishnan, S.: COAST: a convex optimization approach to stress-based embedding. In: Wismath, S., Wolff, A. (eds.) GD 2013. LNCS, vol. 8242, pp. 268–279. Springer, Heidelberg (2013). doi:10.1007/978-3-319-03841-4_24

15. Gansner, E.R., Hu, Y., North, S.C.: A maxent-stress model for graph layout. IEEE Trans. Vis. Comput. Graph. **19**(6), 927–940 (2013)

16. Gansner, E.R., Koren, Y., North, S.: Graph drawing by stress majorization. In: Pach, J. (ed.) GD 2004. LNCS, vol. 3383, pp. 239–250. Springer, Heidelberg (2005). doi:10.1007/978-3-540-31843-9_25

17. Greengard, L.: The Rapid evaluation of potential fields in particle systems. ACM distinguished dissertations, MIT Press, Cambridge (1988). http://opac.inria.fr/ record=b1086802

18. Hachul, S., Jünger, M.: Drawing large graphs with a potential-field-based multilevel algorithm. In: Pach, J. (ed.) GD 2004. LNCS, vol. 3383, pp. 285–295. Springer, Heidelberg (2005). doi:10.1007/978-3-540-31843-9_29

19. Hall, K.M.: An r-dimensional quadratic placement algorithm. Manag. Sci. **17**(3), 219–229 (1970). http://dx.doi.org/10.1287/mnsc.17.3.219

20. Hu, Y., Shi, L.: Visualizing large graphs. Wiley Interdiscip. Rev. Comput. Stat. **7**(2), 115–136 (2015). http://dx.doi.org/10.1002/wics.1343

21. Hu, Y.F.: Efficient and high quality force-directed graph drawing. Mathematica J. **10**, 37–71 (2005).http://www.mathematica-journal.com/issue/v10i1/contents/ graph_draw/graph_draw.pdf

22. Ingram, S., Munzner, T.: Glint: An MDS framework for costly distance functions. In: Kerren, A., Seipel, S. (eds.) Proceedings of SIGRAD 2012, Interactive Visual Analysis of Data. Linköping Electronic Conference Proceedings, vol. 81, pp. 29–38. Linköping University Electronic Press (2012). http://www.ep.liu.se/ecp_article/ index.en.aspx?issue=081;article=005

23. Khoury, M., Hu, Y., Krishnan, S., Scheidegger, C.E.: Drawing large graphs by low-rank stress majorization. Comput. Graph. Forum **31**(3), 975–984 (2012). http://dx.doi.org/10.1111/j.1467-8659.2012.03090.x

24. Klimenta, M., Brandes, U.: Graph drawing by classical multidimensional scaling: new perspectives. In: Didimo, W., Patrignani, M. (eds.) GD 2012. LNCS, vol. 704, pp. 55–66. Springer, Heidelberg (2013). doi:10.1007/978-3-642-36763-2_6

25. Kobourov, S.G.: Force-directed drawing algorithms. In: Tamassia, R. (ed.) Handbook of Graph Drawing and Visualization, pp. 383–408. CRC Press, Boca Raton (2013)

26. Koren, Y., Carmel, L., Harel, D.: ACE: A fast multiscale eigenvectors computation for drawing huge graphs. In: Wong, P.C., Andrews, K. (eds.) InfoVis 2002, pp. 137–144. IEEE Computer Society (2002). http://dx.doi.org/10.1109/INFVIS.2002. 1173159

27. McGee, V.E.: The multidimensional analysis of "elastic" distances. Br. J. Math. Stat. Psychol. **19**(2), 181–196 (1966). http://dx.doi.org/10.1111/j.2044-8317.1966. tb00367.x

28. Meyerhenke, H., Nöllenburg, M., Schulz, C.: Drawing large graphs by multilevel maxent-stress optimization. In: Di Giacomo, E., Lubiw, A. (eds.) GD 2015. LNCS, vol. 9411, pp. 30–43. Springer, Heidelberg (2015). doi:10.1007/978-3-319-27261-0_3

29. Ortmann, M., Klimenta, M., Brandes, U.: A sparse stress model. CoRR abs/1608.08909 (2016). http://arxiv.org/abs/1608.08909

30. Quigley, A.J.: Large scale relational information visualization, clustering, and abstraction. Ph.D. thesis, University of Newcastle (2000)

31. Sibson, R.: Studies in the robustness of multidimensional scaling: procrustes statistics. J. Roy. Stat. Soc. Ser. B (Methodol.) 40(2), 234–238 (1978). http://www.jstor.org/stable/2984761

32. de Silva, V., Tenenbaum, J.B.: Global versus local methods in nonlinear dimensionality reduction. In: Becker, S., Thrun, S., Obermayer, K. (eds.) NIPS 2002, pp. 705–712. MIT Press, Cambridge (2002). http://papers.nips.cc/paper/2141-global-versus-local-methods-in-nonlinear-dimensionality-reduction

33. Tunkelang, D.: JIGGLE: java interactive graph layout environment. In: Whitesides, S.H. (ed.) GD 1998. LNCS, vol. 1547, pp. 413–422. Springer, Heidelberg (1998). doi:10.1007/3-540-37623-2_33

34. Tunkelang, D.: A numerical optimization approach to general graph drawing. Ph.D. thesis, Carnegie Mellon University (1999)

35. Walshaw, C.: A multilevel algorithm for force-directed graph-drawing. J. Graph Algorithms Appl. 7(3), 253–285 (2003). http://www.cs.brown.edu/publications/jgaa/accepted/2003/Walshaw2003.7.3.pdf

Node Overlap Removal by Growing a Tree

Lev Nachmanson[1], Arlind Nocaj[2], Sergey Bereg[3(✉)], Leishi Zhang[4],
and Alexander Holroyd[1]

[1] Microsoft Research, Redmond, USA
{levnach,holroyd}@microsoft.com
[2] University of Konstanz, Konstanz, Germany
arlind.nocaj@uni-konstanz.de
[3] The University of Texas at Dallas, Richardson, USA
besp@utdallas.edu
[4] Middlesex University, London, UK
L.X.Zhang@mdx.ac.uk

Abstract. Node overlap removal is a necessary step in many scenarios including laying out a graph, or visualizing a tag cloud. Our contribution is a new overlap removal algorithm that iteratively builds a Minimum Spanning Tree on a Delaunay triangulation of the node centers and removes the node overlaps by "growing" the tree. The algorithm is simple to implement yet produces high quality layouts. According to our experiments it runs several times faster than the current state-of-the-art methods.

1 Introduction

Removing node overlap after laying out a graph is a common task in network visualization. Most graph layout algorithms [23] consider nodes as points that do not occupy any geometrical space. In practice, nodes often have shapes, labels, and so on. These shapes and labels may overlap in the visualization and affect the visual readability. To remove such overlaps a specialized algorithm is usually applied.

The main contribution of this paper is a new node overlap removal algorithm that we call Growing Tree, or GTree further on. The basic idea is to first capture most of the overlap and the local structure with a specific spanning tree on top of a proximity graph, and then resolve the overlap by letting the tree "grow".

We compare GTree with PRISM [6], which is widely used for the same purpose. Needing more area than PRISM, our method preserves the original layout well and is up to eight times faster than PRISM. To compare the two algorithms we implemented GTree in the open source graph visualization software Graphviz [1], where PRISM is the default overlap removal algorithm. On the other side, GTree is the default in MSAGL[2], where we also have an implementation of PRISM. We ran comparisons by using both tools.

---

[1] http://www.graphviz.org/.
[2] https://github.com/Microsoft/automatic-graph-layout.

© Springer International Publishing AG 2016
Y. Hu and M. Nöllenburg (Eds.): GD 2016, LNCS 9801, pp. 33–43, 2016.
DOI: 10.1007/978-3-319-50106-2_3

2 Related Work

There is vast research on node overlap removal. Some methods, including hierarchical layouts [4], incorporate the overlap removal with the layout step. Likewise, force-directed methods [5] have been extended to take the node sizes into account [16,17,24], but it is difficult to guarantee overlap-free layouts without increasing the repulsive forces extensively. Dwyer et al. [2] show how to avoid node overlaps with Stress Majorization [7]. The method can remove node overlaps during the layout step, but it needs an initial state that is overlap free; sometimes such a state is not given.

Another approach, which we also choose, is to use a post-processing step. In Cluster Busting [8,18] the nodes are iteratively moved towards the centers of their Voronoi cells. The process has the disadvantage of distributing the nodes uniformly in a given bounding box.

Imamichi et al. [15] approximate the node shapes by circles and minimize a function penalizing the circle overlaps.

Starting from the center of a node, RWorldle [22] removes the overlaps by discovering the free space around a node by using a spiral curve and then utilizing this space. The approach requires a large number of intersection queries that are time consuming. This idea is extended by Strobelt et al. [21] to discover available space by scanning the plane with a line or a circle.

Another set of node overlap removal algorithms focus on the idea of defining pairwise node constraints and translating the nodes to satisfy the constraints [11,13,19,20]. These methods consider horizontal and vertical problems separately, which often leads to a distorted aspect ratio [6]. A Force-transfer-algorithm is introduced by Huang et al. [14]; horizontal and vertical scans of overlapped nodes create forces moving nodes vertically and horizontally; the algorithm takes $\mathcal{O}(n^2)$ steps, where n is the number of the nodes. Gomez et al. [9] develop Mixed Integer Optimization for Layout Arrangement to remove overlaps in a set of rectangles. The paper discusses the quality of the layout, which seems to be high, but not the effectiveness of the method, which relies on a mixed integer problem solver. Dwyer et al. [3] reduce the overlap removal to a quadratic problem and solve it efficiently in $\mathcal{O}(n \log n)$ steps. According to Gansner and Hu [6], the quality and the speed of the method of Dwyer et al. [3] is very similar to the ones of PRISM.

The ProjSnippet method [10] generates good quality layouts. The method requires $\mathcal{O}(n^2)$ amount of memory, at least if applied directly as described in the paper, and the usage of a nonlinear problem solver.

In PRISM [6,12], a Delaunay triangulation on the node centers is used as the starting point of an iterative step. Then a stress model for node overlap removal is built on the edges of the triangulation and the stress function of the model is minimized. GTree also starts with building this Delaunay triangulation, but then the algorithms diverge.

3   GTree Algorithm

An input to GTree is a set of nodes V, where each node $i \in V$ is represented by an axis-aligned rectangle B_i with the center p_i. We assume that for different $i, j \in V$ the centers p_i, p_j are different too. If this is not the case, we randomly shift the nodes by tiny offsets. We denote by D a Delaunay triangulation of the set $\{p_i : i \in V\}$, and let E be the set of edges of D.

On a high level, our method proceeds as follows. First we calculate the triangulation D, then we define a cost function on E and build a minimum cost spanning tree on D for this cost function. Finally, we let the tree "grow". The steps are repeated until there are no more overlaps. The last several steps are slightly modified. Now we explain the algorithm in more detail.

We define the cost function c on E in such a way that the larger the overlap on an edge becomes, the smaller the cost of this edge comes to be. Let $(i, j) \in E$. If the rectangles B_i and B_j do not overlap then $c(i, j) = dist(B_i, B_j)$, that is the distance between B_i and B_j. Otherwise, for a real number t let us denote by $B_j(t)$ a rectangle with the same dimensions as B_j and with the same orientation, but with the center at $p_i + t(p_j - p_i)$. We find $t_{ij} > 1$ such that the rectangles B_i and $B_j(t_{ij})$ touch each other. Let $s = \|p_j - p_i\|$, where $\|\|$ denotes the Euclidean norm. We set $c(i, j) = -(t_{ij} - 1)s$. See Fig. 1 for an illustration.

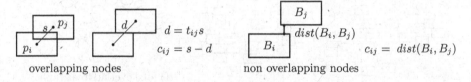

overlapping nodes   non overlapping nodes

Fig. 1. Cost function c_{ij} for edges of the Delaunay triangulation. For overlapping nodes $-c_{ij}$ is equal to the minimal distance that is necessary to shift the boxes along the edge direction so they touch each other.

Having the cost function ready, we compute a minimum spanning tree T on D. Remember that it is a tree with the set of vertices V for which the cost, $\sum_{e \in E'} c(e)$, is minimal, where E' is the set of edges of the tree. We use Prim's algorithm to find T.

The algorithm proceeds by growing T, similar to the growth of a tree in nature. It starts from the root of T. For each child of the root overlapping with the root, it extends the edge connecting the root and the child to remove the overlap. To achieve this, it keeps the root fixed but translates the sub-tree of the child. The edges between the root and other children remain unchanged. The algorithm recursively processes the children of the root in the same manner. This process is described in Algorithm 1.

The number t_{ij} in line 5 of Algorithm 1 is the same as in the definition of the cost of the edge (i, j) when B_i and B_j overlap, and is 1 otherwise.

Algorithm 1: Growing T

Input: Current center positions p and root r
Output: New center positions p'

1   $p'_r = p_r$
2   GrowAtNode (r)
3   **function** GrowAtNode (i)
4   **foreach** $j \in Children(i)$ **do**
5          $p'_j = p'_i + t_{ij}(p_j - p_i)$
6   GrowAtNode (j);

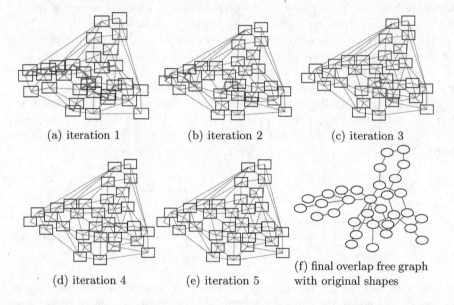

(a) iteration 1   (b) iteration 2   (c) iteration 3

(d) iteration 4   (e) iteration 5   (f) final overlap free graph with original shapes

Fig. 2. Overlap removal process with the minimum spanning tree on the proximity graph, where the latter here is a Delaunay triangulation on rectangle centers. The blue edges form a tree; there are four different trees in the figure. The tree edges connecting overlapped nodes are thick and solid. In each iteration the thick edges are elongated and the dashed tree edges shift accordingly. Overlap is completely resolved in four iterations.

The algorithm does not update all positions for the child sub-tree nodes immediately, but updates only the root of the sub-tree. Using the initial positions of a parent and a child, and the new position of the parent, the algorithm obtains the new position of the child in line 5. In total, Algorithm 1 works in $O(|V|)$ steps. The choice of the root of the tree does not matter. Different roots produce the same results modulo a translation of the plane by a vector. Indeed it can be shown that after applying the algorithm, for any $i, j \in V$ the vector $p'_j - p'_i$ is defined uniquely by the path from i to j in T.

While an overlap along any edge of the triangulation exists, we iterate, starting from finding a Delaunay triangulation, then building a minimum spanning tree on it, and finally running Algorithm 1. See Fig. 2 for an example.

When there are no overlaps on the edges of the triangulation, as noticed by Gansner and Hu [6], overlaps are still possible. We follow the same idea as PRISM and modify the iteration step. In addition to calculating the Delaunay triangulation we run a sweep-line algorithm to find all overlapping node pairs and augment the Delaunay graph D with each such a pair. As a consequence, the resulting minimum spanning tree contains non-Delaunay edges catching the overlaps, and the rest of the overlaps are removed. This stage usually requires much less time than the previous one.

It is possible to create an example where the algorithm will not remove all overlaps. However, such examples are extremely rare and have not been seen yet in practice of using MSAGL or in our experiments. MSAGL applies random tiny changes to the initial layout which prevents GTree from cycling.

4 Comparing PRISM and GTree by Measuring Layout Similarity, Quality, and Run Time

Our data includes the same set of graphs that was used by the authors of PRISM to compare it with other algorithms [6]. The set is available in the Graphviz open source package[3]. We also used a small collection of random graphs and a collection of about 10,000 files[4]. For the experiments we use a modified version of Dot, where we can invoke either GTree or Prism for the overlap removal step, and we also used MSAGL, where we implemented PRISM and GTree. MSAGL was used only to obtain the quality measures. We ran the experiments on a PC with Linux, 64bit and an Intel Core i7-2600K CPU@3.40 GHz with 16 GB RAM.

Some of resulting layouts can be seen in Figs. 3, 5, 6.

One can try to resolve overlap by scaling the node centers of the original layout. If there are no two coincident node centers this will work, but the resulting layout may require a huge area if some centers are close to each other. We consider the *area* of the final layout as one of the quality measures, and usually PRISM produces a smaller area than GTree, see Table 1.

In addition to comparing the areas, we compare some other layout properties. Following Gansner and Hu [6], we look at *edge length dissimilarity*, denoted as σ_{edge}. This measure reflects the relative change of the edge lengths of a Delaunay Triangulation on the node centers of the original layout.

The other measure, which is denoted by σ_{disp}, is the Procrustean similarity [1]. It shows how close the transformation of the original graph is to a combination of a scale, a rotation, and a shift transformation. PRISM and GTree performs similar in the last two measures as Table 1 shows.

---

[3] http://www.graphviz.org.

[4] https://www.dropbox.com/sh/4q0k89yrv4x3ae3/AAA3xyKFRhLyyHXcG9jpcgata?dl=0.

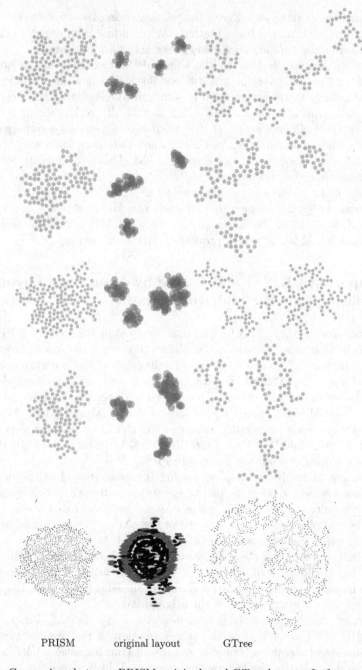

PRISM   original layout   GTree

Fig. 3. Comparison between PRISM, original, and GTree layouts. In four top rows the initial layouts were generated randomly. At the bottom are the drawings of nodes of graph "root" which was initially laid out by the Multi Dimensional Scaling algorithm of MSAGL. In our opinion, the initial structure is more preserved in the right column, containing the results of GTree.

Table 1. Similarity to the initial layout (left) and number of iterations for different graph sizes and different initialization methods (right). PR stands for PRISM

Graph	σ_{edge} PR	GTree	σ_{disp} PR	GTree	area PR	GTree
dpd	0.34	0.28	0.37	0.36	0.82	0.84
unix	0.22	0.19	0.24	0.20	2.38	2.38
rowe	0.29	0.26	0.23	0.24	0.68	0.73
size	0.39	0.37	0.24	0.26	1.09	1.28
ngk10_4	0.30	0.30	0.27	0.30	0.00	0.00
NaN	0.56	0.44	0.73	0.51	4.03	4.34
b124	0.55	0.53	0.97	0.83	5.52	6.22
b143	0.67	0.70	1.12	0.93	3.62	3.88
mode	0.54	0.50	0.59	0.53	1.53	2.29
b102	0.71	0.77	1.43	1.27	4.50	6.62
xx	0.75	0.70	1.65	1.42	6.21	9.57
root	1.09	1.19	2.89	2.45	34.58	91.87
badvoro	0.88	0.92	2.27	2.42	25.68	47.43
b100	0.84	0.98	3.08	3.14	20.64	37.38

init. layout: Graph	$|V|$	$|E|$	neato PR	GTree	SFDP PR	GTree
dpd	36	108	4	7	3	6
unix	41	49	3	4	12	5
rowe	43	68	5	4	13	7
size	47	55	7	3	9	5
ngk10_4	50	100	6	3	14	7
NaN	76	121	8	3	24	6
b124	79	281	14	4	30	12
b143	135	366	21	6	37	12
mode	213	269	37	8	11	6
b102	302	611	60	24	113	19
xx	302	611	83	18	50	19
root	1054	1083	95	18	99	22
badvoro	1235	1616	40	20	50	23
b100	1463	5806	80	24	136	28

Table 2. k closest neighbors error, the Multi Dimensional Scaling algorithm of MSAGL was used for the initial layout. PR stands for PRISM.

Graph	$k=8$ PR	GTree	$k=9$ PR	GTree	$k=10$ PR	GTree	$k=11$ PR	GTree	$k=12$ PR	GTree
dpd	7.75	6.06	9.61	7.36	9.5	8	10.14	8.5	9.97	7.64
unix	8.56	7.05	10.51	8.8	10.95	10.02	11.66	10.54	13	11.41
rowe	6.28	8.09	7.09	9.95	7.49	10.49	9.12	11.4	11.05	12.51
size	4.68	6.09	5.47	6.47	6.28	7.57	6.89	8.13	8.26	10.02
ngk10_4	6.76	7.4	7.52	9.26	8.28	11.38	10.72	13.74	11.92	14.66
NaN	11.83	8.95	14.46	11.5	17.32	13.88	19.88	16.37	22.17	19.7
b124	11.03	11.44	13.22	13.56	14.76	15.54	15.91	17.32	18.23	20.04
b143	13.49	12.39	16.31	14.99	19.49	17.93	23.11	21.04	26.53	24.43
mode	16.91	11.46	20.58	13.95	24.68	16.85	29.54	19.92	34.48	22.56
b102	15.99	14.62	19.61	18.78	23.38	22.77	27.28	26.77	32.15	31.45
xx	15.68	15.65	19.01	19.45	23.05	23.37	26.98	27.35	31.29	32.47
root	17.09	15.7	20.89	19.36	25.48	23.3	30.48	27.66	35.74	32.83
badvoro	16.18	15.15	20.16	18.98	24.37	23.28	29.18	28.03	34.29	33.29
b100	18	19.25	22.11	23.65	26.79	28.69	32.03	34.46	37.44	40.5

To distinguish the methods further, we measure the change in the set of k closest neighbors of the nodes. Namely, let $p_1, \ldots, p_n$ be the positions of the node centers, and let k be an integer such that $0 < k \leq n$. Let $I = \{1, \ldots, n\}$ be the set of node indices. For each $i \in I$ we define $N_k(i) \subset I \setminus \{i\}$, such that $|N_k(p, i)| = k$, and for every $j \in I \setminus N_k(p, i)$ and for every $j' \in N_k(p, i)$ holds $\|p_j - p_i\| \geq \|p_{j'} - p_i\|$. In other words, $N_k(p, i)$ represents a set of k closest neighbors of i, excluding i. Let $p'_1, \ldots, p'_n$ be transformed node centers. To see how much the layout is distorted nearby node i, we intersect $N_k(p, i)$ and $N_k(p', i)$. We measure the distortion as $(k - m)^2$, where m is the number of elements in the intersection. One can see that if the node preserves its k closest neighbors then the distortion is zero.

Table 3. Statistics on collection A. Here k-cn stands for k-closest neighbors, and "iters" stands for the number of iterations. Each cell contains the number of graphs for the measure on which the method performed better. We can see that PRISM produced a layout of smaller area than the one of GTree on 8498 graph, against 1579 graphs where GTree required less area. From the other side, GTree gives better results on all other measures. The columns of k-cn and "iters" do not sum to 10077, the number of graphs in A, because some of the results were equal for PRISM and GTree.

Method	k-cn	σ_{edge}	σ_{disp}	area	iters	time
PRISM	3237	4741	4114	8498	46	7
GTree	4088	5336	5963	1579	9986	10070

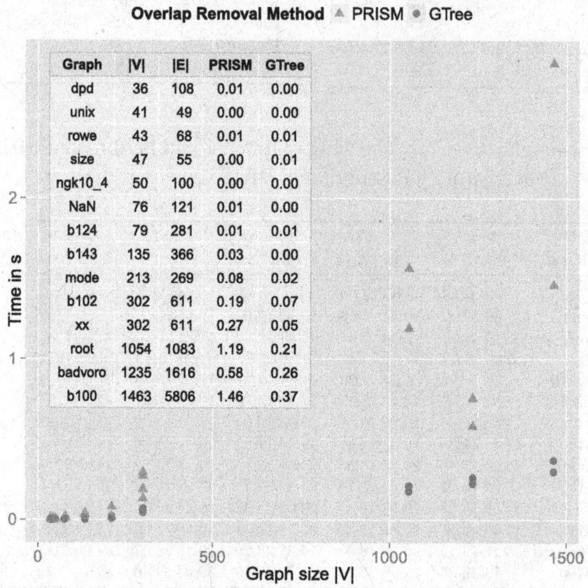

Overlap Removal Method ▲ PRISM ● GTree

Graph	|V|	|E|	PRISM	GTree
dpd	36	108	0.01	0.00
unix	41	49	0.00	0.00
rowe	43	68	0.01	0.01
size	47	55	0.00	0.01
ngk10_4	50	100	0.00	0.00
NaN	76	121	0.01	0.00
b124	79	281	0.01	0.01
b143	135	366	0.03	0.00
mode	213	269	0.08	0.02
b102	302	611	0.19	0.07
xx	302	611	0.27	0.05
root	1054	1083	1.19	0.21
badvoro	1235	1616	0.58	0.26
b100	1463	5806	1.46	0.37

Fig. 4. Runtimes for PRISM and GTree.

Our experiments for k from 8 to 12 show that under this measure GTree produced a smaller error, showing less distortion, on 8 graphs from 14, and on the rest PRISM produced a better result, see Table 2. GTree produced a smaller error on all small random graphs from other collections[5].

We ran tests on the graphs from a subdirectory of the same site called "dot_files", let us call this set of graphs collection A. Each graph from A represents the control flow of a method from a version of the .NET framework. A contains 10077 graphs. The graph sizes do not exceed several thousands. We

---

[5] https://www.dropbox.com/sh/4q0k89yrv4x3ae3/AAA3xyKFRhLyyHXcG9jpcgata ?dl=0.

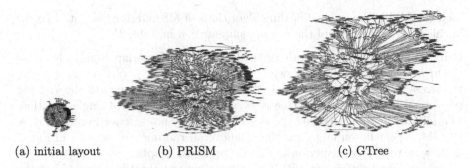

(a) initial layout   (b) PRISM   (c) GTree

Fig. 5. Root graph with 1054 nodes and 1083 edges. (a) initial layout with NEATO, (b) applying PRISM, (c) applying GTree.

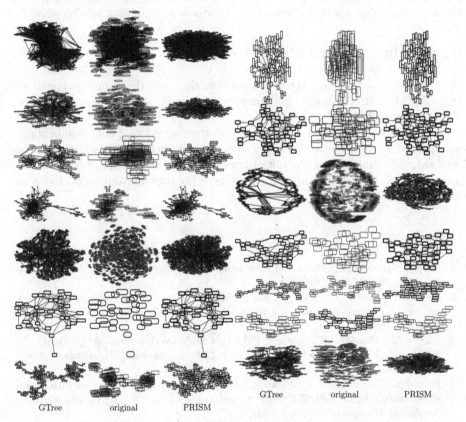

GTree   original   PRISM

GTree   original   PRISM

Fig. 6. Results for GTree and PRISM initialized with SFDP. From top to bottom and left to right: b100, b102, b124, b143, badvoro, dpd, mode, - NaN, ngk10_4, root, rowe, size, unix, and xx. To make the original drawings more readable they have been changed; In most cases the nodes were diminished and the edges removed. The drawings were scaled differently.

used the Multi Dimensional Scaling algorithms of MSAGL for the initial layout in this test. The results of the run are summarized in Table 3.

Runtime Comparison. Both methods remove the overlap iteratively using the proximity graph. However, while PRISM needs $\mathcal{O}(|V| \cdot \sqrt{|V|})$ time to solve the stress model, GTree needs only $\mathcal{O}(|V|)$ time per iteration with the growing tree procedure. Therefore, GTree is asymptotically faster in a single iteration. In addition, as Table 1 (right) shows, GTree usually needs fewer iterations than PRISM, especially on larger graphs. The overall runtime can be seen in Fig. 4. It shows that GTree outperforms PRISM on larger graphs.

In Fig. 5 we experiment with the way we expand the edges. Instead of the formula $p'_j = p'_i + t_{ij}(p_j - p_i)$, which resolves the overlap between the nodes i and j immediately, we use the update $p'_j = p'_i + \min(t_{ij}, 1.5)(p_j - p_i)$. As a result, the algorithm runs a little bit slower but produces layouts with smaller area.

5 Conclusion and Future Work

We proposed a new overlap removal algorithm that uses the minimum spanning tree. The algorithm is simple and easy to implement, and yet it preserves the initial layout well and is efficient.

Although we introduced our approach in the context of graph visualization, our method can also be used for any other purpose where overlap needs to be resolved while maintaining the initial layout. Finding a measure of how well an overlap removal algorithm preserves clusters of the initial layout seems to be an interesting challenge.

References

1. Borg, I., Groenen, P.: Modern Multidimensional Scaling: Theory and Applications. Springer, New York (2005)
2. Dwyer, T., Koren, Y., Marriott, K.: IPSEP-COLA: an incremental procedure for separation constraint layout of graphs. IEEE Trans. Vis. Comput. Graph. **12**(5), 821–828 (2006)
3. Dwyer, T., Marriott, K., Stuckey, P.J.: Fast node overlap removal. In: Healy, P., Nikolov, N.S. (eds.) GD 2006. LNCS, pp. 153–164. Springer, Heidelberg (2006). doi:10.1007/11618058_15
4. Friedrich, C., Schreiber, F.: Flexible layering in hierarchical drawings with nodes of arbitrary size. In: Estivill-Castro, V., (ed.), ACSC, vol. 26, CRPIT, pp. 369–376. Australian Computer Society (2004)
5. Fruchterman, T.M.J., Reingold, E.M.: Graph drawing by force-directed placement. Softw. Pract. Exp. **21**(11), 1129–1164 (1991)
6. Gansner, E.R., Hu, Y.: Efficient, proximity-preserving node overlap removal. J. Graph Algorithms Appl. **14**(1), 53–74 (2010)
7. Gansner, E.R., Koren, Y., North, S.C.: Graph drawing by stress majorization. In: Pach, J. (ed.) GD 2004. LNCS, vol. 3383, pp. 239–250. Springer, Heidelberg (2004)
8. Gansner, E.R., North, S.C.: Improved force-directed layouts. In: Whitesides, S. (ed.) GD 1998. LNCS, vol. 1547, pp. 364–373. Springer, Heidelberg (1998)

9. Gomez-Nieto, E., Casaca, W., Nonato, L.G., Taubin, G.: Mixed integer optimization for layout arrangement. In: 2013 26th SIBGRAPI-Conference on Graphics, Patterns and Images (SIBGRAPI), pp. 115–122. IEEE (2013)

10. Gomez-Nieto, E., San Roman, F., Pagliosa, P., Casaca, W., Helou, E., de Oliveira, M.F., Nonato, L.: Similarity preserving snippet-based visualization of web search results. IEEE Trans. Vis. Comput. Graph. **20**, 457–470 (2013)

11. Hayashi, K., Inoue, M., Masuzawa, T., Fujiwara, H.: A layout adjustment problem for disjoint rectangles preserving orthogonal order. Syst. Comput. Japan **33**(2), 31–42 (2002)

12. Hu, Y.: Visualizing graphs with node and edge labels. CoRR, abs/0911.0626 (2009)

13. Huang, X., Lai, W.: Force-transfer: a new approach to removing overlapping nodes in graph layout. In: Oudshoorn, M.J. (ed.), ACSC, vol. 16, CRPIT, pp. 349–358. Australian Computer Society (2003)

14. Huang, X., Lai, W., Sajeev, A., Gao, J.: A new algorithm for removing node overlapping in graph visualization. Inf. Sci. **177**(14), 2821–2844 (2007)

15. Imamichi, T., Arahori, Y., Gim, J., Hong, S.-H., Nagamochi, H.: Removing node overlaps using multi-sphere scheme. In: Tollis, I.G., Patrignani, M. (eds.) GD 2008. LNCS, vol. 5417, pp. 296–301. Springer, Heidelberg (2008). doi:10.1007/978-3-642-00219-9_28

16. Li, W., Eades, P., Nikolov, N.S.: Using spring algorithms to remove node overlapping. In: Hong, S.-H. (ed.), APVIS, vol. 45, CRPIT, pp. 131–140. Australian Computer Society (2005)

17. Lin, C.-C., Yen, H.-C., Chuang, J.-H.: Drawing graphs with nonuniform nodes using potential fields. J. Vis. Lang. Comput. **20**(6), 385–402 (2009)

18. Lyons, K.A., Meijer, H., Rappaport, D.: Algorithms for cluster busting in anchored graph drawing. J. Graph Algorithms Appl. **2**(1), 1–24 (1998)

19. Marriott, K., Stuckey, P.J., Tam, V., He, W.: Removing node overlapping in graph layout using constrained optimization. Constraints **8**(2), 143–171 (2003)

20. Misue, K., Eades, P., Lai, W., Sugiyama, K.: Layout adjustment and the mental map. J. Vis. Lang. Comput. **6**(2), 183–210 (1995)

21. Strobelt, H., Spicker, M., Stoffel, A., Keim, D., Deussen, O.: Rolled-out wordles: a heuristic method for overlap removal of 2d data representatives. In: Computer Graphics Forum, vol. 31, pp. 1135–1144. Wiley Online Library (2012)

22. Strobelt, H., Spicker, M., Stoffel, A., Keim, D.A., Deussen, O.: Rolled-out wordles: a heuristic method for overlap removal of 2d data representatives. Comput. Graph. Forum **31**(3), 1135–1144 (2012)

23. Tamassia, R.: Handbook of Graph Drawing and Visualization (Discrete Mathematics and Its Applications). Chapman & Hall/CRC, Boca Raton (2007)

24. Wang, X., Miyamoto, I.: Generating customized layouts. In: Brandenburg, F.-J. (ed.) GD 1995. LNCS, vol. 1027, pp. 504–515. Springer, Heidelberg (1995). doi:10.1007/BFb0021835

Placing Arrows in Directed Graph Drawings

Carla Binucci[1]([⊠]), Markus Chimani[2], Walter Didimo[1], Giuseppe Liotta[1],
and Fabrizio Montecchiani[1]

[1] Università degli Studi di Perugia, Perugia, Italy
{carla.binucci,walter.didimo,giuseppe.liotta,
fabrizio.montecchiani}@unipg.it
[2] Osnabrück University, Osnabrück, Germany
markus.chimani@uni-osnabrueck.de

Abstract. We consider the problem of placing arrow heads in directed graph drawings without them overlapping other drawn objects. This gives drawings where edge directions can be deduced unambiguously. We show hardness of the problem, present exact and heuristic algorithms, and report on a practical study.

1 Introduction

The default way of drawing a directed edge is to draw it as a line with an arrow head at its target. While there also exist other models (placing arrows at the middle, drawing edges in a "tapered" fashion, etc.; cf. [7,8]) the former is prevailing in virtually all software systems. However, this simple model becomes problematic when several edges attach to a vertex on a similar trajectory: it may be hard to see whether a specific edge is in- or outgoing, cf. Fig. 1 and [1].

We try to solve this issue by looking for a placement of the arrow heads such that (a) they do not overlap other edges or arrow heads, and (b) still retain the property of being at—or at least close to—the target vertices of the edges. In the following, we show NP-hardness of the problem, propose exact and heuristic algorithms for its discretized variant, and evaluate their practical performance in a brief exploratory study. We remark that our problem is related to map labeling and in particular to edge labeling problems [4,5,9–11,13,15–18].

For space reasons, some proofs and technical details are omitted in this extended abstract, and can be found in the appendix of the ArXiv version [1].

2 The Arrow Placement Problem

We first formally define our arrow placement problem and establish its theoretical time complexity. Let $G = (V, E)$ be a digraph and let Γ be a straight-line drawing of G. We assume that in Γ each vertex $v \in V$ is drawn as a circle

Work is partially supported by the MIUR project AMANDA "Algorithmics for MAssive and Networked DAta", prot. 2012C4E3KT_001.

Y. Hu and M. Nöllenburg (Eds.): GD 2016, LNCS 9801, pp. 44–51, 2016.
DOI: 10.1007/978-3-319-50106-2_4

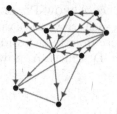

Fig. 1. Layouts of a digraph with 10 vertices and 21 edges. (left) The arrows are placed by a common editor; several arrows overlap and the direction of, e.g., the thick red edge is not clear. (right) The arrows are placed by our exact method. (Color figure online)

(possibly a point) C_v. We also assume that, for each edge $e \in E$, the arrow of e is modeled as a circle C_e of positive radius, centered in a point along the segment that represents e: when Γ is displayed, the arrow of e is drawn as a triangle inscribed in C_e, suitably rotated according to the direction of e. We assume that all circles representing a vertex (arrow) have a common radius r_V (r_E, respectively). We say that two arrows—or an arrow and a vertex—*overlap* if their corresponding circles intersect in two points. An arrow and an edge *overlap* if the segment representing the edge intersects the circle representing the arrow in two points. For the sake of simplicity, we reuse terms of theoretical concepts also for their visual representation: "arrow" and "vertex" also refer to their corresponding circle in Γ; "edge" also refers to its corresponding segment in Γ.

Definition 1. *Let a_e denote the arrow of an edge $e \in E$. A valid position for a_e in Γ is such that: (P1) for every vertex $v \in V$, a_e and v do not overlap; (P2) for every edge $g \in E$, $g \neq e$, a_e and g do not overlap. An assignment of a valid position to each arrow is called a* valid placement *of the arrows, denoted by P_Γ.*

Definition 2. *Given a valid placement P_Γ, the* overlap number *of P_Γ is the number of pairs of overlapping arrows, and is denoted as $\mathrm{ov}(P_\Gamma)$.*

Given a straight-line drawing Γ of a digraph $G = (V, E)$, and constants r_V, r_E, we ask for a valid placement P_Γ of the arrows (if one exists) such that $\mathrm{ov}(P_\Gamma)$ is minimum. This optimization problem is NP-hard; we prove this by showing the hardness of the following decision problem ARROW-PLACEMENT.

Problem: ARROW-PLACEMENT
INSTANCE: $\langle G = (V, E), \Gamma, r_V, r_E \rangle$.
QUESTION: Does there exist a valid placement P_Γ of the arrows with $\mathrm{ov}(P_\Gamma) = 0$?

Theorem 1. *The* ARROW-PLACEMENT *problem is NP-hard.*

The proof of Theorem 1 uses a reduction from PLANAR 3-SAT [12], and is similar to those used in the context of edge labeling [9,13,16,18]. It yields an instance of ARROW-PLACEMENT where the search of a valid placement P_Γ with $\mathrm{ov}(P_\Gamma) = 0$ can be restricted to a finite number of valid positions for each arrow. Hence, ARROW-PLACEMENT remains NP-hard even if we fix a finite set of positions for each arrow, and a valid placement with overlap number zero (if any) may only choose from these positions. As this variant of ARROW-PLACEMENT, which we call DISCRETE-ARROW-PLACEMENT, clearly belongs to NP, it is NP-complete.

3   Algorithms

We describe algorithms for the optimization version of DISCRETE-ARROW-PLACEMENT. We assume that a set of valid positions for each arrow is given, based on $\{\Gamma, r_V, r_E\}$, and look for a valid placement P_Γ that minimizes $\mathrm{ov}(P_\Gamma)$ over this set of positions. We give both an exact algorithm and two variants of a heuristic, which we experimentally compare in Sect. 4. Given an edge $e \in E$, let A_e denote the set of valid positions for the arrow of edge e, and let $A := \bigcup_{e \in E} A_e$ be the set of all valid positions. Our algorithms are based on an *arrow conflict graph* C_A, depending on A, Γ, and r_E. The positions A form the node set of C_A. Two positions are *conflicting*, and connected by an (undirected) edge in C_A, if they correspond to positions of different edges and the arrows would overlap when placed on these positions. Finding a valid placement P_Γ with $\mathrm{ov}(P_\Gamma) = 0$ means to select one element from each A_e such that they form an independent set in C_A. More general, finding a valid placement P_Γ with $\mathrm{ov}(P_\Gamma) = k$ ($k \geq 0$) means to select one element from each A_e such that they induce a subgraph with k edges in C_A. Our exact algorithm minimizes k using an ILP formulation, while our heuristic adopts a greedy strategy. Both techniques try to minimize the distance of each arrow from its target vertex as a secondary objective. However, our algorithms can be easily adapted to privilege other positions (e.g., close to the source vertices, in the middle of the edges, etc.), or to consider bidirected edges.

ILP Formulation. For each position $p_e \in A_e$ of an edge $e = (v, u)$, we have a binary variable x_{p_e}. We define a distance $d(p_e) \in \{1, \ldots, |A_e|\}$, from p_e to u: $d(p_e) = 1$ ($d(p_e) = |A_e|$) means that p_e is the position closest (farthest, respectively) to u. Let $E_A := E(C_A)$ be the pairs of conflicting positions. For every $(p_e, p_g) \in E_A$, we define a binary variable $y_{p_e p_g}$. The total number of variables is $O(|A|^2)$, and we write:

$$\min \sum_{(p_e, p_g) \in E_A} y_{p_e p_g} + \frac{1}{M} \cdot \sum_{e \in E} \sum_{p_e \in A_e} d(p_e) x_{p_e} \tag{1}$$

$$\sum_{p_e \in A_e} x_{p_e} = 1 \qquad\qquad \forall e \in E \tag{2}$$

$$x_{p_e} + x_{p_g} \leq y_{p_e p_g} + 1 \qquad\qquad \forall (p_e, p_g) \in E_A \tag{3}$$

The objective function minimizes the overlap number and, secondly, the sum of the distances of the arrows from their target vertices. To do this, the second term is divided for a sufficiently large constant M. For example, one can set $M = |E| \max_{e \in E} \{|A_e|\}$. Equation 2 guarantee that exactly one valid position per edge is selected. Constraint 3 enforces $y_{p_e p_g} = 1$ if both conflicting positions x_{p_e} and x_{p_g} are chosen. In the following, the exact technique will be referred to as Opt. We remark that optimization problems and ILP formulations similar to above have been given in the context of edge and map labeling [4,5,9,11,15,16].

Heuristics. Our heuristics follow a greedy strategy, again based on C_A. Let $p_e \in A_e \subset V(C_A)$ as above. We initially assigns cost $c(p_e)$ to each position p_e, and then execute $|E|$ iterations. In each iteration, we select a position p_e of minimum cost (over all $e \in E$) and place the arrow of the corresponding edge there; then, we remove all positions A_e from C_A (including p_e), and update the costs of the remaining positions. We define $c(p_e) := \delta(p_e) + \frac{1}{M}d(p_e) + T\sigma_{p_e}$, where: $\delta(p_e)$ is the degree of p_e in C_A (i.e., the number of positions conflicting with p_e); constant M and "distance" $d(p_e)$ are defined as in the ILP; σ_{p_e} is the number of already chosen positions conflicting with p_e (0 in the first iteration); T is equal to the maximum initial cost of a valid position. This cost function guarantees that: (i) positions conflicting with already selected positions are chosen only if necessary; (ii) the algorithm prefers positions with the minimum number of conflicts with the remaining positions and, among them, those closer to the target vertex. Since constructing C_A may be time-consuming in practice (we compare all pairs of valid positions), we also consider using only a subset of the edges of C_A; we may consider only those conflicts arising from positions of adjacent edges in the input graph. In the following, HeurGlobal is the heuristic that considers full C_A, while HeurLocal is the variant based on this simplified version of C_A.

4   Experimental Analysis

Experimental Setting. We use three different sets of graph: PLANAR are biconnected planar digraphs with edge density 1.5–2.5, randomly generated with the OGDF [3]. RANDOM are digraphs generated with uniform probability distribution with edge density 1.4–1.6. Both sets contain 30 instances each; 6 graphs for each number of vertices $n \in \{100, 200, \ldots, 500\}$. We did not generate denser graphs, as they give rise to cluttered drawings with few valid positions for the arrows—there, the arrow placement problem seems less relevant. Finally, NORTH is a popular set of 1,275 real-world digraphs with 10–100 vertices and average density 1.4 [14]. We draw each instance of the three sets with straight-line edges using OGDF's FM3 algorithm [6]. The layouts of the PLANAR may contain edge crossings, as they are generated by a force-directed approach.

Value r_E is chosen as the minimum of (a) 40% of the shortest edge length, (b) 25% of the average edge length, and (c) 10 pixels, but enforced to be at least 3 pixels. We set $r_V := r_E$. For each edge $e = (w, u)$ we compute positions A_e as follows. The i-th position, $i \geq 1$, has its center at distance $r_V + i \cdot r_E$ from target u. We generate positions as long as they have distance at least $r_V + r_E$ from source vertex w. We then remove positions that overlap with edges or vertices in Γ. If no valid positions remain, we choose the one closest to u as e's unique arrow position. Thus, in the final placements there might be some conflicts between an arrow and a vertex or edge of the drawing. We call such conflicts *crossings* and observe that a single invalid position may result in several crossings.

We apply Opt, HeurGlobal, and HeurLocal to each of the drawings. The algorithms are implemented in C# and run on an Intel Core i7-3630QM notebook with 8 GB RAM under Windows 10. For the ILP we use CPLEX 12.6.1 with default settings. For each computation, we measure total running time, overlap number, and number of crossings (due to invalid positions, see above). From the qualitative point of view, we also compare the algorithmic results with a trivial placement, called Editor, which simply places each arrow close to its target vertex, similarly as most graph editors do. We also measure *placement time*, i.e., the time spent by an algorithm to find a placement *after* C_A has been computed.

Results. For PLANAR, the average numbers of positions in C_A range from 640 to 7,150. Figures 2(a) and (b) show that for PLANAR all the algorithms are very applicable, although Opt is of course significantly slower. While the pure placement time for HeurGlobal is not much longer than that of HeurLocal, it suffers from the fact that generating the full C_A constitutes roughly 1/3 of its overall runtime, whereas the generation time of the reduced conflict graph is rather neglectable. On the other hand, Fig. 2(c) shows that HeurGlobal practically coincides with the optimum w.r.t. the number of overlaps (its average gap is below 3%; the worst gap is 6.76%). HeurLocal still gives very good solutions, with gaps about half that of Editor. Figure 2(d) shows that our algorithms reduce the number of invalid positions by 33−77% compared to Editor. The number of

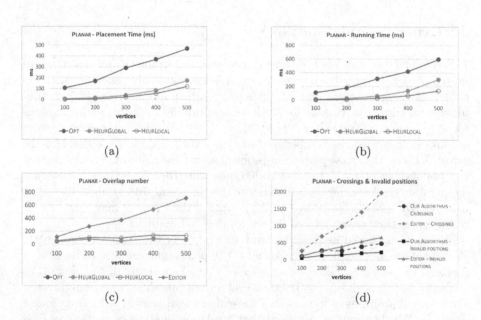

Fig. 2. PLANAR: (a) Placement time; (b) Total running time; (c) Number of overlaps; (d) Number of crossings (edge/vertex with arrow) and of invalid positions.

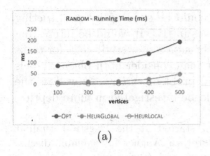

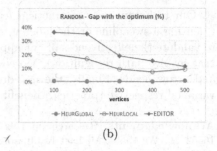

(a)   (b)

Fig. 3. RANDOM: (a) Total running time; (b) Number of overlaps, relative to Opt.

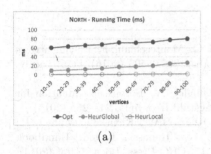

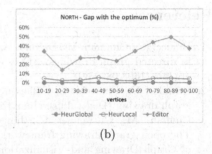

(a)   (b)

Fig. 4. NORTH: (a) Total running time; (b) Number of overlaps, relative to to Opt.

crossings is the same for all our algorithms, as they occur when we cannot find any valid position for arrows during the generation procedure. Figure 2(d) shows that our algorithms cause significantly less crossings than Editor.

For RANDOM, average numbers of positions in C_A range from 640 to 4,377. The general behavior for RANDOM is similar to that of PLANAR but the difference between the running time of Opt and the heuristics is slightly more pronounced (Fig. 3(a)). Again, constructing C_A constitutes roughly 1/3 of HeurGlobal's running time. Still, the quality of HeurGlobal's solutions again essentially coincide with Opt; the other heuristics are now closer than before, see Fig. 3(b).

For NORTH, the average $|V(C_A)|$ range from 62 to 311. We observe the same patterns, see Figs. 4: HeurLocal requires nearly no time, while HeurGlobal is very competitive at just above 20ms for the large graphs (a third of which is the construction of full C_A). Again, Opt always finds a solution very quickly, in fact within roughly 80ms. HeurGlobal again gives essentially optimal solutions, while HeurLocal exhibits 5–10% gaps. Editor requires 30–50% more overlaps than Opt.

5   Conclusions and Future Work

We discussed optimizing arrow head placement in directed graph drawings, to improve readability. As mentioned, this is very related to studies in map and graph labeling, but its specifics seem to make a more focused study worthwhile.

Our techniques are of practical use, and could be sped-up by constructing C_A using a sweepline or the labeling techniques in [17]. It would be interesting to validate the effectiveness of our approach through a user study (e.g. for tasks that involve path recognition). Moreover, one may consider both placing labels and arrow heads. Finally, the non-discretized problem variant, as well as the variants' respective (practical) benefits, should be investigated in more depth.

Acknowledgments. Research on this problem started at the Dagstuhl seminar 15052 [2]. We thank Michael Kaufmann and Dorothea Wagner for valuable discussions, and the anonymous referees for their comments and suggestions.

References

1. Binucci, C., Chimani, M., Didimo, W., Liotta, G., Montecchiani, F.: Placing arrows in directed graph drawings. ArXiv e-prints abs/1608.08505 (2016). http://arxiv.org/abs/1608.08505v1
2. Brandes, U., Finocchi, I., Nöllenburg, M., Quigley, A.: Empirical evaluation for graph drawing (Dagstuhl seminar 15052). Dagstuhl Rep. **5**(1), 243–258 (2015)
3. Chimani, M., Gutwenger, C., Jünger, M., Klau, G.W., Klein, K., Mutzel, P.: The open graph drawing framework (OGDF). In: Tamassia, R. (ed.) Handbook of Graph Drawing and Visualization, chap. 17. CRC Press, Boca Raton (2014). www.ogdf.net
4. Gemsa, A., Niedermann, B., Nöllenburg, M.: Trajectory-based dynamic map labeling. In: Cai, L., Cheng, S.-W., Lam, T.-W. (eds.) ISAAC 2013. LNCS, vol. 8283, pp. 413–423. Springer, Heidelberg (2013). doi:10.1007/978-3-642-45030-3_39
5. Gemsa, A., Nöllenburg, M., Rutter, I.: Evaluation of labeling strategies for rotating maps. In: Gudmundsson, J., Katajainen, J. (eds.) SEA 2014. LNCS, vol. 8504, pp. 235–246. Springer, Heidelberg (2014). doi:10.1007/978-3-319-07959-2_20
6. Hachul, S., Jünger, M.: Drawing large graphs with a potential-field-based multilevel algorithm. In: Pach, J. (ed.) GD 2004. LNCS, vol. 3383, pp. 285–295. Springer, Heidelberg (2005). doi:10.1007/978-3-540-31843-9_29
7. Holten, D., Isenberg, P., van Wijk, J.J., Fekete, J.: An extended evaluation of the readability of tapered, animated, and textured directed-edge representations in node-link graphs. In: IEEE PacificVis 2011, pp. 195–202. IEEE (2011)
8. Holten, D., van Wijk, J.J.: A user study on visualizing directed edges in graphs. In: CHI 2009, pp. 2299–2308. ACM (2009)
9. Kakoulis, K.G., Tollis, I.G.: On the complexity of the edge label placement problem. Comput. Geom. **18**(1), 1–17 (2001)
10. Kakoulis, K.G., Tollis, I.G.: Labeling algorithms. In: Tamassia, R. (ed.) Handbook on Graph Drawing and Visualization, pp. 489–515. Chapman and Hall/CRC, New York (2013)
11. van Kreveld, M.J., Strijk, T., Wolff, A.: Point labeling with sliding labels. Comput. Geom. **13**(1), 21–47 (1999)
12. Lichtenstein, D.: Planar formulae and their uses. SIAM J. Comput. **11**(2), 329–343 (1982)
13. Marks, J., Shieber, S.: The computational complexity of cartographic label placement. Technical Report 05-91, Harvard University (1991)
14. North graphs. http://www.graphdrawing.org/data.html

15. Strijk, T., van Kreveld, M.J.: Practical extensions of point labeling in the slider model. GeoInformatica **6**(2), 181–197 (2002)
16. Strijk, T., Wolff, A.: Labeling points with circles. Int. J. Comput. Geom. Appl. **11**(2), 181–195 (2001)
17. Wagner, F., Wolff, A., Kapoor, V., Strijk, T.: Three rules suffice for good label placement. Algorithmica **30**(2), 334–349 (2001)
18. Wolff, A.: A simple proof for the NP-hardness of edge labeling. Technical Report 11/2000, Institute of Mathematics and Computer Science, Ernst Moritz Arndt University Greifswald (2000)

Peacock Bundles: Bundle Coloring for Graphs with Globality-Locality Trade-Off

Jaakko Peltonen[1,2(✉)] and Ziyuan Lin[1]

[1] Helsinki Institute for Information Technology HIIT,
Department of Computer Science, Aalto University, Espoo, Finland
ziyuan.lin@aalto.fi
[2] School of Information Sciences, University of Tampere, Tampere, Finland
jaakko.peltonen@aalto.fi

Abstract. Bundling of graph edges (node-to-node connections) is a common technique to enhance visibility of overall trends in the edge structure of a large graph layout, and a large variety of bundling algorithms have been proposed. However, with strong bundling, it becomes hard to identify origins and destinations of individual edges. We propose a solution: we optimize edge coloring to differentiate bundled edges. We quantify strength of bundling in a flexible pairwise fashion between edges, and among bundled edges, we quantify how dissimilar their colors should be by dissimilarity of their origins and destinations. We solve the resulting nonlinear optimization, which is also interpretable as a novel dimensionality reduction task. In large graphs the necessary compromise is whether to differentiate colors sharply between locally occurring strongly bundled edges ("local bundles"), or also between the weakly bundled edges occurring globally over the graph ("global bundles"); we allow a user-set global-local tradeoff. We call the technique "peacock bundles". Experiments show the coloring clearly enhances comprehensibility of graph layouts with edge bundling.

Keywords: Graph visualization · Network data · Machine learning · Dimensionality reduction

1 Introduction

Graphs are a prominent type of data in visual analytics. Prominent graph types include for instance hyperlinks of webpages, social networks, citation networks between publications, interaction networks between genes, variable dependency networks of probabilistic graphical models, message citations and replies in discussion forums, traces of eye fixations, and many others. 2D or 3D visualization of graphs is a common need in data analysis systems. If node coordinates are not available from the data, several node layout methods have been developed, from constrained layouts such as circular layouts ordered by node degree to unconstrained layouts optimized by various criteria; the latter methods can be based on the node and edge set (node adjacency matrix) alone, or can make use of

© Springer International Publishing AG 2016
Y. Hu and M. Nöllenburg (Eds.): GD 2016, LNCS 9801, pp. 52–64, 2016.
DOI: 10.1007/978-3-319-50106-2_5

multivariate node and edge features, typically aiming to reduce edge crossings and place nodes close-by if they are similar by some criterion.

In layouts with numerous edges it may be hard to see trends in node-to-node connections. Edge bundling draws multiple edges as curves that are close-by and parallel for at least part of their length. Bundling simplifies the appearance of the graph, and bundles also summarize connection trends between areas of the layout. However, when edges are drawn close together, the ability to visually follow edges and discover their start and end points is lost. Interactive systems [10] can allow inspection of edges, but inspecting numerous edges is laborious.

Comprehensibility of edges can be enhanced by distinguishing them by visual properties, such as line style, line width, markers along the curve, or color. Following an edge by its color can allow an analyst to see where each edge goes, but poorly assigned colors can make this task hard to do at a glance. **We present a machine learning method that optimizes edge colors in graphs with edge bundling, to keep bundled edges maximally distinguishable.** We focus on edge color as it has several degrees of freedom suitable for optimization (up to three continuous-valued color channels if using RGB color space), but our method is easily applicable to other continuous-valued edge properties. We call our solution **peacock bundles** as it is inspired by the plumage of a peacock; our method results in a fan of colors, reminiscent of a peacock tail, at fan-in locations of edges arriving into a bundle and fan-out locations of edges departing from a bundle. Figure 1 (middle) illustrates the concept and how it can help follow edges. We next review related works and then present the method and experiments.

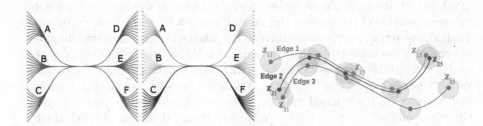

Fig. 1. Illustration of peacock bundle coloring. **Left:** A graph with node groups A–F, drawn with hierarchical edge bundling. With plain gray coloring finding the connecting vertex pairs is not possible. **Middle:** Peacock bundle coloring reveals that nodes in group A connect to nodes in group E in order, and similarly B to D in order, and C to F in reverse order. The connections are easily seen from the optimized coloring produced by our Peacock Bundles method: bundled edges traveling from and to close-by nodes get close-by colors. **Right:** Pairwise bundling detection as described in Sect. 3.1, for three edges, control points z_{ij} shown as circles, distance threshold T as the radius of light gray circles (small threshold used for illustration). Control points $z_{12}, \ldots, z_{15}$ of edge 1 are near control points of edge 2, but only z_{13} is near a control point of edge 3. If, e.g., $K_{ij} = 2$ nearby control points are required between edges, edge 1 is considered bundled with edge 2 but not edge 3; edge 2 is considered bundled with edges 1 and 3, and edge 3 with edge 2 but not edge 1. Since edges 1 and 3 are not considered bundled they could be assigned a similar color. (Color figure online)

2   Background: Node Layout, Edge Bundling, and Coloring

Node layouts of graphs have been optimized by many approaches, see [9] for a survey. Our approach is not specific to any node layout approach and can be run for any resulting layout. Several methods have been proposed for edge bundling [4,6–8,13,16,17,19]. For example, Cui et al. [4] generate a mesh covering the graph on the display based on node positions and edge distribution. The mesh helps cluster edges spatially; edges within a cluster are bundled. Hierarchical Edge Bundling [13] embeds a tree representation for data with hierarchy onto the 2D display. Tree nodes are used as spline control points for edges; bundles come from reusing control points. See Zhou et al. [22] for a recent review and taxonomy.

Unlike node layout and edge bundling, relatively little attention has been paid to practical edge coloring; while graph theory papers exist about the "edge coloring problem" of setting distinct colors to adjacent edges with a minimum number of colors, that combinatorial problem does not reflect real-life graph visual analytics where a continuous edge color space exists and the task is to set colors to be informative about graph properties. Simple coloring approaches exist. A naive coloring sets a random color to each edge: such coloring is unrelated to spatial positions of nodes and edges and is chaotic, making it hard to grasp an overview of edge origins and destinations at a glance. Edge colors are sometimes reserved to show discrete or multivariate annotations such as edge strengths; such coloring relies on external data and may not help gain an overview of the graph layout itself. A simple layout-driven solution is to color each edge by onscreen position of the start or end node. If edges have been clustered by some method, one often sets the same color to the whole cluster [7,21]; this simplifies coloring, but prevents telling apart origins and destinations of individual edges.

Hu and Shi [15] create edge colorings with a maximal distinguishability motivation related to ours, but their method does not consider actual edge bundling and operates on the original graph; we operate on bundled graphs and quantify edge bundling. Also, instead of only using binary detection of bundled edge pairs (a hard criterion whether two edges are bundled) we differentiate all edges, emphasizing each pair by a weight that is high between strongly bundled local edges and smaller between others, with a user-set global-local tradeoff. Lastly, their method tries to set maximally distinct colors between all bundled edges, needing harder compromises for larger bundles: we quantify which bundled edges need the most distinct colors by comparing their origin and destination coordinates in the layout, and thus can devote color resources efficiently even in large graphs.

Our algorithm is related to nonlinear dimensionality reduction. Although dimensionality reduction has been used in colorization for other domains [3,5], to our knowledge ours is the first method to optimize local graph coloring with edge bundling as a dimensionality reduction task.

3   The Method: Peacock Bundles

Bundle coloring has several challenges. **1.** Efficient coloring should depend not only on high-dimensional graph properties on the low-dimensional graph layout: if two edges are spatially distinct they do not need different colors. **2.** *Bundles are typically not clearly defined*: the curve corresponding to an individual edge may become *locally* bundled with several other edges at different places along the curve between its start and end node, and edges cannot be cleanly separated into groups that would correspond to some *globally* nonoverlapping bundles. Solutions requiring nonoverlapping bundles would be suboptimal: they would either not be applicable to real-life edge-bundled graphs or would need to artificially approximate the bundle structure of such graphs as nonoverlapping subsets. **3.** The solution should scale up to large graphs with large bundles. In large bundles it is typically not feasible to assign strongly distinct colors between all edges; it is then crucial to quantify how to make the compromise, that is, which edges should have the most distinct colors within the bundle.

Our coloring solution neatly solves these challenges, by posing the coloring as an optimization task defined based on local bundling between *each individual pair of edges*. Our solution is applicable to all graphs and takes into account the full bundle structure in a graph layout without approximations. For any two edges it is easy to define whether their curves are bundled (close enough and parallel) for some part of their length, without requiring a notion of a globally defined bundle; we optimize the coloring to *tell each edge apart from the ones it has been bundled with*. Such optimization makes maximally efficient use of the colors: two edges need distinct colors only if they are bundled together, whereas two edges that are not bundled can share the same color or very similar colors. Moreover, even between two bundled edges, how distinct their colors need to be can be quantified in a natural way based on the node layout: the more dissimilar their origins and destinations are, the more dissimilar their colors should be. Differentiating origins and destinations helps analysts assuming the node layout is meaningful. Computation of peacock bundles requires two steps:

1. *Detection of which pairs of edges are bundled together* at some location along their curve. We solve this by a well-defined closeness threshold of consecutive curve segments. An edge may participate in multiple bundles along its curve.
2. *Definition of the color optimization task.* We formalize the color assignment task as a dimensionality reduction task from two input matrices, a pairwise edge-to-edge bundling matrix and a dissimilarity matrix that quantifies how dissimilar colors of bundled edges should be, to a continuous-valued low-dimensional colorspace, which can be one-dimensional (1D) to achieve a color gradient, or 2D or 3D for greater variety. (Properties like width or continuous line-style attributes could be included in a higher than 3D output space; here we use color only.) We define the color assignment as an edge dissimilarity preservation task: colors are optimized to preserve spatial dissimilarities of start and end nodes among each pair of bundled edges, whereas no constraint is placed between colors of non-bundled edge pairs.

Peacock bundle coloring can be integrated into edge bundling algorithms, but can be also run as standalone postprocessing for graphs with edge bundling, regardless of which algorithms yielded the node layout and edge curves. Peacock bundles optimize colors taking both the graph and its visualization (node and edge layout) into account: color separation needs to be emphasized only for edges that appear spatially bundled. We demonstrate the result on several graphs with different node layouts and a popular edge bundling technique.

3.1   Detection of Bundled Pairs of Edges

Let the graph contain M edges $i = 1, \ldots, M$, each represented by a curve. If the curves are spline curves, let each curve be generated by C_i control points; if the curves are piecewise linear, let each curve be divided into C_i segments represented e.g. by the midpoint of a segment. For brevity we use the terminology of control points in the following, but the algorithm can be used just as well for other definitions of a curve, such as midpoints of piecewise linear curves or equidistributed points on the curves if getting such locations is convenient.

Let B_{ij} be a variable in $[0, 1]$ denoting whether edge i is bundled together with edge j. If the edge bundling has been created by an algorithm that explicitly defines bundle memberships for edges, B_{ij} can simply be set to 1 for edges assigned to the same bundle and zero otherwise. However, for several situations this is insufficient: (i) sometimes the bundling algorithm is not available or the bundling has e.g. been created interactively; (ii) some bundling algorithms only e.g. attract edge segments and do not define which edges are bundled; (iii) an edge may be close to several different other edges, so that no single bundle membership is sufficient to describe its relationship to other edges. For these reasons we provide a way to define pairwise edge bundling variables B_{ij} that does not require availability of any previous bundling algorithm.

We set $B_{ij} = 1$ if at least K_{ij} consecutive control points of edge i are each close enough to one or more control points of edge j. Intuitively, if several consecutive control points of edge i are close to edge j, the edges travel close and parallel (as a bundle) at least between those control points. Since our choice of control points does not allow the curves to change drastically between two consecutive control points, the defined B_{ij} is stable when the control point densities between curve i and curve j do not differ too much. In practice, we set K_{ij} to an integer at least 1, separately for each pair of edges, as a fraction of the number of available control points as detailed later in this section.

Formally, for edge i denote the on-screen coordinates of the C_i control points by $\mathbf{z}_{i1}, \ldots, \mathbf{z}_{iC_i}$, and similarly for edge j. Let $d(\cdot, \cdot)$ denote the Euclidean distance between two control points, and let T be a distance threshold. Then

$$B_{ij} = \max_{r_0 = 1, \ldots, C_i - K_{ij} + 1} \prod_{r=r_0}^{r_0 + K_{ij} - 1} 1\left(\min_{s=1, \ldots, C_j} d(\mathbf{z}_{ir}, \mathbf{z}_{js}) \leq T \right) \tag{1}$$

where $r = r_0, \ldots, r_0 + K_{ij} - 1$ are indices of consecutive control points in edge i. The term $1(\cdot)$ is 1 if the statement inside is true and zero otherwise: that is, the term is 1 if the rth control point of edge i is close to edge j (to some control point s of edge j). The whole product term is 1 if the K_{ij} consecutive control points of i from r_0 onwards are all close to edge j. Finally, the whole term B_{ij} is 1 if edge i has K_{ij} consecutive points (from any r_0 onwards) that are all close to edge j. Figure 1 (right) illustrates the pairwise bundling detection.

The distance threshold T should be set to a value below which line segments appear very similar; a rule of thumb is to set T to a fraction of the total diameter (or larger dimension) of the screen area of the graph. Similarly, a convenient way to set the required number of close-by control points K_{ij} is to set it to a fraction of the maximum number of control points in the two edges, requiring at least 1 control point, so that for each pair of edges i and j we set $K_{ij} = \max(1, \lfloor \max(C_i, C_j) K_{min} \rfloor)$ where $K_{min} \in (0, 1]$ is the desired fraction.

Detected pairwise bundles match ground truth in all simple examples we tried (e.g. Fig. 1 left); in experiments of Sect. 4 where no ground truth is available the bundling is visually good; edges bundled with any edge of interest can be interactively checked at http://ziyuang.github.io/peacock-examples/.

3.2   Optimization of Edge Colors by Dimensionality Reduction

Our coloring is based on dimensionality reduction of bundled edges from an original dissimilarity (distance) matrix to a color space; we thus need to define how dissimilar two bundled edges are. We aim to help analysts differentiate where in the graph layout each edge goes; we thus use the node locations of edges to define the similarity. Denote the two on-screen node layout coordinates of edge i by $\mathbf{v}_i^1$ and $\mathbf{v}_i^2$. We first define

$$d_{ij}^{original} = \min(\|\mathbf{v}_i^1 - \mathbf{v}_j^1\| + \|\mathbf{v}_i^2 - \mathbf{v}_j^2\|, \ \|\mathbf{v}_i^1 - \mathbf{v}_j^2\| + \|\mathbf{v}_i^2 - \mathbf{v}_j^1\|) . \quad (2)$$

Denote the set of p features for edge i as a vector $\mathbf{x}_i = [x_{i1}, \ldots, x_{ip}]$, and denote the low-dimensional output features for edge i as a vector $\mathbf{y}_i = [y_{i1}, \ldots, y_{iq}]$ where $q \in \{1, 2, 3\}$ is the output dimensionality. We define the dimensionality reduction task as minimizing the difference between the endpoint dissimilarity of bundled edges and dissimilarity of their optimized colors. This yields the cost function

$$\min_{\{\mathbf{y}_1, \ldots, \mathbf{y}_M\}} \sum_i \sum_j B_{ij} (d_{ij}^{original} - d^{out}(\mathbf{y}_i, \mathbf{y}_j))^2 \quad (3)$$

where $d^{out}(\mathbf{y}_i, \mathbf{y}_j)$ is the Euclidean distance between the output features. The terms B_{ij} are large for only those pairs of edges that are bundled, thus minimizing the cost assigns colors to preserve dissimilarity within bundled edges, but allows freedom of color assignment between non-bundled edges. The cost encapsulates that greater difference of edge destinations should yield greater color difference, and that color differentiation is most needed for strongly bundled edges. While alternative formulations are possible, (3) is simple and works well.

From Local to Global Color Differentiation. The weights B_{ij} detect edges according to thresholds T and K_{ij}. Some edge pairs that fail the detection might still visually appear nearly bundled: instead of differentiating only within detected bundles, it is meaningful to differentiate other edges too. The simplest way is to encode a *tradeoff* between local (within-bundle) and global differentiation in the B_{ij}: we set $B_{ij} = 1$ if edges i and j are bundled, otherwise $B_{ij} = \epsilon$ where $\epsilon \in [0, 1]$ is a user-set parameter for the preferred global-local tradeoff. When $0 < \epsilon < 1$, the cost emphasizes achieving desired color differences between bundled edges (where $B_{ij} = 1$) according to their dissimilarity of origins and destinations, but also aims to achieve color differences between other edges ($B_{ij} = \epsilon$) according to the same dissimilarity. As the optimization is based on desired dissimilarities between edges, it intelligently optimizes colors even when all edge pairs can have nonzero weight: $\epsilon = 0$ means a pure local coloring where only bundled edge pairs matter, and $\epsilon = 1$ means a pure global coloring that aims to show dissimilarity of origin and destination for all edges regardless of bundling. In our tests coloring changes gradually with respect to ϵ. In experiments, when emphasizing local color differences, we set $\epsilon = 0.001$ which achieved local differentiation and formed color gradients for bundles in most cases.

A way to set a more nuanced tradeoff is to run edge detection with multiple settings and set weaker B_{ij} for edges detected with weaker thresholds; in practice the above simple tradeoff already worked well.

Relationship to Nonlinear Multidimensional Scaling. Interestingly, minimizing (3) can be seen as a specialized weighted form of *nonlinear multidimensional scaling*, with several differences: unlike traditional multidimensional scaling we treat edges (not data items or nodes) as input items whose dissimilarities are preserved; our output is not a spatial layout but a color scheme; and most importantly, the cost function does not aim to preserve all "distances" but weights each pairwise distance according to how strongly that pair of edges is bundled. The theoretical connection lets us make use of optimization approaches previously developed for multidimensional scaling, here we choose to use the popular stress majorization algorithm (SMACOF) [1] to minimize the cost function.

Color Range Normalization. After optimization, output features $\mathbf{y}_i$ of each edge must be normalized to the range of the color channels (or positions along a color gradient). Simple ideas like applying an affine transform to the output matrix $Y = (\mathbf{y}_1, \ldots, \mathbf{y}_M)$ would give different amounts of color space to different bundles, thus colors within bundles would not be well differentiated. We propose a normalization to maximally distinguish edges within each bundle. Let Col denote the color matrix to be obtained from normalization. For each $\mathbf{y}_i$, let $\{\mathbf{y}_{i_l}\}_{l=1}^{M_i}$ be the set of output features where each edge l is bundled with edge i. We assemble $\mathbf{y}_i$ and $\{\mathbf{y}_{i_l}\}_{l=1}^{M_i}$ into a matrix $Y^i = (\mathbf{y}_i, \mathbf{y}_{i_1}, \ldots, \mathbf{y}_{i_{M_i}})$, affinely transform Y^i to $\tilde{Y}^i = (\tilde{\mathbf{y}}_i, \tilde{\mathbf{y}}_{i_1}, \ldots, \tilde{\mathbf{y}}_{i_{M_i}})$ so that each entry in $\tilde{Y}^i$ is within the allowed range (say, $[0, 1]$), then set Col_i, the i-th column of Col (color vector for edge i) as $\tilde{\mathbf{y}}_i$. This normalization expands the color range within bundles.

Where to Show Colors. The optimized colors can be shown along the whole edge, or at "fan-in" segments where the edge enters a bundle and "fan-out" segments where it departs a bundle. Edge i is bundled with j if several consecutive curve segments of i are close to j; the last segment before the close-by ones is the fan-in segment; the first segment after the close-by ones is the fan-out segment. In experiments we show color along the whole edge for simplicity. Note that, as with any edge coloring, colors of close-by edges may perceptually blend, but our optimized colors then remain visible at fan-in and fan-out locations.

4   Experiments

We demonstrate the Peacock bundles method on five graphs (Figs. 2, 3 and 4): two graphs with hierarchical edge bundling [13], and three with force-directed edge bundling [14]. The two graphs with hierarchical edge bundling are created

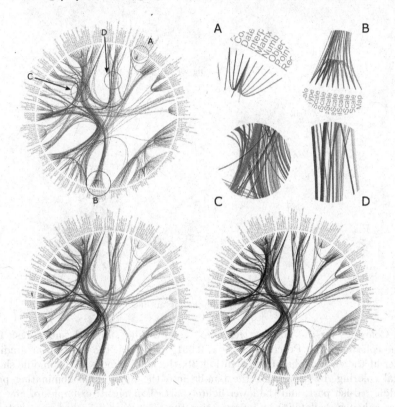

Fig. 2. Colorings for the graph "radial". **Top left:** the coloring from Peacock with $\epsilon = 0.001$. **Top right:** zoomed-in versions of the parts within dashed-line circles in the top-left figure as examples of local coloring. In the four zoomed-in parts, colors show a linear gradient and vary in yellow-red-blue, thus (1) local colors are differentiated, and (2) they span roughly the same full color range. The local colors also help follow edges at bottom right of the graph, where colors are homogeneous in the baseline coloring. **Bottom left:** coloring from Peacock, $\epsilon = 1$. The bundles are colored into 3 parts: the blue-ish upper half, green-ish lower-left part, and yellow-ish lower-right part. There are also red bundles joining the blue and green parts, differentiating itself from other bundles. **Bottom right:** the baseline coloring. The bundles are colored into the red-ish upper half and blue-ish lower half. Compared with the coloring with $\epsilon = 1$ from Peacock, the bundle from left to right, and the bundle at the top-left corner are less distinguishable. (Color figure online)

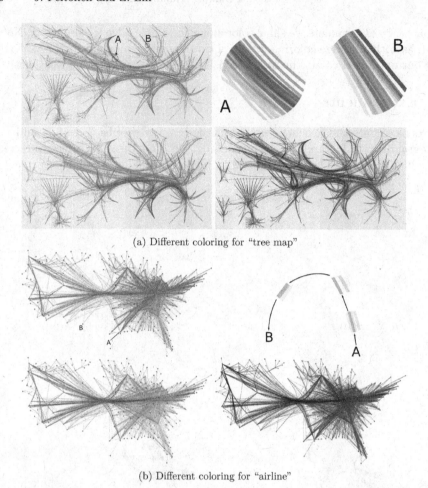

(a) Different coloring for "tree map"

(b) Different coloring for "airline"

Fig. 3. Colorings for the graphs "tree map" and "airline". In both subfigures: **Top left:** the colorings from Peacock with $\epsilon = 0.001$. In Fig. 3a, the local linear gradient is clearer at the ends of the bundles. In Fig. 3b, the large bundle in the middle shows the local coloring, by separating the bundle into the upper blue dominating part, the middle red-ish part, and the lower lighter part. **Top right:** examples of how the colorings enhance readability by investigating the parts within dashed line circles in both top-left subfigures. In Fig. 3a, the colors in bundle A help the user to recognize, for example, 1) the blue-ish part in bundle A leads to the blue-ish part of the top-right "claw" or the right "claw"; 2) the red-ish pa0rt in bundle A leads to the red-ish part of the "claw" at the right of bundle B, or the top right "claw"; 3) the yellow-ish half that joins in the middle leads to bundle B or to the "claw" at the right of bundle B. Figure 3b shows how the coloring help a pink edge from A to B stand out against other edges in the same bundle. **Bottom left:** the coloring from Peacock with $\epsilon = 1$. In Fig. 3a, bundles are globally differentiated. In Fig. 3b, for nodes of large degrees, the edges connecting to them have distinct colors for different directions. **Bottom right:** the baseline coloring. In Fig. 3a, the user may mis-recognize that there are edges from bundle A to bundle B. In Fig. 3b, unlike the bottom-left subfigure, edges connecting to the same node tend to have similar colors. (Color figure online)

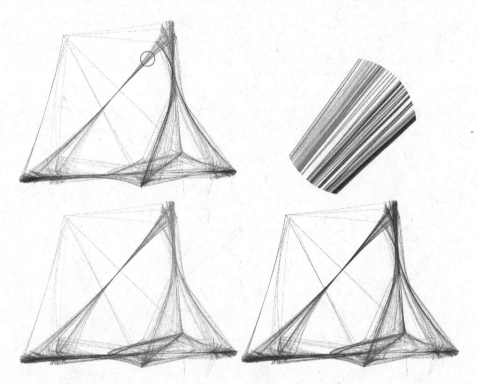

Fig. 4. Colorings for graph "Jane Austen". **Top left:** the coloring from Peacock with $\epsilon = 0.001$. The "crossing" at the right half of the figure shows a clear differentiation: red edges go from upper-right to lower-left; green edges go from upper to lower; blue edges go from upper-left to lower-right. Without the local coloring, it is difficulty to tell whether the bundles or the edges are crossing or just tangental to one another. **Top right:** another example from the zoomed-in version of the part within the dashed line circle in the top-left figure. Edge colors change from green to purple-ish. The colors help the user follow the edges after the heavily bundled part in the middle: red edges mostly go leftwards, green edges are scattered, and blue edges mostly go rightwards. **Bottom left:** the coloring from Peacock with $\epsilon = 1$. Less locality but more globality. The earlier blue edges in the right "crossing" become purple, but still distinguishable from the other two bundles. **Bottom right:** the baseline coloring, which loses the distinguishablity shown in the Peacock coloring. (Color figure online)

from the class hierarchy of the visualization toolkit Flare [11], with the built-in radial layout (graph named "radial"; Fig. 2) and tree map layout ("tree map"; Fig. 3a) in d3.js [2] respectively. The four graphs with force-directed edge bundling are: a spatial graph of US flight connections ("airline"; Fig. 3b); a graph of consecutive word-to-word appearances in novels of Jane Austen ("Jane Austen"; Fig. 4); and a graph of matches between US college football teams ("football"; Fig. 5). The last three graphs are laid out as an unconstrained 2D graph by a recent node-neighborhood preserving layout method [18]. For all

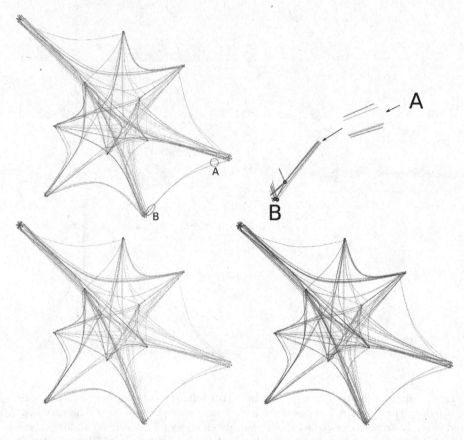

Fig. 5. Colorings for graph "football". **Top left:** the coloring from Peacock with $\epsilon =$ 0.001. The uncertainty in this graph is mostly from the small clusters of nodes at the end of the edges. **Top right:** an example showing how the coloring help distinguish heavily bundled edges from A to B. The shown segments are the zoomed-in version of the parts within the dashed line circle or ellipse in the top-left figure. We can see, for example, that the blue edge leads to the top node in B, while the yellow edge leads to leftmost node in B. This is also noticeable in the top-left figure, particularly for the blue edge. However, it will be a difficult task with the baseline coloring since the part between A and B is heavily bundled. **Bottom left:** the coloring from Peacock with $\epsilon = 1$. Colors of edges from the same cluster are differentiated (e.g., at the top-left cluster, edge colors vary from red to blue). **Bottom right:** the baseline coloring, which only reflects the locations of the bundles. (Color figure online)

graphs, edge bundles were created by a d3.js plugin implementing the algorithm [20] adapted to splines. All coloring are compared with a baseline coloring from end point positions.

The Baseline. We compare our method with a baseline coloring that directly encodes end point positions into color channels. We choose channel red and blue for the encoding in the experiments. Let $\mathbf{v}_i^1 = (x_i^1, y_i^1)$ and $\mathbf{v}_i^2 = (x_i^2, y_i^2)$ be the onscreen coordinates of edge i's two end points as in (2). We first create a 3-dimensional vector $\widetilde{Col}_i^{\text{baseline}}$ as the "unnormalized" color for edge i as

$$\widetilde{Col}_i^{\text{baseline}} = (\min(x_{i,1}, x_{i,2}), 0, \min(y_{i,1}, y_{i,2}))^{\text{T}} \qquad (4)$$

then we affinely normalize the matrix $\widetilde{Col}^{\text{baseline}}$ into $[0, 1]$ to obtain the final baseline colors Col^{baseline}.

Choices of Peacock Parameters. The parameters T and K_{min} in (1) must be chosen to determine B_{ij}. We set T to $2 - 4\%$ of max(graph width, graph height), and fix K_{min} as 0.4. Experiments show the choices give good results empirically.

Figures 2, 3 and 4 show the results from the proposed method and the baseline. The top-left subfigures are with the tradeoff parameter set to prefer locality in the coloring. The top-right subfigures provide zoomed-in views detailing the local color variation ("peacock fans") and demonstrating how the coloring improves readability and helps follow edges. The bottom-left figures are optimized to differentiate origins and destinations globally (tradeoff parameter $\epsilon = 1$), hence colors indicate overall trends of connections between areas of the graph layout, at the expense of less color variability within bundles. The bottom-right figures are from the baseline, also aiming to show variability of endpoint positions the coloring but not optimized by machine learning; the simple baseline coloring leaves bundles and within-bundle variation less distinguishable.

5 Conclusions

We introduced "peacock bundles", a novel edge coloring algorithm for graphs with edge bundling. Colors are optimized both to preserve differences between bundle locations, and differentiate edges within bundles. The algorithm is based on dimensionality reduction without need to explicitly define bundles. Experiments show the method outperforms the baseline coloring with several graphs and bundling algorithms, greatly improving the comprehensibility of graphs with edge bundling. Potential future work includes incorporating color perception models [12], and more nuanced weighting schemes for global-local tradeoffs.

We acknowledge computational resources from the Aalto Science-IT project. Authors belong to the COIN centre of excellence. The work was supported by Academy of Finland grants 252845 and 256233.

References

1. Borg, I., Groenen, P.J.F.: Modern Multidimensional Scaling: Theory and Applications. Springer Series in Statistics, 2nd edn. Springer, New York (2005)
2. Bostock, M., Ogievetsky, V., Heer, J.: D3 data-driven documents. IEEE Trans. Vis. Comput. Graph. **17**(12), 2301–2309 (2011)

3. Casaca, W., et al.: Colorization by multidimensional projection. In: Proceedings of SIBGRAPI 2012, pp. 32–38. IEEE (2012)

4. Cui, W., Zhou, H., Qu, H., Wong, P.C., Li, X.: Geometry-based edge clustering for graph visualization. IEEE Trans. Vis. Comput. Graph. **14**(6), 1277–1284 (2008)

5. Daniels, J., Anderson, E.W., Nonato, L.G., Silva, C.T., et al.: Interactive vector field feature identification. IEEE Trans. Vis. Comput. Graph. **16**(6), 1560–1568 (2010)

6. Dwyer, T., Marriott, K., Wybrow, M.: Integrating edge routing into force-directed layout. In: Kaufmann, M., Wagner, D. (eds.) GD 2006. LNCS, vol. 4372, pp. 8–19. Springer, Heidelberg (2007). doi:10.1007/978-3-540-70904-6_3

7. Ersoy, O., Hurter, C., Paulovich, F.V., Cantaneira, G., Telea, A.: Skeleton-based edge bundling for graph visualization. IEEE Trans. Vis. Comput. Graph. **17**(12), 2364–2373 (2011)

8. Gansner, E.R., Hu, Y., North, S., Scheidegger, C.: Multilevel agglomerative edge bundling for visualizing large graphs. In: Proceedings of PacificVis 2011, pp. 187–194. IEEE (2011)

9. Gibson, H., Faith, J., Vickers, P.: A survey of two-dimensional graph layout techniques for information visualisation. Info. Vis. **12**(3–4), 324–357 (2013)

10. Grossman, T., Balakrishnan, R.: The bubble cursor: enhancing target acquisition by dynamic resizing of the cursor's activation area. In: Proceedings of CHI 2005, pp. 281–290. ACM (2005)

11. Heer, J.: Flare (2009). https://git.io/v6buH

12. Heer, J., Stone, M.: Color naming models for color selection, image editing and palette design. In: Proceedings of CHI 2012 (2012)

13. Holten, D.: Hierarchical edge bundles: visualization of adjacency relations in hierarchical data. IEEE Trans. Vis. Comput. Graph. **12**(5), 741–748 (2006)

14. Holten, D., Van Wijk, J.J.: Force-directed edge bundling for graph visualization. Comput. Graph. Forum **28**(3), 983–990 (2009)

15. Hu, Y., Shi, L.: A coloring algorithm for disambiguating graph and map drawings. In: Duncan, C., Symvonis, A. (eds.) GD 2014. LNCS, vol. 8871, pp. 89–100. Springer, Heidelberg (2014). doi:10.1007/978-3-662-45803-7_8

16. Hurter, C., Ersoy, O., Telea, A.: Graph bundling by kernel density estimation. Comput. Graph. Forum **31**(3pt1), 865–874 (2012)

17. Luo, S.J., Liu, C.L., Chen, B.Y., Ma, K.L.: Ambiguity-free edge-bundling for interactive graph visualization. IEEE Trans. Vis. Comput. Graph. **18**(5), 810–821 (2012)

18. Parkkinen, J., Nybo, K., Peltonen, J., Kaski, S.: Graph visualization with latent variable models. In: Proceedings of MLG 2010, pp. 94–101. ACM (2010)

19. Pupyrev, S., Nachmanson, L., Kaufmann, M.: Improving layered graph layouts with edge bundling. In: Brandes, U., Cornelsen, S. (eds.) GD 2010. LNCS, vol. 6502, pp. 329–340. Springer, Heidelberg (2011). doi:10.1007/978-3-642-18469-7_30

20. Sugar, C.: d3.forcebundle (2016). https://git.io/v6GgL

21. Telea, A., Ersoy, O.: Image-based edge bundles: simplified visualization of large graphs. Comput. Graph. Forum **29**(3), 843–852 (2010)

22. Zhou, H., Xu, P., Yuan, X., Qu, H.: Edge bundling in information visualization. Tsinghua Sci. Technol. **18**(2), 145–156 (2013)

Clustered Graphs

Twins in Subdivision Drawings of Hypergraphs

René van Bevern[1,2], Iyad Kanj[3], Christian Komusiewicz[4], Rolf Niedermeier[5],
and Manuel Sorge[5(✉)]

[1] Novosibirsk State University, Novosibirsk, Russian Federation
rvb@nsu.ru
[2] Sobolev Institute of Mathematics, Siberian Branch of the Russian Academy
of Sciences, Novosibirsk, Russian Federation
[3] DePaul University, Chicago, USA
ikanj@cs.depaul.edu
[4] Friedrich-Schiller-Universität Jena, Jena, Germany
christian.komusiewicz@uni-jena.de
[5] TU Berlin, Berlin, Germany
{rolf.niedermeier,manuel.sorge}@tu-berlin.de

Abstract. Visualizing hypergraphs, systems of subsets of some universe, has continuously attracted research interest in the last decades. We study a natural kind of hypergraph visualization called *subdivision drawings*. Dinkla et al. [Comput. Graph. Forum '12] claimed that only few hypergraphs have a subdivision drawing. However, this statement seems to be based on the assumption (also used in previous work) that the input hypergraph does not contain *twins*, pairs of vertices which are in precisely the same hyperedges (subsets of the universe). We show that such vertices may be necessary for a hypergraph to admit a subdivision drawing. As a counterpart, we show that the number of such "necessary twins" is upper-bounded by a function of the number m of hyperedges and a further parameter r of the desired drawing related to its number of layers. This leads to a linear-time algorithm for determining such subdivision drawings if m and r are constant; in other words, the problem is linear-time fixed-parameter tractable with respect to the parameters m and r.

1 Introduction

Hypergraph drawings are useful as visual aid in diverse applications [1], among them electronic circuit design [11] and relational databases [2,18]. There are several methods for embedding hypergraphs in the plane. The combinatorial problem studied in this work stems from obtaining *subdivision drawings* [14,15].

Significant parts of the work were done while all authors were with TU Berlin and while RvB, IK, and MS were supported by the DFG, project NI 369/12.

R. van Bevern—Supported by grant 16-31-60007 mol_a_dk of the Russian Foundation for Basic Research.

C. Komusiewicz—Supported by the DFG, project KO 3669/4-1.

© Springer International Publishing AG 2016
Y. Hu and M. Nöllenburg (Eds.): GD 2016, LNCS 9801, pp. 67–80, 2016.
DOI: 10.1007/978-3-319-50106-2_6

Herein, given a hypergraph $\mathcal{H}$, we divide the plane into closed regions that one-to-one correspond to the vertices of $\mathcal{H}$ in such a way that, for each hyperedge F, the union of the regions corresponding to the vertices in F is connected. Subdivision drawings have also been called *vertex-based Venn diagrams* [14]. Figure 1 shows an example for such a drawing.

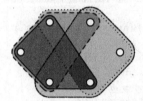 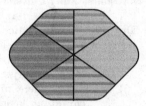

Fig. 1. Two drawings of the same hypergraph. On the left, we see a drawing in the *subset standard* in which the vertices (white circles) are enclosed by curves that correspond to hyperedges. On the right, we see a *subdivision drawing* in which we assign vertices to regions (enclosed by black lines) and we color these regions with colors that one-to-one correspond to the hyperedges; for each hyperedge, the union of the regions of the vertices in that hyperedge is connected. (Color figure online)

Subdivision drawings are a natural extension of planarity for ordinary graphs: A graph is planar if and only if it has a subdivision drawing when viewed as a hypergraph. For hypergraphs, having a subdivision drawing is a rather general concept of planar embeddings, as, for example, each Zykov planar hypergraph (meaning that the incidence graph is planar) and each hypergraph with a well-formed Euler diagram (see Flower et al. [12]) has a subdivision drawing. Still, Dinkla et al. [10] claimed that "most hypergraphs do not have [subdivision drawings]". However, this claim might have been based on the fact that several works on subdivision drawings assumed that the input hypergraph is twinless, that is, there are no two vertices contained in precisely the same hyperedges (see Mäkinen [18, p. 179], Buchin et al. [6, p. 535], and Kaufmann et al. [15, p. 399]). Twins do not seem useful at first glance: whatever role one vertex can play to obtain a subdivision drawing, its twin can also fulfill. One of our contributions is disproving the general validity of this assumption in Sect. 3. More specifically, we give a hypergraph with two twins that has a subdivision drawing but, removing one twin, it ceases to have one. Thus, twins may indeed be helpful to find a solution.

More generally, we can construct hypergraphs with ℓ twins that allow for subdivision drawings but cease doing so when removing one of the twins. However, the number of hyperedges in the construction grows with ℓ. It is thus natural to ask whether there is a function $\psi\colon \mathbb{N} \to \mathbb{N}$ such that, in each hypergraph with m hyperedges, we can forget all but $\psi(m)$ twins while maintaining the property of having a subdivision drawing. Using well-quasi orderings, one can relatively easily prove the *existence* of such a function ψ, yet finding a closed form for ψ turned out to be surprisingly difficult: so far we could only compute a concrete upper bound when considering a second parameter r measuring the number of "layers" in the drawing. A small number r of layers, however, is a relevant special case [4,6].

We study subdivision drawings from a combinatorial point of view, exploiting the fact that it is equivalent for a hypergraph to have a subdivision drawing and to have a support that is planar [14]. Herein, a *support* for a hypergraph $\mathcal{H} = (V, \mathcal{E})$ is a graph G on the same vertex set as $\mathcal{H}$ such that each hyperedge $F \in \mathcal{E}$ induces a connected subgraph $G[F]$. The outerplanarity number r of the support roughly translates to the number of layers in a corresponding drawing:[1] An r-outerplanar graph admits a planar embedding (without edge crossings) which has the property that, after removing r times all vertices on the outer face, we obtain an empty graph. Similar restrictions were studied before [4,6]. Formally, we study the following problem.

Problem (r-Outerplanar Support).
Input: A connected hypergraph $\mathcal{H}$ with n vertices and m hyperedges, and $r \in \mathbb{N}$.
Question: Does $\mathcal{H}$ admit an r-outerplanar support?

Our main result is a concrete upper bound on the number $\psi(m, r)$ of twins that might be necessary to obtain an r-outerplanar support. Since superfluous twins can then be removed in linear time, this gives the following algorithmic result.

Theorem 1. *There is an algorithm solving r-OUTERPLANAR SUPPORT which, for constant r and m, has linear running time.*

In contrast to Theorem 1, r-OUTERPLANAR SUPPORT remains NP-complete for $r = \infty$ [14] and even for every fixed $r > 1$ [6] (see below). The constants in the running time of the algorithm in Theorem 1 have a large dependence on m and r. However, it is conceivable that the parameters m and r are small in practical instances: for a large number m of hyperedges, it is plausible that we obtain only hardly legible drawings unless the hyperedges adhere to some special structure. Thus, it makes sense to design algorithms particularly for hypergraphs with a small number of hyperedges, as done by Verroust and Viaud [21]. Moreover, a small outerplanarity number r leads to few layers in the drawing which may lead to aesthetically pleasing drawings, similarly to path- or cycle-supports [6].

Related Work. For specifics on the relations of some different planar embeddings for hypergraphs, see Kaufmann et al. [15], Brandes et al. [4].

As mentioned before, Johnson and Pollak [14] showed that finding a *planar* support is NP-complete. Buchin et al. [6] proved that r-OUTERPLANAR SUPPORT is NP-complete for $r = 2, 3$. From their proof it follows that r-OUTERPLANAR SUPPORT is also NP-complete for every $r > 3$. This is due to a property of the reduction that Buchin et al. use. Given a formula ϕ

---

[1] We refer to Kaufmann et al. [15] for a method to obtain a subdivision drawing from a planar support.

in 3CNF, they construct a hypergraph $\mathcal{H}$ that has a *planar* support if and only if ϕ is satisfiable. Due to the way in which $\mathcal{H}$ is constructed, if there is any planar support, then it is 3-outerplanar. Thus, deciding whether an r-outerplanar support, $r \geq 3$, exists also decides the satisfiability of the corresponding formula.

Towards determining the computational complexity of finding an *outerplanar* hypergraph support, Buchin et al. [4] gave a polynomial-time algorithm for cactus supports (graphs in which each edge is contained in at most one cycle). They also showed that finding an outerplanar support (or planar support) can be done in polynomial time if, in the input hypergraph, each intersection or difference of two hyperedges is either a singleton or again a hyperedge in the hypergraph. Getting even more special, a tree support can be found in linear time [2,19]. Buchin et al. [6] gave a polynomial-time algorithm that can deal with an additional upper bound on the vertex degrees in the tree support. Klemz et al. [16] studied so-called area-proportional Euler diagrams, for which the corresponding computational problem reduces to finding a minimum-weight tree support. Such supports can also be found in polynomial time [16,17].

In a wider scope, motivated by drawing metro maps and metro map-like diagrams, Brandes et al. [5] studied the problem of finding *path-based* planar hypergraph supports, that is, planar supports that fulfill the additional constraint that the subgraph induced by each hyperedge contains a Hamiltonian path, giving NP-hardness and tractability results. Finding path-based tree supports is also known as the GRAPH REALIZATION problem, for which several polynomial-time algorithms were already known [3].

Chen et al. [7] showed that for obtaining minimum-edge supports (not necessarily planar), twins show a similar behavior as for r-outerplanar supports: Removing a twin can increase the minimum number of edges needed for a support and finding a minimum-edge support is linear-time solvable for a constant number of hyperedges via removing superfluous twins. The proof is quite different, however.

Organization. In Sect. 2 we provide some technical preliminaries used throughout the work. In Sect. 3 we give an example that shows that twins can be crucial for a hypergraph to have a planar support. As mentioned, for each $m \in \mathbb{N}$, there is a number $\psi(m)$ such that in each hypergraph with a planar support we can safely forget all but $\psi(m)$ twins (a proof is deferred to a full version). In Sect. 4 we give a concrete upper bound for $\psi(m)$ in the case of r-outerplanar supports and derive the linear-time algorithm for r-OUTERPLANAR SUPPORT promised in Theorem 1. We conclude and give some directions for future research in Sect. 5.

2   Preliminaries

General Notation. By $A \uplus B$ we denote the union of two disjoint sets A and B. For a family of sets $\mathcal{F}$, we write $\bigcup \mathcal{F}$ in place of $\bigcup_{S \in \mathcal{F}} S$. For equivalence relations ρ over some set S and $v \in S$ we use $[v]_\rho$ to denote the equivalence class of v in ρ.

Hypergraphs. A *hypergraph* $\mathcal{H}$ is a tuple $(V, \mathcal{E})$ consisting of a *vertex set* V, also denoted $V(\mathcal{H})$, and a *hyperedge set* $\mathcal{E}$, also denoted $\mathcal{E}(\mathcal{H})$. The hyperedge set $\mathcal{E}$ is a family of subsets of V, that is, $F \subseteq V$ for every hyperedge $F \in \mathcal{E}$. Where it is not ambiguous, we use $n := |V|$ and $m := |\mathcal{E}|$. When specifying running times, we use $|\mathcal{H}|$ to denote $|V(\mathcal{H})| + \sum_{F \in \mathcal{E}(\mathcal{H})} |F|$. The *size* $|F|$ of a hyperedge F is the number of vertices in it. Unless stated otherwise, we assume that hypergraphs do not contain hyperedges of size at most one or multiple copies of the same hyperedge. (These do not play any role for the problem under consideration, and removing them can be done easily and efficiently.)

A vertex $v \in V$ and a hyperedge $F \in \mathcal{E}$ are *incident* with one another if $v \in F$. For a vertex $v \in V(\mathcal{H})$, we let $\mathcal{E}_{\mathcal{H}}(v) := \{F \in \mathcal{H} \mid v \in F\}$. If it is not ambiguous, then we omit the subscript $\mathcal{H}$ from $\mathcal{E}_{\mathcal{H}}$. A vertex v *covers* a vertex u if $\mathcal{E}(u) \subseteq \mathcal{E}(v)$. Two vertices $u, v \in V$ are *twins* if $\mathcal{E}(v) = \mathcal{E}(u)$. Clearly, the relation τ on V defined by $\forall u, v \in V \colon (u, v) \in \tau \iff \mathcal{E}(u) = \mathcal{E}(v)$ is an equivalence relation. The equivalence classes $[u]_\tau$, $u \in V$, are called *twin classes*.

Removing a vertex subset $S \subseteq V(\mathcal{H})$ from a hypergraph $\mathcal{H} = (V, \mathcal{E})$ results in the hypergraph $\mathcal{H} - S := (V \setminus S, \mathcal{E}')$ where $\mathcal{E}'$ is obtained from $\{F \setminus S \mid F \in \mathcal{E}\}$ by removing empty and singleton sets. For brevity, we also write $\mathcal{H} - v$ instead of $\mathcal{H} - \{v\}$. The *subhypergraph shrunken to* $V' \subseteq V$ is the hypergraph $\mathcal{H}|_{V'} := \mathcal{H} - (V \setminus V')$.

Graphs. Our notation related to graphs is basically standard and heavily borrows from Diestel's book [9]. In particular, a *bridge* of a graph is an edge whose removal increases its number of connected components. Analogously, a *cut-vertex* is a vertex whose removal increases its number of connected components. Some special notation including the gluing of graphs is given below. We use the usual notation for planar and plane graphs. An r-*outerplanar* graph admits a planar embedding which has the property that, after r times of removing all vertices on the outer face, we obtain an empty graph. The ith layer L_i of a plane graph is defined as the set of vertices on the outer face, after having $i - 1$ times removed all vertices on the outer face.

Boundaried Graphs and Gluing. For a nonnegative integer $b \in \mathbb{N}$, a b-*boundaried graph* is a tuple (G, B, β) where G is a graph, $B \subseteq V(G)$ such that $|B| = b$, and $\beta \colon B \to \{1, \ldots, b\}$ is a bijection. Vertex subset B is called the *boundary* and β the *boundary labeling*. For ease of notation we also refer to (G, B, β) as the b-*boundaried* graph G with boundary B and boundary labeling β. For brevity, we also denote by β-*boundaried* graph G that b-boundaried graph G whose boundary is the domain of β and whose boundary labeling is β.

For a nonnegative integer b, the *gluing* operation $\circ_b$ maps two b-boundaried graphs to an ordinary graph as follows: Given two b-boundaried graphs G_1, G_2 with corresponding boundaries B_1, B_2 and boundary labelings β_1, β_2, to obtain the graph $G_1 \circ_b G_2$ take the disjoint union of G_1 and G_2, and identify each $v \in B_1$ with $\beta_2^{-1}(\beta_1(v)) \in B_2$. We omit the index b in $\circ_b$ if it is clear from the context.

3   Beware of Removing Twins

In Fig. 2, we provide a concrete example that shows that twins can be necessary
to obtain a 2-outer-planar support:

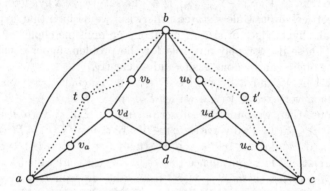

Fig. 2. A hypergraph $\mathcal{H}$ and its support, showing that twins can be essential for obtaining a 2-outer-planar support. The set of hyperedges consists of size-two hyperedges
that are drawn as solid lines between the corresponding vertices and, additionally,
$\{a, v_a, t, t', c\}$, $\{a, v_b, t, t', c\}$, $\{b, v_a, t, t', c\}$, $\{b, v_b, t, t', c\}$, $\{b, u_b, t, t', a\}$, $\{b, u_c, t, t', a\}$,
$\{c, u_b, t, t', a\}$, and $\{c, u_c, t, t', a\}$. Note that the vertices t and t' are twins. The hyper-
graph $\mathcal{H}$ has a (2-outer)planar support whose edges are indicated by the solid and the
dotted lines. However, $\mathcal{H} - t$ does not have a planar support.

The vertex set of the hypergraph $\mathcal{H}$ shown in Fig. 2 is $V := \{a, b, c, d, v_a, v_b,$
$v_d, u_b, u_c, u_d, t, t'\}$. We choose the hyperedges in such a way that t and t' are twins
and $\mathcal{H}$ has a planar (more precisely, 2-outerplanar) support but $\mathcal{H} - t$ does not.
First, we add to the set of hyperedges $\mathcal{E}$ of $\mathcal{H}$ the size-two hyperedges represented
by solid lines between the corresponding vertices in Fig. 2. The corresponding
"solid" hyperedges incident with (and only with) a, b, c, d form a K_4 and have the
purpose of essentially fixing the embedding of each support G: Since the complete
graph on four vertices, K_4, is 3-connected, it has only one planar embedding up to
the choice of the outer face [20, p. 747]. The remaining solid hyperedges (incident
with v_a, v_b, v_d and u_a, u_c, u_d) have the purpose of anchoring the u- and v-vertices
within two different faces of the embedding of the K_4: These hyperedges form
two connected components that are adjacent to a, b, d and b, c, d, respectively.
Hence, these connected components reside in those (unique) faces of the K_4 that
are incident with a, b, d and b, c, d, respectively.

With the following additional hyperedges, our goal is to enforce that t and
t' are used as conduits to connect the v-vertices to c via both a and b, and to
connect the u-vertices to a via both b and c. As we explain below, this is achieved
by the following hyperedges:

$$\{a, v_a, t, t', c\}, \qquad \{a, v_b, t, t', c\}, \qquad \{b, v_a, t, t', c\}, \qquad \{b, v_b, t, t', c\},$$
$$\{b, u_b, t, t', a\}, \qquad \{b, u_c, t, t', a\}, \qquad \{c, u_b, t, t', a\}, \qquad \{c, u_c, t, t', a\}.$$

Clearly, t and t' are twins. As can easily be verified, adding t and t' and the dotted edges in Fig. 2 to the graph induced by the solid edges gives a planar support for $\mathcal{H}$.

We now show that t and t' have to reside in different faces for each planar support G for $\mathcal{H}$. First, observe that, in G, either v_a is not adjacent to b or v_b is not adjacent to a. Moreover, neither of v_a and v_b is adjacent to c. Thus, to connect the subgraphs induced by the hyperedges that contain v_a or v_b, either vertex t or its twin t' must be adjacent to one of the two vertices in G. For the same reason, one of t and t' must be adjacent to one of u_b and u_c. Since there is no face in G that is simultaneously incident with one of v_a or v_b and one of u_b or u_c, the twins t and t' thus have to be in different faces. This implies that it is impossible to obtain a planar support if t or t' is missing. Consequently, removing one vertex of a twin class can transform a yes-instance of r-OUTERPLANAR SUPPORT into a no-instance.

The example above is not a pathology of having only one pair of twins, in a full version, we extend it so that an arbitrarily large set of twins is required for the existence of a planar support.

4   Relevant Twins for r-Outerplanar Supports

In this section, we show that there is an explicit function ψ such that, out of each twin class of a given hypergraph $\mathcal{H}$, we can remove all but $\psi(m, r)$ twins such that the resulting hypergraph has an r-outerplanar support if and only if $\mathcal{H}$ has. In other words, we prove that the following data reduction rule is correct.

Rule 1. Let $\mathcal{H}$ be a hypergraph with m edges. If there is a twin class with more than $\psi(m, r) = 2^{6r \cdot 2^{m \cdot (2r^2 + r + 1)} \cdot (r+1)^{32r^2 + 8r}}$ vertices, then remove one vertex out of this class.

Assuming that Rule 1 is correct, Theorem 1 follows.

Proof (Theorem 1). Rule 1 can be applied exhaustively in linear time because the twin classes can be computed in linear time [13]. After this, each twin class contains at most $\psi(m, r)$ vertices, meaning that, overall, at most $2^m \psi(m, r)$ vertices remain. Testing all possible planar graphs for whether they are a support for the resulting hypergraph thus takes constant time if m and r are constant. Hence, the overall running time is linear in the input size.   □

We mention in passing that, in the terms of parameterized algorithmics, exhaustive application of Rule 1 can be seen as a problem kernel (see [8], for example).

The correctness proof for Rule 1 consists of two parts. First, in Theorem 2, we show that each r-outerplanar graph has a long sequence of nested separators. Here, *nested* means that each separator separates the graph into a *left* side and a *right* side, and each left side contains all previous left sides. Furthermore, the sequence of separators has the additional property that, for any pair of separators S_1, S_2, we can glue the left side of S_1 and the right side of S_2, obtaining another r-outerplanar graph.

In the second part of the proof, we fix an initial support for our input hypergraph. We then show that, in a long sequence of nested separators for this support as above, there are two separators such that we can carry out the following procedure. We discard all vertices between the separators, glue their left and right sides, and reattach the vertices which we discarded as degree-one vertices. Furthermore, we can do this in such a way that the resulting graph is an r-outerplanar support. The reattached degree-one vertices hence are not crucial to obtain an r-outerplanar support. We will show that if our input hypergraph is large enough, that is, larger than some function of m and r, then there is always at least one non-crucial vertex which can be removed.

We now formalize our approach. Theorem 2 will guarantee the existence of a long sequence of gluable separators. To formally state it, we need the following notation.

Definition 1. *For an edge bipartition $A \uplus B = E(G)$ of a graph G, let $M(A, B)$ be the set of vertices in G which are incident with both an edge in A and in B, that is,*
$$M(A, B) := \{v \in V(G) \mid \exists a \in A \exists b \in B : v \in a \cap b\}.$$

We call $M(A, B)$ the middle set *of A, B. For an edge set $A \subseteq E(G)$, denote by $G\langle A \rangle := (\bigcup_{e \in A} e, A)$ the subgraph induced by A.*

Recall from Sect. 2 the definitions of graph gluing, boundary, and boundary labeling.

Theorem 2 ($\star^2$). *For every connected, bridgeless, r-outerplanar graph G with n vertices there is a sequence $((A_i, B_i, \beta_i))_{i=1}^s$ where each pair $A_i, B_i \subseteq E(G)$ is an edge bipartition of G and $\beta_i \colon M(A_i, B_i) \to \{1, \ldots, |M(A_i, B_i)|\}$ such that $s \geq \log(n)/(r+1)^{32r^2+8r}$, and, for every i, j, $1 \leq i < j \leq s$,*

(i) $|M(A_i, B_i)| = |M(A_j, B_j)| \leq 2r$,
(ii) $A_i \subsetneq A_j$, $B_i \supsetneq B_j$, and
(iii) $G\langle A_i \rangle \circ G\langle B_j \rangle$ is r-outerplanar, where $G\langle A_i \rangle$ is understood to be β_i-boundaried and $G\langle B_j \rangle$ is understood to be β_j-boundaried.

To gain some intuition for Theorem 2 note that each $M(A_i, B_i)$ is a separator, separating its left side $G\langle A_i \rangle$ from its right side $G\langle B_i \rangle$ in G. Statement (ii) ensures that each left sides contains all previous left sides, that is, the separators are nested. Statement (iii) ensures that for any two separators in the sequence, we can glue their left and right sides and again obtain an r-outerplanar graph. In this new graph, the vertices inbetween the separators are missing—these will be the vertices which are not crucial to obtain an r-outerplanar support.

The reason why we can prove the lower bound on the length of the sequence is basically because r-outerplanar graphs have a tree-like structure, whence large r-outerplanar graphs have a long "path" in this structure, and a long path in

---

[2] Results labeled by $\star$ are deferred to a full version of the paper.

such a structure induces many nested separators from which we can glean the separators that are amenable to Statement (iii).

We next formalize the crucial vertices for obtaining an r-outerplanar support. These are the vertices in a smallest representative support, defined as follows.

Definition 2 (Representative support). *Let $\mathcal{H}$ be a hypergraph. A graph G is a representative support for $\mathcal{H}$ if $V(G) \subseteq V(\mathcal{H})$, graph G is a support for subhypergraph $\mathcal{H}|_{V(G)}$ shrunken to $V(G)$, and every vertex in $V(\mathcal{H}) \setminus V(G)$ is covered in $\mathcal{H}$ by some vertex in $V(G)$.*

Using the sequence of separators from Theorem 2, we show that the size of a smallest representative r-outerplanar support is upper-bounded by a function of m and r. To this end, we take an initial support, find two separators whose vertices in between we can remove and reattach as non-crucial vertices, that is, vertices not in a representative support. Intuitively, the two separators have to have the same "status" with respect to the hyperedges that cross them. We formalize this as follows.

Definition 3 (Edge-bipartition signature). *Let $\mathcal{H} = (V, \mathcal{E})$ be a hypergraph and let G be a representative planar support for $\mathcal{H}$. Let (A, B, β) be a tuple where (A, B) is an edge bipartition of G, and $\beta \colon M(A, B) \to \{1, \ldots, |M(A, B)|\}$. Let $\ell := |M(A, B)|$. The signature of (A, B, β) is a triple $(\mathcal{T}, \phi, \mathcal{C})$, where*

- $\mathcal{T} := \{[u]_\tau \mid u \in \bigcup A\}$ *is the set of twin classes in $\bigcup A$,*
- $\phi \colon \{1, \ldots, \ell\} \to \{[u]_\tau \mid u \in V\} \colon j \mapsto [\beta^{-1}(j)]_\tau$ *maps each index of a vertex in $M(A, B)$ to the twin class of that vertex, and*
- $\mathcal{C} := \{(F, \gamma_F) \mid F \in \mathcal{E}\}$, *where γ_F is the relation on $\{1, \ldots, \ell\}$ defined by $(i, j) \in \gamma_F$ whenever $\beta^{-1}(i), \beta^{-1}(j) \in F$ and $\beta^{-1}(i)$ is connected to $\beta^{-1}(j)$ in $G\langle B\rangle[F \cap \bigcup B]$. Herein, $G\langle B\rangle[F \cap \bigcup B]$ is the subgraph of $G\langle B\rangle$ induced by $F \cap \bigcup B$.*

We have the following upper bound on the number of different separator states.

Lemma 1 ($\star$). *In a sequence $((A_i, B_i, \beta_i))_{i=1}^s$ as in Theorem 2 the number of distinct edge-bipartition signatures is upper-bounded by $2^{m \cdot (2r^2 + r + 1)}$.*

As before, let $\psi(m, r) := 2^{6r \cdot 2^{m \cdot (2r^2 + r + 1)} \cdot (r+1)^{32r^2 + 8r}}$.

Lemma 2. *If a hypergraph $\mathcal{H} = (V, \mathcal{E})$ has an r-outerplanar support, then it has a representative r-outerplanar support with at most $\psi(m, r)$ vertices.*

Proof. Let $G = (W, E)$ be a representative r-outerplanar support for $\mathcal{H}$ with the minimum number of vertices and fix a corresponding planar embedding. Assume towards a contradiction that $|W| > \psi(m, r)$. We show that there is a representative support for $\mathcal{H}$ with less than $\psi(m, r)$ vertices.

We aim to apply Theorem 2 to G. For this we need that G is connected and does not contain any bridges. Indeed, if G is not connected, then add edges between its connected components in a tree-like fashion. This does not affect the outerplanarity number of G (although it adds bridges). If G has a bridge $\{u, v\}$, then proceed as follows. At least one of the ends of the bridge, say v, has degree at least two because $|W| > \psi(m, r) \geq 2$. One neighbor $w \neq u$ of v is incident with the same face as u, because $\{u, v\}$ is a bridge. Add the edge $\{u, w\}$. Thus, edge $\{u, v\}$ ceases to be a bridge. We can embed $\{u, w\}$ in such a way that the face $\mathfrak{F}$ incident with u, v, and w is split into one face that is incident with only $\{u, v, w\}$ and devoid of any other vertex, and one face $\mathfrak{F}'$ that is incident with all the vertices that are incident with $\mathfrak{F}$ including u, v, and w. This implies that each vertex retains its layer L_i, meaning that G remains r-outerplanar. Thus, we may assume that G is connected, bridgeless, and r-outerplanar.

Since G contains more than $\psi(m, r)$ vertices, there is a sequence $\mathcal{S} = ((A_i, B_i, \beta_i))_{i=1}^s$ as in Theorem 2 of length at least

$$s \geq \frac{\log(\psi(m, r))}{(r+1)^{32r^2+8r}} = \frac{6r \cdot 2^{m \cdot (2r^2+r+1)} \cdot (r+1)^{32r^2+8r}}{(r+1)^{32r^2+8r}} = 6r \cdot 2^{m \cdot (2r^2+r+1)}.$$

Since there are less than $2^{m \cdot (2r^2+r+1)}$ different signatures in $\mathcal{S}$ (Lemma 1), there are $6r$ elements of $\mathcal{S}$ with the same signature. Note that each middle set $M(A_i, B_i)$ induces a plane graph in G and, since $|M(A_i, B_i)| \leq 2r$, induces at most $\max\{1, 3|M(A_i, B_i)| - 6\} \leq \max\{1, 6r - 6\}$ edges. Thus, there are two edge bipartitions (A_i, B_i, β_i) and (A_j, B_j, β_j), $i < j$, in $\mathcal{S}$ with the same signature such that the middle sets $M(A_i, B_i)$, $M(A_j, B_j)$ differ in at least one vertex.

Let $G_{ij} := G\langle A_i\rangle \circ G\langle B_j\rangle$, wherein $G\langle A_i\rangle$ is β_i-boundaried and $G\langle B_j\rangle$ is β_j-boundaried. Let $W' := V(G_{ij})$, where we assume that $W' \cap M(A_j, B_j) \subseteq M(A_i, B_i)$ for the sake of a simpler notation. Note that $W \setminus W' \neq \emptyset$ since the middle sets of the two edge bipartitions differ in at least one vertex and since $A_i \subsetneq A_j$.

We prove that G_{ij} is a representative support for $\mathcal{H}$, that is, each vertex $V \setminus W'$ is covered by some vertex in W' in $\mathcal{H}$ and that G_{ij} is a support for $\mathcal{H}|_{W'}$. Since G_{ij} is r-outerplanar by Theorem 2, Statement (iii), this contradicts the choice of G according to the minimum number of vertices, thus proving the lemma.

To prove that each vertex $V \setminus W'$ is covered by some vertex in W', we show that $\{[u]_\tau \mid u \in V\} = \{[u]_\tau \mid u \in W'\}$. Since $G = (W, E)$ is a representative support, $\{[u]_\tau \mid u \in V\} = \{[u]_\tau \mid u \in W\}$. Furthermore, by the definition of signature, we have $\{[u]_\tau \mid u \in \bigcup A_i\} = \{[u]_\tau \mid u \in \bigcup A_j\}$. Thus, for each vertex $u \in W \setminus W'$, there is a vertex $v \in W'$ with $[u]_\tau = [v]_\tau$, meaning that, indeed, $\{[u]_\tau \mid u \in V\} = \{[u]_\tau \mid u \in W'\}$.

To show that G_{ij} is a representative support it remains to show that it is a support for $\mathcal{H}|_{W'}$, that is, each hyperedge F' of $\mathcal{H}|_{W'}$ induces a connected graph $G_{ij}[F']$. Let F be a hyperedge of $\mathcal{H}$ such that $F \cap W' = F'$. Observe that such a hyperedge F exists and that $G[F \cap W]$ is connected since G is a representative support of $\mathcal{H}$. Denote by S_k the middle set $M(A_k, B_k)$ of (A_k, B_k) in G for $k \in \{i, j\}$ and by S the middle set $M(A_i, B_j) = S_i = S_j$ of (A_i, B_j) in G_{ij}.

To show that $G_{ij}[F']$ is connected, consider first the case that $F \cap (S_i \cup S_j) = \emptyset$. Since each vertex in $V \setminus W'$ is covered by a vertex in W' we have that each vertex in F is contained in either $G\langle A_i \rangle$ or $G\langle B_j \rangle$ along with all edges of $G[F]$. All these edges are also present in G_{ij} whence $G_{ij}[F']$ is connected.

Now consider the case that $F \cap (S_i \cup S_j) \neq \emptyset$. Since S_i and S_j are separators in G, each vertex in $F \setminus (S_i \cup S_j)$ is connected in $G[F]$ to some vertex in S_i or S_j via a path with internal vertices in $F \setminus (S_i \cup S_j)$. We consider the connectivity relation of their corresponding vertices in S. To this end, for a graph H and $T \subseteq V(H)$ use $\gamma(T, H)$ for the equivalence relation on T of connectivity in H. That is, for $u, v \in T$ we have $(u, v) \in \gamma(T, H)$ if u and v are connected in H. Using this terminology, since both S_i and S_j equal S in G_{ij}, to show that $G_{ij}[F']$ is connected, it is enough to prove that the transitive closure δ of $\gamma(F' \cap S, G_{ij}\langle A_i \rangle) \cup \gamma(F' \cap S, G_{ij}\langle B_j \rangle)$ contains only one equivalence class.

Denote by $\hat{G}$ the graph obtained from G by identifying each $v \in S_i$ with $\beta_j^{-1}(\beta_i(v)) \in S_j$, hence, identifying S_i and S_j, resulting in the set S. Relation $\alpha := \gamma(F \cap S, \hat{G})$ has only one equivalence class and, moreover, it is the transitive closure of $\gamma(F \cap S_i, G\langle A_i \rangle) \cup \gamma(F \cap S, \hat{G}\langle B_i \setminus B_j \rangle) \cup \gamma(F \cap S_j, G\langle B_j \rangle)$, wherein we identify each $v \in S_i$ with $\beta_j^{-1}(\beta_i(v)) \in S_j$ as above and, thus, $S_i = S_j = S$. We have $\gamma(F' \cap S, G_{ij}\langle A_i \rangle) = \gamma(F \cap S_i, G\langle A_i \rangle)$ and $\gamma(F' \cap S, G_{ij}\langle B_j \rangle) = \gamma(F \cap S_j, G\langle B_j \rangle)$. Thus for $\alpha = \delta$ it suffices to prove that $\gamma(F \cap S, \hat{G}\langle B_i \setminus B_j \rangle) \subseteq \gamma(F' \cap S_j, G_{ij}\langle B_j \rangle)$. Indeed, the left-hand side $\gamma(F \cap S, \hat{G}\langle B_i \setminus B_j \rangle)$ is contained in $\gamma(F \cap S_i, G\langle B_i \rangle)$. Let $(\mathcal{T}, \phi, \mathcal{C})$ be the signature of (A_i, B_i, β_i) and (A_j, B_j, β_j) and $(F, \gamma_F) \in \mathcal{C}$. Note that $\gamma(F \cap S_i, G\langle B_i \rangle) = \gamma_F = \gamma(F \cap S_j, G\langle B_j \rangle)$ where we abuse notation and set $u = \beta_i(u)$ for $u \in S_i$ and $v = \beta_j(v)$ for $v \in S_j$. Hence, $\gamma(F \cap S, \hat{G}\langle B_i \setminus B_j \rangle) \subseteq \gamma(F \cap S_j, G\langle B_j \rangle) = \gamma(F' \cap S_j, G\langle B_j \rangle) = \gamma(F' \cap S_j, G_{ij}\langle B_i \rangle)$. Thus, indeed, $\delta = \alpha$, that is, F' is connected. $\square$

We now use the upper bound on the number of vertices in representative supports to get rid of superfluous twins. First, we show that representative supports can be extended to obtain a support.

Lemma 3. *Let $G = (W, E)$ be a representative r-outerplanar support for a hypergraph $\mathcal{H} = (V, \mathcal{E})$. Then $\mathcal{H}$ has an r-outerplanar support in which all vertices of $V \setminus W$ have degree one.*

Proof. Let G' be the graph obtained from G by making each vertex v of $V \setminus W$ a degree-one neighbor of a vertex in W that covers v (such a vertex exists by the definition of representative support). Clearly, the resulting graph is planar. It is also r-outerplanar, which can be seen by adapting an r-outerplanar embedding of G for G': If the neighbor v of a new degree-one vertex u is in L_1, then place u in the outer face. If $v \in L_i$, $i > 1$, then place u in a face which is incident with v and a vertex in L_{i-1} (such a face exists by the definition of L_i).

It remains to show that G' is a support for $\mathcal{H}$. Consider a hyperedge $F \in \mathcal{E}$. Since G is a representative support for $\mathcal{H}$, we have that $F \cap W$ is nonempty and that $G[F \cap W]$ is connected. In G', each vertex $u \in F \setminus W$ is adjacent to some vertex $v \in W$ that covers u. Hence $v \in F$. Thus, $G'[F]$ is connected as $G'[F \cap W]$ is connected and all vertices in $F \setminus W$ are neighbors of a vertex in $F \cap W$. $\square$

We now use Lemma 3 to show that, if there is a twin class that contains more vertices than a small representative support, then we can safely remove one vertex from this twin class.

Lemma 4. *Let $\ell \in \mathbb{N}$, let $\mathcal{H}$ be a hypergraph, and let $v \in V(\mathcal{H})$ be a vertex such that $|[v]_\tau| \geq \ell$. If $\mathcal{H}$ has a representative r-outerplanar support with less than ℓ vertices, then $\mathcal{H} - v$ has an r-outerplanar support.*

Proof. Let $G = (W, E)$ be a representative r-outerplanar support for $\mathcal{H}$ such that $|W| < \ell$. Then at least one vertex of $[v]_\tau$ is not in W and we can assume that this vertex is v without loss of generality. Thus, $\mathcal{H}$ has an r-outerplanar support G' in which v has degree one by Lemma 3. The graph $G' - v$ is an r-outerplanar support for $\mathcal{H} - v$: For each hyperedge F in $\mathcal{H} - v$, we have that $G'[F \setminus \{v\}]$ is connected because v is not a cut-vertex in $G'[F]$ (since it has degree one).   □

Now we combine the observations above with the fact that there are small r-outerplanar supports to prove that Rule 1 is correct.

Proof (Correctness of Rule 1). Consider an instance $\mathcal{H} = (V, \mathcal{E})$ of r-OUTERPLANAR SUPPORT to which Rule 1 is applicable and let $v \in V$ be a vertex to be removed, that is, v is contained in a twin class of size more than $\psi(m, r)$. By Lemma 2, if $\mathcal{H}$ has an r-outerplanar support, then it has a representative r-outerplanar support with at most $\psi(m, r)$ vertices. By Lemma 4, this implies that $\mathcal{H} - v$ has an r-outerplanar support. Moreover, if $\mathcal{H} - v$ has an r-outerplanar support, then this r-outerplanar support is a representative r-outerplanar support for $\mathcal{H}$. By Lemma 3, this implies that $\mathcal{H}$ has an r-outerplanar support. Therefore, $\mathcal{H}$ and $\mathcal{H} - v$ are equivalent instances, and v can be safely removed from $\mathcal{H}$.   □

5   Concluding Remarks

The main contribution of this work is to show that twins may be crucial for instances of r-OUTERPLANAR SUPPORT but the number of crucial twins is upper-bounded in terms of the number m of hyperedges and the outerplanarity number r of a support. As a result, we can safely remove non-crucial twins. More specifically, in linear time we can transform any instance of r-OUTERPLANAR SUPPORT into an equivalent one whose size is upper-bounded by a function of m and r only. In turn, this implies fixed-parameter tractability with respect to $m + r$. It is fair to say, however, that due to the strong exponential growth in m and r this result is mainly of classification nature. Improved bounds (perhaps based on further data reduction rules) are highly desirable for practical applications.

Two further directions for future research are as follows. First, above we only showed how to reduce the size of the input instance. We also need an efficient algorithm to construct an r-outerplanar support for such an instance. As a first step, it would be interesting to improve on the $n^{O(n)}$-time brute-force algorithm that simply enumerates all n-vertex planar graphs and tests whether one of them is an r-outerplanar support.[3]

Second, it is interesting to gear the parameters under consideration more towards practice. In Sect. 4 above we attached signatures to each edge bipartition in a sequence of edge bipartitions of a support and we could reduce our input only if there were sufficiently many edge bipartitions with the same signature. This signature contained, among other information, the twin class of each vertex of the separator induced by the edge bipartition. Clearly, if all of these at least 2^{mr} different types of signatures are present, this will lead to an illegible drawing of the hypergraph (and still, in absence of better upper bounds, we cannot reduce our input). It seems thus worthwhile to contemplate parameters that capture legibility of the hypergraph drawing by restricting further the number of possible signatures.

Finally, an obvious open question is whether finding a *planar* support is (linear-time) fixed-parameter tractable with respect to the number m of hyperedges only. A promising direction might be to show that there is a planar representative support (as in Definition 2) which has treewidth upper-bounded by a function of m. From this, we would get a sequence of gluable subgraphs similarly to the one we have used here, amenable to the same approach as in Sect. 4.

References

1. Alsallakh, B., Micallef, L., Aigner, W., Hauser, H., Miksch, S., Rodgers, P.J.: The state-of-the-art of set visualization. Comput. Graph. Forum **35**(1), 234–260 (2016)
2. Beeri, C., Fagin, R., Maier, D., Yannakakis, M.: On the desirability of acyclic database schemes. J. ACM **30**(3), 479–513 (1983)
3. Bixby, R.E., Wagner, D.K.: An almost linear-time algorithm for graph realization. Math. Oper. Res. **13**(1), 99–123 (1988)
4. Brandes, U., Cornelsen, S., Pampel, B., Sallaberry, A.: Blocks of hypergraphs—applied to hypergraphs and outerplanarity. In: Iliopoulos, C.S., Smyth, W.F. (eds.) IWOCA 2010. LNCS, vol. 6460, pp. 201–211. Springer, Heidelberg (2011). doi:10.1007/978-3-642-19222-7_21
5. Brandes, U., Cornelsen, S., Pampel, B., Sallaberry, A.: Path-based supports for hypergraphs. J. Discrete Algorithms **14**, 248–261 (2012)
6. Buchin, K., van Kreveld, M.J., Meijer, H., Speckmann, B., Verbeek, K.: On planar supports for hypergraphs. J. Graph Algorithms Appl. **15**(4), 533–549 (2011)
7. Chen, J., Komusiewicz, C., Niedermeier, R., Sorge, M., Suchý, O., Weller, M.: Polynomial-time data reduction for the subset interconnection design problem. SIAM J. Discrete Math. **29**(1), 1–25 (2015)
8. Cygan, M., Fomin, F.V., Kowalik, L., Lokshtanov, D., Marx, D., Pilipczuk, M., Pilipczuk, M., Saurabh, S.: Parameterized Algorithms. Springer, Berlin (2015)
9. Diestel, R.: Graph Theory. Graduate Texts in Mathematics, vol. 173, 4th edn. Springer, Berlin (2010)

---

[3] Each planar graph has an ordering of the vertices such that each vertex has at most five neighbors later in the ordering. To achieve $n^{O(n)}$ enumeration time we simply guess such an ordering and then for each vertex its at most five later neighbors.

10. Dinkla, K., van Kreveld, M.J., Speckmann, B., Westenberg, M.A.: Kelp diagrams: point set membership visualization. Comput. Graph. Forum **31**(3), 875–884 (2012)
11. Eschbach, T., Günther, W., Becker, B.: Orthogonal hypergraph drawing for improved visibility. J. Graph Algorithms Appl. **10**(2), 141–157 (2006)
12. Flower, J., Fish, A., Howse, J.: Euler diagram generation. J. Vis. Lang. Comput. **19**(6), 675–694 (2008)
13. Habib, M., Paul, C., Viennot, L.: Partition refinement techniques: an interesting algorithmic tool kit. Int. J. Found. Comput. Sci. **10**(2), 147–170 (1999)
14. Johnson, D.S., Pollak, H.O.: Hypergraph planarity and the complexity of drawing Venn diagrams. J. Graph Theory **11**(3), 309–325 (1987)
15. Kaufmann, M., Kreveld, M., Speckmann, B.: Subdivision drawings of hypergraphs. In: Tollis, I.G., Patrignani, M. (eds.) GD 2008. LNCS, vol. 5417, pp. 396–407. Springer, Heidelberg (2009). doi:10.1007/978-3-642-00219-9_39
16. Klemz, B., Mchedlidze, T., Nöllenburg, M.: Minimum tree supports for hypergraphs and low-concurrency Euler diagrams. In: Ravi, R., Gørtz, I.L. (eds.) SWAT 2014. LNCS, vol. 8503, pp. 265–276. Springer, Heidelberg (2014). doi:10.1007/978-3-319-08404-6_23
17. Korach, E., Stern, M.: The clustering matroid and the optimal clustering tree. Math. Program. **98**(1–3), 385–414 (2003)
18. Mäkinen, E.: How to draw a hypergraph. Int. J. Comput. Math. **34**, 178–185 (1990)
19. Tarjan, R., Yannakakis, M.: Simple linear-time algorithms to test chordality of graphs, test acyclicity of hypergraphs, and selectively reduce acyclic hypergraphs. SIAM J. Comput. **13**(3), 566–579 (1984)
20. Tutte, W.T.: How to draw a graph. Proc. London Math. Soc. **13**(3), 743–768 (1963)
21. Verroust, A., Viaud, M.-L.: Ensuring the drawability of extended Euler diagrams for up to 8 sets. In: Blackwell, A.F., Marriott, K., Shimojima, A. (eds.) Diagrams 2004. LNCS (LNAI), vol. 2980, pp. 128–141. Springer, Heidelberg (2004). doi:10.1007/978-3-540-25931-2_13

Multi-colored Spanning Graphs

Hugo A. Akitaya[1], Maarten Löffler[2], and Csaba D. Tóth[1,3(✉)]

[1] Tufts University, Medford, MA, USA
hugo.alves_akitaya@tufts.edu
[2] Utrecht University, Utrecht, The Netherlands
m.loffler@uu.nl
[3] California State University Northridge, Los Angeles, CA, USA
csaba.toth@csun.edu

Abstract. We study a problem proposed by Hurtado et al. [10] motivated by sparse set visualization. Given n points in the plane, each labeled with one or more primary colors, a *colored spanning graph* (CSG) is a graph such that for each primary color, the vertices of that color induce a connected subgraph. The MIN-CSG problem asks for the minimum sum of edge lengths in a colored spanning graph. We show that the problem is NP-hard for k primary colors when $k \geq 3$ and provide a $(2 - \frac{1}{3+2\varrho})$-approximation algorithm for $k = 3$ that runs in polynomial time, where ϱ is the Steiner ratio. Further, we give a $O(n)$ time algorithm in the special case that the input points are collinear and k is constant.

1 Introduction

Visualizing set systems is a basic problem in data visualization. Among the oldest and most popular set visualization tools are the Venn and Euler diagrams. However, other methods are preferred when the data involves a large number of sets with complex intersection relations [2]. In particular, a variety of tools have been proposed for set systems where the elements are associated with location data. Many of these methods use geometric graphs to represent set membership, motivated by reducing the amount of ink used in the representation, including LineSets [1], Kelp Diagrams [7] and KelpFusion [11].

Hurtado et al. [10] recently proposed a method for drawing sets using outlines that minimise the total visual clutter. The underlying combinatorial problem is to compute a *minimum colored spanning graph*; see Fig. 1. They studied the problem for n points in a plane and two sets (each point is a member of one or both sets). The output is a graph with the minimum sum of edge lengths such that the subgraph induced by each set is connected. They gave an algorithm that runs in $O(n^6)$-time,[1] and a $(\frac{1}{2}\varrho + 1)$-approximation in $O(n \log n)$ time, where ϱ is the Steiner ratio (the ratio between the length of a minimum spanning tree and the length of a minimum Steiner tree). Efficient algorithms are known in two special cases: One runs in $O(n)$ time for collinear points that are already sorted [10]; the other runs in $O(m^2 + n)$ time for cocircular points, where m is

---

[1] An earlier claim that the problem was NP-hard [9] turned out to be incorrect [10].

© Springer International Publishing AG 2016
Y. Hu and M. Nöllenburg (Eds.): GD 2016, LNCS 9801, pp. 81–93, 2016.
DOI: 10.1007/978-3-319-50106-2_7

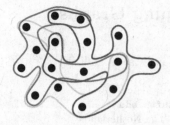 

Fig. 1. Left: A set of points and three subsets, S_1, S_2, and S_3, drawn as outlines in different colors. Right: The corresponding (minimum) colored spanning graph. Refer to Sect. 2 for an explanation of color use. (Color figure online)

the number of points that are elements of both sets [5]. This problem also has applications for connecting different networks with minimum cost, provided that edges whose endpoints belong to both networks can be shared.

Results and Organization. We study the minimum colored spanning graph problem for n points in a plane and k sets, $k \geq 3$. The formal definition and some properties of the optimal solution are in Sect. 2. In Sect. 3, we show that MIN-kCSG is NP-complete for all $k \geq 3$, and in Sect. 4 we provide an $(2 - \frac{2}{2+2\varrho})$-approximation algorithm for $k = 3$ that runs in $O(n \log n + m^6)$ time, where m is the number of multichromatic points. This improves the previous $(2 + \frac{\varrho}{2})$-approximation from [10]. Section 5 describes an algorithm for the special case of collinear points that runs in $2^{O(k^2 2^k)} \cdot n$ time. Due to space constraints, some proofs are omitted; they can be found in the full version of this paper.

2   Preliminaries

In this section, we define the problem and show a property of the optimal solution related to the minimum spanning trees, which is used in Sects. 3 and 4.

Definitions. Given a set of n points in the plane $S = \{p_1, \ldots, p_n\}$ and subsets $S_1, \ldots, S_k \subseteq S$, we represent set membership with a function $\alpha : S \to 2^{\{1,\ldots,k\}}$, where $p \in S_c$ iff $c \in \alpha(p)$ for every *primary color* $c \in \{1, \ldots, k\}$. We call $\alpha(p)$ the *color* of point p. A point p is *monochromatic* if it is a member of a single set S_i, that is, $|\alpha(p)| = 1$, and *multi-chromatic* if $|\alpha(p)| > 1$. For an edge $\{p_i, p_j\} \in E$ in a graph $G = (S, E)$, we use the shorthand notation $\alpha(\{p_i, p_j\}) = \alpha(p_i) \cap \alpha(p_j)$ for the shared primary colors of the two vertices. For every $c \in \{1, \ldots, k\}$, we let $G_c = (S_c, E_c)$ denote the subgraph of $G = (S, E)$ induced by S_c. All figures in this paper depict only three primary colors: r, b, and y for red, blue, and yellow respectively. Multi-chromatic points and edges are shown green, orange, purple, or black if their color is $\{b, y\}$, $\{r, y\}$ or $\{r, b\}$, or $\{r, b, y\}$, respectively. See, for example, Fig. 1 (b).

A *colored spanning graph* for the pair (S, α), denoted CSG(S, α), is a graph $G = (S, E)$ such that (S_c, E_c) is connected for every primary color $c \in \{1, \ldots, k\}$. The *minimum colored spanning graph* problem (MIN-CSG), for a given pair

(S, α), asks for the minimum cost $\sum_{e \in E} w(e)$ of a $\mathrm{CSG}(S, \alpha)$, where $w(e)$ is the Euclidean length of e. When we wish to emphasize the number k of primary colors, we talk about the MIN-kCSG problem.

Monochromatic Edges in a Minimum CSG. The following lemma shows that we can efficiently compute some of the monochromatic edges of a minimum CSG for an instance (S, α) using the *minimum spanning tree* (*MST*) of S_c for every primary color $c \in \{1, \ldots, k\}$.

Lemma 1. *Let (S, α) be an instance of MIN-CSG and $c \in \{1, \ldots, k\}$. Let $E(MST(S_c))$ be the edge set of an MST of S_c, and let S'_c be the set of multi-chromatic points in S_c. Then there exists a minimum CSG that contains at least $|E(MST(S_c))| - |S'_c| + 1$ edges of $E(MST(S_c))$. The common edges of $E(MST(S_c))$ and of such a minimum CSG can be computed in $O(n \log n)$ time.*

Proof. Construct a monochromatic subset $E'_c \subset E(MST(S_c))$ by successively removing a longest edge from the path in $MST(S_c)$ between any two points in S'_c. An $MST(S_c)$ can be computed in $O(n \log n)$ time, and E'_c can be obtained in $O(n)$ time. The graph (S_c, E'_c) has $|S'_c|$ components, each containing one element of S'_c, hence $|E'_c| = |E(MST(S_c))| - |S'_c| + 1$.

Let (S, E^{OPT}) be a minimum CSG. While there is an edge $e \in E'_c \setminus E^{OPT}$, we can find an edge $e^* \in E^{OPT} \setminus E'_c$ such that exchanging e^* for e yields another minimum CSG. Indeed, since (S_c, E_c^{OPT}) is connected, the insertion of the edge e creates a cycle C that contains e. Consider the longest (open or closed) path $P \subseteq C$ that is monochromatic and contains e. Note that at least one of the endpoints of e is monochromatic, therefore P contains at least two monochromatic edges. Since every component of (S_c, E'_c) is a tree and contains only one multi-chromatic point, there is a monochromatic edge $e^* \in E^{OPT} \setminus E'_c$ in P. We have $w(e) \leq w(e^*)$, because there is a cut of the complete graph on S_c that contains both e and e^*, and $e \in E(MST(S_c))$. Since $\alpha(e^*) = c$, the deletion of e^* can only influence the connectivity of the induced subgraph (S_c, E_c^{OPT}). Consequently, $(S, E^{OPT} \cup \{e\} \setminus \{e^*\})$ is a CSG with equal or lower cost than (S, E^{OPT}). By successively exchanging the edges in $E'_c \setminus E^{OPT}$, we obtain a minimal CSG containing E'_c. $\square$

Hurtado et al. [10] gave an $O(n^6)$-time algorithm for MIN-2CSG, by a reduction to a matroid intersection problem on the set of all possible edges on S, which has $O(n^2)$ elements. Their algorithm for matroid intersection finds $O(n^2)$ single source shortest paths in a bipartite graph with $O(n^2)$ vertices and $O(n^4)$ edges, which leads to an overall running time of $O(n^6)$. We improve the runtime to $O(n \log n + m^6)$, where m is the number of multi-chromatic points.

Corollary 1. *An instance (S, α) of MIN-2CSG can be solved in $O(n \log n + m^6)$ time, where m is the number of multi-chromatic points in S.*

Proof. By Lemma 1, we can compute two spanning forests on S_1 and S_2, respectively, each with m components, that are subgraphs of a minimum CSG in $O(n \log n)$ time. It remains to find edges of minimum total length that connect these components in each color, for which we can use the same matroid intersection algorithm as in [10], but with a ground set of size $O(m^2)$. $\square$

3   General Case

We show that the decision version of MIN-CSG is NP-complete. We define the decision version of MIN-CSG as follows: given an instance (S, α) and $W > 0$, is there a CSG (S, E) such that $\sum_{e \in E} w(e) < W$?

Lemma 2. MIN-kCSG *is in NP.*

Proof. Given a set of edges E, we can verify if (S, E) is a $CSG(S, \alpha)$ in $O(k|S|)$ time by testing connectivity in (S_c, E_c) for each primary color $c \in \{1, \ldots, k\}$, and then check whether $\sum_{e \in E} w(e) \leq W$ in $O(|E|)$ time.   □

We reduce MIN-3CSG from PLANAR-MONOTONE-3SAT, which is known to be NP-complete [4]. For every instance A of PLANAR-MONOTONE-3SAT, we construct an instance $f(A)$ of MIN-3CSG. An instance A consists of a plane bipartite graph between n variable and m clause vertices such that every clause has degree three or two, all variables lie on the x-axis and edges do not cross the x-axis. Clauses are called *positive* if they are in the upper half-plane or *negative* otherwise. The problem asks for an assignment from the variable set to $\{$true, false$\}$ such that each positive (negative) clause is adjacent to a true (false) variable.

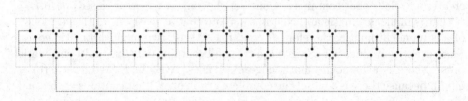

Fig. 2. Construction for an instance A equivalent to the boolean formula $(x_1 \vee x_3 \vee x_5) \wedge (\neg x_1 \vee \neg x_5) \wedge (\neg x_2 \vee \neg x_3 \vee \neg x_4)$. (Color figure online)

Given an instance A of PLANAR-MONOTONE-3SAT, we construct $f(A)$ as shown in Fig. 2 (refer to the full paper for a figure of a single variable gadget). The points marked with small disks are called *active* and they are the only multi-chromatic points in the construction. The dashed lines in a primary color represent a chain of equidistant monochromatic points, where the gap between consecutive points is ε. A purple (resp., black) dashed line represents a red and a blue (resp., a red, a blue, and a yellow) dashed line that run ε close to each other. Informally, the value of ε is set small enough such that every point in the interior of a dashed line is adjacent to its neighbors in any minimum CSG. The boolean assignment of A is encoded in the edges connecting active points. We break the construction down to gadgets and explain their behavior individually.

The long horizontal purple dashed line is called *spine* and the set of yellow dashed lines (shown in Fig. 3(a)) is called *cage*. The rest of the construction consists of variable and clause gadgets (shown in Figs. 3(b) and (c)). The width

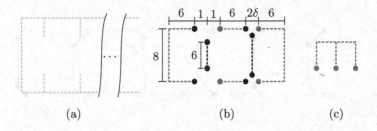

Fig. 3. (a) Cage. (b) Variable gadget. (c) Clause gadget. (Color figure online)

of a variable gadget depends on the degree of the corresponding variable in the bipartite graph given by the instance A. For every edge incident to the variable, we repeat the middle part of the gadget as shown in Fig. 3(b) (see Fig. 2 for variables of degree 1 and 2). The vertical black dashed lines are called *ribs* and the set of three or four active points close to an endpoint of a rib is called *switch*. The variable gadget contains switches of two different sizes alternately from left to right. A *2-switch* (resp., *2δ-switch*) is a switch in which active points are at most 2 (resp., 2δ) apart. The clause gadgets are positioned as the embedding of clauses in A; refer to Fig. 2. Each active point of a positive (negative) clause is assigned to a 2δ-switch and positioned vertically above (below) the active point of the rib, at distance 2δ from it.

Let E' be the set of all monochromatic edges of a minimum CSG computable by Lemma 1. Let r be the number of edges in the bipartite graph of A. The instance $f(A)$ contains $13r$ active points, so (S, E') contains $13r$ connected components. By construction, the number of ε-edges in a solution of $f(A)$ between components of (S, E') is upper bounded by $39r$ (one edge per color per component). Finally, we set $W = (\sum_{e \in E'} w(e)) + 39r\varepsilon + r(2 + 2\sqrt{2}) + r\delta(2 + 2\sqrt{2}) + m\delta(2\sqrt{2} - 2)$ and we choose $\varepsilon = \frac{1}{500r^2}$ and $\delta = \frac{1}{10r}$. This particular choice of ε and δ is justified by the proofs of Corollaries 2 and 3. By construction, $f(A)$ has the following property:

(I) For every partition of the components of (S_c, E'_c) into two sets C_1, C_2, where c is a primary color, let $\{p_1, p_2\}$ be the shortest edge between C_1 and C_2. Then either $w(\{p_1, p_2\}) = \varepsilon$ or p_1 and p_2 are active points in the same switch.

Definition 1. *A* standard solution *of* MIN-3CSG *is a solution that contains E' and in which every edge longer than ε is between two active points of the same switch.*

Lemma 3. *Let A be a positive instance of* PLANAR-MONOTONE-3SAT. *Then $f(A)$ is a positive instance of* MIN-3CSG.

To prove the lemma, we construct a standard solution for $f(A)$ based on the solution for A. This proof, and subsequent proofs, argues about all possible ways to connect the vertices in a switch of $f(A)$. The most efficient ones are shown in Fig. 4; these may appear in an optimal solution. Refer to the full paper for a complete list.

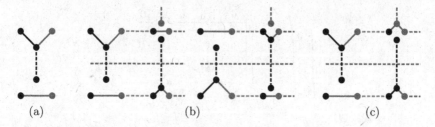

Fig. 4. Possible ways to connect the vertices in a switch of $f(A)$. (a) One of the two states of a 2-switch, encoding the truth value of the variable. (b) The two possible states of a 2δ-switch if the incident clause is not satisfied through this variable. (c) The only possible state of a 2δ-switch if the incident clause is satisfied through this variable. (Color figure online)

Lemma 4. *If $f(A)$ is a positive instance of* MIN-3CSG, *there exists a standard solution for this instance.*

Before proving the other direction of the reduction, we show some properties of a standard solution. The active points in a switch impose some local constraints. The black and purple points attached to horizontal dashed lines determine the *switch constraint*: since these points have more colors than their incident dashed lines, they each are incident to at least one edge in the switch. Each rib determines a *rib constraint* to a pair of switches that contain its endpoints: at least one of these switches must contain an edge between its black active points or else there is no yellow path between this rib and the cage. The following lemmas establish some bounds on the length of the edges used to satisfy local constraints of a pair of switches adjacent to a rib. We refer to this pair as a 2-pair or 2δ-pair according to the type of the switch.

Lemma 5. *In a standard solution, the minimum length required to satisfy the local constraints of a 2-pair (resp., 2δ-pair) is $2(1 + \sqrt{2})$ (resp., $2\delta(1 + \sqrt{2})$).*

Corollary 2. *In a standard solution, every 2-pair is connected minimally.*

Lemma 6. *In a standard solution, for each clause gadget, there exists a 2δ-pair with local cost at least $4\delta\sqrt{2}$.*

Corollary 3. *In a standard solution, for each clause gadget, there exists a 2δ-pair connected as Fig. 4(c). All other 2δ-pairs are connected minimally as shown in Fig. 4(b).*

Lemma 7. *Let $f(A)$ be a positive instance of* MIN-3CSG. *Then A is a positive instance of* PLANAR-MONOTONE-3SAT.

The following theorem is a direct consequence of Lemmata 2, 3, and 7.

Theorem 1. MIN-kCSG *is NP-complete for $k \geq 3$.*

4   Approximation

Hurtado et al. [10] gave an approximation algorithm for MIN-kCSG that runs in $O(n \log n)$ time and achieves a ratio of $\lceil k/2 \rceil + \lfloor k/2 \rfloor \varrho/2$, where ϱ is the Steiner ratio. The value of ϱ is not known and the current best upper bound is $\varrho \leq 1.21$ by Chung and Graham [6] (Gilbert and Pollack [8] conjectured $\varrho = \frac{2}{\sqrt{3}} \approx 1.15$). For the special case $k = 3$, the previous best approximation ratio is $2 + \varrho/2 \leq 2.6$. We improve the approximation ratio to 2, and then further to 1.81. Our first algorithm immediately generalises to $k \geq 3$, and yields an $\lceil k/2 \rceil$-approximation, improving on the general result by Hurtado et al.; our second algorithm also generalizes to $k > 3$, however, we do not know whether it achieves a good ratio.

Suppose we are given an instance of MIN-3CSG defined by (S, α) where $|S| = n$ and the set of primary colors is $\{\mathtt{r}, \mathtt{b}, \mathtt{y}\}$. We define $\alpha_{\mathtt{rb}} : S_{\mathtt{r}} \cup S_{\mathtt{b}} \to 2^{\{r,b\}} \setminus \{\emptyset\}$ where $\alpha_{\mathtt{rb}}(p) = \alpha(p) \setminus \{\mathtt{y}\}$. Let G^* be an optimal solution for MIN-3CSG, and put OPT $= \|G^*\|$. Algorithm A1 computes a minimum red-blue-purple graph $G_{\mathtt{rb}} = CSG(S_{\mathtt{r}} \cup S_{\mathtt{b}}, \alpha_{\mathtt{rb}})$ in $O(n \log n + m^6)$ time, where $m = |S_{\mathtt{r}} \cap S_{\mathtt{b}}|$ by Corollary 1; then computes a minimum spanning tree $G_{\mathtt{y}}$ of $S_{\mathtt{y}}$, and returns the union $G_{\mathtt{rb}} \cup G_{\mathtt{y}}$. Since G^* contains a red, a blue, and a yellow spanning tree, we have $\|G_{\mathtt{rb}}\| \leq$ OPT and $\|G_{\mathtt{y}}\| \leq$ OPT; that is, Algorithm A1 returns a solution to MIN-3CSG whose length is at most 2OPT.

Theorem 2. *Algorithm A1 returns a 2-approximation for* MIN-3CSG*; it runs in* $O(n \log n + m^6)$ *time on* n *points,* m *of which are multi-chromatic.*

Algorithm A1 can be extended to k colors by partitioning the primary colors into $\lceil \frac{k}{2} \rceil$ groups of at most two and computing the minimum CSG for each group. The union of these graphs is a $\lceil \frac{k}{2} \rceil$-approximation that can be computed in $O(kn^6)$ time.

Algorithm A2 computes six solutions for a given instance of MIN-3CSG, $G_1, \ldots, G_6$, and returns one with minimum weight. Graph G_1 is the union of $G_{\mathtt{rb}}$ and $G_{\mathtt{y}}$ defined above. Graphs G_2 and G_3 are defined analogously: $G_2 = G_{\mathtt{ry}} \cup G_{\mathtt{b}}$ and $G_3 = G_{\mathtt{by}} \cup G_{\mathtt{r}}$, each of which can be computed in $O(n^6)$ time by [10]. Let $S_{\mathtt{rby}} \subseteq S$ be the set of "black" points that have all three colors, and let H be an MST of $S_{\mathtt{rby}}$, which can be computed in $O(n \log n)$ time. We augment H into a solution of MIN-3CSG in three different ways as follows. First, let $G_{\mathtt{rb}:H}$ be the minimum forest such that $H \cup G_{\mathtt{rb}:H}$ is a minimum red-blue-purple spanning graph on $S_{\mathtt{r}} \cup S_{\mathtt{b}}$. $G_{\mathtt{rb}:H}$ can be computed in $O(n \log n + m^6)$ time by the same matroid intersection algorithm as in Corollary 1, by setting the weight of any edge between components containing black points to zero. Similarly, let $G_{\mathtt{y}:H}$ be the minimum forest such that $H \cup G_{\mathtt{y}:H}$ is a spanning tree on $S_{\mathtt{y}}$, which can be computed in $O(n \log n)$ time by Prim's algorithm. Now we let $G_4 = H \cup G_{\mathtt{rb}:H} \cup G_{\mathtt{y}:H}$. Similarly, let $G_5 = H \cup G_{\mathtt{ry}:H} \cup G_{\mathtt{b}:H}$ and $G_6 = H \cup G_{\mathtt{by}:H} \cup G_{\mathtt{r}:H}$.

Theorem 3. *Algorithm A2 returns a* $(2 - \frac{1}{3 + 2\varrho})$*-approximation for* MIN-3CSG*; it runs in* $O(n \log n + m^6)$ *time on an input of* n *points,* m *of which are multi-chromatic.*

Proof. Consider an instance (S, α) of MIN-3CSG, and let $G^* = (S, E^*)$ be an optimal solution with $\|E^*\| = \text{OPT}$. Partition E^* into 7 subsets: for every color $\gamma \in 2^{\{r,b,y\}} \setminus \emptyset$, let $E^*_\gamma = \{e \in E^* : \alpha(e) = \gamma\}$, that is E^*_γ is the set of edges of color γ in G^*. Put $\beta = \|E^*_{rby}\|/\text{OPT}$. Then we have $2(1 - \beta)\text{OPT} = (2\|E^*_r\| + \|E^*_{rb}\| + \|E^*_{ry}\|) + (2\|E^*_b\| + \|E^*_{rb}\| + \|E^*_{by}\|) + (2\|E^*_y\| + \|E^*_{ry}\| + \|E^*_{by}\|)$. Without loss of generality, we may assume $2\|E^*_y\| + \|E^*_{ry}\| + \|E^*_{by}\| \le \frac{2}{3}(1-\beta)\text{OPT}$.

First, consider $G_1 = G_{rb} \cup G_y$. The edges of G^* whose colors include red or blue (resp., yellow) form a connected graph on $S_r \cup S_b$ (resp., S_y). Consequently,

$$\|G_{rb}\| \le \|E^*_r\| + \|E^*_b\| + \|E^*_{rb}\| + \|E^*_{ry}\| + \|E^*_{by}\| + \|E^*_{rby}\|. \tag{1}$$

$$\|G_y\| \le \|E^*_y\| + \|E^*_{ry}\| + \|E^*_{by}\| + \|E^*_{rby}\|. \tag{2}$$

The combination of (1) and (2) yields

$$\|G_1\| \le \|G_{rb}\| + \|G_y\| \le \text{OPT} + \|E^*_{ry}\| + \|E^*_{by}\| + \|E^*_{rby}\|$$

$$\le \text{OPT} + \frac{2}{3}(1 - \beta) \cdot \text{OPT} + \beta \cdot \text{OPT} = \frac{5 + \beta}{3}\text{OPT}. \tag{3}$$

Next, consider $G_4 = H \cup G_{rb:H} \cup G_{y:H}$. The edges of G^* whose colors include yellow contain a spanning tree on S_y, hence a Steiner tree on the black points S_{rby}. Specifically, the black edges in E^*_{rby} form a black spanning forest, which is completed to a Steiner tree by some of the edges of $E^*_y \cup E^*_{by} \cup E^*_{ry}$. This implies

$$\|H\| \le \|E^*_{rby}\| + \varrho \cdot (\|E^*_y\| + \|E^*_{by}\| + \|E^*_{ry}\|)$$

$$\le \beta \cdot \text{OPT} + \varrho\frac{2}{3}(1 - \beta) \cdot \text{OPT} = \left(\beta + \frac{2}{3}\varrho - \frac{2}{3}\beta\varrho\right)\text{OPT}.$$

Since H is a spanning tree on the black vertices S_{rby}, (1) and (2) reduce to

$$\|G_{rb:H}\| \le \|E^*_r\| + \|E^*_b\| + \|E^*_{rb}\| + \|E^*_{ry}\| + \|E^*_{by}\|, \tag{4}$$

$$\|G_{y:H}\| \le \|E^*_y\| + \|E^*_{ry}\| + \|E^*_{by}\|. \tag{5}$$

The combination of (4) and (5) yields

$$\|G_{rb:H}\| + \|G_{y:H}\| \le (\text{OPT} - \|E^*_{rby}\|) + \|E^*_{ry}\| + \|E^*_{by}\|$$

$$\le (1 - \beta) \cdot \text{OPT} + \frac{2}{3}(1 - \beta) \cdot \text{OPT} = \frac{5}{3}(1 - \beta) \cdot \text{OPT}.$$

Therefore,

$$\|G_4\| = \|H\| + \|G_{rb:H}\| + \|G_{y:H}\| \le \left(\frac{5}{3} + \frac{2}{3}(\varrho - \beta - \beta\varrho)\right)\text{OPT}. \tag{6}$$

If we set $\beta = \frac{2\varrho}{3+2\varrho}$, then both (3) and (6) give the same upper bound

$$\frac{\min(\|G_1\|, \|G_4\|)}{\text{OPT}} \le \frac{5 + \beta}{3} = 2 - \frac{1}{3 + 2\varrho} \le 1.816,$$

where we used the current best upper bound for the Steiner ratio $\varrho \le 1.21$ from [6]. $\qquad\square$

5   Collinear Points

In this section we consider instances of MIN-kCSG, (S, α), where $k \geq 3$ and S consists of collinear points. An example is shown in Fig. 5. Without loss of generality, $S = \{p_1, \ldots, p_n\}$ and the points p_i, $1 \leq i \leq n$, lie on the x-axis sorted by x-coordinates. We present a dynamic programming algorithm that solves MIN-kCSG in $2^{O(k^2 2^k)} \cdot n$ time.

Fig. 5. An example with optimal solution for collinear points. (Color figure online)

Our first observation is that if the points in S are collinear, we may assume that every edge satisfies the following property.

If $\{p_a, p_b\}$, $a < b$, is an edge, then there is no r, $a < r < b$, such that $\alpha(\{p_a, p_b\}) \subseteq \alpha(p_r)$. $\hspace{3em} (\star)$

Lemma 8. *For every graph $G = (S, E)$, there exists a graph $G' = (S, E')$ of the same cost that satisfies $(\star)$ and for each color $c \in \{1, \ldots, k\}$, every component of (S_c, E_c) is contained in some component of (S_c, E'_c). In particular, MIN-kCSG has a solution with property $(\star)$.*

Proof. Let $G = (S, E)$ be a graph, and let X_G denote the set of triples $(i, j; r)$ such that $1 \leq i < r < j \leq n$, $\{p_i, p_j\} \in E$, and $\alpha(\{p_i, p_j\}) \subseteq \alpha(p_r)$. If $X_G = \emptyset$, then G satisfies $(\star)$. Suppose $X_G \neq \emptyset$. For every triple $(i, j; r) \in X_G$, successively, replace the edge $\{p_i, p_j\}$ by two edges $\{p_i, p_r\}$ and $\{p_r, p_j\}$ (i.e., subdivide edge $\{p_i, p_j\}$ at p_r). Note that $\alpha(\{p_h, p_i\}), \alpha(\{p_i, p_j\}) \subseteq \alpha(p_i)$, consequently p_i and p_j remain in the same component for each primary color $c \in \alpha(\{p_i, p_j\})$. Each step maintains the total edge length of the graph and strictly decreases X_G. After $|X_G|$ subdivision steps, we obtain a graph $G' = (S, E')$ as required. $\hspace{1em}\square$

In the remainder of this section we assume that every edge has property $(\star)$. Furthermore, all graphs considered in this section are defined on an interval of consecutive vertices of S.

Corollary 4. *Let $G = (S, E)$ be a graph and let $i \in \{1, \ldots, n\}$.*

1. *If $e \in E$ is an edge between $\{p_1, \ldots, p_i\}$ and $\{p_{i+1}, \ldots, p_n\}$ and $\alpha(e) = \gamma$, then the endpoints of e are uniquely determined. Specifically, if $e = \{p_a, p_b\}$ with $1 \leq a \leq i < b \leq n$, then $a \in \{1, \ldots, i\}$ is the largest index such that $\gamma \subset \alpha(p_a)$, and $b \in \{i+1, \ldots, n\}$ is the smallest index such that $\gamma \subset \alpha(p_b)$.*
2. *If two edges $e_1, e_2 \in E$ overlap, then $\alpha(e_1) \neq \alpha(e_2)$.*

Proof.

(1) Suppose, to the contrary, that there is index j, $a < j \leq i$, such that $\gamma \subset \alpha(p_j)$. Then edge $\{p_a, p_b\}$ and point p_r violate $(\star)$. The case that there is some j, $i + 1 \leq j < b$, leads to the same contradiction.
(2) Without loss of generality $e_1 = \{p_a, p_b\}$ and $e_2 = \{p_i, p_j\}$ with $a \leq i < b \leq j$. Then both edges e_1 and e_2 are between $\{p_1, \ldots, p_i\}$ and $\{p_{i+1}, \ldots, p_n\}$, contradicting part (1). $\qquad\square$

The basis for our dynamic programming algorithm is that MIN-kCSG has the *optimal substructure* and *overlapping substructures* properties when the points in S are collinear. We introduce some notation for defining the subproblems. For indices $1 \leq a \leq b \leq n$, let $S[a, b] = \{p_a, \ldots, p_b\}$. For every graph $G = (S, E)$ and index $i \in \{1, \ldots, n\}$, we partition the edge set E into three subsets as follows: let E_i^- be the set of edges induced by $S[1, i]$, E_i^+ the set of edges induced by $S[i + 1, n]$, and E_i^0 the set of edges between $S[1, i]$ and $S[i + 1, n]$. With this notation, MIN-kCSG has the following optimal substructure property.

Lemma 9. *Let $G = (S, E)$ be a minimum CSG, $i \in \{1, \ldots, n\}$, and $\mathcal{X}$ be the family of edge sets X_i^- on $S[1, i]$ such that $(S, X_i^- \cup E_i^0 \cup E_i^+)$ is a CSG. Then $(S, X_i^- \cup E_i^0 \cup E_i^+)$ is a minimum CSG iff $X_i^- \in \mathcal{X}$ has minimum cost.*

Proof. If $(S, X_i^- \cup E_i^0 \cup E_i^+)$ is a minimum CSG, but some $Y_i^- \in \mathcal{X}$ costs less than E_i^-, then $(S, Y_i^- \cup E_i^0 \cup E_i^+)$ would be a CSG that costs less than $G = (S, E)$, contradicting the minimality of $(S, X_i^- \cup E_i^0 \cup E_i^+)$. If $X_i^- \in \mathcal{X}$ has minimum cost, but $G = (S, E)$ costs less than $(S, X_i^- \cup E_i^0 \cup E_i^+)$, then $E_i^- \in \mathcal{X}$ would costs less than X_i^-, contradicting the minimality of $X_i^- \in \mathcal{X}$. $\qquad\square$

Lemma 9 immediately suggests a naïve algorithm for MIN-kCSG: Guess the edge set $E_i^0 \cup E_i^+$ of a minimum CSG $G = (S, E)$, and compute a minimum-cost set X_i^- on $S[1, i]$ such that $(S, X_i^- \cup E_i^0 \cup E_i^+)$ is a CSG. However, all possible edge sets $E_i^0 \cup E_i^+$ could generate $2^{\Theta(n)}$ subproblems. We reduce the number of subproblems using the overlapping subproblem property. Instead of guessing $E_i^0 \cup E_i^+$, it is enough to guess the information relevant for finding the minimal cost X_i^- on $S[1, i]$. First, the edges in E_i^0 can be uniquely determined by the set of their colors (using Corollary 4 (1)). Second, the only useful information from E_i^+ is to tell which points in $S[1, i]$ are adjacent to the same component of $(S[i + 1, n]_c, (E_i^+)_c)$, for each primary color $c \in \{1, \ldots, k\}$. This information can be summarized by k equivalence relations on the sets $(E_i^0)_1, \ldots, (E_i^0)_k$. We continue with the details.

We can encode E_i^0 by the set of its colors $\Gamma_i = \{\alpha(e) : e \in E_i^0\}$. For $i \in \{1, \ldots, n\}$, a set of edges X_i^0 between $S[1, i]$ and $S[i + 1, n]$ is *valid* if there exists a CSG $G = (S, E)$ such that $X_i^0 = E_i^0$.

Lemma 10. *For $i \in \{1, \ldots, n\}$, an edge set X_i^0 between $S[1, i]$ and $S[i + 1, n]$ is valid iff for every primary color $c \in \{1, \ldots, k\}$, there is an edge $e \in X_i^0$ such that $c \in \alpha(e)$ whenever both $S[1, i]_c$ and $S[i + 1, n]_c$ are nonempty.*

We encode the relevant information from E_i^+ using k equivalence relations as follows. For every $c \in \{1, \ldots, k\}$, the components of $(S[i + 1, n]_c, (E_i^+)_c)$ define an equivalence relation on $(E_i^0)_c$, which we denote by π_i^c: two edges in $(E_i^0)_c$ are related iff they are incident to the same component of $(S[i+1,n]_c, (E_i^+)_c)$. Let $\Pi_i = (\pi_i^1, \ldots, \pi_i^k)$. The equivalence relation π_i^c, in turn, determines a graph $(S[1,i]_c, E(\pi_i^c))$: two distinct vertices in $S[1,i]_c$ are adjacent iff they are incident to equivalent edges in $(E_i^0)_c$ (that is, two distinct vertices in $S[1,i]_c$ are adjacent iff they both are adjacent to the same component of $(S[i+1,n]_c, (E_i^+)_c))$. See Fig. 6 for examples of E_i^0 and Π_i. The condition that $(S, X_i^- \cup E_i^0 \cup E_i^+)$ is a CSG can now be formulated in terms of E_i^0 and Π_i (without using E_i^+ directly).

Lemma 11. *Let $G = (S, E)$ be a CSG, $i \in \{1, \ldots, n\}$, and X_i^- an edge set on $S[1, i]$. The graph $(S, X_i^- \cup E_i^0 \cup E_i^+)$ is a CSG iff the graph $(S[1,i]_c, (X_i^-)_c \cup E(\pi_i^c))$ is connected for every $c \in \{1 \ldots, k\}$.*

We can now define subproblems for MIN-kCSG. For an index $i \in \{1, \ldots, n\}$, a valid set E_i^0, and equivalence relations $\Pi_i = (\pi_i^1, \ldots, \pi_i^k)$, let $\mathcal{X}(E_i^0, \Pi_i)$ be the family of edge sets X_i^- on $S[1, i]$ such that for every $c \in \{1 \ldots, k\}$, the graph $(S[1,i]_c, (X_i^-)_c \cup E(\pi_i^c))$ is connected. The subproblem $\mathbf{A}[i, E_i^0, \Pi_i]$ is to find the minimum cost of an edge set $X_i^- \in \mathcal{X}(E_i^0, \Pi_i)$.

Note that for $i = n$, $\mathbf{A}[n, \emptyset, (\emptyset, \ldots, \emptyset)]$ is the minimum cost of a CSG for an instance (S, α) of MIN-kCSG. Next, we establish a recurrence relation for $\mathbf{A}[i, E_i^0, \Pi_i]$, which will allow computing $\mathbf{A}[n, \emptyset, (\emptyset, \ldots, \emptyset)]$ by dynamic programming. For $i = 1$, we have $\mathbf{A}[1, E_1^0, \Pi_1] = 0$ for any valid E_1^0 and Π_1. For all i, $1 < i \leq n$, we wish to express $\mathbf{A}[i, E_i^0, \Pi_i]$ in terms of $\mathbf{A}[i-1, E_{i-1}^0, \Pi_{i-1}]$'s for suitable E_{i-1}^0 and Π_{i-1}.

We say that two valid edge sets E_{i-1}^0 and E_i^0 are *compatible* if there exists an $X_i^- \in \mathcal{X}(E_i^0, \Pi_i)$ for some Π_i such that $E_{i-1}^0 = (X_i^- \cup E_i^0)_{i-1}^0$. We can characterize compatible edge sets as follows.

Lemma 12. *Two valid edge sets E_{i-1}^0 and E_i^0 are compatible iff every edge e in the symmetric difference of E_{i-1}^0 and E_i^0 is incident to p_i.*

For two valid compatible edge sets, E_{i-1} and E_i, and a sequence of equivalence relations Π_i, we define equivalence relations $\widehat{\Pi}_{i-1} = (\hat{\pi}_{i-1}^1, \ldots, \hat{\pi}_{i-1}^k)$ as follows. For every primary color $c \in \{1, \ldots, k\}$, let the equivalence relation $\hat{\pi}_{i-1}^c$ on $(E_{i-1}^0)_c$ be the transitive closure of the union of four equivalence relations: two edges in $(E_{i-1}^0)_c$ are related if (1) they both incident to p_i; (2) they both are in $(E_i^0)_c$ and π_i^c-equivalent; (3) they are both in $(E_i^0)_c$ and each are equivalent to some edge in $(E_i^0)_c$ that are π_i^c-equivalent; (4) one is incident to p_i and the other is in $(E_i^0)_c$ and π_i^c-equivalent to some edge in $(E_i^0)_c$ incident to π_i^c.

Lemma 13. *Let E_{i-1}^0 and E_i^0 be two valid compatible edge sets, and $\Pi_i = (\pi_i^1, \ldots, \pi_i^c)$. Let E_{i-1}^- be a set of edges on $S[1, i-1]$, and put $E = E_{i-1}^- \cup E_{i-1}^0 \cup E_i^0$. Then, $\widehat{\Pi}_{i-1}$ has the following property: $E_i^- \in \mathcal{X}(E_i^0, \Pi_i)$ if and only if*

(d1) $E_{i-1}^- \in \mathcal{X}(E_{i-1}^0, \widehat{\Pi}_{i-1})$; *and*

(d2) if $c \in \alpha(p_i)$ and $S[1, i]_c \neq \{p_i\}$, then p_i is incident to an edge in $(E_{i-1}^0)_c$ or an edge in $(E_i^0)_c$ that is π_i^c-equivalent to some edge incident to $S[1, i-1]_c$.

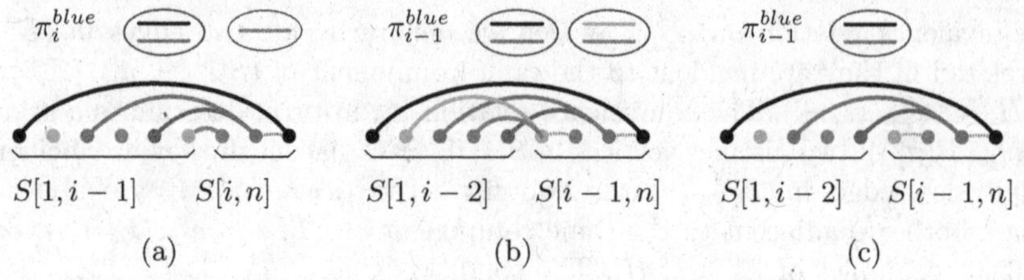

$$\pi_i^{blue} \qquad\qquad \pi_{i-1}^{blue} \qquad\qquad \pi_{i-1}^{blue}$$

$$S[1,i-1]\quad S[i,n] \qquad S[1,i-2]\quad S[i-1,n] \qquad S[1,i-2]\quad S[i-1,n]$$

$$\text{(a)} \qquad\qquad\qquad \text{(b)} \qquad\qquad\qquad \text{(c)}$$

Fig. 6. (a) E_i^0 and π_i^{blue}. (b) E_{i+1}^0 and π_{i+1}^{blue}, where E_{i+1}^0 and E_i^0 are compatible. (c) E_{i+1}^0 and π_{i+1}^{blue} violate condition (**d2**). (Color figure online)

Lemma 14. *For all $i \in \{2,\dots,n\}$, we have the following recurrence:*

$$A[i, E_i^0, \Pi_i] = \sum_{\{p_h,p_i\}\in E_i^0} w(\{p_h,p_i\}) + \min_{E_{i-1}^0 \ compatible} A[i-1, E_{i-1}^0, \widehat{\Pi}_{i-1}]. \quad (7)$$

Theorem 4. *For every constant $k \geq 1$, MIN-kCSG can be solved in $O(n)$ time when the input points are collinear.*

Proof. We determine the number of subproblems. By Corollary 4, every valid E_i^0 contains at most $|2^{\{1,\dots,k\}} \setminus \{\emptyset\}| = 2^k - 1$ edges. We have $|(E_i^0)_c| \leq 2^{k-1}$, since 2^{k-1} different colors contain any primary color $c \in \{1,\dots,k\}$. The number of equivalence relations of a set of size t is known as the t-th *Bell number*, denoted $B(t)$. It is known [3] that $B(t) \leq (0.792t/\ln(t+1))^t < 2^{O(t \log t)}$. Consequently, the number of possible Π_i is at most $(B(2^{k-1}))^k$. The total number of subproblems is $O(n2^k(B(2^{k-1}))^k)$, which is $O(n)$ for any constant k. We solve the subproblems $A[i, E_i^0, \Pi_i]$, $1 < i \leq n$, by dynamic programming, using the recursive formula (7). The time required to evaluate (7) is $O(2^k)$ for the sum of edge weights and $O(2^k(B(2^{k-1}))^k)$ to compare all compatible subproblems $A[i-1, E_{i-1}^0, \widehat{\Pi}_{i-1}]$, that is, $O(1)$ time when k is a constant. Therefore, the dynamic programming can be implemented in $O(n)$ time. $\qquad\square$

6 Conclusions

We have shown that MIN-3CSG is NP-complete in general and given a $O(n)$ time algorithm for MIN-kCSG in the special case that all points are collinear and k is a constant. We also improved the approximation factor of a polynomial time algorithm from $(2 + \frac{1}{2}\varrho)$ [10] to $(2 - \frac{2}{2+2\varrho})$ when $k = 3$. It remains open whether there exists a PTAS for MIN-kCSG, $k \geq 3$. Several other special cases are open for MIN-3CSG, such as when the points in S are on a circle or in convex position. We can generalize MIN-kCSG so that the edge weights need not be Euclidean distances. Given an arbitrary graph (V, E) and a coloring $\alpha : V \to \mathcal{P}(\{1,\dots,k\})$, what is the minimum set $E' \subseteq E$ such that (V, E') is a colored spanning graph? Since the 2-approximation algorithm presented here did not rely on the geometry of the problem, it extends to the generalization; however, this problem may be harder to approximate than its Euclidean counterpart.

Acknowledgements. Research on this paper was supported in part by the NSF awards CCF-1422311 and CCF-1423615. Akitaya was supported by the Science Without Borders program. Löffler was partially supported by the Netherlands Organisation for Scientific Research (NWO) projects 639.021.123 and 614.001.504.

References

1. Alper, B., Riche, N., Ramos, G., Czerwinski, M.: Design study of linesets, a novel set visualization technique. IEEE Trans. Vis. Comput. Graph. **17**(12), 2259–2267 (2011)
2. Alsallakh, B., Micallef, L., Aigner, W., Hauser, H., Miksch, S., Rodgers, P.: Visualizing sets and set-typed data: state-of-the-art and future challenges. In: Proceedings of Eurographics Conference on Visualization (EuroVis), pp. 1–21 (2014)
3. Berend, D., Tassa, T.: Improved bounds on Bell numbers and on moments of sums of random variables. Prob. Math. Stat. **30**(2), 185–205 (2010)
4. de Berg, M., Khosravi, A.: Optimal binary space partitions for segments in the plane. Int. J. Comput. Geom. Appl. **22**(3), 187–206 (2012)
5. Biniaz, A., Bose, P., van Duijn, I., Maheshwari, A., Smid, M.: A faster algorithm for the minimum red-blue-purple spanning graph problem for points on a circle. In: Proceedings of 28th Canadian Conference on Computational Geometry, Vancouver, BC, pp. 140–146 (2016)
6. Chung, F., Graham, R.: A new bound for Euclidean Steiner minimum trees. Ann. N.Y. Acad. Sci. **440**, 328–346 (1986)
7. Dinkla, K., van Kreveld, M.J., Speckmann, B., Westenberg, M.A.: Kelp diagrams: point set membership visualization. Comput. Graph. Forum **31**(3), 875–884 (2012)
8. Gilbert, E., Pollak, H.: Steiner minimal trees. SIAM J. Appl. Math. **16**, 1–29 (1968)
9. Hurtado, F., Korman, M., van Kreveld, M., Löffler, M., Sacristán, V., Silveira, R.I., Speckmann, B.: Colored spanning graphs for set visualization. In: Wismath, S., Wolff, A. (eds.) GD 2013. LNCS, vol. 8242, pp. 280–291. Springer, Heidelberg (2013). doi:10.1007/978-3-319-03841-4_25
10. Hurtado, F., Korman, M., van Kreveld, M., Löffler, M., Sacristan, V., Shioura, A., Silveira, R.I., Speckmann, B., Tokuyama, T.: Colored spanning graphs for set visualization (2016). Preprint arXiv:1603.00580
11. Meulemans, W., Riche, N.H., Speckmann, B., Alper, B., Dwyer, T.: KelpFusion: a hybrid set visualization technique. IEEE Trans. Vis. Comput. Graph. **19**(11), 1846–1858 (2013)

C-Planarity of Embedded Cyclic c-Graphs

Radoslav Fulek[(✉)]

IST Austria, Am Campus 1, 3400 Klosterneuburg, Austria
radoslav.fulek@gmail.com

Abstract. We show that c-planarity is solvable in quadratic time for flat clustered graphs with three clusters if the combinatorial embedding of the underlying graph is fixed. In simpler graph-theoretical terms our result can be viewed as follows. Given a graph G with the vertex set partitioned into three parts embedded on a 2-sphere, our algorithm decides if we can augment G by adding edges without creating an edge-crossing so that in the resulting spherical graph the vertices of each part induce a connected sub-graph. We proceed by a reduction to the problem of testing the existence of a perfect matching in planar bipartite graphs. We formulate our result in a slightly more general setting of cyclic clustered graphs, i.e., the simple graph obtained by contracting each cluster, where we disregard loops and multi-edges, is a cycle.

1 Introduction

Testing planarity of graphs with additional constraints is a popular theme in the area of graph visualizations. One of most the prominent such planarity variants, c-planarity, raised in 1995 by Feng, Cohen and Eades [12,13] asks for a given planar graph G equipped with a hierarchical structure on its vertex set, i.e., clusters, to decide if a planar embedding G with the following property exists: the vertices in each cluster are drawn inside a disc so that the discs form a laminar set family corresponding to the given hierarchical structure and the embedding has the least possible number of edge-crossings with the boundaries of the discs. Shortly after, several groups of researchers tried to settle the main open problem formulated by Feng et al. asking to decide its complexity status, i.e., either provide a polynomial/sub-exponential-time algorithm for c-planarity or show its NP-hardness. First, Biedl [5] gave a polynomial-time algorithm for c-planarity with two clusters. A different approach for two clusters was considered by Hong and Nagamochi [19] and quite recently in [15]. The result also follows from a work by Gutwenger et al. [17]. Beyond two clusters a polynomial time algorithm for c-planarity was obtained only in special cases, e.g., [8,16,17,20,21], and most recently in [6,7]. Cortese et al. [9] shows that c-planarity is solvable in polynomial time if the underlying graph is a cycle and the number of clusters is at most three.

R. Fulek—The research leading to these results has received funding from the People Programme (Marie Curie Actions) of the European Union's Seventh Framework Programme (FP7/2007-2013) under REA grant agreement no [291734].

Y. Hu and M. Nöllenburg (Eds.): GD 2016, LNCS 9801, pp. 94–106, 2016.
DOI: 10.1007/978-3-319-50106-2_8

In the present work we generalize the result of Cortese et al. to the class of all planar graphs with a given combinatorial embedding. In a recent pre-print [14] we established a strengthening for trees, where we do not fix the embedding. In the general case (including already the case of three clusters) of so-called flat clustered graphs a similar result was obtained only in very limited cases. Specifically, either when every face of G is incident to at most five vertices [10,15], or when there exist at most two vertices of a cluster incident to a single face [7]. We remark that the techniques of the previously mentioned papers do not give a polynomial-time algorithm for the case of three clusters, and also do not seem to be adaptable to this setting. Our result and the technique used to achieve it suggest that, for a fairly general class of clustered graphs, c-planarity could be tractable/solvable in sub-exponential time at least with a fixed combinatorial embedding.

Notation. Let $G = (V, E)$ denote a connected planar graph possibly with multi-edges. For standard graph theoretical definitions such as path, cycle, walk etc., we refer reader to [11, Sect. 1]. A *drawing* of G is a representation of G in the plane where every vertex in V is represented by a unique point and every edge $e = uv$ in E is represented by a Jordan arc joining the two points that represent u and v. We assume that in a drawing no edge passes through a vertex, no two edges touch and every pair of edges cross in finitely many points. An *embedding* of G is an edge-crossing free drawing. If it leads to no confusion, we do not distinguish between a vertex or an edge and its representation in the drawing and we use the words "vertex" and "edge" in both contexts. A *face* in an embedding is a connected component of the complement of the embedding of G (as a topological space) in the plane. The *facial walk* of f is the closed walk in G with a fixed orientation that we obtain by traversing the boundary of f counter-clockwise. In order to simplify the notation we sometimes denote the facial walk of a face f by f. A pair of consecutive edges e and e' in a facial walk f creates a *wedge* incident to f at their common vertex. A vertex or an edge is *incident* to a face f, if it appears on its facial walk. The *rotation* at a vertex is the counter-clockwise cyclic order of the end pieces of its incident edges in a drawing of G. An embedding of G is up to an isotopy and the choice of an *outer* (unbounded) face described by the rotations at its vertices. We call such a description of an embedding of G a *combinatorial embedding*. Remaining faces are *inner faces*. The *interior* and *exterior* of a cycle in an embedded graph is the bounded and unbounded, respectively, connected component of its complement in the plane. Similarly, the *interior* and *exterior* of an inner face in an embedded graph is the bounded and unbounded, respectively, connected component of the complement of its facial walk in the plane, and vice-versa for the outer face. When talking about interior/exterior or area of a cycle in a graph G with a combinatorial embedding and a *designated* outer face we mean it with respect to an embedding in the isotopy class that G defines. For $V' \subseteq V$ we denote by $G[V']$ the sub-graph of G induced by V'.

A *flat clustered graph*, shortly *c-graph*, is a pair (G, T), where $G = (V, E)$ is a graph and $T = \{V_0, \ldots, V_{c-1}\}$, $\biguplus_i V_i = V$, is a partition of the vertex set into *clusters.* See Fig. 1 for an illustration. A c-graph (G, T) is *clustered planar* (or briefly *c-planar*) if G has an embedding in the plane such that (i) for every $V_i \in T$ there is a topological disc $D(V_i)$, where interior$(D(V_i)) \cap$ interior$(D(V_j)) = \emptyset$, if $i \neq j$, containing all the vertices of V_i in its interior, and (ii) every edge of G intersects the boundary of $D(V_i)$ at most once for every $D(V_i)$. A c-graph (G, T) with a given combinatorial embedding of G is *c-planar* if additionally the embedding is combinatorially described as given. A *clustered drawing and embedding* of a flat clustered graph (G, T) is a drawing and embedding, respectively, of G satisfying (i) and (ii). In 1995 Feng, Cohen and Eades [12,13] introduced the notion of clustered planarity for clustered graphs, shortly c-planarity, (using, a more general, hierarchical clustering) as a natural generalization of graph planarity. (Under a different name Lengauer [22] studied a similar concept in 1989.)

By slightly abusing the notation for the rest of the paper G denotes a flat c-graph $(G, T) = (V_0 \uplus V_1 \uplus \ldots \uplus V_{c-1}, E)$ with c clusters $V_0, V_1, \ldots$ and V_{c-1}, and a given combinatorial embedding, and we assume that G is *cyclic* [15, Sect. 6]. Thus, every $e = uv$ of G is such that $u \in V_i$ and $v \in V_j$ where $j, - i \mod c \leq 1$ and for every i there exists an edge in G between V_i and $V_{i+1 \mod c}$. In the case of three clusters, the first condition is redundant.

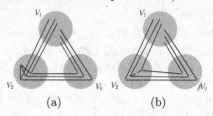

Fig. 1. A c-graph that is not c-planar (left); and a c-planar c-graph (right).

If the second condition is violated, the problem was essentially solved for three clusters as discussed in Sect. 2.3. We assume that G is connected, since in the problem that we are studying, the connected components of G can be treated separately. Indeed, without loss of generality we assume throughout the paper that in a clustered embedding of G the clusters are unbounded wedges defined by pairs of rays emanating from the origin (see Fig. 2a) that is disjoint from all the edges. We call such a clustered drawing a *fan drawing*.

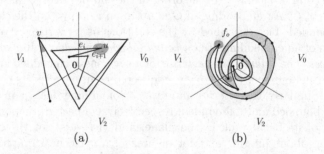

Fig. 2. (a) A clustered graph $G = (V_0 \uplus V_1 \uplus V_2, E)$ with clusters represented by wedges bounded by rays meeting at the origin. The highlighted wedge at u is concave and at v convex. (b) A semi-simple face f and the outer face f_o with an incident concave wedge.

Thus, a connected component in a clustered embedding can be drawn so that it is disjoint from a ball B centered at the origin of radius $\epsilon > 0$ for any ϵ. The rest of the graph is then embedded inductively inside B. The aim of the present work is to prove the following.

Theorem 1. *There exists a quadratic-time algorithm in $|V(G)|$ to test if a cyclic c-graph (G, T) is c-planar.*

Further Research Directions. We think that our technique should be extendable by means of Euler's formula to resolve the c-planarity in more general situations than the one treated in the present paper. In particular, we suspect that the technique should yield a generalization of the characterization of strip planar clustered graphs [14, Sect. 5]. That would allow us to work with graphs without a fixed embedding. We mention that the tractability in a special case of our problem known as cyclic level planarity, when the embedding is not fixed, follows from a recent work of Angelini et al. [2].

Organization. In Sect. 2 we introduce concepts used in the proof of our result. We give an outline of our approach in Sect. 2.1. A more detailed description and a proof of correctness of our algorithm is in Sect. 3.

2   Preliminaries

2.1   Outline of the Approach

By [13, Theorem 1] deciding c-planarity of instances G in which all $G[V_i]$'s are connected amounts to checking if an outer face of G can be chosen so that every V_i is embedded in the outer face of $G[V \setminus V_i]$. On the other hand, once we have a clustered embedding of G we can augment G by adding edges drawn inside clusters without creating an edge-crossing so that clusters become connected. These observations suggest that c-planarity of G could be viewed as a connectivity augmentation problem, for example as in [7,15], in which we want to decide if it is possible to make clusters connected while maintaining the planarity of G. One minor problem with this viewpoint is the fact that if G is c-planar we do not allow a cluster V_i to induce a cycle such that clusters V_j and $V_{j'}$, $i \neq j, j'$, are drawn on its opposite sides. However, this cannot happen if G is cyclic. Following the above line of thought our algorithm tries to augment G by subdividing its faces with paths and edges. We proceed in two steps. In the first step, Sect. 3.2, we either detect that G is not c-planar or similarly as in [1] and [14] by turning clusters into independent sets and adding certain paths we normalize the instance. In the second step, Sect. 3.1, we decide if the normalized instance can be further augmented by edges as desired.

In order to prove the correctness of the second step of the algorithm we use the notion of the *winding number* $\text{wn}(W) \in \mathbb{Z}$ of a walk W of G, as defined in Sect. 2.3. The parameter $\text{wn}(W)$ says how many times and in which sense a walk W of G winds around the origin in a clustered drawing of G. Thus, G is not c-planar if there exists a face f such that for its facial walk $|\text{wn}(f)| > 1$ or if there exists at least two inner faces f with $|\text{wn}(f)| > 0$. However, it can be easily seen that this necessary condition of c-planarity is not sufficient except when G is a cycle [9]. The necessary condition allows us to reduce the c-planarity testing problem of a normalized instance to the problem of finding a perfect matching in an auxiliary face-vertex incidence graph which is polynomially solvable. The novelty of our work lies in the use of the winding number in the context of connectivity augmentation guided by the flow and matching in the auxiliary face-vertex incidence graph à la [1,14], respectively.

We remark that the approach of [1] via a variant of upward embeddings for directed graphs in our settings has several problems that seem quite hard to overcome, the main one being the fact that the result of Bertolazzi et al. [4] does not extend, at least not in a natural way, to the drawings on the rolling cylinder, see e.g., Auer et al. [3] for the definition of these drawings. We are not aware of a polynomial-time algorithm for the corresponding problem, nor a corresponding NP-hardness result, and find the corresponding algorithmic question interesting and related to our problem.

2.2   Winding Number

We define the winding number $\text{wn}(W)$ of a closed oriented walk W in a drawing disjoint from the origin of a graph G (possibly with crossings). In what follows facial walks are understood with the orientations as in an embedding of G with the given rotations and a face f_o being a designated outer face. By viewing a closed walk W in the drawing as a continuous function w from the unit circle S^1 to $\mathbb{R}^2 \setminus \mathbf{0}$, the winding number $\text{wn}(W) \in \mathbb{Z}$ corresponds to the element of the fundamental group of S^1 [18, Chap. 1.1] represented by $\frac{w(x)}{\|w(x)\|_2}$. Let W_1 and W_2 denote a pair of oriented closed walks meeting in a vertex v. Let W denote the closed oriented walk from v to v obtained by concatenating W_1 and W_2. By the definition of wn we have $\text{wn}(W) = \text{wn}(W_1) + \text{wn}(W_2)$. Let f_1 and f_2, $f_o \neq f_1, f_2$, denote a pair of faces of G whose walks intersect in a single walk. Let G' denote a graph we get from G by deleting edges incident to both f_1 and f_2. Let f denote the new face thereby obtained. Since f_1 and f_2 intersect in a single walk, the boundary of f is connected. In the drawing of G' inherited from the drawing of G we have $\text{wn}(f) = \text{wn}(f_1) + \text{wn}(f_2)$, since common edges of f_1 and f_2 are traversed in opposite directions by f_1 and f_2. A face or a vertex is in the interior of a closed walk W in G if it is in the interior of a cycle induced by the edges of W in an embedding of G with the given rotations and f_o as the outer face. The previous observation is easily generalized by a simple inductive argument as follows

$(*)$ $\quad \sum_f \text{wn}(f) = \text{wn}(W)$

where we sum over all faces f of G in the interior of the closed walk W in G. In particular, $\sum_f \text{wn}(f) = \text{wn}(f_o)$, where we sum over all faces $f \neq f_o$ of G.

2.3   Labeling Vertices

Let $\gamma : V \to \{0, 1, \ldots c - 1\}$ be a labeling of the vertices V by integers such that $\gamma(v) = i$ if $v \in V_i$. Let W denote an oriented closed walk in a clustered drawing of G. We put height$(W) = \sum_{v'u' \in E(W)} g(\gamma(u') - \gamma(v'))$, where $g(0) = 0$, $g(1) = g(1 - c) = 1$ and $g(c - 1) = g(-1) = -1$. We have the following.

Lemma 1. *For a walk W in a fan drawing of G we have* $\mathrm{wn}(W) = \mathrm{height}(W)/c$.

Proof. The number of times the walk W crosses the ray between V_i and $V_{i+1 \bmod c}$ from right to left w.r.t. to the direction of the ray is $\mathrm{wn}_i^+(W) = \sum_{v'u'} g(\gamma(u') - \gamma(v'))$, where we sum over the edges $v'u'$ in the walk W, where $v' \in V_i$ immediately precedes $u' \in V_{i+1 \bmod c}$ in the walk. Similarly, we define $\mathrm{wn}_i^-(W) = \sum_{v'u'} g(\gamma(u') - \gamma(v'))$, where we sum over the edges $v'u'$ in W, where $v' \in V_{i+1 \bmod c}$ immediately precedes $u' \in V_i$ in the walk. We have, $\mathrm{wn}(W) = \mathrm{wn}_i^+(W) + \mathrm{wn}_i^-(W)$ which in turn implies $c \cdot \mathrm{wn}(W) = \sum_i (\mathrm{wn}_i^+(W) + \mathrm{wn}_i^-(W)) = \mathrm{height}(W)$. ∎

By the previous lemma $\mathrm{wn}(W)$ is determined already by the c-graph G and is the same in all clustered drawings of G, and hence, putting $\mathrm{wn}(W) := \mathrm{height}(W)/c$, for a walk W with a fixed orientation, allows us to speak about $\mathrm{wn}(W)$ without referring to a particular drawing of G. Thus, $\mathrm{wn}(W)$ tells us the winding number of W in any clustered drawing. By Jordan-Schönflies theorem G the following holds.

Lemma 2. *G is not c-planar if there exists a face f such that $|\mathrm{wn}(f)| > 1$ or if there exists more than one inner face f' with $|\mathrm{wn}(f')| = 1$.*

Proof. In a crossing free drawing $|\mathrm{wn}(f)| \leq 1$ for every face f. If $|\mathrm{wn}(f')| = 1$ the origin $\mathbf{0}$ lies in the interior of f' since otherwise the facial walk is null-homotopic, i.e., homotopic to a constant map, in $\mathbb{R}^2 \setminus \mathbf{0}$ (contradiction). However, interiors of faces are disjoint. ∎

If $\mathrm{wn}(f) = 0$ for all faces f, [14, Lemma 1.2] extends easily to this case, reducing the problem to the work of Angelini et al. [1]. Thus, by Lemma 2 and for the sake of simplicity of the presentation, throughout the paper we assume that there exists a pair of faces f_o, f_o', $\mathrm{wn}(f_o) = \mathrm{wn}(f_o') \neq 0$ (by ($*$) there cannot be just one such face) one of which, let's say f_o, we designate as an *outer face*. The roles of f_o and f_o' are, in fact, interchangeable. Also such a restriction is by no means crucial in our problem, and alternatively, it is always possible to choose and subdivide the outer face in the normalized instance (defined later) by a path so that the restriction is satisfied.

Viewing a facial walk f as a sequence of vertices and edges $w_0 e_0 w_1 e_2 \ldots e_m w_m$, where $e_{i-1} = w_{i-1} w_i$, let V_f be the set $\{w_0, \ldots, w_m\}$ of *vertex occurrences* along f. We treat V_f also as a multi-set of vertices, and thus, γ is defined on its elements. Let $\gamma_f : V_f \to \mathbb{N}$, for $f \neq f_o, f_o'$, be a labeling of the elements of V_f by integers defined as follows. We mark all the vertex occurrences in V_f as unprocessed. We pick an arbitrary vertex occurrence $v \in V_f$, set $\gamma_f(v) := \gamma(v)$ and mark v as processed. We repeatedly pick an unprocessed vertex occurrence $u \in V_f$ that has its predecessor or successor v along the boundary walk of f in V_f processed. We put $\gamma_f(u) := \gamma_f(v) + g(\gamma(u) - \gamma(v))$. Intuitively, γ_f records the distance in terms of "winding around origin" of vertex occurrences along the boundary walk of f from a single chosen vertex occurrence. Since $\mathrm{wn}(f) = 0$ the function $\gamma_f(u)$ is completely determined by the choice of the first occurrence of a vertex we processed. This choice is irrelevant for our use of γ_f as we see later. Also notice that $\gamma(v) = \gamma_f(v) \mod c$ for all vertices incident to f.

A normalized instance allows only the faces of the types defined next. An element v in V_f is a *local minimum* (*maximum*) of a face f if in the facial walk f the value of $\gamma(v)$ is not bigger (not smaller) with respect to the relation $0 < 1 < \ldots < c - 1 < 0$ than the value of its successor and predecessor. A walk W in G is *(strictly) monotone with respect to* γ if the labels of the occurrences of vertices on W form a (strictly) monotone sequence with respect to the relation $0 < 1 < \ldots < c - 1 < 0$ when ordered in the correspondence with their appearance on W. The face f is *simple* if f has at most one local minimum. It follows that a simple face f has also at most one local maximum. The inner face $f \neq f'_o$ is *semi-simple* if f has exactly two local minima and maxima and these minima and maxima, respectively, have the same γ_f value.

3 Algorithm

A cyclic c-graph G is *normalized* if

(i) G is connected;
(ii) each cluster V_i induces an independent set; and
(iii) each face of G is simple or semi-simple, and f_o and f'_o are both simple.

Suppose that (i)–(iii) are satisfied. By (ii) we put directions on all the edges in G as follows. Let $\overrightarrow{G}$ denote the directed c-graph obtained from G by orienting every edge uv from the vertex with the smaller label $\min(\gamma(u), \gamma(v))$ to the vertex with the bigger label $\max(\gamma(u), \gamma(v))$ with respect to the relation $0 < 1 < \ldots < c - 1 < 0$. A *sink* and *source* of $\overrightarrow{G}$ is a vertex with no outgoing and incoming, respectively, edges.

Let e denote an edge of G not contained in a single cluster. Given a clustered embedding $\mathcal{D}$ of G let $\mathbf{p_e} := \mathbf{p_e}(\mathcal{D})$ denote the intersection point of e with a ray separating a pair of clusters. Let $e_0, \ldots, e_{k-1}$ be the edges incident to a sink or source u. By Jordan curve theorem it is not hard to see that (i)–(iii) imply that a clustered embedding $\mathcal{D}$ of G is "combinatorially" determined once we order the set of intersection points $\mathbf{p_{e_0}}, \ldots, \mathbf{p_{e_{k-1}}}$ along rays separating clusters for every sink and sources u in G. Moreover, the set of intersection points corresponding to a sink or source u admits in an embedding only orders that are cyclic shifts of one another, since we have the rotations at vertices of G fixed. The wedge in $\mathcal{D}$ formed by a pair of edges e_i and e_{i+1} incident to a face f at its local extreme u is *concave* (see Fig. 2a for an illustration) if u is a sink or source of $\overrightarrow{G}$ and the line segment $\mathbf{p}_{e_i}\mathbf{p}_{e_{i+1}}$ contains all the other points $\mathbf{p}_{e_j}$ or in other words the order of intersection points corresponding to u is $\mathbf{p}_{e_{i+1}}, \mathbf{p}_{e_{i+2}}, \ldots, \mathbf{p}_{e_{k-1}}, \mathbf{p}_{e_0}, \ldots, \mathbf{p}_{e_i}$. A non-concave wedge is *convex*. Note that in $\mathcal{D}$ every sink or source is incident to exactly one concave wedge that in turn determines the order of intersection points. Thus, combinatorially $\mathcal{D}$ is also determined by a prescription of concave wedges at sink and sources.

Let S be the set of sinks and sources of $\overrightarrow{G}$. Let F denote the union of the set of semi-simple faces of G with a subset of $\{f_o, f'_o\}$ containing faces incident to a sink and a source. We construct a planar bipartite graph $I = (S \cup F, E(I))$ with parts S and F, where $s \in S$ and $f \in F$ is joined by an edge if s is incident to f. Given that (i)–(iii) are satisfied, the existence of a perfect matching M in I is a necessary condition for G being c-planar. Indeed, as we just said, in a clustered embedding, each source or sink has exactly one of its wedges concave. On the other hand, by Jordan curve theorem it can be easily checked that in the clustered embedding

(A) every semi-simple face is incident to exactly one concave wedge
(B) faces f_o and f'_o are incident to one concave wedge if they are incident to a sink and source, and
(C) all the other faces are not incident to any concave wedges at the minimum and maximum.

This is fairly easy to see if G is vertex two-connected, see Fig. 2b for an illustration. The cycle C *corresponding to a closed walk* is obtained by traversing the walk and introducing a new vertex for each vertex occurrence in the walk. For a face f incident to cut-vertices, **(A)**–**(C)** follows by considering the cycle corresponding to the facial walk of f (treated as a face) embedded in a close vicinity of the boundary of f. Thus, a desired matching M is obtained by matching each source or sink with the face incident to its concave wedge.

We show in Sect. 3.1 that if M exists G is c-planar by augmenting G with edges as described in Sect. 2.1. Testing the existence, but even counting perfect matchings in a planar bipartite graph can be carried out in a polynomial time [23, Sect. 8].

The running time of our algorithm is $O(|V|^2)$ since finding the perfect matching can be done in $O(|V|^2)$ time, due to $|E(I)| = O(|V|)$, and the pre-processing step including the construction of I and the normalization will be easily seen to have this time complexity. Also computing the winding number for all the faces can be performed in a linear time by Lemma 1. First, we explain and prove the correctness of the algorithm for instances satisfying (i)–(iii). In Sect. 3.2, we show a polynomial-time reduction of the general case to instances satisfying (i)–(iii). We often use Jordan–Schönflies theorem without explicitly mentioning it.

3.1 Constructing a Clustered Embedding

Given a normalized instance G and a matching M between sources and sinks in S, and faces in F of G we construct a clustered embedding of G as follows. Recall that we assume that G does not have a face f with $|\mathrm{wn}(f)| > 0$ besides f_o and f'_o. We start with $\overrightarrow{G}$ defined above and add edges to it thereby eliminating all the sinks

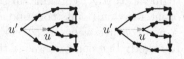

Fig. 3. Subdividing a semi-simple face (left). Subdividing a simple face f'_o (right).

and sources, see Fig. 3. Let $u \in S$ be a source matched in M with f. If f is a semi-simple inner face let u' denote another local minimum incident to f. We add to $\overrightarrow{G}$ an edge $\overrightarrow{u'u}$ embedded in the interior of f. If $f = f_o$ or $f = f'_o$ we join u by $\overrightarrow{u'u}$ with the vertex in the same cluster u' so that we subdivide f into two simple faces f' and f'' such that $\mathrm{wn}(f') = 0$ and $\mathrm{wn}(f'') = \mathrm{wn}(f)$. If $f = f_o$ face f'' is the new outer face. By Lemma 1, such a vertex u' exists and it is unique.

We proceed with $u \in S$ that are sinks analogously thereby eliminating all the sinks and source in the resulting graph $\overrightarrow{G'}$, where by G' we denote its underlying undirected graph. By Lemma 1, there still exists exactly one inner face f'_o with a non-zero winding number in the resulting graph G'.

Lemma 3. G' has exactly one inner face f'_o such that $|\mathrm{wn}(f'_o)| = 1$.

Since $\gamma(v) = \gamma_f(v) \mod c$ for every face $f \neq f_o, f'_o$ and v incident to f, every edge we added joins a pair of vertices in the same cluster.

Lemma 4. The induced sub-graph $G'[V_i]$ of (undirected) G' does not contain a cycle for $i = 0, 1, \ldots, c - 1$.

Proof. For the sake of contradiction suppose that a cycle C is contained in $G'[V_{j'}]$. Let us choose C such that the area of its interior is minimized. Since $G[V_{j'}]$ is an independent set all the edges of C are newly added. Thus, by looking at the rotation of an arbitrary vertex v' of C we see that v' is incident to a vertex v from V_j, $j \neq j'$, in the interior of C. Indeed, no two edges of C subdivide the same face of G.

Using the fact that $\overrightarrow{G'}$ does not contain any source or sink, we show that a vertex w in the interior of C belongs to an oriented cycle C' (by chance also directed in $\overrightarrow{G'}$), whose interior is contained in the interior of C such that $\mathrm{wn}(C') > 0$. The cycle C' is obtained by following a directed path in $\overrightarrow{G'}$ (from which it inherits its orientation) passing through v. Either both ends of the path meet each other, they both meet C, or the path meet itself in the interior. In the first two cases we can take $w := v$ in the last case it can happen that the directed path gives rise to a cycle C' not containing v. However, C' is not induced by a single cluster by the choice of C, and thus, $\mathrm{wn}(C') > 0$ by Lemma 1 and C' contains a vertex w from V_j. Let F' denote the set of faces in the interior of C and not in the interior of C'. In all cases it can be seen by Lemma 1 that $\mathrm{wn}(C') > 0$.

Indeed, as we proved in the proof of Lemma 1 $\mathrm{wn}(C') = \mathrm{wn}_j^+(C') + \mathrm{wn}_j^-(C')$. Since C' follows a directed path and is not induced by a single cluster we have $\mathrm{wn}_j^+(W) > 0$ and $\mathrm{wn}_j^-(W) = 0$. Hence, $\mathrm{wn}(C') = \mathrm{wn}_j^+(C') + \mathrm{wn}_j^-(C') > 0$.

By (*) it follows that C' contains the unique inner face with a non-zero winding number in its interior. Then Lemma 3 with (*) yields the following contradiction

$$0 = \mathrm{wn}(C) = \mathrm{wn}(C') + \sum_{f \in F'} \mathrm{wn}(f) = \mathrm{wn}(C') \neq 0$$

■

Let $E' \subseteq \bigcup_i \binom{V_i}{2} \setminus E(G')$ such that each edge in E' can be added to the embedding of G' without creating a crossing or increasing the number of inner faces with a non-zero winding number. We do not put any direction on the edges in E'. Since every inner face $\neq f_o'$ in G' is simple, and its outer face and the face f_o' are not adjacent to a source or sink, all the edges in E' can be introduced simultaneously without creating a crossing. In particular, no edge of E' subdivides f_o' or the outer face. Let E'' denote a maximal subset of E' that does not introduce a cycle in $(G' \cup E'')[V_i]$ for every $i = 0, 1, \ldots, c-1$ (see Fig. 4), where $G' \cup E'' = (V(G'), E(G') \cup E'')$. By Lemma 4, E'' is well-defined.

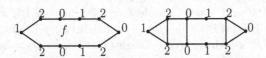

Fig. 4. A simple face f of G' (left). The face f subdivided with edges of E'' (right). Labels at vertices are their γ values (or indices of their clusters).

Lemma 5. $(G' \cup E'')[V_i]$ *is a tree for* $i = 0, 1, \ldots, c-1$.

Proof. Suppose for the sake of contradiction that $(G' \cup E'')[V_i]$ for some i is not a tree, and thus, it is just a forest with more than one connected component. It follows that either (1) there exists a cycle in $(G' \cup E'')[V \setminus V_i]$ containing a vertex v of V_i in its interior or (2) a pair of vertices of V_i in different connected components of $(G' \cup E'')[V_i]$ are incident to the same face of $(G' \cup E'')$. The claim (1) or (2) implies that there exists a cycle C in $(G' \cup E')[V \setminus V_i]$ containing a vertex w of V_i in its interior. Similarly as in the proof of Lemma 4, by following a directed path through v we obtain an oriented cycle C' (this time not necessarily directed) in G, whose interior is contained in the interior of C with $\mathrm{wn}(C') > 0$ yielding a contradiction.

Indeed, as we proved in the proof of Lemma 1 $\mathrm{wn}(C') = \mathrm{wn}_i^+(C') + \mathrm{wn}_i^-(C')$. Since C' is not induced by a single cluster and follows in the interior of C a directed path, and C does not have any vertex in V_i we have $\mathrm{wn}_i^+(W) > 0$ and $\mathrm{wn}_i^-(W) = 0$. Hence, $\mathrm{wn}(C') = \mathrm{wn}_i^+(C') + \mathrm{wn}_i^-(C') > 0$.   ■

By Lemma 5, every F_i is a tree. Taking a close neighborhood of each such F_i as a disc representing the cluster V_i we obtain a desired clustered embedding of $(G' \cup E'')$. In the obtained embedding we just delete edges not belonging to G and that concludes the proof of the correctness of our algorithm.

3.2 Normalization

In the present section we normalize the instance so that (i)–(iii) are satisfied. We argued the connectedness in Introduction, and hence, (i) is taken care of. To achieve (ii) is fairly standard by contracting components induced by clusters to vertices. Thus, it remains to satisfy (iii).

We want to sub-divide a non-simple face f into a pair of faces one of which is semi-simple by a monotone path P' w.r.t. γ. Let uPv denote an oriented monotone sub-walk of f w.r.t. γ joining a local minimum u and maximum v of f minimizing $|\text{height}(P)|$. Let vQv' denote the oriented monotone walk with $|\text{height}(P)| = |\text{height}(Q)|$ immediately following P on the facial walk of f, and let $u'Q'u$ be such walk immediately preceding P on the facial walk of f. Note that Q and Q' exists due to the minimality of P and that we have $\text{height}(Q) = \text{height}(Q') = -\text{height}(P)$. Similarly as in [14] we subdivide f into two faces f' and f'' by a strictly monotone path $v'P'u'$ w.r.t. γ. Hence, $\text{height}(P) = \text{height}(P')$. We have $\text{height}(Q) = \text{height}(Q') = -\text{height}(P) = -\text{height}(P')$. Thus, by Lemma 1 if f with $\text{wn}(f) \neq 0$ is semi-simple we obtain a simple face f' with $\text{wn}(f') \neq 0$ and a semi-simple face f'' with $\text{wn}(f'') = 0$ as desired. Indeed, $\text{wn}(f'') = \text{height}(P') + \text{height}(Q') + \text{height}(P) + \text{height}(Q) = 0$ and $c \cdot \text{wn}(f) = \text{height}(v'P''u') + \text{height}(Q') + \text{height}(P) + \text{height}(Q) = \text{height}(v'P''u') - \text{height}(P') = c \cdot \text{wn}(f')$. It remains to show the following lemma, since both f' and f'' are incident to less local minima and maxima than f if f is not semi-simple. Hence, after $O(|V|)$ facial subdivisions we obtain a desired instance, since $|E(I)| = O(|V|)$.

Lemma 6. *If the c-graph G is c-planar then by subdividing f of G by P' into a pair of faces f' and f'', where f'' is semi-simple we obtain a c-planar c-graph. Moreover, $\text{wn}(f') = \text{wn}(f)$ and $\text{wn}(f'') = 0$.*

Proof. The second statement is proved above. Hence, we deal just with the first one. Let e_u and e'_u denote the first edge on P and the last edge on Q', respectively. Let e_v and e'_v denote the last edge on P and the first edge on Q, respectively. Let $e_{v'}$ and $e_{u'}$ denote the last edge on Q and the first edge on Q'. Let $\mathbf{p_u} = \mathbf{p_{e_u}}$, and $\mathbf{p_{v'}} = \mathbf{p_{e_{v'}}}$ denote the intersection of the edges e_u, and $e_{v'}$, respectively, with a ray separating a pair of clusters. Let ω_u and ω_v denote the wedge between e_u, e'_u and e_v, e'_v, respectively, in f.

We presently show that subdividing f with P' preserves c-planarity, since a clustered embedding without P' can be deformed so that P' can be added to a clustered planar embedding without creating a crossing, while keeping the embedding clustered. This is not hard to see if, let's say ω_v, is convex and the line segment $\mathbf{p_u}\mathbf{p_{v'}}$ is not crossed by an edge. Since ω_v is convex, the relative

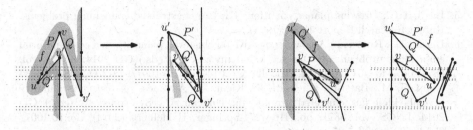

Fig. 5. A pair of deformations of the clustered embedding of G so that f can be subdivided by P'. For the sake of clarity clusters are drawn as horiznotal strips rather than wedges.

interior of $\mathbf{p_u p_{v'}}$ is contained in the interior of f. Note that $u'Q'PQv'$ is a subwalk of f since f is not simple. We draw a curve C joining u' with v' following the walk $u'Q'PQv'$ in its small neighborhood in the interior f; we cut C at its (two) intersection points with $\mathbf{p_u p_{v'}}$ and reconnected the severed ends on both sides by a curve following $\mathbf{p_u p_{v'}}$ in its small neighborhood thereby obtaining a closed curve, and a curve C' joining v' and u'. Finally, C' can be subdivided by vertices thereby yielding a desired embedding of $G \cup P'$. Otherwise, if ω_v is concave or $\mathbf{p_u p_{v'}}$ is crossed by an edge of G we need to deform the clustered embedding of G so that this is not longer the case.

By a *spur* with the *tip* u we understand a closed curve obtained as a concatenation of a line segment contained in a ray separating clusters and a curve contained in the boundary of f passing through exactly one extreme u of f such that the curve is longest possible. The *length* is the spur is one plus the number of its crossings with rays separating clusters divided by two. If ω_u is concave, the vertex u is a tip of a spur whose length is the distance of u to a closest other extreme along the face. Note that both P and Q' must be paths in this case. The rough idea in the omitted part of the proof is that shortest spurs have room around them to be deformed while maintaining the embedding clustered such that P' can be added. Spurs are deformed as illustrated in Fig. 5. ∎

Acknowledgment. I would like to thank Jan Kynčl and Dömötör Pálvölgyi for many comments and suggestions that helped to improve the presentation of the result.

References

1. Angelini, P., Lozzo, G., Battista, G., Frati, F.: Strip planarity testing. In: Wismath, S., Wolff, A. (eds.) GD 2013. LNCS, vol. 8242, pp. 37–48. Springer, Heidelberg (2013). doi:10.1007/978-3-319-03841-4_4
2. Angelini, P., Da Lozzo, G., Di Battista, G., Frati, F., Patrignani, M., Rutter, I., Beyond level planarity. arXiv preprint arXiv: 1510.08274 (2015)
3. Auer, C., Bachmaier, C., Brandenburg, F.J., Gleissner, A., Hanauer, K.: Upward planar graphs and their duals. Theoret. Comput. Sci. **571**, 36–49 (2015)
4. Bertolazzi, P., Di Battista, G., Liotta, G., Mannino, C.: Upward drawings of triconnected digraphs. Algorithmica **12**(6), 476–497 (1994)

5. Biedl, T.C.: Drawing planar partitions III: two constrained embedding problems. Rutcor Research Report 13-98 (1998)

6. Bläsius, T., Rutter, I.: A new perspective on clustered planarity as a combinatorial embedding problem. In: Duncan, C., Symvonis, A. (eds.) GD 2014. LNCS, vol. 8871, pp. 440–451. Springer, Heidelberg (2014). doi:10.1007/978-3-662-45803-7_37

7. Chimani, M., Battista, G., Frati, F., Klein, K.: Advances on testing c-planarity of embedded flat clustered graphs. In: Duncan, C., Symvonis, A. (eds.) GD 2014. LNCS, vol. 8871, pp. 416–427. Springer, Heidelberg (2014). doi:10.1007/978-3-662-45803-7_35

8. Cortese, P.F., Di Battista, G., Frati, F., Patrignani, M., Pizzonia, M.: C-planarity of c-connected clustered graphs. J. Graph Algorithms Appl. **12**(2), 225–262 (2008)

9. Cortese, P.F., Di Battista, G., Patrignani, M., Pizzonia, M.: Clustering cycles into cycles of clusters. J. Graph Algorithms Appl. **9**(3), 391–413 (2005)

10. Battista, G., Frati, F.: Efficient c-planarity testing for embedded flat clustered graphs with small faces. In: Hong, S.-H., Nishizeki, T., Quan, W. (eds.) GD 2007. LNCS, vol. 4875, pp. 291–302. Springer, Heidelberg (2008). doi:10.1007/978-3-540-77537-9_29

11. Diestel, R.: Graph Theory. Graduate Texts in Mathematics, vol. 173, 3rd edn. Springer, Heidelberg (2005)

12. Feng, Q.-W., Cohen, R.F., Eades, P.: How to draw a planar clustered graph. In: Du, D.-Z., Li, M. (eds.) COCOON 1995. LNCS, vol. 959, pp. 21–30. Springer, Heidelberg (1995). doi:10.1007/BFb0030816

13. Feng, Q.-W., Cohen, R.F., Eades, P.: Planarity for clustered graphs. In: Spirakis, P. (ed.) ESA 1995. LNCS, vol. 979, pp. 213–226. Springer, Heidelberg (1995). doi:10.1007/3-540-60313-1_145

14. Fulek, R.: Toward the Hanani-Tutte theorem for clustered graphs (2014). arXiv:1410.3022v2

15. Fulek, R., Kynčl, J., Malinovic, I., Pálvölgyi, D.: Clustered planarity testing revisited. Electron. J. Comb. **22**, 22 (2015)

16. Goodrich, M.T., Lueker, G.S., Sun, J.Z.: C-planarity of extrovert clustered graphs. In: Healy, P., Nikolov, N.S. (eds.) GD 2005. LNCS, vol. 3843, pp. 211–222. Springer, Heidelberg (2006). doi:10.1007/11618058_20

17. Gutwenger, C., Jünger, M., Leipert, S., Mutzel, P., Percan, M., Weiskircher, R.: Advances in C-planarity testing of clustered graphs. In: Goodrich, M.T., Kobourov, S.G. (eds.) GD 2002. LNCS, vol. 2528, pp. 220–236. Springer, Heidelberg (2002). doi:10.1007/3-540-36151-0_21

18. Hatcher, A.: Algebraic Topology. Cambridge University Press, Cambridge (2002)

19. Hong, S.-H., Nagamochi, H.: Simpler algorithms for testing two-page book embedding of partitioned graphs. Theor. Comput. Sci. (2016)

20. Jelínek, V., Jelínková, E., Kratochvíl, J., Lidický, B.: Clustered planarity: embedded clustered graphs with two-component clusters. In: Tollis, I.G., Patrignani, M. (eds.) GD 2008. LNCS, vol. 5417, pp. 121–132. Springer, Heidelberg (2009). doi:10.1007/978-3-642-00219-9_13

21. Jelínková, E., Kára, J., Kratochvíl, J., Pergel, M., Suchý, O., Vyskočil, T.: Clustered planarity: small clusters in cycles and Eulerian graphs. J. Graph Algorithms Appl. **13**(3), 379–422 (2009)

22. Lengauer, T.: Hierarchical planarity testing algorithms. J. ACM **36**(3), 474–509 (1989)

23. Lovász, L., Plummer, M.D.: Matching Theory. AMS Chelsea Publishing Series. American Mathematical Soc., Providence (2009)

Computing NodeTrix Representations
of Clustered Graphs

Giordano Da Lozzo, Giuseppe Di Battista, Fabrizio Frati[(✉)],
and Maurizio Patrignani

Roma Tre University, Rome, Italy
{dalozzo,gdb,frati,patrigna}@dia.uniroma3.it

Abstract. NodeTrix representations are a popular way to visualize clustered graphs; they represent clusters as adjacency matrices and inter-cluster edges as curves connecting the matrix boundaries. We study the complexity of constructing NodeTrix representations focusing on planarity testing problems, and we show several $\mathbb{NP}$-completeness results and some polynomial-time algorithms.

1 Introduction and Overview

NodeTrix representations have been introduced by Henry et al. [17] in one of the most cited papers of the InfoVis conference [1]. A NodeTrix representation is a hybrid representation for the visualization of social networks where the node-link paradigm is used to visualize the overall structure of the network, within which adjacency matrices show communities.

Formally, a NodeTrix (NT for short) representation is defined as follows. A *flat clustered graph* $(V, E, \mathcal{C})$ is a graph (V, E) with a partition $\mathcal{C}$ of V into sets $V_1, \ldots, V_k$, called *clusters*, that can be defined according to the application needs. The word "flat" is used to underline that clusters are not arranged in a multi-level hierarchy [10,13]. An edge $(u, v) \in E$ with $u \in V_i$ and $v \in V_j$ is an *intra-cluster edge* if $i = j$ and is an *inter-cluster edge* if $i \neq j$. In an *NT representation* clusters $V_1, \ldots, V_k$ are represented by non-overlapping symmetric adjacency matrices $M_1, \ldots, M_k$, where M_i is drawn in the plane so that its boundary is a square Q_i with sides parallel to the coordinate axes. Thus, the matrices $M_1, \ldots, M_k$ convey the information about the intra-cluster edges of $(V, E, \mathcal{C})$, while each inter-cluster edge (u, v) with $u \in V_i$ and $v \in V_j$ is represented by a curve connecting a point on Q_i with a point on Q_j, where the point on Q_i (on Q_j) belongs to the column or to the row of M_i (resp. of M_j) associated with u (resp. with v).

Several papers aimed at improving the readability of NT representations by reducing the number of crossings between inter-cluster edges. For this purpose, vertices can have duplicates in different matrices [16] or clusters can be computed so to have dense intra-cluster graphs and a planar inter-cluster graph [9].

Research partially supported by MIUR project AMANDA, prot. 2012C4E3KT_001.

Y. Hu and M. Nöllenburg (Eds.): GD 2016, LNCS 9801, pp. 107–120, 2016.
DOI: 10.1007/978-3-319-50106-2_9

In this paper we study the problem of automatically constructing an NT representation of a given flat clustered graph. This problem combines traditional graph drawing issues, like the placement of a set of geometric objects in the plane (here the squares $Q_1, \ldots, Q_k$) and the routing of the graph edges (here the inter-cluster edges), with a novel algorithmic challenge: To handle the degrees of freedom given by the choice of the *order* for the rows and the columns of the matrices and by the choice of the *side* of the matrices to which the inter-cluster edges attach to. Indeed, the order of the rows and columns of a matrix M_i is arbitrary, as long as M_i is symmetric; further, an inter-cluster edge incident to M_i can arbitrarily exit M_i from four sides: left or right if it exits M_i from its associated row, or top or bottom if it exits M_i from its associated column.

When working on a new model for graph representations, the very first step is usually to study the complexity of testing if a graph admits a planar representation within that model. Hence, in Sect. 2 we deal with the problem of testing if a flat clustered graph admits a planar NT representation. An NT representation is *planar* if no inter-cluster edge e intersects any matrix M_i, except possibly at an end-point of e on Q_i, and no two inter-cluster edges e and e' cross each other, except possibly at a common end-point. The NT PLANARITY problem asks if a flat clustered graph admits a planar NT representation.

Our findings show how tough the problem is (see Table 1). Namely, we show that NT PLANARITY is NP-complete even if the order of the rows and of the columns of the matrices is fixed (i.e., it is part of the input), or if the exit sides of the inter-cluster edges are fixed. It is easy to show that NT PLANARITY becomes linear-time solvable if both the order and the sides are fixed. But this is probably too restrictive for practical applications since all the degrees of freedom that are representation-specific are lost.

Table 1. Complexity results for NT PLANARITY. The result marked † assumes that the number of clusters is constant.

		General model		Monotone model	
		Free sides	Fixed sides	Free sides	Fixed sides
Row/column order	Free	NPC [Theorem 1]	NPC [Theorem 2]	NPC [Theorem 5]	NPC [Theorem 6]
	Fixed	NPC [Theorem 3]	P [Theorem 4]	P [Theorem 8]†	P [Theorem 7]

Motivated by such complexity results, in Sect. 3 we study a more constrained model that is still useful for practical applications. A *monotone NT representation* is an NT representation in which the matrices have prescribed positions and the inter-cluster edges are represented by xy-monotone curves inside the convex hull of their incident matrices. We require that this convex hull does not intersect any other matrix. We study this model for two reasons. First, in most of (although not in all) the available examples of NT representations the inter-cluster edges are represented by xy-monotone curves (see, e.g., NodeTrix clips and prototype [2]). Second, we are interested in supporting a visualization system where the position of the matrices is decided by the user and the inter-cluster edges are automatically drawn with "few" crossings. Therefore, the crossings between inter-cluster edges not incident to a common matrix are some-how unavoidable, as they depend on the positions of the matrices selected by the users, and we are only interested in reducing the number of *local* crossings, that are the crossings between pairs of edges incident to the same matrix.

We say that an NT representation is *locally planar* if no two inter-cluster edges incident to the same matrix cross. While testing if a flat clustered graph admits a monotone NT locally planar representation is $\mathbb{NP}$-complete even if the sides are fixed (see Table 1), the problem becomes polynomial-time solvable in the reasonable scenario in which the number of matrices is constant, the order of the rows and columns is fixed, and the sides of the matrices to which the inter-cluster edges attach is variable.

Conclusions and open problems are discussed in Sect. 4 where NT PLA-NARITY is related to graph drawing problems of theoretical interest.

Before proceeding to prove our results, we establish formal definitions and notation. An NT representation consists of: (a) A *row-column order* σ_i for each cluster V_i, that is, a bijection $\sigma_i : V_i \leftrightarrow \{1, \ldots, |V_i|\}$. (b) A *side assignment* s_i for each inter-cluster edge incident to V_i, that is, an injective mapping $s_i : \bigcup_{j \neq i} E_{i,j} \rightarrow \{\text{T}, \text{B}, \text{L}, \text{R}\}$, where $E_{i,j}$ is the set of inter-cluster edges between the clusters V_i and V_j (V_i and V_j are *adjacent* if $E_{i,j} \neq \emptyset$). (c) A *matrix* M_i for each cluster V_i, that is, a representation of V_i as a symmetric adjacency matrix such that: (i) the boundary of M_i is a square Q_i with sides parallel to the coordinate axes; let $\min_x(Q_i)$ be the minimum x-coordinate of a point on Q_i; $\min_y(Q_i)$, $\max_x(Q_i)$, and $\max_y(Q_i)$ are defined analogously; (ii) the left-to-right order of the columns and the top-to-bottom order of the rows in M_i is σ_i; and (iii) every two distinct matrices are disjoint; if V_i has only one vertex, we often talk about the matrix representing that vertex, rather than the matrix representing V_i. (d) An *edge drawing* for each inter-cluster edge $e = (u, v)$ with $u \in V_i$ and $v \in V_j$, that is, a representation of e as a Jordan curve between two points p_u and p_v defined as follows. Let m_T^u be the mid-point of the line segment that is the intersection of the top side of Q_i with the column associated to u in M_i; points m_B^u, m_L^u, and m_R^u are defined analogously. Then p_u coincides with m_T^u, m_B^u, m_L^u, or m_R^u if $s_i(e) = \text{T}$, $s_i(e) = \text{B}$, $s_i(e) = \text{L}$, or $s_i(e) = \text{R}$, respectively. Point p_v is defined analogously. The full versi on of the paper [12] contains complete proofs.

2   Testing NodeTrix Planarity

In this section we study the time complexity of testing NODETRIX PLANARITY.

Theorem 1. NODETRIX PLANARITY *is* $\mathbb{NP}$-*complete even if at most three clusters contain more than one vertex.*

Proof Sketch: Lemma 1 will prove that NT PLANARITY is in $\mathbb{NP}$. For the $\mathbb{NP}$-hardness we give a reduction from the $\mathbb{NP}$-complete problem PARTITIONED 3-PAGE BOOK EMBEDDING [7] that, given a graph $\langle V, E = E_1 \cup E_2 \cup E_3 \rangle$, asks whether a total ordering $\mathcal{O}$ of V exists such that no two edges e and e' in the same set E_i have alternating end-vertices in $\mathcal{O}$. We construct an instance $\langle V', E', \mathcal{C}' \rangle$ of NT PLANARITY from $\langle V, E = E_1 \cup E_2 \cup E_3 \rangle$ as follows; see Fig. 1.

The instance $\langle V', E', \mathcal{C}' \rangle$ has a cycle D composed of vertices u_l, t_j^i, u_r, b_j^i, and u_b^i, where $i = 1, \ldots, 3$ and $j = 1, \ldots, 7$ (each in a distinct cluster containing that vertex only) and of inter-cluster edges called *bounding edges*. The instance contains three "big" clusters $V_i'' = V_i' \cup \{x_i, y_i, w_i, z_i\}$ with $i = 1, 2, 3$, where V_i' is in bijection with V; these are the only clusters with more than one vertex. Any two vertices, one in V_i' and one in V_{i+1}', that are in bijection (via a vertex in V) are connected by an *order-preserving* edge. Further, the vertices x_i, y_i, w_i, z_i of V_i'' are connected to the vertices in D via *corner edges*, and *side-filling edges* connect u_l with every vertex in V_1', u_r with every vertex in V_3', and u_b^i with every vertex in V_i'. Finally, $\langle V', E', \mathcal{C}' \rangle$ contains paths corresponding to the edges in E; namely, for every $e = (r, s) \in E_i$, $\langle V', E', \mathcal{C}' \rangle$ contains a cluster $\{u_e'\}$ and two *equivalence edges* (u_e', r_i') and (u_e', s_i'), where r_i' and s_i' are the vertices in V_i' in bijection with r and s, respectively. The construction can be easily performed in polynomial time. We now prove the equivalence between the two instances.

For the direction ($\Longrightarrow$), consider a total order $\mathcal{O}$ of V which solves instance $\langle V, E \rangle$. An order σ_i' of V_i' is constructed from $\mathcal{O}$ via the bijection between V_i' and V; then define an order σ_i of V_i'' as $x_i, y_i, \sigma_i', w_i, z_i$. Embed D in the plane and embed each matrix M_i representing V_i'' inside D with row-column order σ_i. The corner, order-preserving, and side-filling edges are routed inside D so that their end-vertices are not assigned to the top side of M_i; for example, the side-filling edges incident to u_l (to u_b^1) are assigned to the left (resp. bottom) side of M_1 and the order-preserving edges incident to V_1' are assigned to the right side of M_1; the order-preserving edges can be routed without crossings since σ_i' and σ_{i+1}' coincide (via the bijection of V_i' and V_{i+1}' with V). Finally, each path (r_i', u_e', s_i') corresponding to an edge $(r, s) \in E_i$ is incident to the top side of M_i; no two of these paths have alternating end-vertices since σ_i' coincides with $\mathcal{O}$ (via the bijection between V_i' and V) and since no two edges in E_i have alternating end-vertices in $\mathcal{O}$. This results in a planar NT representation of $\langle V', E', \mathcal{C}' \rangle$.

The direction ($\Longleftarrow$) is more involved. Consider a planar NT representation Γ of $\langle V', E', \mathcal{C}' \rangle$. First, the matrices representing clusters not in D induce a connected part of Γ, hence they are all on the same side of D, say they are all inside D. Second, the boundary Q_i of M_i and the corner edges incident to it subdivide the interior of D into five regions, namely one containing M_i, and four incident to the sides of Q_i. All the vertices in V_i' are incident to each of the latter four regions; this is proved by arguing that the first two and the last two vertices in the row-column order σ_i of M_i are among $\{x_i, y_i, w_i, z_i\}$, and by arguing about how the corner edges are incident to the sides of Q_i. Third, the side-filling edges incident to a same vertex in D "fill" one of such regions and so do the order-preserving edges connecting M_i with M_{i+1} (or with M_{i-1}). Hence, all the equivalence edges are incident to the same side of Q_i. Finally, the order-preserving edges between vertices in V_i' and in V_{i+1}' are in the region shared by M_i and M_{i+1}; these regions are gray in Fig. 1. This implies that σ_i' and σ_{i+1}' are

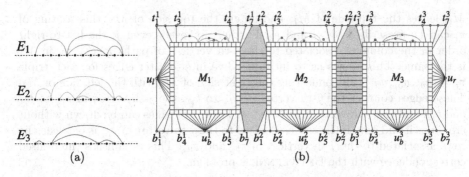

E_1
E_2
E_3

(a)   (b)

Fig. 1. (a) An instance $\langle V, E = E_1 \cup E_2 \cup E_3 \rangle$ of PARTITIONED 3-PAGE BOOK EMBED-DING and (b) the corresponding instance $\langle V', E', \mathcal{C}' \rangle$ of NT PLANARITY.

either the same or the reverse of each other, via the bijection with the vertices in V. Hence, we define an order $\mathcal{O}$ of V according to the bijection with the order σ_i' of V_i'; then no two edges in E_1, in E_2, or in E_3 have alternating end-vertices, as otherwise the corresponding paths would cross in Γ.   □

Let $G = (V, E, \mathcal{C})$ be a flat clustered graph with a given side assignment s_i, for each $V_i \in \mathcal{C}$. We say that G is *NT planar with fixed side* if G admits a NT planar representation Γ such that, $\forall e = (u, v) \in E$ with $u \in V_i$ and $v \in V_j$, the incidence points of e with the matrices M_i and M_j representing V_i and V_j in Γ, respectively, lie on the segments corresponding to the $s_i(u)$ side of M_i and to the $s_j(v)$ side of M_j, respectively.

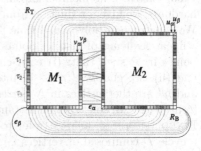

Fig. 2. Illustration for Theorem 2.

Theorem 2. NODETRIX PLANARITY WITH FIXED SIDE *is* NP-*complete even for instances with two clusters.*

Proof Sketch: Lemma 1 will prove that NT PLANARITY WITH FIXED SIDE is in NP. For the NP-hardness we give a reduction from BETWEENNESS [18], whose input is a set of items $\{a_1, \ldots a_h\}$ and a collection of t ordered triples $\tau_j = \langle a_{b_j}, a_{c_j}, a_{d_j} \rangle$. The goal is to find a total order of the items such that, for each $\tau_j = \langle a_{b_j}, a_{c_j}, a_{d_j} \rangle$, item a_{c_j} is between a_{b_j} and a_{d_j}. We construct the corresponding instance of NT PLANARITY WITH FIXED SIDE by defining a flat clustered graph $(V, E, \mathcal{C} = \{V_1, V_2\})$ and side assignments s_1 and s_2 as follows; refer to Fig. 2. Cluster V_1 (V_2) contains t vertices for each a_i plus two vertices v_α and v_β (plus two vertices u_α and u_β and $2t$ vertices t_{j1} and t_{j2} for $j = 1, \ldots, t$). Let M_1 and M_2 be matrices representing V_1 and V_2, respectively; also, let $e_\alpha = (u_\alpha, v_\alpha)$ ($e_\beta = (u_\beta, v_\beta)$) be assigned to the right (left) side of M_1 and to the left (right) side of M_2. We associate to each element a_i a $(2t+1)$-vertex path π_i that starts at u_β, and repeatedly leaves the bottom side of M_2, enters the bottom side of

M_1, leaves the top side of M_1, and enters the top side of M_2; this routing of π_i can be prescribed by s_1 and s_2. Further, for every even j, the left-to-right order of the columns associated to the j-th vertices of paths $\pi_1, \ldots, \pi_h$ in M_1 is the same. This allows us to introduce five inter-cluster edges for each triple $\tau_j = \langle a_{b_j}, a_{c_j}, a_{d_j} \rangle$, all connecting the right side of M_1 with the left side of M_2. These edges connect the j-th vertex in π_{b_j} to t_{j1}, the j-th vertex in π_{c_j} to t_{j1} and t_{j2}, and the j-th vertex in π_{d_j} to t_{j2}. These five edges can be drawn without crossings if and only if the row associated to the j-th vertex in π_{c_j} is between the rows associated to the j-th vertices in π_{b_j} and π_{d_j}. This establishes the desired correspondence with the BETWEENNESS problem.   □

Let $G = (V, E, \mathcal{C})$ be a flat clustered graph with a given row-column order σ_i, for each $V_i \in \mathcal{C}$. We say that G is *NT planar with fixed order* if it admits a NT planar representation Γ where, for each $V_i \in \mathcal{C}$, each vertex $v \in V_i$ is associated with the $\sigma_i(v)$-th row and column of the matrix M_i representing V_i in Γ.

Theorem 3. NODETRIX PLANARITY WITH FIXED ORDER *is* NP-*complete even if at most one cluster contains more than one vertex.*

Proof Sketch: The membership in NP will be proved in Lemma 1. For the NP-hardness, we give a reduction from the 4-coloring problem for *circle graphs* [20]. We construct in polynomial time [19] a representation $\langle \mathcal{P}, \mathcal{O} \rangle$ of G, where $\mathcal{P}$ is a linear sequence of distinct points on a circle and $\mathcal{O}$ is a set of chords between points in $\mathcal{P}$ such that: (i) each chord $c \in \mathcal{O}$ corresponds to a vertex $n \in N$ and (ii) two chords $c', c'' \in \mathcal{O}$ intersect if and only if $(n', n'') \in A$, where n' and n'' are the vertices in N corresponding to c' and c'', respectively. Starting from $\langle \mathcal{P}, \mathcal{O} \rangle$ we construct an instance $(V, E, \mathcal{C})$ of NODETRIX PLANARITY WITH FIXED ORDER as follows (refer to Fig. 3). The instance $(V, E, \mathcal{C})$ contains: (i) a cycle D composed of vertices v_{tl}, v'_{tr}, v''_{tr}, v_{br}, v'_{bl}, and v''_{bl} (each in a distinct cluster containing that vertex only) and of *bounding edges*; (ii) a cluster V_* containing a vertex v_i for each point $p_i \in \mathcal{P}$, plus vertices v_α and v_ω; (iii) *corner edges* connecting v_{tl}, v'_{tr}, v''_{tr}, v_{br}, v'_{bl}, and v''_{bl} with either v_α or v_ω; and (iv) for every chord $c = (p_i, p_j) \in \mathcal{O}$, a path corresponding to c composed of a cluster

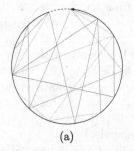

(a)

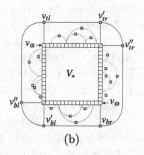

(b)

Fig. 3. (a) An intersection representation $\langle \mathcal{P}, \mathcal{O} \rangle$ of a circle graph $G = (N, A)$. (b) Instance $(V, E, \mathcal{C})$ of NODETRIX PLANARITY corresponding to $\langle \mathcal{P}, \mathcal{O} \rangle$.eps

$\{v_c\}$ and of two *chord edges* (v_i, v_c) and (v_c, v_j). Let the row-column order σ_* of V_* be $v_\alpha, \mathcal{P}, v_\omega$. We now prove the equivalence between the instances.

($\Longrightarrow$) Suppose that the chords of $\langle \mathcal{P}, \mathcal{O} \rangle$ can be assigned colors $1, 2, 3, 4$ so that no two chords with the same color cross. Embed D in the plane and embed the matrix M_* representing V_* inside D with row-column order σ_*. Route the corner edges inside D, subdividing the region inside D and outside M_* into four regions, each incident to a distinct side of M_*. Arbitrarily color these four regions with colors $1, 2, 3, 4$; embed a path (v_i, v_c, v_j) inside a region if the chord (p_i, p_j) corresponding to (v_i, v_c, v_j) has the color of the region. Then only paths in the same region might intersect, however if they do then they correspond to chords with the same color that cross in $\langle \mathcal{P}, \mathcal{O} \rangle$, given that the order of the vertices in σ_* is the same as the order of the corresponding points in $\mathcal{P}$.

($\Longleftarrow$) Suppose that $(V, E, \mathcal{C})$ has a planar NT representation Γ with row-column order σ_* for the matrix M_* representing V_*. Since v_α and v_ω are the first and last vertex in σ_*, the corner edges subdivide the region inside D and outside M_* into four regions, each incident to a distinct side of M_*, which we arbitrarily color $1, 2, 3, 4$. By the planarity of Γ, each path (v_i, v_c, v_j) is in one of such regions; then we color each chord (p_i, p_j) with the color of the region path (v_i, v_c, v_j) is embedded into. If two chords with the same color cross in $\langle \mathcal{P}, \mathcal{O} \rangle$, then the corresponding paths cross in Γ, as the order of the vertices in σ_* is the same as the order of the corresponding points in $\mathcal{P}$. □

Let $G = (V, E, \mathcal{C})$ be a flat clustered graph with a given row-column order σ_i and side assignment s_i, for each $V_i \in \mathcal{C}$. Then G is *NT planar with fixed order and fixed side* if it is simultaneously planar with fixed order and with fixed side.

Theorem 4. NODETRIX PLANARITY WITH FIXED ORDER AND FIXED SIDE *can be solved in linear time.*

Proof Sketch: Consider the graph G' obtained from an instance $G = (V, E, \mathcal{C})$ by collapsing each cluster V_i into a vertex v_i. Instance G is NT planar with fixed order and fixed side if and only if G' is planar with the additional constraint that the clockwise order of the edges incident to each vertex v_i is compatible with the order of the rows of the matrix representing V_i and the side assignment for each inter-cluster edge incident to V_i. We obtain an instance of constrained planarity that can be tested in linear time with known techniques [15]. □

We conclude the section with the following lemma.

Lemma 1. NODETRIX PLANARITY, NODETRIX PLANARITY WITH FIXED SIDE, *and* NODETRIX PLANARITY WITH FIXED ORDER *are in* NP.

Proof Sketch: The number of distinct row-column orders and side assignments for an instance $(V, E, \mathcal{C})$ is a function of $|V| + |E|$. The statement follows from Theorem 4. □

3   Monotone NodeTrix Representations

Let $G = (V, E, C)$ be a flat clustered graph and γ be a square assignment for G that maps each cluster in C to an axis-aligned square in the plane. A curve is x-*monotone* (resp. y-*monotone*) if no two of its points have the same projection on the x-axis (resp. on the y-axis) and is xy-*monotone* if it is either a horizontal or a vertical segment or it is both x- and y-monotone. A *monotone* NT representation Γ of $\langle G, \gamma \rangle$ is a NT representation such that: (i) all the inter-cluster edges are represented by xy-monotone curves; (ii) the boundary of the matrix M_i representing cluster $V_i \in C$ is $Q_i = \gamma(V_i)$; (iii) for each pair of adjacent clusters V_i and V_j, with $i \neq j$, the convex hull of Q_i and Q_j does not intersect any other Q_k, with $k \neq i, j$ – we call this convex hull the *pipe* of Q_i and Q_j; and (iv) all the inter-cluster edges between vertices in V_i and vertices in V_j lie inside the pipe of Q_i and Q_j. In a monotone NT representation Γ of G let $\chi_i(\Gamma)$ denote the number of edge crossings between pairs of inter-cluster edges incident to V_i. Let $\chi(\Gamma) = \sum_{i=1}^{k} \chi_i(\Gamma)$; we say that Γ is *locally planar* if $\chi(\Gamma) = 0$ and no inter-cluster edge intersects any matrix except at its incidence points. The notions of *fixed order* and *fixed side* easily extend to monotone NT representations.

We study the complexity of testing if a flat clustered graph with fixed square assignment admits a monotone locally-planar NT representation, a problem which we call MONOTONE NT LOCAL PLANARITY (MNTLP). The next two theorems show the NP-hardness of MNTLP and of its variant with fixed side.

Theorem 5. MNTLP *is* NP-*complete.*

Theorem 6. MNTLP WITH FIXED SIDE *is* NP-*complete.*

Since the instances of MNTLP used in the proof of Theorem 5 are planar whenever they are locally planar, testing the existence of a planar monotone NT representation with fixed square assignment is also NP-complete. Further, the instances of NT PLANARITY used in the proof of Theorem 1 can be drawn planarly with straight-line (i.e., monotone) edges, whenever they are planar. Hence, testing whether a flat clustered graph admits a monotone planar NT representation – without square assignment – is NP-complete.

Consider now a flat clustered graph $G = (V, E, C)$ and a monotone NT representation Γ of G with fixed square assignment γ. Consider two clusters $V_a, V_b \in C$ and let $Q_a = \gamma(V_a)$ and $Q_b = \gamma(V_b)$. Since Q_a and Q_b are disjoint, there exists either a vertical or a horizontal line separating them. Suppose that the former holds, the other case being analogous. Also suppose that $\max_x(Q_a) < \min_x(Q_b)$ and $\max_y(Q_a) \geq \max_y(Q_b)$, the other cases being analogous up to reflections of the Cartesian axes (refer to Fig. 4). Also, consider an inter-cluster edge $e = (u, v) \in E_{a,b}$. Depending on the relative positions of Q_a and Q_b in Γ, not all the possible combinations of side assignments for e might be allowed, as described in the following property.

Property 1. Let y_u and y_v be the y-coordinate of points m_R^u and m_L^v, respectively. The following three arrangements are possible for Q_a and Q_b in Γ.

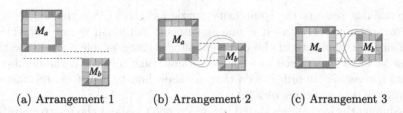

(a) Arrangement 1   (b) Arrangement 2   (c) Arrangement 3

Fig. 4. Possible arrangements of squares Q_a and Q_b. Thick red segments represent sides of Q_a and Q_b edge (u, v) cannot be assigned to. Red curves show further forbidden side assignment pairs for edge (u, v).eps (Color figure online)

Arrangement 1: $\max_y(Q_b) < \min_y(Q_a)$. Then $s_b(e) \neq$ B and all other four side assignments $\langle s_a(e) = \text{R}, s_b(e) = \text{T} \rangle$, $\langle s_a(e) = \text{R}, s_b(e) = \text{L} \rangle$, $\langle s_a(e) = \text{B}, s_b(e) = \text{T} \rangle$, and $\langle s_a(e) = \text{B}, s_b(e) = \text{L} \rangle$ are allowed for e.

Arrangement 2: $\min_y(Q_b) < \min_y(Q_a) \leq \max_y(Q_b)$. Then $s_b(e) \neq$ B; also, pair $\langle s_a(e) = \text{B}, s_b(e) = \text{T} \rangle$ is not allowed, while pair $\langle s_a(e) = \text{R}, s_b(e) = \text{L} \rangle$ is allowed. The remaining two possible pairs $\langle s_a(e) = \text{R}, s_b(e) = \text{T} \rangle$ and $\langle s_a(e) = \text{B}, s_b(e) = \text{L} \rangle$ are or are not allowed, depending on y_u and y_v. In particular, if $y_u \leq \max_y(Q_b)$, then $\langle s_a(e) = \text{R}, s_b(e) = \text{T} \rangle$ is not allow+ed, otherwise it is; also, if $y_v \geq \min_y(Q_a)$, then $\langle s_a(e) = \text{B}, s_b(e) = \text{L} \rangle$ is not allowed, otherwise it is.

Arrangement 3: $\min_y(Q_a) \leq \min_y(Q_b)$. Then $s_a(e) \neq$ B; also, pair $\langle s_a(e) = \text{R}, s_b(e) = \text{L} \rangle$ is allowed. The remaining two possible pairs $\langle s_a(e) = \text{R}, s_b(e) = \text{T} \rangle$ and $\langle s_a(e) = \text{R}, s_b(e) = \text{B} \rangle$ are or are not allowed, depending on y_u. In particular, if $y_u \leq \max_y(Q_b)$, then $\langle s_a(e) = \text{R}, s_b(e) = \text{T} \rangle$ is not allowed, otherwise it is, and if $y_u \geq \min_y(Q_b)$, then $\langle s_a(e) = \text{R}, s_b(e) = \text{B} \rangle$ is not allowed, otherwise it is.

Note that if an edge e can be drawn as an xy-monotone curve not crossing any matrix then it can also be drawn as a straight-line segment not crossing any matrix, since the pipe of Q_a and Q_b does not intersect any matrix other than M_a and M_b. The next lemma extends this observation by arguing that the xy-monotonicity constraint can be replaced by a straight-line requirement also for what concerns crossings between inter-cluster edges incident to the same matrix.

Lemma 2. *An instance* $\langle G = (V, E, \mathcal{C}), \gamma \rangle$ *of* MNTLP WITH FIXED ORDER AND FIXED SIDE *is locally planar if and only if it admits a locally planar monotone NT representation where all the inter-cluster edges are drawn as straight-line segments.*

In contrast to the negative results of Theorems 5 and 6, we show that MNTLP WITH FIXED ORDER AND FIXED SIDE is solvable in polynomial time.

Theorem 7. MNTLP WITH FIXED ORDER AND FIXED SIDE *can be solved in polynomial time.*

Proof: We check whether every edge can be represented as an xy-monotone curve by Property 1. Further, we check whether all pairs of inter-cluster edges incident to the same cluster admit a non-crossing straight-line drawing by Lemma 2.   □

The remaining piece of the complexity puzzle for MNTLP is the setting with fixed row-column order and free side assignment. Although we are not able to establish the complexity of the corresponding decision problem, we show that testing MNTLP with fixed order is polynomial-time solvable if the number of clusters is constant. In order to do that, we show how to transform instances of our problem into instances of 2-SAT.

Assuming the hypotheses stated before Property 1 about the relative position of Q_a and Q_b, we say that an inter-cluster edge $e = (u \in V_a, v \in V_b)$ is *S-drawn* in Γ if: (i) Q_a and Q_b are arranged as in Arrangement 1 of Property 1 and either $\langle s_a(e) = \text{R}, s_b(e) = \text{L} \rangle$ or $\langle s_a(e) = \text{B}, s_b(e) = \text{T} \rangle$; or (ii) Q_a and Q_b are arranged as in Arrangement 2 of Property 1 and it holds that (a) $\langle s_a(e) = \text{R}, s_b(e) = \text{L} \rangle$, (b) $y_u > \max_y(Q_b)$, and (c) $y_v < min_y(Q_a)$. Note that if Q_a and Q_b are arranged as in Arrangement 3 of Property 1, then e is not S-drawn in Γ, by definition. The representation of an S-drawn edge is an *S-drawing*. We have the following.

Lemma 3. *Let* $\langle G = (V, E, \mathcal{C} = \{V_a, V_b\}), \gamma, \sigma \rangle$ *be an instance of* MNTLP WITH FIXED ORDER. *Consider the following two cases: an inter-cluster edge* $e^* \in E$ *has a given S-drawing* Γ_e *(Case 1), or no inter-cluster edge in* E *has an S-drawing (Case 2). Both in Case 1 and in Case 2, we can construct in* $O(|E|^2)$ *time a 2-SAT formula* $\phi(a, b, \Gamma_e)$ *and* $\phi(a, b)$, *respectively, with length* $O(|E|^2)$ *that is satisfiable if and only if* $\langle G, \gamma, \sigma \rangle$ *admits a locally planar monotone NT representation with fixed order satisfying the constraint of the corresponding case.*

Proof Sketch: If $Q_a = \gamma(V_a)$ and $Q_b = \gamma(V_b)$ are not disjoint, no NT representation of G exists, hence the statement is trivially true. Otherwise, there exists either a vertical or a horizontal line separating them. Suppose that the former holds and that $\max_x(Q_a) < \min_x(Q_b)$ and $\max_y(Q_a) \geq \max_y(Q_b)$, the other cases being analogous. Suppose that an inter-cluster edge e^* is required to have a drawing Γ_e as in Case 1. By the definition of an S-drawn edge, if Q_a and Q_b are arranged as in Arrangement 3 of Property 1, the required NT representation does not exist, thus the statement trivially holds. Hence, we can assume that Q_a and Q_b are arranged as in Arrangement 1 or 2 of Property 1. Let $e \neq e^* \in E$ be any inter-cluster edge not adjacent to e.

Consider Arrangement 1 and suppose $s_a(e^*) = \text{R}$ and $s_b(e^*) = \text{L}$. The end-vertices of e and e^* in V_a (in V_b) have two possible relative positions in σ_a (resp. in σ_b). This leads to four possible combinations for these relative positions.

If $\sigma_a(e^*) < \sigma_a(e)$ and $\sigma_b(e) < \sigma_b(e^*)$, then any xy-monotone curve representing e crosses e^* independently of the side assignment for e and the statement trivially holds. See Fig. 5a. For each of the three remaining combinations, *exactly two* side assignments for e create no crossing with e^*. If $\sigma_a(e) < \sigma_a(e^*)$ and $\sigma_b(e) < \sigma_b(e^*)$, then it holds that either $s_a(e) = \text{R}$ and $s_b(e) = \text{T}$ or $s_a(e) = \text{R}$ and $s_b(e) = \text{L}$. See Fig. 5b. If $\sigma_a(e) < \sigma_a(e^*)$ and $\sigma_b(e^*) < \sigma_b(e)$, then it holds that either $s_a(e) = \text{R}$ and $s_b(e) = \text{T}$ or $s_a(e) = \text{B}$ and $s_b(e) = \text{L}$. See Fig. 5c. If $\sigma_a(e^*) < \sigma_a(e)$ and $\sigma_b(e^*) < \sigma_b(e)$, then it holds that either $s_a(e) = \text{R}$ and $s_b(e) = \text{L}$ or $s_a(e) = \text{B}$ and $s_b(e) = \text{L}$. See Fig. 5d.

The case in which Q_a and Q_b are arranged as in Arrangement 1, $s_a(e^*) = \text{B}$, and $s_b(e^*) = \text{T}$ is analogous. The proof for Arrangement 2 is also analogous.

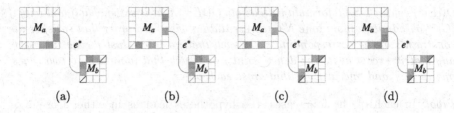

Fig. 5. Illustrations for the proof of Lemma 3, Case 1, Arrangement 1.

We are now ready to show, for Case 1 of the lemma, that a locally planar monotone NT representation of $\langle G = (V, E, \mathcal{C} = \{V_a, V_b\}), \gamma \rangle$ with $s_a(e^*) = $ R and $s_b(e^*) = $ L exists if and only if a suitable 2-SAT formula ϕ is satisfiable. For each inter-cluster edge $e \neq e^* \in E$ not adjacent to e^*, we define a Boolean variable x_e. The above discussion shows that, if a trivially false formula cannot be associated with the instance $\langle G, \gamma, \sigma \rangle$, then there are exactly two distinct side assignments for e. We select one arbitrarily, which we call *canonical side assignment*, and associate $x_e = $ TRUE to it and $x_e = $ FALSE to the other. For each pair of non-adjacent inter-cluster edges $e_1, e_2 \neq e^* \in E$, consider the four possible side assignments for them. We add to ϕ at most four clauses defined as follows. If the canonical side assignment for e_1 and the one for e_2 generate a crossing between e_1 and e_2, then we add clause $\{\overline{x_{e_1}} \vee \overline{x_{e_2}}\}$ to ϕ. If the canonical side assignment for e_1 and the non-canonical side assignment for e_2 generate a crossing between e_1 and e_2, then we add clause $\{\overline{x_{e_1}} \vee x_{e_2}\}$ to ϕ. If the non-canonical side assignment for e_1 and the canonical side assignment for e_2 generate a crossing between e_1 and e_2, then we add clause $\{x_{e_1} \vee \overline{x_{e_2}}\}$ to ϕ. If the non-canonical side assignment for e_1 and the one for e_2 generate a crossing between e_1 and e_2, then we add clause $\{x_{e_1} \vee x_{e_2}\}$ to ϕ.

As a consequence of the above discussion $\langle G = (V, E, \mathcal{C} = \{V_a, V_b\}), \gamma \rangle$ admits a locally planar monotone NT representation such that $s_a(e^*) = $ R and $s_b(e^*) = $ L if and only if ϕ is satisfiable. Further, since the number of clauses in ϕ is upper-bounded by $O(|E|^2)$ and since it can be determined in constant time whether a side assignment for any two edges produces a crossing, then formula ϕ can be constructed in $O(|E|^2)$ time and has $O(|E|^2)$ size. Since 2-SAT formulae can be tested for satisfiability in linear time [8], the statement of Case 1 follows if $s_a(e^*) = $ R and $s_b(e^*) = $ L; a 2-SAT formula can be analogously constructed if $s_a(e^*) = $ B and $s_b(e^*) = $ T.

The discussion of Case 2 and the corresponding construction of the Boolean formula are analogous to those of Case 1. □

We now turn to the study of flat clustered graphs with three clusters.

Lemma 4. *Let* $\langle G = (V, E, \mathcal{C} = \{V_a, V_b, V_c\}), \gamma, \sigma \rangle$ *be an instance of* MNTLP WITH FIXED ORDER. *Consider the four cases that are generated by assuming that an edge* $e^* \in E_{a,b}$ *has a prescribed S-drawing or not and that an edge* $f^* \in E_{a,c}$ *has a prescribed S-drawing or not. In each case, we can construct in*

$O(|E|^2)$ time a 2-SAT formula with length $O(|E|^2)$ that is satisfiable if and only if $\langle G, \gamma, \sigma \rangle$ admits a monotone NT representation with fixed order that satisfies the constraints of the corresponding case, such that no inter-cluster edge intersects any matrix except at its incidence points, and such that there are no two edges, one in $E_{a,b}$ and one in $E_{a,c}$, that cross each other.

Proof: In each of the four cases, the hypotheses lead us in either Case 1 or Case 2 of Lemma 3 for the edges in $E_{a,b}$ and the same holds for the edges in $E_{a,c}$. Hence, by Lemma 3, each of these edges admits at most two side assignments in each case. Moreover, each of these side assignments corresponds to a directed or negated literal. For each pair of edges $e \in E_{a,b}$ and $f \in E_{a,c}$ and for each of the at most four side assignments for them, we exploit Lemma 2 to test whether a side assignment for e and f leads to a crossing and in the case of a crossing we introduce suitable clauses to rule out that side assignment. □

We finally get the following.

Theorem 8. MNTLP WITH FIXED ORDERING *can be tested in* $O(\binom{|2E+1|}{|C|^2} |E|^2)$ *time for an instance* $\langle G = (V, E, C), \gamma, \sigma \rangle$.

Proof Sketch: The proof is based on guessing, for each pair of adjacent clusters, whether they are connected by an S-drawn edge or not. For each guess we exploit Lemmata 3 and 4 to construct a 2-SAT formula that is checked for satisfiability. For each pair of adjacent clusters V_a, V_b we have to guess among $2|E_{a,b}| + 1$ possibilities, corresponding to the choice of $|E_{a,b}|$ edges to be S-drawn, each in two possible ways, plus the possibility of not having any S-drawn edge. This leads to $O(\binom{|2E+1|}{|C|^2})$ guesses. □

Observe that the computational complexity of the algorithm described in the proof of Theorem 8 is polynomial if the number of clusters is constant.

4   Conclusions and Open Problems

We have shown that testing NodeTrix (NT) representations of clustered graphs for planarity is NP-complete even if the order of the rows and columns is fixed or if the sides where the inter-cluster edges attach to the matrices is fixed. We have also studied the setting where matrices have fixed positions and inter-cluster edges are xy-monotone curves. In this case we established negative and positive results; leveraging on the latter, we developed a library that computes a layout of the inter-cluster edges with few crossings. A demo [3] shows that the computation allows the user to move matrices without any slowdown of the interaction.

Several theoretical problems are related to the planarity of NT representations. First, the NP-completeness of NT planarity can be interpreted as a proof of the NP-completeness of clustered planarity (see, for example, [4,6,11,14]) when a specific type of representation is required. Observe, though, that a flat clustered graph may be NT planar even if its underlying graph is not planar.

Second, planarity of hybrid representations have been recently studied [5] in the setting in which clusters are represented as the intersections of geometric objects. Our results can be viewed as a further progress in this area. Third, consider a clustered graph with two clusters represented as matrices aligned along their principal diagonal. Computing a locally planar NT representation is equivalent to solve the 2-page bipartite book embedding with spine crossings problem [5]. Interestingly, if the two matrices are aligned along their secondary diagonal this equivalence is not evident anymore.

Among the future research directions, we mention the one of automatically embedding the matrices to minimize crossings in monotone NT representations.

References

1. CiteVis: Visualizing citations among InfoVis conference papers. http://www.cc.gatech.edu/gvu/ii/citevis
2. NodeTrix. http://www.aviz.fr/Research/Nodetrix
3. NodeTrix Representations: a proof-of-concept editor. http://www.dia.uniroma3.it/~dalozzo/projects/matrix
4. Angelini, P., Da Lozzo, G., Di Battista, G., Frati, F., Patrignani, M., Roselli, V.: Relaxing the constraints of clustered planarity. Comput. Geom.: Theory Appl. **48**(2), 42–75 (2015)
5. Angelini, P., Da Lozzo, G., Di Battista, G., Frati, F., Patrignani, M., Rutter, I.: Intersection-link representations of graphs. In: Di Giacomo, E., Lubiw, A. (eds.) GD 2015. LNCS, vol. 9411, pp. 217–230. Springer, Heidelberg (2015). doi:10.1007/978-3-319-27261-0_19
6. Angelini, P., Da Lozzo, G., Di Battista, G., Frati, F., Roselli, V.: The importance of being proper (in clustered-level planarity and T-level planarity). Theor. Comput. Sci. **571**, 1–9 (2015)
7. Angelini, P., Da Lozzo, G., Neuwirth, D.: Advancements on SEFE and partitioned book embedding problems. Theor. Comput. Sci. **575**, 71–89 (2015)
8. Aspvall, B., Plass, M.F., Tarjan, R.E.: A linear-time algorithm for testing the truth of certain quantified boolean formulas. Inf. Process. Lett. **8**(3), 121–123 (1979)
9. Batagelj, V., Brandenburg, F.J., Didimo, W., Liotta, G., Palladino, P., Patrignani, M.: Visual analysis of large graphs using (x, y)-clustering and hybrid visualizations. IEEE TVCG **17**(11), 1587–1598 (2011)
10. Chimani, M., Battista, G., Frati, F., Klein, K.: Advances on testing c-planarity of embedded flat clustered graphs. In: Duncan, C., Symvonis, A. (eds.) GD 2014. LNCS, vol. 8871, pp. 416–427. Springer, Heidelberg (2014). doi:10.1007/978-3-662-45803-7_35
11. Cortese, P.F., Di Battista, G., Frati, F., Patrignani, M., Pizzonia, M.: C-planarity of c-connected clustered graphs. J. Graph Algorithms Appl. **12**(2), 225–262 (2008)
12. Da Lozzo, G., Di Battista, G., Frati, F., Patrignani, M.: Computing nodetrix representations of clustered graphs. CoRR abs/1608.08952v1 (2016). http://arxiv.org/abs/1608.08952v1
13. Di Battista, G., Frati, F.: Efficient c-planarity testing for embedded flat clustered graphs with small faces. J. Graph Algorithms Appl. **13**(3), 349–378 (2009)
14. Feng, Q.-W., Cohen, R.F., Eades, P.: Planarity for clustered graphs. In: Spirakis, P. (ed.) ESA 1995. LNCS, vol. 979, pp. 213–226. Springer, Heidelberg (1995). doi:10.1007/3-540-60313-1_145

15. Gutwenger, C., Klein, K., Mutzel, P.: Planarity testing and optimal edge insertion with embedding constraints. J. Graph Algorithms Appl. **12**(1), 73–95 (2008)
16. Henry, N., Bezerianos, A., Fekete, J.D.: Improving the readability of clustered social networks using node duplication. IEEE TVCG **14**(6), 1317–1324 (2008)
17. Henry, N., Fekete, J.D., McGuffin, M.J.: NodeTrix: a hybrid visualization of social networks. IEEE Trans. Vis. Comput. Graph. **13**(6), 1302–1309 (2007)
18. Opatrny, J.: Total ordering problem. SIAM J. Comput. **8**(1), 111–114 (1979)
19. Spinrad, J.P.: Recognition of circle graphs. J. Algorithms **16**(2), 264–282 (1994)
20. Unger, W.: On the k-colouring of circle-graphs. In: Cori, R., Wirsing, M. (eds.) STACS 1988. LNCS, vol. 294, pp. 61–72. Springer, Heidelberg (1988). doi:10.1007/BFb0035832

Planar Graphs

1-Bend Upward Planar Drawings of SP-Digraphs

Emilio Di Giacomo, Giuseppe Liotta, and Fabrizio Montecchiani[✉]

Dip. di Ingegneria, Università Degli Studi di Perugia, Perugia, Italy
{emilio.digiacomo,giuseppe.liotta,fabrizio.montecchiani}@unipg.it

Abstract. It is proved that every series-parallel digraph whose maximum vertex-degree is Δ admits an upward planar drawing with at most one bend per edge such that each edge segment has one of Δ distinct slopes. This is shown to be worst-case optimal in terms of the number of slopes. Furthermore, our construction gives rise to drawings with optimal angular resolution $\frac{\pi}{\Delta}$. A variant of the proof technique is used to show that (non-directed) reduced series-parallel graphs and flat series-parallel graphs have a (non-upward) one-bend planar drawing with $\lceil \frac{\Delta}{2} \rceil$ distinct slopes if biconnected, and with $\lceil \frac{\Delta}{2} \rceil + 1$ distinct slopes if connected.

1 Introduction

The *k-bend planar slope number* of a family of planar graphs with maximum vertex-degree Δ is the minimum number of distinct slopes used for the edges when computing a crossing-free drawing with at most $k > 0$ bends per edge of any graph in the family. For example, if $\Delta = 4$, a classic result is that every planar graph has a crossing-free drawing such that every edge segment is either horizontal or vertical and each edge has at most two bends (see, e.g., [2]). Clearly, this is an optimal bound on the number of slopes. This result has been extended to values of Δ larger than four by Keszegh et al. [15], who prove that $\lceil \frac{\Delta}{2} \rceil$ slopes suffice to construct a planar drawing with at most two bends per edge for any planar graph. However, if additional geometric constraints are imposed on the crossing-free drawing, only a few tight bounds on the planar slope number are known. For example, if one requires that the edges cannot have bends, the best known upper bound on the planar slope number is $O(c^\Delta)$ (for a constant $c > 1$) while a general lower bound of just $3\Delta - 6$ has been proved [15]. Tight bounds are only known for outerplanar graphs [17] and subcubic planar graphs [9], while the gap between upper and lower bound has been reduced for planar graphs with treewidth two [18] or three [10,14]. If one bend per edge is allowed, Keszegh et al. [15] show an upper bound of 2Δ and a lower bound of $\frac{3}{4}(\Delta - 1)$ on the planar slope number of the planar graphs with maximum vertex-degree Δ. In a recent paper, Knauer and Walczak [16] improve the upper bound to $\frac{3}{2}(\Delta - 1)$; in the same paper, it is also proved that a tight bound of $\lceil \frac{\Delta}{2} \rceil$ can be achieved for the outerplanar graphs.

Research supported in part by the MIUR project AMANDA "Algorithmics for MAssive and Networked DAta", prot. 2012C4E3KT_001.

Y. Hu and M. Nöllenburg (Eds.): GD 2016, LNCS 9801, pp. 123–130, 2016.
DOI: 10.1007/978-3-319-50106-2_10

In this paper we focus on the 1-bend planar slope number of directed graphs with the additional requirement that the computed drawing be *upward*, i.e., each edge is drawn as a curve monotonically increasing in the y-direction. We recall that upward drawings are a classic research topic in graph drawing, see, e.g., [1,3,11–13] for a limited list of references. Also, upward drawings of ordered sets with no bends and few slopes have been studied by Czyzowicz [4,5]. We show that every series-parallel digraph (SP-digraph for short) G whose maximum vertex-degree is Δ has *1-bend upward planar slope number* Δ. That is, G admits an upward planar drawing with at most one bend per edge where at most Δ distinct slopes are used for the edges. This is shown to be worst-case optimal in terms of the number of slopes. An implication of this result is that the general $\frac{3}{2}(\Delta - 1)$ upper bound for the (undirected) 1-bend planar slope number [16] can be lowered to Δ when the graph is series-parallel. We then extend our drawing technique to undirected graphs and hence look at non-upward drawings. We show a tight bound of $\lceil \frac{\Delta}{2} \rceil$ for the 1-bend planar slope number of biconnected reduced SP-graphs and biconnected flat SP-graphs (see Sect. 2 for definitions). The biconnectivity requirement can be dropped at the expenses of one more slope. To prove the above results, we construct a suitable contact representation γ of an SP-digraph where each vertex is represented as a cross, i.e. a horizontal segment intersected by a vertical segment (Sect. 3); then, we transform γ into a 1-bend upward planar drawing Γ optimizing the number of slopes used in such transformation (Sect. 4). Our algorithm runs in linear time and gives rise to drawings with angular resolution at least $\frac{\pi}{\Delta}$, which is worst-case optimal. Some proofs and technicalities are omitted and can be found in [7].

2   Preliminaries

A *series-parallel digraph* (*SP-diagraph* for short) [6] is a simple planar digraph that has one source and one sink, called *poles*, and it is recursively defined as follows. A single edge is an SP-digraph. The digraph obtained by identifying the sources and the sinks of two SP-digraphs is an SP-digraph (*parallel composition*). The digraph obtained by identifying the sink of one SP-digraph with the source of a second SP-digraph is an SP-digraph (*series composition*). A *reduced* SP-digraph is an SP-digraph with no transitive edges. An SP-digraph G is associated with a binary tree T, called the *decomposition tree* of G. The nodes of T are of three types, *Q-nodes*, *S-nodes*, and *P-nodes*, representing single edges, series compositions, and parallel compositions, respectively. An example is shown in Fig. 1(a). The decomposition tree of G has $O(n)$ nodes and can be constructed in $O(n)$ time [6]. An SP-digraph is *flat* if its decomposition tree does not contain two P-nodes that share only one pole and that are not in a series composition (see, e.g., [8]). The underlying undirected graph of an SP-digraph is called an *SP-graph* , and the definitions of reduced and flat SP-digraphs translate to it.

The *slope* s of a line ℓ is the angle that a horizontal line needs to be rotated counter-clockwise in order to make it overlap with ℓ. The slope of a segment is the slope of its supporting line. We denote by $\mathcal{S}_k$ the set of slopes: $s_i = \frac{\pi}{2} + i\frac{\pi}{k}$

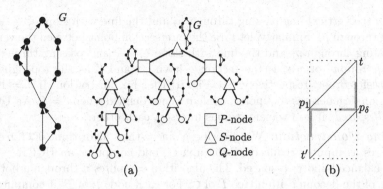

Fig. 1. (a) An SP-digraph G and its decomposition tree. (b) The safe-region (dotted) of a cross.

$(i = 0, \ldots, k - 1)$. Note that $\mathcal{S}_k$ contains the slope $\frac{\pi}{2}$ for any value of k. Also, any polyline drawing using only slopes in $\mathcal{S}_k$ has angular resolution (i.e. the minimum angle between any two consecutive edges around a vertex) at least $\frac{\pi}{k}$.

3   Cross Contact Representations

Basic Definitions. A *cross* consists of one horizontal and one vertical segment that share an interior point, called *center* of the cross. A cross is *degenerate* if either its horizontal or its vertical segment has zero length. The center of a degenerate cross is its midpoint. A point p of a cross c is an *end-point* (*interior point*) of c if it is an end-point (interior point) of the horizontal or vertical segment of c. Two crosses c_1 and c_2 *touch* if they share a point p, called *contact*, such that p is an end-point of the vertical (horizontal) segment of c_1 and an interior point of the horizontal (vertical) segment of c_2. A *cross-contact representation (CCR)* of a graph G is a drawing γ such that: (i) Every vertex v of G is represented by a cross $c(v)$; (ii) All intersections of crosses are contacts; and (iii) Two crosses $c(u)$ and $c(v)$ touch if and only if the edge (u, v) is in G.

We now consider CCRs of digraphs, and define properties that will be useful to transform a CCR into a 1-bend upward planar drawing with few slopes and good angular resolution. Let γ be a CCR of a digraph G with maximum vertex-degree Δ. Let (u, v) be an edge of G oriented from u to v. Let p be the contact between $c(u)$ and $c(v)$. The point p is an *upward contact* if the following two conditions hold: (a) p is an end-point of the vertical segment of one of the two crosses and an interior point of the other cross, and (b) the center of $c(v)$ is above the center of $c(u)$. A CCR of a digraph G such that all its contacts are upward is an *upward CCR (UCCR)*. An UCCR γ is *balanced* if for every non-degenerate cross $c(u)$ of γ, we have that $|n_l(u) - n_r(u)| \leq 1$, where $n_l(u)$ $(n_r(u))$ is the number of contacts to the left (right) of the center of $c(u)$. Let $\{p_1, p_2, \ldots, p_\delta\}$ be the $\delta \geq 0$ contacts along the horizontal segment of $c(u)$, in this order from the leftmost one (p_1) to the rightmost one (p_δ). Let t be the intersection point between the vertical line passing through p_δ and the line with slope $\frac{\pi}{2} - \frac{\pi}{\Delta}$ and passing through p_1. Similarly, let t' be the intersection point between the vertical line passing through p_1 and the line with slope $\frac{\pi}{2} - \frac{\pi}{\Delta}$ and passing through p_δ. The *safe-region* of $c(u)$ is the rectangle having t and t' as the top-right and bottom-left corner, respectively. See Fig. 1(b) for an illustration. If $\delta = 1$, the safe-region degenerates to a point, while it is not defined when $\delta = 0$. An UCCR γ is *well-spaced* if no two safe-regions intersect each other.

Drawing Construction. We describe a linear-time algorithm, `UCCRDrawer`, that takes as input a reduced SP-digraph G, and computes an UCCR γ of G that is balanced and well-spaced. The algorithm computes γ through a bottom-up visit of the decomposition tree T of G. For each node μ of T, it computes an UCCR γ_μ of the graph G_μ associated with μ satisfying the following properties: **P1.** γ_μ is balanced; **P2.** γ_μ is well-spaced; **P3.** Let s_μ and t_μ be the two poles of G_μ. If μ is not a Q-node, then both $c(s_\mu)$ and $c(t_\mu)$ are degenerate, with $c(s_\mu)$ at the bottom side of a rectangle R_μ that contains γ_μ, and $c(t_\mu)$ at the top side of R_μ.

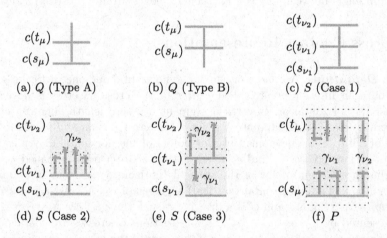

(a) Q (Type A)   (b) Q (Type B)   (c) S (Case 1)

(d) S (Case 2)   (e) S (Case 3)   (f) P

Fig. 2. Illustration for `UCCRDrawer`. The safe-regions are dotted (and not in scale).

For each leaf node μ (which is a Q-node) the associated graph G_μ consists of a single edge (s_μ, t_μ). We define two possible types of UCCR, γ_μ^A (type A) and γ_μ^B (type B), of G_μ, which are shown in Fig. 2(a) and (b), respectively. Properties **P1** – **P2** trivially hold in this case, while property **P3** does not apply.

For each non-leaf node μ of T, `UCCRDrawer` computes the UCCR γ_μ by suitably combining the (already) computed UCCRs γ_{ν_1} and γ_{ν_2} of the two graphs associated with the children ν_1 and ν_2 of μ. If μ is an S-node of T, we distinguish between the following cases, where $t_{\nu_1} = s_{\nu_2}$ is the pole shared by ν_1 and ν_2.

Case 1. Both ν_1 and ν_2 are Q-nodes. Then an UCCR of G_μ is computed by combining $\gamma_{\nu_1}^A$ and $\gamma_{\nu_2}^B$ as in Fig. 2(c). Properties **P1** – **P3** trivially hold.

Case 2. ν_1 is a Q-node, while ν_2 is not (the case when ν_2 is a Q-node and ν_1 is not is symmetric). We combine the drawing $\gamma_{\nu_1}^A$ of G_{ν_1} and the drawing γ_{ν_2} of G_{ν_2} as in Fig. 2(d). Notice that to combine the two drawings we may need to scale one of them so that their widths are the same. To ensure **P1**, we move the vertical segment of $c(t_{\nu_1}) = c(s_{\nu_2})$ so that $|n_l(t_{\nu_1}) - n_r(t_{\nu_1})| \leq 1$. We may also need to shorten its upper part in order to avoid crossings with other segments, and to extend its lower part so that $c(s_{\nu_1})$ is outside the safe-region of $c(t_{\nu_1}) = c(s_{\nu_2})$, thus guaranteeing property **P2**. Property **P3** holds by construction.

Case 3. If none of ν_1 and ν_2 is a Q-node, then we combine γ_{ν_1} and γ_{ν_2} as in Fig. 2(e). We may need to scale one of the two drawings so that their widths are the same. Property **P1** holds, as it holds for γ_{ν_1} and γ_{ν_2}. Furthermore, we ensure **P2** by performing the following stretching operation. Let ℓ_a and ℓ_b be two horizontal lines slightly above and slightly below the horizontal segment of $c(t_{\nu_1}) = c(s_{\nu_2})$, respectively. We extend all the vertical segments intersected by ℓ_a or ℓ_b until the safe-region of $c(t_{\nu_1}) = c(s_{\nu_2})$ does not intersect any other safe-region. Property **P3** holds by construction.

Let μ be a P-node of T, having ν_1 and ν_2 as children (recall that neither ν_1 nor ν_2 is a Q-node, since G is a reduced SP-digraph). We combine γ_{ν_1} and γ_{ν_2} as in Fig. 2(f). We may need to scale one of the two drawings so that their heights are the same. Property **P1** holds, as it holds for γ_{ν_1} and γ_{ν_2}. To ensure **P2**, a stretching operation similar to the one described in Case 3 is possibly performed by using a horizontal line slightly above (below) the horizontal segment of $c(s_\mu)$ $(c(t_\mu))$. Property **P3** holds by construction.

To deal with the time complexity of algorithm UCCRDrawer, we represent each cross with the coordinates of its four end-points. To obtain linear time complexity, for each drawing γ_μ of a node μ, we avoid moving all the crosses of its children. Instead, for each child of μ, we only store the offset of the top-left corner of the bounding box of its drawing. Afterwards, we fix the final coordinates of each cross through a top-down visit of T. The above discussion can be summarized as follows.

Lemma 1. *Let G be an n-vertex reduced SP-digraph. Algorithm UCCRDrawer computes a balanced and well-spaced UCCR γ of G in $O(n)$ time.*

4   1-Bend Drawings

We start by describing how to transform an UCCR of a reduced SP-digraph into a 1-bend upward planar drawing that uses the slope-set $\mathcal{S}_\Delta$. Let γ be an UCCR of a reduced SP-digraph G and let $c(u)$ be the cross representing a vertex u of G in γ. Let $p_1, \ldots, p_\delta$ ($\delta \geq 1$) be the contacts along the horizontal segment of $c(u)$, in this order from the leftmost one (p_1) to the rightmost one (p_δ). Let c be either the center of $c(u)$, if $c(u)$ is non-degenerate, or $p_{\lfloor \delta/2 \rfloor + 1}$ if $c(u)$ is degenerate. Consider the set of lines $\ell_0, \ldots, \ell_{\Delta-1}$, such that ℓ_i passes through c

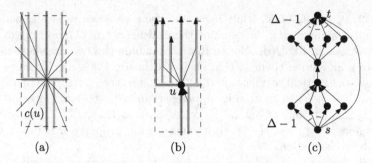

Fig. 3. (a)–(b) Transforming an UCCR into a 1-bend drawing. (c) An SP-digraph requiring at least Δ slopes in any 1-bend upward planar drawing.

and has slope $s_i \in \mathcal{S}_\Delta$ (for $i = 0, \ldots, \Delta - 1$). These lines, except for ℓ_0, intersect all the vertical segments forming a contact with the horizontal segment of $c(u)$. If $c(u)$ is not degenerate, then ℓ_0 coincides with the vertical segment, which has at least one contact. In particular, each quadrant of $c(u)$ contains a number of lines that is at least the number of vertical segments touching $c(u)$ in that quadrant. Since γ is well-spaced, these intersections are inside the safe-region of $c(u)$. Hence we can replace each contact of $c(u)$ with two segments having slope in $\mathcal{S}_\Delta$ as shown in Fig. 3(a) and (b). More precisely, each contact p_i of $c(u)$ is replaced with two segments that are both in the quadrant of $c(u)$ that contains the vertical segment defining p_i. This guarantees the upwardness of the drawing. Also, each edge has one bend, since it is represented by a single contact between a horizontal and a vertical segment and we introduce one bend only when dealing with the cross containing the horizontal segment. Finally, Γ is planar, because there is no crossing in γ and each cross is only modified inside its safe-region which, by the well-spaced property, is disjoint by any other safe-region. Thus, every reduced SP-digraph admits a 1-bend upward planar drawing with at most Δ slopes. To deal with a general SP-digraph, we subdivide each transitive edge and compute a drawing of the obtained reduced SP-digraph. We then modify this drawing to remove subdivision vertices (see also [7]).

Figure 3(c) shows a family of SP-digraphs such that, for every value of Δ, there exists a graph in this family with maximum vertex-degree Δ and that requires at least Δ slopes in any 1-bend upward planar drawing. Namely, if a digraph G has a source (or a sink) of degree Δ, then it requires at least $\Delta - 1$ slopes in any upward drawing because each slope, with the only possible exception of the horizontal one, can be used for a single edge. In the digraph of Fig. 3(c) however, the edge (s, t) must be either the leftmost or the rightmost edge of s and t in any upward planar drawing. Therefore, if only $\Delta - 1$ slopes are allowed, such edge cannot be drawn planarly and with one bend. Thus, the following theorem holds.

Theorem 1. *Every n-vertex SP-digraph G with maximum vertex-degree Δ admits a 1-bend upward planar drawing Γ with at most Δ slopes and angular* *resolution at least $\frac{\pi}{\Delta}$. These bounds are worst-case optimal. Also, Γ can be computed in $O(n)$ time.*

Since every SP-graph can be oriented to an SP-digraph (by computing a so-called *bipolar orientation* [19,20]), the next corollary is implied by Theorem 1 and improves the upper bound of $\frac{3}{2}(\Delta - 1)$ [16] for the case of SP-graphs.

Corollary 1. *The 1-bend planar slope number of SP-graphs with maximum vertex-degree Δ is at most Δ.*

Our drawing technique can be naturally extended to construct 1-bend planar drawings of two sub-families of biconnected SP-graphs using $\lceil \frac{\Delta}{2} \rceil$ slopes. Intuitively, if the drawing does not need to be upward, then for each cross $c(u)$ (see e.g. Fig. 3(a)), one can use the same slope for two distinct edges incident to u. Also, the biconnectivity requirement can be dropped by using one more slope.

Theorem 2. *Let G be a 2-connected SP-graph with maximum vertex-degree Δ and n vertices. If G is reduced or flat, then G admits a 1-bend planar drawing Γ with at most $\lceil \frac{\Delta}{2} \rceil$ slopes and angular resolution at least $\frac{2\pi}{\Delta}$. Also, Γ can be computed in $O(n)$ time.*

Corollary 2. *Let G be an SP-graph with maximum vertex-degree Δ and n vertices. If G is reduced or flat, then G admits a 1-bend planar drawing Γ with at most $\lceil \frac{\Delta}{2} \rceil + 1$ slopes and angular resolution at least $\frac{2\pi}{\Delta+1}$. Also, Γ can be computed in $O(n)$ time.*

5   Open Problems

We proved that the 1-bend upward planar slope number of SP-digraphs with maximum vertex-degree Δ is at most Δ and this is a tight bound. Is the bound of Corollary 1 also tight? Moreover, can it be extended to any partial 2-tree?

References

1. Bertolazzi, P., Di Battista, G., Mannino, C., Tamassia, R.: Optimal upward planarity testing of single-source digraphs. SIAM J. Comput. **27**(1), 132–169 (1998)
2. Biedl, T.C., Kant, G.: A better heuristic for orthogonal graph drawings. Comput. Geom. **9**(3), 159–180 (1998)
3. Binucci, C., Didimo, W., Giordano, F.: Maximum upward planar subgraphs of embedded planar digraphs. Comput. Geom. **41**(3), 230–246 (2008)
4. Czyzowicz, J.: Lattice diagrams with few slopes. J. Comb. Theory Ser. A **56**(1), 96–108 (1991). http://dx.doi.org/10.1016/0097-3165(91)90025-C
5. Czyzowicz, J., Pelc, A., Rival, I.: Drawing orders with few slopes. Discret. Math. **82**(3), 233–250 (1990). http://dx.doi.org/10.1016/0012-365X(90)90201-R
6. Di Battista, G., Eades, P., Tamassia, R., Tollis, I.G.: Graph Drawing. Prentice-Hall, Upper Saddle River (1999)

7. Di Giacomo, E., Liotta, G., Montecchiani, F.: 1-Bend Upward Planar Drawings of SP-Digraphs. ArXiv e-prints abs/1608.08425 (2016). http://arxiv.org/abs/1608.08425v1
8. Di Giacomo, E.: Drawing series-parallel graphs on restricted integer 3D grids. In: Liotta, G. (ed.) GD 2003. LNCS, vol. 2912, pp. 238–246. Springer, Heidelberg (2004). doi:10.1007/978-3-540-24595-7_22
9. Di Giacomo, E., Liotta, G., Montecchiani, F.: The planar slope number of subcubic graphs. In: Pardo, A., Viola, A. (eds.) LATIN 2014. LNCS, vol. 8392, pp. 132–143. Springer, Heidelberg (2014). doi:10.1007/978-3-642-54423-1_12
10. Di Giacomo, E., Liotta, G., Montecchiani, F.: Drawing outer 1-planar graphs with few slopes. J. Graph. Algorithms Appl. **19**(2), 707–741 (2015). http://dx.doi.org/10.7155/jgaa.00376
11. Didimo, W.: Upward planar drawings and switch-regularity heuristics. J. Graph Algorithms Appl. **10**(2), 259–285 (2006)
12. Didimo, W., Giordano, F., Liotta, G.: Upward spirality and upward planarity testing. SIAM J. Discret. Math. **23**(4), 1842–1899 (2009)
13. Garg, A., Tamassia, R.: On the computational complexity of upward and rectilinear planarity testing. SIAM J. Comput. **31**(2), 601–625 (2001)
14. Jelínek, V., Jelínková, E., Kratochvíl, J., Lidický, B., Tesar, M., Vyskocil, T.: The planar slope number of planar partial 3-trees of bounded degree. Gr. Combin. **29**(4), 981–1005 (2013)
15. Keszegh, B., Pach, J., Pálvölgyi, D.: Drawing planar graphs of bounded degree with few slopes. SIAM J. Discret. Math. **27**(2), 1171–1183 (2013)
16. Knauer, K., Walczak, B.: Graph drawings with one bend and few slopes. In: Kranakis, E., Navarro, G., Chávez, E. (eds.) LATIN 2016. LNCS, vol. 9644, pp. 549–561. Springer, Heidelberg (2016). doi:10.1007/978-3-662-49529-2_41
17. Knauer, K.B., Micek, P., Walczak, B.: Outerplanar graph drawings with few slopes. Comput. Geom. **47**(5), 614–624 (2014)
18. Lenhart, W., Liotta, G., Mondal, D., Nishat, R.I.: Planar and plane slope number of partial 2-trees. In: Wismath, S., Wolff, A. (eds.) GD 2013. LNCS, vol. 8242, pp. 412–423. Springer, Heidelberg (2013). doi:10.1007/978-3-319-03841-4_36
19. Rosenstiehl, P., Tarjan, R.E.: Rectilinear planar layouts and bipolar orientations of planar graphs. Discr. Comput. Geom. **1**, 343–353 (1986)
20. Tamassia, R., Tollis, I.G.: A unified approach a visibility representation of planar graphs. Discr. Comput. Geom. **1**, 321–341 (1986)

Non-aligned Drawings of Planar Graphs

Therese Biedl[1] and Claire Pennarun[2(✉)]

[1] David R. Cheriton School of Computer Science, University of Waterloo, Waterloo,
Canada
biedl@uwaterloo.ca
[2] University of Bordeaux, CNRS, LaBRI, UMR 5800, 33400 Talence, France
claire.pennarun@labri.fr

Abstract. A *non-aligned* drawing of a graph is a drawing where no two
vertices are in the same row or column. Auber et al. showed that not
all planar graphs have a non-aligned planar straight-line drawing in the
$n \times n$-grid. They also showed that such a drawing exists if up to $n-3$
edges may have a bend.

In this paper, we give algorithms for non-aligned planar drawings
that improve on the results by Auber et al. In particular, we give such
drawings in an $n \times n$-grid with at most $\frac{2n-5}{3}$ bends, and we study what
grid-size can be achieved if we insist on having straight-line drawings.

1 Introduction

At last year's GD conference, Auber et al. [2] introduced the concept of *rook-drawings*: these are drawings of a graph in an $n \times n$-grid such that no two
vertices are in the same row or the same column (thus, if the vertices were
rooks on a chessboard, then no vertex could beat any other). They showed that
not all planar graphs have a planar straight-line rook-drawing, and then gave a
construction of planar rook-drawings with at most $n-3$ bends. From now on,
all drawings are required to be planar.

In this paper, we continue the study of rook-drawings. Note that if a graph has
no straight-line rook-drawing, then we can relax the restrictions in two possible
ways. We could either, as Auber et al. did, allow to use bends for some of the
edges, and try to keep the number of bends small. Or we could increase the
grid-size and ask what size of grid can be achieved for straight-line drawings in
which no two vertices share a row or a column; this type of drawing is known
as *non-aligned drawing* [1]. A rook-drawing is then a non-aligned drawing on an
$n \times n$-grid.

Existing Results: Apart from the paper by Auber et al., non-aligned drawings
have arisen in a few other contexts. Alamdari and Biedl showed that every
graph that has an inner rectangular drawing also has a non-aligned drawing
[1]. These drawings are so-called rectangle-of-influence drawings and can hence
be assumed to be in an $n \times n$-grid. In particular, every 4-connected planar
graph with at most $3n-7$ edges therefore has a rook-drawing (see Sect. 3.1 for
details). Non-aligned drawings were also created by Di Giacomo et al. [9] in the

© Springer International Publishing AG 2016
Y. Hu and M. Nöllenburg (Eds.): GD 2016, LNCS 9801, pp. 131–143, 2016.
DOI: 10.1007/978-3-319-50106-2_11

context of upward-rightward drawings. They showed that every planar graph has a non-aligned drawing in an $O(n^4) \times O(n^4)$-grid. Finally, there have been studies about drawing graphs with the opposite goal, namely, creating as many collinear vertices as possible [15].

Our Results: In this paper, we show the following (the bounds listed here are upper bounds; see the sections for tighter bounds):

- Every planar graph has a non-aligned straight-line drawing in an $n^2 \times n^2$-grid. This is achieved by taking any weak barycentric representation (for example, the one by Schnyder [16]), scaling it by a big enough factor, and then moving vertices slightly so that they have distinct coordinates while maintaining a weak barycentric representation.
- Every planar graph has a non-aligned straight-line drawing in an $n \times \frac{1}{2}n^3$-grid. This is achieved by creating drawings with the canonical ordering [11] in a standard fashion (similar to [8]). However, we pre-compute all the x-coordinates (and in particular, make them a permutation of $\{1, \ldots, n\}$), and then argue that with the standard construction the slopes do not get too big, and hence the height is quadratic. Modifying the construction a bit, we can also achieve that all y-coordinates are distinct and that the height is cubic.
- Every planar graph has a rook-drawing with at most $\frac{2n-5}{3}$ bends. This is achieved via creating a so-called rectangle-of-influence drawing of a modification of G, and arguing that each modification can be undone while adding only one bend.

Our bounds are even better for 4-connected planar graphs. In particular, every 4-connected planar graph has a rook-drawing with at most 1 bend (and more generally, the number of bends is no more than the number of so-called filled triangles). We also show that any so-called nested-triangle graph has a non-aligned straight-line drawing in an $n \times (\frac{4}{3}n - 1)$-grid.

2   Non-aligned Straight-Line Drawings

In this section, all drawings are required to be straight-line drawings.

2.1   Non-aligned Drawings on an $n^2 \times n^2$-Grid

We first show how to construct non-aligned drawings in an $n^2 \times n^2$-grid by scaling and perturbing a so-called *weak barycentric representation* (reviewed below). In the following, a vertex v is assigned to a triplet of non-negative integer coordinates $(p_0(v), p_1(v), p_2(v))$. For two vertices u, v and $i = 0, 1, 2$, we say that $p_i(u) <_{\text{lex}} p_i(v)$ if either $p_i(u) < p_i(v)$ or $p_i(u) = p_i(v)$ and $p_{i+1}(u) < p_{i+1}(v)$. Note that in this section, addition on the subscripts is done modulo 3.

Definition 1 ([16]). *A* weak barycentric representation *of a graph G is an injective function mapping each $v \in V(G)$ to a point $(p_0(v), p_1(v), p_2(v)) \in \mathbb{N}_0^3$ such that:*

- $p_0(v) + p_1(v) + p_2(v) = c$ for every vertex v, where c is a constant independent of the vertex,
- for each edge (u, v) and each vertex $w \neq \{u, v\}$, there is some $k \in \{0, 1, 2\}$ such that $p_k(u) <_{lex} p_k(z)$ and $p_k(v) <_{lex} p_k(z)$.

Theorem 1 ([16]). *Every planar graph with n vertices has a weak barycentric representation with $c = n - 1$. Furthermore, $0 \leq p_i(v) \leq n - 2$ for all vertices $v \in V$ and all $i \in \{0, 1, 2\}$.*

Observe that weak barycentric representations are preserved under scaling, i.e., if we have a weak barycentric representation $\mathcal{P}$ (say with constant c), then we can scale all assigned coordinates by the same factor N and obtain another weak barycentric representation (with constant $c \cdot N$). We need to do slightly more, namely scale and "twist", as detailed in the following lemma (not proved here).

Lemma 1. *Let G be a graph with a weak barycentric representations $\mathcal{P} = ((p_0(v), p_1(v), p_2(v))_{v \in V})$. Let $N \geq 1 + \max_{v \in V}\{\max_{i=0,1,2} p_i(v)\}$ be a positive integer. Define $\mathcal{P}'$ to be the assignment $p_i'(v) := N \cdot p_i(v) + p_{i+1}(v)$ for $i = 0, 1, 2$. Then $\mathcal{P}'$ is also a weak barycentric representation.*

Applying this to Schnyder's weak barycentric representation, we now have:

Theorem 2. *Every planar graph has a non-aligned straight-line planar drawing in an $(n(n-2)) \times (n(n-2))$-grid.*

Proof. Let $\mathcal{P} = ((p_0(v), p_1(v), p_2(v))_{v \in V})$ be the weak barycentric representation of Theorem 1; we know that $0 \leq p_i(v) \leq n - 2$ for all v and all i. Now apply Lemma 1 with $N = n - 1$ to obtain the weak barycentric representation $\mathcal{P}'$ with $p_i'(v) = (n - 1)p_i(v) + p_{i+1}(v)$. Observe that $p_i'(v) \leq (n - 1)(n - 2) + (n - 2) = n(n - 2)$. Also, $p_i'(v) \geq 1$ since not both $p_i(v)$ and $p_{i+1}(v)$ can be 0. (More precisely, $p_i(v) = 0 = p_{i+1}(v)$ would imply $p_{i+2}(v) = n - 1$, contradicting $p_{i+2}(v) \leq n - 2$.)

As shown by Schnyder [16], mapping each vertex v to point $(p_0'(v), p_1'(v))$ gives a planar straight-line drawing of G. By the above, this drawing has the desired grid-size. It remains to show that it is non-aligned, i.e., for any two vertices u, v and any $i \in \{0, 1\}$, we have $p_i'(u) \neq p_i'(v)$. Assume after possible renaming that $p_i(u) \leq p_i(v)$. We have two cases:

- If $p_i(u) < p_i(v)$, then $p_i(u) \leq p_i(v) - 1$ since $\mathcal{P}$ assigns integers. Thus $N \cdot p_i(u) \leq N \cdot p_i(v) - N < N \cdot p_i(v) - p_{i+1}(u) + p_{i+1}(v)$ since $p_{i+1}(u) < N$ and $p_{i+1}(v) \geq 0$. Therefore $p_i'(u) < p_i'(v)$.
- If $p_i(u) = p_i(v)$, then $p_{i+1}(u) \neq p_{i+1}(v)$ (else the three coordinates of u and v would be the same, which is impossible since $\mathcal{P}$ is an injective function). Then $p_i'(u) = N \cdot p_i(u) + p_{i+1}(u) \neq N \cdot p_i(v) + p_{i+1}(v) = p_i'(v)$. $\square$

2.2 Non-aligned Drawings on an $n \times f(n)$-Grid

We now show how to build non-aligned drawings for which the width is the minimum-possible n, and the height is $\approx \frac{1}{2}n^3$. We use the well-known canonical ordering for *triangulated plane graphs*, i.e., graphs for which the planar embedding is fixed and all faces (including the outer-face) are triangles. We hence assume throughout that G is triangulated; we can achieve this by adding edges and delete them in the obtained drawing.

The *canonical ordering* [11] of such a graph is a vertex order $v_1, \ldots, v_n$ such that $\{v_1, v_2, v_n\}$ is the outer-face, and for any $3 \leq k \leq n$, the graph G_k induced by $v_1, \ldots, v_k$ is 2-connected. This implies that v_k has at least 2 *predecessors* (i.e., neighbours in G_{k-1}), and its predecessors form an interval on the outer-face of G_{k-1}. We assume (after possible renaming) that v_1 is the neighbor of v_2 found in clockwise order on the outer-face, and enumerate the outer-face of graph G_{k-1} in clockwise order as $c_1, \ldots, c_L$ with $c_1 = v_1$ and $c_L = v_2$. Then the predecessors of v_k consist of $c_\ell, \ldots, c_r$ for some $1 \leq \ell < r \leq L$; we call c_ℓ and c_r the *leftmost* and *rightmost* predecessor of v_k (see also Fig. 1). In this section, $x(v)$ and $y(v)$ denote the x- and y-coordinates of a vertex v, respectively.

Distinct x-Coordinates. We first give a construction that achieves distinct x-coordinates in $\{1, \ldots, n\}$ (but y-coordinates may coincide). Let $v_1, \ldots, v_n$ be a canonical ordering. The goal is to build a straight-line drawing of the graph G_k induced by $v_1, \ldots, v_k$ using induction on k. The key idea is to define *all* x-coordinates beforehand. Define an orientation of the edges of G as follows. Direct (v_1, v_2) as $v_1 \to v_2$. For $k \geq 3$, if c_r is the rightmost predecessor of v_k, then direct all edges from predecessors of v_k towards v_k, with the exception of (v_k, c_r), which is directed $v_k \to c_r$.

By induction on k, one easily shows that the orientation of G_k is acyclic, with unique source v_1 and unique sink v_2, and the outer-face directed $c_1 \to \cdots \to c_L$. Find a topological order $x : V \to \{1, \ldots, n\}$ of the vertices, i.e., if $u \to v$ then $x(u) < x(v)$. We use this topological order as our x-coordinates, and hence have $x(v_1) = 1$ and $x(v_2) = n$. (We thus use two distinct vertex-orderings: one defined by the canonical ordering, which is used to compute y-coordinates, and one defined by the topological ordering derived from the canonical ordering, which directly gives the x-coordinates.)

Now construct a drawing of G_k that respects these x-coordinates by induction on k; see also Fig. 1. Start with v_1 at $(1, 2)$, v_3 at $(x(v_3), 2)$ and v_2 at $(n, 1)$.

For $k \geq 3$, let c_ℓ and c_r be the leftmost and rightmost predecessor of v_{k+1}. Notice that $x(c_\ell) < \cdots < x(c_r)$ due to our orientation. Let y^* be the smallest integer value such that any c_j, for $\ell \leq j \leq r$, can "see" the point $p = (x(v_{k+1}), y^*)$ in the sense that that the line segment from c_j to p intersects no other vertices or edges. Since c_j is on the outer-face, then c_j can see all points on the ray upward from c_j. Furthermore, by tilting this ray slightly, c_j can also see all sufficiently high points on the vertical line $\{x = x(v_{k+1})\}$. Thus, such a y^* exists. Placing v_{k+1} at $(x(v_{k+1}), y^*)$ hence gives a planar drawing of G_{k+1}, and we continue until we get a drawing of $G_n = G$.

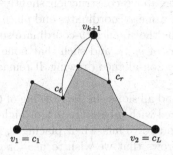

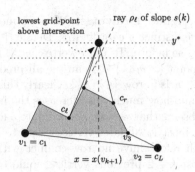

Fig. 1. (Left) Illustration of a canonical order. (Right) Finding a y-coordinate for v_{k+1}.

To analyze the height of this construction, we bound the slopes.

Lemma 2. *Define $s(k) := k - 3$ for $k \geq 3$. All edges on the outer-face of the constructed drawing of G_k have slope at most $s(k)$ for $k \geq 3$.*

Proof. (Sketch) In the drawing of G_k, let ρ_ℓ be the ray of slope $s(k)$ starting at c_ℓ. Consider the place where this ray intersects the vertical line $\{x = x(v_{k+1})\}$, and let y' be the smallest y-coordinate of a grid point vertically above this intersection point. Hence

$$y' \leq y(c_\ell) + (x(v_{k+1}) - x(c_\ell)) \cdot s(k) + 1. \tag{1}$$

Since all edges on the outer-face of G_k have slope at most $s(k)$, one can easily verify that point $(x(v_{k+1}), y')$ can see all of $c_\ell, \ldots, c_r$, so we have $y^* \leq y'$. Also, the worst new slope among the edges of the outer-face of G_k occurs at edge (c_ℓ, v_{k+1}), which by Eq. (1) has slope at most

$$\frac{y' - y(c_\ell)}{x(v_{k+1}) - x(c_\ell)} \leq s(k) + \frac{1}{x(v_{k+1}) - x(c_\ell)} \leq s(k) + 1 = s(k+1). \qquad \square$$

Vertex v_n has x-coordinate at most $n - 1$, and the edge from v_1 to v_n has slope at most $s(n) = n - 3$. This shows that the y-coordinate of v_n is at most $2 + (n - 2) \cdot (n - 3)$. Since triangle $\{v_1, v_2, v_n\}$ bounds the drawing, this gives:

Theorem 3. *Every planar graph has a planar straight-line drawing in an $n \times (2 + (n - 2)(n - 3))$-grid such that all vertices have distinct x-coordinates.*

While this theorem per se is not useful for non-aligned drawings, we find it interesting from a didactic point of view: It proves that polynomial coordinates can be achieved for straight-line drawings of planar graphs, and requires for this only the canonical ordering, but neither the properties of Schnyder trees [16] nor the details of how to "shift" that is needed for other methods using the canonical ordering (e.g. [8,11]). We believe that our bound on the height is much too big, and that the true height is $o(n^2)$ and possibly $O(n)$.

Non-aligned Drawings. We now modify the above construction slightly to achieve distinct y-coordinates. Define the exact same x-coordinates and place v_1 and v_2 as before. To place vertex v_{k+1}, let y^* be the smallest y-coordinate such that point $(x(v_{k+1}), y^*)$ can see all predecessors of v_{k+1}, and such that none of $v_1, \ldots, v_k$ is in row $\{y = y^*\}$. Clearly this gives a non-aligned drawing. It remains to bound how much this increases the height.

Observe that we use y-coordinate 3 for v_3, and all slopes in the drawing of G_3 are at most 1. All later vertices want to be placed with y-coordinate 3 or higher, and therefore have no row-conflict with v_1 and v_2. Thus when placing v_{k+1}, at most $k - 2$ previous vertices could occupy rows that we wish to use for y^*. In consequence, Eq. (1) is now replaced by

$$y^* \leq y(c_\ell) + (x(v_{k+1}) - x(c_\ell)) \cdot s'(k) + (k - 1) \tag{2}$$

where $s'(k)$ is the bound on the slope of the drawing of G_k. A proof similar to the one of Lemma 2 now shows:

Lemma 3. *Define $s'(k) := \sum_{i=1}^{k-2} i = \frac{1}{2}(k-1)(k-2)$ for $k \geq 3$. All edges on the outer-face of the constructed non-aligned drawing of G_k have slope at most $s'(k)$ for $k \geq 3$.*

The maximal slope is hence at most $\frac{1}{2}(n-1)(n-2)$, and if applicable, achieved at edge (v_1, v_n). Since $x(v_n) - x(v_1) \leq n - 2$ and $y(v_1) = 2$, therefore the height is at most $2 + \frac{1}{2}(n-1)(n-2)^2$.

Theorem 4. *Every planar graph has a non-aligned straight-line drawing in an $n \times \left(2 + \frac{1}{2}(n-1)(n-2)^2\right)$-grid.*

Comparing this to Theorem 2, we see that the aspect ratio is much worse, but the area is smaller. We suspect that the method results in a smaller height than the proven upper bound: Eq. (2) is generally not tight, and so a smaller slope-bound (implying a smaller height) is likely to hold.

2.3   The Special Case of Nested Triangles

We now turn to non-aligned drawings of a special graph class. Define a *nested-triangle graph* G as follows. G has $3k$ vertices for some $k \geq 1$, say $\{u_i, v_i, w_i\}$ for $i = 1, \ldots, k$. Vertices $\{u_i, v_i, w_i\}$ form a triangle (for $i = 1, \ldots, k$). We also have paths $u_1, u_2, \ldots, u_k$ as well as $v_1, v_2, \ldots, v_k$ and $w_1, w_2, \ldots, w_k$. With this the graph is 3-connected; we assume that its outer-face is $\{u_1, v_1, w_1\}$. All interior faces that are not triangles may or may not have a diagonal in them, and there are no restrictions on which diagonal (if any). Nested-triangle graphs are of interest in graph drawing because they are the natural lower-bound graphs for the area of straight-line drawings [10].

Theorem 5. *Any nested-triangle graph with $n = 3k$ vertices has a non-aligned straight-line drawing in an $n \times (\frac{4}{3}n - 1)$-grid.*

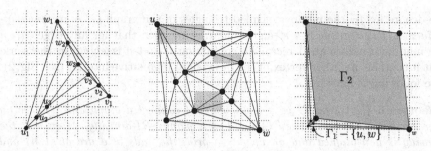

Fig. 2. A non-aligned straight-line drawing of a nested-triangle graph with $k = 3$ on an 9×11-grid, an RI-drawing satisfying the conditions of Theorem 6, and combining two RI-drawings if all separating triangles contain (u, w).

Proof. The 4-cycle $\{w_k, v_k, v_{k-1}, w_{k-1}\}$ may or may not have a diagonal in it; after possible exchange of $w_1, \ldots, w_k$ and $v_1, \ldots, v_k$ we assume that there is no edge between v_{k-1} and w_k. For $i = 1, \ldots, k$, place u_i at (i, i), vertex v_i at $(3k + 1 - i, k + i)$, and w_i at $(k + i, 4k + 1 - 2i)$ (see Fig. 2). The x- and y-coordinates are all distinct. The x-coordinates range from 1 to n, and the maximal y-coordinate is $4k - 1 = \frac{4}{3}n - 1$. It is easy to check that all interior faces are drawn strictly convex, with the exception of $\{v_k, v_{k-1}, w_{k-1}, w_k\}$ which has a 180° angle at v_k, but our choice of naming ensured that there is no edge (v_{k-1}, w_k). Thus any diagonal inside these 4-cycles can be drawn without overlap. Since G is planar, two edges joining vertices of different triangles cannot cross. Thus G is drawn without crossing in an $n \times \left(\frac{4}{3}n - 1\right)$-grid. □

In particular, notice that the octahedron is a nested-triangle graph (for $k = 2$) and this construction gives a non-aligned straight-line drawing in a 6×7-grid. This is clearly optimal since it has no straight-line rook-drawing [2].

We conjecture that this construction gives the minimum-possible height for nested-triangle graphs among all non-aligned straight-line drawings.

3   Rook-Drawings with Bends

We now construct rook-drawings with bends; as before we do this only for triangulated graphs. The main idea is to find rook-drawings with only 1 bend for 4-connected triangulated graphs. Then convert any graph into a 4-connected triangulated graph by subdividing few edges and re-triangulating, and argue that the drawing for it, modified suitably, gives a rook-drawing with few bends.

We need a few definitions first. Fix a triangulated graph G. A *separating triangle* is a triangle that has vertices both strictly inside and strictly outside the triangle. G is *4-connected* (i.e., cannot be made disconnected by removing 3 vertices) if and only if it has no separating triangle. A *filled triangle* [5] of G is a triangle that has vertices strictly inside. Graph G has at least one filled triangle (namely, the outer-face) and every separating triangle is also a filled triangle.

We denote by f_G the number of filled triangles of the graph G. A *rectangle-of-influence (RI) drawing* is a straight-line drawing such that for any edge (u,v), the minimum axis-aligned rectangle containing u and v is *empty* in the sense that it contains no other vertex of the drawing in its relative interior (see Fig. 2). The following is known:

Theorem 6 ([5]). *Let G be a triangulated 4-connected graph and let e be an edge on the outer-face. Then $G - e$ has a planar RI-drawing. Moreover, the drawing is non-aligned and on an $n \times n$-grid, the ends of e are at $(1,n)$ and $(n,1)$, and the other two vertices on the outer-face are at $(2,2)$ and $(n-1,n-1)$.*

The latter part of this claim is not specifically stated in [5], but can easily be inferred from the construction (see also a simpler exposition in [4]). RI-drawings are useful because they can be deformed (within limits) without introducing crossings. We say that two drawings Γ and Γ' of a graph have *the same relative coordinates* if for any two vertices v and w, we have $x_\Gamma(v) < x_\Gamma(w)$ if and only if $x_{\Gamma'}(v) < x_{\Gamma'}(w)$, and $y_\Gamma(v) < y_\Gamma(w)$ if and only if $y_{\Gamma'}(v) < y_{\Gamma'}(w)$, where $x_\Gamma(v)$ denotes the x-coordinate of v in Γ, etc. The following result appears to be folklore; we sketch a proof for completeness.

Observation 1. *Let Γ be an RI-drawing. If Γ' is a straight-line drawing with the same relative coordinates as Γ, then Γ' is an RI-drawing, and it is planar if and only if Γ is.*

Proof. The claim on the RI-drawing was shown by Liotta et al. [14]. It remains to argue planarity. Assume that edge (u,v) crosses edge (w,z) in an RI-drawing. Since all rectangles-of-influence are empty, this happens if and only if (up to renaming) we have $x(w) \leq x(u) \leq x(v) \leq x(z)$ and $y(u) \leq y(w) \leq y(z) \leq y(v)$. This only depends on the relative orders of u,v,w,z, and hence a transformation that maintains relative coordinates maintains planarity. □

We need a slight strengthening of Theorem 6.

Lemma 4. *Let G be a triangulated graph, let $e \in E$ be an edge on the outer-face, and assume all separating triangles contain e. Then $G - e$ has a planar RI-drawing. Moreover, the drawing is non-aligned and on an $n \times n$-grid, the ends of e are at $(1,n)$ and $(n,1)$, and the other two vertices on the outer-face are at $(2,2)$ and $(n-1,n-1)$.*

Proof. We proceed by induction on the number of separating triangles of G. In the base case, G is 4-connected and the claim holds by Theorem 6. For the inductive step, assume that $T = \{u,x,w\}$ is a separating triangle. By assumption it contains e, say $e = (u,w)$. Let G_1 be the graph consisting of T and all vertices inside T, and let G_2 be the graph obtained from G by removing all vertices inside T. Apply induction to both graphs. In drawing Γ_2 of $G_2 - e$, vertex x is on the outer-face and hence placed at $(2,2)$. Now insert a (scaled-down) copy of the drawing Γ_1 of G_1, minus vertices u and w, in the square $(1,2] \times (1,2]$ (see Fig. 2). Since x was in the top-right corner of $\Gamma_1 - \{u,w\}$, the two copies of x can be identified. One easily verifies that this gives an RI-drawing, because within each drawing the relative coordinates are unchanged, and the two drawings are disjoint except at u and w. Finally, re-assign coordinates to the vertices while keeping relative coordinates intact so that we have an $n \times n$-grid; by Observation 1 this gives a planar RI-drawing.   □

3.1   4-Connected Planar Graphs

Combining Theorem 6 with Observation 1, we immediately obtain:

Theorem 7. *Let G be a triangulated 4-connected planar graph. Then G has a planar rook-drawing with at most one bend.*

Proof. Fix an arbitrary edge e on the outer-face, and apply Theorem 6 to obtain an RI-rook-drawing Γ of $G-e$ (see Fig. 3(a)). It remains to add in edge $e = (u, v)$. One end u of e is in the top-left corner, and the leftmost column contains no other vertex. The other end v is in the bottom-right corner, and the bottommost row contains no other vertex. We can hence route (u, v) by going vertically from u and horizontally from v, with the bend in the bottom-left corner.   □

Corollary 1. *Let G be a 4-connected planar graph. Then G has a rook-drawing with at most one bend, and with no bend if G is not triangulated.*

Proof. If G is triangulated then the result was shown above, so assume G has at least one face of degree 4 or more. Since G is 4-connected, one can add edges to G such that the result G' is triangulated and 4-connected [6]. Pick a face incident to an added edge e as outer-face of G', and apply Theorem 6 to obtain an RI-drawing of $G' - e$. Deleting all edges in $G' - G$ gives the result.   □

Since we have only one bend, and the ends of the edge (u, v) that contain it are the top-left and bottom-right corner, we can remove the bend by stretching.

Theorem 8. *Every 4-connected planar graph has a non-aligned planar drawing in an $n \times (n^2 - 3n + 4)$-grid and in a $(2n - 2) \times (2n - 2)$-grid.*

Proof. Let Γ be the RI-drawing of $G - (u, v)$ with u at $(1, n)$ and v at $(n, 1)$. Relocate u to point $(1, n^2 - 3n + 4)$. The resulting drawing is still a planar RI-drawing by Observation 1. Now $y(u) - y(v) = (n - 2)(n - 1) + 1$, hence the line segment from u to v has slope less than $-(n - 2)$, and is therefore above point $(n - 1, n - 1)$ (and hence above all other vertices of the drawing). So we can add this edge without violating planarity, and obtain a non-aligned straight-line drawing of G (see Fig. 3(b)). For the other result, start with the same drawing Γ. Relocate u to $(1, 2n - 2)$ and v to $(2n - 2, 1)$. The line segment from u to v has slope -1 and crosses Γ only in the top right grid-square, which was empty. So we obtain a non-aligned planar straight-line drawing. (see Fig. 3(c)).   □

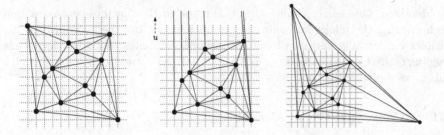

Fig. 3. An RI-rook-drawing of $G-e$, transformation into a non-aligned drawing of G of width n (where the top-left corner moved sufficiently high), and a second non-aligned drawing of G.

3.2   Constructing Rook-Drawings with Few Bends

We now explain the construction of a (poly-line) rook-drawing for a triangulated graph G with at least 5 vertices. We proceed as follows:

1. Find a small independent-filled-hitting set E_f. Here, an *independent-filled-hitting set* E' is a set of edges such that (i) every filled triangle has at least one edge in E' (we say that E' *hits* all filled triangles), and (ii) every face of G has at most one edge in E' (we say that E' is *independent*). We can show the following bound:

Lemma 5. *Any triangulated graph G of order n has an independent-filled-hitting set of size at most*

- f_G *(where f_G is the number of filled triangles of G), and it can be found in $O(n)$ time,*
- $\frac{2n-5}{3}$, *and it can be found in $O((n \log n)^{1.5}\alpha(n,n))$ time (where α denotes the inverse Ackermann function) or approximated arbitrarily close in $O(n)$ time.*

Proof. (Sketch) Compute a 4-coloring of the planar graph G, defining three perfect matchings in the dual graph. Let M_1, M_2, M_3 be the corresponding edge sets in G, and let E_i (for $i = 1, 2, 3$) be M_i with all edges removed that do not belong to a filled triangle. Since M_i stems from a 4-coloring, it contains exactly one edge of each triangle, hence E_i hits all filled triangles and $|E_i| \leq f_G$. Since M_i is the dual of a matching, E_i is independent. Cardinal et al. [7] showed that any planar graph has at most $2n - 7$ edges that belong to a separating triangle, and similarly one can show that at most $2n - 5$ edges belong to a filled triangle (there are $2n-5$ interior faces, and we can assign to each edge in a filled triangle the face on its "interior" side without double-counting faces). Therefore the best of E_1, E_2, E_3 contains at most $(2n-5)/3$ edges. To find it, we can either compute a minimum-weight perfect matching for suitable weights [12], or approximate it with a linear-time PTAS based on Baker's technique [3].   □

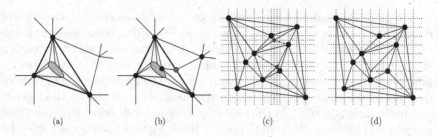

Fig. 4. (a, b) A separating triangle removed by subdividing and re-triangulating. (c) The RI-drawing of Fig. 2 reordered such that subdivision vertices (grey) are not at integer coordinates. (d) The drawing of (c), with subdivison vertices shifted to integer gridpoints.

2. Since the outer-face is a filled triangle, there exists one edge $e_o \in E_f$ that belongs to the outer-face. Define $E_s := E_f - \{e_o\}$ and notice that E_s contains no outer-face edges since E_f is independent.

3. As done in some previous papers [7,13], remove separating triangles by subdividing all edges $e \in E_s$, and re-triangulate by adding edges from the subdivision vertex (see Fig. 4(a, b)). Let V_x be the new set of vertices, and let G_1 be the new graph. Observe that G_1 may still have separating triangles, but all those separating triangles contain e_o since E_f hits all filled triangles.

4. By Lemma 4, $G_1 - e_o$ has a non-aligned RI-drawing Γ where the ends of e_o are at the top-left and bottom-right corner.

5. Transform Γ into drawing Γ' so that the relative orders stay intact, the original vertices (i.e., vertices of G) are on an $n \times n$-grid and the subdivision vertices (i.e., vertices in V_x) are inbetween.

   This can be done by enumerating the vertices in x-order, and assigning new x-coordinates in this order, increasing to the next integer for each original vertex and increasing by $\frac{1}{|V_x|+1}$ for each subdivision vertex. Similarly update the y-coordinates (see Fig. 4(c)). Drawing Γ' is still a non-aligned RI-drawing, and the ends of e_o are on the top-left and bottom-right corner.

6. Let e be an edge in E_s with subdivision vertex x_e. Since e is an interior edge of G, x_e is an interior vertex of G_1. Now move x_e to some integer gridpoint nearby. This is possible due to the following (not proved here):

Lemma 6. *Let Γ be a planar RI-drawing. Let x be an interior vertex of degree 4 with neighbours u_1, u_2, u_3, u_4 that form a 4-cycle. Assume that none of x, u_1, u_2, u_3, u_4 share a grid-line. Then we can move x to a point on grid-lines of its neighbours and obtain a planar RI-drawing.*

Note that the 4-cycle among the neighbours of x_e contains no other vertices in V_x, since E_s was independent. So any two subdivision vertices are separated via such a 4-cycle, and we can apply this operation to any subdivision-vertex independently.

7. Now replace each subdivision-vertex x_e by a bend, connected to the ends of e along the corresponding edges from x_e, see Fig. 4(d). (Sometimes, as is the case in the example, we could also simply delete the bend and draw edge e straight-line.) None of the shifting changed positions for vertices of G, so we now have a rook-drawing of $G - e_o$ with bends. The above shifting of vertices does not affect outer-face vertices, so the ends of e_o are still in the top-left and bottom-right corner. As the final step draw e_o by drawing vertically from one end and horizontally from the other; these segments are not occupied by the rook-drawing.

We added one bend for each edge in E_f. By Lemma 5, we can find an E_f with $|E_f| \leq f_G$ and $|E_f| \leq \frac{2n-5}{3}$ (neither bound is necessarily smaller than the other), and hence have:

Theorem 9. *Any planar graph G of order n has a planar rook-drawing with at most b bends, with $b \leq \min\{\frac{2n-5}{3}, f_G\}$.*

4   Conclusion

In this paper, we continued the work on planar rook-drawings initiated by Auber et al. [2]. We constructed planar rook-drawings with at most $\frac{2n-5}{3}$ bends; the number of bends can also be bounded by the number of filled triangles. We also considered drawings that allow more rows and columns while keeping vertices on distinct rows and columns; we proved that such non-aligned planar straight-line drawings always exist and have area $O(n^4)$. As for open problems, the most interesting question is lower bounds. No planar graph is known that needs more than one bend in a planar rook-drawing, and no planar graph is known that needs more than $2n+1$ grid-lines in a planar non-aligned drawing. The "obvious" approach of taking multiple copies of the octahedron fails because the property of having a rook-drawing is not closed under taking subgraphs: if vertices are added, then they could "use up" extraneous grid-lines in the drawing of a subgraph. We conjecture that the $n \times (\frac{4}{3}n - 1)$-grid achieved for nested-triangle graphs is optimal for planar straight-line non-aligned drawings with width n.

References

1. Alamdari, S., Biedl, T.: Planar open rectangle-of-influence drawings with non-aligned frames. In: Kreveld, M., Speckmann, B. (eds.) GD 2011. LNCS, vol. 7034, pp. 14–25. Springer, Heidelberg (2012). doi:10.1007/978-3-642-25878-7_3

2. Auber, D., Bonichon, N., Dorbec, P., Pennarun, C.: Rook-drawing for plane graphs. In: Di Giacomo, E., Lubiw, A. (eds.) GD 2015. LNCS, vol. 9411, pp. 180–191. Springer, Heidelberg (2015). doi:10.1007/978-3-319-27261-0_15

3. Baker, B.: Approximation algorithms for NP-complete problems on planar graphs. J. ACM **41**(1), 153–180 (1994)

4. Biedl, T., Derka, M.: The (3,1)-canonical order. J. Gr. Algorithms Appl. **20**(2), 347–362 (2016)

5. Biedl, T., Bretscher, A., Meijer, H.: Rectangle of influence drawings of graphs without filled 3-cycles. In: Kratochvíl, J. (ed.) GD 1999. LNCS, vol. 1731, pp. 359–368. Springer, Heidelberg (1999). doi:10.1007/3-540-46648-7_37
6. Biedl, T., Kant, G., Kaufmann, M.: On triangulating planar graphs under the four-connectivity constraint. Algorithmica 19(4), 427–446 (1997)
7. Cardinal, J., Hoffmann, M., Kusters, V., Tóth, C.D., Wettstein, M.: Arc diagrams, flip distances, and Hamiltonian triangulations. In: Symposium on Theoretical Aspects of Computer Science, STACS 2015. LIPIcs, vol. 30, pp. 197–210 (2015)
8. Chrobak, M., Kant, G.: Convex grid drawings of 3-connected planar graphs. Int. J. Comput. Geom. Appl. 7(3), 211–223 (1997)
9. Di Giacomo, E., Didimo, W., Kaufmann, M., Liotta, G., Montecchiani, F.: Upward-rightward planar drawings. In: Information, Intelligence, Systems and Applications (IISA 2014), pp. 145–150. IEEE (2014)
10. Dolev, D., Leighton, F.T., Trickey, H.W.: Planar embedding of planar graphs. Adv. Comput. Res. 2, 147–161 (1984)
11. de Fraysseix, H., Pach, J., Pollack, R.: How to draw a planar graph on a grid. Combinatorica 10, 41–51 (1990)
12. Gabow, H.N., Tarjan, R.E.: Faster scaling agorithms for general graph matching problems. J. ACM 38, 815–853 (1991)
13. Kaufmann, M., Wiese, R.: Embedding vertices at points: few bends suffice for planar graphs. J. Gr. Algorithms Appl. 6(1), 115–129 (2002)
14. Liotta, G., Lubiw, A., Meijer, H., Whitesides, S.: The rectangle of influence drawability problem. Comput. Geom. 10(1), 1–22 (1998)
15. Da Lozzo, G., Dujmović, V., Frati, F., Mchedlidze, T., Roselli, V.: Drawing planar graphs with many collinear vertices. In: Nöllenberg, M., Hu, Y. (eds.) GD 2016. LNCS, vol. 9801, pp. 152–165. Springer, Heidelberg (2016)
16. Schnyder, W.: Embedding planar graphs on the grid. In: ACM-SIAM Symposium on Discrete Algorithms (SODA 1990), pp. 138–148 (1990)

Snapping Graph Drawings to the Grid Optimally

Andre Löffler[✉], Thomas C. van Dijk, and Alexander Wolff[✉]

Lehrstuhl Für Informatik I, Universität Würzburg, Würzburg, Germany
andre.loeffler@uni-wuerzburg.de
http://www1.informatik.uni-wuerzburg.de/en/staff

Abstract. In geographic information systems and in the production of digital maps for small devices with restricted computational resources one often wants to round coordinates to a rougher grid. This removes unnecessary detail and reduces space consumption as well as computation time. This process is called *snapping to the grid* and has been investigated thoroughly from a computational-geometry perspective. In this paper we investigate the same problem for given drawings of planar graphs under the restriction that their combinatorial embedding must be kept and edges are drawn straight-line. We show that the problem is NP-hard for several objectives and provide an integer linear programming formulation. Given a plane graph G and a positive integer w, our ILP can also be used to draw G straight-line on a grid of width w and minimum height (if possible).

1 Introduction

When compressing geographic data, for example in order to ship it to devices with small memory, small screens and slow CPUs, the main objective is to reduce unnecessary detail. One way to do this is to round data points to a grid.

In the computational geometry community, a process called *snap rounding* has been proposed and has since become well-established: given an arrangement of line segments, each grid cell that contains vertices or intersections is "hot". Then every segment becomes a polygonal chain whose edges (*fragments*) connect center points of hot cells, namely those that the original segment (*ursegment*) intersects. Guibas and Marimont [7] showed that during snap rounding, vertices of the arrangement never cross a polygonal chain, so after snapping no two fragments cross. Moreover, the circular order of the fragments around an output vertex is the same as the order in which the corresponding ursegments intersect the boundary of its grid cell. The resulting arrangement approximates the original one in the sense that any fragment lies within the Minkowski sum of the corresponding ursegments and a unit square centered at the origin. However, the structure of the graph can be affected (vertices merge, faces disappear, edges bend). Further work in this direction includes that of De Berg et al. [3].

Motivated by the above GIS application, we investigate the problem of moving the drawing of a graph to a given grid. Since we still want to be able to

---

The full version of this paper is available at http://arxiv.org/abs/1608.08844.

© Springer International Publishing AG 2016
Y. Hu and M. Nöllenburg (Eds.): GD 2016, LNCS 9801, pp. 144–151, 2016.
DOI: 10.1007/978-3-319-50106-2_12

recognize the original graph, we do not tolerate new incidences. Then we must accept the possibility that a vertex does not go to the nearest grid point, but we still want to minimize change. This can by measured, for example, by the sum of the distances or the maximum distance in the Euclidean (L_2-) or Manhattan (L_1-) metric. Apparently, this problem, which we call TOPOLOGICALLY-SAFE SNAPPING, has not been studied yet. (Note that we carry over the term "snapping," although we don't necessarily snap to the *nearest* grid point.)

From a graph-drawing perspective, restricting to the grid has a (relatively) long history. Motivated by the fact that Tutte's barycenter method [15] for drawing planar graphs yields drawings that need precision linear in the size of the graph, Schnyder [14] and, independently, de Fraysseix et al. [5] have shown that any planar graph with n vertices admits a straight-line drawing on a grid of size $O(n) \times O(n)$. This is asymptotically optimal in the worst case [5]. Chrobak and Nakano [2] have investigated drawing planar graphs on grids of smaller width, at the expense of a larger height. Grid-snapping techniques can be found in any diagram creation tool. Aesthetic properties of force-directed drawing algorithms are widely researched, see e.g. Kieffer et al. [8] for grid layouts of diagrams.

Although minimizing the area of straight-line grid drawings has been the topic of several graph drawing contests, there has been rather little previous work. It is known that the problem is $\mathcal{NP}$-hard [9], but not even for special cases exact or approximation algorithms have been proposed.

Our Contribution. We show that optimal snapping is $\mathcal{NP}$-hard, with a reduction that asks for compressing each coordinate by just a single bit (Sect. 2). The proof is somewhat similar in concept to the proof of the $\mathcal{NP}$-hardness of Metro-Map Layout [12,16], but new constructions are required since the snapping problem does not easily allow the construction of "rigid" gadgets. Second, we give an integer linear program (ILP) for optimal snapping (Sect. 3). This ILP generalizes the one for Metro-Map Layout [13]. Where that ILP assumes a constant number of possible edge directions (namely 8), we have to cope with a number that is quadratic in the size of the grid. The numbers of variables and constraints of our ILP are polynomial in grid and graph size, but are quite large in practice. In fact, on a grid of size $k \times k$, there are $\Theta(k^2)$ edge directions. Thus, for an n-vertex planar graph, we must generate $O(k^2 n^2)$ constraints, among others, to preserve planarity and the cyclic order of edges around the vertices. To ameliorate this, we apply delayed constraint generation, a technique that adds certain constraints only when needed. Still, runtime is prohibitive for graphs with more than about 15 vertices. Our techniques can be adapted to draw (small) graphs with minimal area. This is interesting even for small graphs since minimum-area drawings can be useful for validating (counter) examples in graph drawing theory.

2   NP-Hardness

We start with a formal definition of TOPOLOGICALLYSAFESNAPPING – or TSS for short. To measure the cost of rounding a graph, we utilize Manhattan distance and the total cost of rounding a graph is the sum over the individual costs of the vertices. As input we take a plane graph $G = (V, E)$ with vertex positions and a bounding box $[0, X_{\max}] \times [0, Y_{\max}]$. The TSS problem is then to minimize the cost of rounding the vertices of G to the integer grid within the box without altering the topology with respect to the given plane straight-line drawing of G.

We prove $\mathcal{NP}$-hardness of TSS by considering the decision variant: is there a rounding that does not exceed a given cost bound c? We reduce from PLANAR MONOTONE 3-SAT (which is $\mathcal{NP}$-hard [4]): given a formula F in 3-CNF that is monotone and whose graph $H(F)$ is planar, is F satisfiable? The graph $H(F)$ has a vertex for each variable and each clause of F and an edge between a variable vertex v_X and a clause vertex v_C if X is part of C. We will only consider formulae whose graphs are planar and that are *monotone* in the usual sense: for any clause C, variables in C either are all negated or all unnegated. We can assume that the graph $H(F)$ can be laid out as in Fig. 1 (a): all variable vertices lie on the x-axis, the vertices of all-negated clauses lie above the x-axis, and the vertices of all-unnegated clauses lie below the x-axis [4].

Theorem 1. TOPOLOGICALLYSAFESNAPPING *is $\mathcal{NP}$-hard.*

Proof. For a given monotone, planar 3-CNF formula F, we construct a cost bound $c_{\min}$ and a plane graph G with vertices at half-integer coordinates. The sum of all vertex movements induced by rounding G to integer coordinates is exactly $c_{\min}$ if and only if F is satisfiable. To achieve this, we introduce gadgets for the elements of $H(F)$ – variables, clauses, edges and bends – and construct G and $c_{\min}$ in polynomial time.

For exposition, we consider two types of vertices. Black vertices start on integer grid points and do not need to be rounded. Moving a black vertex to another integer grid point is allowed, but we will show that this is not optimal if F is satisfiable. White vertices start at grid cell centers and thus will always move at least one unit by rounding. Let $W \subseteq V(G)$ be the set of white vertices. Now we give the construction of the various gadgets.

First, we introduce the line and bend gadgets. These ensure consistency between variable and clause gadgets. Every segment of the line gadget consists of four black vertices and two edges forming a *tunnel*, and a single white vertex inside; see Fig. 1 (b). The white vertex can be rounded most cheaply to exactly two possible integer grid points, depicted by the red and blue arrows. By rounding a white vertex in one direction, we prohibit the neighbor in that direction to go the opposite way – as both vertices would end up on the same integer grid point (which violates topological safety). So, if the white vertex at one end of the line is rounded inward (blue arrow) the white vertex at the other end of that line must be rounded outward – we say it is *pushed*. The same holds for the bend gadgets, as can be seen in Fig. 1 (c).

Next, consider the variable gadget depicted in Fig. 1 (d). It has tunnels for vertical line gadgets for every negated and unnegated occurrence at the top and bottom respectively. At the center of this gadget, there is a white vertex that is connected to the gadget's walls by two triangles. Call this the *assignment*

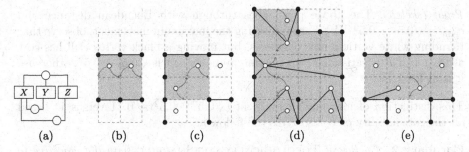

Fig. 1. (a) Graph $H(F)$ for formula $F = (\overline{X} \vee \overline{Y} \vee \overline{Z}) \wedge (X \vee Y) \wedge (X \vee Z)$, (b) Horizontal line gadget, (c) Bottom-to-right bend gadget, both with possible roundings, (d) Gadget for variable with two negated and one unnegated occurrences, (e) All-negated clause gadget with three negated variables. Inner area of each gadget highlighted gray. (Color figure online)

vertex and note that it can be rounded up or down, which makes the edges of the triangles block grid points on the top or bottom tunnels, respectively. The tunnels of that direction are then all forced to push into the connected clause gadgets. This represents the truth assignment of the corresponding variable.

Finally, the clause gadget is shown in Fig. 1 (e). We describe the all-negated degree-3 version; the degree-2 version can be constructed similarly. There is a white *satisfaction* vertex that can go to any of three possible integer grid points at equal cost. These grid points belong to line gadgets and are only available if the line does not "push". Then the satisfaction vertex can be rounded at cost 1 if and only if the clause is satisfied. Gadgets for all-unnegated clauses can be obtained by mirroring the construction of Fig. 1 (e) at a horizontal line.

The rounding cost of G is bounded from below by $c_{\min} = |W|$ since every white vertex must be rounded at cost at least 1. If F is satisfiable, there is a rounding that achieves this because then we can round the assignment vertices such that the satisfaction vertices can be rounded at cost 1. In the other direction, a satisfying assignment can be read off from the assignment vertices if rounding occurred at cost $c_{\min}$.

If none of the three candidate grid points for the satisfaction vertex are available, a topologically correct rounding must move a black vertex associated with that clause (of either the clause itself, the connected variables or the edges and bends connecting them). This adds at least 1 to the rounding cost without reducing the movement of any white vertex and thus such solutions cost strictly more than $c_{\min}$. That is, if $c_{\min}$ is exceeded, then F is unsatisfiable: any rounding corresponding to a satisfying truth assignment is cheaper. This concludes our Karp reduction and the claim follows.   □

Corollary 1. TopologicallySafeSnapping *is also $\mathcal{NP}$-hard when using Euclidean distance. In this case it is also $\mathcal{NP}$-hard to minimize the maximum movement instead of the sum.*

Proof (sketch). The above proof goes through with Euclidean distance and $c_{min} = \sqrt{0.5^2 + 0.5^2} \cdot |W|$. For minimizing the maximum movement, observe that rounding white vertices now costs less, but moving a black vertex still has cost at least 1: if F is satisfiable, the maximum movement is $\sqrt{0.5^2 + 0.5^2}$, otherwise it is at least 1.   □

This distinction of maximum movement ($\sqrt{0.5^2 + 0.5^2} \approx 0.71$ versus 1) based on the satisfiability of F also gives the following.

Corollary 2. *Euclidean* TOPOLOGICALLYSAFESNAPPING *with the objective to minimize maximum movement is $\mathcal{APX}$-hard.*

3   Exact Solution Using Integer Linear Programming

In this section we provide an ILP-based exact algorithm for TSS. Recall that an instance is a graph $G = (V, E)$ with vertex coordinates. For all $v \in V$, call these (X_v, Y_v) and introduce integer decision variables $0 \leq x_v \leq X_{max}$ and $0 \leq y_v \leq Y_{max}$ to represent the "rounded" output position. This leads to the following objective function.

$$\text{Minimize} \sum_{v \in V} |x_v - X_v| + |y_v - Y_v| \qquad (1)$$

This formula is itself not linear, but can be made so with standard transformations [11]. Note that without any further constraints, this would just move every vertex to the nearest integer grid point. We will now introduce constraints to ensure topological safety, that is, in the output no two points are on same grid point, no two edges intersect, and the edges at every node have the same cyclic order as in the input.

Vertices do not Coincide. This can be ensured by adding the following constraints. They too are not linear as stated, but can be readily linearized.

$$(x_v \neq x_w) \vee (y_v \neq y_w) \qquad \forall v, w \in V, v \neq w \qquad (2)$$

Possible Directions. The most important departure from the metro-map drawing ILP is that, clearly, more than eight different directions are allowed. A priori we have no further constraints than that every rounded vertex lies somewhere within the given bounding box. Let $\mathcal{D}$ be the set of unique directions $D = (D_X, D_Y)$ in $[-X_{max}, X_{max}] \times [-Y_{max}, Y_{max}]$. Considering the Farey sequence [6], we know that $|\mathcal{D}|$ is $\Theta(X_{max} \cdot Y_{max})$. In the following, we let the set $\mathcal{D}$ be ordered counterclockwise, starting at the positive x-axis, allowing comparison of directions.

No Two Edges Cross. The following constraints ensure that nonincident edges do not cross. (Incident edges are allowed to touch in the shared vertex.) We will follow the idea of Nöllenburg and Wolff [13]. While producing octilinear drawings of metro maps, they ensured planarity by forcing every pair of nonincident edges to be separated by at least some distance $D_{\min}$ in at least one of the eight octilinear directions. This minimum distance was partly an aesthetic guideline, but also guarantees planarity. We are only interested in the latter and therefore pick $D_{\min}$ such that all planar realizations on the grid are allowed.

The separation distance $D_{\min}$ has to be small enough to separate any non-intersecting pair of edges in the output. Here the bounding box leads to a bound since it bounds the slope of the edges; it suffices to choose $D_{\min} = 1/(\max\{X_{\max}, Y_{\max}\} + 1)$.

For every pair of nonincident edges $e_1, e_2 \in E$ and every direction $D \in \mathcal{D}$, we introduce a binary decision variable $\gamma_D(e_1, e_2) \in \{0, 1\}$ indicating that e_1 and e_2 are apart by $D_{\min}$ in direction D. Every such pair must be separated in some direction (following the idea of [12]).

$$\sum_{D \in \mathcal{D}} \gamma_D(e_1, e_2) = 1 \qquad \forall e_1, e_2 \in E, e_1, e_2 \text{ nonincident} \qquad (3)$$

Let $L_\gamma = 2 \cdot \max\{X_{\max}, Y_{\max}\} + 1$. Then, for any direction $D \in \mathcal{D}$, any pair of nonincident edges e_1, e_2 and any $v \in e_1, w \in e_2$, we require the following.

$$D_X \cdot (x_v - x_w) + D_Y \cdot (y_v - y_w) + (1 - \gamma_D(e_1, e_2))L_\gamma \geq D_{\min} \qquad (4)$$

Constraint (3) yields a unique direction D with $\gamma_D = 1$. By choice of L_γ, any constraint (4) that involves a direction D with $\gamma_D = 0$ is trivially fulfilled.

Determine Direction of Incident Edges. For incident edges $e_1, e_2 \in E$, we have to ensure that the directions of e_1 and e_2 differ. Again, we generalize the metro-map drawing ILP – dropping the "relative position rule" – allowing edges to have any direction $D \in \mathcal{D}$.

To keep track of this, we introduce a binary decision variable $\alpha_D(v, w) \in \{0, 1\}$ for every vertex $v \in V$, every neighbor $w \in N(v)$ and every direction $D \in \mathcal{D}$. The meaning of $\alpha_D(v, w) = 1$ is that the direction of edge (v, w) is D.

$$\sum_{D \in \mathcal{D}} \alpha_D(v, w) = 1 \qquad \forall v \in V \ \forall w \in N(v) \qquad (5)$$

For any vertex $v \in V$, any neighbor $w \in N(v)$, and any direction $D \in \mathcal{D}$, the following ensures that edge (v, w) indeed has direction D. Let $L_\alpha = 2 \cdot \max\{X_{\max}, Y_{\max}\} + 1$.

$$\begin{aligned} x_w \cdot D_Y + y_v \cdot D_X - x_v \cdot D_Y \pm (1 - \alpha_D(v, w))L_\alpha &\gtrless y_w \cdot D_X \\ (1 - \alpha_D(v, w))L_\alpha + (x_w - x_v) \cdot D_X + (y_w - y_v) \cdot D_Y &\geq 0 \end{aligned} \qquad (6)$$

From constraint (5) we get that for every vertex-neighbor pair one α has to be set to 1. This α again enables one subset—as L_α dominates all other terms—of constraints from (6), forcing comparison between edge slope and direction. This gives us the direction of edge (v, w) with the correct sign.

Preserve Cyclic Order of Outgoing Edges. We use a binary decision variable $\beta(v, w) \in \{0, 1\}$ for every vertex-neighbor pair, indicating if w is the "last" neighbor of v according to the order of $\mathcal{D}$. The following preserves cyclic order.

$$\sum_{w \in N(v)} \beta(v, w) = 1 \quad \forall v \in V \text{ with } \deg(v) > 1 \tag{7}$$

$$\alpha_{D_1}(v, w_i) \leq \beta(v, w_i) + \sum_{D_w \in \mathcal{D}: \, D_w > D_1} \alpha_{D_w}(v, w_{i+1})$$
$$\forall D_1 \in \mathcal{D} \; \forall v \in V, N(v) = \{w_1, w_2, \ldots, w_k\} \; (k = \deg v > 1) \tag{8}$$

For notational convenience, we let $w_{k+1} = w_1$, as $N(v)$ is conceptually circular. For any α set to 0, the inequalities of (8) are trivially satisfied. Otherwise, there has to be a neighbor whose connecting edge has a later direction (and thus the corresponding α set to 1), unless it is the last neighbor in the embedding of v. To ensure that there is only one "last neighbor"-violation of the constraints from (8), we introduce the constraints of (7). Adding β to every constraint of (8) also allows for the whole neighborhood of v to be rotated around it. This describes the full ILP and gives to the following.

Theorem 2. *The above ILP solves* TOPOLOGICALLYSAFESNAPPING.

Graph Drawing. Replacing the objective function with Minimize $\max_{v \in V} y_v$, the ILP computes a straight-line grid drawing with the given embedding, width at most $X_{\max}$, and minimum height. This allows us to find minimum-area drawings of small graphs.

Delayed Constraint Generation. We can apply a delayed constraint generation approach (see for example Cinneck [1]) to the above ILP as follows. First we run the ILP without any constraints, which snaps each vertex to the nearest grid point. (This takes practically no time.) We then test the result for topological validity, adding constraints corresponding to any violations. Then we repeat until no violations occur. This improves the runtime when few iterations suffice for a particular instance, but the approach should still be considered practically infeasible, especially for large bounding boxes: the set of possible directions $\mathcal{D}$ still results in a large program. Future work could focus on reducing the brute-force inclusion of all possible directions. Experimental results are found in the full version of this paper [10].

Acknowledgments. We thank Gergely Mincsovics for suggesting this problem to us.

References

1. Chinneck, J.W.: Feasibility and Infeasibility in Optimization: Algorithms and Computational Methods. International Series in Operations Research and Management Science, vol. 118. Springer, Heidelberg (2008)
2. Chrobak, M., Nakano, S.-I.: Minimum-width grid drawings of plane graphs. Comput. Geom. **11**(1), 29–54 (1998)

3. de Berg, M., Halperin, D., Overmars, M.: An intersection-sensitive algorithm for snap rounding. Comput. Geom. **36**(3), 159–165 (2007)
4. de Berg, M., Khosravi, A.: Optimal binary space partitions for segments in the plane. Int. J. Comput. Geom. Appl. **22**(03), 187–205 (2012)
5. De Fraysseix, H., Pach, J., Pollack, R.: How to draw a planar graph on a grid. Combinatorica **10**(1), 41–51 (1990)
6. Graham, R.L., Knuth, D.E., Patashnik, O.: Concrete Mathematics–A Foundation for Computer Science. Addison-Wesley, Reading (1994)
7. Guibas, L.J., Marimont, D.H.: Rounding arrangements dynamically. Int. J. Comput. Geom. Appl. **8**(2), 157–178 (1998)
8. Kieffer, S., Dwyer, T., Marriott, K., Wybrow, M.: Incremental grid-like layout using soft and hard constraints. In: Wismath, S., Wolff, A. (eds.) GD 2013. LNCS, vol. 8242, pp. 448–459. Springer, Heidelberg (2013). doi:10.1007/978-3-319-03841-4_39
9. Krug, M., Wagner, D.: Minimizing the area for planar straight-line grid drawings. In: Hong, S.-H., Nishizeki, T., Quan, W. (eds.) GD 2007. LNCS, vol. 4875, pp. 207–212. Springer, Heidelberg (2008). doi:10.1007/978-3-540-77537-9_21
10. Löffler, A., van Dijk, T.C., Wolff, A.: Snapping graph drawings to the grid optimally. Arxiv report arXiv.org/abs/1608.08844 (2016)
11. McCarl, B.A., Spreen, T.H.: Applied mathematical programming using algebraic systems. Texas A&M University (1997)
12. Nöllenburg, M.: Automated drawing of metro maps. Master's thesis, Fakultät für Informatik, Universität Karlsruhe (2005). http://www.ubka.uni-karlsruhe.de/indexer-vvv/ira/2005/25
13. Nöllenburg, M., Wolff, A.: Drawing and labeling high-quality metro maps by mixed-integer programming. IEEE Trans. Visual. Comput. Graphics **17**(5), 626–641 (2011)
14. Schnyder, W.: Embedding planar graphs on the grid. In: Proceedings of 1st ACM-SIAM Symposium on Discrete Algorithms (SODA 1990), pp. 138–148 (1990)
15. Tutte, W.T.: How to draw a graph. Proc. London Math. Soc. **13**(52), 743–768 (1963)
16. Wolff, A.: Drawing subway maps: A survey. Informatik - Forschung & Entwicklung **22**(1), 23–44 (2007)

Drawing Planar Graphs with Many Collinear Vertices

Giordano Da Lozzo[1], Vida Dujmović[2], Fabrizio Frati[1(⊠)],
Tamara Mchedlidze[3], and Vincenzo Roselli[1]

[1] University Roma Tre, Rome, Italy
{dalozzo,frati,roselli}@dia.uniroma3.it
[2] University of Ottawa, Ottawa, Canada
vida.dujmovic@uottawa.ca
[3] Karlsruhe Institute of Technology, Karlsruhe, Germany
mched@iti.uka.de

Abstract. Given a planar graph G, what is the maximum number of collinear vertices in a planar straight-line drawing of G? This problem resides at the core of several graph drawing problems, including universal point subsets, untangling, and column planarity. The following results are known: Every n-vertex planar graph has a planar straight-line drawing with $\Omega(\sqrt{n})$ collinear vertices; for every n, there is an n-vertex planar graph whose every planar straight-line drawing has $O(n^{0.986})$ collinear vertices; every n-vertex planar graph of treewidth at most two has a planar straight-line drawing with $\Theta(n)$ collinear vertices. We extend the linear bound to planar graphs of treewidth at most three and to triconnected cubic planar graphs, partially answering two problems posed by Ravsky and Verbitsky. Similar results are not possible for all bounded treewidth or bounded degree planar graphs. For planar graphs of treewidth at most three, our results also imply asymptotically tight bounds for all of the other above mentioned graph drawing problems.

1 Introduction

A set S of vertices in a planar graph G is *collinear* if G has a planar straight-line drawing where all the vertices in S are collinear. Ravsky and Verbitsky [20] considered the problem of determining the maximum cardinality of collinear sets in planar graphs. A collinear set S is *free* if a total order $<_S$ of S exists such that, for any $|S|$ points on a straight line ℓ, G has a planar straight-line drawing where the vertices in S are mapped to the $|S|$ points and their order on ℓ matches $<_S$. Free collinear sets were first used (but not named) by Bose *et al.* [3] and then formally introduced by Ravsky and Verbitsky [20]. Collinear and free collinear sets relate to several graph drawings problems, as will be discussed later.

By exploiting the results in [3], Dujmović [9] showed that every n-vertex planar graph has a free collinear set with size $\sqrt{n/2}$. Ravsky and Verbitsky [20] negatively answered the question whether this bound can be improved to linear. Namely, they noted that if a planar triangulation has a collinear set S, then its

© Springer International Publishing AG 2016
Y. Hu and M. Nöllenburg (Eds.): GD 2016, LNCS 9801, pp. 152–165, 2016.
DOI: 10.1007/978-3-319-50106-2_13

dual has a cycle of length $\Omega(|S|)$. Since there are m-vertex triconnected cubic planar graphs whose longest cycle has length $O(m^\sigma)$ [14], then there are n-vertex planar graphs in which every collinear set has size $O(n^\sigma)$. Here σ is a graph-theoretic constant called *shortness exponent*; it is known that $\sigma < 0.986$.

Which classes of planar graphs have (free) collinear sets with linear size? Goaoc *et al.* [12] proved (implicitly) that n-vertex outerplanar graphs have free collinear sets with size $(n + 1)/2$. Ravsky and Verbitsky [20] proved that n-vertex planar graphs of treewidth at most two have free collinear sets with size $n/30$; they also asked for other classes of graphs with (free) collinear sets with linear size, calling special attention to planar graphs of bounded treewidth and to planar graphs of bounded degree. In this paper we prove the following results.

Theorem 1. *Every n-vertex planar graph of treewidth at most three has a free collinear set with size at least $\lceil \frac{n-3}{8} \rceil$.*

Theorem 2. *Every n-vertex triconnected cubic planar graph has a collinear set with size at least $\lceil \frac{n}{4} \rceil$.*

Theorem 3. *Every planar graph of treewidth k has a collinear set with size $\Omega(k^2)$.*

Theorem 1 generalizes the result on planar graphs of treewidth 2 [20]. Ravsky and Verbitsky [21, Corollary 3.5] constructed n-vertex planar graphs of treewidth 8 whose largest collinear set has size $o(n)$; by using the dual of Tutte's graph rather than the dual of the Barnette-Bosák-Lederberg's graph in that construction, it is readily seen that the sub-linear bound holds true for planar graphs of treewidth at most 5. Thus, the question whether planar graphs of treewidth k admit (free) collinear sets with linear size remains open only for $k = 4$. Theorem 2 provides the first linear lower bound on the size of collinear sets for a wide class of bounded-degree planar graphs. The result cannot be extended to planar graphs of degree at most 7, since there are n-vertex planar triangulations of maximum degree 7 whose dual graph has a longest cycle of length $o(n)$ [17]. Finally, Theorem 3 improves the $\Omega(\sqrt{n})$ bound on the size of collinear sets in general planar graphs for all planar graphs with treewidth $\omega(\sqrt[4]{n})$. We now discuss implications of Theorems 1–3 for other graph drawing problems.

A *column planar set* in a graph G is a set $Q \subseteq V(G)$ satisfying the following property: there is a function $\gamma : Q \to \mathbb{R}$ such that, for any function $\lambda : Q \to \mathbb{R}$, there is a planar straight-line drawing of G where each vertex $v \in Q$ lies at point $(\gamma(v), \lambda(v))$. Column planar sets were defined by Evans *et al.* [11] motivated by applications to partial simultaneous geometric embeddings. They proved that n-vertex trees have column planar sets of size $14n/17$. The bounds in Theorems 1–3 carry over to the size of column planar sets for the corresponding graph classes.

A *universal point subset* for the family $\mathcal{G}_n$ of n-vertex planar graphs is a set P of points in the plane such that, for every $G \in \mathcal{G}_n$, there is a planar straight-line drawing of G in which $|P|$ vertices lie at the points in P. Universal point subsets were introduced by Angelini *et al.* [1]. Every n points in general position form a universal point subset for the n-vertex outerplanar graphs [2,5,13] and every

$\sqrt{n/2}$ points in the plane form a universal point subset for $\mathcal{G}_n$ [9]. By Theorem 1, we obtain that every $\lceil\frac{n-3}{8}\rceil$ points in the plane form a universal point subset for the n-vertex planar graphs of treewidth at most three.

Given a straight-line drawing of a planar graph, possibly with crossings, to *untangle* it means to assign new locations to some vertices so that the resulting straight-line drawing is planar. The goal is to do so while keeping as many vertices as possible fixed [3,4,7,12,15,18,20]. General n-vertex planar graphs can be untangled while keeping $\Omega(n^{0.25})$ vertices fixed [3]; this bound cannot be improved above $O(n^{0.4948})$ [4]. Asymptotically tight bounds are known for paths [7], trees [12], outerplanar graphs [12], and planar graphs of treewidth 2 [20]. By Theorem 1, we obtain that every n-vertex planar graph of treewidth at most 3 can be untangled while keeping $\Omega(\sqrt{n})$ vertices fixed. This bound is the best possible [3] and generalizes most of the mentioned previous results [12,20].

Complete proofs can be found in the full version of the paper [8].

2 Preliminaries

A *k-tree* is either K_{k+1} or can be obtained from a smaller k-tree G by the insertion of a vertex adjacent to all the vertices in a k-clique of G. The *treewidth* of a graph G is the minimum k such that G is a subgraph of a k-tree.

A connected *plane graph* G is a connected planar graph with a *plane embedding* – an equivalence class of planar drawings of G, where two drawings are *equivalent* if each vertex has the same clockwise order of its incident edges and the outer faces are delimited by the same walk. We think about any plane graph G as drawn according to its plane embedding; also, when we talk about a planar drawing of G, we mean that it respects its plane embedding. The *interior* of G is the closure of the union of its internal faces. A subgraph H of G has the plane embedding obtained from the one of G by deleting vertices and edges not in H.

We denote the degree of a vertex v in a graph G by $\delta_G(v)$. A graph is *cubic* (*subcubic*) if every vertex has degree 3 (resp. at most 3). If $U \subseteq V(G)$, we denote by $G - U$ the graph $(V(G) - U, \{(u,v) \in E(G)|u,v \notin U\})$; the subgraph of G *induced* by U is $(U, \{(u,v) \in E(G)|u,v \in U\})$. If H is a subgraph of G and $v \in V(G)-V(H)$, we let $H\cup\{v\}$ be the graph $(V(H)\cup\{v\}, E(H))$. An *H-bridge* B is either *trivial* – it is an edge of G not in H with both end-vertices in H – or *non-trivial* – it is a connected component of $G - V(H)$ together with the edges from that component to H. The vertices in $V(H)\cap V(B)$ are called *attachments*.

Let G be a connected graph. If G has no *cut-vertex* – a vertex whose removal disconnects G – and it is not an edge, then it is *biconnected*. A *biconnected component* of G is a maximal biconnected subgraph of G. If G is biconnected, then a *separation pair* is a pair of vertices $\{a,b\}$ whose removal disconnects G; also, an $\{a,b\}$-*component* is either *trivial* – it is edge (a,b) – or *non-trivial* – it is the subgraph of G induced by a, b, and the vertices of a connected component of $G - \{a,b\}$. If G has no separation pair, then it is *triconnected*.

3   From a Geometric to a Topological Problem

In this section we show that the problem of determining a large collinear set in a planar graph, which is geometric by definition, can be turned into a purely topological problem. This may be useful to obtain bounds for the size of collinear sets in classes of planar graphs different from the ones we studied in this paper.

An open simple curve λ is *good* for a planar drawing Γ of a plane graph G if each edge e of G is either contained in λ or has at most one point in common with λ (if λ passes through an end-vertex of e, that counts as a common point). Clearly, the existence of a good curve passing through a certain sequence of vertices, edges, and faces of G does not depend on the actual drawing Γ, but only on the plane embedding of G. Hence, we often talk about the existence of good curves in plane graphs, rather than in their planar drawings. We denote by $R_{G,\lambda}$ the only unbounded region of the plane defined by G and λ. Curve λ is *proper* if both its end-points are incident to $R_{G,\lambda}$. We have the following.

Theorem 4. *A plane graph G has a planar straight-line drawing with x collinear vertices if and only if G has a proper good curve that passes through x vertices.*

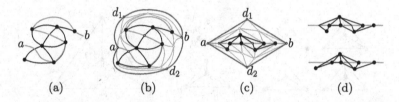

Fig. 1. (a) A proper good curve for a plane graph G. (b) Augmentation of G. (c) A planar straight-line drawing of the augmented graph G. (d) Planar polyline (top) and straight-line (bottom) drawings of the original G.

Proof Sketch. The necessity is readily proved. For the sufficiency, let λ be a proper good curve through x vertices of G; refer to Fig. 1. Add dummy vertices at two points d_1 and d_2 in $R_{G,\lambda}$, at the end-points a and b of λ, and at each crossing between an edge and λ; also, add dummy edges (d_1, a), (d_1, b), (d_2, a), (d_2, b) and between any two consecutive vertices along λ (the latter edges form a path P_λ); finally, triangulate the internal faces of G with dummy vertices and edges that do not connect non-consecutive vertices on λ. Let C_1 (C_2) be the cycle composed of P_λ and of the edges (d_1, a) and (d_1, b) (resp. (d_2, a) and (d_2, b)). Represent C_1 (C_2) as a convex polygon Q_1 (resp. Q_2), with P_λ on a horizontal line ℓ; since the subgraphs of G inside C_1 and C_2 are triconnected, they have planar straight-line drawings with C_1 and C_2 represented by Q_1 and Q_2, respectively [22]. Removing the dummy vertices and edges results in a planar drawing Γ of the original graph G where each edge is y-monotone. A planar straight-line drawing Γ' of G in which the y-coordinate of each vertex is the same as in Γ always exists [10, 19]. Then the x vertices of G curve λ passes through lie on ℓ in Γ'.   □

4   Planar Graphs with Treewidth at Most Three

In this section we prove Theorem 1. We regard a plane 3-cycle as a plane 3-tree; then every plane graph G with $n \geq 3$ vertices and treewidth at most 3 can be augmented with dummy edges to a plane 3-tree G' [16] and every free collinear set in G' is also a free collinear set in G. Thus, in the following, we assume that G is a plane 3-tree. We first prove that G has a collinear set with size $\lceil \frac{n-3}{8} \rceil$; by Theorem 4 it suffices to prove that G has a proper good curve through $\lceil \frac{n-3}{8} \rceil$ vertices. Let u, v, and z be the external vertices of G. If $n = 3$, then G is *empty*. Otherwise, the *central vertex* of G is the unique internal vertex w adjacent to u, v, and z. The plane 3-trees G_1, G_2, and G_3 which are the subgraphs of G inside cycles (u, v, w), (u, z, w), and (v, z, w) are the *children* of G and of w. We associate to each internal vertex x of G a plane 3-tree $G(x)$ as follows. We associate G to w and we use recursion on the children of G; then x is the central vertex of $G(x)$. An internal vertex x of G is of *type A, B, C,* or *D* if, respectively, 3, 2, 1, or 0 of the children of $G(x)$ are empty (see Fig. 2(a)). Let $a(G)$, $b(G)$, $c(G)$, and $d(G)$ be the number of internal vertices of G of type A, B, C, and D, respectively, and let $m = n - 3$ be the number of internal vertices of G.

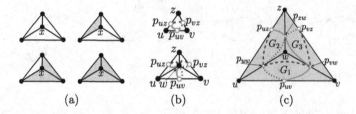

Fig. 2. (a) A vertex x of type A (top-left), B (top-right), C (bottom-left), and D (bottom-right). (b) $\lambda_u(G)$ (solid), $\lambda_v(G)$ (dotted), and $\lambda_z(G)$ (dashed) if $m = 0$ (top) and $m = 1$ (bottom). (c) $\lambda_u(G)$, $\lambda_v(G)$, and $\lambda_z(G)$ if w is of type C or D.

In the following we present an algorithm that computes three proper good curves $\lambda_u(G)$, $\lambda_v(G)$, and $\lambda_z(G)$ lying in the interior of G. For every edge (x, y) of G, let p_{xy} be an arbitrary internal point of (x, y). The end-points of $\lambda_u(G)$ are p_{uv} and p_{uz}, those of $\lambda_v(G)$ are p_{uv} and p_{vz}, and those of $\lambda_z(G)$ are p_{uz} and p_{vz}. Although each of $\lambda_u(G)$, $\lambda_v(G)$, and $\lambda_z(G)$ is a good curve, any two of these curves might cross each other and pass through the same vertices of G. Each of these curves passes through all the internal vertices of type A, through no vertex of type C or D, and through "some" vertices of type B. We will prove that the total number of internal vertices of G these curves pass through is at least $3m/8$, hence one of them passes through at least $\lceil m/8 \rceil$ internal vertices.

The curves $\lambda_u(G)$, $\lambda_v(G)$, and $\lambda_z(G)$ are constructed by induction on m. If $m = 0$, then $\lambda_u(G)$ traverses the internal face (u, v, z) from p_{uv} to p_{uz}, while if $m = 1$, then $\lambda_u(G)$ traverses the internal face (u, v, w) from p_{uv} to the central vertex w of G and the internal face (u, z, w) from w to p_{uz} (see Fig. 2(b)). Curves $\lambda_v(G)$ and $\lambda_z(G)$ are defined analogously.

If $m > 1$, then we distinguish the case in which w is of type C or D from the one in which w is of type B. In the former case (see Fig. 2(c)), the curves are constructed by composing the curves inductively constructed for the children of G. In the latter case (see Fig. 3), a sequence of vertices of type B, called *B-chain*, is recovered; its arrangement in G is exploited in order to ensure that $\lambda_u(G)$, $\lambda_v(G)$, and $\lambda_z(G)$ pass through many vertices of type B.

Assume first that w is of type C or D. Inductively construct curves $\lambda_u(G_1)$, $\lambda_v(G_1)$, and $\lambda_w(G_1)$ for G_1, curves $\lambda_u(G_2)$, $\lambda_z(G_2)$, and $\lambda_w(G_2)$ for G_2, and curves $\lambda_v(G_3)$, $\lambda_z(G_3)$, and $\lambda_w(G_3)$ for G_3. Let $\lambda_u(G) = \lambda_v(G_1) \cup \lambda_w(G_3) \cup \lambda_z(G_2)$, $\lambda_v(G) = \lambda_u(G_1) \cup \lambda_w(G_2) \cup \lambda_z(G_3)$, and $\lambda_z(G) = \lambda_u(G_2) \cup \lambda_w(G_1) \cup \lambda_v(G_3)$.

Assume next that w is of type B. Let $H_0 = G$, let $w_1 = w$, and let H_1 be the only non-empty child of G. If cycle (v, w, z) delimits the outer face of H_1, define three paths $P_u = (u, w)$, $P_v = (v)$, and $P_z = (z)$; analogously, if cycle (u, w, z) delimits the outer face of H_1, let $P_u = (u)$, $P_v = (v, w)$, and $P_z = (z)$; finally, if cycle (u, v, w) delimits the outer face of H_1, let $P_u = (u)$, $P_v = (v)$, and $P_z = (z, w)$.

Now suppose that, for $i \geq 1$, a sequence $w_1, \ldots, w_i$ of vertices of type B, a sequence $H_0, \ldots, H_i$ of plane 3-trees, and paths P_u, P_v, and P_z have been defined satisfying the following properties: (1) for $1 \leq j \leq i$, w_j is the central vertex of H_{j-1} and H_j is the only non-empty child of H_{j-1}; (2) P_u, P_v, and P_z are vertex-disjoint and each of them is induced in G; and (3) P_u, P_v, and P_z connect u, v, and z with the three external vertices u', v', and z' of H_i, where $u' \in P_u$, $v' \in P_v$, and $z' \in P_z$. Properties (1)–(3) are indeed satisfied with $i = 1$. Consider the central vertex w_{i+1} of H_i.

If w_{i+1} is of type B, then let H_{i+1} be the unique non-empty child of H_i. If cycle (v', z', w_{i+1}) delimits the outer face of H_{i+1}, add edge (u', w_{i+1}) to P_u and leave P_v and P_z unaltered; the other cases are analogous. Properties (1)–(3) are clearly satisfied by this construction.

If w_{i+1} is not of type B, then we call the sequence $w_1, \ldots, w_i$ a *B-chain* of G. Let $H = H_i$, let $P_u = (u = u_1, \ldots, u_U = u')$, let $P_v = (v = v_1, \ldots, v_V = v')$, and let $P_z = (z = z_1, \ldots, z_Z = z')$; also, define cycles $C_{uv} = P_u \cup (u, v) \cup P_v \cup (u', v')$, $C_{uz} = P_u \cup (u, z) \cup P_z \cup (u', z')$, and $C_{vz} = P_v \cup (v, z) \cup P_z \cup (v', z')$. Each of these cycles has no vertex of G inside, and every edge of G inside one of them connects two vertices on distinct paths among P_u, P_v, and P_z, by Property (2). We are going to use the following (a similar lemma can be stated for C_{uz} and C_{vz}).

Lemma 1. *Let p_1 and p_2 be points on C_{uv}, possibly coinciding with vertices of C_{uv}, and not both on the same edge of G. A good curve exists that connects p_1 and p_2, that lies inside C_{uv}, except at p_1 and p_2, and that intersects each edge of G inside C_{uv} at most once.*

We now construct $\lambda_u(G)$, $\lambda_v(G)$, and $\lambda_z(G)$. Inductively construct curves $\lambda_{u'}(H)$, $\lambda_{v'}(H)$, and $\lambda_{z'}(H)$ for H. We distinguish three cases based on how many

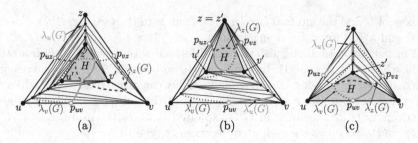

Fig. 3. $\lambda_u(G)$, $\lambda_v(G)$, and $\lambda_z(G)$ if w is of type B. (a) None of P_u, P_v, and P_z is a single vertex. (b) Only P_z is a single vertex. (c) P_u and P_v are single vertices.

among P_u, P_v, and P_z are single vertices (not all of them are, since $w_1 \neq u, v, z$). We discuss here the case in which none of them is a single vertex, as in Fig. 3(a); the other cases, which are illustrated in Figs. 3(b)–(c), are similar. We show how to construct $\lambda_u(G)$; the construction of $\lambda_v(G)$ and $\lambda_z(G)$ is analogous.

If $Z > 2$, then $\lambda_u(G)$ consists of curves $\lambda_u^0, \ldots, \lambda_u^4$; curve λ_u^0 lies inside C_{uz} and connects p_{uz} with z_2, which is internal to P_z since $Z > 2$; λ_u^1 coincides with path $(z_2, \ldots, z_{Z-1})$ (which is a single vertex if $Z = 3$); λ_u^2 lies inside C_{vz} and connects z_{Z-1} with $p_{v'z'}$; λ_u^3 coincides with $\lambda_{v'}(H)$; finally, λ_u^4 lies inside C_{uv} and connects $p_{u'v'}$ with p_{uv}. Curves λ_u^0, λ_u^2, and λ_u^4 are constructed as in Lemma 1. If $Z = 2$, then $\lambda_u(G)$ consists of curves $\lambda_u^1, \ldots, \lambda_u^4$; curve λ_u^1 lies inside C_{uz} and connects p_{uz} with $p_{zz'}$; λ_u^2 lies inside C_{vz} and connects $p_{zz'}$ with $p_{v'z'}$; λ_u^3 and λ_u^4 are defined as in the case $Z > 2$. Curves λ_u^1, λ_u^2, and λ_u^4 are constructed as in Lemma 1. This completes the construction of $\lambda_u(G)$, $\lambda_v(G)$, and $\lambda_z(G)$.

The curves $\lambda_u(G)$, $\lambda_v(G)$, and $\lambda_z(G)$ are clearly proper. Lemmas 2 and 3 prove that they are good and pass through many vertices. We introduce three parameters for the latter proof: $s(G)$ is the number of vertices the curves pass through (counting each vertex with a multiplicity equal to the number of curves that pass through it), $x(G)$ is the number of internal vertices of type B none of the curves passes through, and $h(G)$ is the number of B-chains of G.

Lemma 2. *Curves $\lambda_u(G)$, $\lambda_v(G)$, and $\lambda_z(G)$ are good.*

Lemma 3. *The following hold true if $m \geq 1$:* **(1)** $a(G)+b(G)+c(G)+d(G) = m$; **(2)** $a(G) = c(G) + 2d(G) + 1$; **(3)** $h(G) \leq 2c(G) + 3d(G) + 1$; **(4)** $x(G) \leq b(G)$; **(5)** $x(G) \leq 3h(G)$; *and* **(6)** $s(G) \geq 3a(G) + b(G) - x(G)$.

Proof Sketch. **(1)** is true since every internal vertex is of one of types A–D. **(4)** follows by definition of $x(G)$. **(5)** is true since every internal vertex of type B is in a B-chain, and for every B-chain the three curves pass through all but at most three of its vertices. **(2)**, **(3)**, and **(6)** can be proved by induction on m, by distinguishing four cases based on the type of w. In particular, **(6)** exploits the fact that, in each case, the three curves contain all the inductively constructed curves and pass through all but at most three vertices of a B-chain.   □

We use Lemma 3 as follows. Let $k = 1/8$. If $a(G) \geq km$, then by **(4)** and **(6)** we get $s(G) \geq 3a(G) \geq 3km$. If $a(G) < km$, by **(1)** and **(6)** we get $s(G) \geq$

$3a(G) + (m - a(G) - c(G) - d(G)) - x(G)$, which by (5) becomes $s(G) \geq m + 2a(G) - c(G) - d(G) - 3h(G)$. Using (2) and (3) we get $s(G) \geq m + 2(c(G) + 2d(G) + 1) - c(G) - d(G) - 3(2c(G) + 3d(G) + 1) = m - 5c(G) - 6d(G) - 1$. Again by (2) and by hypothesis we get $c(G) + 2d(G) + 1 < km$, thus $5c(G) + 6d(G) + 1 < 5c(G) + 10d(G) + 5 < 5km$. Hence, $s(G) \geq m - 5km$. Since $k = 1/8$, we have $s(G) \geq 3m/8$ both if $a(G) \geq m/8$ and if $a(G) < m/8$. Thus one of $\lambda_u(G)$, $\lambda_v(G)$, and $\lambda_z(G)$ is a proper good curve passing through $\lceil \frac{n-3}{8} \rceil$ internal vertices of G. This concludes the proof that G has a collinear set with size $\lceil \frac{n-3}{8} \rceil$.

We now strengthen this result by proving that G has a *free* collinear set with the same size. This is accomplished by means of the following lemma, which concludes the proof of Theorem 1.

Theorem 5. *Every collinear set in a plane 3-tree is also a free collinear set.*

Proof Sketch. Let G be a plane 3-tree and Ψ be a planar straight-line drawing of G with a set S of vertices on a straight line ℓ. Let $<_\Psi$ be the order of the vertices in S along ℓ in Ψ. Our proof shows that, for any set X_S of $|S|$ points on ℓ, there is a planar straight-line drawing Γ of G such that: (1) every vertex is above, below, or on ℓ in Γ if and only the same holds in Ψ; and (2) the i-th vertex in $<_\Psi$ is at the i-th point in X_S in left-to-right order along ℓ. This is proved by assuming an arbitrary drawing Δ of (u, v, z), by drawing w so as to split Δ into three triangles with a suitable number of points of X_S in their interior, and by then using recursion on the children of G.   □

5   Triconnected Cubic Planar Graphs

In this section we prove Theorem 2. By Theorem 4 it suffices to prove that every n-vertex triconnected cubic plane graph has a proper good curve λ through $\lceil \frac{n}{4} \rceil$ vertices. The proof is by induction on n; Lemma 4 below states our inductive hypothesis. In order to split the graph into subgraphs on which induction can be applied, we use a structural decomposition that is derived from a paper by Chen and Yu [6] and that applies to a class of graphs, called *strong circuit graphs* in [6], wider than triconnected cubic plane graphs. We introduce the concept of *well-formed quadruple* in order to point out some properties of the graphs in this class. In particular, the inductive hypothesis handles carefully the set X of degree-2 vertices of the graph, which have neighbors that are not in the graph at the current level of the induction; since λ might pass through these neighbors, it has to avoid the vertices in X, in order to be good. Special conditions are ensured for two vertices u and v which work as link to the rest of the graph.

We introduce some definitions. Given two external vertices u and v of a biconnected plane graph G, let $\tau_{uv}(G)$ and $\beta_{uv}(G)$ be the paths delimiting the outer face of G in clockwise and counter-clockwise direction from u to v, respectively. Let π be one of $\tau_{uv}(G)$ and $\beta_{uv}(G)$. An *intersection point* (a *proper intersection point*) between an open curve λ and π is a point p belonging to both λ and π such that, for every $\epsilon > 0$, the part of λ in the disk centered at p with radius ϵ

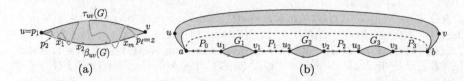

Fig. 4. (a) Illustration for Lemma 4. The gray region is the interior of G. The vertices in X are squares, the intersection points between λ and $\beta_{uv}(G)$ are circles, and u and v are disks. (b) Illustration for Lemma 5 with $k = 3$.

contains points not in π (resp. points in the outer face of G); if the end-vertices of λ are in π, then we regard them as intersection points.

A quadruple (G, u, v, X) is *well-formed* if: **(a)** G is a biconnected subcubic plane graph; **(b)** u and v are two external vertices of G; **(c)** $\delta_G(u) = \delta_G(v) = 2$; **(d)** if edge (u, v) exists, it coincides with $\tau_{uv}(G)$; **(e)** for every separation pair $\{a, b\}$ of G, a and b are external vertices of G and at least one of them is internal to $\beta_{uv}(G)$; further, every non-trivial $\{a, b\}$-component of G contains an external vertex of G different from a and b; and **(f)** $X = (x_1, \ldots, x_m)$ is a (possibly empty) sequence of degree-2 vertices of G in $\beta_{uv}(G)$, different from u and v, and in this order along $\beta_{uv}(G)$ from u to v. We have the following main lemma.

Lemma 4. *Let (G, u, v, X) be a well-formed quadruple. There exists a proper good curve λ such that (see Fig. 4(a)):*

(1) λ starts at u, does not pass through v, and ends at a point z of $\beta_{uv}(G)$;

(2) z is between x_m and v on $\beta_{uv}(G)$ (if $X = \emptyset$, this condition is vacuous);

(3) the intersection points between λ and $\beta_{uv}(G)$ occur along λ from u to z and occur along $\beta_{uv}(G)$ from u to v in the same order $u = p_1, \ldots, p_\ell = z$;

(4) the vertices in X are incident to $R_{G,\lambda}$ and are not on λ; if p_i, x_j and p_{i+1} are in this order along $\beta_{uv}(G)$, then the part of λ between p_i and p_{i+1} is in the interior of G;

(5) λ and $\tau_{uv}(G)$ have no proper intersection point; and

(6) let L_λ (N_λ) be the subset of vertices in $V(G) - X$ that are (resp. are not) on λ; each vertex in N_λ can be charged to a vertex in L_λ so that each vertex in L_λ is charged with at most 3 vertices and u is charged with at most 1 vertex.

Before proving Lemma 4 we state the following (see Fig. 4(b)).

Lemma 5. *Let (G, u, v, X) be a well-formed quadruple and $\{a, b\}$ be a separation pair of G with $a, b \in \beta_{uv}(G)$. The $\{a, b\}$-component G_{ab} of G containing $\beta_{ab}(G)$ either coincides with $\beta_{ab}(G)$ or consists of: **(i)** a path $P_0 = (a, \ldots, u_1)$ (possibly a single vertex); **(ii)** for $i = 1, \ldots, k$ with $k \geq 1$, a biconnected component G_i of G_{ab} containing vertices u_i and v_i, where (G_i, u_i, v_i, X_i) is a well-formed quadruple with $X_i = X \cap V(G_i)$; **(iii)** for $i = 1, \ldots, k-1$, a path $P_i = (v_i, \ldots, u_{i+1})$, where $u_{i+1} \neq v_i$; and **(iv)** a path $P_k = (v_k, \ldots, b)$ (possibly a single vertex).*

We outline the proof of Lemma 4, which is by induction on the size of G.

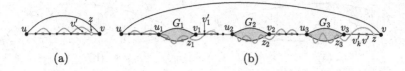

Fig. 5. Base case (a) and Case 1 with $k = 3$ (b) for the proof of Lemma 4.

Base Case: G is a cycle; see Fig. 5(a). By Property (e) of (G, u, v, X), $\{u, v\}$ is not a separation pair of G, hence edge (u, v) exists and coincides with $\tau_{uv}(G)$. Curve λ starts at u; it then passes through the vertices in $V(G) - (X \cup \{v\})$ in the order as they appear along $\beta_{uv}(G)$ from u to v; if two vertices in $V(G) - (X \cup \{v\})$ are consecutive in $\beta_{uv}(G)$, then λ contains the edge between them. If the neighbor v' of v in $\beta_{uv}(G)$ is not in X, then λ ends at v', otherwise λ ends at a point z in the interior of edge (v, v'). Finally, charge v to u.

Next we describe the inductive cases. In the description of each case, we implicitly assume that none of the previously described cases applies.

Case 1: edge (u, v) exists; see Fig. 5(b). By Property (d) of (G, u, v, X), edge (u, v) coincides with $\tau_{uv}(G)$. By Property (c), v has a unique neighbor $v' \neq u$, hence $\{u, v'\}$ is a separation pair to which Lemma 5 applies. For $i = 1, \ldots k$, use induction to construct a proper good curve λ_i satisfying the properties of Lemma 4 for the well-formed quadruple (G_i, u_i, v_i, X_i), defined as in Lemma 5.

Curve λ starts at u and passes through the vertices in $V(P_0) \backslash X$ until reaching u_1; this part of λ lies in the internal face of G incident to edge (u, v) and is constructed similarly to the base case. Curve λ continues with λ_1, which ends at a point z_1. Then λ traverses the outer face of G to reach the neighbor v_1' of v_1 in P_1 (if $v_1' \notin X$) or a point in the interior of edge (v_1, v_1') (if $v_1' \in X$); this part of λ can be drawn without causing self-intersections since λ_1 satisfies Properties (2), (3), and (5) of Lemma 4 – these properties ensure that z_1 and v_1' are both incident to R_{G, λ_1}. Curve λ continues similarly until a point z_k in $\beta_{u_k v_k}(G_k)$ is reached. If the neighbor v_k' of v_k in P_k is v, then λ stops at $z = z_k$; otherwise, it traverses the outer face of G from z_k to a point on edge (v_k, v_k') – this point is v_k' if $v_k' \notin X$ – and it ends by passing through the vertices in $V(P_k) \backslash (X \cup \{v\})$, similarly to the base case. Inductively compute a charge of the vertices in $(N_\lambda \cap V(G_i))$ to the vertices in $L_\lambda \cap V(G_i)$; finally, charge v to u.

If Case 1 does not apply, by Property (e) of (G, u, v, X), $\{u, v\}$ is not a separation pair of G, hence u is not a cut-vertex of graph $G - \{v\}$. Let H be the biconnected component of $G - \{v\}$ containing u. Graph G is composed of H, of a trivial $H \cup \{v\}$-bridge $B_1 = (y_1, v)$, which is an edge in $\tau_{uv}(G)$, and of an $H \cup \{v\}$-bridge B_2 with attachments v and y_2, where y_1 and y_2 are in H. Let $X' = \{y_2\} \cup (X \cap V(H))$. Then (H, u, y_1, X') is a well-formed quadruple.

Case 2: B_2 contains a vertex not in $X \cup \{v, y_2\}$. Refer to Fig. 6(a). Curve λ is composed of curves λ_1, λ_2, and λ_3. Curve λ_1 is inductively constructed for (H, u, y_1, X'). Since $y_2 \in X'$, λ_1 ends at a point z_0 in $\beta_{y_2 y_1}(H)$. Curve λ_2 lies in the internal face of G incident to edge (y_1, v) and connects z_0 with the first

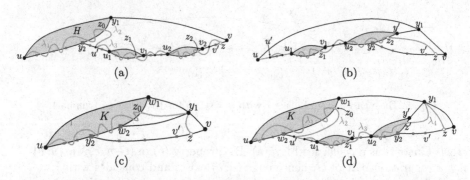

Fig. 6. (a) Case 2, (b) Case 3, (c) Case 4, and (d) Case 5 of the proof of Lemma 4.

vertex $u' \neq y_2$ not in X encountered when traversing $\beta_{y_2 v}(G)$ from y_2 to v; u' exists by the hypothesis of Case 2 and by Property (f) of (G, u, v, X). Properties (3)–(5) of λ_1 ensure that y_2 is not on λ_1 and is incident to R_{G,λ_1}. Thus, even if u' is adjacent to y_2, still λ intersects (y_2, u') only once. Finally, λ_3 connects u' with a point $z \neq y_2, v$ on $\beta_{y_2 v}(G)$; since $\{y_2, v\}$ is a separation pair of G, Lemma 5 applies and curve λ_3 is constructed as in Case 1. Inductively determine the charge of the vertices in $(N_\lambda \cap V(H)) - \{y_2\}$ to the vertices in $L_\lambda \cap V(H)$, and the charge of the vertices in N_λ in each biconnected component G_i of B_2 to the vertices in $L_\lambda \cap V(G_i)$. Finally, charge y_2 and v to u'.

If Case 2 does not apply, B_2 is a path $\beta_{y_2 v}$ whose internal vertices are in X.

Case 3: edge (u, y_1) exists. By Property (d) of (H, u, y_1, X'), edge (u, y_1) coincides with $\tau_{uy_1}(H)$. Let y' be the neighbor of y_1 in $\beta_{uy_1}(H)$. If H has a vertex not in $X' \cup \{u, y_1\}$ as in Fig. 6(b) – otherwise λ is easily constructed – then $\{u, y'\}$ is a separation pair of H and Lemma 5 applies. Construct a curve λ_1 between u and a point $z_k \neq y_1$ on $\beta_{y_2 y_1}(H)$ as in Case 1. Curve λ consists of λ_1 and of a curve λ_2 in the internal face of G incident to edge (v, y_1) between z_k and a point z on edge (v, v'). Inductively charge the vertices in the biconnected components on which induction is applied. Charge v to u, and y_1 and y_2 to the first vertex $u' \neq u$ not in X' encountered when traversing $\beta_{uy_1}(H)$ from u to y_1.

If Case 3 does not apply, then u is not a cut-vertex of graph $H - \{y_1\}$, since $\{u, y_1\}$ is not a separation pair of H. Graph H is composed of the biconnected component K of $H - \{y_1\}$ containing u, of a trivial $K \cup \{y_1\}$-bridge $D_1 = (w_1, y_1)$, and of a $K \cup \{y_1\}$-bridge D_2 with attachments y_1 and w_2, where $w_1, w_2 \in V(K)$.

Case 4: $y_2 \in K$. Refer to Fig. 6(c). Since $\delta_G(y_2) \leq 3$, y_2 and w_2 are distinct. Also, w_2 is an internal vertex of G; hence, D_2 is a trivial $K \cup \{y_1\}$-bridge. Let $X'' = (X \cap V(K)) \cup \{y_2, w_2\}$; inductively construct a curve λ_1 connecting u with a point $z_0 \neq w_1$ in $\beta_{w_2 w_1}(K)$ for the well-formed quadruple (K, u, w_1, X''). Curve λ consists of λ_1 and of a curve λ_2 from z_0 to a point z on edge (v, v') passing through y_1. Curve λ_2 lies in the internal faces of G incident to edges (w_1, y_1) and (y_1, v). Inductively charge the vertices in $(N_\lambda \cap V(K)) - \{y_2, w_2\}$ to the vertices in $L_\lambda \cap V(K)$; charge v, y_2, and w_2 to y_1.

Case 5: $y_2 \notin K$. Let $X'' = \{w_2\} \cup (X \cap V(K))$. Curve λ consists of four curves $\lambda_1, \ldots, \lambda_4$. Inductively construct λ_1 for the well-formed quadruple (K, u, w_1, X'') between u and a point $z_0 \neq w_1$ in $\beta_{w_2 w_1}(K)$. If D_2 has a vertex not in $X' \cup \{y_1, w_2\}$, as in Fig. 6(d) – otherwise λ is constructed similarly to Case 4 – then λ_2 connects z_0 with the first vertex $u' \neq w_2$ not in X' encountered while traversing $\beta_{w_2 y_1}(H)$ from w_2 to y_1; λ_2 is in the internal face of G incident to edge (w_1, y_1). Curve λ_3 connects u' with a point z' in $\beta_{y_2 y_1}(H)$; $\{w_2, y_1\}$ is a separation pair of H, hence Lemma 5 applies and curve λ_3 is constructed as in Case 1. Finally, λ_4 connects z' with a point z on edge (v, v') passing through y_1. Inductively charge the vertices in the biconnected components on which induction is applied. Charge v, y_2, and w_2 to y_1. This concludes the proof of Lemma 4.

We now prove Theorem 2. Let G be an n-vertex triconnected cubic plane graph. Let H be the plane graph obtained from G by removing any edge (u, v) incident to the outer face of G. Then $(H, u, v, \emptyset)$ is a well-formed quadruple and H has a proper good curve λ as in Lemma 4. Insert (u, v) in the outer face of H, restoring the plane embedding of G. By Properties (1)–(5) of λ edge (u, v) does not intersect λ other than at u, hence λ remains proper and good. By Property (6) with $X = \emptyset$, λ passes through $\lceil \frac{n}{4} \rceil$ vertices of G. This concludes the proof.

6   Implications for Other Graph Drawing Problems

In this section we present corollaries of our results to other graph drawing problems. The key tool to establish these connections is a lemma that appeared in [3, Lemma 1], which we explicitly state here in two more readily applicable versions.

Lemma 6. [3] *Let G be a planar graph that has a planar straight-line drawing Γ with a set S of vertices on the x-axis. For any assignment of y-coordinates to the vertices in S, there exists a planar straight-line drawing of G such that each vertex in S has the same x-coordinate as in Γ and has the assigned y-coordinate.*

Lemma 7. [3] *Let G be a planar graph, S be a free collinear set, and $<_S$ be the total order associated with S. Consider any assignment of x- and y-coordinates to the vertices in S such that the assigned x-coordinates are distinct and the order of the vertices in S by increasing or decreasing x-coordinates is $<_S$. There exists a planar straight-line drawing of G such that each vertex in S has the assigned x- and y-coordinates.*

Lemma 7 and the fact that planar graphs of treewidth at most 3 have free collinear sets with linear size, established in Theorem 1, imply the following.

Corollary 1. *Every set of at most $\lceil \frac{n-3}{8} \rceil$ points in the plane is a universal point subset for all n-vertex plane graphs of treewidth at most three.*

As noted in [3,20], Lemmas 6 and 7 imply that every straight-line drawing of a planar graph G with a free collinear set of size x can be untangled while keeping $\sqrt{x}$ vertices fixed. Together with Theorem 1 this implies the following.

Corollary 2. *Any straight-line drawing of an n-vertex planar graph of treewidth at most three can be untangled while keeping at least $\sqrt[3]{\lceil (n-3)/8 \rceil}$ vertices fixed.*

Finally, Lemma 6 implies that every collinear set is a column planar set. That and our three main results imply our final corollary.

Corollary 3. *Triconnected cubic planar graphs and planar graphs of treewidth at most three have column planar sets of linear size. Further, planar graphs of treewidth at least k have column planar sets of size $\Omega(k^2)$.*

7   Conclusions

We studied the problems of determining the maximum cardinality of collinear sets and free collinear sets in planar graphs; it would be interesting to close the gap between the best bounds of $\Omega(n^{0.5})$ and $O(n^{0.986})$ known for these problems.

We proved that triconnected cubic plane graphs have collinear sets with linear size. Generalizing the bound to subcubic plane graphs seems like a plausible goal.

We proved that plane graphs with treewidth at most 3 have free collinear sets with linear size. In order to do that, we proved that every collinear set is free in a plane 3-tree, which brings us to a question posed in [20]: is every collinear set free, and if not, how close are the sizes of these two sets in a planar graph?

Finally, the *maximum* number of collinear vertices in any planar straight-line drawing of a plane 3-tree can be determined by dynamic programming. An implementation of the algorithm has shown that, for $m \leq 50$ and for every plane 3-tree G with m internal vertices, the maximum number of collinear internal vertices in any planar straight-line drawing of G is at least $\lceil \frac{m+2}{3} \rceil$ (this bound is the best possible for every $m \leq 50$). Is this the case for every $m \geq 1$?

Acknowledgments. Research partially supported by MIUR Project "AMANDA" under PRIN 2012C4E3KT, by NSERC, and by Ontario Ministry of Research and Innovation. The authors thank Giuseppe Liotta, Sue Whitesides, and Stephen Wismath for stimulating discussions.

References

1. Angelini, P., Binucci, C., Evans, W., Hurtado, F., Liotta, G., Mchedlidze, T., Meijer, H., Okamoto, Y.: Universal point subsets for planar graphs. In: Chao, K.-M., Hsu, T., Lee, D.-T. (eds.) ISAAC 2012. LNCS, vol. 7676, pp. 423–432. Springer, Heidelberg (2012). doi:10.1007/978-3-642-35261-4_45

2. Bose, P.: On embedding an outer-planar graph in a point set. Comput. Geom. **23**(3), 303–312 (2002)

3. Bose, P., Dujmović, V., Hurtado, F., Langerman, S., Morin, P., Wood, D.R.: A polynomial bound for untangling geometric planar graphs. Discrete Comput. Geom. **42**(4), 570–585 (2009)

4. Cano, J., Tóth, C.D., Urrutia, J.: Upper bound constructions for untangling planar geometric graphs. SIAM J. Discrete Math. **28**(4), 1935–1943 (2014)

5. Castañeda, N., Urrutia, J.: Straight line embeddings of planar graphs on point sets. In: Fiala, F., Kranakis, E., Sack, J. (eds.) 8th Canadian Conference on Computational Geometry (CCCG 1996), pp. 312–318. Carleton University Press (1996)
6. Chen, G., Yu, X.: Long cycles in 3-connected graphs. J. Comb. Theory Ser. B **86**(1), 80–99 (2002)
7. Cibulka, J.: Untangling polygons and graphs. Discrete Comput. Geom. **43**(2), 402–411 (2010)
8. Da Lozzo, G., Dujmović, V., Frati, F., Mchedlidze, T., Roselli, V.: Drawing planar graphs with many collinear vertices. CoRR abs/1606.03890v3 (2016). http://arxiv.org/abs/1606.03890v3
9. Dujmović, V.: The utility of untangling. In: Di Giacomo, E., Lubiw, A. (eds.) GD 2015. LNCS, vol. 9411, pp. 321–332. Springer, Heidelberg (2015). doi:10.1007/978-3-319-27261-0_27
10. Eades, P., Feng, Q., Lin, X., Nagamochi, H.: Straight-line drawing algorithms for hierarchical graphs and clustered graphs. Algorithmica **44**(1), 1–32 (2006)
11. Evans, W., Kusters, V., Saumell, M., Speckmann, B.: Column planarity and partial simultaneous geometric embedding. In: Duncan, C., Symvonis, A. (eds.) GD 2014. LNCS, vol. 8871, pp. 259–271. Springer, Heidelberg (2014). doi:10.1007/978-3-662-45803-7_22
12. Goaoc, X., Kratochvíl, J., Okamoto, Y., Shin, C., Spillner, A., Wolff, A.: Untangling a planar graph. Discrete Comput. Geom. **42**(4), 542–569 (2009)
13. Gritzmann, P., Mohar, B., Pach, J., Pollack, R.: Embedding a planar triangulation with vertices at specified points (solution to problem e3341). Am. Math. Mon. **98**, 165–166 (1991)
14. Grünbaum, B., Walther, H.: Shortness exponents of families of graphs. J. Comb. Theory Ser. A **14**(3), 364–385 (1973)
15. Kang, M., Pikhurko, O., Ravsky, A., Schacht, M., Verbitsky, O.: Untangling planar graphs from a specified vertex position - hard cases. Discrete Appl. Math. **159**(8), 789–799 (2011)
16. Kratochvíl, J., Vaner, M.: A note on planar partial 3-trees. CoRR abs/1210.8113 (2012). http://arxiv.org/abs/1210.8113
17. Owens, P.J.: Non-hamiltonian simple 3-polytopes whose faces are all 5-gons or 7-gons. Discrete Math. **36**(2), 227–230 (1981)
18. Pach, J., Tardos, G.: Untangling a polygon. Discrete Comput. Geom. **28**(4), 585–592 (2002)
19. Pach, J., Tóth, G.: Monotone drawings of planar graphs. J. Graph Theory **46**(1), 39–47 (2004)
20. Ravsky, A., Verbitsky, O.: On collinear sets in straight-line drawings. In: Kolman, P., Kratochvíl, J. (eds.) WG 2011. LNCS, vol. 6986, pp. 295–306. Springer, Heidelberg (2011). doi:10.1007/978-3-642-25870-1_27
21. Ravsky, A., Verbitsky, O.: On collinear sets in straight line drawings. CoRR abs/0806.0253 (2011). http://arxiv.org/abs/0806.0253
22. Tutte, W.T.: How to draw a graph. Proc. London Math. Soc. **13**(52), 743–768 (1963)

Drawing Graphs on Few Lines and Few Planes

Steven Chaplick[1], Krzysztof Fleszar[1], Fabian Lipp[1], Alexander Ravsky[2(✉)],
Oleg Verbitsky[3], and Alexander Wolff[1]

[1] Lehrstuhl für Informatik I, Universität Würzburg, Würzburg, Germany
{steven.chaplick,krzysztof.fleszar,fabian.lipp,
alexander.wolff}@uni-wuerzburg.de
[2] Pidstryhach Institute for Applied Problems of Mechanics and Mathematics,
National Academy of Sciences of Ukraine, Lviv, Ukraine
oravsky@mail.ru
[3] Institut für Informatik, Humboldt-Universität zu Berlin, Berlin, Germany
verbitsk@informatik.hu-berlin.de
http://www1.informatik.uni-wuerzburg.de

Abstract. We investigate the problem of drawing graphs in 2D and 3D
such that their edges (or only their vertices) can be covered by few lines
or planes. We insist on straight-line edges and crossing-free drawings.
This problem has many connections to other challenging graph-drawing
problems such as small-area or small-volume drawings, layered or track
drawings, and drawing graphs with low visual complexity. While some
facts about our problem are implicit in previous work, this is the first
treatment of the problem in its full generality. Our contribution is as
follows.

- We show lower and upper bounds for the numbers of lines and planes
  needed for covering drawings of graphs in certain graph classes. In
  some cases our bounds are asymptotically tight; in some cases we
  are able to determine exact values.
- We relate our parameters to standard combinatorial characteristics of graphs (such as the chromatic number, treewidth, maximum
  degree, or arboricity) and to parameters that have been studied in
  graph drawing (such as the track number or the number of segments
  appearing in a drawing).
- We pay special attention to planar graphs. For example, we show
  that there are planar graphs that can be drawn in 3-space on a lot
  fewer lines than in the plane.

1 Introduction

It is well known that any graph admits a straight-line drawing in 3-space. Suppose that we are allowed to draw edges only on a limited number of planes. How
many planes do we need for a given graph G? For example, K_6 needs four planes;

The full version of this paper is available on arXiv [10]. Whenever we refer to the
Appendix, we mean the appendix of arXiV:1607.01196v2

O. Verbitsky was supported by DFG grant VE 652/1-2.

Y. Hu and M. Nöllenburg (Eds.): GD 2016, LNCS 9801, pp. 166–180, 2016.
DOI: 10.1007/978-3-319-50106-2_14

see Fig. 1. Note that this question is different from the well-known concept of a *book embedding* where all vertices lie on one line (the spine) and edges lie on a limited number of adjacent half-planes (the pages). In contrast, we put no restriction on the mutual position of planes, the vertices can be located in the planes arbitrarily, and the edges must be straight-line.

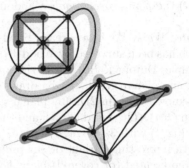

Fig. 1. K_6 can be drawn straight-line and crossing-free on four planes. This is optimal, that is, $\rho_3^2(K_6) = 4$.

Fig. 2. Planar 9-vertex graph G with $\pi_3^1(G) = 3$, 3D-drawing on three lines.

In a weaker setting, we require only the vertices to be located on a limited number of planes (or lines). For example, the graph in Fig. 2 can be drawn in 2D such that its vertices are contained in three lines; we conjecture that it is the smallest planar graph that needs more than two lines even in 3D. This version of our problem is related to the well-studied problem of drawing a graph straight-line in a 3D grid of bounded volume [16,37]: If a graph can be drawn with all vertices on a grid of volume v, then $v^{1/3}$ planes and $v^{2/3}$ lines suffice. We now formalize the problem.

Definition 1. Let $1 \leq l < d$, and let G be a graph. We define the *l-dimensional affine cover number* of G in $\mathbb{R}^d$, denoted by $\rho_d^l(G)$, as the minimum number of *l*-dimensional planes in $\mathbb{R}^d$ such that G has a drawing that is contained in the union of these planes. We define $\pi_d^l(G)$, the *weak l*-dimensional affine cover number of G in $\mathbb{R}^d$, similarly to $\rho_d^l(G)$, but under the weaker restriction that the vertices (and not necessarily the edges) of G are contained in the union of the planes. Finally, the *parallel* affine cover number, $\bar{\pi}_d^l(G)$, is a restricted version of $\pi_d^l(G)$, in which we insist that the planes are parallel. We consider only straight-line and crossing-free drawings. Note: $\rho_d^l(G)$, $\pi_d^l(G)$, and $\bar{\pi}_d^l(G)$ are only undefined when $d = 2$ and G is non-planar.

Clearly, for any combination of l and d, it holds that $\pi_d^l(G) \leq \bar{\pi}_d^l(G)$ and $\pi_d^l(G) \leq \rho_d^l(G)$. Larger values of l and d give us more freedom for drawing graphs and, therefore, smaller π- and ρ-values. Formally, for any graph G, if $l' \leq l$ and $d' \leq d$ then $\pi_d^l(G) \leq \pi_{d'}^{l'}(G)$, $\rho_d^l(G) \leq \rho_{d'}^{l'}(G)$, and $\bar{\pi}_d^l(G) \leq \bar{\pi}_{d'}^{l'}(G)$.

But in most cases this freedom is not essential. For example, it suffices to consider $l \leq 2$ because otherwise $\rho_d^l(G) = 1$. More interestingly, we can actually focus on $d \leq 3$ because every graph can be drawn in 3-space as effectively as in high dimensional spaces, i.e., for any integers $1 \leq l \leq d$, $d \geq 3$, and for any graph G, it holds that $\pi_d^l(G) = \pi_3^l(G)$, $\bar{\pi}_d^l(G) = \bar{\pi}_3^l(G)$, and $\rho_d^l(G) = \rho_3^l(G)$. We prove this important fact in Appendix A. Thus, our task is to investigate the cases $1 \leq l < d \leq 3$. We call $\rho_2^1(G)$ and $\rho_3^1(G)$ the *line cover numbers* in 2D and 3D, $\rho_3^2(G)$ the *plane cover number*, and analogously for the weak versions.

Related Work. We have already briefly mentioned 3D graph drawing on the grid, which has been surveyed by Wood [37] and by Dujmović and Whitesides [16]. For example, Dujmović [13], improving on a result of Di Battista et al. [4], showed that any planar graph can be drawn into a 3D-grid of volume $O(n \log n)$. It is well-known that, in 2D, any planar graph admits a plane straight-line drawing on an $O(n) \times O(n)$ grid [20,33] and that the nested-triangles graph $T_k = K_3 \times P_k$ (see Fig. 4) with $3k$ vertices needs $\Omega(k^2)$ area [20].

An interesting variant of our problem is to study drawings whose edge sets are represented (or covered) by as few objects as possible. The type of objects that have been used are straight-line segments [14,17] and circular arcs [34]. The idea behind this objective is to keep the visual complexity of a drawing low for the observer. For example, Schulz [34] showed how to draw the dodecahedron by using 10 arcs, which is optimal.

Our Contribution. Our research goes into three directions.

First, we show lower and upper bounds for the numbers of lines and planes needed for covering drawings of graphs in certain graph classes such as graphs of bounded degree or subclasses of planar graphs. The most natural graph families to start with are the complete graphs and the complete bipartite graphs. Most versions of the affine cover numbers of these graphs can be determined easily. Two cases are much more subtle: We determine $\rho_3^2(K_n)$ and $\rho_3^1(K_{n,n})$ only asymptotically, up to a factor of 2 (see Theorem 12 and Example 10). Some efforts are made to compute the exact values of $\rho_3^2(K_n)$ for small n (see Theorem 15). As another result in this direction, we prove that $\rho_3^1(G) > n/5$ for almost all cubic graphs on n vertices (Theorem 9(b)).

Second, we relate the affine cover numbers to standard combinatorial characteristics of graphs and to parameters that have been studied in graph drawing. In Sect. 2.1, we characterize $\pi_3^1(G)$ and $\pi_3^2(G)$ in terms of the *linear vertex arboricity* and the *vertex thickness*, respectively. This characterization implies that both $\pi_3^1(G)$ and $\pi_3^2(G)$ are linearly related to the chromatic number of the graph G. Along the way, we refine a result of Pach et al. [28] concerning the volume of 3D grid drawings (Theorem 2). We also prove that any graph G has balanced separators of size at most $\rho_3^1(G)$ and conclude from this that $\rho_3^1(G) \geq \text{tw}(G)/3$, where $\text{tw}(G)$ denotes the treewidth of G (Theorem 9). In Sect. 3.2, we analyze the relationship between $\rho_2^1(G)$ and the segment number $\text{segm}(G)$ of a graph, which was introduced by Dujmović et al. [14]. We prove that $\text{segm}(G) = O(\rho_2^1(G)^2)$

for any connected G and show that this bound is optimal (see Theorem 23 and Example 22).

Third, we pay special attention to planar graphs (Sect. 3). Among other results, we show examples of planar graphs with a large gap between the parameters $\rho_3^1(G)$ and $\rho_2^1(G)$ (see Theorem 24).

We also investigate the parallel affine cover numbers $\bar{\pi}_2^1$ and $\bar{\pi}_3^1$. Observe that for any graph G, $\bar{\pi}_3^1(G)$ equals the *improper track number* of G, which was introduced by Dujmović et al. [15].

Due to lack of space, our results for the parallel affine cover numbers (along with a survey of known related results) appear in Appendix B. We defer some other proofs to Appendices C and D and list some open problems in Appendix E.

Remark on the Computational Complexity. In a follow-up paper [9], we investigate the computational complexity of computing the ρ- and π-numbers. We argue that it is NP-hard to decide whether a given graph has a π_3^1- or π_3^2-value of 2 and that both values are even hard to approximate. This result is based on Theorems 2 and 4 and Corollaries 3 and 5 in the present paper. While the graphs with ρ_3^2-value 1 are exactly the planar graphs (and hence, can be recognized in linear time), it turns out that recognizing graphs with a ρ_3^2-value of 2 is already NP-hard. In contrast to this, the problems of deciding whether $\rho_3^1(G) \leq k$ or $\rho_2^1(G) \leq k$ are solvable in polynomial time for any fixed k. However, the versions of these problems with k being part of the input are complete for the complexity class $\exists\mathbb{R}$ which is based on the *existential theory of the reals* and that plays an important role in computational geometry [32].

Notation. For a graph $G = (V, E)$, we use n and m to denote the numbers of vertices and edges of G, respectively. Let $\Delta(G) = \max_{v \in V} \deg(v)$ denote the maximum degree of G. Furthermore, we will use the standard notation $\chi(G)$ for the chromatic number, $\text{tw}(G)$ for the treewidth, and $\text{diam}(G)$ for the diameter of G. The Cartesian product of graphs G and H is denoted by $G \times H$.

2 The Affine Cover Numbers in $\mathbb{R}^3$

2.1 Placing Vertices on Few Lines or Planes (π_3^1 and π_3^2)

A *linear forest* is a forest whose connected components are paths. The *linear vertex arboricity* $\text{lva}(G)$ of a graph G equals the smallest size r of a partition $V(G) = V_1 \cup \cdots \cup V_r$ such that every V_i induces a linear forest. This notion, which is an induced version of the fruitful concept of *linear arboricity* (see Remark 8 below), appears very relevant to our topic. The following result is based on a construction of Pach et al. [28]; see Appendix C for the proof.

Theorem 2. *For any graph G, it holds that $\pi_3^1(G) = \text{lva}(G)$. Moreover, any graph G can be drawn with vertices on r lines in the 3D integer grid of size $r \times 4rn \times 4r^2n$, where $r = \text{lva}(G)$.*

Corollary 3. $\chi(G)/2 \le \pi_3^1(G) \le \chi(G)$.

Corollary 3 readily implies that $\pi_3^1(G) \le \Delta(G) + 1$ [7]. This can be considerably improved using a relationship between the linear vertex arboricity and the maximum degree that is established by Matsumoto [27]. Matsumoto's result implies that $\pi_3^1(G) \le \Delta(G)/2 + 1$ for any connected graph G. Moreover, if $\Delta(G) = 2d$, then $\pi_3^1(G) = d + 1$ if and only if G is a cycle or the complete graph K_{2d+1}.

We now turn to the weak plane cover numbers. The *vertex thickness* $\text{vt}(G)$ of a graph G is the smallest size r of a partition $V(G) = V_1 \cup \cdots \cup V_r$ such that $G[V_1], \ldots, G[V_r]$ are all planar. We prove the following theorem in Appendix C.

Theorem 4. *For any graph G, it holds that $\pi_3^2(G) = \bar{\pi}_3^2(G) = \text{vt}(G)$ and that G can be drawn such that all vertices lie on a 3D integer grid of size $\text{vt}(G) \times O(m^2) \times O(m^2)$, where m is the number of edges of G. Note that this drawing occupies $\text{vt}(G)$ planes.*

Corollary 5. $\chi(G)/4 \le \pi_3^2(G) \le \chi(G)$.

Example 6. (a) $\pi_3^1(K_n) = \lceil n/2 \rceil$.
(b) $\pi_3^1(K_{p,q}) = 2$ for any $1 \le p \le q$; except for $\pi_3^1(K_{1,1}) = \pi_3^1(K_{1,2}) = 1$.
(c) $\pi_3^2(K_n) = \lceil n/4 \rceil$; therefore, $\pi_3^2(G) \le \lceil n/4 \rceil$ for every graph G.

2.2 Placing Edges on Few Lines or Planes (ρ_3^1 and ρ_3^2)

Clearly, $\Delta(G)/2 \le \rho_3^1(G) \le m$ for any graph G. Call a vertex v of a graph G *essential* if $\deg v \ge 3$ or if v belongs to a K_3 subgraph of G. Denote the number of essential vertices in G by $\text{es}(G)$.

Lemma 7. (a) $\rho_3^1(G) > (1 + \sqrt{1 + 8\,\text{es}(G)})/2$.
(b) $\rho_3^1(G) > \sqrt{m^2/n - m}$ for any graph G with $m \ge n \ge 1$.

Proof. (a) In any drawing of a graph G, any essential vertex is shared by two edges not lying on the same line. Therefore, each such vertex is an intersection point of at least two lines, which implies that $\text{es}(G) \le \binom{\rho_3^1(G)}{2}$. Hence, $\rho_3^1(G) \ge (1 + \sqrt{1 + 8\,\text{es}(G)})/2 > \sqrt{2\,\text{es}(G)}$.

(b) Taking into account multiplicity of intersection points (that is, each vertex v requires at least $\lceil \deg v/2 \rceil(\lceil \deg v/2 \rceil - 1)/2$ intersecting line pairs), we obtain

$$\binom{\rho_3^1(G)}{2} \ge \frac{1}{2}\sum_{v \in V(G)} \left\lceil \frac{\deg v}{2} \right\rceil \left(\left\lceil \frac{\deg v}{2} \right\rceil - 1\right) \ge \sum \frac{\deg v(\deg v - 2)}{8} =$$

$$= \frac{1}{8}\sum(\deg v)^2 - \frac{1}{4}\sum \deg v \ge \frac{1}{8n}\left(\sum \deg v\right)^2 - \frac{1}{4}2m = \frac{m^2}{2n} - \frac{m}{2}.$$

The last inequality follows by the inequality between arithmetic and quadratic means. Hence, $\rho_3^1(G) > \sqrt{m^2/n - m}$.   □

Part (a) of Lemma 7 implies that $\rho_3^1(G) > \sqrt{2n}$ if a graph G has no vertices of degree 1 and 2, while Part (b) yields $\rho_3^1(G) > \sqrt{m/2}$ for all such G. Note that a disjoint union of k cycles can have no essential vertices, but each cycle will need 3 intersection points of lines, i.e., such a graph has $\rho_3^1 \in \Omega(\sqrt{k})$. Thus, ρ_3^1 cannot be bounded from above by a function of essential vertices.

Remark 8. The *linear arboricity* $\mathrm{la}(G)$ of a graph G is the minimum number of linear forests which partition the edge set of G; see [24]. Clearly, we have $\rho_3^1(G) \geq \mathrm{la}(G)$. There is no function of $\mathrm{la}(G)$ that is an upper bound for $\rho_3^1(G)$. Indeed, let G be an arbitrary cubic graph. Akiyama et al. [2] showed that $\mathrm{la}(G) = 2$. On the other hand, any vertex of G is essential, so $\rho_3^1(G) > \sqrt{2n}$ by Lemma 7(a). Theorem 9 below shows an even larger gap.

We now prove a general lower bound for $\rho_3^1(G)$ in terms of the treewidth of G. Note for comparison that $\pi_3^1(G) \leq \chi(G) \leq \mathrm{tw}(G) + 1$ (the last inequality holds because the graphs of treewidth at most k are exactly partial k-trees and the construction of a k-tree easily implies that it is $k + 1$-vertex-chromatic). The relationship between $\rho_3^1(G)$ and $\mathrm{tw}(G)$ follows from the fact that graphs with low parameter $\rho_3^1(G)$ have small separators. This fact is interesting by itself and has yet another consequence: Graphs with bounded vertex degree can have linearly large value of $\rho_3^1(G)$ (hence, the factor of n in the trivial bound $\rho_3^1(G) \leq m \leq \frac{1}{2} n \Delta(G)$ is best possible).

We need the following definitions. Let $W \subseteq V(G)$. A set of vertices $S \subset V(G)$ is a *balanced* W-*separator* of the graph G if $|W \cap C| \leq |W|/2$ for every connected component C of $G \backslash S$. Moreover, S is a *strongly balanced* W-*separator* if there is a partition $W \backslash S = W_1 \cup W_2$ such that $|W_i| \leq |W|/2$ for both $i = 1, 2$ and there is no path between W_1 and W_2 avoiding S. Let $\mathrm{sep}_W(G)$ (resp. $\mathrm{sep}_W^*(G)$) denote the minimum k such that G has a (resp. strongly) balanced W-separator S with $|S| = k$. Furthermore, let $\mathrm{sep}(G) = \mathrm{sep}_{V(G)}(G)$ and $\mathrm{sep}^*(G) = \mathrm{sep}_{V(G)}^*(G)$. Note that $\mathrm{sep}_W(G) \leq \mathrm{sep}_W^*(G)$ for any W and, in particular, $\mathrm{sep}(G) \leq \mathrm{sep}^*(G)$.

It is known [19, Theorem 11.17] that $\mathrm{sep}_W(G) \leq \mathrm{tw}(G) + 1$ for every $W \subseteq V(G)$. On the other hand, if $\mathrm{sep}_W(G) \leq k$ for all W with $|W| = 2k + 1$, then $\mathrm{tw}(G) \leq 3k$.

The *bisection width* $\mathrm{bw}(G)$ of a graph G is the minimum possible number of edges between two sets of vertices W_1 and W_2 with $|W_1| = \lceil n/2 \rceil$ and $|W_2| = \lfloor n/2 \rfloor$ partitioning $V(G)$. Note that $\mathrm{sep}^*(G) \leq \mathrm{bw}(G) + 1$.

Theorem 9. *(a)* $\rho_3^1(G) \geq \mathrm{bw}(G)$.
(b) $\rho_3^1(G) > n/5$ *for almost all cubic graphs with n vertices.*
(c) $\rho_3^1(G) \geq \mathrm{sep}_W^*(G)$ *for every $W \subseteq V(G)$.*
(d) $\rho_3^1(G) \geq \mathrm{tw}(G)/3$.

Proof. (a) Fix a drawing of the graph G on $r = \rho_3^1(G)$ lines in $\mathbb{R}^3$. Choose a plane L that is not parallel to any of the at most $\binom{n}{2}$ lines passing through two vertices of the drawing. Let us move L along the orthogonal direction until it separates the vertex set of G into two almost equal parts W_1 and W_2. The plane L can intersect at most r edges of G, which implies that $\mathrm{bw}(G) \leq r$.

(b) follows from Part (a) and the fact that a random cubic graph on n vertices has bisection width at least $n/4.95$ with probability $1 - o(1)$ (Kostochka and Melnikov [25]).

(c) Given $W \subseteq V(G)$, we have to prove that $\mathrm{sep}_W^*(G) \leq \rho_3^1(G)$. Choose a plane L as in the proof of Part (a) and move it until it separates W into two equal parts W_1' and W_2'; if $|W|$ is odd, then L should contain one vertex w of W. If $|W|$ is even, we can ensure that L does not contain any vertex of G. We now construct a set S as follows. If L contains a vertex $w \in W$, i.e., $|W|$ is odd, we put w in S. Let E be the set of those edges which are intersected by L but are not incident to the vertex w (if it exists). Note that $|E| < r$ if $|W|$ is odd and $|E| \leq r$ if $|W|$ is even. Each of the edges in E contributes one of its incident vertices into S. Note that $|S| \leq r$. Set $W_1 = W_1' \backslash S$ and $W_2 = W_2' \backslash S$ and note that there is no edge between these sets of vertices. Thus, S is a strongly balanced W-separator.

(d) follows from (c) by the relationship between treewidth and balanced separators. □

On the other hand, note that $\rho_3^1(G)$ cannot be bounded from above by any function of $\mathrm{tw}(G)$. Indeed, by Lemma 7(a) we have $\rho_3^1(T) = \Omega(\sqrt{n})$ for every caterpillar T with linearly many vertices of degree 3. The best possible relation in this direction is $\rho_3^1(G) \leq m < n\,\mathrm{tw}(G)$. The factor n cannot be improved here (take $G = K_n$).

Example 10. (a) $\rho_3^1(K_n) = \binom{n}{2}$ for any $n \geq 2$.
(b) $pq/2 \leq \rho_3^1(K_{p,q}) \leq pq$ for any $1 \leq p \leq q$.

We now turn to the plane cover number.

Example 11. For any integers $1 \leq p \leq q$, it holds that $\rho_3^2(K_{p,q}) = \lceil p/2 \rceil$.

Determining the parameter $\rho_3^2(G)$ for complete graphs $G = K_n$ is a much more subtle issue. We are able to determine the asymptotics of $\rho_3^2(K_n)$ up to a factor of 2.

By a *combinatorial cover* of a graph G we mean a set of subgraphs $\{G_i\}$ such that every edge of G belongs to G_i for some i. A *geometric cover* of a crossing-free drawing $d \colon V(K_n) \to \mathbb{R}^3$ of a complete graph K_n is a set $\mathcal{L}$ of planes in $\mathbb{R}^3$ so that for each pair of vertices $v_i, v_j \in V(K_n)$ there is a plane $\ell \in \mathcal{L}$ containing both points $d(v_i)$ and $d(v_j)$. This geometric cover $\mathcal{L}$ induces a combinatorial cover $\mathcal{K}_{\mathcal{L}} = \{G_\ell \mid \ell \in \mathcal{L}\}$ of the graph K_n, where G_ℓ is the subgraph of K_n induced by the set $d^{-1}(\ell)$. Note that each G_ℓ is a K_s subgraph with $s \leq 4$ (because K_5 is not planar).

Let $c(K_n, K_s)$ denote the minimum size of a combinatorial cover of K_n by K_s subgraphs ($c(K_n, K_s) = 0$ if $s > n$). The asymptotics of the numbers $c(K_n, K_s)$ for $s = 3, 4$ can be determined via the results about *Steiner systems* by Kirkman and Hanani [5,23]. This yields the following bounds for $\rho_3^2(K_n)$ (see Appendix C).

Theorem 12. *For all $n \geq 3$,*

$$(1/2 + o(1))\,n^2 = c(K_n, K_4) \leq \rho_3^2(K_n) \leq c(K_n, K_3) = (1/6 + o(1))\,n^2.$$

Table 1. Lower and upper bounds for $\rho_3^2(K_n)$ for small values of n.

n	4	5	6	7	8	9
$\geq$	1	3	4	6	6	7
$\leq$	1	3	4	6	7	

Note that we cannot always realize a combinatorial cover of K_n by copies of K_4 geometrically. For example, $c(K_6, K_4) = 3 < 4 = \rho_3^2(K_6)$ (see Theorem 15).

In order to determine $\rho_3^2(K_n)$ for particular values of n, we need some properties of geometric and combinatorial covers of K_n.

Lemma 13. *Let $d: V(K_n) \rightarrow \mathbb{R}^3$ be a crossing-free drawing of K_n and $\mathcal{L}$ a geometric cover of d. For each 4-vertex graph $G_\ell \in \mathcal{K}_\mathcal{L}$, the set $d(G_\ell)$ not only belongs to a plane ℓ, but also defines a triangle with an additional vertex in its interior.*

Lemma 14. *Let $d: V(K_n) \rightarrow \mathbb{R}^3$ be a crossing-free drawing of K_n and $\mathcal{L}$ a geometric cover of d. No two different 4-vertex graphs $G_\ell, G_{\ell'} \in \mathcal{K}_\mathcal{L}$ can have three common vertices.*

Theorem 15. *For $n \leq 9$, the value of $\rho_3^2(K_n)$ is bounded by the numbers in Table 1.*

Proof. Here, we show only the bounds for $n = 6$. For the remaining proofs, see Appendix C. Figure 1 shows that $\rho_3^2(K_6) \leq 4$. Now we show that $\rho_3^2(K_6) \geq 4$. Assume that $\rho_3^2(K_6) < 4$. Consider a combinatorial cover $\mathcal{K}_\mathcal{L}$ of K_6 by its complete planar subgraphs corresponding to a geometric cover $\mathcal{L}$ of its drawing by 3 planes. Graph K_6 has 15 edges, so to cover it by complete planar graphs we have to use at least two copies of K_4 and, additionally, a copy of K_k for $3 \leq k \leq 4$. But, since each two copies of K_4 in K_6 have a common edge (and by Lemma 14 this edge is unique), the cover $\mathcal{K}_\mathcal{L}$ consists of three copies of K_4. Denote these copies by K_4^1, K_4^2, and K_4^3. By Lemma 13, for each i, $d(K_4^i)$ is a triangle with an additional vertex $d(v_i)$ in its interior. Let $V_0 = \{v_1, v_2, v_3\}$. By the Krein–Milman theorem [26,36], the convex hull $\mathrm{Conv}(d(K_6))$ is the convex hull $\mathrm{Conv}(d(V(K_6)) \backslash d(V_0))$. If all the vertices v_i are mutually distinct then the set $d(V(K_6)) \backslash d(V_0)$ is a triangle, so the drawing d is planar, a contradiction. Hence, $v_i = v_j$ for some $i \neq j$. Let k be the third index that is distinct from both i and j. Since graphs K_4^i and K_4^j have exactly one common edge, this is an edge (v_i, v) for some vertex v of K_6 (see Fig. 1 with u_4 for v_i and u_1 for v). Let $V(K_4^i) = \{v, v_i, v_i^1, v_i^2\}$ and $V(K_4^j) = \{v, v_j, v_j^1, v_j^2\}$. Since the union $K_4^1 \cup K_4^2 \cup K_4^3$ covers all edges of K_6, all edges (v_i^1, v_j^1), (v_i^1, v_j^2), (v_i^2, v_j^1), and (v_i^2, v_j^2) belong to K_4^k. Thus $V(K_4^k) = \{v_i^1, v_i^2, v_j^1, v_j^2\}$. But vertices v_i^1, v_i^2, v_j^1, and v_j^2 are in convex position (see Fig. 1), a contradiction to Lemma 13. $\qquad \square$

3   The Affine Cover Numbers of Planar Graphs ($\mathbb{R}^2$ and $\mathbb{R}^3$)

3.1   Placing Vertices on Few Lines (π_2^1 and π_3^1)

Combining Corollary 3 with the 4-color theorem yields $\pi_3^1(G) \leq 4$ for planar graphs. Given that outerplanar graphs are 3-colorable (they are partial 2-trees), we obtain $\pi_3^1(G) \leq 3$ for these graphs. These bounds can be improved using the equality $\pi_3^1(G) = \mathrm{lva}(G)$ of Theorem 2 and known results on the linear vertex arboricity:

(a) For any planar graph G, it holds that $\pi_3^1(G) \leq 3$ [21,29].
(b) There is a planar graph G with $\pi_3^1(G) = 3$ [11].
(c) For any outerplanar graph G, $\pi_3^1(G) \leq 2$ [1,6,35].

According to Chen and He [12], the upper bound $\mathrm{lva}(G) \leq 3$ for planar graphs by Poh [29] is constructive and yields a polynomial-time algorithm for partitioning the vertex set of a given planar graph into three parts, each inducing a linear forest. By combining this with the construction given in Theorem 2, we obtain a polynomial-time algorithm that draws a given planar graph such that the vertex set "sits" on three lines.

The example of Chartrand and Kronk [11] is a 21-vertex planar graph whose *vertex arboricity* is 3, which means that the vertex set of this graph cannot even be split into two parts both inducing (not necessarily linear) forests. Raspaud and Wang [30] showed that all 20-vertex planar graphs have vertex arboricity at most 2. We now observe that a smaller example of a planar graph attaining the extremal value $\pi_3^1(G) = 3$ can be found by examining the *linear* vertex arboricity.

Example 16. The planar 9-vertex graph G in Fig. 2 has $\pi_3^1(G) = \mathrm{lva}(G) = 3$. (See a proof in Appendix D.)

Now we show lower bounds for the parameter $\pi_2^1(G)$.

Recall that the *circumference* of a graph G, denoted by $c(G)$, is the length of a longest cycle in G. For a planar graph G, let $\bar{v}(G)$ denote the maximum k such that G has a straight-line plane drawing with k collinear vertices.

Lemma 17. *Let G be a planar graph. Then $\pi_2^1(G) \geq n/\bar{v}(G)$. If G is a triangulation then $\pi_2^1(G) \geq (2n - 4)/c(G^*)$.*

Proof. Since the first claim is obvious, we prove only the second. Let $\gamma(G)$ denote the minimum number of cycles in the dual graph G^* sharing a common vertex and covering every vertex of G^* at least twice. Note that, as G is a triangulation, $\gamma(G) \geq (4n - 8)/c(G^*)$, where $2n - 4$ is the number of vertices in G^* (as a consequence of Euler's formula). We now show $\pi_2^1(G) \geq \gamma(G)/2$, which implies the claimed result.

Given a drawing realizing $\pi_2^1(G)$ with line set $\mathcal{L}$, for every line $\ell \in \mathcal{L}$, draw two parallel lines ℓ', ℓ'' sufficiently close to ℓ such that they together intersect the interiors of all faces touched by ℓ and do not go through any vertex of the drawing. Note that ℓ' and ℓ'' cross boundaries of faces only via inner points of edges. Each such crossing corresponds to a transition from one vertex to another along an edge in the dual graph G^*. Since all the faces of G are triangles, each of them is visited by each of ℓ' and ℓ'' at most once. Therefore, the faces crossed along ℓ' and the faces crossed along ℓ'', among them the outer face of G, each form a cycle in G^*. It remains to note that every face f of the graph G is crossed at least twice, because f is intersected by at least two different lines from $\mathcal{L}$ and each of these two lines has a parallel copy that crosses f. □

An infinite family of triangulations G with $\bar{v}(G) \leq n^{0.99}$ is constructed in [31]. By the first part Lemma 17 this implies that there are infinitely many triangulations G with $\pi_2^1(G) \geq n^{0.01}$. The second part of Lemma 17 along with an estimate of Grünbaum and Walther [22] (that was used also in [31]) yields a stronger result.

Theorem 18. *There are infinitely many triangulations G with $\Delta(G) \leq 12$ and $\pi_2^1(G) \geq n^{0.01}$.*

Proof. The *shortness exponent* $\sigma_{\mathcal{G}}$ of a class $\mathcal{G}$ of graphs is the infimum of the set of the reals $\liminf_{i \to \infty} \log c(H_i)/\log |V(H_i)|$ for all sequences of $H_i \in \mathcal{G}$ such that $|V(H_i)| < |V(H_{i+1})|$. Thus, for each $\epsilon > 0$, there are infinitely many graphs $H \in \mathcal{G}$ with $c(H) < |V(H)|^{\sigma_{\mathcal{G}}+\epsilon}$. The dual graphs of triangulations with maximum vertex degree at most 12 are exactly the cubic 3-connected planar graphs with each face incident to at most 12 edges (this parameter is well defined by the Whitney theorem). Let σ denote the shortness exponent for this class of graphs. It is known [22] that $\sigma \leq \frac{\log 26}{\log 27} = 0.988\ldots$. The theorem follows from this bound by the second part of Lemma 17. □

Problem 19. Does $\pi_2^1(G) = o(n)$ hold for all planar graphs G?

A *track drawing* [18] of a graph is a plane drawing for which there are parallel lines, called *tracks*, such that every edge either lies on a track or its endpoints lie on two consecutive tracks. We call a graph *track drawable* if it has a track drawing. Let $\mathrm{tn}(G)$ be the minimum number of tracks of a track drawing of G. Note that $\pi_2^1(G) \leq \bar{\pi}_2^1(G) \leq \mathrm{tn}(G)$.

The following proposition is similar to a lemma of Bannister et al. [3, Lemma 1] who say it is implicit in the earlier work of Felsner et al. [18].

Theorem 20. (cf. [3,18]). *Let G be a track drawable graph. Then $\pi_2^1(G) \leq 2$.*

Proof. Consider a track drawing of G, which we now transform to a drawing on two intersecting lines. Put the tracks consecutively along a spiral so that they correspond to disjoint intervals on the half-lines as depicted on the right. Tracks whose indices are equal modulo 4 are placed on the same half-line; for more details see Fig. 8 in Appendix D on page 26. (Bannister et al. [3, Fig. 1] use three half-lines meeting in a point.) □

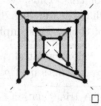

Observe that any tree is track drawable: two vertices are aligned on the same track iff they are at the same distance from an arbitrarily assigned root. Moreover, any outerplanar graph is track drawable [18]. This yields an improvement over the bound $\pi_3^1(G) \leq 2$ for outerplanar graphs stated in the beginning of this section.

Corollary 21. *For any outerplanar graph G, it holds that $\pi_2^1(G) \leq 2$.*

3.2 Placing Edges on Few Lines (ρ_2^1 and ρ_3^1)

The parameter $\rho_2^1(G)$ is related to two parameters introduced by Dujmović et al. [14]. They define a *segment* in a straight-line drawing of a graph G as an inclusion-maximal (connected) path of edges of G lying on a line. A *slope* is an inclusion-maximal set of parallel segments. The *segment number* (resp., *slope number*) of a planar graph G is the minimum possible number of segments (resp., slopes) in a straight-line drawing of G. We denote these parameters by $\mathrm{segm}(G)$ (resp., $\mathrm{slop}(G)$). Note that $\mathrm{slop}(G) \leq \rho_2^1(G) \leq \mathrm{segm}(G)$.

These parameters can be far away from each other. Figure 4 shows a graph with $\mathrm{slop}(G) = O(1)$ and $\rho_2^1(G) = \Omega(n)$ (see the proof of Theorem 24). On the other hand, note that $\rho_2^1(mK_2) = 1$ while $\mathrm{segm}(mK_2) = m$ where mK_2 denotes the graph consisting of m isolated edges. The gap between $\rho_2^1(G)$ and $\mathrm{segm}(G)$ can be large even for connected graphs. It is not hard to see that $\mathrm{segm}(G)$ is bounded from below by half the number of odd degree vertices (see [14] for details). Therefore, if we take a caterpillar G with k vertices of degree 3 and $k+2$ leaves, then $\mathrm{segm}(G) \geq n/2$, while $\rho_2^1(G) = O(\sqrt{n})$ because G can easily be drawn in a square grid of area $O(n)$. Note that, for the same G, the gap between $\mathrm{slop}(G)$ and $\rho_2^1(G)$ is also large. Indeed, $\mathrm{slop}(G) = 2$ while $\rho_2^1(G) > \sqrt{n-2}$ by Lemma 7(a).

It turns out that a large gap between $\rho_2^1(G)$ and $\mathrm{segm}(G)$ can be shown also for 3-connected planar graphs and even for triangulations.

Example 22. There are triangulations with $\rho_2^1(G) = O(\sqrt{n})$ and $\mathrm{segm}(G) = \Omega(n)$.[1] Note that this gap is the best possible because any 3-connected graph G has minimum vertex degree 3 and, hence, $\rho_2^1(G) \geq \rho_3^1(G) > \sqrt{2n}$ by Lemma 7(a). Consider the graph shown in Fig. 3. Its vertices are placed on the standard orthogonal grid and two slanted grids, which implies that at most $O(\sqrt{n})$ lines are involved. The pattern can be completed to a triangulation by adding three vertices around it and connecting them to the vertices on the pattern boundary. Since the pattern boundary contains $O(\sqrt{n})$ vertices, $O(\sqrt{n})$ new lines suffice for this. Thus, we have $\rho_2^1(G) = O(\sqrt{n})$ for the resulting triangulation G. Note that the vertices drawn fat in Fig. 3 have degree 5, and there are linearly many of them. This implies that $\mathrm{segm}(G) = \Omega(n)$.

Somewhat surprisingly, the parameter $\mathrm{segm}(G)$ can be bounded from above by a function of $\rho_2^1(G)$ for all connected graphs.

---

[1] A triangulation G with $\mathrm{segm}(G) = O(\sqrt{n})$ has been found by Dujmović et al. [14, Fig. 12].

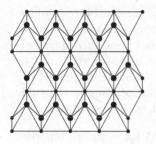
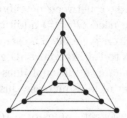
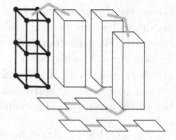

Fig. 3. The main body of a triangulation G with $\rho_2^1(G) = O(\sqrt{n})$ and segm $(G) = \Omega(n)$.

Fig. 4. The nested-triangles graph T_k.

Fig. 5. Sketch of the construction in the proof of Theorem 24(b).

Theorem 23. *For any connected planar graph G, $\mathrm{segm}(G) = O(\rho_2^1(G)^2)$.*

Note that $\Delta(G)/2 \leq \rho_3^1(G) \leq \rho_2^1(G) \leq \mathrm{segm}(G) \leq m$ for any planar graph G. For all inequalities here except the second one, we already know that the gap between the respective pair of parameters can be very large (by considering a caterpillar with linearly many degree 3 vertices and applying Lemma 7(a), by Example 22, and by considering the path graph P_n, for which $\mathrm{segm}(P_n) = 1$). Part (b) of the following theorem shows a large gap also between the parameters $\rho_3^1(G)$ and $\rho_2^1(G)$, that is, some planar graphs can be drawn much more efficiently, with respect to the line cover number, in 3-space than in the plane.

Theorem 24. *(a) There are infinitely many planar graphs with constant maximum degree, constant treewidth, and linear ρ_2^1-value.*
(b) For infinitely many n there is a planar graph G on n vertices with $\rho_2^1(G) = \Omega(n)$ and $\rho_3^1(G) = O(n^{2/3})$.

Proof. Consider the nested-triangles graph $T_k = C_3 \times P_k$ shown in Fig. 4. To prove statements (a) and (b), it suffices to establish the following bounds:

(i) $\rho_2^1(T_k) \geq n/2$ and
(ii) $\rho_3^1(T_k) = O(n^{2/3})$.

To see the linear lower bound (i), note that T_k is 3-connected. Hence, Whitney's theorem implies that, in any plane drawing of T_k, there is a sequence of nested triangles of length at least $k/2$. The sides of the triangles in this sequence must belong to pairwise different lines. Therefore, $\rho_2^1(T_k) \geq 3k/2 = n/2$.

For the sublinear upper bound (ii), first consider the graph $C_4 \times P_k$. We build wireframe rectangular prisms that are stacks of $O(\sqrt[3]{n})$ squares each. These prisms are placed onto the base plane in an $O(\sqrt[3]{n}) \times O(\sqrt[3]{n})$ grid; see Fig. 5. So far we can place the edges on the $O(n^{2/3})$ lines of the 3D cubic grid of volume $O(n)$. Next, we construct a path that traverses all squares by passing through the prisms from top to bottom (resp., vice versa) and connecting neighboring prims. We rotate and move some of the squares at the top (resp., bottom) of the prisms to be able to draw the edges between neighboring prisms according to this path. For this "bending" we need $O(n^{2/3})$ additional lines. In Appendix D we provide a drawing; see Fig. 11 on page 30. The same approach works for the graph $T_k = C_3 \times P_k$. In addition to the standard 3D grid, here we need also its slanted, diagonal version (and, again, additional lines for bending in the cubic box of volume $O(n)$). The number of lines increases just by a constant factor. □

We are able to determine the exact values of $\rho_2^1(G)$ for complete bipartite graphs $K_{p,q}$ that are planar.

Example 25. $\rho_2^1(K_{1,q}) = \lceil m/2 \rceil$ and $\rho_2^1(K_{2,q}) = \lceil (3n-7)/2 \rceil = \lceil (3m-2)/4 \rceil$. See Appendix D for details.   ·

Motivated by Example 25, we ask:

Problem 26. What is the smallest c such that $\rho_2^1(G) \leq (c+o(1))m$ for any planar graph G? Example 25 shows that $c \geq 3/4$. Durocher and Mondal [17], improving on an earlier bound of Dujmović et al. [14], showed that $\text{segm}(G) < \frac{7}{3}n$ for any planar graph G. This implies that $c \leq 7/9$.

For any binary tree T, it holds that $\rho_2^1(T) = O(\sqrt{n \log n})$. This follows from the known fact [8] that T has an orthogonal drawing on a grid of size $O(\sqrt{n \log n}) \times O(\sqrt{n \log n})$. For complete binary trees lower and upper bounds are described in Example 37 in Appendix D.

References

1. Akiyama, J., Era, H., Gervacio, S.V., Watanabe, M.: Path chromatic numbers of graphs. J. Graph Theory **13**(5), 571–573 (1989)
2. Akiyama, J., Exoo, G., Harary, F.: Covering and packing ingraphs III: cyclic and acyclic invariants. Math. Slovaca **30**, 405–417 (1980)
3. Bannister, M.J., Devanny, W.E., Dujmović, V., Eppstein, D., Wood, D.R.: Track layouts, layered path decompositions, and leveled planarity (2015). http://arxiv.org/abs/1506.09145
4. Battista, G.D., Frati, F., Pach, J.: On the queue number of planar graphs. SIAM J. Comput. **42**(6), 2243–2285 (2013)
5. Bollobás, B.: Combinatorics: Set Systems, Hypergraphs, Families of Vectors and Combinatorial Probability, 1st edn. Cambridge University Press, Cambridge (1986)
6. Broere, I., Mynhardt, C.M.: Generalized colorings of outerplanar and planar graphs. In: Proceedings of the 5th International Conference Graph Theory Applications Algorithms and Computer Science, Kalamazoo, MI, pp. 151–161 (1984, 1985)
7. Brooks, R.L.: On colouring the nodes of a network. Math. Proc. Camb. Philos. Soc. **37**, 194–197 (1941)
8. Chan, T.M., Goodrich, M.T., Kosaraju, S.R., Tamassia, R.: Optimizing area and aspect ratio in straight-line orthogonal tree drawings. Comput. Geom. Theory Appl. **23**, 153–162 (2002)

9. Chaplick, S., Fleszar, K., Lipp, F., Ravsky, A., Verbitsky, O., Wolff, A.: The complexity of drawing graphs on few lines and few planes (2016). http://arxiv.org/abs/1607.06444

10. Chaplick, S., Fleszar, K., Lipp, F., Ravsky, A., Verbitsky, O., Wolff, A.: Drawing graphs on few lines and few planes (2016). http://arxiv.org/abs/1607.01196

11. Chartrand, G., Kronk, H.V.: The point-arboricity of planar graphs. J. Lond. Math. Soc. **44**, 612–616 (1969)

12. Chen, Z., He, X.: Parallel complexity of partitioning a planar graph into vertex-induced forests. Discrete Appl. Math. **69**(1–2), 183–198 (1996)

13. Dujmović, V.: Graph layouts via layered separators. J. Comb. Theory Ser. B **110**, 79–89 (2015)

14. Dujmović, V., Eppstein, D., Suderman, M., Wood, D.R.: Drawings of planar graphs with few slopes and segments. Comput. Geom. Theory Appl. **38**(3), 194–212 (2007)

15. Dujmović, V., Morin, P., Wood, D.R.: Layout of graphs with bounded tree-width. SIAM J. Comput. **34**(3), 553–579 (2005)

16. Dujmović, V., Whitesides, S.: Three-dimensional drawings. In: Tamassia, R. (ed.) Handbook of Graph Drawing and Visualization, chap. 14, pp. 455–488. CRC Press, Boca Raton (2013)

17. Durocher, S., Mondal, D.: Drawing plane triangulations with few segments. In: Proceedings of the Canadian Conference Computational Geometry (CCCG 2014), pp. 40–45 (2014). http://cccg.ca/proceedings/2014/papers/paper06.pdf

18. Felsner, S., Liotta, G., Wismath, S.: Straight-line drawings on restricted integer grids in two and three dimensions. J. Graph Algorithms Appl. **7**(4), 363–398 (2003)

19. Flum, J., Grohe, M.: Parametrized Complexity Theory. Springer, Berlin (2006)

20. de Fraysseix, H., Pach, J., Pollack, R.: How to draw a planar graph on a grid. Combinatorica **10**(1), 41–51 (1990)

21. Goddard, W.: Acyclic colorings of planar graphs. Discrete Math. **91**(1), 91–94 (1991)

22. Grünbaum, B., Walther, H.: Shortness exponents of families of graphs. J. Comb. Theory Ser. A **14**(3), 364–385 (1973)

23. Hanani, H.: The existence and construction of balanced incomplete block designs. Ann. Math. Stat. **32**, 361–386 (1961). http://www.jstor.org/stable/2237750

24. Harary, F.: Covering and packing in graphs I. Ann. N.Y. Acad. Sci. **175**, 198–205 (1970)

25. Kostochka, A., Melnikov, L.: On a lower bound for the isoperimetric number of cubic graphs. In: Proceedings of the 3rd International Petrozavodsk Conference Probabilistic Methods in Discrete Mathematics, pp. 251–265. TVP, Moskva, VSP, Utrecht (1993)

26. Krein, M., Milman, D.: On extreme points of regular convex sets. Studia Math. **9**, 133–138 (1940)

27. Matsumoto, M.: Bounds for the vertex linear arboricity. J. Graph Theory **14**(1), 117–126 (1990)

28. Pach, J., Thiele, T., Tóth, G.: Three-dimensional grid drawings of graphs. In: DiBattista, G. (ed.) GD 1997. LNCS, vol. 1353, pp. 47–51. Springer, Heidelberg (1997). doi:10.1007/3-540-63938-1_49

29. Poh, K.S.: On the linear vertex-arboricity of a planar graph. J. Graph Theory **14**(1), 73–75 (1990)

30. Raspaud, A., Wang, W.: On the vertex-arboricity of planar graphs. Eur. J. Comb. **29**(4), 1064–1075 (2008)

31. Ravsky, A., Verbitsky, O.: On collinear sets in straight-line drawings. In: Kolman, P., Kratochvíl, J. (eds.) WG 2011. LNCS, vol. 6986, pp. 295–306. Springer, Heidelberg (2011). doi:10.1007/978-3-642-25870-1_27. https://arXiv.org/abs/0806.0253

32. Schaefer, M., Štefankovič, D.: Fixed points, Nash equilibria, and the existential theory of the reals. Theory Comput. Syst., 1–22 (2015, has appeared online)

33. Schnyder, W.: Embedding planar graphs on the grid. In: Proceedings of the 1st ACM-SIAM Symposium Discrete Algorithms (SODA 1990), pp. 138–148 (1990)

34. Schulz, A.: Drawing graphs with few arcs. J. Graph Algorithms Appl. **19**(1), 393–412 (2015)

35. Wang, J.: On point-linear arboricity of planar graphs. Discrete Math. **72**(1–3), 381–384 (1988)

36. Wikipedia: Krein-Milman theorem. https://en.wikipedia.org/wiki/Krein-Milman_theorem. Accessed 21 Apr 2016

37. Wood, D.R.: Three-dimensional graph drawing. In: Kao, M.Y. (ed.) Encyclopedia of Algorithms, pp. 1–7. Springer, Boston (2008)

Layered and Tree Drawings

Algorithms for Visualizing Phylogenetic Networks

Ioannis G. Tollis[1] and Konstantinos G. Kakoulis[2]($\boxtimes$)

[1] Department of Computer Science, University of Crete, Heraklion, Greece
tollis@csd.uoc.gr
[2] Department of Mechanical and Industrial Design Engineering,
T.E.I. of West Macedonia, Kozani, Greece
kkakoulis@teiwm.gr

Abstract. We study the problem of visualizing phylogenetic networks, which are extensions of the *Tree of Life* in biology. We use a space filling visualization method, called DAGmaps, in order to obtain clear visualizations using limited space. In this paper, we restrict our attention to galled trees and galled networks and present linear time algorithms for visualizing them as DAGmaps.

1 Introduction

The quest of the *Tree of Life* arose centuries ago, and one of the first illustrations of an evolutionary tree was produced by Charles Darwin in 1859, in his book *"The Origin of Species"*. Over a century later, evolutionary biologists still used *phylogenetic trees* to depict evolution. A *phylogenetic tree T on X* is obtained by labeling the leaves of a tree by the set of taxa $X = \{x_1, x_2, \ldots, x_n\}$. Each taxon x_i represents a species or an organism.

The branches of the phylogenetic trees represent the evolution of species, and sometimes the length of their edges is scaled in order to represent the time.

As pointed out in [4], molecular phylogeneticists were failing to find the true tree of life, not because their methods were inadequate or because they had chosen the wrong genes, but perhaps because the history of life cannot be properly represented as a tree. Indeed, the mechanisms of horizontal gene transfer, hybridization and genetic recombination necessitate the use of *phylogenetic* network models to illustrate them.

There are many different types of phylogenetic networks which can be separated in two main classes according to [8]: *implicit* phylogenetic networks that provide tools to visualize and analyze incompatible phylogenetic signals, such as split networks [7], and *explicit* phylogenetic networks that provide explicit scenarios of reticulate evolution, such as hybridization networks [16,17], horizontal gene transfer networks [6] and recombination networks [5,10].

Visualization of phylogenetic trees and networks is an important part of this area, since most of these graphs are huge. Furthermore, the usual node-link representation leads to visual clutter. Thus, alternative visualization of phylogenetic trees, such as treemaps, may be preferable.

© Springer International Publishing AG 2016
Y. Hu and M. Nöllenburg (Eds.): GD 2016, LNCS 9801, pp. 183–195, 2016.
DOI: 10.1007/978-3-319-50106-2_15

Treemaps [14], a space filling technique for visualizing large hierarchical data sets, display trees as a set of nested rectangles. The (root of the) tree is the initial rectangle. Each subtree is assigned to a subrectangle, which is then tiled into smaller rectangles representing further subtrees. Space filling visualizations, such as treemaps, have the capacity to display thousands of items legibly in limited space via a two dimensional map. Treemaps have been used in bioinformatics to visualize phylogenetic trees [1], gene expression data [18], gene ontologies [2,20,21], and the Encyclopedia of Life [1]. An extension of treemaps is presented in [22], which manages to visualize not only trees, but also Directed Acyclic Graphs (DAGs). As shown in [22], it is not always possible to visualize a DAG with a DAGmap without having node duplications.

In this paper we present space filling techniques that use DAGmap drawings for the visualization of two categories of phylogenetic networks, galled trees and planar galled networks. No node duplications appear in both visualization algorithms that we present. In Sect. 2 we introduce an algorithm which locates the galls of a graph and examines whether this graph is a galled tree or a galled network. In Sect. 3 we describe how to draw the DAGmaps of galled trees, and we examine whether the galled trees and galled networks can be one-dimensionally DAGmap drawn. Finally, in Sect. 4 we present an algorithm for producing DAGmap drawings of planar galled networks.

2   Preliminaries

Let $G = (V, E)$ be a directed graph (digraph) with $n = |V|$ nodes and $m = |V|$ edges. If $e = (u, v) \in E$ is a directed edge, we say that e is incident from u (or outgoing from u) and incident to v (or incoming to v); edge u is the origin of e and node v is the destination of e. A directed acyclic graph (DAG) is a digraph that contains no cycles. A source of digraph G is a node without incoming edges. A sink of G is a node without outgoing edges. An internal node of G has both incoming and outgoing edges.

A drawing of a graph G maps each node v to a distinct point of the plane and each edge (u, v) to a simple open Jordan curve, with endpoints u and v. A drawing is planar if no two edges intersect except, possibly, at common endpoints. A graph is planar if it admits a planar drawing. Two planar drawings of a graph are equivalent if they determine the same circular ordering of the edges around each node. An equivalence class of planar drawings is a (combinatorial) embedding of G. An embedded graph is a graph with a specified embedding. A planar drawing partitions the plane into topologically connected regions that are called faces.

An upward drawing of a digraph is such that all the edges are represented by directed curves increasing monotonically in the vertical direction. A digraph has an upward drawing if and only if it is acyclic. A digraph is upward planar if it admits a planar upward drawing. Note that a planar acyclic digraph does not necessarily have a planar upward drawing. A graph is layered planar if it can be drawn such that the nodes are placed in horizontal rows or layers, the edges are drawn as polygonal chains connecting their end nodes, and there are no edge crossings.

In a phylogenetic network there can be three kind of nodes: *root*, *tree*, and *reticulation* nodes. A root node has no incoming edges. There is only one root node in every rooted phylogenetic network. Tree nodes have exactly one ancestor. Reticulation nodes have more than one ancestors. It is easy to realize that a phylogenetic tree is a phylogenetic network without reticulation nodes.

In addition, there can be two kind of edges: *tree*, and *reticulation* edges. A tree edge leads to a node that has exactly one incoming edge. A reticulation edge leads to a node that has more than one incoming edges.

Reticulation cycles are defined as follows. Since there is only one root node in every rooted phylogenetic network, in the corresponding undirected graph every reticulation node belongs to a cycle. This cycle, in the directed graph, is called reticulation cycle.

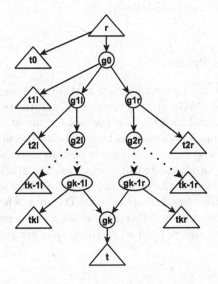

Fig. 1. The structure of a gall.

A *gall* is a reticulation cycle in a phylogenetic network that shares no nodes with any other reticulation cycle. It consists of a beginning node g_0, two chains (the left and the right one) and a reticulation node g_k, as shown in Fig. 1. The beginning node g_0 is on level 1 of this subgraph, the reticulation node on level $k + 1$, and the chain nodes are on the i levels, $i \in \{2, \ldots, k\}$. Every level i may contain either one or two chain nodes. Every node g_i, $i \in \{0, \ldots, k\}$, of the gall may have a subtree t_{i+1} as a descendant. These subtrees do not have more connections with this gall, because in that case a reticulation cycle would be created, which would share a node with the gall, and this is not allowed according to the definition of a gall.

A *galled* tree is a phylogenetic network whose reticulation cycles are galls [5,23]. This is called the *galled tree condition*. Considering the definition of a gall, it is easy to realize that the reticulation nodes of a galled tree have indegree two.

A *Galled* network is a rooted phylogenetic network in which every reticulation cycle shares no reticulation nodes with any other reticulation cycle [9]. This is called the *galled network condition*.

In contrast to the galled trees, galled networks allow the reticulation cycles to share nodes, as long as they are not reticulation nodes. These reticulation cycles are called *loose galls*. In the rest of the paper, whenever we refer to loose galls of a galled network, we will use the term *galls* for simplification.

Galled trees [5,12,19,23] and galled networks [8,9,11,13] have received much attention in recent years. They are important types of phylogenetic networks due to their biological significance and their simple, almost treelike, structure. A galled tree or network may suffice to accurately describe an evolutionary process when the number of recombination events is limited and most of them have occurred recently [5].

2.1 The DAGmap Problem

DAGmaps are space filling visualizations of DAGs that generalize treemaps [22]. The main properties of DAGmaps are shown in Fig. 2. In treemaps the rectangle of a child node is included into the rectangle of its parent node (see Fig. 2(a)). In DAGmaps the rectangle of a node is included into the union of rectangles of its ancestors. Also the rectangle of an edge is contained in the intersection of the rectangles of its source and destination nodes (see Fig. 2(b)).

The DAGmap problem is the problem of deciding whether a graph admits a DAGmap drawing without node duplications. Deciding whether or not a DAG admits a DAGmap drawing is NP-complete [22]. Furthermore, the DAGmap problem remains NP-complete even when the graphs are restricted to be galled networks:

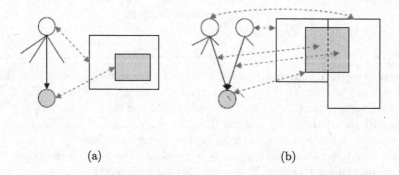

(a)   (b)

Fig. 2. (a) A treemap drawing. (b) A DAGmap drawing.

Theorem 1. *The DAGmap problem for galled networks is NP-complete.*

Proof. Omitted due to space limitations.   □

2.2   Locating the Galls

The first task is to recognize whether a given phylogenetic network is a galled tree or a galled network. Since they both contain galls, we will need to locate the galls of the given phylogenetic network. This will allow us to check whether our network is a galled tree, a galled network, or none of them. This can be accomplished by the following algorithm:

Algorithm 1. Locating the galls of a graph
Input: A Graph G.
Output: The set of galls and the characterization of G as a galled tree or a galled network, or null if the graph is neither of them.

```
1. Perform a simple graph traversal in order to locate the
   reticulation nodes.
2. If a node with more than two incoming edges is found, then
   return null.
3. For every reticulation node find its two parents. Each of these
   parents belongs to a chain of the gall.
4.  For every parent find its parent and assign it to the same
    chain. (At each step discover one node from each chain.)
5.  Continue this process until a node is found which already
    belongs to the other chain. This is the beginning node of the
    gall. If no such node is found, return null.
6. After locating all the galls, test the galled tree and the
   galled network condition.
7. If the galled tree condition holds then characterize the graph
   as a"galled tree".
8. Else if the galled network condition holds then characterize
   the graph as a "galled network".
9. Else return null.
10.Return the located galls.
```

This process will discover all the galls of the graph, since every reticulation node corresponds to exactly one gall. In addition, every chain node will be visited a constant number of times if we use a hash table to store the chain nodes. Also, the property that every gall has exactly one reticulation node guarantees that this algorithm will neither leave any gall undiscovered, nor claim to discover a gall that does not exist. Thus, it is straightforward to show that Algorithm 1 runs in $O(n + m)$ time.

3   DAGmaps for Galled Trees

In this section we present techniques for drawing galled trees as DAGmaps.

3.1   Drawing Galled Trees as DAGmaps

Next, we present a three step algorithm for drawing galled trees as DAGmaps. First, we transform the input galled tree into a tree by collapsing the two chains of each gall into a single chain. Then, we use treemap techniques to draw the tree. Finally, we expand the collapsed galls. Next, we make some interesting observations:

Fact 1. *Any node of a galled tree has indegree at most two.*

If there were a node with indegree more than two in a galled tree, then this node would belong to more than one reticulation cycles, which means that there would be (more than one) reticulation cycles.

Fact 2. *Every galled tree is planar.*

This is easy to realize considering that galled trees are almost like trees, but with some branches being made of two parallel chains, instead of one (see Fig. 3). Furthermore, this implies that the number of edges of a galled tree is $O(n)$.

We now present an algorithm for constructing a DAGmap of a galled tree:

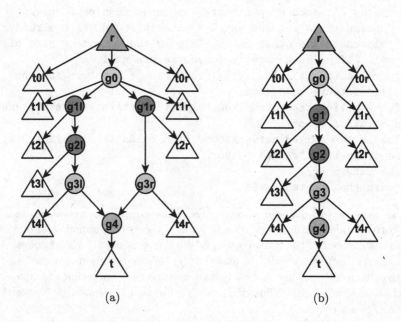

(a)   (b)

Fig. 3. Transformation of a galled tree (a) into a tree (b).

Algorithm 2. DAGmap drawing of galled trees

Input: A galled tree G.

Output: A DAGmap drawing of G.

1. Transform the galled tree G into a tree T, by unifying the two chains of each gall.
2. Draw the treemap of T, according to the chosen treemap technique.
3. Split the rectangles, corresponding to the nodes of the unified chains of the galls, to obtain the initial parallel chains.

Step 1 of the above algorithm is illustrated in Fig. 3. The parallel chains have been united, and nodes g_{i_l}, g_{i_r} have been replaced by node g_i while the subtrees t_{i_l} and t_{i_r} remain unchanged, $i \in \{1, \ldots, k\}$.

The treemap of T, in Step 2, is drawn under the constraint that the g_i nodes (which represent the union of nodes g_{i_l} and g_{i_r} of the DAG) must always touch both t_{i_l} and t_{i_r}, in the same direction. This means that if we choose to place g_{i_l} on the left and g_{i_r} on the right, where $i \in \{1, \ldots, k-1\}$, then we will follow this convention for every $i \in \{1, \ldots, k-1\}$ (see Fig. 4(a)). Drawing the treemap of T needs $O(n)$ time, if we choose a linear time layout algorithm like the slice and dice layout.

Slice and dice [14] is a treemap drawing technique, where the initial rectangle is recursively divided. The direction of each subdivision changes in each level, from horizontal to vertical.

The output of Step 3 is shown in Fig. 4(b), where the unified nodes are split. Note that the reticulation node g_k lies on both g_{k-1_l} and g_{k-1_r}. This step needs $O(n)$ time, because in the worst case it traverses all the nodes of the graph.

From the above we conclude that:

Theorem 2. *Every galled tree admits a DAGmap drawing, which can be computed in $O(n)$ time.*

In the next section we show that galled trees can be drawn as one-dimensional DAGmaps.

3.2   Drawing Galled Trees as One-Dimensional DAGmaps

A DAGmap is called one-dimensional if the initial rectangle is sliced only along the vertical (horizontal) direction. Since the height (width) of all the rectangles is constant and equal to the height (width) of the initial drawing rectangle, the problem is one-dimensional.

Next, we show that galled trees can be drawn as one-dimensional DAGmaps.

Theorem 3. *Every galled tree can be drawn as a one-dimensional DAGmap.*

Sketch of Proof. Let $G = (V, E)$ be a proper layered DAG with vertex partition $V = L_1 \cup L_2 \cup \ldots \cup L_h$, where $h > 1$, such that the source (root) is in L_h

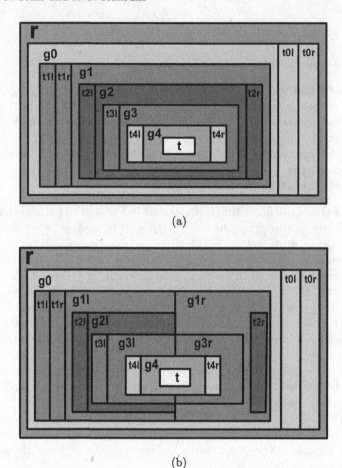

Fig. 4. (a) The treemap drawing of the tree shown in Fig. 3(b). (b) The DAGmap drawing of the gall shown in Fig. 3(a).

and the sinks are in L_1. Tsiaras *et al.* [22] have shown that a DAG G admits a one-dimensional DAGmap if and only if it is layered planar. We will show that every galled tree is layered planar, using its tree-like structure.

We transform the galled tree G into a tree T, as shown in Fig. 3. We take the vertex partition of T: $V_T = L_1 \cup L_2 \cup \ldots \cup L_h$, where $h > 1$, such that the source (root) is in L_h and the sinks are in L_1. Then, we define the vertex partition of the galled tree $V_G = L_1 \cup L_2 \cup \ldots \cup L_h$, where $h > 1$, such that every node of T which also belongs to G remains at the same layer. Moreover, for every node g_i of T which belongs to layer L_j of the partition, and is originated from the union of the nodes g_{i_l} and g_{i_r} of G, it is concluded that g_{i_l} and g_{i_r} will belong to layer L_j of the partition V_G.

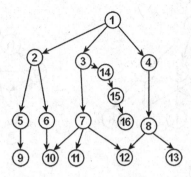

Fig. 5. An example of a galled network that does not admit a one-dimensional DAGmap.

Since every tree is layered planar and we obtained the vertex partition of G from the vertex partition of T, we conclude that every galled tree admits a one-dimensional DAGmap.   □

However, not every planar galled network admits a one-dimensional DAGmap.

Lemma 1. *Not every planar galled network admits a one-dimensional DAGmap.*

Sketch of Proof. In Fig. 5 an example of such a planar galled network is shown, that does not admit a one-dimensional DAGmap. Node 16 will not be able to be drawn in the line of level 4 without edge crossings. However, as it will be shown in the next section, this graph can be DAGmap drawn.   □

4   DAGmaps for Galled Networks

In this section we investigate how to draw galled networks as DAGmaps. From Theorem 1 we have that this problem is NP-complete. Therefore, it is worth examining the problem of drawing planar galled networks as DAGmaps. In the following lemma we show that planar galled networks are a subset of the set of galled networks.

Lemma 2. *Not every galled network is planar.*

Sketch of Proof. This lemma can be proved by creating a family of galled networks that contain a subgraph homeomorphic to K_5 [15]. Figure 6(a) depicts a Galled network. This is a non planar galled network since it is topologically the same as the network shown in Fig. 6 (b), which is homeomorphic to K_5.   □

Since planar galled networks represent phylogenetic networks, it is clear that all edges flow in the same direction monotonically. This means that planar galled networks are upward (downward) planar graphs. Therefore, we have the following:

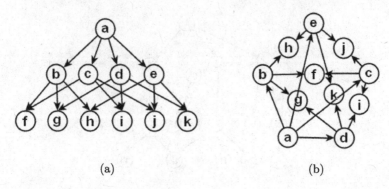

(a)   (b)

Fig. 6. An example of a non planar galled network. As we can see the network (a) is the same with the network (b), which is topologically equivalent to K_5.

Fact 3. *Planar galled networks are upward planar.*

By definition, the phylogenetic networks are single source directed acyclic graphs. Therefore, we have the following:

Fact 4. *Each planar galled network is a single source upward planar directed acyclic graph.*

In order to draw planar galled networks as DAGmaps, without node duplication, we will relax the rule for drawing DAGmaps, which states that every node is drawn as a rectangle. Specifically, we will allow nodes to be drawn as rectilinear cohesive polygons. Next, we present an algorithm that produces DAGmaps of planar galled networks.

Algorithm 3. DAGmap drawing of planar galled networks
Input: A planar galled network G.
Output: A DAGmap drawing of G.

1. Transform the galled network G into a galled tree GT, by splitting the nodes that belong to more than one galls, so as no gall shares its nodes with other galls.
2. Order all subtrees of GT such that:
3.   The nodes created by the splitting of nodes of G are moved so that they are adjacent to each other.
4. Draw the DAGmaps of the galls of GT. Nested galls are drawn recursively.
5.   Unify the split nodes and remove unused space.

Step 1 of the above algorithm is illustrated in Fig. 7. As shown, every node u that participates in k galls ($k > 1$) is being replaced by k nodes u_i, $i \in \{1, \ldots, k\}$. Each node u_i participates in only one gall. Consequently GT is a galled tree because there is no gall that shares nodes with any other gall. This step needs $O(n)$ time, because in the worst case it traverses all the nodes of the graph, and the number of edges of a planar graph is $O(n)$.

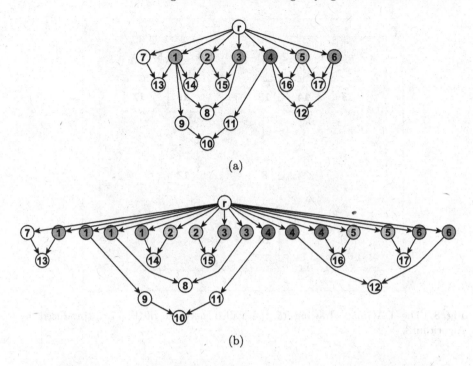

Fig. 7. Transformation of a galled network (a) to a galled tree (b).

In Steps 2 and 3 we define the order of all subtrees of the galled tree GT. The goal is to find an ordering such that all splitted nodes are neighbors. We observe that a proper nesting of the galls produces a planar embedding of G. Thus, given a planar embedding Γ of G, it is easy to find the correct order of all subtrees. Specifically, the order of the subtrees of GT is determined by the clockwise order of the incoming and outgoing edges of each node (to be splitted) in Γ. Bertolazzi *et al.* [3] have shown that if a single source digraph is upward planar, then its drawing can be constructed in $O(n)$ time. Thus, given Fact 4, we can produce an upward planar drawing of a planar galled network in linear time.

The drawings of the DAGmaps of the galls of GT (Step 4) are obtained by executing Algorithm 2. The running time of this algorithm is $O(n)$. Finally, the unification of Step 5 needs $O(n)$ time in the worst case, since it is the reverse procedure of Step 1. The output is shown in Fig. 8.

Generally speaking, the node splitting process triggers the duplication of all of its out-neighbors. Therefore, the transformation of a DAG into a tree leads to trees with (potentially exponentially) many more nodes than the original DAG. However, the node splitting of Step 1 does not have the exponential effects of the ordinary node duplication, since all the duplicated nodes of this case are neighbors. From the above, we realise that Algorithm 4 takes $O(n)$ time, and combining this with Algorithm 3, we conclude that:

Theorem 4. *Every planar galled network admits a DAGmap drawing, which can be computed in $O(n)$ time.*

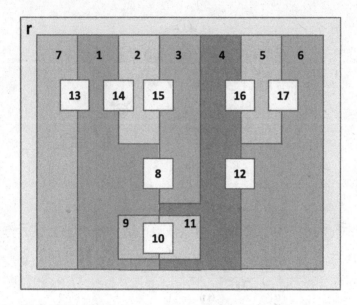

Fig. 8. The DAGmap drawing of the galled network of Fig. 7(a) produced by Algorithm 3.

5   Conclusions and Future Work

DAGmaps, an extension of Treemaps, represent an effective space filling visualization method to display and analyze hierarchical data. In this paper we have presented algorithms that use DAGmap drawings for the visualization of two categories of phylogenetic networks, galled trees and planar galled networks. Future work will cover the study of more categories of phylogenetic networks, in addition to answering the question whether one could minimize the number of node duplications performed during Step 1 of Algorithm 3 in the case of non planar galled networks. Furthermore, we intend to develop a visualization tool for processing phylogenetic networks and displaying them as DAGmaps.

Acknowledgments. We thank Irini Koutaki-Pantermaki who contributed some ideas in an earlier version of the paper, and Vassilis Tsiaras for useful discussions.

References

1. Arvelakis, A., Reczko, M., Stamatakis, A., Symeonidis, A., Tollis, I.G.: Using treemaps to visualize phylogenetic trees. In: Oliveira, J.L., Maojo, V., Martín-Sánchez, F., Pereira, A.S. (eds.) ISBMDA 2005. LNCS, vol. 3745, pp. 283–293. Springer, Heidelberg (2005). doi:10.1007/11573067_29
2. Baehrecke, E.H., Dang, N., Babaria, K., Shneiderman, B.: Visualization and analysis of microarray and gene ontology data with treemaps. BMC Bioinform. **5**(1), 1–12 (2004)

3. Bertolazzi, P., Di Battista, G., Mannino, C., Tamassia, R.: Optimal upward planarity testing of single-source digraphs. SIAM J. Comput. **27**(1), 132–169 (1998)
4. Doolittle, W.F.: Phylogenetic classification and the universal tree. Science **284**(5423), 2124–2129 (1999)
5. Gusfield, D., Eddhu, S., Langley, C.H.: Efficient reconstruction of phylogenetic networks with constrained recombination. In: Computational Systems Bioinformatics Conference (Proceedings of the CSB 2003), pp. 363–374 (2003)
6. Hallett, M.T., Lagergren, J., Tofigh, A.: Simultaneous identification of duplications and lateral transfers. In: 8th Annual International Conference on Research in Computational Molecular Biology (Proceedings of the RECOMB 2004), pp. 347–356. Springer (2004)
7. Huson, D.H., Bryant, D.: Application of phylogenetic networks in evolutionary studies. Mol. Biol. Evol. **23**(2), 254–267 (2006)
8. Huson, D.H., Klöpper, T.H.: Beyond galled trees - decomposition and computation of galled networks. In: Speed, T., Huang, H. (eds.) RECOMB 2007. LNCS, vol. 4453, pp. 211–225. Springer, Heidelberg (2007). doi:10.1007/978-3-540-71681-5_15
9. Huson, D.H., Rupp, R., Berry, V., Gambette, P., Paul, C.: Computing galled networks from real data. Bioinformatics **25**(12), i85–i93 (2009)
10. Huson, D.H., Rupp, R., Scornavacca, C.: Phylogenetic Networks: Concepts, Algorithms and Applications. Cambridge University Press, Cambridge (2011)
11. Huson, D.H., Scornavacca, C.: A survey of combinatorial methods for phylogenetic networks. Genome Biol. Evol. **3**, 23–35 (2011)
12. Jansson, J., Lingas, A.: Computing the rooted triplet distance between galled trees by counting triangles. J. Discrete Algorithms **25**, 66–78 (2014)
13. Jansson, J., Nguyen, N.B., Sung, W.K.: Algorithms for combining rooted triplets into a galled phylogenetic network. SIAM J. Comput. **35**(5), 1098–1121 (2006)
14. Johnson, B., Shneiderman, B., Tree-Maps : a space-filling approach to the visualization of hierarchical information structures. In: 2nd Conference on Visualization (Proceedings of the VIS 1991), pp. 284–291. IEEE Computer Society Press (1991)
15. Kuratowski, C.: Sur le problème des courbes gauches en topologie. Fundamenta Mathematica **16**, 271–283 (1930)
16. Linder, R.C., Rieseberg, L.H.: Reconstructing patterns of reticulate evolution in plants. Am. J. Bot. **91**, 1700–1708 (2004)
17. Maddison, W.P.: Gene trees in species trees. Syst. Biol. **46**(3), 523–536 (1997)
18. McConnell, P., Johnson, K., Lin, S.: Applications of tree-maps to hierarchical biological data. Bioinformatics **18**(9), 1278–1279 (2002)
19. Nakhleh, L., Warnow, T., Linder, R.C., Reconstructing reticulate evolution in species: theory and practice. In: 8th Annual International Conference on Research in Computational Molecular Biology (Proceedings of the RECOMB 2004), pp. 337–346 (2004)
20. Symeonidis, A., Tollis, I.G., Reczko, M.: Visualization of functional aspects of microRNA regulatory networks using the gene ontology. In: Maglaveras, N., Chouvarda, I., Koutkias, V., Brause, R. (eds.) ISBMDA 2006. LNCS, vol. 4345, pp. 13–24. Springer, Heidelberg (2006). doi:10.1007/11946465_2
21. Tao, Y., Liu, Y., Friedman, C., Lussier, Y.A.: Information visualization techniques in bioinformatics during the postgenomic era. Drug Discov. Today BIOSILICO **2**(6), 237–245 (2004)
22. Tsiaras, V., Triantafilou, S., Tollis, I.G.: DAGmaps: space filling visualization of directed acyclic graphs. Graph Algorithms Appl. **13**(3), 319–347 (2009)
23. Wang, L., Zhang, K., Zhang, L.: Perfect phylogenetic networks with recombination. J. Comput. Biol. **8**(1), 69–78 (2001)

A Generalization of the Directed Graph Layering Problem

Ulf Rüegg[1]([✉]), Thorsten Ehlers[1], Miro Spönemann[2],
and Reinhard von Hanxleden[1]

[1] Department of Computer Science, Kiel University, Kiel, Germany
{uru,the,rvh}@informatik.uni-kiel.de
[2] TypeFox GmbH, Kiel, Germany
miro.spoenemann@typefox.io

Abstract. The Directed Layering Problem (DLP) solves a step of the widely used layer-based approach to automatically draw directed acyclic graphs. To cater for cyclic graphs, usually a preprocessing step is used that solves the Feedback Arc Set Problem (FASP) to make the graph acyclic before a layering is determined.

Here we present the Generalized Layering Problem (GLP), which solves the combination of DLP and FASP simultaneously, allowing general graphs as input. We present an integer programming model and a heuristic to solve the NP-completeGLP and perform thorough evaluations on different sets of graphs and with different implementations for the steps of the layer-based approach.

We observe that GLP reduces the number of dummy nodes significantly, can produce more compact drawings, and improves on graphs where DLP yields poor aspect ratios.

Keywords: Layer-based layout · Layer assignment · Linear arrangement · Feedback arc set · Integer programming

1 Introduction

The layer-based approach is a well-established and widely used method to automatically draw directed graphs. It is based on the idea to assign nodes to subsequent *layers* that show the inherent direction of the graph, see Fig. 1a for an example. The approach was introduced by Sugiyama et al. [20] and remains a subject of ongoing research.

Given a directed graph, the layer-based approach was originally defined for acyclic graphs as a pipeline of three phases. However, two additional phases are necessary to allow practical usage, which are marked with asterisks:

1. *Cycle removal**: Eliminate all cycles by reversing a preferably small subset of the graph's edges. This phase adds support for cyclic graphs as input.
2. *Layer assignment*: Assign all nodes to numbered *layers* such that edges point
   . from layers of lower index to layers of higher index. Edges connecting nodes that are not on consecutive layers are split by so-called *dummy nodes*.

© Springer International Publishing AG 2016
Y. Hu and M. Nöllenburg (Eds.): GD 2016, LNCS 9801, pp. 196–208, 2016.
DOI: 10.1007/978-3-319-50106-2_16

3. *Crossing reduction:* Find an ordering of the nodes within each layer such that the number of crossings is minimized.
4. *Coordinate assignment:* Determine explicit node coordinates with the goal to minimize the distance of edge endpoints.
5. *Edge routing*:* Compute bend points for edges, e.g. with an orthogonal style.

While state-of-the-art methods produce drawings that are often satisfying, there are graph instances where the results show bad *compactness* and unfavorable *aspect ratio* [8]. In particular, the number of layers is bound from below by the longest path of the input graph after the first phase. When placing the layers vertically one above the other, this affects the height of the drawing, see Fig. 1a. Following these observations, we present new methods to overcome current limitations.

Contributions. The focus of this paper is on the first two phases stated above. They determine the initial topology of the drawing and thus directly impact the compactness and the aspect ratio of the drawing.

We introduce a new layer assignment method which is able to handle cyclic graphs and to consider compactness properties for selecting an edge reversal set. Specifically, (1) it can overcome the previously mentioned lower bound on the number of layers arising from the longest path of a graph, (2) it can be flexibly configured to either favor elongated or narrow drawings, thus improving on aspect ratio, and (3) compared to previous methods it is able to reduce both the number of dummy nodes and reversed edges for certain graphs. See Figs. 1 and 2 for examples.

We discuss how to solve the new method to optimality using an integer programming model as well as heuristically, and evaluate both.

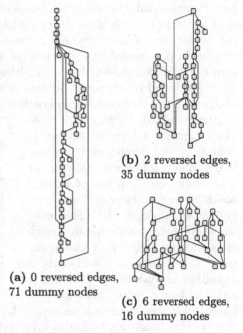

(b) 2 reversed edges, 35 dummy nodes

(a) 0 reversed edges, 71 dummy nodes

(c) 6 reversed edges, 16 dummy nodes

Fig. 1. Different drawings of the g.39.29 graph from the North graphs collection [3]. (a) is drawn with known methods [7], (b) and (c) are results of the methods presented here. Backward edges are drawn bold and dashed.

Outline. The next section presents related work. We introduce problems and definitions in Sect. 3, and present methods to solve the newly introduced problems in Sects. 4 and 5. Section 6 discusses thorough evaluations before we conclude in Sect. 7.

2 Related Work

The cycle removal phase targets the NP-complete Feedback Arc Set Problem (FASP). Several approaches have been proposed to solve FASP either to optimality or heuristically [11]. In the context of layered graph drawing, reversing a minimal number of edges does not necessarily yield the best results, and application-inherent information might make certain edges better candidates to be reversed [7]. Moreover, the decision which edges to reverse in order to make a graph acyclic has a big impact on the results of the subsequent layering phase. Nevertheless the two phases are executed separately until today.

To solve the second phase, i.e. the layer assignment problem, several approaches with different optimization goals have emerged. Eades and Sugyiama employ a longest path layering, which requires linear time, and the resulting number of layers equals the number of nodes of the graph's longest path [6]. Gansner et al. solve the layering phase by minimizing the sum of the edge lengths regarding the number of necessary dummy nodes [7]. They show that the problem is solvable in polynomial time and present a network simplex algorithm which in turn is not proven to be polynomial, although it runs fast in practice. This approach was found to inherently produce compact drawings and performed best in comparison to other layering approaches [10].

Healy and Nikolov tackle the problem of finding a layering subject to bounds on the number of layers and the maximum number of nodes in any layer with consideration of dummy nodes using an integer linear programming approach [10]. The problem is NP-hard, even without considering dummy nodes. In a subsequent paper they present a branch-and-cut algorithm to solve the problem faster and for larger graph instances [9]. Later, Nikolov et al. propose and evaluate several heuristics to find a layering with a restricted number of nodes in each layer [14]. Nachmanson et al. present an iterative algorithm to produce drawings with an aspect ratio close to a previously specified value [13].

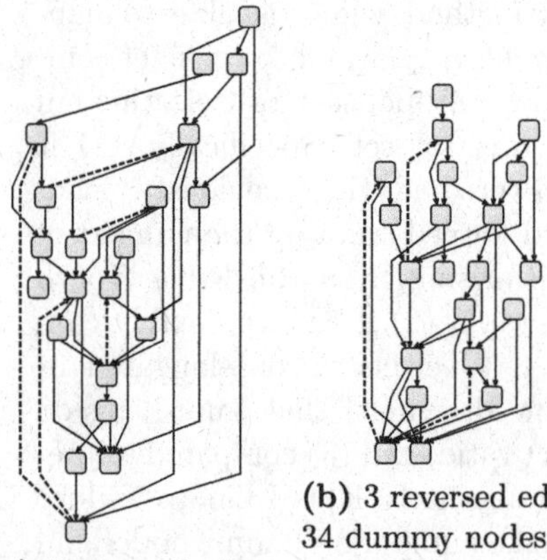

All of the previously mentioned layering methods have two major

(b) 3 reversed edges, 34 dummy nodes

(a) 5 reversed edges, 55 dummy nodes

Fig. 2. A graph drawn with (a) EaGa (known methods as described in Sect. 2) and (b) 1-30-GLP (this work). This example illustrates that GLP-IP can perform better in both metrics: reversed edges (dashed) and dummy nodes.

drawbacks. (1) They require the input graph to be acyclic upfront, and (2) they are bound to a minimum number of layers equal to the longest path of the graph. In particular this means that the bound on the number of layers in the methods of Nikolov et al. cannot be smaller than the longest path.

In the context of force-directed layout, Dwyer and Koren presented a method that can incorporate constraints enforcing all directed edges to point in the same direction [4]. They explored the possibility to relax some of the constraints, i.e. let some of the edges point backwards, and found that this improves the readability of the drawing. In particular, it reduced the number of edge crossings.

3   Definitions and Problem Classification

Let $G = (V, E)$ denote a graph with a set of nodes V and a set of edges E. We write an edge between nodes u and v as (u, v) if we care about direction, as $\{u, v\}$ otherwise. A *layering* of a directed graph G is a mapping $L : V \to \mathbb{N}$. A layering L is *valid* if $\forall (u, v) \in E$: $L(v) - L(u) \geq 1$.

Problem 1 (Directed Layering (DLP)). Let $G = (V, E)$ be an acyclic directed graph. The problem is to find a minimum k and a valid layering L such that $\sum_{(v,w) \in E}(L(w) - L(v)) = k$.

As mentioned in Sect. 2, DLP was originally introduced by Gansner et al. [7]. We extend the idea of a layering for directed acyclic graphs to general graphs, i.e. graphs that are either directed or undirected and that can possibly be cyclic. Undirected graphs can be handled by assigning an arbitrary direction to each edge, thus converting it into a directed one, and by hardly penalizing reversed edges. We call a layering L of a general graph G *feasible* if $\forall \{u, v\} \in E : |L(u) - L(v)| \geq 1$.

Problem 2 (Generalized Layering (GLP)). Let $G = (V, E)$ be a possibly cyclic directed graph and let $\omega_{\text{len}}, \omega_{\text{rev}} \in \mathbb{N}$ be weighting constants. The problem is to find a minimum k and a feasible layering L such that

$$\omega_{\text{len}} \left(\sum_{(v,w) \in E} |L(w) - L(v)| \right) + \omega_{\text{rev}} |\{(v, w) \in E : L(v) > L(w)\}| \ = \ k.$$

Intuitively, the left part of the sum represents the overall edge length (i.e. the number of dummy nodes) and the right part represents the number of reversed edges (i.e. the FAS). After reversing all edges in this FAS, the feasible layering becomes a valid layering. Compared to the standard cycle removal phase combined with DLP, the generalized layering problem allows more flexible decisions on which edges to reverse. Also note that GLP with $\omega_{\text{len}} = 1, \omega_{\text{rev}} = \infty$ is equivalent to DLP for acyclic input graphs and that while DLP is solvable in polynomial time, both parts of GLP are NP-complete [16].

4   The IP Approach

In the following, we describe how to solve GLP using integer programming. The rough idea of this model is to assign integer values to the nodes of the given graph that represents the layer in which a node is to be placed.

Input and Parameters. Let $G = (V, E)$ be a graph with node set $V = \{1, \ldots, n\}$. Let e be the adjacency matrix, i.e. $e(u, v) = 1$ if $(u, v) \in E$ and $e(u, v) = 0$ otherwise. ω_{len} and ω_{rev} are weighting constants.

Integer Decision Variables. $l(v)$ takes a value in $\{1, \ldots, n\}$ indicating that node v is placed in layer $l(v)$, for all $v \in V$.

Boolean Decision Variables. $r(u, v) = 1$ if and only if edge $e = (u, v) \in E$ and e is reversed, i.e. $l(u) > l(v)$, for all $u, v \in V$. Otherwise, $r(u, v) = 0$.

$$\text{Minimize} \quad \omega_{\text{len}} \sum_{(u,v) \in E} |l(u) - l(v)| + \omega_{\text{rev}} \sum_{(u,v) \in E} r(u, v).$$

The sums represent the edge lengths, i.e. the number of dummy nodes, and the number of reversed edges, respectively. Constraints are defined as follows:

$$1 \leq l(v) \leq n \quad \forall v \in V \tag{A}$$
$$|l(u) - l(v)| \geq 1 \quad \forall(u, v) \in E \tag{B}$$
$$n \cdot r(u, v) + l(v) \geq l(u) + 1 \quad \forall(u, v) \in E \tag{C}$$

Constraint (A) restricts the range of possible layers. (B) ensures that the resulting layering is feasible. (C) binds the decision variables in r to the layering, i.e. because r is part of the objective, and $\omega_{\text{rev}} > 0$, $r(u, v)$ gets assigned 0 unless $l(v) < l(u)$, for all $(u, v) \in E$.

Variations. The model can easily be extended to restrict the number of layers by replacing the n in constraint (1) by a desired bound $b \leq n$.

The edge matrix can be extended to contain a weight $w_{u,v}$ for each edge $(u, v) \in E$. This can be helpful if further semantic information is available, i.e. about feedback edges that lend themselves well to be reversed.

5   The Heuristic Approach

Interactive modeling tools providing automatic layout facilities require execution times significantly shorter than one second. As the IP formulation discussed in the previous section rarely meets this requirement, we present a heuristic to solve GLP. It proceeds as follows. (1) Leaf nodes are removed iteratively, since it is trivial to place them with minimum edge length and desired edge direction. Note that therefore the heuristic is not yet able to improve on trees that yield a poor compactness. We leave this for future research. (2) For the (possibly cyclic) input graph an initial feasible layering is constructed which is used to deduce edge directions yielding an acyclic graph. (3) Using the network simplex method presented by Gansner et al. [7], a solution with minimal edge length is created. (4) We execute a greedy improvement procedure after which we again deduce edge directions and re-attach the leaves. (5) We apply the network simplex algorithm a second time to get a valid layering with minimal

---

Algorithm 1. constructLayering

Input: directed graph $G = (V, E)$
Data: Sets U, C. For all $v \in V$ $score[v]$, $incAs[v]$, $outAs[v]$
$lIndex \leftarrow -1$, $rIndex \leftarrow 0$
Output: index[v]: feasible layering of G

```
1  for v ∈ V do
2  │   score[v] ← |{w | {v, w} ∈ E}|; incAs[v] ← 0; outAs[v] ← 0
3  │   add v to U

4  remove random v from U
5  c ← v
6  while U not empty do
7  │   if incAs[c] < outAs[c] then
8  │   │   index[c] ← lIndex--
9  │   else
10 │   │   index[c] ← rIndex++

11 │   remove c from U and C; cScore ← ∞
12 │   for v ∈ {w | {c, w} ∈ E ∧ w ∈ U} do
13 │   │   add v to C; score[v]--
14 │   │   if (c, v) ∈ E then incAs[v]++ else outAs[v]++

15 │   for v ∈ C do
16 │   │   if score[v] < cScore then cScore ← score[v]; c ← v
```

---

edge lengths for the next steps of the layer-based approach. In the following we will discuss steps 2 and 4 in further detail.

Step 2: Layering Construction. To construct an initial feasible solution we follow an idea that was first presented by McAllister as part of a greedy heuristic for the Linear Arrangement Problem (LAP) [12] and later extended by Pantrigo et al. [15].

Nodes are assigned to distinct indexes, where as a start, a node is selected randomly, assigned to the first index, and added to a set of assigned nodes. Based on the set of assigned nodes a candidate list is formed, and the most promising node is assigned to the next index. As decision criterion we use the difference between the number of edges incident to unassigned nodes and the number of edges incident to assigned nodes. This procedure is repeated until all nodes are assigned to distinct indices (see Algorithm 1).

In contrast to McAllister, for GLP we allow nodes to be added to either side of the set of assigned nodes, and decide the side based on the number of reversed edges that would emerge from placing a certain node on that side. For this we use a decreasing left index variable and an increasing right index variable.

Step 4: Layering Improvement. At this point a feasible layering with a minimum number of dummy nodes w. r. t. the chosen FAS is given since we execute the network simplex method of Gansner et al. beforehand. Thus we can only improve on the number of reversed edges. We determine possible *moves* and decide whether to take the move based on a *profit* value. Let a graph $G = (V, E)$ and a feasible layering L be given. For ease of presentation, we define the following notions. An example: For a node v, *topSuc* are the nodes connected to v via an outgoing edge of v and are currently assigned to a layer with lower index than v's index.

Intuitively, *topSuc* (just as *botPre*) are nodes connected by an edge pointing into the "wrong" direction.

$$
\begin{aligned}
&v.topSuc = \{w : (v,w) \in E \wedge L(v) > L(w)\} \quad &&v.botSuc = \{w : (v,w) \in E \wedge L(v) < L(w)\} \\
&v.topPre = \{w : (w,v) \in E \wedge L(w) < L(v)\} \quad &&v.botPre = \{w : (w,v) \in E \wedge L(w) > L(v)\} \\
&v.topAdj = v.topSuc \cup v.topPre \quad &&v.botAdj = v.botSuc \cup v.botPre
\end{aligned}
$$

For all these functions we define suffixes that allow to query for a certain set of nodes before or after a certain index. For instance, for all top successors of v before index i we write $v.topSucBefore(i) = \{w : w \in v.topSuc \wedge L(w) < i\}$.

Let $move : V \mapsto \mathbb{N}$ denote a function assigning to each node a natural value. The function describes whether it is possible to move a node without violating the layering's feasibility as well as how far the node should be moved. For instance, let for a node v *topPre* be empty but *topSuc* be not empty. Thus, we can move v to an arbitrary layer with lower index than $L(v)$. A good choice would be one layer before any of v's *topSuc* since this would alter the connected edges to point downwards.

$$
move(v) = \begin{cases}
0 & \text{if } v.topSuc = \emptyset, \\
L(v) - \min(\{L(w) : w \in v.topSuc\}) + 1 & \text{if } v.topPre = \emptyset, \\
L(v) - \max(\{L(w) : w \in v.topPre\}) - 1 & \text{otherwise.}
\end{cases}
$$

Let $profit : V \times \mathbb{N} \times \mathbb{N} \mapsto \mathbb{Z}$ denote a function assigning a quality score to each node v if it were moved by $m \in \mathbb{N}$ to a different layer x, i.e. if it is worth to increase some edges' lengths for a subset of them to point downwards. Note that we reuse ω_{len} and ω_{rev} here but do not expect them to have an impact as strong as for the IP. For the rest of the paper we fix them to 1 and 5.

$$
profit(v,m,x) = \begin{cases}
0 & \text{if } m \leq 1, \\
\omega_{\text{len}}(m|v.topAdjBefore(x)| - m|v.botAdj|) & \\
\quad + \omega_{\text{rev}}|v.topSucAfter(x)| & \text{otherwise.}
\end{cases}
$$

As seen in Algorithm 2, the *move* and *profit* functions are determined initially for a given feasible layering. A queue, sorted based on profit values, is then used to successively perform moves that yield a profit. After a move of node n, both functions can be updated for all nodes in the adjacency of n.

Time Complexity. Removing leaf nodes requires linear time, $O(|V| + |E|)$. Algorithm 1 is quadratic in the number of nodes, $O(|V|^2)$. The while loop has to assign an index to every node and the two inner for loops are, for a complete graph, iterated $\frac{|V|}{2}$ times on average. Determining the next candidate (lines 15–16) could be accelerated using dedicated data structures. The improvement step strongly depends on the input graph. The network simplex method runs reportedly fast in practice [7], although it has not been proven to be polynomial. Our evaluations showed that the heuristic's overall execution time is clearly dominated by the network simplex method (cf. Sect. 6).

Algorithm 2. improveLayering

Input: feasible layering of $G = (V, E)$ in $index[v]$
Data: priority queue PQ
For all $v \in V$ $move[v]$, $profit[v]$
Output: index[v]: feasible layering of G

1 **for** $v \in V$ **do**
2 $\quad$ $move[v] \leftarrow move(v)$
3 $\quad$ $profit[v] \leftarrow profit(v, \ move[v], \ index[v] - move[v])$
4 $\quad$ **if** $profit[v] > 0$ **then** enqueue v to PQ

5 **while** PQ not empty **do**
6 $\quad$ $v \leftarrow$ dequeue PQ
7 $\quad$ $index[v]$ $-=$ $move[v]$
8 $\quad$ **for** $w \in \{w \mid \{v, w\} \in E\}$ **do**
9 $\quad\quad$ update $move[w]$ and $profit[w]$
10 $\quad\quad$ **if** $profit[w] > 0$ **then** enqueue w to PQ **else** possibly dequeue w from PQ

---

6 Evaluation

In this section we evaluate three points: (1) the general feasibility of GLP to improve the compactness of drawings, (2) the quality of metric estimations for area and aspect ratio, and (3) the performance of the presented IP and heuristic. Our main metrics of interest here are height, area, and aspect ratio, as defined in an earlier paper [8]. Remember that the layer-based approach is defined as a pipeline of several independent steps. After the layering phase, which is the focus of our research here, these latter two metrics can only be estimated using the number of dummy nodes, the number of layers, and the maximal number of nodes in a layer. Results can be seen in Table 1 and Table 2, which we will discuss in more detail in the remainder of this section.

Obtaining a Final Drawing. To collect all metrics we desire, we have to create a final drawing of a graph. Over time numerous strategies have been presented for each step of the layer-based approach, we thus present several alternatives. To break cycles we use a popular heuristic by Eades et al. [5]. To determine a layering we use our newly presented approach GLP (both the IP method and heuristic, denoted by GLP-IP and GLP-H) and alternatively the network simplex method presented by Gansner et al. [7]. We denote the combination of the cycle breaking of Eades et al. and the layering of Gansner et al. as EaGa and consider it to be an alternative to GLP. Crossings between pairs of layers are minimized using a layer sweep method in conjunction with the barycenter heuristic, as originally proposed by Sugiyama et al. [20]. We employ two different strategies to determine fixed coordinates for nodes within the layers. First, we consider a method introduced by Buchheim et al. that was extended by Brandes and Köpf [1,2], which we denote as BK. Second, we use a method inspired by Sander [18] that we call LS. Edges are routed either using polylines (Poly) or orthogonal segments (Orth). The orthogonal router is based on the methods presented by Sander [19]. Overall, this gives twelve setups of the algorithm: three layering methods, two node placement algorithms, and two edge routing

Table 1. Average values for different layering strategies employed to the test graphs. Different weights are used for GLP-IP as specified in the column head and final drawings were created using BK and Poly. For GLP-H* no improvement was performed. A detailed version of these results can be found in [17].

	1–10	1–20	1–30	1–40	1–50	EaGa	GLP-H	GLP-H*
Reversed edges	3.71	2.89	2.64	2.54	2.44	2.93	8.67	10.36
Dummy nodes	34.45	46.73	52.79	56.14	60.53	72.64	48.48	58.21
Height	843	943	980	1,004	1,025	1,084	930	1,027
Area	631,737	672,717	691,216	700,385	708,361	737,159	656,070	720,798
Aspect ratio	0.77	0.65	0.63	0.61	0.59	0.55	0.67	0.60

(a) Random graphs

	1–10	1–20	1–30	1–40	1–50	EaGa	GLP-H	GLP-H*
Reversed edges	2.74	1.47	1.02	0.72	0.56	0	7.07	8.55
Dummy nodes	39.91	55.47	65.73	75.66	82.47	141.30	53.53	68.91
Height	1,068	1,224	1,334	1,409	1,469	1,727	1,137	1,216
Area	587,727	622,838	641,581	660,842	695,494	874,374	629,778	691,372
Aspect ratio	0.34	0.28	0.24	0.23	0.22	0.20	0.33	0.32

(b) North graphs

procedures. In the following, let ω_{len}-ω_{rev}-GLP denote the used weights. If we do not further qualify GLP, we refer to the IP model.

Test Graphs. Our new approach is intended to improve the drawings of graphs with a large height and relatively small width, hence unfavorable aspect ratio. Nevertheless, we also evaluate the generality of the approach using a set of 160 randomly generated graphs with 17 to 60 nodes and an average of 1.5 edges per node. The graphs were generated by creating a number of nodes, assigning out-degrees to each node such that the sum of outgoing edges is 1.5 times the nodes, and finally creating the outgoing edges with a randomly chosen target node. Unconnected nodes were removed. Second, we filtered the graph set provided by North[1] [3] based on the aspect ratio and selected 146 graphs that have at least 20 nodes and a drawing[2] with an aspect ratio below 0.5, i.e. are at least twice as high as wide. We also removed plain paths, that is, pairs of nodes connected by exactly one edge, and trees. For these special cases GLP in its current form would not change the resulting number of reversed edges as all edges can be drawn with length 1. This is also true for any bipartite graph. Note however that GLP can easily incorporate a bound on the number of layers which can straightforwardly be used to force more edges to be reversed, resulting in a drawing with better aspect ratio.

General Feasibility of GLP. An exemplary result of the GLP approach compared to EaGa can be seen in Fig. 2. For that specific drawing, GLP produces fewer reversed edges, fewer dummy nodes, and less area (both in width and height). For all tested setups the average effective height and area (normalized by the

---

[1] http://www.graphdrawing.org/data/.
[2] Created using BK and Poly.

Table 2. Results for final drawings of the set of random graphs, when applying different layout strategies. For GLP-IP $\omega_{len} = 1$ and $\omega_{rev} = 30$ were used. Area is normalized by a graph's node count. The most interesting comparisons are between columns where EaGa and GLP use the same strategies for the remaining steps. Detailed results can be found in [17].

Edge routing	Poly						Orth					
Node coord.	BK			LS			BK			LS		
Layering	EaGa	GLP-IP	GLP-H	EaGa	GLP-IP	GLP-H	EaGa	GLP-IP	GLP-H	EaGa	GLP-IP	GLP-H
Height	1,165	1,043	898	943	824	732	790	711	652	817	746	678
Area	20,194	18,683	15,575	12,383	11,035	10,075	13,582	12,642	11,272	10,666	9,917	9,295
Aspect ratio	0.59	0.67	0.67	0.55	0.64	0.64	0.84	0.96	0.90	0.63	0.70	0.68

number of nodes) of GLP and the heuristic are smaller than EaGa's, see Table 2. The average aspect ratios come closer to 1.0. For simplicity, in this paper we desire aspect ratios closer to 1.0. For a more detailed discussion on this topic see Gutwenger et al. [8].

Furthermore, we found that by altering the weights ω_{rev} and ω_{len} a trade-off between reversed edges and resulting dummy nodes (and thus area and aspect ratio) can be achieved, which can be seen in Table 1a.

The results for the North graphs are similar. Since the North graphs are acyclic, the cycle breaking phase is not required and current layering algorithms cannot improve the height. The GLP approach, however, can freely reverse edges and hereby change the height and aspect ratio. Results can be seen in Table 2. Clearly, EaGa has no reversed edges as all graphs are acyclic. 1-10-GLP starts with an average of 2.7 reversed edges and the value constantly decreases with an increased weight on reversed edges. The number of dummy nodes on the other hand constantly decreases from 141.3 for EaGa to 39.9 for 1-10-GLP.

The average height and average area of the final drawings decrease with an increasing number of reversed edges. For 1-10-GLP the average height and area are 38.2 % and 33.8 % smaller than EaGa. The aspect ratio changes from an average of 0.20 for EaGa to 0.34 for 1-10-GLP.

The results show that for the selected graphs, for which current methods cannot improve on height, the weights of the new approach allow to find a satisfying trade-off between reversed edges and dummy nodes. Furthermore, the improvements in compactness stem solely from the selection of weights, not from an upper bound on the number of layers. Naturally, such a bound can further improve the aspect ratio and height.

Metric Estimations. Table 2 presents results that were measured on the final drawing of a graph. As mentioned earlier, after the layering step these values are not available and estimations are commonly used to deduce the quality of a result. For our example graphs, the estimated area reduced from 222.9 (EaGa) to 187.4 (1-30-GLP) on average. The estimated aspect ratios increase on average from 0.74 to 0.84. Both tendencies conform to the averaged effective values in Table 2, i.e. GLP-IP and the GLP-H perform better. However, we observed that for 64 % of the graphs the tendency of the estimated area contradicts the tendency of the effective area.[3] 54% when not considering dummy nodes. In other words, for a specific graph the estimated area might be decreased for GLP compared to EaGa but the effective area is increased for GLP (or vice versa). This clearly indicates that an estimation can be misleading. Besides, node placement and edge routing can have a non-negligible impact on the aspect ratio and compactness of the final drawing.

Performance of the Heuristic. Results for final drawings using the presented heuristic are included in Table 2 and are comparable to 1-30-GLP, i.e. the heuristic performs better than EaGa w. r. t. the desired metrics.

Tables 1a and 1b underline this result and show that the improvement step of the heuristic clearly improves on all measured metrics. Further, more detailed results, can be found in [17]. Nevertheless, the heuristic yields significantly more reversed edges. When aiming for compactness, we consider this to be acceptable.

Execution Times. To solve the IP model we used CPLEX 12.6 and executed the evaluations on a server with an Intel Xeon E5540 CPU and 24 GB memory. The execution times for GLP-IP vary between 476 ms for a graph with 19 nodes and 541 s for a graph with 58 nodes and exponentially increase with the graph's node count. This is impracticable for interactive tools that rely on automatic layout, but is fast enough to collect optimal results for medium sized graphs.

The execution time of the heuristic is compared to EaGa and was measured on a laptop with an Intel i7-3537U CPU and 8 GB memory. The reported time includes only the first two steps of the layer-based approach. It turns out that the execution time of the heuristic is on average 2.3 times longer than EaGa. This seems reasonable, as it involves two executions of the network simplex layering method. For the tested graphs, the construction and improvement steps of the heuristic hardly contribute to its overall execution time. The effective execution time ranges between 0.1 ms and 10.0 ms for EaGa and 0.3 ms and 19.7 ms for the heuristic. Hence, the heuristic is fast enough to be used in interactive tools.

We also ran the algorithm five times for five randomly generated graphs with 1000 nodes and 1500 edges. EaGa required an average of 374 ms, the heuristic 666 ms with about 4 ms for construction an 2 ms for improvement. This shows that the time contribution of the latter two is negligible even for larger graphs.

7   Conclusion

In this paper we address problems with current methods for the first two phases of the layer-based layout approach. We argue that separately performing cycle breaking and layering is disadvantageous when aiming for compactness.

We present a configurable method for the layering phase that, compared to other state-of-the-art methods, shows on average improved performance on compactness. That is, the number of dummy nodes is reduced significantly for most graphs and can never increase. While the number of dummy nodes only allows for an estimation of the area, the effective area of the final drawing is

---

[3] Using BK and Poly.

reduced as well. Furthermore, graph instances for which current methods yield unfavorable aspect ratios can easily be improved. Also, the presented heuristic clearly improves on the desired metrics. Depending on the application, a slight increase in the number of reversed edges is often acceptable.

We want to stress that the common practice to determine the quality of methods developed for certain phases of the layer-based approach based on metrics that represent estimations of the properties of the final drawing is error-prone. For instance, estimations of the area and aspect ratio after the layering phase can vary significantly from the effective values of the final drawing and strongly depend on the used strategies for computing node and edge coordinates.

Future work will include improving the heuristic, e.g. selecting the initial node based on a certain criterion instead of randomly in Algorithm 1 should improve the results. We also plan to incorporate hard bounds on the width of a drawing. It is important that methods support to prevent, or at least to strongly penalize, the reversal of certain edges, since certain diagram types demand several edges to be drawn forwards. Also, user studies could help understand which edges are natural candidates to be reversed from a human's perspective.

Furthermore, in an accompanying technical report we present a variation of GLP where we fix the size of the FAS while remaining free in the choice of which edges to reverse [16], which so far has only been evaluated using integer programming.

Acknowledgements. We thank Chris Mears for support regarding the initial IP formulation. We further thank Tim Dwyer and Petra Mutzel for valuable discussions. This work was supported by the German Research Foundation under the project *Compact Graph Drawing with Port Constraints* (ComDraPor, DFG HA 4407/8-1).

References

1. Brandes, U., Köpf, B.: Fast and simple horizontal coordinate assignment. In: Mutzel, P., Jünger, M., Leipert, S. (eds.) GD 2001. LNCS, vol. 2265, pp. 31–44. Springer, Heidelberg (2002). doi:10.1007/3-540-45848-4_3
2. Buchheim, C., Jünger, M., Leipert, S.: A fast layout algorithm for k-level graphs. In: Marks, J. (ed.) GD 2000. LNCS, vol. 1984, pp. 229–240. Springer, Heidelberg (2001). doi:10.1007/3-540-44541-2_22
3. Battista, G., Garg, A., Liotta, G., Parise, A., Tamassia, R., Tassinari, E., Vargiu, F., Vismara, L.: Drawing directed acyclic graphs: an experimental study. In: North, S. (ed.) GD 1996. LNCS, vol. 1190, pp. 76–91. Springer, Heidelberg (1997). doi:10.1007/3-540-62495-3_39
4. Dwyer, T., Koren, Y.: DIG-COLA: directed graph layout through constrained energy minimization. In: Proceedings of the IEEE Symposium on Information Visualization (INFOVIS 2005), pp. 65–72, October 2005
5. Eades, P., Lin, X., Smyth, W.F.: A fast and effective heuristic for the feedback arc set problem. Inf. Process. Lett. **47**(6), 319–323 (1993)
6. Eades, P., Sugiyama, K.: How to draw a directed graph. J. Inf. Process. **13**(4), 424–437 (1990)

7. Gansner, E.R., Koutsofios, E., North, S.C., Vo, K.P.: A technique for drawing directed graphs. Softw. Eng. **19**(3), 214–230 (1993)

8. Gutwenger, C., von Hanxleden, R., Mutzel, P., Rüegg, U., Spönemann, M.: Examining the compactness of automatic layout algorithms for practical diagrams. In: Proceedings of the Workshop on Graph Visualization in Practice (GraphViP 2014), Melbourne, Australia, July 2014

9. Healy, P., Nikolov, N.S.: A branch-and-cut approach to the directed acyclic graph layering problem. In: Goodrich, M.T., Kobourov, S.G. (eds.) GD 2002. LNCS, vol. 2528, pp. 98–109. Springer, Heidelberg (2002). doi:10.1007/3-540-36151-0_10

10. Healy, P., Nikolov, N.S.: How to layer a directed acyclic graph. In: Mutzel, P., Jünger, M., Leipert, S. (eds.) GD 2001. LNCS, vol. 2265, pp. 16–30. Springer, Heidelberg (2002). doi:10.1007/3-540-45848-4_2

11. Healy, P., Nikolov, N.S.: Hierarchical drawing algorithms. In: Tamassia, R. (ed.) Handbook of Graph Drawing and Visualization, pp. 409–453. CRC Press, Boca Raton (2013)

12. McAllister, A.J.: A new heuristic algorithm for the linear arrangement problem. University of New Brunswick, Technical report (1999)

13. Nachmanson, L., Robertson, G., Lee, B.: Drawing graphs with GLEE. In: Hong, S.-H., Nishizeki, T., Quan, W. (eds.) GD 2007. LNCS, vol. 4875, pp. 389–394. Springer, Heidelberg (2008). doi:10.1007/978-3-540-77537-9_38

14. Nikolov, N.S., Tarassov, A., Branke, J.: In search for efficient heuristics for minimum-width graph layering with consideration of dummy nodes. J. Exp. Algorithmics **10** (2005)

15. Pantrigo, J., Martí, R., Duarte, A., Pardo, E.: Scatter search for the cutwidth minimization problem. Ann. Oper. Res. **199**(1), 285–304 (2012)

16. Rüegg, U., Ehlers, T., Spönemann, M., von Hanxleden, R.: A generalization of the directed graph layering problem. Technical report 1501, Kiel University, Department of Computer Science, ISSN 2192–6247, February 2015

17. Rüegg, U., Ehlers, T., Spönemann, M., von Hanxleden, R.: A generalization of the directed graph layering problem, August 2016. arXiv:1608.07809 [cs.OH]

18. Sander, G.: A fast heuristic for hierarchical Manhattan layout. In: Brandenburg, F.J. (ed.) GD 1995. LNCS, vol. 1027, pp. 447–458. Springer, Heidelberg (1996). doi:10.1007/BFb0021828

19. Sander, G.: Layout of directed hypergraphs with orthogonal hyperedges. In: Liotta, G. (ed.) GD 2003. LNCS, vol. 2912, pp. 381–386. Springer, Heidelberg (2004). doi:10.1007/978-3-540-24595-7_35

20. Sugiyama, K., Tagawa, S., Toda, M.: Methods for visual understanding of hierarchical system structures. IEEE Trans. Syst. Man Cybern. **11**(2), 109–125 (1981)

Compact Layered Drawings of General Directed Graphs

Adalat Jabrayilov[1]([⊠]), Sven Mallach[3], Petra Mutzel[1], Ulf Rüegg[2],
and Reinhard von Hanxleden[2]

[1] Department of Computer Science, Technische Universität Dortmund,
Dortmund, Germany
`{adalat.jabrayilov,petra.mutzel}@tu-dortmund.de`
[2] Department of Computer Science, Kiel University, Kiel, Germany
`{uru,rvh}@informatik.uni-kiel.de`
[3] Department of Computer Science, Universität zu Köln, Köln, Germany
`mallach@informatik.uni-koeln.de`

Abstract. We consider the problem of layering general directed graphs under height and possibly also width constraints. Given a directed graph $G = (V, A)$ and a maximal height, we propose a layering approach that minimizes a weighted sum of the number of reversed arcs, the arc lengths, and the width of the drawing. We call this the *Compact Generalized Layering Problem (CGLP)*. Here, the width of a drawing is defined as the maximum sum of the number of vertices placed on a layer and the number of dummy vertices caused by arcs traversing the layer. The CGLP is $\mathcal{NP}$-hard. We present two MIP models for this problem. The first one (EXT) is our extension of a natural formulation for directed *acyclic* graphs as suggested by Healy and Nikolov. The second one (CGL) is a new formulation based on partial orderings. Our computational experiments on two benchmark sets show that the CGL formulation can be solved much faster than EXT using standard commercial MIP solvers. Moreover, we suggest a variant of CGL, called MML, that can be seen as a heuristic approach. In our experiments, MML clearly improves on CGL in terms of running time while it does not considerably increase the average arc lengths and widths of the layouts although it solves a slightly different problem where the dummy vertices are not taken into account.

Keywords: Layer-based layout · Layer assignment · Mixed integer programming

1 Introduction

A widely used hierarchical drawing style for directed graphs is the method proposed by Sugiyama et al. [11] that involves the following steps in this order: (i) cycle removal, (ii) layering phase, (iii) crossing minimization, and (iv) coordinate assignment and arc routing. One of the consequences of this workflow is

© Springer International Publishing AG 2016
Y. Hu and M. Nöllenburg (Eds.): GD 2016, LNCS 9801, pp. 209–221, 2016.
DOI: 10.1007/978-3-319-50106-2_17

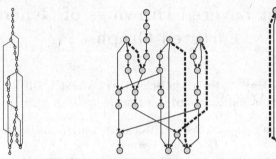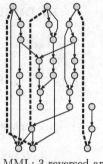

(a) Classic: 0 reversed arcs, 44 dummy vertices   **(b)** CGL: 3 reversed arcs, 20 dummy vertices   **(c)** MML: 3 reversed arcs, 37 dummy vertices

Fig. 1. (a) A graph drawn with traditional methods [2,4] where every arc has to point downwards, resulting in a poor aspect ratio – here emphasized by scaling down the image to fit with the right ones. The methods CGL and MML, presented in this paper, are able to impose a bound on the height of the drawing, allowing some arcs to point upwards. The created drawings (b) and (c) are significantly more compact and improve readability. Reversed arcs are drawn bold and dashed.

that it is hard to control the aspect ratio of the final layout. Phase (ii) requires an acyclic graph as input and the height of the produced layering inherently depends on its longest path. So if phase (i) breaks cycles inappropriately or if an acyclic graph whose longest path is much larger than its width is already given as initial input, it is impossible to construct a compact layering. However, the readability and compactness of a drawing might be considerably improved if arcs to be reversed are chosen carefully and in an integrated fashion, see Fig. 1 for an example. If it is required that all arcs point downward, then the layering will have a poor aspect ratio as shown in Fig. 1a. If we allow reversing some arcs so that they point upward, the aspect ratio can be improved drastically (see Figs. 1b and 1c).

In order to achieve this, Rüegg et al. [10] suggested to investigate the *Generalized Layering Problem (GLP)* that combines the first two interdependent phases of the Sugiyama approach. As the problem is $\mathcal{NP}$-hard, they proposed an integer linear programming (ILP) formulation in which the weighted sum of the number of reversed arcs and the arcs lengths is minimized, and where also the height of the drawing can be restricted. Their approach improved the compactness of the derived layouts. However, it permits only limited control over the width since it does not take the dummy vertices into account that are caused by arcs connecting vertices on non-adjacent layers. On the other hand, Healy and Nikolov [5] have considered the layering problem with dummy vertices, but only for acyclic directed graphs (DAGs) and for the case that both the desired maximum height and the width of the drawing are given as inputs.

Contributions. The purpose of this paper is to close this gap, i.e., to derive a model which is capable of computing a layering of a general digraph by minimizing a weighted sum of the number of reversed arcs, the total arc length, and the width W of the resulting drawing taking the dummy vertices into account. We call this the *Compact Generalized Layering Problem (CGLP)*. The only input to our approach, besides the graph and the weights for the objective function, is the desired maximum height H of the drawing. We will discuss how H can be chosen such that the existence of a feasible layering is always guaranteed by our model — which is in contrast to a setting where the user has to specify both H and W. Nevertheless, an upper bound on W can still be specified.

We present two mixed integer linear programming (MIP) formulations for the CGLP. The first one (EXT) is a natural extension of the already mentioned model by Healy and Nikolov for DAGs, and based on assignment variables. The second one (CGL) is a completely new formulation based on partial orderings. Our computational experiments on two benchmark sets show that the CGL formulation can be solved much faster than EXT when using a standard commercial MIP solver (Gurobi[1]) for both of the models. Moreover, CGL is able to compute optimal solutions for each of the tested problem instances in less than seven minutes of computation time on a standard PC while taking only a few seconds for most of them.

While in general we try to keep edges short, we may want to emphasize reversed arcs, for example in a flow diagram with cycles, by drawing them rather long; see also Fig. 1c. For this reason, we propose a variant of CGL, called MML (Min+Max Length), for the problem of minimizing the weighted sum of reversed arcs, positive forward arc lengths, negative backward arc lengths, and the maximum number of real vertices on a layer. Within our experiments, MML is faster than both EXT and CGL without considerably increasing the average arc lengths and widths of the layouts.

As mentioned by Rüegg et al., the aspect ratio of a final drawing (i.e. in pixels) does not only depend on the layering but also on the final phases of the Sugiyama approach, i.e. coordinate assignment and arc routing. Even more, it can deviate significantly from the aspect ratio estimated after the layering phase, as described in [10]. Hence, in practice, there is a high demand for methods that can be quickly adjusted to the specific graph instance and use case. Here, the models presented in this paper provide more control over the produced layering compared to existing approaches to the GLP.

Outline. The paper is organized as follows. First, we discuss related research in Sect. 2. Definitions, preliminaries and motivations for our studies and models are given in Sect. 3. Section 4 presents our newly developed MIP models which are finally evaluated experimentally in Sect. 5. We conclude with Sect. 6.

2   Related Work

Over the years several approaches have been proposed for the layering phase of the Sugiyama approach. Eades and Sugiyama proposed a method that is known as longest path layering [3]. This approach guarantees to produce a layering with

---

[1] http://www.gurobi.com/.

a minimum number of layers, i.e. minimum height, but the width can become arbitrarily large. Gansner et al. [4] use an ILP formulation to create layerings with minimum total arc length that can be solved efficiently using a network simplex algorithm. The Coffman-Graham algorithm [1] delivers approximate solutions to the precedence-constrained multi-processor scheduling problem that can be used to calculate a layering with a maximum number of real vertices per layer. However, dummy vertices are not taken into account but can have a significant impact on the actual width of the' drawing. Still, Nachmanson et al. [8] use the Coffman-Graham algorithm as part of an iterative heuristic procedure to produce drawings with a certain aspect ratio.

The layering problem with restricted width and consideration of dummy vertices has been studied in the literature as well. Healy and Nikolov found that minimizing the number of dummy vertices during layering inherently produces compact drawings [6]. Following this observation, they target dummy vertex minimization with a branch and cut algorithm that is able to incorporate bounds on both width and height [5]. Nikolov et al. discuss heuristics to find layerings with small width when considering dummy vertices [9], however, no explicit bound on the width and height can be used.

Since all of the above methods rely on the input graph to be acyclic, the minimum height of their layerings directly relates to the longest path of the graph. Rüegg et al. [10] integrate the first two phases of the Sugiyama approach to allow arbitrary graphs as input by minimizing a weighted sum of the number of reversed arcs and the number of dummy vertices. They showed that this overcomes the previously mentioned problem regarding a graph's longest path and also allows more compact drawings in general. Still, they did not consider hard bounds on the width and height of the drawing and did not consider the contribution of dummy vertices to the width.

3   Preliminaries

As we are concerned with the layering phase of the layout method suggested by Sugiyama et al. [11], we briefly recall the definition of layerings of directed graphs at the beginning of this section. Afterwards, we state our new layering problem definition and discuss how to guarantee the existence of feasible drawings under this model.

Layerings of Directed Graphs. A *layering* of a digraph $G = (V, A)$ with vertex set V and arc set A is a function $\ell : V \to \mathbb{N}^+$ assigning a layer $\ell(v)$ to each vertex $v \in V$. In our context, for a feasible layering, it is necessary that no two adjacent vertices are placed on the same layer, i.e., $\ell(u) \neq \ell(v)$ for all $(u, v) \in A$. The layering determines the height, the width, the arc lengths, and the number of reversed arcs, which will be defined in the following. The *height* H of a layering ℓ is the maximum layer used by ℓ. The *width* $W_k(\ell)$ of a layer k in a layering ℓ is the sum of the number of vertices assigned to layer k and the number of dummy vertices caused by arcs traversing the layer k. More formally, let $V_k = \{v \in V \mid \ell(v) = k\}$ and $D_k = \{(u, v) \in A \mid \ell(u) < k$ and $\ell(v) > k$ or $\ell(v) < k$ and $\ell(u) > k\}$. Then $W_k(\ell) = |V_k| + |D_k|$ and the width of ℓ is defined as the maximum width of all layers in ℓ, i.e., $W = \max_{1 \leq k \leq H} W_k(\ell)$. The total arc length $\mathrm{len}(\ell)$ is the sum of the arc lengths $|\ell(u) - \ell(v)|$ of all arcs $(u, v) \in A$ in the layering ℓ. An arc $(u, v) \in A$ in a layering ℓ is called a *reverse arc* if $\ell(u) > \ell(v)$ and the total number of reversed arcs in ℓ is denoted by $\mathrm{rev}(\ell)$. The *estimated aspect ratio* of a layering with width W and height H is defined as W/H. In contrast, the *aspect ratio*, considers the width and height of a final layout after all of the Sugiyama phases. So far, we assumed that real as well as dummy vertices have unit width. Nevertheless, each of our MIP models presented in Sect. 4 can easily be extended to deal with varying vertex widths.

Asking the user to specify bounds on both the height H and the width W of a hierarchical drawing a priori can easily lead to infeasible problem settings or require several iterations to fit the parameters to the graph structure. We circumvent these issues in our subsequently defined variation of the problem by requiring only H as an input parameter while making W a subject of optimization. We also discuss how H can be chosen safely.

The Compact Generalized Layering Problem (CGLP). As an extension to the Generalized Layering Problem (GLP) described in [10], we define the Compact Generalized Layering Problem as follows: Given a (not necessarily acyclic) directed graph $G = (V, A)$ and a maximum layering height H, compute a layering ℓ such that the end-vertices of each arc are assigned to different layers and the following objective function is minimized: the weighted sum of the number of reversed arcs $\mathrm{rev}(\ell)$, the total arc length $\mathrm{len}(\ell)$, and the width $W(\ell)$.

Lower Bounds on the Height H of Feasible Layerings. Since, in our generalized setting, the direction of the arcs can be arbitrary, we can think of undirected graphs to determine a lower bound on H. To assign layers to vertices such that no two adjacent vertices are on the same layer is an equivalent problem as to assign colors to vertices such that no two adjacent vertices have the same color. Hence, the minimum number of layers, i.e., height, necessary for a feasible layering of an undirected graph $G = (V, E)$ is equal to its chromatic number $\chi(G)$.

Since it is $\mathcal{NP}$-hard to compute $\chi(G)$, we suggest to approximate it. A valid upper bound for $\chi(G)$ that can be computed in linear time is the maximum vertex degree $\max_{v \in V} \deg(v)$ of G plus one. This bound is tight if G is the complete graph or an odd cycle, otherwise $\chi(G) \leq \max_{v \in V} \deg(v)$ for any connected graph G. A better approximation can be achieved by using the largest eigenvalue λ^* of the adjacency matrix of G. Wilf showed that $\chi(G) \leq 1 + \lambda^*$ [12] and together with the Perron-Frobenius Theorem we have that $2\frac{|E|}{|V|} \leq \lambda^* \leq \max_{v \in V} \deg(v)$. Summing up, we have $H \geq \chi(G) \leq \lambda^* + 1 \leq \max_{v \in V} \deg(v) + 1$.

4 Description of the MIP Models

4.1 A Generalization of the Model by Healy and Nikolov (EXT)

As a reference, we consider a natural extension of the model by Healy and Nikolov [5] that minimizes the total arc length of layered drawings of *acyclic* digraphs while restricting the height as well as the width and taking dummy vertices into account. In their model, the height H and the width W are fixed input parameters. In our extension for general digraphs $G = (V, A)$, we only require H as an input, and incorporate the width of the drawing into the optimization process. The model describes an assignment problem (AP) with variables $x_{v,k}$ to decide whether vertex $v \in V$ is placed on layer $1 \le k \le H$ ($x_{v,k} = 1$) or not ($x_{v,k} = 0$). If an arc $(u, v) \in A$ is reversed, then this is expressed by a variable $r_{u,v} = 1$, otherwise $r_{u,v} = 0$. The variables $z_{uv,k}$ model whether arc (u, v) causes a dummy vertex on layer $2 \le k \le H - 1$, which is important to formulate a proper width constraint. To save extra arc length variables, we exploit the fact that the length of an arc is exactly the number of dummy vertices it causes plus one. The full model is:

$$\min \left(\omega_{rev} \sum_{(u,v) \in A} r_{u,v} \right) + \left(\omega_{len} \sum_{(u,v) \in A} \sum_{k=2}^{H-1} z_{uv,k} \right) + \omega_{wid} W$$

$$s.t. \quad \sum_{k=1}^{H} x_{v,k} = 1 \qquad \text{for all } v \in V \tag{1}$$

$$x_{u,k} + x_{v,k} \le 1 \qquad \text{for all } (u,v) \in A, 1 \le k \le H \tag{2}$$

$$x_{u,k} - \sum_{l=k}^{H} x_{v,l} \le r_{uv} \qquad \text{for all } (u,v) \in A, 1 \le k \le H \tag{3}$$

$$\sum_{v \in V} x_{v,k} \le W \qquad \text{for all } k \in \{1, H\} \tag{4}$$

$$\sum_{v \in V} x_{v,k} + \sum_{(u,v) \in A} z_{uv,k} \le W \qquad \text{for all } 2 \le k \le H - 1 \tag{5}$$

$$\sum_{l<k} x_{v,l} - \sum_{l \le k} x_{u,l} \le z_{uv,k} \qquad \text{for all } (u,v) \in A, 2 \le k \le H - 1 \tag{6}$$

$$\sum_{l<k} x_{u,l} - \sum_{l \le k} x_{v,l} \le z_{uv,k} \qquad \text{for all } (u,v) \in A, 2 \le k \le H - 1 \tag{7}$$

$$x_{v,k} \in \{0, 1\} \quad \text{for all } v \in V, 1 \le k \le H$$

$$r_{u,v} \in [0, 1] \quad \text{for all } (u,v) \in A$$

$$z_{uv,k} \in [0, 1] \quad \text{for all } (u,v) \in A, 2 \le k \le H - 1$$

$$W \in \mathbb{R}_{\ge 0}$$

The objective function minimizes the weighted sum of the number of reversed arcs, the total arc length, and the width W of the drawing. Equations (1) ensure that exactly one layer is assigned to each $v \in V$. Inequalities (2) enforce adjacent vertices to be placed on different layers. If an arc (u, v) is reversed due to the positions of its end vertices, then inequality (3) makes sure that $r_{u,v}$ is equal to 1 (otherwise, it will be 0 due to the objective function). The total number of vertices and dummy vertices assigned to one layer must never exceed W which is ensured by (4) and (5). Finally, inequalities (6) and (7) enforce a variable $z_{uv,k}$ to be 1 if $\ell(u) < k$ and $\ell(v) > k$ or vice versa. The integrality of all continuous variables is implied by the integrality of the x-variables due to the constraints and the objective function. Model EXT has $\mathcal{O}(|V| \cdot H + |A| \cdot H)$ variables and $\mathcal{O}(|V| + |A| \cdot H)$ constraints.

4.2 Our New Ordering-Based MIP Model (CGL)

The CGL model is based on the observation that the layering problem is a partial ordering problem (POP) in the sense that a vertex u is smaller than v (i.e., $u < v$) in the partial order if $\ell(u) < \ell(v)$.

Following this idea, we introduce, for each $v \in V$ and for each $1 \leq k \leq H$, the variables $y_{v,k}$ that are equal to 1 if and only if $\ell(v) < k$. Conceptually, we also have the reverse variables $y_{k,v}$ (equal to 1 if and only if $k < \ell(v)$), but we will see soon that these can be discarded. However, with the reverse variables at hand, it is easy to see that $\ell(v) = k$ if and only if $y_{k,v} = y_{v,k} = 0$. In addition, the new model also comprises the variables $r_{u,v}$ and $z_{uv,k}$ as already introduced in Sect. 4.1. The interplay between the y- and the r-variables as described in the following will lead to the desired partial ordering of V. The full model is:

$$\min \left(\omega_{rev} \sum_{(u,v) \in A} r_{u,v} \right) + \left(\omega_{len} \sum_{(u,v) \in A} \sum_{k=2}^{H-1} z_{uv,k} \right) + \omega_{wid} W$$

$$
\begin{array}{lll}
s.t. \quad y_{v,1} & = 0 & \text{for all } v \in V \qquad (8) \\
y_{H,v} & = 0 & \text{for all } v \in V \qquad (9) \\
y_{k,v} + y_{v,k+1} & = 1 & \text{for all } v \in V, 1 \leq k \leq H-1 \quad (10) \\
y_{k+1,v} - y_{k,v} & \leq 0 & \text{for all } v \in V, 1 \leq k \leq H-2 \quad (11) \\
-y_{u,k} - y_{k,v} - r_{u,v} & \leq -1 & \text{for all } (u,v) \in A, 1 \leq k \leq H \quad (12) \\
-y_{k,u} - y_{v,k} + r_{u,v} & \leq 0 & \text{for all } (u,v) \in A, 1 \leq k \leq H \quad (13) \\
y_{k,u} + y_{v,k} - z_{uv,k} & \leq 1 & \text{for all } (u,v) \in A, 2 \leq k \leq H-1 \\
& & \qquad\qquad\qquad\qquad\qquad (14) \\
y_{k,v} + y_{u,k} - z_{uv,k} & \leq 1 & \text{for all } (u,v) \in A, 2 \leq k \leq H-1 \\
& & \qquad\qquad\qquad\qquad\qquad (15) \\
\displaystyle\sum_{u \in V} (1 - y_{u,k} - y_{k,u}) & \leq W & \text{for all } k \in \{1, H\} \qquad (16) \\
\displaystyle\sum_{u \in V} (1 - y_{u,k} - y_{k,u}) + \sum_{(u,v) \in A} z_{uv,k} & \leq W & \text{for all } 2 \leq k \leq H-1 \quad (17) \\
y_{v,k}, \ y_{k,v} & \in \{0,1\} & \text{for all } v \in V, 1 \leq k \leq H \\
r_{u,v} & \in [0,1] & \text{for all } (u,v) \in A \\
z_{uv,k} & \in [0,1] & \text{for all } (u,v) \in A, 2 \leq k \leq H-1 \\
W & \in \mathbb{R}_{\geq 0} &
\end{array}
$$

The switch from an AP to a POP requires a more involved approach to yield consistency of the model. The first four constraints enforce the graph to be embedded into the layers $1, \ldots, H$. Equations (8) and (9) make sure no vertex is assigned a layer smaller than one or larger than H. For each layer $1 \leq k \leq H$, each vertex v is either assigned a layer larger than k (in which case $y_{k,v} = 1$) or not (in which case $y_{v,k+1} = 1$) as is enforced by (10). These equations can be used to eliminate one half of the y-variables (and then be eliminated themselves) as mentioned before. If $\ell(v) > k + 1$, then this implies $\ell(v) > k$ as well and this is expressed in the transitivity inequalities (11). It remains to show that arc directions and layer assignments will be consistent and no two adjacent vertices can be on the same layer. This is achieved by inequalities (12) and (13). Suppose that $r_{u,v} = 0$, i.e., the arc $(u, v) \in A$ shall be a forward arc. Then the inequalities (12) enforce that, for each layer $1 \leq k \leq H$, either $\ell(u) < k$ or $\ell(v) > k$ (or both). In this case, the inequalities (13) are inactive, but they take the equivalent role in the reversed-arc case where $r_{u,v} = 1$ and then inequalities (12) are inactive.

As already discussed for the previous model, a dummy vertex on layer k is caused by arc $(u, v) \in A$ if either $\ell(u) > k$ and $\ell(v) < k$ $(y_{k,u} + y_{v,k} - 1 = 1)$, or vice versa $(y_{k,v} + y_{u,k} - 1 = 1)$. In the first case, inequality (14) will force $z_{uv,k}$ to be 1, in the second case, inequality (15) will do so. In any other case, the variable will be zero due to the objective function. Finally, inequalities (16) and (17) count the vertices and dummy vertices placed on each layer k and make sure that W is a proper upper bound on the width of the layering. The CGL formulation has $\mathcal{O}(|V| \cdot H + |A| \cdot H)$ variables and constraints.

4.3   A Min+Max Length Variant Without Dummy Vertices (MML)

We shortly describe a variant of CGL, called MML, that can produce appealing results usually faster when dummy vertices need not be taken into account in terms of the width. As opposed to W, let W_r be the width of a layering where only the real vertices are counted.

The idea is to remove the dummy vertex variables $z_{uv,k}$ together with the constraints (14), (15) and to replace inequalities (16) and (17) simply by:

$$\sum_{u \in V} (1 - y_{u,k} - y_{k,u}) \qquad \leq W_r \qquad \text{for all } 1 \leq k \leq H$$

We now need to count arc lengths in an ordinary fashion. The usual way to do this is to introduce length variables $l_{u,v} \in \mathbb{R}$ and the following two inequalities per arc in order to capture the absolute length depending on the arc direction.

$$\sum_{k=1}^{H} (y_{k,v} - y_{k,u}) \qquad \leq l_{u,v} \qquad \text{for all } (u, v) \in A \qquad (18)$$

$$\sum_{k=1}^{H} (y_{k,u} - y_{k,v}) \qquad \leq l_{u,v} \qquad \text{for all } (u, v) \in A \qquad (19)$$

However, for certain use cases, e.g., when feedback should be emphasized, it can be desirable to draw forward arcs as short as possible and maximize the length of the reversed arcs. Therefore, we propose not to introduce the l-variables but to directly incorporate the terms used on the left hand side of inequalities (18) into the objective function, which results in the desired minimization of the backward arcs' negative lengths. MML's objective function is:

$$\min \left(\omega_{rev} \sum_{(u,v) \in A} r_{u,v} \right) + \left(\omega_{len} \sum_{(u,v) \in A} \sum_{k=1}^{H} (y_{k,v} - y_{k,u}) \right) + \omega_{wid} \, W_r$$

This model has only $\mathcal{O}(|V| \cdot H + |A|)$ variables and $\mathcal{O}(|V| \cdot H + |A| \cdot H)$ constraints.

5   Evaluation

Setup. The experiments were performed single-threadedly on an Intel Core i7-4790, 3.6 GHz, with 32 GB of memory and running Ubuntu Linux 14.04. For solving the MIPs, we used Gurobi 6.5. In our implementation, the CGL model has been reduced in terms of its variables as is indicated in Sect. 4.2. Further, in all the models we enforce at least one vertex to be placed on layer $k = 1$ to eliminate some symmetries. The parameters were set to $H = \lceil 1.6 * \sqrt{|V|} \rceil$, $w_{rev} = |E| \cdot H$, $w_{len} = 1$, and $w_{wid} = 1$. This choice of H delivered feasible problems for all of our instances and emphasizes our target to have a good aspect ratio and a drawing that adheres to standard forms such as flat screens following the golden ratio. Due to our choice of w_{rev}, arcs are reversed only if this is unavoidable due to the specified height or because they are part of a cycle.

We are interested in answering the following questions[2]:

(H1) Does the POP-oriented CGL model dominate the AP-based EXT model in terms of running times?

(H2) Is MML a good alternative concerning the running times and the metrics (arc length, W, and estimated aspect ratio) of the generated layerings?

(H3) How do CGL and MML influence the aspect ratio of the *final* drawings?

We used two benchmark sets[3]. The first set *ATTar*, the same as used by Rüegg et al. [10], is an extraction of 146 acyclic AT&T graphs with at least 20 vertices having aspect ratio smaller than 0.5 when drawn with the classic Sugiyama approach. These graphs have between 20 and 99 vertices and between 20 and 168 arcs. Their arc to vertex density is about 1.5 on average but varies significantly. Especially, the density of some graphs with about 60 vertices is up to 4.7 which is why the results displayed in our figures and boxplots stand out for these instances.

The second benchmark set *Random* consists of 340 randomly generated, not necessarily acyclic graphs with 17 to 100 vertices, 30 to 158 arcs, and 1.5 arcs per vertex. We used these graphs in order to analyze our approach also for cyclic sparse digraphs. First, a number of vertices was created. Afterwards, for each vertex, a random number of outgoing arcs (with arbitrary target) is created such that the overall number of arcs is 1.5 times the number of vertices.

(H1): Comparison of CGL and EXT. First, we look at the ATTar instances and model EXT. While small instances with up to 25 vertices can be solved within at most seven seconds, there is already one instance with 29 vertices that cannot be solved within the time limit of 10 minutes of CPU and system time. In total, eleven of 66 instances with $25 < |V| < 50$ time out, while for the others the running times highly deviate within the full spectrum between a second and about nine minutes. However, none of the 33 instances with more

---

[2] Additional experimental results and example drawings can be found in [7].

[3] Both benchmark sets are available on https://ls11-www.cs.tu-dortmund.de/mutzel/gdbenchmarks.

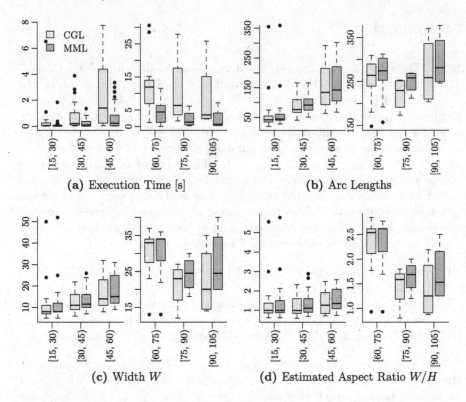

Fig. 2. Summary of the results of the ATTar graphs. For each of the four metrics the graphs were binned based on their vertex counts (x axis) and the y axis represents the metric's value. For each bin the left box represents CGL's result and the right box MML's result.

than 50 vertices can be solved within the time limit. The picture for the random instances is similar. The first instance remaining unsolved within the time limit has 37 vertices and those with 50 or more vertices can be solved only sporadically (33 of 201).

With the CGL model, however, we were able to solve all the instances (ATTar and Random) to optimality. The running times are shown in Figs. 2a and 3a. All but ten of the ATTar instances were solved in less than 10 seconds of CPU and system time, and the highest running time observed was 30 seconds. Concerning the random instances, we observed that 314 of the 340 instances were solved within 30 seconds. However, for $|V| \geq 80$, a higher dispersion of running times could be observed. The largest observed running time was 388 seconds for an instance with 92 vertices. Since both models solve the same problem to optimality, we can conclude that the EXT model is clearly dominated by the CGL model when a state-of-the-art commercial MIP solver is used.

(H2): Alternative MML. In Sect. 4.3 we introduced the model MML that, as opposed to the models EXT and CGL, maximizes the length of reversed arcs

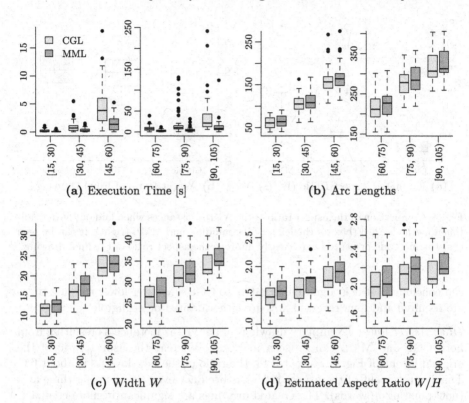

Fig. 3. Summary of the results of the random graphs. For each of the four metrics, the graphs were binned based on their vertex counts (x axis) and the y axis represents the metric's value. For each bin, the left box represents CGL's result and the right box MML's result. To improve presentation, we removed one outlier in (a) with 92 vertices that took CGL 388*s* of computation time.

and does not regard the contribution of dummy vertices to a layer's width. Our hope was that MML is much faster than CGL and EXT without sacrificing the quality of the generated layouts too much.

In Figs. 2a and 3a, one can see that the MML model could almost always (except for 14 instances in total) be solved much faster than the CGL model and hence also the EXT model. Especially, with MML all but ten instances of the ATTar benchmark set were solved within three seconds. The boxplots also show that the running times for solving the MML model are more robust for both benchmark sets.

As can be seen in Figs. 2 and 3, the average arc lengths and widths of layerings created by MML increase only moderately when compared to CGL. On average, the increase in the total arc lengths is only about 6% and the increase in the width is about 7% for each of the benchmark sets. The displayed widths of MML also include dummy vertices. Since MML is significantly faster than CGL, we

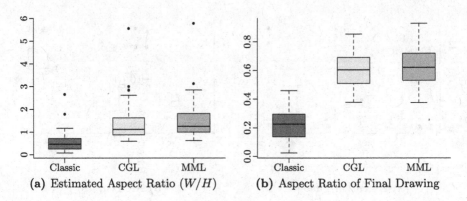

(a) Estimated Aspect Ratio (W/H) **(b)** Aspect Ratio of Final Drawing

Fig. 4. Comparison of the aspect ratio of the ATTar instances when laid out with traditional methods [2,4] (box on the left), CGL (middle), and MML (right). It can be seen that the methods presented here clearly improve the aspect ratio of the final drawing.

can conclude that it is a good alternative to CGL whenever lower running times are required and long reversed arcs are either desired or negligible.

(H3): Aspect Ratio. Exemplary drawings of an ATTar instance as resulting from both CGL and MML can be seen in Fig. 1. Whereas the aspect ratio of the original layout of Fig. 1 is about 0.14, the ratio of the new layouts is about 0.6. This improvement of the aspect ratio has been achieved by reversing three arcs (now pointing upwards). The created drawings are significantly more compact. In Fig. 1c, generated with the MML model, the reversed arcs can be found easily, since the model tries to maximize their length.

An average of about 3.13 arcs needed to be reversed on the ATTar graphs and 3.82 on the random graphs to adhere to the selected H. Also the maximum number of reversed arcs in both benchmark sets is similar; it is 8 for the ATTar instances and 9 for the random instances. The reversed arcs changed the estimated aspect ratio of the ATTar graphs from an average of 0.51 to an average of 1.36, see Fig. 4a. As mentioned earlier, the estimated aspect ratio must not necessarily coincide with the final drawing's aspect ratio. To further inspect this, we produced final drawings using the same strategies of the Sugiyama approach as discussed in [10]. In Fig. 4b, one can see that the average aspect ratio improves from 0.22 to about 0.61 for CGL and to about 0.63 for MML. From this we conclude that both models lead to compact layouts with improved aspect ratio.

6   Conclusion

This paper introduces the CGLP, which can be seen as an extension of the DAG Layering Problem suggested by Healy and Nikolov [5,6], and the GLP suggested by Rüegg et al. [10]. The CGLP gives more control over the desired layering by integrating the reversal of arcs and taking the contribution of dummy vertices to a layering's width into account.

We suggest two MIP models for CGLP, one of which is based on partial orderings and show that the model can be solved to optimality within a short computation time for typical instances with up to 100 vertices. In addition, we suggest an alternative MIP model (MML) for a slightly different problem which can be solved even faster while the widths and arc lengths of the generated layerings do not increase significantly. Our experiments have shown that using the CGLP, indeed, the aspect ratio of the generated final drawings can be influenced.

Acknowledgements. This work was supported by the German Research Foundation under the project *Compact Graph Drawing with Port Constraints* (ComDraPor, DFG HA 4407/8-1 and MU 1129/9-1).

References

1. Coffman, E.G., Graham, R.L.: Optimal scheduling for two-processor systems. Acta Informatica **1**(3), 200–213 (1972)
2. Eades, P., Lin, X., Smyth, W.F.: A fast and effective heuristic for the feedback arc set problem. Inf. Process. Lett. **47**(6), 319–323 (1993)
3. Eades, P., Sugiyama, K.: How to draw a directed graph. J. Inf. Process. **13**(4), 424–437 (1990)
4. Gansner, E.R., Koutsofios, E., North, S.C., Vo, K.P.: A technique for drawing directed graphs. Softw. Eng. **19**(3), 214–230 (1993)
5. Healy, P., Nikolov, N.S.: A branch-and-cut approach to the directed acyclic graph layering problem. In: Goodrich, M.T., Kobourov, S.G. (eds.) GD 2002. LNCS, vol. 2528, pp. 98–109. Springer, Heidelberg (2002). doi:10.1007/3-540-36151-0_10
6. Healy, P., Nikolov, N.S.: How to layer a directed acyclic graph. In: Mutzel, P., Jünger, M., Leipert, S. (eds.) GD 2001. LNCS, vol. 2265, pp. 16–30. Springer, Heidelberg (2002). doi:10.1007/3-540-45848-4_2
7. Jabrayilov, A., Mallach, S., Mutzel, P., Rüegg, U., von Hanxleden, R.: Compact layered drawings of general directed graphs. arXiv:1609.01755 [cs.DS] (2016)
8. Nachmanson, L., Robertson, G., Lee, B.: Drawing graphs with GLEE. In: Hong, S.-H., Nishizeki, T., Quan, W. (eds.) GD 2007. LNCS, vol. 4875, pp. 389–394. Springer, Heidelberg (2008). doi:10.1007/978-3-540-77537-9_38
9. Nikolov, N.S., Tarassov, A., Branke, J.: In search for efficient heuristics for minimum-width graph layering with consideration of dummy nodes. J. Exp. Algorithmics **10**, 1–27 (2005). Article No. 2.7
10. Rüegg, U., Ehlers, T., Spönemann, M., von Hanxleden, R.: A generalization of the directed graph layering problem. In: Proceedings of the 24th International Symposium on Graph Drawing and Network Visualization (GD 2016) (2016)
11. Sugiyama, K., Tagawa, S., Toda, M.: Methods for visual understanding of hierarchical system structures. IEEE Trans. Syst. Man Cybern. **11**(2), 109–125 (1981)
12. Wilf, H.S.: The eigenvalues of a graph and its chromatic number. J. Lond. Math. Soc. **42**, 330–332 (1967)

Bitonic *st*-orderings for Upward Planar Graphs

Martin Gronemann[✉]

University of Cologne, Cologne, Germany
`gronemann@informatik.uni-koeln.de`

Abstract. Canonical orderings serve as the basis for many incremental planar drawing algorithms. All these techniques, however, have in common that they are limited to undirected graphs. While *st*-orderings do extend to directed graphs, especially planar *st*-graphs, they do not offer the same properties as canonical orderings. In this work we extend the so called bitonic *st*-orderings to directed graphs. We fully characterize planar *st*-graphs that admit such an ordering and provide a linear-time algorithm for recognition and ordering. If for a graph no bitonic *st*-ordering exists, we show how to find in linear time a minimum set of edges to split such that the resulting graph admits one. With this new technique we are able to draw every upward planar graph on n vertices by using at most one bend per edge, at most $n - 3$ bends in total and within quadratic area.

1 Introduction

Drawing directed graphs is a fundamental problem in graph drawing and has therefore received a considerable amount of attention in the past. Especially the so called *upward planar drawings*, a planar drawing in which the curve representing an edge has to be strictly y-monotone from its source to target. The directed graphs that admit such a drawing are called the *upward planar* graphs. Deciding if a directed graph is upward planar turned out to be NP-complete in the general case [11], but there exist special cases for which the problem is polynomial-time solvable [1,2,8,16,19,20]. An important result in our context is from Di Battista and Tamassia [6]. They show that every upward planar graph is the spanning subgraph of a planar *st*-graph, that is, a planar directed acyclic graph with a single source and a single sink. They also show that every such graph has an upward planar straight-line drawing [6], but it may require exponential area which for some graphs cannot be avoided [5,7].

If one allows bends on the edges, then every upward planar graph can be drawn within quadratic area. Di Battista and Tamassia [6] describe an approach that is based on the visibility representation of a planar *st*-graph. Every edge has at most two bends, therefore, the resulting drawing has at most $6n - 12$ bends with n being the number of vertices. With a more careful choice of the vertex positions and by employing a special visibility representation, the authors manage to improve this bound to $(10n - 31)/3$. Moreover, the drawing requires only quadratic area and can be obtained in linear time. Another approach by Di

© Springer International Publishing AG 2016
Y. Hu and M. Nöllenburg (Eds.): GD 2016, LNCS 9801, pp. 222–235, 2016.
DOI: 10.1007/978-3-319-50106-2_18

Battista et al. [7] uses an algorithm that creates a straight-line dominance drawing as an intermediate step. A dominance drawing, however, has much stronger requirements than an upward planar drawing. Therefore, the presented algorithm in [7] cannot handle planar *st*-graphs directly. Instead it requires a *reduced planar st-graph*, that is, a planar *st*-graph without *transitive edges*. In order to obtain such a graph, Di Battista et al. [7] split every transitive edge by replacing it with a path of length two. The result is a reduced planar *st*-graph for which a straight-line dominance drawing is obtained that requires only quadratic area and can be computed in linear time. Then they reverse the procedure of splitting the edges by using the coordinates of the inserted dummy vertices as bend points. Since a planar *st*-graph has at most $2n - 5$ transitive edges, the resulting layout has not more than $2n - 5$ bends and at most one bend per edge. To our knowledge, this bound is the best achieved so far.

These techniques are very different to the ones used in the undirected case. One major reason is the availability of *canonical orderings* for undirected graphs, introduced by de Fraysseix et al. [9] to draw every (maximal) planar graph straight-line within quadratic area. From there on this concept has been further improved and generalized [15,17,18]. Biedl and Derka [3] discuss various variants and their relation. Another similar concept that extends to non-planar graphs is the Mondshein sequence [21]. However, all these orderings have in common that they do not extend to directed graphs, that is, for every edge (u, v), it holds that u precedes v in the ordering. An exception are *st*-orderings. While they are easy to compute for planar *st*-graphs, they lack a certain property compared to canonical orderings. In [13] we introduced for undirected biconnected planar graphs the *bitonic st-ordering*, a special *st*-ordering which has properties similar to canonical orderings. However, the algorithm in [13] uses canonical orderings for the triconnected case as a subroutine. Since finding a canonical ordering is in general not a trivial task, respecting the orientation of edges makes it even harder. Nevertheless, such an ordering is desirable, since one would be able to use incremental drawing approaches for directed graphs that are usually limited to the undirected case.

In this paper we extend the bitonic *st*-ordering to directed graphs, namely planar *st*-graphs. We start by discussing the consequences of having such an ordering available. Based on the observation that the algorithm of de Fraysseix et al. [9] can easily be modified to obtain an upward planar straight-line drawing, we show that for good reasons not every planar *st*-graph admits such an ordering. After deriving a full characterization of the planar *st*-graphs that do admit a bitonic *st*-ordering, we provide a linear-time algorithm that recognizes these and computes a corresponding ordering. For a planar *st*-graph that does not admit a bitonic *st*-ordering, we show that splitting at most $n - 3$ edges is sufficient to transform it into one for which then an ordering can be found. Furthermore, a linear-time algorithm is described that determines the smallest set of edges to split. By combining these results, we are able to draw every planar *st*-graph with at most one bend per edge, $n - 3$ bends in total within quadratic area in linear time. This improves the upper bound on the total number of bends considerably. Some proofs have been omitted and can be found in the full version [12] or in [14].

2   Preliminaries

In this work we are solely concerned with a special type of directed graph, the so-called *planar st-graph*, that is, a planar acyclic directed graph $G = (V, E)$ with a single source $s \in V$, a single sink $t \in V$ and no parallel edges. It should be noted that some definitions assume that $(s, t) \in E$, we explicitly do not require this edge to be present. However, we assume a fixed embedding scenario such that s and t are on the outer face. Under such constraints, planar *st*-graphs possess the property of being *bimodal*, that is, the incoming and outgoing edges appear as a consecutive sequence around a vertex in the embedding. Given an edge $(u, v) \in E$, we refer to v as a *successor* of u and call u a *predecessor* of v. Similar to [13], we define for every vertex $u \in V$ a list of successors $S(u) = \{v_1, \ldots, v_m\}$, ordered by the outgoing edges $(u, v_1), \ldots, (u, v_m)$ of u as they appear in the embedding clockwise around u. For $S(s)$ we choose v_1 and v_m such that v_m, s, v_1 appear clockwise on the outer face. A central problem will be the existence of paths between vertices. Therefore, we refer to a path from u to v and its existence with $u \rightsquigarrow v \in G$. With a few exceptions, G is clear from the context, thus, we omit it. If there exists no path $u \rightsquigarrow v$, we may abbreviate it by writing $u \not\rightsquigarrow v$.

Let $G = (V, E)$ be a planar *st*-graph and $\pi : V \mapsto \{1, \ldots, |V|\}$ be the rank of the vertices in an ordering $s = v_1, \ldots, v_n = t$. π is said to be an *st-ordering*, if for all edges $(u, v) \in E$, $\pi(u) < \pi(v)$ holds. In case of a (planar) *st*-graph such an ordering can be obtained in linear time by using a simple topological sorting algorithm [4]. We are interested in a special type of *st*-ordering, the so called bitonic *st*-ordering introduced in [13]. We say an ordered sequence $A = \{a_1, \ldots, a_n\}$ is *bitonic increasing*, if there exists $1 \leq h \leq n$ such that $a_1 \leq \cdots \leq a_h \geq \cdots \geq a_n$ and *bitonic decreasing*, if $a_1 \geq \cdots \geq a_h \leq \cdots \leq a_n$. Moreover, we say A is bitonic increasing (decreasing) with respect to a function f, if $A' = \{f(a_1), \ldots, f(a_n)\}$ is bitonic increasing (decreasing). In the following, we restrict ourselves to bitonic increasing sequences and abbreviate it by just referring to it as being bitonic. An *st*-ordering π for G is a *bitonic st-ordering* for G, if at every vertex $u \in V$ the ordered sequence of successors $S(u) = \{v_1, \ldots, v_m\}$ as implied by the embedding is bitonic with respect to π, that is, there exists $1 \leq h \leq m$ with $\pi(v_1) < \cdots < \pi(v_h) > \cdots > \pi(v_m)$. Notice that the successors of a vertex are distinct and so are their labels in an *st*-ordering.

3   Upward Planar Straight-Line Drawings and Bitonic *st*-orderings

We start by assuming that we are given a planar *st*-graph $G = (V, E)$ together with a bitonic *st*-ordering π. The idea is to use the straight-line algorithm from [13] which is based on the one in [15] to produce an upward planar straight-line layout. Due to space constraints, we omit details here and only sketch the two modifications that are necessary. For a full pseudocode listing, an example and a detailed description, see the full version [12] or [14]. When using a bitonic *st*-ordering to drive the planar straight-line algorithm of de Fraysseix et al. [9],

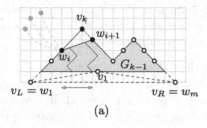
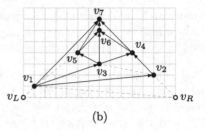

(a)   (b)

Fig. 1. (a) A vertex v_k with only one predecessor w_i using the vertex w_{i+1} as second neighbor. Vertices in grey have not been drawn yet. The two dummy vertices v_L, v_R remain the left- and rightmost ones. (b) Example of an upward planar straight-line drawing on seven vertices.

the only critical case is the one in which a vertex v_k must be placed that has only one neighbor, say w_i, in the subgraph drawn so far. In [13] we use the idea of Harel and Sardas [15] who guarantee with their ordering that the edges preceding or following (w_i, v_k) in the embedding around w_i have already been drawn. Hence one may just pretend that v_k has a second neighbor either to the right or left of w_i. The idea is illustrated in Fig. 1a where v_k uses w_{i+1}, the successor of w_i on the contour, as second neighbor. The following lemma captures the required property and shows that a bitonic *st*-ordering complies with it.

Lemma 1. *Let $G = (V, E)$ be an embedded planar st-graph with a corresponding bitonic st-ordering π. Moreover, let v_k be the k-th vertex in π and $G_k = (V_k, E_k)$ the subgraph induced by $v_1, \ldots, v_k$. For every $1 < k \leq |V|$ the following holds:*

1. *G_k and $G - G_k$ are connected,*
2. *v_k is in the outer face of G_{k-1},*
3. *For every vertex $v \in V_k$, the neighbors of v that are not in G_k appear consecutively in the embedding around v.*

Sketch of Proof. The first two properties hold for all *st*-orderings. For the third, assume to the contrary, contradicting that $S(v)$ is bitonic with respect to π. □

Due to the third statement we can always choose a second neighbor either to the left or right, since otherwise the grey vertices in Fig. 1a would not be consecutive in the embedding around w_i. The second modification solves a problem that arises in the initialization phase of the drawing algorithm. Recall that in [9] the first three vertices are drawn as a triangle. This of course works in the case of a canonical ordering, but requires extra care when using a bitonic *st*-ordering. In order to avoid subcases and keep things simple, we add two isolated dummy vertices v_L and v_R that take the roles of the first two vertices and pretend to form a triangle with $v_1 = s$. This has another side effect: It avoids distinguishing between subcases when we have to find a second neighbor at the boundary of the contour, because v_L is always the first, and v_R always the last vertex on every contour during the incremental construction. See the example in Fig. 1b.

Theorem 1. *Given an embedded planar st-graph $G = (V, E)$ and a corresponding bitonic st-ordering π for G. An upward planar straight-line drawing for G of size $(2|V| - 2) \times (|V| - 1)$ can be obtained from π in linear time.*

Proof. The upward property is obtained by the following observation: The original planar straight-line algorithm installs every vertex v_k with $k > 2$ above its predecessors. Since we start with v_L, v_R, v_1, the drawing is upward. It remains to bound the area. Notice that the input consists of the two additional vertices v_L, v_R. The original algorithm, without any area improvements, produces a drawing with a size of $2((|V| + 2) - 4) \times (|V| + 2) - 2 = 2|V| \times |V|$. However, v_L and v_R are dummy vertices and can be removed anyway. Moreover, every other vertex is located above them. Hence, their removal yields a smaller drawing of size $(2|V| - 2) \times (|V| - 1)$. □

Now the first question that comes to mind is, if we can always find a bitonic st-ordering. Although every planar st-graph admits an upward planar straight-line drawing [6], there exist some classes for which it is known that they require exponential area [5,7]. Since Theorem 1 clearly states that the drawing requires only polynomial area, these graphs cannot admit a bitonic st-ordering.

Corollary 1. *Not every planar st-graph admits a bitonic st-ordering.*

While this had to be expected, we now have to solve an additional problem. Before we think about how to compute a bitonic st-ordering, we must first be able to recognize planar st-graphs that admit such an ordering.

4   Characterization, Recognition and Ordering

We proceed as follows: As a first step, we identify a necessary condition that a planar st-graph has to meet for admitting a bitonic st-ordering. Then we exploit this condition to compute a bitonic st-ordering which proves sufficiency. We start with an alternative characterization of bitonic sequences. Since we will use the labels of an st-ordering, we can assume that the elements are pairwise distinct.

Lemma 2. *An ordered sequence $A = \{a_1, \ldots, a_n\}$ of pairwise distinct elements is bitonic increasing if and only if the following holds:*

$$\forall 1 \leq i < j < n : a_i < a_{i+1} \vee a_j > a_{j+1}.$$

Sketch of Proof. For "$\Rightarrow$", assume to the contrary which yields $i \geq j$. For "$\Leftarrow$", we choose, if exists, $h = \min\{j \mid a_j > a_{j+1}\}$, otherwise we set $h = n$. □

In general a planar st-graph may have many st-orderings, some of them being bitonic while others are not. To deal with this in a more formal manner, we introduce some additional notation. Given an embedded planar st-graph $G = (V, E)$, we refer with $\Pi(G)$ to all feasible st-orderings of G, that is,

$$\Pi(G) = \{\pi : V \mapsto \{1, \ldots, |V|\} \mid \pi \text{ is an } st - \text{ordering for } G\}.$$

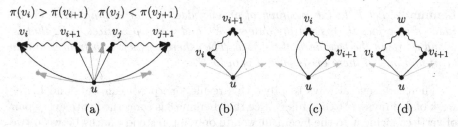

$$\pi(v_i) > \pi(v_{i+1}) \quad \pi(v_j) < \pi(v_{j+1})$$

(a)   (b)   (c)   (d)

Fig. 2. (a) A successor list $S(u) = \{\ldots, v_i, v_{i+1}, \ldots, v_j, v_{j+1}, \ldots\}$ with $i < j$ and a forbidden configuration of paths $v_{i+1} \rightsquigarrow v_i$ and $v_j \rightsquigarrow v_{j+1}$. (b)–(d) The three cases at a face between two successors v_i and v_{i+1} of the face-source u: (b) v_{i+1} is the sink of the face indicating the existence of a path from v_i to v_{i+1}. (c) A path from v_{i+1} to v_i results in a face having v_i as sink. (d) There exists no path between v_i and v_{i+1}, if and only if neither v_i nor v_{i+1} is the face-sink.

Furthermore, let $\Pi_b(G)$ be the subset of $\Pi(G)$ that contains all bitonic st-orderings. By definition, we can describe $\Pi_b(G)$ by

$$\Pi_b(G) = \{\pi \in \Pi(G) \mid \forall u \in V : S(u) \text{ is bitonic with respect to } \pi\}.$$

Applying the alternative characterization of bitonicity from Lemma 2 to the bitonic property of the successor lists $S(u)$ yields the following expression for the existence of a bitonic st-ordering:

$$\exists \pi \in \Pi_b(G) \Leftrightarrow \exists \pi \in \Pi(G) \quad \forall u \in V \text{ with } S(u) = \{v_1, \ldots, v_m\}$$
$$\forall 1 \le i < j < m : \pi(v_i) < \pi(v_{i+1}) \vee \pi(v_j) > \pi(v_{j+1}). \tag{1}$$

Next we translate this expression from st-orderings to the existence of paths. Consider a path from some vertex u to some other vertex v in G, then for every $\pi \in \Pi(G)$, by the definition of st-orderings, $\pi(u) < \pi(v)$ holds. Now it is not hard to imagine that if there exists $\pi \in \Pi_b(G)$, then there must exist configurations of paths that are forbidden. To clarify this, let us rewrite the last part of the condition in Eq. 1, that is, $\pi(v_i) < \pi(v_{i+1}) \vee \pi(v_j) > \pi(v_{j+1})$, using a simple boolean transformation, which yields $\neg(\pi(v_i) > \pi(v_{i+1}) \wedge \pi(v_j) < \pi(v_{j+1}))$. So if there exists a path from v_{i+1} to v_i and one from v_j to v_{j+1} with $i < j$, then this expression evaluates to false for every $\pi \in \Pi(G)$. Therefore, we may refer to the pair of paths $v_{i+1} \rightsquigarrow v_i$ and $v_j \rightsquigarrow v_{j+1}$ with $i < j$ as a *forbidden configuration* of paths. See Fig. 2a for an illustration.

We may state now that in case there exists a bitonic st-ordering, the aforementioned configuration of paths cannot exist:

$$\exists \pi \in \Pi_b(G) \Rightarrow \forall u \in V \text{ with } S(u) = \{v_1, \ldots, v_m\}$$
$$\forall 1 \le i < j < m : v_{i+1} \not\rightsquigarrow v_i \vee v_j \not\rightsquigarrow v_{j+1}.$$

Conversely, if we find an u with v_i and v_j in a graph for which these paths exist, then we can safely reject it as one that does not admit a bitonic st-ordering. The following well-known property of planar st-graphs will prove itself useful when it comes to testing for the existence of a path between two vertices.

Lemma 3. *Let F be the subgraph of an embedded planar st-graph $G = (V, E)$ induced by a face that is not the outer face[1], and u, v two vertices of F, that is, u and v are on the boundary of the face. Then there exists a path from u to v in G, if and only if there exists such a path in F.*

There are several ways to prove this result, one proof can be found in the work of de Fraysseix et al. [10]. Notice that Lemma 3 is concerned with every pair of vertices incident to the face. But we are only interested in paths between two consecutive successors v_i and v_{i+1} of a vertex u. Notice that v_i, v_{i+1} and u share a common face which is not the outer face and in which u is the face-source. Figure 2b–d illustrates all three possible cases: $v_i \rightsquigarrow v_{i+1}$ (b), $v_{i+1} \rightsquigarrow v_i$ (c), and no path at all (d). Hence, we can decide the existence of a path based on the sink of the common face.

To prove that the absence of forbidden configurations is sufficient for the existence of a bitonic st-ordering, we require the following technical proposition.

Proposition 1. *Given an embedded planar st-graph $G = (V, E)$ and a vertex $u \in V$ with successor list $S(u) = \{v_1, \dots, v_m\}$. If it holds that*

$$\forall\, 1 \leq i < j < m \,:\, v_{i+1} \not\rightsquigarrow v_i \vee v_j \not\rightsquigarrow v_{j+1},$$

then there exists $1 \leq h \leq m$ such that

$$(\forall\, 1 \leq i < h : v_{i+1} \not\rightsquigarrow v_i) \wedge (\forall\, h \leq i < m : v_i \not\rightsquigarrow v_{i+1})$$

holds. In other words, there exists at least one v_h in $S(u)$ whose preceding vertices in $S(u)$ are only connected by paths in clockwise direction, whereas paths between following vertices are directed counterclockwise.

Sketch of Proof. If exists, set $h = \min\{i \mid v_{i+1} \rightsquigarrow v_i\}$, otherwise set $h = m$. □

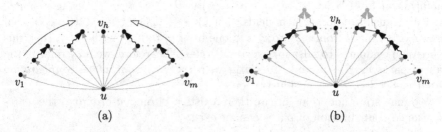

(a)   (b)

Fig. 3. (a) Paths orientations between consecutive successors of u. All of them directed towards v_h as described by Proposition 1. (b) The augmented graph G' in the proof of Lemma 4 obtained by adding edges between consecutive successors of u such that they are oriented towards v_h.

---

[1] This restriction is necessary due to the possible absence of the st-edge which is allowed by our definition of planar st-graphs.

The idea is now the following: If we have a graph that satisfies our necessary condition, then we can find for every $u \in V$ with $u \neq t$ a successor v_h with the property as described in Proposition 1. The intuition behind this property is that all paths that exist between successors of u, are directed in some way towards v_h. See Fig. 3a for an illustration. The next lemma exploits this property to obtain a bitonic *st*-ordering, which proves that this condition is indeed sufficient for the existence of a bitonic *st*-ordering.

Lemma 4. *Given a planar st-graph $G = (V, E)$ with a fixed embedding. If at every vertex $u \in V$ with successor list $S(u) = \{v_1, \ldots, v_m\}$ the following holds:*

$$\forall\, 1 \leq i < j < m \,:\, v_{i+1} \not\rightsquigarrow v_i \vee v_j \not\rightsquigarrow v_{j+1},$$

then G admits a bitonic st-ordering $\pi \in \Pi_b(G)$.

Proof. To show that there exists $\pi \in \Pi_b(G)$, we augment G into a new graph G' by inserting additional edges that we refer to as E'. These edges ensure that between every pair of consecutive successors in G, there exists a path in $G' = (V, E \cup E')$. Afterwards, we show that every *st*-ordering $\pi \in \Pi(G')$ for G' is a bitonic *st*-ordering for G.

For every vertex u with successor list $S(u) = \{v_1, \ldots, v_m\}$, we may assume by Proposition 1 that there exists $1 \leq h \leq m$ such that for every $1 \leq i < h$ there exists no path from v_{i+1} to v_i, and for every $h \leq i < m$ no path from v_i to v_{i+1} in G. Our goal is to add specific edges to fill the gaps such that there exist two paths in G', $v_1 \rightsquigarrow v_2 \rightsquigarrow \cdots \rightsquigarrow v_h \in G'$ and $v_m \rightsquigarrow v_{m-1} \rightsquigarrow \cdots \rightsquigarrow v_h \in G'$. Figure 3b illustrates the idea. More specifically, for every $1 \leq i < m$, there are three cases to consider: (i) There already exists a path between v_i and v_{i+1} in G, that is, $v_i \rightsquigarrow v_{i+1} \in G$ or $v_{i+1} \rightsquigarrow v_i \in G$. Proposition 1 ensures that the path is directed towards v_h, thus, we just skip the pair. (ii) If there exists no path between v_i and v_{i+1} in G and $i < h$ holds, we add an edge from v_i to v_{i+1}. (iii) When there also exists no path between v_i and v_{i+1}, but now $h \leq i < m$ holds, we add the reverse edge (v_{i+1}, v_i) to E'.

Before we continue, we show that $G' = (V, E \cup E')$ is *st*-planar. Consider a single edge in E' which has been added either by case (ii) or (iii) while traversing the successors $S(u)$ of some vertex $u \in V$. This edge will be added to a face in which u is the source, and since every face has only one source, only one edge will be added to the corresponding face, hence, planarity is preserved. Since case (ii) and (iii) only apply, when there exists no path between the two vertices, adding this edge will not generate a cycle. Induction on the number of added edges yields then *st*-planarity for G'.

Consider now an *st*-ordering $\pi \in \Pi(G')$. Since clearly $E' \subseteq E \cup E'$ holds, π is also an *st*-ordering for G, that is, $\Pi(G') \subseteq \Pi(G)$ holds. Recall that we constructed G' such that for every $u \in V$ with $S(u) = \{v_1, \ldots, v_m\}$, there exists $v_1 \rightsquigarrow v_2 \rightsquigarrow \cdots \rightsquigarrow v_h \in G'$ and $v_m \rightsquigarrow v_{m-1} \rightsquigarrow \cdots \rightsquigarrow v_h \in G'$. It follows that for every $\pi \in \Pi(G')$

$$\forall\, 1 \leq i < h : \pi(v_i) < \pi(v_{i+1}) \wedge \forall\, h \leq i < m : \pi(v_i) > \pi(v_{i+1})$$

Algorithm 1. Recognition and ordering algorithm for planar-st graphs

---

input : Embedded planar st-graph $G = (V, E)$ with $S(u)$ for every $u \in V$.
output: If exists, a bitonic st-ordering π for G.
begin
    $E' \leftarrow \emptyset$;
    for $u \in V$ *with* $S(u) = \{v_1, \ldots, v_m\}$ **do**
        decreasing $\leftarrow$ **false**;
        for $i = 1$ **to** $m - 1$ **do**
            $w \leftarrow$ FACESINK(u, v_i, v_{i+1});
            if $w = v_{i+1}$ **and** *decreasing* **then return** REJECT;
            if $w = v_i$ **then** *decreasing* $\leftarrow$ **true**;
            if $v_i \neq w \neq v_{i+1}$ **then**
                if *decreasing* **then** $E' \leftarrow E' \cup (v_{i+1}, v_i)$ **else** $E' \leftarrow E' \cup (v_i, v_{i+1})$;

    compute $\pi \in \Pi(V, E \cup E')$;
    return π

---

holds, which implies that $S(u)$ is bitonic with respect to π. Since this holds for all $u \in V$, it follows that $\Pi(G') \subseteq \Pi_b(G)$. Moreover, G' has at least one st-ordering, that is, $\Pi(G') \neq \emptyset$, thus, there exists $\pi \in \Pi_b(G)$. □

Let us summarize the implications of the lemma. The only requirement is that the graph complies with our necessary condition, that is, the absence of forbidden configurations. If this is the case, then Lemma 4 provides us with a bitonic st-ordering, which in turn proves that this condition is sufficient.

$$\exists \pi \in \Pi_b(G) \Leftrightarrow \forall u \in V \text{ with } S(u) = \{v_1, \ldots, v_m\}$$
$$\forall\, 1 \leq i < j < m\, :\, v_{i+1} \not\rightsquigarrow v_i \vee v_j \not\rightsquigarrow v_{j+1}$$

With a full characterization now at our disposal and in combination with Lemma 3, we are able to describe a simple linear-time algorithm (Algorithm 1) which tests a given graph and in case it admits a bitonic st-ordering, computes one. We iterate over $S(u)$ and as long as there is no path $v_{i+1} \rightsquigarrow v_i$, we assume $i < h$ and fill possible gaps. Once we encounter a path $v_{i+1} \rightsquigarrow v_i$ for the first time, we implicitly set $h = i$ via the flag and continue to add edges, but now the reverse ones. But in case we find a path $v_i \rightsquigarrow v_{i+1}$, then it forms with $v_{h+1} \rightsquigarrow v_h$ a forbidden configuration and the graph can be rejected. If we succeed in all successor list, an st-ordering for G' is computed, which is a bitonic one for G. Since G' is st-planar and has the same vertex set as G, we can claim that the overall runtime is linear. Let us state this as the main result of this section.

Theorem 2. *Deciding whether an embedded planar st-graph G admits a bitonic st-ordering π or not is linear-time solvable. Moreover, if G admits such an ordering, π can be found in linear time.*

Next we will consider the case in which no bitonic st-ordering exists. Although our initial motivation was to create upward planar straight-line drawings, we now allow bends and shift our efforts to upward planar poly-line drawings.

5   Upward Planar Poly-line Drawings with Few Bends

We start with a simple observation. Consider a forbidden configuration consisting of two paths $v_{i+1} \leadsto v_i$ and $v_j \leadsto v_{j+1}$ with $i < j$ between successors of a vertex u as shown in Fig. 2a. Notice that (u, v_i) and (u, v_{j+1}) are transitive edges. Since a reduced planar *st*-graph has no transitive edges, we can argue the following.

Corollary 2. *Every reduced planar st-graph admits a bitonic st-ordering.*

This leads to the idea to use the same transformation as Di Battista et al. [7] in their dominance-based approach. We can split every transitive edge to obtain a reduced planar *st*-graph and draw it upward planar straight-line. Replacing the dummy vertices with bends results in an upward planar poly-line drawing with at most $2|V| - 5$ bends, at most one bend per edge and quadratic area.

But we can do better using the following idea: If we have a single forbidden configuration, it suffices to split only one of the two transitive edges. More specifically, if we split in Fig. 2a the edge (u, v_i) into two new edges (u, v_i') and (v_i', v_i) with v_i' being the dummy vertex, then v_i' replaces v_i in $S(u)$. But now there exists no path from v_{i+1} to v_i', hence, the forbidden configuration has been destroyed at the cost of one split. Moreover, a pair of transitive edges does not necessarily induce a forbidden configuration. At this point the question arises how such a split affects other successor lists and if it may even create new forbidden configurations. The following trivial observation is helpful in this regard.

Lemma 5. *Let $G' = (V', E')$ be the graph obtained from splitting an edge (u, v) of a graph $G = (V, E)$ by inserting a dummy vertex v'. More specifically, let $V' = V \cup \{v'\}$ and $E' = (E - (u, v)) \cup \{(u, v'), (v', v)\}$. Then for all $w, x \in V$ there exists a path $w \leadsto x \in G$, if and only if there exists a path $w \leadsto x \in G'$.*

Since a forbidden configuration is solely defined by the existence of paths, we can argue now with Lemma 5 that a split does not create nor resolves forbidden configurations in other successor lists. However, one vertex that is not covered by the lemma is the dummy vertex itself, but it has only one successor which is insufficient for a forbidden configuration. This locality is of great value, because it enables us to focus on one successor list, instead of having to deal with a bigger picture. Next we prove an upper bound on the number of edges to split in order to resolve all forbidden configurations.

Lemma 6. *Every embedded planar st-graph $G = (V, E)$ can be transformed into a new one that admits a bitonic st-ordering by splitting at most $|V| - 3$ edges.*

Proof. Consider a vertex u and its successor list $S(u) = \{v_1, \ldots, v_m\}$ that contains multiple forbidden configurations of paths. Instead of arguing by means of forbidden configurations, we use our second condition from Proposition 1, that is, the existence of a vertex v_h such that every path that exists between two consecutive successors v_i and v_{i+1}, is directed from v_i towards v_{i+1} for $i < h$, or from v_{i+1} towards v_i if $i \le h$ holds. Of course h does not exist due to the forbidden configurations. But we can enforce its existence by splitting some edges.

Assume that we want v_h to be the first successor, that is, $h = 1$. Then every path from v_i to v_{i+1} with $1 \leq i < m$ is in conflict with this choice. We can resolve this by splitting every edge (u, v_{i+1}) for which a path $v_i \rightsquigarrow v_{i+1}$ exists. Clearly, the maximum number of edges to split is at most $m - 1$, that is the case in which for every $1 \leq i < m$, there exists a path from v_i to v_{i+1}. However, there do not exist paths $v_i \rightsquigarrow v_{i+1}$ and $v_{i+1} \rightsquigarrow v_i$ at the same time, because G is acyclic. So, if the number of edges to split is more than $\frac{m-1}{2}$, then there are less than $\frac{m-1}{2}$ paths of the form $v_{i+1} \rightsquigarrow v_i$. In that case, we may choose in a symmetric manner v_h to be the last successor ($h = m$), instead of being the first. Or in other words, we choose v_h to be the first or the last successor, depending on the direction of the majority of paths. And as a result, at most $\frac{m-1}{2}$ edges have to be split. Notice that the overall length of all successor lists is exactly the number of edges in the graph. Hence, with $m = |S(u)|$ we get $\sum_{u \in V} |S(u)| = |E| \leq 3|V| - 6$, and the claimed upper bound can be derived by

$$\sum_{u \in V} \frac{|S(u)| - 1}{2} \leq \frac{3|V| - 6 - |V|}{2} = |V| - 3.$$

Moreover, the split procedure preserves st-planarity of G.   □

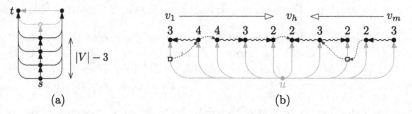

Fig. 4. (a) Example of a graph with $|V| - 3$ forbidden configurations, each requiring one split to be resolved. (b) Example for finding the smallest set of edges to split. The numbers indicate how many splits are necessary when choosing the corresponding vertex to be v_h. For v_5, v_6, v_8 and v_9 only two splits are necessary. Choosing $h = 6$ results in $E_{\text{split}} = \{(u, v_1), (u, v_8)\}$. The squares indicate the result of the two splits, whereas the dotted edges represent E' in Algorithm 1.

One may wonder now if this bound can be improved. Unfortunately, the graph shown in Fig. 4a is an example that requires $|V| - 3$ splits, hence, the bound is tight. It also shows that there exist graphs that can be drawn upward planar straight-line in polynomial area but do not admit a bitonic st-ordering. But we will push the idea of splitting edges a bit further from a practical point of view, and focus on the problem of finding a minimum set of edges to split.

In the following we describe an algorithm that solves this problem in linear time. To do so, we introduce some more notation. Let $u \in V$ be a vertex with successor list $S(u) = \{v_1, \ldots, v_m\}$. We define $L(u, h) = |\{i < h : v_{i+1} \rightsquigarrow v_i\}|$ and $R(u, h) = |\{i < h : v_i \rightsquigarrow v_{i+1}\}|$. If we choose now a particular $1 \leq h \leq m$ at

---

Algorithm 2. Algorithm for computing the minimum set of edges to split.

---

input : Embedded planar *st*-graph $G = (V, E)$ with $S(u)$ for every $u \in V$.
output: Minimum set $E_{\text{split}} \subset E$ to split for admitting a bitonic *st*-ordering.
begin

> $E_{\text{split}} \leftarrow \emptyset$;
> **for** $u \in V$ with $S(u) = \{v_1, \ldots, v_m\}$ **do**
>> $h \leftarrow 1$;
>> $c_{\min} \leftarrow c \leftarrow 0$;
>> **for** $i = 2$ **to** m **do**
>>> $w \leftarrow \text{FACESINK}(u, v_{i-1}, v_i)$;
>>> **if** $w = v_{i-1}$ **then** $c \leftarrow c + 1$;
>>> **if** $w = v_i$ **then** $c \leftarrow c - 1$;
>>> **if** $c < c_{\min}$ **then**
>>>> $c_{\min} \leftarrow c$;
>>>> $h \leftarrow i$;
>>
>> **for** $i = 1$ **to** $h - 1$ **do**
>>> **if** $v_i = \text{FACESINK}(u, v_i, v_{i+1})$ **then** $E_{\text{split}} \leftarrow E_{\text{split}} \cup (u, v_i)$;
>>
>> **for** $i = h$ **to** $m - 1$ **do**
>>> **if** $v_{i+1} = \text{FACESINK}(u, v_i, v_{i+1})$ **then** $E_{\text{split}} \leftarrow E_{\text{split}} \cup (u, v_{i+1})$;
>
> **return** E_{split}

---

u, then we have to split every edge (u, v_{i+1}) with $i < h$ for which there exists a path $v_{i+1} \rightsquigarrow v_i$, and every edge (u, v_i) with $h \leq i$ for which G contains a path $v_i \rightsquigarrow v_{i+1}$, that is, we have to split $L(u, h) + R(u, m) - R(u, h)$ edges. See Fig. 4b for an example. When now considering all successor lists, the minimum number of edge splits is

$$\sum_{u \in V} \left(R(u, m) + \min_{1 \leq h \leq m} \{ L(u, h) - R(u, h) \} \right).$$

Notice that the locality of a split allows us to minimize the number of edge splits for every successor list independently. From an algorithmic point of view, we are interested in the value of h and not in the number of splits, hence, we may drop $R(u, m)$ and consider the problem of finding h for which $L(u, h) - R(u, h)$ is minimum. Since this is now only a matter of counting paths for which we can again exploit Lemma 3, a linear-time algorithm becomes straightforward (see Algorithm 2). And as a result, we may state the following lemma without proof.

Lemma 7. *Every embedded planar st-graph $G = (V, E)$ can be transformed into a planar st-graph that admits a bitonic st-ordering by splitting every edge at most once. Moreover, the minimum number of edges to split is at most $|V| - 3$ and they can be found in linear time.*

Now we may use this to create upward planar poly-line drawings with few bends.

Theorem 3. *Every embedded planar st-graph $G = (V, E)$ admits an upward planar poly-line drawing within quadratic area having at most one bend per edge, at most $|V| - 3$ bends in total, and such a drawing can be obtained in linear time.*

Proof. We use Lemma 7 to obtain a new planar st-graph $G' = (V', E')$ with $|V'| \leq 2|V| - 3$ and a corresponding bitonic st-ordering π with Algorithm 1. With Theorem 1, an upward planar straight-line layout of size $(2|V'| - 2) \times (|V'| - 1)$ for G' is computed. Replacement of the dummy vertices by bends, yields an upward planar poly-line drawing for G of size at most $(4|V| - 8) \times (2|V| - 4)$. □

Recall that every upward planar graph is a spanning subgraph of a planar st-graph [6]. Therefore, the bound of $|V| - 3$ translates to all upward planar graphs.

Corollary 3. *Every upward planar graph $G = (V, E)$ admits an upward planar poly-line drawing within quadratic area having at most one bend per edge and at most $|V| - 3$ bends in total.*

6   Conclusion

In this work we have introduced the bitonic st-ordering for planar st-graphs. Although this technique has its limitations, it provides the properties of canonical orderings for the directed case. We have shown that this concept is viable by using a classic undirected incremental drawing algorithm for creating upward planar drawings with few bends.

References

1. Abbasi, S., Healy, P., Rextin, A.: Improving the running time of embedded upward planarity testing. Inf. Process. Lett. **110**(7), 274–278 (2010)
2. Bertolazzi, P., Di Battista, G., Liotta, G., Mannino, C.: Upward drawings of tri-connected digraphs. Algorithmica **6**(12), 476–497 (1994)
3. Biedl, T.C., Derka, M.: The (3, 1)-ordering for 4-connected planar triangulations. CoRR abs/1511.00873 (2015)
4. Cormen, T.H., Leiserson, C.E., Rivest, R.L., Stein, C.: Introduction to Algorithms, 3rd edn. The MIT Press, Cambridge (2009)
5. Di Battista, G., Frati, F.: A survey on small-area planar graph drawing. CoRR abs/1410.1006 (2014)
6. Di Battista, G., Tamassia, R.: Algorithms for plane representations of acyclic digraphs. Theoret. Comput. Sci. **61**(2–3), 175–198 (1988)
7. Di Battista, G., Tamassia, R., Tollis, I.: Area requirement and symmetry display of planar upward drawings. Discrete Comput. Geom. **7**(1), 381–401 (1992)
8. Didimo, W., Giordano, F., Liotta, G.: Upward spirality and upward planarity testing. SIAM J. Discrete Math. **23**(4), 1842–1899 (2009)
9. de Fraysseix, H., Pach, J., Pollack, R.: How to draw a planar graph on a grid. Combinatorica **10**(1), 41–51 (1990)
10. de Fraysseix, H., de Mendez, P.O., Rosenstiehl, P.: Bipolar orientations revisited. Discrete Appl. Math. **56**(2–3), 157–179 (1995). 5th Franco-Japanese Days

11. Garg, A., Tamassia, R.: On the computational complexity of upward and rectilinear planarity testing. In: Tamassia, R., Tollis, I.G. (eds.) GD 1994. LNCS, vol. 894, pp. 286–297. Springer, Heidelberg (1995). doi:10.1007/3-540-58950-3_384

12. Gronemann, M.: Bitonic st-orderings for Upward Planar Graphs. arXiv e-prints, August 2016. http://arxiv.org/abs/1608.08578v1

13. Gronemann, M.: Bitonic *st*-orderings of biconnected planar graphs. In: Duncan, C., Symvonis, A. (eds.) GD 2014. LNCS, vol. 8871, pp. 162–173. Springer, Heidelberg (2014). doi:10.1007/978-3-662-45803-7_14

14. Gronemann, M.: Algorithms for incremental planar graph drawing and two-page book embeddings. Ph.D. thesis, University of Cologne (2015)

15. Harel, D., Sardas, M.: An algorithm for straight-line drawing of planar graphs. Algorithmica **20**(2), 119–135 (1998)

16. Hutton, M.D., Lubiw, A.: Upward planarity testing of single-source acyclic digraphs. SIAM J. Comput. **25**(2), 291–311 (1996)

17. Kant, G.: Drawing planar graphs using the canonical ordering. Algorithmica **16**, 4–32 (1996)

18. Kant, G., He, X.: Regular edge labeling of 4-connected plane graphs and its applications in graph drawing problems. Theoret. Comput. Sci. **172**(1), 175–193 (1997)

19. Papakostas, A.: Upward planarity testing of outerplanar dags (extended abstract). In: Tamassia, R., Tollis, I.G. (eds.) GD 1994. LNCS, vol. 894, pp. 298–306. Springer, Heidelberg (1995). doi:10.1007/3-540-58950-3_385

20. Samee, M.A.H., Rahman, M.S.: Upward planar drawings of series-parallel digraphs with maximum degree three. In: WALCOM 2012, pp. 28–45. Bangladesh Academy of Sciences (BAS) (2007)

21. Schmidt, J.M.: The mondshein sequence. In: Esparza, J., Fraigniaud, P., Husfeldt, T., Koutsoupias, E. (eds.) ICALP 2014. LNCS, vol. 8572, pp. 967–978. Springer, Heidelberg (2014). doi:10.1007/978-3-662-43948-7_80

Low Ply Drawings of Trees

Patrizio Angelini[1], Michael A. Bekos[1(✉)], Till Bruckdorfer[1],
Jaroslav Hančl Jr.[2], Michael Kaufmann[1], Stephen Kobourov[3],
and Antonios Symvonis[4], and Pavel Valtr[2]

[1] Institut für Informatik, Universität Tübingen, Tübingen, Germany
{angelini,bekos,bruckdor,mk}@informatik.uni-tuebingen.de
[2] Department of Applied Mathematics, Charles University (KAM),
Prague, Czech Republic
{jaroslav,valtr}@kam.mff.cuni.cz
[3] Department for Computer Science, University of Arizona, Tucson, USA
kobourov@cs.arizona.edu
[4] School of Applied Mathematical and Physical Sciences, NTUA, Athens, Greece
symvonis@math.ntua.gr

Abstract. We consider the recently introduced model of *low ply graph drawing*, in which the ply-disks of the vertices do not have many common overlaps, which results in a good distribution of the vertices in the plane. The *ply-disk* of a vertex in a straight-line drawing is the disk centered at it whose radius is half the length of its longest incident edge. The largest number of ply-disks having a common overlap is called the *ply-number* of the drawing.

We focus on trees. We first consider drawings of trees with constant ply-number, proving that they may require exponential area, even for stars, and that they may not even exist for bounded-degree trees. Then, we turn our attention to drawings with logarithmic ply-number and show that trees with maximum degree 6 always admit such drawings in polynomial area.

1 Introduction

Let Γ be a straight-line drawing of a graph G. For a vertex $v \in G$, let the *ply-disk* D_v of v be the open disk with center v and radius r_v that is half of the length of the longest incident edge of v. For a point $q \in \mathbb{R}^2$ in the plane, denote by S_q the set of disks with q in their interior, i.e., $S_q = \{D_v \mid \|v - q\| < r_v\}$.

The *ply-number* of a straight-line drawing Γ is $\mathsf{pn}(\Gamma) = \max_{q \in \mathbb{R}^2} |S_q|$. In other words, it describes the maximum number of ply-disks that have a common non-empty intersection. The *ply-number* of a graph G is $\mathsf{pn}(G) = \min_{\Gamma \text{ of } G} \mathsf{pn}(\Gamma)$.

The ply-number is one of the most recent quality measures for graph layouts [5]. While traditional measures, such as edge crossings [4] and symmetries [6], have been studied for decades, the notion of optimizing a graph layout

An open access version of the full-text of the paper is also available [2]. Research partially supported by DFG grant Ka812/17-1. The research by Pavel Valtr was supported by the grant GAČR 14-14179S of the Czech Science Foundation.

© Springer International Publishing AG 2016
Y. Hu and M. Nöllenburg (Eds.): GD 2016, LNCS 9801, pp. 236–248, 2016.
DOI: 10.1007/978-3-319-50106-2_19

so that the spheres of influence of each vertex (see [12] for different variants) are well distributed is new. Goodrich and Eppstein [7] observed that real-word geographic networks usually have only constant sphere-of-influence overlap, or in the terminology of this paper, constant ply-number.

The problem of computing graph drawings with low ply-number is related to *circle-contact representations* of graphs, where vertices are interior-disjoint circles in the plane and two vertices are adjacent if the corresponding pair of circles touch each other [9,10]. Every maximal planar graph has a circle-contact representation [11]. A drawback of such representations is that the sizes of the circles may vary exponentially, making the resulting drawings difficult to read. In *balanced* circle packings and circle-contact representations, the ratio of the maximum and minimum diameters for the set of circles is polynomial in the number of vertices in the graph. Such drawings could be drawn with polynomial area, for instance, where the smallest circle determines the minimum resolution. It is known that trees and planar graphs with bounded tree-depth have balanced circle-contact representation [1]. Breu and Kirkpatrick [3] show that it is NP-complete to test whether a graph has a perfectly-balanced circle-contact representation, in which all circles have the same size, i.e., they are unit disks.

Very recently, Di Giacomo et al. [5] showed that binary trees, stars, and caterpillars have drawings with ply-number 2 (with exponential area, that is, the ratio of the longest to the shortest edge is exponential in the number of vertices), while general trees with height h admit drawings with ply-number $h + 1$. Also, they showed that the class of graphs with ply-number 1 coincides with the class of graphs that have a weak contact representation with unit disks, which makes the recognition problem NP-hard for general graphs [8]. On the other hand, testing whether an internally triangulated biconnected planar graph has ply-number 1 can be done in $O(n \log n)$ time. This paper left several natural questions open. Of particular interest are the following two questions:

(i) *Is it possible to draw a binary tree, a star, or a caterpillar in polynomial area with ply-number 2?*

(ii) *While binary trees have constant ply-number, is this true also for trees with larger bounded degree?*

In this paper we provide answers to the two above questions (Sect. 3). For the first question, we prove an exponential lower bound on the area requirements of drawings with constant ply-number of stars, and hence of caterpillars. For the second question, we prove that there exist trees with maximum degree 11 that do not have constant ply-number. Motivated by these two negative results, we consider in Sect. 4 drawings of trees with logarithmic ply-number. In this case, we present an algorithm to construct a drawing of every tree with maximum degree 6 in polynomial area[1]. We give preliminary definitions in Sect. 2 and discuss some open problems in Sect. 5.

---

[1] The *area* of a drawing is the area of the smallest axis-aligned rectangle containing it, under the resolution rule that each edge has length at least 1.

2 Preliminaries

Let G be a graph. We denote by $\ell(e)$ (by $\ell(u, v)$) the length of an edge $e \in G$ (an edge $(u, v) \in G$) in a straight-line drawing of G. Also, for a path $P = v_1, \ldots, v_m$, we denote by $\ell(P) = \sum_{i=1}^{m-1} \ell(v_i, v_{i+1})$ the total length of its edges. Further, we denote by D_v the ply-disk in Γ of a vertex $v \in G$ and by r_v the radius of D_v. Finally, we call *constant-ply drawing* (or *log-ply drawing*) a straight-line drawing Γ such that $\mathsf{pn}(\Gamma) = O(1)$ (such that $\mathsf{pn}(\Gamma) = O(\log n)$).

Let T be a tree rooted at a vertex r. The *depth* d_v of a vertex $v \in T$ is the length of the path between v and r; note that $d_r = 0$. The *height* h of T is the maximum depth of a vertex of T.

3 Constant-Ply Drawings of Trees

In this section we provide negative answers to two open questions [5] about constant-ply drawings of trees. In Subsect. 3.1 we prove that drawings of this type may require exponential area, even for stars, while in Subsect. 3.2 we prove that there exist bounded-degree trees not admitting any of such drawings.

3.1 Area Lower Bound for Stars

In the original paper on the topic [5], it has been shown that a star admits a drawing with ply-number 1 if and only if it has at most six leaves, and that every star admits a drawing with ply-number 2, independently of the number of leaves. The algorithm for the latter result is based on a placement of the leaves at exponentially-increasing distances from the central vertex, which results in a drawing with exponential area; see Fig. 1a. In this subsection we prove that this is in fact unavoidable, as we give an exponential lower bound for the area requirements of any drawing of a star with constant ply-number.

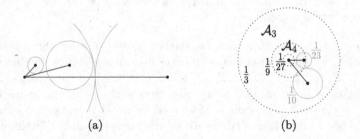

(a)   (b)

Fig. 1. (a) An exponential-area drawing with ply-number 2 of a star [5]. (b) Illustration for the proof of Theorem 1; two disks belonging to class $\mathcal{T}_3$ are entirely contained inside annuli $\mathcal{A}_3$ and $\mathcal{A}_4$. The number close to each disk is its radius.

Theorem 1. *Any constant-ply drawing of an n-vertex star has exponential area.*

Proof. Let $K_{1,n-1}$ be an n-vertex star with central vertex v, and let Γ be a straight-line drawing of $K_{1,n-1}$ with ply-number p, where $p = O(1)$. We prove the statement by showing that the ratio of the longest to the shortest edge in Γ is exponential in n. Assume that the longest edge e of Γ has length $\ell(e) = 2$, after a possible scaling of Γ; thus, the largest ply-disk in Γ has radius 1.

For any $i \in \mathbb{N}$ we define $\mathcal{A}_i$ to be the annulus delimited by two circles centered at v with radius 3^{-i+2} and 3^{-i+1}, respectively. Refer to Fig. 1b. Then, we partition the ply-disks of the $n - 1$ leaves of $K_{1,n-1}$ into the classes $\mathcal{T}_1, \ldots, \mathcal{T}_k$ in such a way that all the disks with radius in $(3^{-j}, 3^{-j+1}]$ belong to $\mathcal{T}_j$, with $1 \leq j \leq k$. We observe that every disk in class $\mathcal{T}_j$ is entirely contained inside the annulus $\mathcal{A}_j \cup \mathcal{A}_{j+1}$; see Fig. 1b. However, there can be at most

$$\frac{p|\mathcal{A}_j \cup \mathcal{A}_{j+1}|}{\min_{D \in \mathcal{T}_j} |D|} \leq \frac{p(\pi 3^{-2j+4} - \pi 3^{-2j})}{\pi 3^{-2j}} = 80p$$

disks in any $\mathcal{A}_j \cup \mathcal{A}_{j+1}$, and hence at most $80p$ disks belong to class $\mathcal{T}_j$. Therefore, $n = 1 + \sum_{j=1}^{k} |T_j| \leq 80pk$ implies that the smallest radius of the ply-disk of a vertex in a drawing is at most 3^{-k}. This implies that the ratio between the largest and the smallest ply-disk radii in Γ, and hence between the longest and the shortest edge, is at least $3^k \geq 3^{n/(80p)}$. This concludes the proof. □

3.2   Large Bounded-Degree Trees

In this section we consider the question posed in [5] on whether bounded-degree trees admit constant-ply drawings. While the answer is positive for binary trees [5], as they admit drawings with ply-number 2, we prove that this positive result cannot be extended to all bounded-degree trees, and in particular to 10-*ary trees*, that is, rooted trees with maximum degree 11.

In the following we denote a complete 10-ary tree of height h by T_{10}^h; note that T_{10}^h has 10^h leaves and 10^d vertices with depth $d \leq h$. The root of a tree T is denoted by $\text{root}(T)$. In the rest of the section we prove the following theorem.

Theorem 2. *For every $M > 0$ there is an integer $h > 0$ such that $\text{pn}(T_{10}^h) \geq M$.*

A *branch* of T_{10}^h is a path in T_{10}^h connecting the root with a leaf. Let e and f be two edges of T_{10}^h. Refer to Fig. 2a. We say that e *dominates* f and write $e >_D f$, if e and f lie on a common branch and $\ell(e) \geq 3^{s+1}\ell(f)$, where s is the number of edges on the path between e and f different from e and f. Observe that on each branch of T_{10}^h the relation $>_D$ is transitive. We say that e *first-hand dominates* f and write $e >_{\text{FD}} f$, if the following three conditions are satisfied: (i) f lies on the path connecting e with the root of T_{10}^h, (ii) e dominates f, and (iii) no other edge on the path between e and f dominates f.

Lemma 1. *Let P be a path with edges $f_0, f_1, \ldots, f_p$. Suppose that f_0 dominates each of the edges $f_1, \ldots, f_p$. Let v be the common vertex of the edges f_0 and f_1. Then the edges $f_1, \ldots, f_p$ lie entirely inside the ply-disk D_v of v.*

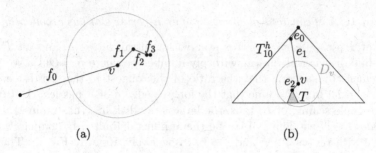

Fig. 2. (a) A path P with $\ell(f_0) = 28$, $\ell(f_1) = 6$, $\ell(f_2) = 4$, and $\ell(f_3) = 1$. Edge f_0 dominates each of f_1, f_2, and f_3, which in fact lie inside D_v. Also, f_0 first-hand dominates f_1 and f_2, but does not first-hand dominate f_3, since f_2 dominates f_3. (b) Illustration for the proof of Lemma 3.

Proof. See Fig. 2a. Since the radius of D_v is at least $\frac{\ell(f_0)}{2}$, it suffices to prove $\ell(f_1) + \cdots + \ell(f_p) < \frac{\ell(f_0)}{2}$. Let $i \in \{1, \ldots, p\}$. Since f_0 dominates f_i, we have $\ell(f_i) \leq \frac{\ell(f_0)}{3^i}$. Thus, $\ell(f_1) + \cdots + \ell(f_p) \leq \ell(f_0)(\frac{1}{3} + \frac{1}{3^2} + \cdots + \frac{1}{3^p}) < \frac{\ell(f_0)}{2}$.   □

Lemma 2. *Let* $e_1, \ldots, e_M$ *be* M *edges in* T_{10}^h *such that* $e_1 >_{FD} e_2 >_{FD} \cdots >_{FD} e_M$. *Then,* $\mathsf{pn}(T_{10}^h) \geq M$.

Proof. By definition, $e_1, \ldots, e_M$ appear in this order, possibly not consecutively, along the same branch of T_{10}^h. Let $\overrightarrow{P}$ be the oriented path that is the subpath of this branch from e_1 to e_M. Since $e_i >_{FD} e_{i+1}$, edge e_i dominates all the edges between e_i and e_{i+1}. Due to the transitivity of $>_D$, each edge e_i dominates all the edges $e_{i+1}, \ldots, e_M$, and hence all the edges appearing after it along $\overrightarrow{P}$.

By Lemma 1, the endvertex v_M of e_M lies inside the ply-disk D_{v_i} of v_i, for each $i = 1, \ldots, M$, where v_i is the last vertex of e_i along $\overrightarrow{P}$. Thus, the M disks $D_{v_1}, \ldots, D_{v_M}$ have a non-empty intersection, and the statement follows.   □

Consider a vertex v with depth d in T_{10}^h. We say that a vertex $u \neq v$ is a *descendant* of v if the path from $\mathrm{root}(T_{10}(h))$ to u contains v. For any $i = 1, \ldots, h-d$, we denote by $T_{10}^i(v)$ the subtree of T_{10}^h rooted at v induced by v and by all the descendants of v with depth $d+1, d+2, \ldots, d+i$. Note that $T_{10}^i(v)$ is a 10-ary tree of height i, thus it has 10^i leaves. We have the following.

Lemma 3. *Let* T' *be a subtree of a rooted 10-ary tree* T *and let* P *be the path from* $\mathrm{root}(T)$ *to* $\mathrm{root}(T')$. *If every edge of* T' *is dominated by at least one edge of* P, *then there exists a vertex* $v \in P$ *such that* T' *lies completely inside* D_v.
    Consequently, $\mathsf{pn}(T) \geq \mathsf{pn}(T') + 1$.

Proof. Refer to Fig. 2b. Let $e_0, e_1, \ldots, e_t$ be the edges of P in the order in which they appear along P, when P is oriented from $\mathrm{root}(T)$ to $\mathrm{root}(T')$. Let i be an index maximizing the value of $3^i \cdot \ell(e_i)$. Then e_i dominates all the edges $e_{i+1}, e_{i+2}, \ldots, e_t$. Also, due to the choice of i and since any edge of T' is dominated by some edge of P, any edge of T' is dominated by e_i. Let v be the root of

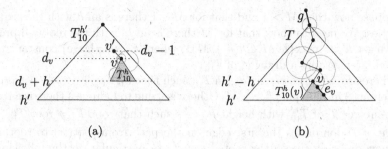

Fig. 3. (a) Illustration for Lemma 4; no edge of $T_{10}^h(v)$ dominates edge (v, v'). (b) Illustration for **Case 2**: for each vertex v with depth $h' - h$, there exists an edge e_v in $T_{10}^h(v)$ that is not dominated by any edge of path from v to the root of T.

T, if $i = 0$, or the common vertex of e_{i-1} and e_i otherwise. Then Lemma 1 can be applied on the path from v to any leaf of T' to show that its subpath from $\mathrm{root}(T')$ to the leaf lies inside D_v, which proves the statement.

As a consequence, we have $\mathsf{pn}(T) \geq \mathsf{pn}(T') + 1$.   □

Lemma 4. *Let h, h', M be three positive integers such that $h' \geq h(M-1) + 1$. If there exists a drawing Γ of $T_{10}^{h'}$ that contains no M edges $e_1, \ldots, e_M$ such that $e_1 >_{FD} e_2 >_{FD} \cdots >_{FD} e_M$, then there exists a vertex v in $T_{10}^{h'}$ with depth $1 \leq d_v \leq h' - h$ such that no edge of $T_{10}^h(v)$ in Γ dominates the edge (v, v'), where v' is the neighbor of v with depth $d_v - 1$. Refer to Fig. 3a.*

Proof. We fix h and proceed by induction on M. If $M = 1$, then there exists no drawing Γ of $T_{10}^{h'}$ satisfying the conditions of the lemma, and thus the statement holds. Suppose now that $M > 1$ and that the lemma holds for $M - 1$. We want to show that the lemma holds for M. Let $h' \geq h(M-1) + 1$. Suppose that a drawing of $T_{10}^{h'}$ contains no M edges $e_1, \ldots, e_M$ such that $e_1 >_{FD} e_2 >_{FD} \cdots >_{FD} e_M$.

Consider the subtree $T' := T_{10}^{h'-h}(\mathrm{root}(T_{10}^{h'}))$ of $T_{10}^{h'}$, with the same root as $T_{10}^{h'}$, that is induced by the vertices with depth at most $h' - h$. If T' does not contain $M - 1$ edges $e_1, \ldots, e_{M-1}$ such that $e_1 >_{FD} e_2 >_{FD} \cdots >_{FD} e_{M-1}$, then the required vertex v exists by induction. Otherwise, consider $M - 1$ edges $e_2, \ldots, e_M$ in T' such that $e_2 >_{FD} e_3 >_{FD} \cdots >_{FD} e_M$. Let d and $d + 1$, with $d < h' - h$, be the depth of the endvertices v' and v of e_2, respectively, in T' (it is the same depth as they have in $T_{10}^{h'}$). Consider the subtree $T_{10}^h(v)$ of $T_{10}^{h'}$ rooted at v. Suppose, for a contradiction, that there exists an edge in $T_{10}^h(v)$ dominating e_2. Then, consider the edge e_1 in $T_{10}^h(v)$ dominating e_2 with the property that no other edge on the path from e_1 to e_2 dominates e_2, that is, e_1 first-hand dominates e_2. Thus, $e_1 >_{FD} e_2 >_{FD} \cdots >_{FD} e_M$, a contradiction. This implies that no edge of $T_{10}^h(v)$ dominates edge e_2, and the statement follows.   □

We are now ready to complete the proof of the main result of the section.

Proof (of Theorem 2). We proceed by induction on M. For $M = 1$ the statement trivially holds.

Suppose now that $M > 1$ and that for $M - 1$ there is an h with the required properties. We need to show that for M there is an h' with the required properties. We set $h' := \max\{h^2 M, Ch(h + M)\}$, where C is a (large) constant to be specified later. We fix a drawing of $T_{10}^{h'}$.

If there are M edges $e_1, \ldots, e_M$ in $T_{10}^{h'}$ such that $e_1 >_{\mathrm{FD}} e_2 >_{\mathrm{FD}} \cdots >_{\mathrm{FD}} e_M$, then Lemma 2 implies $\mathsf{pn}(T_{10}^{h'}) \geq M$. Otherwise, due to Lemma 4 there is a rooted 10-ary subtree T of $T_{10}^{h'}$ with height $\overline{h} \geq \frac{h'}{M}$ such that $\mathrm{root}(T) \neq \mathrm{root}(T_{10}^{h'})$ and no edge of T dominates the first edge on the path from $\mathrm{root}(T)$ to $\mathrm{root}(T_{10}^{h'})$. From now on, we focus on the rooted tree T. In particular, in the following we refer to the depth of a vertex as its depth in T. We distinguish two cases.

In **Case 1** there exists a vertex v with depth $\overline{h} - h$ in T such that every edge of the tree $T_{10}^h(v)$ is dominated by at least one edge of the path from v to $\mathrm{root}(T)$. In this case, Lemma 3 (applied on tree $T_{10}^h(v)$) and the inductive hypothesis show that $\mathsf{pn}(T_{10}^{h'}) \geq \mathsf{pn}(T_{10}^h(v)) + 1 \geq M$.

In **Case 2** there exists no vertex in T with the above properties. Refer to Fig. 3b. Thus, for any vertex v with depth $\overline{h} - h$ in T, the subtree $T_{10}^h(v)$ rooted at v contains at least one edge that is not dominated by any edge of the path from v to $\mathrm{root}(T)$; among these edges of $T_{10}^h(v)$ we choose one, denoted by $e(v)$, whose endvertices have the smallest possible depth. This implies that $e(v)$ is not dominated by any edge of the path P_v from its endvertex u_v to $\mathrm{root}(T)$.

Let g be the first edge from $\mathrm{root}(T)$ to $\mathrm{root}(T_{10}^{h'})$. Note that edges of P_v dominate neither g nor $e(v)$. W.l.o.g., assume $\ell(g) = 1$. Since edges g and $e(v)$ do not dominate each other, we have $1/3^{\overline{h}} < \ell(e(v)) < 3^{\overline{h}}$. Thus, there is a unique integer $k(v) \in \{-\overline{h}, -\overline{h} + 1, \ldots, \overline{h} - 1\}$ such that $\ell(e(v)) \in [3^{k(v)}, 3^{k(v)+1})$.

Let k be a most frequent value of $k(v)$ over all the vertices v with depth $\overline{h} - h$. Since $k(v)$ may have $2\overline{h}$ different values, the set V_k of vertices v at level $\overline{h} - h$ with $k(v) = k$ has size at least $10^{\overline{h}-h}/(2\overline{h})$. Consider now a vertex $v \in V_k$ and the path P_v from $\mathrm{root}(T)$ to u_v. Since no edge of this path dominates g or $e(v)$, we have the following two upper bounds on the length of the i-th edge e_i of the path P_v oriented from $\mathrm{root}(T)$ to u_v:

$$\ell(e_i) \leq 3^i \quad \text{and}$$

$$\ell(e_i) \leq 3^{\overline{h}-i} \cdot \ell(e(v)) < 3^{\overline{h}-i+k+1}.$$

For the latter, we use $\ell(e(v)) < 3^{k+1}$, which follows from the fact that $v \in V_k$.

The edges e_i with $i \leq (\overline{h} + k)/2$ have total length at most $\sum_{i=0}^{\lfloor(\overline{h}+k)/2\rfloor} 3^i \leq 3^{(\overline{h}+k)/2+1}$, and the total length of the other edges is at most

$$\sum_{i=\lfloor(\overline{h}+k)/2+1\rfloor}^{\overline{h}} 3^{\overline{h}-i+k+1} = 3^{k+1} \cdot \sum_{i=\lfloor(\overline{h}+k)/2\rfloor+1}^{\overline{h}} 3^{\overline{h}-i}$$

$$= 3^{k+1} \cdot \sum_{j=0}^{\overline{h}-\lfloor(\overline{h}+k)/2\rfloor-1} 3^j \leq 3^{k+1} \cdot 3^{(\overline{h}-k)/2+1} = 3^{(\overline{h}+k)/2+2}.$$

It follows that the total length of the path P_v is smaller than $12 \cdot 3^{(\overline{h}+k)/2}$.

Thus all the edges $e(v)$, $v \in V_k$, lie in the disk D of radius $12 \cdot 3^{(\overline{h}+k)/2}$ centered at $\text{root}(T)$. The area of D is $12^2 3^{\overline{h}+k}$. Let $v \in V_k$, and let u'_v be the vertex of the path P_v adjacent to u_v. The ply-disk $D_{u'_v}$ contains the disk of radius $3^k/2$ centered at u'_v, which is entirely contained in D. It follows that the region $D_{u'_v} \cap D$ has area at least $\pi(3^k/2)^2 = (\pi/4)3^{2k}$. Therefore there is a point of D lying in at least

$$\frac{|V_k| \cdot (\pi/4)3^{2k}}{\text{area}(D)} \geq \frac{(10^{\overline{h}-h}/(2\overline{h})) \cdot (\pi/4)3^{2k}}{12^2\pi 3^{\overline{h}+k}} = \frac{(10/3)^{\overline{h}}/\overline{h} \cdot 3^k}{12^2 \cdot 8 \cdot 10^h} \geq \frac{(10/9)^{\overline{h}}/\overline{h}}{12^2 \cdot 8 \cdot 10^h}$$

disks $D_{u'_v}$, with $v \in V_k$.

Since $h' \geq CM(h + \log M)$, we have $\overline{h} \geq C(h + \log M)$. If C is a sufficiently large constant then some point of D lies in at least

$$\frac{(10/9)^{\overline{h}}/\overline{h}}{12^2 \cdot 8 \cdot 10^h} \geq M$$

disks $D_{u'_v}$, with $v \in V_k$, which concludes the proof.   □

4   Log-Ply Drawings of Bounded-Degree Trees in Polynomial Area

Motivated by the fact that constant-ply drawings of stars may require exponential area (Theorem 1) and by the fact that not all the bounded-degree trees admit a constant-ply drawing (Theorem 2), in this section we ask whether allowing a logarithmic ply-number makes it possible to always construct drawings of trees, possibly in polynomial area. We give a first answer by proving in Theorem 3 that this is true for 5-ary trees, that is, trees with maximum degree 6.

We start with some definitions. A 2-*drawing* of a path $P = v_1, \ldots, v_m$ is a straight-line drawing of P in which all the vertices lie along the same straight-line segment in the same order as they appear in P and for each $i = 2, \ldots, m$ we have $\frac{\ell(v_{i-1}, v_i)}{2} \leq \ell(v_i, v_{i+1}) \leq 2\ell(v_{i-1}, v_i)$; see Fig. 4a. We have the following.

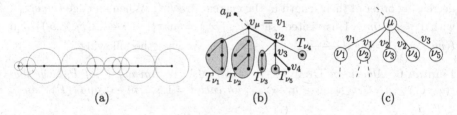

(a)   (b)   (c)

Fig. 4. (a) A 2-drawing of a path. (b) A ternary tree T_μ and the path μ, represented by fat edges, that is a node of the heavy-path tree $\mathcal{T}$; the subtrees $T_{\nu_1}, \ldots, T_{\nu_5}$ obtained when removing μ are inside shaded region. (c) The portion of $\mathcal{T}$ containing nodes μ and its children $\nu_1, \ldots, \nu_5$. The arc of $\mathcal{T}$ between μ and a node ν_i is labeled with the node of μ that is the anchor of ν_i.

Lemma 5. *A 2-drawing of a path $P = (v_1, \ldots, v_n)$ has ply-number at most 2.*

Proof. Refer to Fig. 4a. For each vertex v_i, we have radius $r_{v_i} \leq \ell(v_i, v_{i+1})$ and $r_{v_i} \leq \ell(v_{i-1}, v_i)$, since $\frac{\ell(v_{i-1}, v_i)}{2} \leq \ell(v_i, v_{i+1}) \leq 2\ell(v_{i-1}, v_i)$. This, together with the fact that all the vertices of P lie along the same straight-line segment, implies that the ply-disk D_{v_i} of v_i may only intersect with $D_{v_{i-1}}$ and with $D_{v_{i+1}}$, but not with any of the other disks (note that D_{v_i} may touch $D_{v_{i-2}}$ and $D_{v_{i+2}}$ in a single point, namely the one where vertices v_{i-1} and v_{i+1} lie, respectively), but cannot overlap with them. □

The *heavy-path tree* $\mathcal{T}$ of a rooted tree T is a decomposition tree of T first defined by Sleator and Tarjan [13] as follows; see Figs. 4b–c. Each node $\mu \in \mathcal{T}$ is a path in T between a vertex v_μ of T and a leaf of the subtree T_μ of T rooted at v_μ. At the first step, v_μ is the root of T, T_μ is T, and the path μ we construct is the root of $\mathcal{T}$. To construct μ, we start from v_μ and we always select the child of the current vertex whose subtree contains the largest number of vertices, until a leaf of T_μ is reached. Then, we remove all the vertices of μ from T_μ and their incident edges, hence obtaining a set of subtrees of T_μ. For each of these subtrees T_ν, rooted at a vertex v_ν, we add a new node ν as a child of μ in $\mathcal{T}$ and recursively construct the corresponding path. Since each subtree T_ν has at most half of the vertices of T_μ, the height of the heavy-path tree $\mathcal{T}$ is $O(\log n)$.

Let $\mu = (v_\mu = v_1, \ldots, v_m)$ be any node in $\mathcal{T}$ and let τ be its parent. The vertex of τ that is adjacent to v_μ is the *anchor* a_μ of μ; in order to have an anchor a_μ also when μ is the root of $\mathcal{T}$, we add a dummy vertex to T that is only incident to its root. The proof of the main theorem of this section is based on the following algorithm, which we call DRAWPATH, to construct a special 2-drawing of the path P that is the concatenation of edge (a_μ, v_μ) and of path μ.

Let n_μ be the total number of vertices in the subtrees of T_τ whose corresponding paths have a_μ as an anchor. Since T_μ is one of these subtrees, we have that $n_\mu > \sum_{i=1}^m n_i$, where n_i is the total number of vertices in the subtrees $T_{\nu_1}, \ldots, T_{\nu_h}$ of T_μ such that paths $\nu_1, \ldots, \nu_h$ have v_i as anchor. Also, since μ is a path in a heavy-path tree, we have $n_i \leq n_\mu/2$ for each $1 \leq i \leq m$.

Algorithm DRAWPATH starts by initializing $\ell(a_\mu, v_1) = n_1$ and $\ell(v_i, v_{i+1}) = n_i + n_{i+1}$, for each $i = 1, \ldots, m - 1$. Then, it visits the edges of P one by one in decreasing order of their length in the current drawing. When an edge (v_i, v_{i+1}), with $1 \leq i \leq m - 1$, is visited, set $\ell(v_{i-1}, v_i) = \max\{\frac{\ell(v_i, v_{i+1})}{2}, \ell(v_{i-1}, v_i)\}$ and $\ell(v_{i+1}, v_{i+2}) = \max\{\frac{\ell(v_i, v_{i+1})}{2}, \ell(v_{i+1}, v_{i+2})\}$. We have the following.

Lemma 6. *Algorithm DRAWPATH constructs a 2-drawing Γ of P such that $\ell(a_\mu, v_1) \geq n_1$, $\ell(v_i, v_{i+1}) \geq n_i + n_{i+1}$, for each $i = 1, \ldots, m-1$, and $\ell(P) \leq 6n_\mu$.*

Proof. First observe that $\ell(a_\mu, v_1) \geq n_1$ and $\ell(v_i, v_{i+1}) \geq n_i + n_{i+1}$ for each $i = 1, \ldots, m - 1$, since this is true already after the initialization and since no operation performed by the algorithm reduces the length of any edge.

Also, the fact that Γ is a 2-drawing can be derived from the operations that are performed when an edge is visited. Note that after an edge has been visited by DRAWPATH, its length is not modified any longer, since the edges are visited in decreasing order of edge lengths and since the length of an edge is modified only if this edge is shorter than one of its adjacent edges.

For the same reason, if an edge (v_i, v_{i+1}), with $1 \leq i \leq m - 1$, determines a local maximum in the sequence of edge lengths in Γ (that is, $\ell(v_h, v_{h+1}) \geq \ell(v_{h-1}, v_h)$ and $\ell(v_h, v_{h+1}) \geq \ell(v_{h+1}, v_{h+2})$), then $\ell(v_i, v_{i+1}) = n_i + n_{i+1}$. We use this property to prove that the total length of the edges in Γ is at most $6n_\mu$.

Consider any two edges (v_h, v_{h+1}) and (v_q, v_{q+1}), with $1 \leq h < q \leq m - 1$, such that $\ell(v_h, v_{h+1}) = n_h + n_{h+1}$, $\ell(v_q, v_{q+1}) = n_q + n_{q+1}$, and such that $\ell(v_i, v_{i+1}) > n_i + n_{i+1}$ for each $i = h + 1, \ldots, q - 1$; namely, (v_h, v_{h+1}) and (v_q, v_{q+1}) are two edges that have not been modified by algorithm DRAWPATH after the initialization and such that all edges between them have been modified.

Claim. The total length of the edges in the subpath P' of Γ between v_h and v_{q+1} is at most $2(n_h + n_{h+1}) + 2(n_q + n_{q+1})$.

Proof. Note that there exists no edge in P' different from (v_h, v_{h+1}) and (v_q, v_{q+1}) that determines a local maximum in the sequence of edge lengths, since this would contradict the fact that $\ell(v_i, v_{i+1}) > n_i + n_{i+1}$ for each $i = h+1, \ldots, q-1$. Hence, P' is composed of a sequence of edges starting at (v_h, v_{h+1}) and ending at an edge (v_{j-1}, v_j), with $h < j \leq q-1$, with decreasing edge lengths, and of a sequence of edges starting at (v_j, v_{j+1}) and ending at (v_q, v_{q+1}) with increasing edge lengths. We have $\ell(v_h, v_{h+1}) = n_h + n_{h+1}$ and $\ell(v_q, v_{q+1}) = n_q + n_{q+1}$, by construction. Also, $\sum_{i=h+1}^{j-1} \ell(v_i, v_{i+1}) = \sum_{i=1}^{j-1-h} \frac{n_h + n_{h+1}}{2^i} < n_h + n_{h+1}$, since Γ is a 2-drawing. Analogously, $\sum_{i=j}^{q-1} \ell(v_i, v_{i+1}) < n_q + n_{q+1}$. □

Hence, every edge (v_h, v_{h+1}) such that $\ell(v_h, v_{h+1}) = n_h + n_{h+1}$, together with the possible sequence of edges with increasing (decreasing) edge lengths preceding (following) it, gives a contribution of less than $3(n_h + n_{h+1})$. Since $\sum_{i=1}^m (n_i + n_{i+1}) < 2n_\mu$, the total edge length is at most $6n_\mu$. □

We describe an algorithm to construct a log-ply drawing of any rooted n-vertex 5-ary tree T with polynomial area. To simplify the description, we first give the algorithm for ternary trees; we discuss later the extension to 5-ary trees.

Construct the heavy-path tree $\mathcal{T}$ of T. Then, construct a drawing of $\mathcal{T}$ recursively according to a bottom-up traversal of $\mathcal{T}$. At each step of the traversal, consider a path $\mu \in \mathcal{T}$. We associate μ with a half-disk D_μ of radius $6^{h-d_\mu} n_\mu$, where h is the height of $\mathcal{T}$ and d_μ is the depth of μ. Refer to Fig. 5a. The goal is to construct a drawing with ply-number at most $2(h - d_\mu + 1)$ of the subtree T_μ rooted at v_μ, augmented with the anchor a_μ of μ and with edge (a_μ, v_μ), inside D_μ in such a way that a_μ lies on the center of D_μ and all the vertices of μ lie along the radius of D_μ that is perpendicular to the diameter delimiting D_μ.

If $\mu = (v_1, \ldots, v_m)$ is a leaf of $\mathcal{T}$, place a_μ on the center of D_μ and the vertices of μ along the radius of D_μ perpendicular to the diameter delimiting it, so that each edge has length 1. This drawing has ply-number 1 and satisfies the required properties by construction.

If $\mu = (v_1, \ldots, v_m)$ is not a leaf, let $\nu_1, \ldots, \nu_k$ be its children. Assume inductively that for each child ν_j, with $j = 1, \ldots, k$, there exists a drawing with ply-number at most $2(h - d_{\nu_j} + 1)$ inside the half-disk D_{ν_j} with radius $6^{h-d_{\nu_j}} n_{\nu_j}$

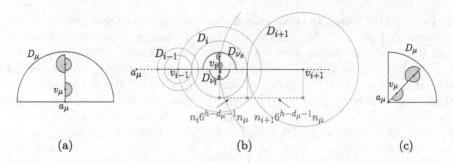

Fig. 5. (a) The half-disk D_μ associated with μ. (b) Illustration for the algorithm for ternary trees. Black dotted circles are the disks D_i; orange solid circles are the ply-disks. (c) Using a quarter-disk instead of a half-disk for 5-ary trees. (Color figure online)

with the required properties. We show how to construct a drawing with ply-number at most $2(h - d_\mu + 1)$ of T_μ inside the half-disk D_μ with radius $6^{h-d_\mu} n_\mu$ with the required properties; recall that $d_\mu = d_{\nu_j} - 1$, for each $j = 1, \ldots, k$.

Refer to Fig. 5b. First, apply algorithm DRAWPATH to construct a 2-drawing of the path P composed of μ and of its anchor a_μ such that $\ell(a_\mu, v_1) \geq n_1$, $\ell(v_i, v_{i+1}) \geq n_i + n_{i+1}$, for $i = 1, \ldots, m-1$, and the total length of the edges in P is at most $6n_\mu$. Then, scale the obtained drawing by a factor of $6^{h-d_\mu-1} n_\mu$, which implies that the total length of the edges in P is at most $6^{h-d_\mu} n_\mu$. Hence, it is possible to place the obtained drawing inside D_μ in such a way that a_μ lies on its center and the vertices of μ lie along the radius that is perpendicular to the diameter delimiting D_μ. Further, for each vertex $v_i \in \mu$, consider a disk D_i centered at v_i of diameter $6^{h-d_\mu-1} n_i$. Due to the scaling performed before, no two disks D_i and D_h, with $1 \leq i, h \leq m$, intersect with each other.

Consider now the at most two children ν_s and ν_t of μ whose anchor is v_i; since $d_\mu = d_{\nu_j} - 1$, for each $j = 1, \ldots, k$, and since $n_i = n_{\nu_s} + n_{\nu_t}$, the diameter of the half-disk D_{ν_s} and the one of the half-disk D_{ν_t} are both not larger than the diameter of disk D_i. Thus, we can plug the drawings of T_{ν_s} and T_{ν_t} lying inside D_{ν_s} and D_{ν_t}, which exist by induction, so that the centers of D_{ν_s} and D_{ν_t} coincide with the center of D_i, and the diameters delimiting D_{ν_s} and D_{ν_t} lie along edges (v_{i-1}, v_i) and (v_i, v_{i+1}); see Fig. 5b. Hence, the constructed drawing of T_μ lies inside D_μ and satisfies all the required properties.

By Lemma 5, the ply-number of the 2-drawing of μ constructed by algorithm DRAWPATH is at most 2, and it remains the same after the scaling. Also, the ply-disk of any vertex in T_{ν_s} (in T_{ν_t}) entirely lies inside half-disk D_{ν_s} (half-disk D_{ν_t}) and hence inside disk D_i; thus, it does not overlap with the ply-disk of any vertex in a different subtree. Since the drawing of T_{ν_j}, for each child ν_j of μ, has ply-number at most $2(h - d_{\nu_j} + 1)$, the drawing of T_μ has ply-number at most $2 + 2(h - d_{\nu_j} + 1) = 2(h - d_{\nu_j} + 2) = 2(h - d_\mu + 1)$, given that $d_\mu = d_{\nu_j} + 1$.

At the end of the traversal, when the root ρ of T has been visited, we have a drawing with ply-number at most $2(h - d_\rho + 1) \leq 2\log n$ of $T_\rho = T$ inside the half-disk D_ρ of radius $6^{h-d_\rho} n_\rho \leq 6^{\log n} n = O(n^{1+\log 6}) = O(n^{3.6})$, and hence area $O(n^{7.2})$.

In order to extend the algorithm to work for 5-ary trees, we have to be able to fit inside the ply-disk D_i of each vertex $v_i \in \mu$ the drawings of the at most four subtrees T_{ν_j} whose anchor is v_i. Hence, we associate with each node μ a *quarter-disk* D_μ (a sector of a disk with internal angle $\frac{\pi}{2}$; see Fig. 5c) instead of a half-disk, still with radius $6^{h-d_\mu} n_\mu$, and we draw T_μ inside D_μ in such a way that the anchor a_μ of μ lies on the center of D_μ and all the vertices of μ lie along the radius of D_μ along the bisector of D_μ. Also in this case, the ply-disk of each vertex of μ entirely lies inside D_μ. We thus have the following.

Theorem 3. *Every n-vertex 5-ary tree has a drawing with ply-number at most $2 \log n$ and $O(n^{7.2})$ area.*

To extend this approach for trees with larger degree, we should use a disk sector D_μ with an internal angle smaller than $\frac{\pi}{2}$. In this case, however, we could not guarantee that the ply-disk of each vertex of μ lies inside D_μ, and thus we could not compute the ply-number of the subtrees independently of each other.

5   Conclusions and Open Problems

In this work we considered drawings of trees with low ply-number. We proved that requiring the ply-number to be bounded by a constant is often a somewhat too strong limitation, as these drawings may not exist, even for bounded-degree trees, or may require exponential area. On the positive side, we showed that relaxing the requirement on the ply-number, allowing it to be bounded by a logarithmic function, makes the problem easier, as we gave an algorithm for constructing polynomial-area drawings with this property for trees with maximum degree 6. Our work leaves several interesting open questions.

First, while it is known that stars, caterpillars, and binary trees admit constant-ply drawings in exponential area [5], we were able to prove that this is unavoidable only for stars and caterpillars; this leaves open the question on the area-requirements of constant-ply drawings of binary trees.

Second, it would be interesting to reduce the gap between binary trees, which always admit constant-ply drawings, and 10-ary trees, which may not admit any of such drawings. More in general, a characterization of the trees admitting these drawings is a fundamental open question.

Finally, in this paper we provided the first results on log-ply drawings of trees. It would be worth studying which trees (or other classes of graphs) always admit this type of drawings, possibly with polynomial area.

References

1. Alam, M.J., Eppstein, D., Goodrich, M.T., Kobourov, S.G., Pupyrev, S.: Balanced circle packings for planar graphs. In: Duncan, C.A., Symvonis, A. (eds.) GD 2014. LNCS, vol. 8871, pp. 125–136. Springer, Heidelberg (2014). doi:10.1007/978-3-662-45803-7_11
2. Angelini, P., Bekos, M.A., Bruckdorfer Jr., T.J.H., Kaufmann, M., Kobourov, S., Symvonis, A., Valtr, P.: Low ply drawings of trees. CoRR abs/1608.08538v2 (2016)

3. Breu, H., Kirkpatrick, D.G.: Unit disk graph recognition is NP-hard. Comput. Geometry **9**(1–2), 3–24 (1998)
4. Buchheim, C., Chimani, M., Gutwenger, C., Jünger, M., Mutzel, P.: Crossings and planarization. In: Tamassia, R. (ed.) Handbook on Graph Drawing and Visualization, pp. 43–85. Chapman and Hall/CRC, Boca Raton (2013)
5. Di Giacomo, E., Didimo, W., Hong, S., Kaufmann, M., Kobourov, S.G., Liotta, G., Misue, K., Symvonis, A., Yen, H.: Low ply graph drawing. In: 6th International Conference on Information, Intelligence, Systems and Applications, IISA 2015, pp. 1–6. IEEE (2015)
6. Eades, P., Hong, S.: Symmetric graph drawing. In: Tamassia, R. (ed.) Handbook on Graph Drawing and Visualization, pp. 87–113. Chapman and Hall/CRC, Boca Raton (2013)
7. Eppstein, D., Goodrich, M.T.: Studying (non-planar) road networks through an algorithmic lens. In: GIS 2008, pp. 1–10. ACM (2008)
8. Fekete, S.P., Houle, M.E., Whitesides, S.: The wobbly logic engine: Proving hardness of non-rigid geometric graph representation problems. In: DiBattista, G. (ed.) GD 1997. LNCS, vol. 1353, pp. 272–283. Springer, Heidelberg (1997). doi:10.1007/3-540-63938-1_69
9. Hliněný, P.: Contact graphs of curves. In: Brandenburg, F. (ed.) Graph Drawing. LNCS, vol. 1027, pp. 312–323. Springer, Heidelberg (1995). doi:10.1007/BFb0021814
10. Hliněný, P.: Classes and recognition of curve contact graphs. J. Combin. Theory Ser. B **74**(1), 87–103 (1998)
11. Koebe, P.: Kontaktprobleme der konformen Abbildung. Berichte über die Verhandlungen der Sächsischen Akad. der Wissenschaften zu Leipzig. Math.-Phys. Klasse **88**, 141–164 (1936)
12. Liotta, G.: Proximity drawings. In: Tamassia, R. (ed.) Handbook on Graph Drawing and Visualization, pp. 115–154. Chapman and Hall/CRC, Boca Raton (2013)
13. Sleator, D.D., Tarjan, R.E.: A data structure for dynamic trees. J. Comput. Syst. Sci. **26**(3), 362–391 (1983)

Visibility Representations

Visibility Representations of Boxes in 2.5 Dimensions

Alessio Arleo[1], Carla Binucci[1], Emilio Di Giacomo[1(✉)], William S. Evans[2],
Luca Grilli[1], Giuseppe Liotta[1], Henk Meijer[3], Fabrizio Montecchiani[1],
Sue Whitesides[4], and Stephen Wismath[5]

[1] Università degli Studi di Perugia, Perugia, Italy
alessio.arleo@studenti.unipg.it,
{carla.binucci,emilio.digiacomo,luca.grilli,giuseppe.liotta,
fabrizio.montecchiani}@unipg.it
[2] University of British Columbia, Vancouver, Canada
will@cs.ubc.ca
[3] University College Roosevelt, Middelburg, The Netherlands
h.meijer@ucr.nl
[4] University of Victoria, Victoria, Canada
sue@uvic.ca
[5] University of Lethbridge, Lethbridge, Canada
wismath@uleth.ca

Abstract. We initiate the study of 2.5D box visibility representations
(2.5D-BR) where vertices are mapped to 3D boxes having the bottom
face in the plane $z = 0$ and edges are unobstructed lines of sight parallel
to the x- or y-axis. We prove that: (i) Every complete bipartite graph
admits a 2.5D-BR; (ii) The complete graph K_n admits a 2.5D-BR if
and only if $n \leqslant 19$; (iii) Every graph with pathwidth at most 7 admits
a 2.5D-BR, which can be computed in linear time. We then turn our
attention to 2.5D grid box representations (2.5D-GBR) which are 2.5D-
BRs such that the bottom face of every box is a unit square at integer
coordinates. We show that an n-vertex graph that admits a 2.5D-GBR
has at most $4n - 6\sqrt{n}$ edges and this bound is tight. Finally, we prove
that deciding whether a given graph G admits a 2.5D-GBR with a given
footprint is NP-complete. The footprint of a 2.5D-BR Γ is the set of
bottom faces of the boxes in Γ.

1 Introduction

A *visibility representation (VR)* of a graph G maps the vertices of G to non-
overlapping geometric objects and the edges of G to *visibilities*, i.e., segments
that do not intersect any geometric object other than at their end-points.
Depending on the type of geometric objects representing the vertices and on
the rules used for the visibilities, different types of representations have been
studied in computational geometry and graph drawing.

Research started at the 2016 Bertinoro workshop on Graph Drawing. Research supported in part by NSERC, and by MIUR project AMANDA prot. 2012C4E3KT_001.

© Springer International Publishing AG 2016
Y. Hu and M. Nöllenburg (Eds.): GD 2016, LNCS 9801, pp. 251–265, 2016.
DOI: 10.1007/978-3-319-50106-2_20

A *bar visibility representation (BVR)* maps the vertices to horizontal segments, called *bars*, while visibilities are vertical segments. BVRs were introduced in the 80 s as a modeling tool for VLSI problems [16,27,28,34–36]. The graphs that admit a BVR are planar and they have been characterized under various models [16,28,34,36].

Extensions and generalizations of BVRs have been proposed in order to enlarge the family of representable graphs. In a *rectangle visibility representation (RVR)* the vertices are axis-aligned rectangles, while visibilities are both horizontal or vertical segments [4,5,10,12,13,23,29,31]. RVRs can exist only for graphs with thickness at most two and with at most $6n - 20$ edges [23]. Recognizing these graphs is NP-hard in general [29] and can be done in polynomial time in some restricted cases [4,31]. Generalizations of RVRs where orthogonal shapes other than rectangles are used to represent the vertices have been recently proposed [15,26]. Another generalization of BVRs are *bar k-visibility representations (k-BVRs)*, where each visibility segment can "see" through at most k bars. Dean et al. [11] proved that the graphs admitting a 1-BVR have at most $6n - 20$ edges. Felsner and Massow [20] showed that there exist graphs with a 1-BVR whose thickness is three. The relationship between 1-BVRs and 1-planar graphs has also been investigated [1,7,17,32].

RVRs are extended to 3D space by *Z-parallel Visibility Representations (ZPR)*, where vertices are axis-aligned rectangles belonging to planes parallel to the xy-plane, while visibilities are parallel to the z-axis. Bose et al. [6] proved that K_{22} admits a ZPR, while K_{56} does not. Štola [30] subsequently reduced the upper bound on the size of the largest representable complete graph by showing that K_{51} does not admits a ZPR. Fekete et al. [18] showed that K_7 is the largest complete graph that admits a ZPR if unit squares are used to represent the vertices. A different extension of RVRs to 3D space are the *box visibility representations (BR)* where vertices are 3D boxes, while visibilities are parallel to the x-, y- and z- axis. This model was studied by Fekete and Meijer [19] who proved that K_{56} admits a BR, while K_{184} does not.

In this paper we introduce *2.5D box visibility representations (2.5D-BR)* where vertices are 3D boxes whose bottom faces lie in the plane $z = 0$ and visibilities are parallel to the x- and y-axis. Like the other 3D models that use the third dimension, 2.5D-BRs overcome some limitations of the 2D models. For example, graphs with arbitrary thickness can be realized. In addition 2.5D-BRs seem to be simpler than other 3D models from a visual complexity point of view and have the advantage that they can be physically realized, for example by 3D printers or by using physical boxes. Furthermore, this type of representation can be used to model visibility between buildings of a urban area [9]. The main results of this paper are as follows.

- We show that every complete bipartite graph admits a 2.5D-BR (Sect. 3). This implies that there exist graphs that admit a 2.5D-BR and have arbitrary thickness.
- We prove that the complete graph K_n admits a 2.5D-BR if and only if $n \leqslant 19$ (Sect. 3). Thus, every graph with $n \leqslant 19$ vertices admits a 2.5D-BR.

- We describe a technique to construct a 2.5D-BR of every graph with pathwidth at most 7, which can be computed in linear time (Sect. 4).
- We then study *2.5D grid box representations* (*2.5D-GBR*) which are 2.5D-BRs such that the bottom face of every box is a unit square with corners at integer coordinates (Sect. 5). We show that an n-vertex graph that admits a 2.5D-GBR has at most $4n - 6\sqrt{n}$ edges and that this bound is tight. It is worth remarking that VRs where vertices are represented with a limited number of shapes have been investigated in the various models of visibility representations. Examples of these shape-restricted VRs are unit bar VRs [14], unit square VRs [10], and unit box VRs [19].
- Finally, we prove that deciding whether a given graph G admits a 2.5D-GBR with a given footprint is NP-complete (Sect. 5). The *footprint* of a 2.5D-BR Γ is the set of bottom faces of the boxes in Γ.

For reasons of space, some proofs and details are omitted and can be found in [3].

2   Preliminaries

A *box* is a six-sided polyhedron of non-zero volume with axis-aligned sides in a 3D Cartesian coordinate system. In a *2.5D box representation* (*2.5D-BR*) the vertices are mapped to boxes that lie in the non-negative half space $z \geqslant 0$ and include one face in the plane $z = 0$, while each edge is mapped to a *visibility* (i.e. a segment whose endpoints lie in faces of distinct boxes and whose interior does not intersect any box) parallel to the x- or to the y-axis. We remark that visibilities between non-adjacent objects may exist, i.e., we adopt the so called *weak visibility model* (in the *strong visibility model* each visibility between two geometric objects corresponds to an edge of the graph). The weak model seems to be the most effective when representing non-planar graphs and it has been adopted in several works (see e.g. [4,7,17]). As in many papers on visibility representations [19,24,31,33,36], we assume the ϵ-*visibility model*, where each segment representing an edge is the axis of a positive-volume cylinder that intersects no box except at its ends; this implies that an intersection point between a visibility and a box belongs to the interior of a box face. In what follows, when this leads to no confusion, we shall use the term *edge* to indicate both an edge and the corresponding visibility, and the term *vertex* for both a vertex and the corresponding geometric object.

Given a box b of a 2.5D-BR, the face that lies in the plane $z = 0$ is called the *footprint* of b. The intersection of the plane $z = 0$ with a 2.5D-BR Γ is called the *footprint* of Γ and is denoted by Γ_0. In other words, the footprint of a 2.5D-BR Γ consists of the footprint of all the boxes in Γ. If Γ is a 2.5D-BR of a complete graph then its footprint Γ_0 satisfies a trivial necessary condition (throughout the paper we will refer to this condition as *NC*): for every pair of boxes b_1 and b_2 of Γ, there must exist a line ℓ (in the plane $z = 0$) such that (i) ℓ passes through the footprints of b_1 and b_2, and (ii) ℓ is either parallel to the x-axis or to the y-axis. A *2.5D grid box representation* (*2.5D-GBR*) is a 2.5D-BR such that every box has a footprint that is a unit square with corners at integer coordinates.

Two boxes *see each other* if there exists a visibility between them; we say that they *see each other above another box* b, if there exists a visibility between them and the projection of this visibility on the plane $z = 0$ intersects the interior of the footprint of b. Notice that this implies that the two boxes are both taller than b. We say that two boxes have a *ground visibility* or are *ground visible* if there exists a visibility between their footprints, i.e. if there exists an unobstructed axis-aligned line segment connecting their footprints. If two boxes are ground visible then they see each other regardless of their heights and the heights of the other boxes. Let G be a graph, let Λ be a collection of boxes each lying in the non-negative half space $z \geqslant 0$ with one face in the plane $z = 0$, such that the boxes of Λ are in bijection with the vertices of G. Note that Λ may not be a 2.5D-BR of G. For a vertex v of G, $\Lambda(v)$ denotes the corresponding box in Λ, while $h(\Lambda(v))$, or simply $h(v)$, indicates the height of this box. For a subset $S \subset V(G)$, $\Lambda(S)$ denotes the subset of boxes associated with S, while $\Lambda_0(S)$ is the footprint of $\Lambda(S)$. Let $G[S]$ be the subgraph of G induced by S. We say that $\Lambda(S)$ is a 2.5D-BR of $G[S]$ in Λ, if for any edge (u,v) of $G[S]$ there exists a visibility in Λ between $\Lambda(u)$ and $\Lambda(v)$; that is, the visibility is not destroyed by the presence of the other boxes in Λ.

3   2.5D Box Representations of Complete Graphs

In this section we consider 2.5D-BRs of complete graphs and complete bipartite graphs.

Theorem 1. *Every complete bipartite graph admits a 2.5D-BR.*

Proof. Let $K_{m,n}$ be a complete bipartite graph. We represent the m vertices in the first partite set with m boxes $a_0, a_1, \ldots, a_{m-1}$ such that box a_i has a footprint with corners at $(2i, 0, 0)$, $(2i + 1, 0, 0)$, $(2i, 2n - 1, 0)$ and $(2i + 1, 2n - 1, 0)$ and height $m - i$. Then we represent the n vertices in the second partite set with n boxes $b_0, b_1, \ldots, b_{n-1}$ such that box b_j has a footprint with corners at $(2m, 2j, 0)$, $(2m + 1, 2j, 0)$, $(2m, 2j + 1, 0)$ and $(2m + 1, 2j + 1, 0)$ and height m. Consider now a box a_i and a box b_j. By construction a_i and b_j see each other above all boxes a_l with $l > i$.   □

A consequence of Theorem 1 is that there exist graphs with unbounded thickness that admit a 2.5D-BR. This contrasts with other models of visibility representations (e.g., k-BVRs, and RVRs), which can only represent graphs with bounded thickness.

We now prove that the largest complete graph that admits a 2.5D-BR is K_{19}. We first show that given a 2.5D-BR of a complete graph there is one line parallel to the x-axis and one line parallel to the y-axis whose union intersect all boxes and such that each of them intersects at most 10 boxes. This implies that there can be at most 20 boxes in a 2.5D-BR of a complete graph. We then show that there must be a box that is intersected by both lines, thus lowering this bound to 19. We finally exhibit a 2.5D-BR of K_{19}. We start with some technical lemmas.

Lemma 1. *Let G be an n-vertex graph that admits a 2.5D-BR Γ'. Then there exists a 2.5D-BR Γ of G such that every box of Γ has a distinct integer height in the range $[1, n]$ and the footprint of Γ is the same as that of Γ'.*

The following lemma is proved in [25, Observation 1]. Given an axis-aligned rectangle r in the plane $z = 0$, we denote by $x(r)$ the x-extent of r and by $y(r)$ the y-extent of r, so $r = x(r) \times y(r)$.

Lemma 2 [25]. *For every arrangement $\mathcal{R}$ of n axis-aligned rectangles in the plane such that for all $a, b \in \mathcal{R}$, either $x(a) \cap x(b) \neq \emptyset$ or $y(a) \cap y(b) \neq \emptyset$, there exists a vertical and a horizontal line whose union intersects all rectangles in $\mathcal{R}$.*

The following lemma is similar to the Erdős–Szekeres lemma and can be proved in a similar manner [18]. A sequence of distinct integers is *unimaximal* if no element of the sequence is smaller than both its predecessor and successor.

Lemma 3 [18]. *For all $m > 1$, in every sequence of $\binom{m}{2} + 1$ distinct integers, there exists at least one unimaximal sequence of length m.*

Given a 2.5D-BR Γ and a line ℓ parallel to the x-axis or to the y-axis, we say that ℓ *stabs* a set of boxes B of Γ if it intersects the interior of the footprints of each box in B. Let $b_1, b_2, \ldots, b_h$ be the boxes of B in the order they are stabbed by ℓ. We say that B has a *staircase layout*, if $h(b_i) > h(b_{i-1})$ for $i = 2, 3, \ldots, h$.

Lemma 4. *In a 2.5D-BR of a complete graph no line parallel to the x-axis or to the y-axis can stab five boxes whose heights, in the order in which the boxes are stabbed, form a unimaximal sequence.*

Proof. Assume, as a contradiction, that there exists a line ℓ parallel to the x-axis or to the y-axis that stabs 5 boxes $b_1, \ldots, b_5$ whose heights form a unimaximal sequence in the order in which the boxes are stabbed by ℓ. Let r_i be the footprint of box b_i (with $1 \leqslant i \leqslant 5$). We claim that there exists a ground visibility between every pair of boxes b_i and b_j (with $1 \leqslant i < j \leqslant 5$). If $j = i+1$ this is clearly true. Suppose then that $j \neq i + 1$. If b_i and b_j do not have a ground visibility, then they must see each other above b_l with $i < l < j$, i.e., the height of b_i and of b_j must be larger than the height of b_l, which is impossible because the sequence of heights is unimaximal. Thus, for every pair of boxes b_i and b_j there must be a ground visibility. Since b_i and b_j are both stabbed by ℓ, this visibility must be parallel to ℓ. This implies that the left sides (if ℓ is parallel to the x-axis) or the bottom sides (if ℓ is parallel to the y-axis) of rectangles r_1, r_2, r_3, r_4, r_5 form a bar visibility representation of K_5, which is impossible because bar visibility representations exist only for planar graphs [21]. □

Lemma 5. *In a 2.5D-BR of a complete graph no line parallel to the x-axis or to the y-axis can stab more than 10 boxes.*

Proof. Let Γ be a 2.5D-BR of a complete graph K_n. By Lemma 1 we can assume that all boxes have distinct integer heights. Suppose, as a contradiction, that there exists a line ℓ parallel to the x-axis or to the y-axis that stabs $k > 10$ boxes. Let $h_1, h_2, \ldots, h_k$ be the heights of the stabbed boxes in the order in which the boxes are stabbed by ℓ. By Lemma 3 this sequence of heights contains a unimaximal sequence of length 5, but this is impossible by Lemma 4.   □

Lemma 6. *A complete graph admits a 2.5D-BR only if it has at most* 19 *vertices.*

Proof. Let Γ be a 2.5D-BR of a complete graph K_n (for some $n > 0$). By Lemma 1 we can assume that all boxes of Γ have distinct heights. The footprint Γ_0 of Γ is an arrangement of rectangles that satisfies Lemma 2. Thus there exist a line ℓ_h parallel to the x-axis and a line ℓ_v parallel to the y-axis that together stab all boxes of Γ. By Lemma 5, both ℓ_h and ℓ_v can stab at most 10 boxes each. This means that the number of boxes (and therefore the number of vertices of K_n) is at most 20. We now prove that if ℓ_h and ℓ_v both stab ten boxes, there must be one box that is stabbed by both ℓ_h and ℓ_v, which implies that the number of boxes in Γ is at most 19.

Suppose, for a contradiction, that $p = \ell_h \cap \ell_v$ does not lie in a box. Refer to Fig. 1(a) for an illustration. Denote by T the set of boxes stabbed by ℓ_v that are above p and by B be the set of boxes stabbed by ℓ_v that are below p. Analogously, denote by L the set of boxes stabbed by ℓ_h that are to the left of p and by R the set of boxes stabbed by ℓ_h that are to the right of p. Each of these sets can be empty but $|T| + |B| = 10$ and $|L| + |R| = 10$. Denote by $l_1, l_2, \ldots, l_{|L|}$ the set of boxes in L from right to left, i.e., l_1 is the box closest to p. Analogously, denote by $r_1, r_2, \ldots, r_{|R|}$ the boxes of R from left to right (r_1 is the closest to p), by $t_1, t_2, \ldots, t_{|T|}$ the boxes of T from bottom to top (t_1 is the closest to p) and by $b_1, b_2, \ldots, b_{|B|}$ the boxes of B from top to bottom (b_1 is the closest to p). Let f_T, f_B, f_L, and f_R be the footprints of t_1, b_1, l_1, and r_1, respectively. Let ℓ_X be the line containing the side of f_X that is closest to p and let ℓ'_X be the line containing the opposite side of f_X (for every $X \in \{T, B, L, R\}$).

We first claim that for each f_X there exists a line ℓ_Y (with $X, Y \in \{T, B, L, R\}$ and $Y \neq X$) that intersects the interior of f_X. Suppose, for a contradiction, that this is not true for at least one f_X, say f_L; that is, the interior of f_L is not intersected by ℓ_T and ℓ_B. If so, there must be a line ℓ parallel to the y-axis that intersects all the rectangles in $T \cup B$ and f_L; otherwise the necessary condition NC does not hold for $T \cup B \cup \{l_1\}$. But then ℓ would stab eleven boxes, which is impossible by Lemma 5. Thus, our claim holds and the four rectangles f_X are placed so that ℓ_T, ℓ_R, ℓ_B, and ℓ_L stab f_R, f_B, f_L, and f_T (or, symmetrically, f_L, f_T, f_R, and f_B, which follows a symmetric argument), respectively, as in Fig. 1(a).

We consider now the sets T, B, L, and R. For each set there are two possible configurations. Consider the set B and the line ℓ'_L. If the set $B' = B \setminus \{b_1\}$ contains a box b_j whose footprint is completely to the right of ℓ'_L, we say that B has *configuration A* (see Fig. 1(b)). In the case of configuration A, the footprint of all boxes in $L' = L \setminus \{l_1\}$ must extend below the line ℓ'_B (otherwise the necessary condition NC does not hold for $L' \cup \{b_j\}$). This implies that $y(f_B)$ is contained

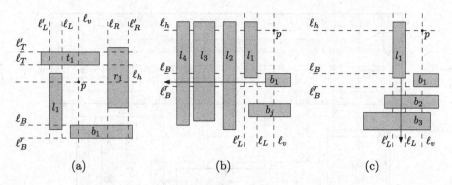

Fig. 1. (a) Placement of the four rectangles f_T, f_R, f_B, and f_L. (b) Configuration A for the boxes of set B. (c) Configuration B for the boxes of set B. The arrow intersects the boxes that must have a staircase layout.

in $y(l_i)$ for all $i \geqslant 2$. The only possibility for b_1 to see all these boxes is that L' has a staircase layout (with l_2 being the shortest box) and b_1 is taller than the second tallest one. So, configuration A for the set B implies that L' has a staircase layout. If all boxes of B' have a footprint that extends to the left of ℓ'_L, we say that B has *configuration B* (see Fig. 1(c)). In this case, $x(f_L)$ is contained in $x(b_i)$ for all $i \geqslant 2$. Again, the only possibility for l_1 to see all these boxes is that B' has a staircase layout and that l_1 is taller than the second tallest one. So, configuration B for the set B implies that B' has a staircase layout. The definitions of configurations A and B for T, L, R are similar to those for B and arise by considering lines ℓ'_R, ℓ'_T, ℓ'_B, respectively.

For any two sets X and Y that are consecutive in the cyclic order T, R, B, L, either X' or Y' has a staircase layout (depending on whether X has configuration A or B). This implies that either B' and T' have both a staircase layout or L' and R' have both a staircase layout. Suppose that B' and T' have a staircase layout (the case when L' and R' have a staircase layout is analogous). If either $|B'| \geqslant 5$ or $|T'| \geqslant 5$, ℓ_v stabs at least five boxes whose heights form a unimaximal sequence, which is impossible by Lemma 4. Thus $|B'| = 4$ and $|T'| = 4$ (recall that $|B'| + |T'| = 8$). Since all boxes of Γ have distinct heights, either $h(b_2) < h(t_2)$ or $h(t_2) < h(b_2)$. In the first case ℓ_v stabs the five boxes t_5, t_4, t_3, t_2, b_2 whose heights form a unimaximal sequence, which is impossible by Lemma 4. In the other case ℓ_v stabs the five boxes b_5, b_4, b_3, b_2, t_2 whose heights form a unimaximal sequence, which is impossible by Lemma 4. □

We conclude this section by exhibiting a 2.5D-BR of K_{19}, illustrated in Fig. 2. To prove the correctness of the drawing the idea is to partition the vertex set of K_{19} into five subsets (shown in Fig. 2) and prove that all boxes in a given set see all other boxes. The following theorem holds.

Theorem 2. *A complete graph K_n admits a 2.5D-BR if and only if $n \leqslant 19$.*

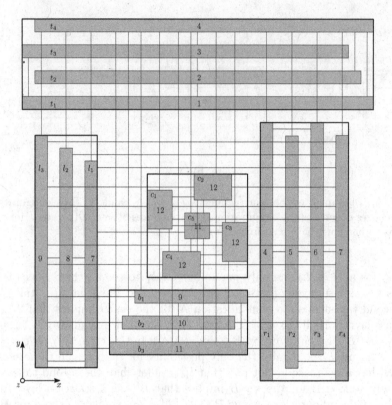

Fig. 2. Illustration of a 2.5D-BR of K_{19}, the footprint is represented by a 2D drawing in the plane $z = 0$, while the heights of boxes are indicated by integer labels. The five rectangles with thick sides represent the partitioning of $V(K_{19})$ into five subsets.

4   2.5D Box Representations of Graphs with Pathwidth at Most 7

A graph G with pathwidth p is a subgraph of a graph that can be constructed as follows. Start with the complete graph K_{p+1} and classify all its vertices as *active*. At each step, a vertex is *deactivated* and a new active vertex is introduced and joined to all the remaining active vertices. The order in which vertices are introduced is given by a *normalized path decomposition*, which can be computed in linear time for a fixed p [22].

Theorem 3. *Every n-vertex graph with pathwidth at most 7 admits a 2.5D-BR, which can be computed in $O(n)$ time.*

Proof. We describe an algorithm to compute a 2.5D-BR of a graph G with pathwidth 7. The algorithm is based on the use of eight groups of rectangles, a subset of which will form the footprint of the 2.5D-BR of G. For graphs with pathwidth $p < 7$, the same algorithm can be applied by considering only $p + 1$ groups, arbitrarily chosen.

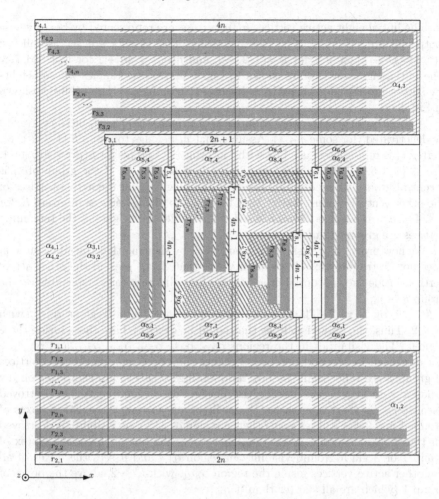

Fig. 3. Construction of a 2.5D-BR for a graph with pathwidth 7.

The eight groups are defined in the plane $z = 0$ and have n rectangles each denoted as $r_{h,1}, r_{h,2}, \ldots, r_{h,n}$ $(1 \leqslant h \leqslant 8)$. The groups are placed as shown in Fig. 3. The groups $h = 5, 6, 7, 8$ will be called *central groups*. A vertex whose footprint is $r_{h,k}$ will be called a *vertex of group* h $(1 \leqslant h \leqslant 8)$.

Let $v_1, v_2, \ldots, v_n$ be the vertices of G in the order given by a normalized path decomposition. We denote by G_i the subgraph of G induced by $\{v_1, v_2, \ldots, v_i\}$. We create a collection of boxes by adding one box per step; at step i we add a box to represent the next vertex v_i to be activated. We denote the collection of the first i boxes as Λ_i and we prove that Λ_i satisfies the following invariant (I1): Λ_i is a 2.5D-BR of G_i such that for any pair of boxes of group j and k $(1 \leqslant j, k \leqslant 8)$ that represent vertices that are adjacent in G_i, there exists a visibility whose projection in the plane $z = 0$ is inside the region $\alpha_{j,k}$. The regions $\alpha_{j,k}$ are highlighted in Fig. 3 as dashed regions.

The initial eight active vertices $v_1, v_2, \ldots, v_8$ are represented by boxes whose footprints are $r_{1,1}, r_{2,1}, \ldots, r_{8,1}$, respectively. The heights are set as follows: $h(v_h) = (h-1) \cdot n + 1$, for $h = 1, 2, 3, 4$, and $h(v_h) = 4n + 1$ for $h = 5, 6, 7, 8$. The initial eight vertices are shown in Fig. 3 as white rectangles whose heights are shown inside them. Λ_8 satisfies invariant I1 thanks to the visibilities shown in Fig. 3.

Assume now that Λ_{i-1} ($i > 8$) satisfies invariant I1 and let v_j be the vertex to be deactivated (for some $j < i$). Assume that v_j belongs to group h ($1 \leqslant h \leqslant 8$). Vertex v_i is represented as a box with footprint $r_{h,i}$ and height $h(v_i) = h(v_j) + 1$, if $h \in \{1, 3, 5, 6, 7, 8\}$, or $h(v_i) = h(v_j) - 1$, if $h \in \{2, 4\}$. If the group of v_i is a central group, we increase by one unit the height of all the active vertices of the other central groups. Notice that the heights of the vertices of group h, for $h \leqslant 4$, are in the range $[(h-1) \cdot n + 1, h \cdot n]$, while the heights of the remaining vertices are greater than $4n$.

We now prove that Λ_i satisfies invariant I1 by showing that the addition of v_i does not destroy any existing visibility and that $\Lambda_i(v_i)$ sees all the other active vertices inside the appropriate regions. We have different cases depending on the group h of v_i.

– $h = 1$ or $h = 2$. The box $\Lambda_i(v_i)$ only intersects the regions $\alpha_{h',2}$, with $h' \neq 2$. Thus, the only visibilities that could be destroyed are those inside these regions. The visibilities in the regions $\alpha_{3,2}$, $\alpha_{4,2}$, $\alpha_{5,2}$, $\alpha_{6,2}$, $\alpha_{7,2}$, and $\alpha_{8,2}$ are not destroyed by the addition of v_i because the boxes representing the vertices of group 2 are taller than the box representing v_i and so are the boxes of any group h' with $h' > 2$. The existing visibilities in the region $\alpha_{1,2}$ are not destroyed because $r_{h,i}$ is short enough (in the x-direction) so that the existing boxes of groups 1 and 2 can still see each other in region $\alpha_{1,2}$. So, no visibility is destroyed for the vertices of group 2. The box $\Lambda_i(v_i)$ sees the box of the active vertex of group 1 or 2 via a ground visibility in region $\alpha_{1,2}$ and it sees the boxes of all the other active vertices inside the region $\alpha_{h',1}$, with $h' > 2$, above the boxes of group 1 (which are all shorter than it).

– $h = 3$ or $h = 4$. The proof of this case is omitted.

– $h = 5$ or $h = 6$. The box $\Lambda_i(v_i)$ only intersect the regions $\alpha_{h,h'}$, with $h' \in \{5, 6, 7, 8\}$ and $h' \neq h$. However, it does not intersect any existing visibility inside these regions and therefore the addition of $\Lambda_i(v_i)$ does not destroy any existing visibility. The box $\Lambda_i(v_i)$ sees the active vertices of groups 1 and 2 inside $\alpha_{h,k}$ (with $h = 5$ or 6, and $k = 1, 2$) and above the boxes of group 1. The active vertices of groups 3 and 4 are seen inside $\alpha_{h,k}$ (with $h = 5$ or 6, and $k = 3, 4$) and above the boxes of group 3. Finally, the active vertices of the central groups are seen inside $\alpha_{h,k}$ (with $h = 5$ or 6, and $k > 4$) and above the boxes of group h. Recall that the active vertices of the central groups have been raised to have the same height as $\Lambda_i(v_i)$ (which is larger than the height of any other box in the central groups).

– $h = 7$ or $h = 8$. The proof of this case is omitted.

The above construction can be done in $O(n)$ time. Since the normalized path decomposition can be computed in $O(n)$ time, the time complexity follows.   □

5   2.5D Grid Box Representations

Next we give a tight bound on the edge density of graphs admitting a 2.5D-GBR. The proof is based on the fact that a set of aligned (unit square) boxes induces an outerplanar graph. A square grid of boxes gives the bound.

Theorem 4. *Every n-vertex graph that admits a 2.5D-GBR has at most $4n - 6\sqrt{n}$ edges, and this bound is tight.*

In the next theorem we prove that deciding whether a given graph admits a 2.5D-GBR with a given footprint is NP-complete. We call this problem 2.5D-GBR-WITH-GIVEN-FOOTPRINT (2.5GBR-WGF). The reduction is from HAMILTONIAN-PATH-FOR-CUBIC-GRAPHS (HPCG), which is the problem of deciding whether a given cubic graph admits a Hamiltonian path [2].

Theorem 5. *Deciding whether a given graph G admits a 2.5D-GBR with a given footprint is NP-complete, even if G is a path.*

Proof sketch: We first prove that 2.5GBR-WGF is in NP. A candidate solution consists of a mapping of the vertices of G to the squares of the given footprint and a choice of the heights of the boxes. By Lemma 1 we can assign to each box an integer height in the set $\{1, 2, \ldots, n\}$. Thus the size of a candidate solution is polynomial in the size of the input graph. Given a candidate solution, we can test in polynomial time whether all edges of G are realized as visibilities. Thus, the problem is in NP.

We now describe a reduction from the HPCG problem. Let G_H be an instance of the HPCG problem, i.e. a cubic graph, with n_H vertices and m_H edges. We compute an orthogonal grid drawing Γ_H of G_H such that every edge has exactly one bend and no two vertices share the same x- or y-coordinate. Such a drawing always exists and can be computed in polynomial time with the algorithm by Bruckdorfer et al. [8]. We now use Γ_H as a trace to construct an instance $\langle G, F \rangle$ of the 2.5GBR-WGF problem, where G is a path and F a footprint, i.e., a set of squares. G is a path with $4n_H + m_H$ vertices and therefore F will contain $4n_H + m_H$ squares. The footprint F is constructed as follows. Γ_H is scaled up by a factor of four. In this way, every two vertices/bends are separated by at least four grid units. Each vertex v of Γ_H is replaced by a set $S(v)$ of four unit squares. In particular if vertex v has coordinates $(4x, 4y)$ in Γ_H, then it is replaced by the following four unit squares: $S_1(v)$ whose bottom-right corner has coordinates $(4x, 4y)$, $S_2(v)$ whose bottom-right corner has coordinates $(4x + 2, 4y)$, $S_3(v)$ whose bottom-right corner has coordinates $(4x, 4y-2)$, and $S_4(v)$ whose bottom-right corner has coordinates $(4x + 2, 4y - 2)$. We associate with each edge e incident to a vertex v, one of the four squares in $S(v)$. If e enters v from West, North, South, or East, the square associated with e is $S_1(v)$, $S_2(v)$, $S_3(v)$, or $S_4(v)$, respectively. Let (u, v) be an edge of Γ_H and let $S_i(u)$ and $S_j(v)$ $(1 \leqslant i, j \leqslant 4)$ be the squares associated with (u, v). The bend of $e = (u, v)$ is replaced by a unit square S_e horizontally/vertically aligned with $S_i(u)$ and $S_j(v)$. The set of squares replacing the vertices of Γ_H, which will be called *vertex squares*

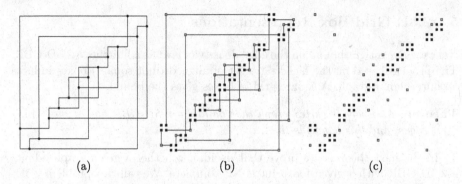

Fig. 4. (a) An orthogonal drawing of a cubic graph. (b) Construction of the footprint. Black (gray) squares are vertex (edge) squares. (c) The constructed footprint.

in the following, together with the set of squares replacing the bends, which will be called *edge squares* in the following, form the footprint F. Figure 4 shows an orthogonal drawing of a cubic graph and the corresponding footprint F. Observe that the footprint F is such that any two squares are separated by at least one unit and in each row/column there are at most three squares. Let F^* be a graph with a vertex for each square in F and an edge between two squares if and only if the two squares are horizontally or vertically aligned. It can be proved that G_H admits a Hamiltonian path if and only if F^* contains a Hamiltonian path.

Consider the instance $\langle G, F \rangle$ of the 2.5GBR-WGF problem, where G is a path. We prove that G admits a 2.5D-GBR with footprint F if and only if F^* admits a Hamiltonian path. Every graph that can be represented by a 2.5D-GBR with footprint F is a spanning subgraph of F^* (because F^* has all possible edges that can be realized as visibilities in a 2.5D-GBR with footprint F). Thus, if G admits a 2.5D-GBR with footprint F, then G is a Hamiltonian path of F^* (recall that G is a path). Suppose now that F^* has a Hamiltonian path H^*. We show that we can choose the heights of the squares in F so that the resulting boxes form a 2.5D-GBR of G. Recall that in each row/column of F there are at most three squares. If an edge connects two squares that are consecutive along a row or column, then any choice of the heights is fine. If an edge connects the first and the last square of a row/column, then the heights of these two squares must be larger than the height of the square in the middle. We assign the heights to one square per step, in the order in which they appear along H^*. We assign to the first square a height equal to the number of squares (i.e., $4n_H + m_H$). Let h be the height assigned to the current square S and let S' be the next square along H^*. If S and S' are consecutive along a row/column then the height assigned to S' is h. If S and S' are the first and the last square of a row/column then the height assigned to S' is h. If S is the first/last square of a row/column and S' is the middle square of the same row/column, then the height assigned to S' is $h - 1$. If S is the middle square of a row/column and S' is the first/last square of the same row/column, then the height assigned to S' is $h + 1$. It is easy to see that all heights are positive and that if an edge connects the first and the last square of a row/column, then the heights of these two squares are greater than the height of the square in the middle. This concludes the proof that G admits a 2.5D-GBR with footprint F if and only if F^* admits a Hamiltonian path. Since F^* has a Hamiltonian path if and only if G_H has a Hamiltonian path, G admits a 2.5D-GBR with footprint F if and only if G_H has a Hamiltonian path, which implies that the 2.5GBR-WGF problem is NP-hard.   □

6   Open Problems

There are several possible directions for further study of 2.5D-BRs. Among them: (*i*) Study the complexity of deciding if a given graph admits a 2.5D-BR. We remark that deciding if a graph admits an RVR is NP-hard. (*ii*) Investigate other classes of graphs that admit a 2.5D-BR. For example, do 1-planar graphs or partial 5-trees always admit a 2.5D-BR? We remark that there are both 1-planar graphs and partial 5-trees not admitting an RVR. (*iii*) Study the 2.5D-BRs under the strong visibility model. For example, which bipartite graphs admit a strong 2.5D-BR?

References

1. Ahmed, M.E., Yusuf, A.B., Polin, M.Z.H.: Bar 1-visibility representation of optimal 1-planar graph. Elect. Inf. Comm. Technol. (EICT) **2013**, 1–5 (2014)
2. Akiyama, T., Nishizeki, T., Saito, N.: NP-completeness of the hamiltonian cycle problem for bipartite graphs. J. Inf. Process. **3**(2), 73–76 (1980)
3. Arleo, A., Binucci, C., Di Giacomo, E, Evans, W.S., Grilli, L., Liotta, G., Meijer, H., Montecchiani, F., Whitesides, S., Wismath, S.: Visibility representations of boxes in 2.5 dimensions. CoRR, abs/1608.08899 (2016)
4. Biedl, T., Liotta, G., Montecchiani, F.: On visibility representations of non-planar graphs. In: Fekete, S., Lubiw, A., (eds.) SoCG 2016, vol. LIPICs, pp. 19:1–19:16. Schloss Dagstuhl - Leibniz-Zentrum fuer Informatik (2016)
5. Bose, P., Dean, A., Hutchinson, J., Shermer, T.: On rectangle visibility graphs. In: North, S. (ed.) GD 1996. LNCS, vol. 1190, pp. 25–44. Springer, Heidelberg (1997). doi:10.1007/3-540-62495-3_35
6. Bose, P., Everett, H., Fekete, S.P., Houle, M.E., Lubiw, A., Meijer, H., Romanik, K., Rote, G., Shermer, T.C., Whitesides, S., Zelle, C.: A visibility representation for graphs in three dimensions. J. Graph Algorithms Appl. **2**(3), 1–16 (1998)
7. Brandenburg, F.: 1-visibility representations of 1-planar graphs. J. Graph Algorithms Appl. **18**(3), 421–438 (2014)
8. Bruckdorfer, T., Kaufmann, M., Montecchiani, F.: 1-bend orthogonal partial edge drawing. J. Graph Algorithms Appl. **18**(1), 111–131 (2014)
9. Carmi, P., Friedman, E., Katz, M.J.: Spiderman graph: visibility in urban regions. Comput. Geometry **48**(3), 251–259 (2015)
10. Dean, A.M., Ellis-Monaghan, J.A., Hamilton, S., Pangborn, G.: Unit rectangle visibility graphs. Electr. J. Comb. **15**(1), 1–24 (2008)
11. Dean, A.M., Evans, W., Gethner, E., Laison, J.D., Safari, M.A., Trotter, W.T.: Bar k-visibility graphs. J. Graph Algorithms Appl. **11**(1), 45–59 (2007)

12. Dean, A.M., Hutchinson, J.P.: Rectangle-visibility representations of bipartite graphs. Discrete Appl. Math. **75**(1), 9–25 (1997)
13. Dean, A.M., Hutchinson, J.P.: Rectangle-visibility layouts of unions and products of trees. J. Graph Algorithms Appl. **2**(8), 1–21 (1998)
14. Dean, A.M., Veytsel, N.: Unit bar-visibility graphs. Congr. Num. **160**, 161–175 (2003)
15. Di Giacomo, E., Didimo, W., Evans, W.S., Liotta, G., Meijer, H., Montecchiani, F., Wismath, S.K.: Ortho-polygon visibility representations of embedded graphs. In: Nöllenburg, M., Hu, Y. (eds.) GD 2016. LNCS, vol. 9801, pp. 280–294. Springer, Heidelberg (2016)
16. Duchet, P., Hamidoune, Y., Las, M., Vergnas, H.M.: Representing a planar graph by vertical lines joining different levels. Discrete Math. **46**(3), 319–321 (1983)
17. Evans, W., Kaufmann, M., Lenhart, W., Mchedlidze, T., Wismath, S.: Bar 1-visibility graphs and their relation to other nearly planar graphs. J. Graph Algorithms Appl. **18**(5), 721–739 (2014)
18. Cobos, F.J., Dana, J.C., Hurtado, F., Márquez, A., Mateos, F.: On a visibility representation of graphs. In: Brandenburg, F.J. (ed.) GD 1995. LNCS, vol. 1027, pp. 152–161. Springer, Heidelberg (1996). doi:10.1007/BFb0021799
19. Fekete, S.P., Meijer, H.: Rectangle and box visibility graphs in 3D. Int. J. Comput. Geometry Appl. **9**(1), 1–28 (1999)
20. Felsner, S., Massow, M.: Parameters of bar k-visibility graphs. J. Graph Algorithms Appl. **12**(1), 5–27 (2008)
21. Garey, M.R., Johnson, D.S., So, H.C.: An application of graph coloring to printed circuit testing. IEEE Trans. Circuits Syst. **CAS–23**(10), 591–599 (1976)
22. Gupta, A., Nishimura, N., Proskurowski, A., Ragde, P.: Embeddings of k-connected graphs of pathwidth k. Discrete Appl. Math. **145**(2), 242–265 (2005)
23. Hutchinson, J.P., Shermer, T., Vince, A.: On representations of some thickness-two graphs. Comp. Geometry **13**(3), 161–171 (1999)
24. Kant, G., Liotta, G., Tamassia, R., Tollis, I.G.: Area requirement of visibility representations of trees. Inf. Process. Lett. **62**(2), 81–88 (1997)
25. Kleitman, J.D., Gyárfás, A., Tóth, G.: Convex sets in the plane with three of every four meeting. Combinatorica **21**(2), 221–232 (2001)
26. Liotta, G., Montecchiani, F.: L-visibility drawings of IC-planar graphs. Inf. Process. Lett. **116**(3), 217–222 (2016)
27. Otten, R.H.J.M., Van Wijk, J.G.: Graph representations in interactive layout design. In: IEEE ISCSS, pp. 91–918. IEEE (1978)
28. Rosenstiehl, P., Tarjan, R.E.: Rectilinear planar layouts and bipolar orientations of planar graphs. Discrete Comput. Geom. **1**, 343–353 (1986)
29. Shermer, T.C.: On rectangle visibility graphs III. external visibility and complexity. In: Canadian Conference on Computational Geometry, pp. 234–239 (1996)
30. Štola, J.: Unimaximal sequences of pairs in rectangle visibility drawing. In: Tollis, I.G., Patrignani, M. (eds.) GD 2008. LNCS, vol. 5417, pp. 61–66. Springer, Heidelberg (2009). doi:10.1007/978-3-642-00219-9_7
31. Streinu, I., Whitesides, S.: Rectangle visibility graphs: characterization, construction, and compaction. In: Alt, H., Habib, M. (eds.) STACS 2003. LNCS, vol. 2607, pp. 26–37. Springer, Heidelberg (2003). doi:10.1007/3-540-36494-3_4
32. Sultana, S., Rahman, M.S., Roy, A., Tairin, S.: Bar 1-visibility drawings of 1-planar graphs. In: Gupta, P., Zaroliagis, C. (eds.) ICAA 2014. LNCS, pp. 62–76. Springer International Publishing, New York (2014). doi:10.1007/978-3-319-04126-1_6
33. Tamassia, R., Tollis, I.G.: A unified approach to visibility representations of planar graphs. Discrete Comput. Geom. **1**(1), 321–341 (1986)

34. Tamassia, R., Tollis, I.G.: Representations of graphs on a cylinder. SIAM J. Discrete Math. **4**(1), 139–149 (1991)
35. Thomassen, C.: Plane representations of graphs. In: Progress in Graph Theory, pp. 43–69. AP (1984)
36. Wismath, S.K.: Characterizing bar line-of-sight graphs. In: Proceedings of 1st Symposium on Computational Geometry, pp. 147–152 (1985)

The Partial Visibility Representation Extension Problem

Steven Chaplick[1]([⊠]), Grzegorz Guśpiel[2], Grzegorz Gutowski[2],
Tomasz Krawczyk[2], and Giuseppe Liotta[3]

[1] Lehrstuhl für Informatik I, Universität Würzburg, Würzburg, Germany
steven.chaplick@uni-wuerzburg.de
[2] Theoretical Computer Science Department,
Faculty of Mathematics and Computer Science, Jagiellonian University,
Kraków, Poland
{guspiel,gutowski,krawczyk}@tcs.uj.edu.pl
[3] Dipartimento di Ingegneria, Università degli Studi di Perugia, Perugia, Italy
giuseppe.liotta@unipg.it

Abstract. For a graph G, a function ψ is called a *bar visibility representation* of G when for each vertex $v \in V(G)$, $\psi(v)$ is a horizontal line segment (*bar*) and $uv \in E(G)$ iff there is an unobstructed, vertical, ε-wide line of sight between $\psi(u)$ and $\psi(v)$. Graphs admitting such representations are well understood (via simple characterizations) and recognizable in linear time. For a directed graph G, a bar visibility representation ψ of G, additionally, for each directed edge (u, v) of G, puts the bar $\psi(u)$ strictly below the bar $\psi(v)$. We study a generalization of the recognition problem where a function ψ' defined on a subset V' of $V(G)$ is given and the question is whether there is a bar visibility representation ψ of G with $\psi|V' = \psi'$. We show that for undirected graphs this problem together with closely related problems are NP-complete, but for certain cases involving directed graphs it is solvable in polynomial time.

1 Introduction

The concept of a visibility representation of a graph is a classic one in computational geometry and graph drawing and the first studies on this concept date back to the early days of these fields (see, e.g. [16,17] and [12] for a recent survey).

This work was partially supported by ESF project EUROGIGA GraDR and preliminary ideas were formed during HOMONOLO 2014. G. Guśpiel was partially supported by the MNiSW grant DI2013 000443. G. Gutowski was partially supported by the Polish National Science Center grant UMO-2011/03/D/ST6/01370. T. Krawczyk was partially supported by the Polish National Science Center grant UMO-2015/17/B/ST6/01873. G. Liotta was partially supported by the MIUR project AMANDA "Algorithmics for MAssive and Networked DAta", prot. 2012C4E3KT_001. The full version of this paper is available on arXiv:1512.00174 [5]. Note: whenever we refer to sections in the *Appendix* we mean the appendix of arXiv:1512.00174v2.

© Springer International Publishing AG 2016
Y. Hu and M. Nöllenburg (Eds.): GD 2016, LNCS 9801, pp. 266–279, 2016.
DOI: 10.1007/978-3-319-50106-2_21

In the most general setting, a visibility representation of a graph is defined as a collection of disjoint sets from an Euclidean space such that the vertices are bijectively mapped to the sets and the edges correspond to unobstructed lines of sight between two such sets. Many different classes of visibility representations have been studied via restricting the space (e.g., to be the plane), the sets (e.g., to be points or line segments) and/or the lines of sight (e.g., to be non-crossing or axis-parallel). In this work we focus on a classic visibility representation setting in which the sets are horizontal line segments (*bars*) in the plane and the lines of sight are vertical. As such, whenever we refer to a visibility representation, we mean one of this type. The study of such representations was inspired by the problems in VLSI design and was conducted by different authors [9,13,14] under variations of the notion of visibility. Tamassia and Tollis [16] gave an elegant unification of different definitions and we follow their approach.

A *horizontal bar* is an open, non-degenerate segment parallel to the x-axis of the coordinate plane. For a set Γ of pairwise disjoint horizontal bars, a *visibility ray* between two bars a and b in Γ is a vertical closed segment spanned between bars a and b that intersects a, b, and no other bar. A *visibility gap* between two bars a and b in Γ is an axis aligned, non-degenerate open rectangle spanned between bars a and b that intersects no other bar.

For a graph G, a *visibility representation* ψ is a function that assigns a distinct horizontal bar to each vertex such that these bars are pairwise disjoint and satisfy additional visibility constraints. There are three standard visibility models:

- *Weak visibility.* In this model, for each edge $\{u, v\}$ of G, there is a visibility ray between $\psi(u)$ and $\psi(v)$ in $\psi(V(G))$.
- *Strong visibility.* In this model, two vertices u, v of G are adjacent if and only if there is a visibility ray between $\psi(u)$ and $\psi(v)$ in $\psi(V(G))$.
- *Bar visibility.* In this model, two vertices u, v of G are adjacent if and only if there is a visibility gap between $\psi(u)$ and $\psi(v)$ in $\psi(V(G))$.

The bar visibility model is also known as the ε-visibility model in the literature.

A graph that admits a visibility representation in any of these models is a planar graph, but the converse does not hold in general. Tamassia and Tollis [16] characterized the graphs that admit a visibility representation in these models as follows. A graph admits a weak visibility representation if and only if it is planar. A graph admits a bar visibility representation if and only if it has a planar embedding with all cut-points on the boundary of the outer face. For both of these models, Tamassia and Tollis [16] presented linear time algorithms for the recognition of representable graphs, and for constructing the appropriate visibility representations. The situation is different for the strong visibility model. Although the planar graphs admitting a strong visibility representation are characterized in [16] (via strong st-numberings), Andreae [1] proved that the recognition of such graphs is NP-complete. Summing up, from a computational point of view, the problems of recognizing graphs that admit visibility representations and of constructing such representations are well understood.

Recently, a lot of attention has been paid to the question of extending partial representations of graphs. In this setting a representation of some vertices of the graph is already fixed and the task is to find a representation of the whole graph that extends the given partial representation. Problems of this kind are often encountered in graph drawing and are sometimes computationally harder than testing for existence of an unconstrained drawing. The problem of extending partial drawings of planar graphs is a good illustration of this phenomenon. On the one hand, by Fáry's theorem, every planar graph can be drawn in the plane so that each vertex is represented as a point, and edges are pairwise non-crossing, straight-line segments joining the corresponding points. Moreover, such a drawing can be constructed in linear time. On the other hand, testing whether a partial drawing of this kind (i.e., an assignment of points to some of the vertices) can be extended to a straight-line drawing of the whole graph is NP-hard [15]. However, an analogous problem in the model that allows the edges to be drawn as arbitrary curves instead of straight-line segments has a linear-time solution [2]. A similar phenomenon occurs when we consider contact representations of planar graphs. Every planar graph is representable as a disk contact graph or a triangle contact graph. Every bipartite planar graph is representable as a contact graph of horizontal and vertical segments in the plane. Although such representations can be constructed in polynomial time, the problems of extending partial representations of these kinds are NP-hard [4].

In this paper we initiate the study of extending partial visibility representations of graphs. From a practical point of view, it may be worth recalling that visibility representations are not only an appealing way of drawing graphs, but they are also typically used as an intermediate step towards constructing visualizations of networks in which all edges are oriented in a common direction and some vertices are aligned (for example to highlight critical activities in a PERT diagram). Visibility representations are also used to construct orthogonal drawings with at most two bends per edge. See, e.g. [6] for more details about these applications. The partial representation extension problem that we study in this paper occurs, for example, when we want to use visibility representations to incrementally draw a large network and we want to preserve the user's mental map in a visual exploration that adds a few vertices and edges per time.

Both for weak visibility and for strong visibility, the partial representation extension problems are easily found to be NP-hard. For weak visibility, the hardness follows from results on contact representations by Chaplick et al. [4]. For strong visibility, it follows trivially from results by Andreae [1]. Our contribution is the study of the partial representation extension problem for bar visibility. Hence, the central problem for this paper is the following:

Bar Visibility Representation Extension:

Input: (G, ψ'); G is a graph; ψ' is a map assigning bars to a $V' \subseteq V(G)$.
Question: Does G admit a bar visibility representation ψ with $\psi|V' = \psi'$?

One of our results is the following.

Theorem 1. *The Bar Visibility Representation Extension Problem is* NP-*complete.*

The proof is a standard reduction from PLANARMONOTONE3SAT problem, which is known to be NP-complete thanks to de Berg and Khosravi [3]. The reduction uses gadgets that simulate logic gates and constructs a planar boolean circuit that encodes the given formula. Theorem 1 is proven in Appendix D. We investigate a few natural modifications of the problem. Most notably, we study the version of the problem for directed graphs. We provide some efficient algorithms for extension problems in this setting. A visibility representation induces a natural orientation on edges of the graph – each edge is oriented from the lower bar to the upper one. This leads to the definition of a visibility representation for a directed graph. The function ψ is a representation of a digraph G if, additionally to satisfying visibility constraints, for each directed edge (u, v) of G, the bar $\psi(u)$ is strictly below the bar $\psi(v)$. Note that a planar digraph that admits a visibility representation also admits an *upward planar drawing* (see e.g., [10]), that is, a drawing in which the edges are represented as non-crossing y-monotone curves.

A *planar st-graph* is a planar acyclic digraph with exactly one source s and exactly one sink t which admits a planar embedding such that s and t are on the outer face. Di Battista and Tamassia [7] proved that a planar digraph admits an upward planar drawing if and only if it is a subgraph of a planar st-graph if and only if it admits a weak visibility representation. Garg and Tamassia [11] showed that the recognition of planar digraphs that admit an upward planar drawing is NP-complete. It follows that the recognition of planar digraphs that admit a weak visibility representation is NP-complete, and so is the corresponding partial representation extension problem. Nevertheless, as is shown in Lemma 1 (see Appendix A for the proof), the situation might be different for bar visibility.

Lemma 1. *Let $st(G)$ be a graph constructed from a planar digraph G by adding two vertices s and t, the edge (s, t), an edge (s, v) for each source vertex v of G, and an edge (v, t) for each sink vertex v of G. A planar digraph G admits a bar visibility representation if and only if the graph $st(G)$ is a planar st-graph.*

As planar st-graphs can be recognized in linear time, the same is true for planar digraphs that admit a bar visibility representation. The natural problem that arises is the following:

Bar Visibility Representation Extension for Digraphs:

Input: (G, ψ'); G is a digraph; ψ' is a map assigning bars to a $V' \subseteq V(G)$.
Question: Does G admit a bar visibility representation ψ with $\psi | V' = \psi'$?

Although we do not provide a solution for this problem, we present an efficient algorithm for an important variant. A bar visibility representation ψ of a directed graph G is called *rectangular* if ψ has a unique bar $\psi(s)$ with the lowest y-coordinate, a unique bar $\psi(t)$ with the highest y-coordinate, $\psi(s)$ and $\psi(t)$ span the same x-interval, and all other bars are inside the rectangle spanned between $\psi(s)$ and $\psi(t)$. See Fig. 1 for an example of a rectangular bar visibility representation of a planar st-graph.

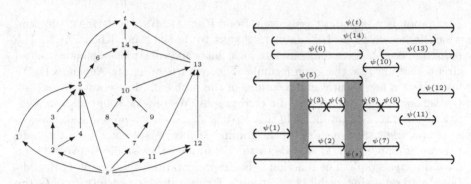

Fig. 1. A planar st-graph G and a rectangular bar visibility representation ψ of G.

Tamassia and Tollis [16] showed that a planar digraph G admits a rectangular bar visibility representation if and only if G is a planar st-graph. In Sect. 3 we give an efficient algorithm for the following problem:

Rectangular Bar Visibility Representation Extension for st-graphs:

Input: (G, ψ'); G is a planar st-graph; ψ' is a map assigning bars to a $V' \subseteq V(G)$.
Question: Does G admit a rectangular bar visibility representation ψ with $\psi|V' = \psi'$?

The main result in this paper is the following.

Theorem 2. *The Rectangular Bar Visibility Representation Extension Problem for an st-graph with n vertices can be solved in $O(n \log^2 n)$ time.*

Our algorithm exploits the correspondence between bar visibility representations and st-orientations of planar graphs, and utilizes the $SPQR$-decomposition.

The rest of the paper is organized as follows. Section 2 contains the necessary definitions and description of the necessary tools. Section 3 contains the general ideas for the proof of Theorem 2. The omitted parts of the proof are reported in Appendix C together with some figures illustrating the ideas behind the proofs. Section 4 mentions further results from the full version and open problems.

2   Preliminaries

For a horizontal bar a, functions $y(a)$, $l(a)$, $r(a)$ give respectively the y-coordinate of a, the x-coordinate of the left end of a, and the x-coordinate of the right end of a. For any bounded object Q in the plane, we use functions $X(Q)$ and $Y(Q)$ to denote the smallest possible, possibly degenerate, closed interval containing the projection of Q on the x-, and on the y-axis respectively. We denote the left end of $X(Q)$ by $l(Q)$ and the right end of $X(Q)$ by $r(Q)$. Let a and b be two horizontal bars with $y(a) < y(b)$. We say that Q is *spanned between a and b* if $X(Q) \subseteq X(a)$, $X(Q) \subseteq X(b)$, and $Y(Q) = [y(a), y(b)]$.

For a graph G, we often describe the visibility representation ψ by providing the values of functions $y_\psi = y(\psi(v))$, $l_\psi = l(\psi(v))$, $r_\psi = r(\psi(v))$ for any vertex v of G. We drop the subscripts when the representation is known from the context.

Let G be a planar st-graph. An st-embedding of G is any planar embedding with s and t on the boundary of the outer face. A planar st-graph together with an st-embedding is called a plane st-graph. Vertices s and t of a planar (plane) st-graph are called the poles of G. We abuse notation and we use the term planar (plane) uv-graph to mean a planar (plane) st-graph with poles u and v. An inner vertex of G is a vertex of G other than the poles of G. A real valued function ξ from $V(G)$ is an st-valuation of G if for each edge (u, v) we have $\xi(u) < \xi(v)$.

Tamassia and Tollis [16] showed that the following properties hold for any plane st-graph:

1. For every inner face f, the boundary of f consists of two directed paths with a common origin and a common destination.
2. The boundary of the outer face consists of two directed paths, with a common origin s and a common destination t.
3. For every inner vertex v, edges from v (to v) are consecutive around v.

Let G be a plane st-graph. We introduce two objects associated with the outer face of G: the left outer face s^* and the right outer face t^*. Properties (1)–(3) allow us to introduce the following standard notions: left/right face of an edge and a vertex, left/right path of a face, and the dual G^* of G – a planar st-graph with vertex set consisting of inner faces of G, s^*, and t^*. For two faces f and g in $V(G^*)$ we say that f is to the left of g, and that g is to the right of f, if there is a directed path from f to g in G^*. See Appendix B.2 for the precise definitions which follow the standard definitions given by Tamassia and Tollis [16].

3   Rectangular Bar Visibility Representations of st-graphs

In this section we provide an efficient algorithm that solves the rectangular bar visibility representation extension problem for st-graphs. Our algorithm employs a specific version of the $SPQR$-decomposition that allows us to describe all st-embeddings of a planar st-graph. See Appendix B.1 for the exact definition which follows the one given by Di Battista and Tamassia [8]. In particular, an $SPQR$-tree T of a planar st-graph G consists of nodes of four different types: S for series nodes, P for parallel nodes, Q for edge nodes, and R for rigid nodes. Each node μ represents a pertinent graph G_μ, a subgraph of G which is an st-graph with poles s_μ and t_μ. Additionally, μ has an associated directed multigraph $skel(\mu)$ called the skeleton of μ. The only difference between our definition of the $SPQR$-tree and the one given in [8] is that we do not add an additional edge between the poles of the skeleton of a node. Our definition ensures that we have a one-to-one correspondence between the edges of $skel(\mu)$ and the children of μ in T. In Sect. 3.1, we use the $SPQR$-tree T of G to describe how a rectangular bar visibility representation is composed of rectangular bar visibility representations of the pertinent graphs of T.

The skeleton of a rigid node has only two st-embeddings, one being the flip of the other around the poles of the node. The skeleton of a parallel node with k children has $k!$ st-embeddings, one for every permutation of the edges of the skeleton. The skeleton of a series node or an edge node has only one st-embedding.

Section 3.1 presents structural properties of bar visibility representations in relation to an $SPQR$-decomposition. In Sect. 3.2 we present an algorithm that solves this extension problem in quadratic time. In Appendix C.6 we give a refined algorithm that works in $O(n \log^2 n)$ time for an st-graph with n vertices.

3.1 Structural Properties

Let Γ be a collection of pairwise disjoint bars. For a pair of bars a, b in Γ with $y(a) < y(b)$ let the *set of visibility rectangles* $R(a, b)$ be the interior of the set of points (x, y) in $\mathbb{R}^2$ where:

1. a is the first bar in Γ on a vertical line downwards from (x, y),
2. b is the first bar in Γ on a vertical line upwards from (x, y).

Figure 1 shows (shaded area) the set of visibility rectangles $R(s, 5)$. Note that there is a visibility gap between a and b in Γ iff $R(a, b)$ is non-empty. If $R(a, b)$ is non-empty, then it is a union of pairwise disjoint open rectangles spanned between a and b.

Let G be a planar st-graph and let T be the $SPQR$-tree for G. Let ψ be a rectangular bar visibility representation of G. For every node μ of T we define the set $B_\psi(\mu)$, called the *bounding box of μ with respect to ψ*, as the closure of the following union:

$$\bigcup \{R(\psi(u), \psi(v)) : (u, v) \text{ is an edge of the pertinent digraph } G_\mu\}.$$

If ψ is clear from the context, then the set $B_\psi(\mu)$ is denoted by $B(\mu)$ and is called the *bounding box of μ*. Let $B(\psi) = X(\psi(V(G))) \times Y(\psi(V(G)))$ be the minimal closed axis-aligned rectangle that contains the representation ψ. It follows that:

1. $B(\psi) = B_\psi(\mu)$, where μ is the root of T,
2. each point in $B(\psi)$ is in the closure of at least one set of visibility rectangles $R(\psi(u), \psi(v))$ for some edge (u, v) of G,
3. each point in $B(\psi)$ is in at most one set of visibility rectangles.

The following two lemmas describe basic properties of a bounding box.

Lemma 2 (Q-Tiling Lemma). *Let μ be a Q-node in T corresponding to an edge (u, v) of G. For any rectangular bar visibility representation ψ of G we have:*

1. *$B(\mu)$ is a union of pairwise disjoint rectangles spanned between $\psi(u)$ and $\psi(v)$.*
2. *If $B(\mu)$ is not a single rectangle, then the parent λ of μ in T is a P-node, and u, v are the poles of the pertinent digraph G_λ.*

The Basic Tiling Lemma presented below describes the relation between the bounding box of an inner node μ and the bounding boxes of the children of μ in any rectangular bar visibility representation of G. The next lemma justifies the name *bounding box* for $B(\mu)$.

Lemma 3 (Basic Tiling Lemma). *Let μ be an inner node in T with children $\mu_1, \ldots, \mu_k$, $k \geqslant 2$. For a rectangular bar visibility representation ψ of G we have:*

1. *$\psi(v) \subseteq B(\mu)$ for every inner vertex v of G_μ.*
2. *$B(\mu)$ is a rectangle that is spanned between $\psi(s_\mu)$ and $\psi(t_\mu)$.*
3. *The sets $B(\mu_1), \ldots, B(\mu_k)$ tile the rectangle $B(\mu)$, i.e., $B(\mu_1), \ldots, B(\mu_k)$ cover $B(\mu)$ and the interiors of $B(\mu_1), \ldots, B(\mu_k)$ are pairwise disjoint.*

In the next three lemmas we specialize the Basic Tiling Lemma depending on whether μ is a P-node, an S-node, or an R-node. These lemmas allow us to describe all tilings of $B(\mu)$ by bounding boxes of μ's children. For Lemmas 4, 5, and 7 we let $\mu_1, \ldots, \mu_k$ be μ's children. The next lemma follows from the Basic Tiling Lemma and the Q-Tiling Lemma.

Lemma 4 (P-Tiling Lemma). *Let μ be a P-node. For any rectangular bar visibility representation ψ of G we have:*

1. *If (s_μ, t_μ) is not an edge of G, then the sets $B(\mu_1), \ldots, B(\mu_k)$ are rectangles spanned between $\psi(s_\mu)$ and $\psi(t_\mu)$.*
2. *If (s_μ, t_μ) is an edge of G, then μ has exactly one child that is a Q-node, say μ_1, and:*
   - *For $i = 2, \ldots, k$, $B(\mu_i)$ is a rectangle spanned between $\psi(s_\mu)$ and $\psi(t_\mu)$.*
   - *$B(\mu_1) \neq \emptyset$ is a union of rectangles spanned between $\psi(s_\mu)$ and $\psi(t_\mu)$.*

When μ is an S-node or an R-node, then there is no edge (s_μ, t_μ). By the Q-Tiling Lemma and by the Basic Tiling Lemma, each set $B(\mu_i)$ is a rectangle that is spanned between the bars representing the poles of G_{μ_i}.

Lemma 5 (S-Tiling Lemma). *Let μ be an S-node. Let $c_1, \ldots, c_{k-1}$ be the cut-vertices of G_μ encountered in this order on a path from s_μ to t_μ. Let $c_0 = s_\mu$, and $c_k = t_\mu$. For any rectangular bar visibility representation ψ of G, for every $i = 1, \ldots, k-1$, we have $X(\psi(c_i)) = X(B(\mu))$. For every $i = 1, \ldots, k$, $B(\mu_i)$ is spanned between $\psi(c_{i-1})$ and $\psi(c_i)$ and $X(B(\mu_i)) = X(B(\mu))$.*

The R-Tiling Lemma should describe all possible tilings of the bounding box of an R-node μ that appear in all representations of G. Since there is a one-to-one correspondence between the edges of $skel(\mu)$ and the children of μ, we abuse notation and write $B(u, v)$ to denote the bounding box of the child of μ that corresponds to the edge (u, v). By the Basic Tilling Lemma, $B(u, v)$ is spanned between the bars representing u and v.

Suppose that ψ is a representation of G. The tiling $\tau = (B_\psi(\mu_1), \ldots, B_\psi(\mu_k))$ of $B_\psi(\mu)$ determines a triple $(\mathcal{E}, \xi, \chi)$, where: $\mathcal{E}$ is an $s_\mu t_\mu$-embedding of $skel(\mu)$, ξ is an st-valuation of $\mathcal{E}$, and χ is an st-valuation of $\mathcal{E}^*$, that are defined as follows. Consider the following planar drawing of the st-graph $skel(\mu)$. Draw every vertex u in the middle of $\psi(u)$, and every edge $e = (u,v)$ as a curve that starts in the middle of $\psi(u)$, goes a little above $\psi(u)$ towards the rectangle $B_\psi(u,v)$, goes inside $B_\psi(u,v)$ towards $\psi(v)$, and a little below $\psi(v)$ to the middle of $\psi(v)$. This way we obtain a plane st-graph $\mathcal{E}$, which is an st-embedding of $skel(\mu)$. The st-valuation ξ of $\mathcal{E}$ is just the restriction of y_ψ to the vertices from $skel(\mu)$, i.e., $\xi = y_\psi|V(skel(\mu))$. To define the st-valuation χ of $\mathcal{E}^*$ we use the following lemma.

Lemma 6 (Face Condition).

1. Let f be a face in $V(\mathcal{E}^*)$ different than t^*, and let $v_0, v_1, \ldots, v_p$ be the right path of f. There is a vertical line $L_r(f)$ that contains the left endpoints of $\psi(v_1), \ldots, \psi(v_{p-1})$ and the left sides of $B_\psi(v_0, v_1), \ldots, B_\psi(v_{p-1}, v_p)$.
2. Let f be a face in $V(\mathcal{E}^*)$ different than s^*, and let $u_0, u_1, \ldots, u_m$ be the left path of f. There is a vertical line $L_l(f)$ that contains the right endpoints of $\psi(u_1), \ldots, \psi(u_{q-1})$ and the right sides of $B_\psi(u_0, u_1), \ldots, B_\psi(u_{q-1}, u_q)$.
3. If f is an inner face of $\mathcal{E}$ then $L_l(f) = L_r(f)$.

The above lemma allows us to introduce the notion of a *splitting line* for every face f in $V(\mathcal{E}^*)$; namely, it is: the line $L_l(f) = L_r(f)$ if f is an inner face of $\mathcal{E}$, $L_r(f)$ if f is the left outer face of $\mathcal{E}$, and $L_l(f)$ if f is the right outer face of $\mathcal{E}$. Now, let $\chi(f)$ be the x-coordinate of the splitting line for a face f in $V(\mathcal{E}^*)$. To show that $\chi(f)$ is an st-valuation of $\mathcal{E}^*$, note that for any edge (f,g) of $\mathcal{E}^*$ there is an edge (u,v) of $\mathcal{E}$ that has f on the left side and g on the right side. It follows that $\chi(f) = l(B_\psi(u,v)) < r(B_\psi(u,v)) = \chi(g)$, proving the claim.

The representation ψ of G determines the triple $(\mathcal{E}, \xi, \chi)$. Note that any other representation with the same tiling $\tau = (B_\psi(\mu_1), \ldots, B_\psi(\mu_k))$ of $B(\mu)$ gives the same triple. To emphasize that the triple $(\mathcal{E}, \xi, \chi)$ is determined by tiling τ, we write $(\mathcal{E}_\tau, \xi_\tau, \chi_\tau)$.

Now, assume that $\mathcal{E}$ is an st-embedding of $skel(\mu)$, ξ is an st-valuation of $\mathcal{E}$, and χ is an st-valuation of the dual of $\mathcal{E}$. Consider the function ϕ that assigns to every vertex v of $skel(\mu)$ the bar $\phi(u)$ defined as follows: $y_\phi(v) = \xi(v)$, $l_\phi(v) = \chi(\text{left face of } v)$, $r_\phi(v) = \chi(\text{right face of } v)$. Firstly, Tamassia and Tollis [16] showed that ϕ is a bar visibility representation of $skel(\mu)$ and that for $\tau = (B_\phi(\mu_1), \ldots, B_\phi(\mu_k))$, we have $(\mathcal{E}_\tau, \xi_\tau, \chi_\tau) = (\mathcal{E}, \xi, \chi)$. Secondly, there is a representation ψ of G that agrees with τ on $skel(\mu)$, i.e., such that $\tau = (B_\psi(\mu_1), \ldots, B_\psi(\mu_k))$. To construct such a representation, we take any representation ψ of G, translate and scale all bars in ψ to get $B_\psi(\mu) = B_\phi(\mu)$, and represent the pertinent digraphs $G_{\mu_1}, \ldots, G_{\mu_k}$ so that the bounding box of μ_i coincides with $B_\phi(\mu_i)$ for $i = 1, \ldots, k$. This leads to the next lemma.

Lemma 7 (R-Tiling Lemma). *Let μ be an R-node. There is a bijection between the set $\{(B_\psi(\mu_1), \ldots, B_\psi(\mu_k)) : \psi$ is a rectangular bar visibility representation of $G\}$ of all possible tilings of the bounding box of μ by the bounding boxes of $\mu_1, \ldots, \mu_k$ in all representations of G, and the set $\{(\mathcal{E}, \xi, \chi) : \mathcal{E}$ is an st-embedding of $skel(\mu)$, ξ is an st-valuation of $\mathcal{E}$, χ is an st-valuation of the dual of $\mathcal{E}\}$.*

3.2   Algorithm

Let G be an n-vertex planar st-graph and let ψ' be a partial representation of G with the set V' of fixed vertices. We present a quadratic time algorithm that tests if there exists a rectangular bar visibility representation ψ of G that extends ψ'. If such a representation exists, the algorithm can construct it in the same time.

In the first step, our algorithm calculates y_ψ. Namely, the algorithm checks whether $y_{\psi'} : V' \to \mathbb{R}$ is extendable to an st-valuation of G. When such an extension does not exist, the algorithm rejects the instance (G, ψ'); otherwise any extension of $y_{\psi'}$ can be used as y_ψ. The next lemma verifies this step's correctness.

Lemma 8. *Let ψ be a rectangular bar visibility representation of G that extends ψ'.*

1. *The function y_ψ is an st-valuation of G that extends $y_{\psi'}$,*
2. *If y is an st-valuation of G that extends $y_{\psi'}$, then a function ϕ that sends every vertex v of G into a bar so that $y_\phi(v) = y(v)$, $l_\phi(v) = l_\psi(v)$, $r_\phi(v) = r_\psi(v)$ is also a rectangular bar visibility representation of G that extends ψ'.*

Clearly, checking whether $y_{\psi'}$ is extendable to an st-valuation of G, and constructing such an extension can be done in $O(n)$ time. In the second step, the algorithm computes the $SPQR$-tree T for G, which also takes linear time.

Before we describe the last step in our algorithm, we need some preparation. For an inner node μ in T we define the sets $V'(\mu)$ and $C(\mu)$ as follows:

$V'(\mu) =$ the set of fixed vertices in $V(G_\mu) \smallsetminus \{s_\mu, t_\mu\}$,

$C(\mu) = \begin{cases} \emptyset, \text{if } V'(\mu) = \emptyset; \\ \text{the smallest closed rectangle containing } \psi'(u) \text{ for all } u \in V'(\mu), \text{otherwise.} \end{cases}$

The set $C(\mu)$ is called the *core* of μ. For a node μ whose core is empty, our algorithm can represent G_μ in any rectangle spanned between the poles of G_μ. Thus, we focus our attention on nodes whose core is non-empty.

Assume that μ is a node whose core is non-empty. We describe the 'possible shapes' the bounding box of μ might have in a representation of G that extends ψ'. The bounding box of μ is a rectangle that is spanned between the bars corresponding to the poles of G_μ. By the Basic Tiling Lemma, if $C(\mu)$ is non-empty then $B(\mu)$ contains $C(\mu)$. For our algorithm it is important to distinguish whether the left (right) side of $B(\mu)$ contains the left (right) side of $C(\mu)$. This criterion leads to four types of representations of μ with respect to the core of μ.

The main idea of the algorithm is to decide for each inner node μ whose core is non-empty, which of the four types of representation of μ are possible and which are not. The algorithm traverses the tree bottom-up and for each node and each type of representation it tries to construct the appropriate tiling using the information about possible representations of its children. The types chosen for different children need to fit together to obtain a tiling of the parent node. In what follows, we present our approach in more detail.

Let μ be an inner node in T. Fix $\phi' = \psi'|V'(\mu)$. Function ϕ' gives a partial representation of the pertinent digraph G_μ obtained by restricting ψ' to the inner vertices of G_μ. Let x, x' be two real values. A rectangular bar visibility representation ϕ of G_μ is called an $[x, x']$-*representation of* μ if ϕ extends ϕ' and $X(\phi(s_\mu)) = X(\phi(t_\mu)) = [x, x']$. We say that an $[x, x']$-representation of μ is:

- *left-loose, right-loose* (LL), when $x < l(C(\mu))$ and $x' > r(C(\mu))$,
- *left-loose, right-fixed* (LF), when $x < l(C(\mu))$ and $x' = r(C(\mu))$,
- *left-fixed, right-loose* (FL), when $x = l(C(\mu))$ and $x' > r(C(\mu))$,
- *left-fixed, right-fixed* (FF), when $x = l(C(\mu))$ and $x' = r(C(\mu))$.

The next lemma justifies this categorization of representations. It says that if a representation of a given type exists, then every representation of the same type is also realisable.

Lemma 9 (Stretching Lemma). *Let μ be an inner node whose core is non-empty. If μ has an LL-representation, then μ has an $[x, x']$-representation for any $x < l(C(\mu))$ and any $x' > r(C(\mu))$. If μ has an LF-representation, then μ has an $[x, x']$-representation for any $x < l(C(\mu))$ and $x' = r(C(\mu))$. If μ has an FL-representation, then μ has an $[x, x']$-representation for $x = l(C(\mu))$ and any $x' > r(C(\mu))$.*

The main task of the algorithm is to verify which representations are feasible for nodes that have non-empty cores. We assume that: μ is an inner node whose core is non-empty; $\mu_1, \ldots, \mu_k$ are the children of μ, $k \geqslant 2$; $\lambda_1, \ldots, \lambda_{k'}$ are the children of μ with $C(\lambda_i) \neq \emptyset$, $0 \leqslant k' \leqslant k$; $\theta(\lambda_i)$ is the set of feasible types of representations for λ_i, $\theta(\lambda_i) \subseteq \{LL, LF, FL, FF\}$. We process the tree bottom-up and assume that $\theta(\lambda_i)$ is already computed and non-empty.

Let x and x' be two real numbers such that $x \leqslant l(C(\mu))$ and $x' \geqslant r(C(\mu))$. We provide an algorithm that tests whether an $[x, x']$-representation of μ exists. We use it to find feasible types for μ by calling it 4 times with appropriate values of x and x'. While searching for an $[x, x']$-representation of μ our algorithm tries to tile the rectangle $[x, x'] \times [y(s_\mu), y(t_\mu)]$ with $B(\mu_1), \ldots, B(\mu_k)$. The tiling procedure is determined by the type of μ. Note that as the core of a Q-node is empty, the algorithm splits into three cases: μ is an S-node, a P-node, and an R-node. The pseudocode for the algorithms is given in Appendix C.3.

Case S. μ **is an S-node.** In this case we attempt to align the left and right side of the bounding box of each child λ of μ to x and x' respectively. For example, if the core of λ is strictly contained in $[x, x']$, then λ must have an LL-representation. The other cases follow similarly. We also must set the x-intervals of the bars of the cut vertices of G_μ to $[x, x']$. The S-Tiling Lemma and the Stretching Lemma imply the correctness of this approach.

Case P. μ **is a P-node.** In this case we attempt to tile the rectangle $[x, x'] \times [y(s_\mu), y(t_\mu)]$ by placing the bounding boxes of the children of μ side by side from left to right. The order of children whose cores are non-empty is determined by the position of those cores. We sort $\lambda_1, \ldots, \lambda_{k'}$ by the left ends of their cores. Let $l_i = l(C(\lambda_i))$ and $r_i = r(C(\lambda_i))$, $r_0 = x$, $l_{k'+1} = x'$, and without loss of generality $l_1 < \ldots < l_{k'}$.

We need to find enough space to place the bounding boxes of children whose cores are empty. Additionally, if (s_μ, t_μ) is an edge of G, then we need to leave at least one visibility gap in the tiling for that edge. Otherwise, if (s_μ, t_μ) is not an edge of G, we need to close all the gaps in the tiling. A more detailed description of the algorithm follows.

If there are λ_i, λ_{i+1} such that the interior of the set $X(C(\lambda_i)) \cap X(C(\lambda_{i+1}))$ is non-empty, then we prove that there is no $[x, x']$-representation of G_μ. Indeed, by the P-Tiling Lemma and by $C(\lambda_i) \subseteq B(\lambda_i)$, the interior of $B(\lambda_i) \cap B(\lambda_{i+1})$ is non-empty and hence tiling of $B(\mu)$ with $B(\mu_1), \ldots, B(\mu_k)$ is not possible. Additionally, if $r(C(\lambda_i)) = l(C(\lambda_{i+1}))$, then neither a right-loose representation of λ_i nor a left-loose representation of λ_{i+1} can be used, so we delete such types of representations from $\theta(\lambda_i)$ and $\theta(\lambda_{i+1})$. If that leaves some $\theta(\lambda_i)$ empty, then an $[x, x']$-representation of μ does not exist. These checks take $O(k')$ time.

Let $Q_i = [r_i, l_{i+1}] \times [y(s_\mu), y(t_\mu)]$ for $i \in [0, k']$. We say that Q_i is an *open gap* (after λ_i, before λ_{i+1}) if Q_i has non-empty interior. In particular, if $x = r_0 < l_1$ ($r_{k'} < l_{k'+1} = x'$) then there is an open gap before λ_1 (after $\lambda_{k'}$). On the one hand, if there is an edge (s_μ, t_μ) or there is at least one μ_i whose core is empty then we need at least one open gap to construct an $[x, x']$-representation. On the other hand, if (s_μ, t_μ) is not an edge of G then we need to close all the gaps in the tiling. There are two ways to close the gaps. Firstly, the representation of each child node whose core is empty can be placed so that it closes a gap. The second way is to use loose representations for children nodes $\lambda_1, \ldots, \lambda_{k'}$.

Suppose that c is a function that assigns to every λ_i a feasible type of representation from the set $\theta(\lambda_i)$. Whenever $c(\lambda_i)$ is right-loose or $c(\lambda_{i+1})$ is left-loose, we can stretch the representation of λ_i or λ_{i+1}, so that it closes the gap Q_i. We describe a simple greedy approach to close the maximum number of gaps in this way. We processes the λ_i's from left to right and for each one: we close both adjacent gaps if we can (i.e. $LL \in \theta(\lambda_i)$); otherwise, we prefer to close the left gap if it is not yet closed rather than the right gap. This is optimal by a simple greedy exchange argument.

If there are still $g > 0$ open gaps left and (s_μ, t_μ) is not an edge of G, then each open gap needs to be closed by placing in this gap a representation of one or more of the children whose core is empty. Thus, it is enough to check that $k - k' \geqslant g$. The correctness of the described algorithm follows by the P-Tiling Lemma, and the Stretching Lemma.

Case R. μ is an R-node. The detailed discussion of this case is reported in Appendix C.4. Here, we sketch our approach. By the R-Tiling Lemma, the set of possible tilings of $B(\mu)$ by $B(\mu_1), \ldots, B(\mu_k)$ is in correspondence with the triples $(\mathcal{E}, \xi, \chi)$, where $\mathcal{E}$ is a planar embedding of $skel(\mu)$, ξ is an st-valuation of $\mathcal{E}$, and χ is an st-valuation of $\mathcal{E}^*$. To find an appropriate tiling of $B(\mu)$ (that yields an $[x, x']$-representation of μ) we search through the set of such triples. Since μ is a rigid node, there are only two st-embeddings of $skel(\mu)$ and we consider both of them separately. Let $\mathcal{E}$ be one of these planar embeddings. Since the y-coordinate for each vertex of G is already fixed, the st-valuation ξ is given by the y-coordinates of the vertices from $skel(\mu)$. It remains to find an st-valuation χ of $\mathcal{E}^*$, i.e., to determine the x-coordinate of the splitting line for every face.

We claim that the existence of an st-valuation χ is equivalent to checking the satisfiability of a carefully designed 2-CNF formula. For every child λ of μ whose core is non-empty, we introduce two boolean variables that indicate which type (LL, LF, FL, FF) of representation is used for λ. Additionally, for every inner face f of $\mathcal{E}$ we introduce two boolean variables: the first (the second) indicates if the splitting line of f is set to the leftmost (rightmost) possible position determined by the bounding boxes of nodes on the left (right) path of f. Now, using those variables, we can express that: feasible representations of the children nodes are used, splitting line of a face f agrees with the choice of representation for the nodes on the boundary of f (see Face Condition Lemma), the choice of splitting lines gives an st-valuation of $\mathcal{E}^*$.

In Appendix C.4 we present a formula construction that uses a quadratic number of clauses and results in a quadratic time algorithm. In Appendix C.6, we present a different, less direct, approach that constructs smaller formulas for R-nodes and leads to the $O(n \log^2 n)$ time algorithm. Therefore, Lemma 8, and the discussion of cases S, P, and R, together with the results in Appendix C imply Theorem 2.

4   Concluding Remarks and Open Problems

We considered the representation extension problem for bar visibility representations and provided an efficient algorithm for st-graphs and showed NP-completeness for planar graphs. An important variant of bar visibility representations is when all bars used in the representation have integral coordinates, i.e., *grid representations*. Any visibility representation can be easily modified into a grid representation. However, this transformation does not preserve coordinates of the given vertex bars. Indeed, we can show (in Appendix D.3) that the (Rectangular) Bar Visibility Representation Extension problem is NP-hard on series-parallel st-graphs when one desires a grid representation.

We conclude with two natural, interesting open problems. The first one is to decide if there exists a polynomial time algorithm that checks whether a partial representation of a directed planar graph is extendable to a bar visibility representation of the whole graph. Although we show an efficient algorithm for an important case of planar st-graphs, it seems that some additional ideas are needed to resolve this problem in general. The second one is to decide if there is an efficient algorithm for recognition of digraphs admitting strong visibility representation, and for the corresponding partial representation extension problem.

References

1. Andreae, T.: Some results on visibility graphs. Discrete Appl. Math. **40**(1), 5–17 (1992)
2. Angelini, P., Di Battista, G., Frati, F., Jelínek, V., Kratochvíl, J., Patrignani, M., Rutter, I.: Testing planarity of partially embedded graphs. ACM Trans. Algorithms **11**(4), 32:1–32:42 (2015)

3. de Berg, M., Khosravi, A.: Optimal binary space partitions in the plane. In: Thai, M.T., Sahni, S. (eds.) COCOON 2010. LNCS, vol. 6196, pp. 216–225. Springer, Heidelberg (2010). doi:10.1007/978-3-642-14031-0_25

4. Chaplick, S., Dorbec, P., Kratochvíl, J., Montassier, M., Stacho, J.: Contact representations of planar graphs: extending a partial representation is hard. In: Kratsch, D., Todinca, I. (eds.) WG 2014. LNCS, vol. 8747, pp. 139–151. Springer, Heidelberg (2014). doi:10.1007/978-3-319-12340-0_12

5. Chaplick, S., Guśpiel, G., Gutowski, G., Krawczyk, T., Liotta, G.: The partial visibility representation extension problem (2015). Pre-print arXiv:1512.00174

6. Di Battista, G., Eades, P., Tamassia, R., Tollis, I.G.: Graph Drawing: Algorithms for the Visualization of Graphs. Prentice-Hall, Upper Saddle River (1999)

7. Di Battista, G., Tamassia, R.: Algorithms for plane representations of acyclic digraphs. Theoret. Comput. Sci. 61(2–3), 175–198 (1988)

8. Di Battista, G., Tamassia, R.: On-line planarity testing. SIAM J. Comput. 25(5), 956–997 (1996)

9. Duchet, P., Hamidoune, Y.O., Vergnas, M.L., Meyniel, H.: Representing a planar graph by vertical lines joining different levels. Discrete Math. 46(3), 319–321 (1983)

10. Garg, A., Tamassia, R.: Upward planarity testing. Order 12(2), 109–133 (1995)

11. Garg, A., Tamassia, R.: On the computational complexity of upward and rectilinear planarity testing. SIAM J. Comput. 31(2), 601–625 (2001)

12. Ghosh, S.K., Goswami, P.P.: Unsolved problems in visibility graphs of points, segments, and polygons. ACM Comput. Surv. 46(2), 22:1–22:29 (2013)

13. Luccio, F., Mazzone, S., Wong, C.K.: A note on visibility graphs. Discrete Math. 64(2–3), 209–219 (1987)

14. Otten, R., van Wijk, J.G.: Graph representations in interactive layout design. In: Proceedings of IEEE International Symposium on Circuits and Systems, New York, NY, USA, May 1978, pp. 914–918 (1978)

15. Patrignani, M.: On extending a partial straight-line drawing. Int. J. Found. Comput. Sci. 17(5), 1061–1070 (2006)

16. Tamassia, R., Tollis, I.G.: A unified approach to visibility representations of planar graphs. Discrete Comput. Geom. 1(4), 321–341 (1986)

17. Wismath, S.K.: Characterizing bar line-of-sight graphs. In: Proceedings of 1st Annual Symposium on Computational Geometry SCG 1985, Baltimore, MD, USA, June 1985, pp. 147–152 (1985)

Ortho-Polygon Visibility Representations
of Embedded Graphs

Emilio Di Giacomo[1], Walter Didimo[1], William S. Evans[2], Giuseppe Liotta[1],
Henk Meijer[3], Fabrizio Montecchiani[1(✉)], and Stephen K. Wismath[4]

[1] Università Degli Studi di Perugia, Perugia, Italy
fabrizio.montecchiani@unipg.it
[2] University of British Columbia, Vancouver, Canada
[3] University College Roosevelt, Middelburg, The Netherlands
[4] University of Lethbridge, Lethbridge, Canada

Abstract. An ortho-polygon visibility representation of an n-vertex
embedded graph G (OPVR of G) is an embedding preserving drawing of
G that maps every vertex to a distinct orthogonal polygon and each edge
to a vertical or horizontal visibility between its end-vertices. The vertex
complexity of an OPVR of G is the minimum k such that every polygon
has at most k reflex corners. We present polynomial time algorithms that
test whether G has an OPVR and, if so, compute one of minimum vertex complexity. We argue that the existence and the vertex complexity
of an OPVR of G are related to its number of crossings per edge and
to its connectivity. Namely, we prove that if G is 1-plane (i.e., it has
at most one crossing per edge) an OPVR of G always exists while this
may not be the case if two crossings per edge are allowed. Also, if G is a
3-connected 1-plane graph, we can compute in $O(n)$ time an OPVR of
G whose vertex complexity is bounded by a constant. However, if G is a
2-connected 1-plane graph, the vertex complexity of any OPVR of G may
be $\Omega(n)$. In contrast, we describe a family of 2-connected 1-plane graphs
for which an embedding that guarantees constant vertex complexity can
be computed. Finally, we present the results of an experimental study
on the vertex complexity of OPVRs of 1-plane graphs.

1 Introduction

Visibility representations are among the oldest and most studied methods to
display graphs. The first papers appeared between the late 70s and the mid 80s,
mostly motivated by VLSI applications (see, e.g., [15,24,25,31,32,34]). These
papers were devoted to *bar visibility representations (BVR)* of planar graphs
where the vertices are modeled as non-overlapping horizontal segments, called
bars, and the edges correspond to vertical visibilities, i.e. vertical segments that
do not intersect any bar other than at their end points. The study of visibility

Research of EDG, WD, GL and FM supported in part by the MIUR project
AMANDA, prot. 2012C4E3KT_001. NSERC funding is gratefully acknowledged for
WE and SW.

Y. Hu and M. Nöllenburg (Eds.): GD 2016, LNCS 9801, pp. 280–294, 2016.
DOI: 10.1007/978-3-319-50106-2_22

representations of non-planar graphs started about ten years later when *rectangle visibility representations (RVR)* were introduced in the computational geometry and graph drawing communities (see e.g., [11,20,21,27]). Every vertex is represented as an axis-aligned rectangle and two vertices are connected by an edge using either horizontal or vertical visibilities. Figure 1(a) is an example of a RVR of the complete graph K_5. RVRs are an attractive way to draw a non-planar graph: Edges are easy to follow because they do not bend and can have only one of two possible slopes, edge crossings are perpendicular, textual labels associated with the vertices can be inserted in the rectangles. Motivated by the NP-hardness of recognizing whether a graph admits an RVR [27], Streinu and Whitesides [28] initiated the study of RVRs that must respect a set of topological constraints. They proved that if a graph G is given together with the cyclic order of the edges around each vertex, the outer face, and a horizontal/vertical direction for each edge, then there exists a polynomial-time algorithm to test whether G admits an RVR that respects these constraints. Biedl *et al.* [5] have shown that testing the representability of G is polynomial also with a different set of constraints, namely when G is given with an embedding that must be preserved in the RVR. In these settings, however, even structurally simple "almost planar" graphs may not admit an RVR. For example, the embedded graph of Fig. 1(b) is 1-plane (i.e., it has at most one crossing per edge), and it does not have an embedding-preserving RVR [5].

In this paper we introduce a generalization of RVRs. We study to what extent such a generalization enlarges the family of graphs that are representable, and we describe testing and drawing algorithms. Let G be an embedded graph. An *ortho-polygon visibility representation* of G (*OPVR* of G) is an embedding-preserving drawing of G that maps each vertex to a distinct orthogonal polygon and each edge to a vertical or horizontal visibility between its end-vertices. For example, Fig. 1(c) is an embedding-preserving OPVR of the graph of Fig. 1(b). In Fig. 1(c) all vertices except two are rectangles: The non-rectangular vertices have a reflex corner each; intuitively, each of them is "away from a rectangle" by one reflex corner. We say that the OPVR of Fig. 1(c) has *vertex complexity* one. More generally, we say that an OPVR has vertex complexity k, if k is the minimum

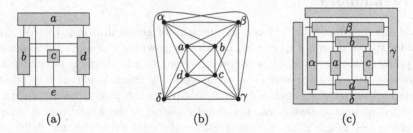

(a)   (b)   (c)

Fig. 1. (a) An RVR of K_5. (b) An embedded graph G that does not admit an embedding preserving RVR. (c) An embedding preserving OPVR of G with vertex complexity one.

integer such that any polygon representing a vertex has at most k reflex corners. We are not only interested in characterizing and testing what graphs admit an OPVR, but we also aim at computing representations of minimum vertex complexity (RVRs if possible). The main results in this paper are as follows.

- In Sect. 3 we present a combinatorial characterization of the graphs that admit an OPVR. This leads to an $O(n^2)$-time algorithm that tests whether an embedded graph G with n vertices admits an embedding-preserving OPVR. If so, an embedding-preserving OPVR of G with minimum vertex complexity is computed in $O(n^{\frac{5}{2}} \log^{\frac{3}{2}} n)$ time. An implication of this characterization is that any 1-plane graph admits an embedding-preserving OPVR. We remark that 1-planar graphs have been widely studied in recent years (see, e.g., [2,4–6,8,16–18,22,29,33]).
- In Sect. 4 we prove that every 3-connected 1-plane graph admits an OPVR whose vertex complexity is bounded by a constant and that this representation can be computed in $O(n)$ time. This implies an $O(n^{\frac{7}{4}} \sqrt{\log n})$-time algorithm to compute OPVRs of minimum vertex complexity for these graphs. Biedl *et al.* [5] proved that not every 3-connected 1-plane graph has a representation with zero vertex complexity, and we show a lower bound of two for infinitely many graphs of this family.
- In Sect. 4 we also study 2-connected 1-plane graphs. Not every 2-connected 1-plane graph can be augmented to become 3-connected (and 1-plane). This has a strong impact on the vertex complexity of the corresponding OPVRs. We prove that an embedding-preserving OPVR of a 2-connected 1-plane graph may require $\Omega(n)$ vertex complexity. Also, we show a sufficient condition that allows to compute an embedding that guarantees constant vertex complexity in $O(n)$ time.
- In Sect. 5 we discuss the results of an experimental study whose aim is to estimate both the vertex complexity of these drawings in practice and the percentage of vertices that are not represented as rectangles.

Some proofs and technicalities are omitted and can be found in [14].

2   Preliminaries

We assume familiarity with basic terminology of graph drawing [13]. We only consider *simple* drawings of graphs, i.e., drawings where two edges have at most one point in common (either a common endpoint or a common interior point where the two edges properly cross each other). A graph is *planar* if it admits a crossing free drawing. Such a drawing subdivides the plane into topologically connected regions, called *faces*. The infinite region is the *outer face*. A *planar embedding* of a graph is an equivalence class of planar drawings that define the same set of faces. A *plane graph* is a planar graph with a given planar embedding. Let f be a face of a plane graph G. The number of vertices encountered in the closed walk along the boundary of f is the *degree* of f, denoted as $\deg(f)$. If G is not 2-connected a vertex may be encountered more than once, thus contributing more than one unit to the degree of the face. The concept of planar embedding can·be extended to non-planar drawings. Given a non-planar drawing, replace each crossing with a dummy vertex. The resulting planarized drawing has a planar embedding. An *embedding* of a graph G is an equivalence class of drawings of G whose planarized versions have the same planar embedding. An *embedded graph* G is a graph with a given embedding: An *embedding-preserving* drawing Γ of G is a drawing of G whose embedding coincides with that of G.

A bar visibility representation (BVR) is *strong* if each visibility between two bars corresponds to an edge of the graph, while it is *weak* when visibilities between non adjacent bars may occur. An *orthogonal polygon* is a polygon whose sides are axis-aligned. A *corner* of an orthogonal polygon is a point of the polygon where a horizontal and a vertical side meet. A corner is a *reflex corner* if it forms a $\frac{3\pi}{2}$ angle inside the polygon. An *ortho-polygon visibility representation* (*OPVR*) of a graph G maps each vertex v of G to a distinct orthogonal polygon $P(v)$ and each edge (u, v) of G to a vertical or horizontal visibility connecting $P(u)$ and $P(v)$ and not intersecting any other polygon $P(w)$, for $w \notin \{u, v\}$. The intersection points between visibilities and polygons are the *attachment points*. We adopt the ϵ-visibility model [21, 28, 31, 34], where the segments representing the edges can be replaced by strips of non-zero width; this implies that an attaching point never coincides with a corner of a polygon. An OPVR is on an *integer grid* if all its corners and attachment points have integer coordinates. Given an OPVR, we can extract a drawing from it as follows. For each vertex v, place a point inside polygon $P(v)$ and connect it to all the attachment points of the boundary of $P(v)$; this can be done without creating any crossing and preserving the circular order of the edges around the vertices. Thus, we refer to an OPVR as a drawing and we extend to OPVRs all the definitions given for drawings. An OPVR γ of an embedded graph is *embedding preserving* if the drawing extracted from γ is embedding preserving. The *vertex complexity* of an OPVR is the maximum number of reflex corners in any polygon representing a vertex. An *optimal OPVR* is an OPVR with minimum vertex complexity.

3   Test and Optimization for Embedded Graphs

Any embedded graph G that admits an OPVR is biplanar, i.e., its edge set can be bicolored so that each color class induces a plane subgraph (use red for the horizontal and blue for the vertical edges of an OPVR of G). However, a biplanar graph G may not have an embedding preserving OPVR. An example is given in Fig. 2 (thin and bold edges define the two colors). The boundary of face f in the figure contains six edge crossings and no vertex. In any OPVR, each crossing forms a $\frac{\pi}{2}$ angle inside f, thus the orthogonal polygon representing f would have six $\frac{\pi}{2}$ corners and no $\frac{3\pi}{2}$ corners in its interior, which is impossible.

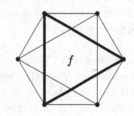

Fig. 2. An embedded graph with no embedding-preserving OPVR.

In the following we first describe an algorithm that, given an embedded graph G that admits an embedding preserving OPVR, computes an optimal OPVR of G (Lemma 2). Then, we describe a characterization of the embedded graphs that admit an embedding preserving OPVR (Lemma 3). This leads to an efficient testing algorithm and it implies that 1-plane graphs always admit an embedding preserving OPVR. Both results extend the *topology-shape-metrics (TSM)* framework to handle OPVRs. The TSM approach, briefly recalled below, was introduced by Tamassia [30] to compute *orthogonal drawings* (see also [13,19]).

The TSM Framework. In an orthogonal drawing of a degree-4 graph each edge is a polyline of horizontal and vertical segments. An angle formed by two consecutive segments incident to the same vertex is a *vertex-angle*; an angle at a bend is a *bend-angle*. The following basic property holds.

Property 1. Let f be a face of an orthogonal drawing and let $N_\alpha(f)$ be the number of vertex-angles of value α inside f, with $\alpha \in \{\frac{\pi}{2}, \frac{3\pi}{2}, 2\pi\}$. Then: $N_{\frac{\pi}{2}}(f) - N_{\frac{3\pi}{2}}(f) - 2N_{2\pi}(f) = 4$ if f is an internal face and $N_{\frac{\pi}{2}}(f) - N_{\frac{3\pi}{2}}(f) - 2N_{2\pi}(f) = -4$ if f is the outer face.

Given a degree-4 graph G, the TSM computes, in three steps, an orthogonal drawing Γ of G with minimum number of bends (see also [13]). The first step, *planarization*, computes an embedding of G and replaces crossing points with dummy vertices. The resulting plane graph G' has $n + c$ vertices, where n and c are the number of vertices and crossings of G, respectively. The second step, *orthogonalization*, computes an *orthogonal representation* H of G', which specifies the values of all vertex-angles and the sequence of bends along each edge. H is computed by means of a flow network N, where each unit of flow corresponds to a $\frac{\pi}{2}$ angle. Each *vertex-node* in N corresponds to a vertex of G' and supplies 4 units of flow; each *face-node* in N corresponds to a face of G' and demands an amount of flow proportional to its degree. Bends along edges correspond to units of flow transferred across adjacent faces of G' through the corresponding arcs of N, and each bend has a unit cost in N. Network N is constructed in $O(n + c)$ time since it has $O(n + c)$ nodes and arcs. Also, it always admits a feasible flow. A feasible flow Φ of cost b of N defines an orthogonal representation H of G' with b bends, and *vice versa*. The third step, *compaction*, computes in $O(n + c + b)$ time an orthogonal drawing preserving the shape of H on an integer grid of size $O(n + c + b) \times O(n + c + b)$.

Our Approach. To exploit the TSM framework, we define a new plane graph $\overline{G}$ obtained from the input embedded graph G as follows (refer to Figs. 3(a) and (b)). Replace each vertex v with a cycle $C(v)$ of $d = \deg(v)$ vertices, so that each of these vertices is incident to one of the edges formerly incident to v, preserving the circular order of the edges around v. If $d = 1$ or $d = 2$, $C(v)$ is a self-loop or a pair of parallel edges, respectively. $C(v)$ is the *expansion cycle* of v; the vertices and the edges of $C(v)$ are the *expansion vertices* and the *expansion edges*, respectively. Also, replace crossings with *dummy vertices*. $\overline{G}$ is called the

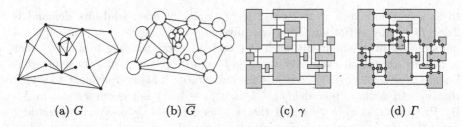

(a) G (b) $\overline{G}$ (c) γ (d) Γ

Fig. 3. (a) An embedded graph G and (b) its planarized expansion $\overline{G}$. (c) An OPVR γ of G and (d) the orthogonal drawing Γ obtained from γ.

planarized expansion of G. The edges of $\overline{G}$ that are not expansion edges are the *original edges*. Each expansion vertex has degree 3 and each dummy vertex has degree 4. The next lemma and properties follow (see also Figs. 3(c) and (d)).

Lemma 1. *An embedded graph* G *admits an embedding preserving OPVR if and only if* $\overline{G}$ *admits an orthogonal representation with the following properties:* **P1.** *Each vertex-angle inside an expansion cycle has value* π. **P2.** *Each original edge has no bend.*

Property 2. If G is biplanar, for each face f of $\overline{G}$ that is not an expansion cycle, $\deg(f) \geq 4$.

Property 3. If G admits an embedding preserving OPVR, then for every internal face f of $\overline{G}$ consisting only of dummy vertices, $\deg(f) = 4$.

Lemma 2. *Let* G *be an* n-*vertex embedded graph that admits an embedding preserving OPVR. There exists an* $O(n^{\frac{5}{2}} \log^{\frac{3}{2}} n)$-*time algorithm that computes an embedding preserving optimal OPVR* γ *of* G. *Also,* γ *has the minimum number of total reflex corners among all embedding preserving optimal OPVRs of* G.

Proof. Since G admits an embedding preserving OPVR, it is biplanar. Hence it has $m \leq 6n - 12$ edges. By Lemma 1, an OPVR of G can be found by computing an orthogonal representation that satisfies **P1** and **P2**. This can be done by computing a feasible flow in the Tamassia's flow network N associated with $\overline{G}$, subject to these constraints: (i) Every arc of N from a vertex-node to a face-node has fixed flow 2 if the face-node corresponds to an expansion cycle (which implies a π angle inside the cycle), and fixed flow 1 otherwise (which implies a $\frac{\pi}{2}$ angle inside the face); (ii) Arcs from two face-nodes such that none of them corresponds to an expansion cycle of $\overline{G}$ are removed (to avoid bends on the original edges). A feasible flow for N may not correspond to an optimal OPVR. To minimize the vertex complexity we construct a different flow network as follows. The amount of flow moved from a vertex-node to an adjacent face-node is fixed *a priori*, and thus we can construct from N an equivalent flow network N', such that all vertex-nodes are removed and their supplies are transferred onto the supply of the adjacent face-nodes. Namely, each face-node v_f corresponding to an expansion cycle f receives $2\deg(f)$ units of flow, while its demand is $2\deg(f) - 4$ by definition. This is equivalent to saying that v_f will supply 4 units of flow in N'. Similarly, each face-node v_f corresponding to a face f that is not an expansion cycle receives $\deg(f)$ units of flow, while its demand is $2\deg(f) - 4$ (or $2\deg(f) + 4$ if f is the outer face). This is equivalent to saying that v_f will demand flow $\deg(f) - 4$ ($\deg(f) + 4$ if f is the outer face) in N'. By Property 2, $\deg(f) \geq 4$ and therefore $\deg(f) - 4 \geq 0$. We now consider every face f of $\overline{G}$ having dummy vertices only (if any), and the corresponding face-node v_f in N'. Note that v_f is an isolated node of N'. Since G admits an embedding preserving OPVR, by Property 3, $\deg(f) = 4$; hence, we can remove v_f from N' and conclude that f must be drawn as a rectangle. Thus, every face-node in N' corresponds to a face of $\overline{G}$ with at least one expansion vertex on its boundary. Since every expansion vertex belongs to at most three faces of $\overline{G}$ and there are $O(n)$ expansion vertices, then N' has $O(n)$ nodes and arcs. We also add gadgets to the network N' in order to impose an upper bound h on the number of reflex corners inside the polygons representing the expansion cycles. Namely, let v_f be a node of N' corresponding to an expansion cycle f. We replace v_f with two face-nodes: a node v_f^{in}, with zero supply and demand; and a node v_f^{out}, with the same supply as v_f (which is 4). The incoming edges of v_f become incoming edges of v_f^{in}, while the outgoing edges of v_f become outgoing edges of v_f^{out}. Finally, we add an edge (v_f^{in}, v_f^{out}) with capacity h. Let N'' be the flow network resulting by applying this transformation to all nodes of N' corresponding to expansion cycles. Since each unit of flow entering in v_f (now in v_f^{in}) corresponds to a $\frac{3\pi}{2}$ angle inside f, a feasible flow of N'' defines an orthogonal representation where each expansion cycle is a polygon with at most h reflex corners, i.e., such a feasible flow defines an OPVR having vertex complexity at most h. N'' is computed in $O(n)$ time and has $O(n)$ nodes and arcs, as N'. In order to guarantee that the OPVR has the minimum number of reflex corners among those with vertex complexity at most h, we compute a feasible flow of minimum cost. Namely, we apply the min-cost flow algorithm of Garg and Tamassia [19], whose complexity is $O(\chi^{\frac{3}{4}} m'' \sqrt{\log n''})$, where n'' and m'' are the number of nodes and arcs of N'', respectively, and χ is the cost of the flow[1]. As already observed, both n'' and m'' are $O(n)$. Also, since the value of the flow is $O(n)$ and in a min-cost flow each unit of flow moved along an augmenting path can traverse each face-node at most once, we have $\chi = O(n^2)$. Hence, a min-cost flow of N'' (if any) is computed in $O(n^{\frac{5}{2}} \sqrt{\log n})$ time.

The supplied flow in N'' is $4n$ (four units for each expansion cycle) and each unit of a min-cost flow can traverse a face-node at most once. Thus, the vertex complexity of an embedding preserving optimal OPVR of G is $k \leq 4n$. We can find the value of k by performing a binary search in the range $[0, 4n]$, testing, for each considered value h, if an OPVR with vertex complexity at most h exists. The number of tests is $O(\log n)$ and each test takes $O(n^{\frac{5}{2}} \sqrt{\log n})$ time, with the algorithm described above. Thus, computing an orthogonal representation H

---

[1] Since N'' may not be planar, we cannot use the faster min-cost flow algorithm in [9].

corresponding to an OPVR with vertex complexity k takes $O(n^{\frac{5}{2}} \log^{\frac{3}{2}} n)$ time. A drawing of H is computed with the compaction step of the TSM. Since H has at most $k \cdot n$ bends, this can be done in $O((k+1)n + c) = O(n^2)$ time.   □

We now introduce a new plane graph associated with the planarized expansion $\overline{G}$ of G. Namely, let $\overline{G}^*$ be the dual graph of $\overline{G}$ where the dual edges associated with the original edges are removed. $\overline{G}^*$ has a vertex for each face of $\overline{G}$ and an edge between two vertices for every edge of an expansion cycle shared by the two corresponding faces. We call $\overline{G}^*$ the *simplified dual* of $\overline{G}$. Given a connected component $\mathcal{C}$ of $\overline{G}^*$, denote by $F_\mathcal{C}$ the set of faces of $\overline{G}$ corresponding to the vertices of $\mathcal{C}$, by $F_\mathcal{C}^{ex}$ the subset of $F_\mathcal{C}$ corresponding to the expansion cycles, and by $F_\mathcal{C}^{nex}$ the set $F_\mathcal{C} \setminus F_\mathcal{C}^{ex}$. Finally, let f_{out} be the outer face of $\overline{G}$. We give the following characterization.

Lemma 3. *An embedded graph G admits an embedding preserving OPVR if and only if for each connected component $\mathcal{C}$ of $\overline{G}^*$ we have $\sum_{f \in F_\mathcal{C}^{nex}} \deg(f) = 4|F_\mathcal{C}| - 8 \cdot \beta$, where $\beta = 1$ if $f_{out} \in F_\mathcal{C}$ and $\beta = 0$ otherwise.*

Lemma 3 leads to an $O(n+c)$-time algorithm that tests whether an embedded graph G with n vertices and c crossings admits an embedding preserving OPVR. Indeed, the size of $\overline{G}^*$ is $O(n + c)$ and thus the condition of Lemma 3 can be checked in $O(n + c)$ time. If G is biplanar it has at most $6n - 12$ edges, and $O(n+c) = O(n^2)$. The next theorem summarizes the contribution of this section.

Theorem 1. *Let G be an n-vertex embedded graph. There exists an $O(n^2)$-time algorithm that tests if G admits an embedding preserving OPVR and, if so, it computes an embedding preserving optimal OPVR γ in $O(n^{\frac{5}{2}} \log^{\frac{3}{2}} n)$ time. Also, γ has the minimum number of reflex corners among all embedding preserving optimal OPVRs of G.*

We remark that an alternative algorithm to test whether G admits an embedding preserving OPVR can be derived from the result in [3]. Namely, Alam et al. [3] showed an algorithm to test whether an n-vertex biconnected plane graph G admits an orthogonal drawing such that edges have no bends, and each face f has most k_f reflex corners. The time complexity of this algorithm is $O((nk)^{\frac{3}{2}})$-time, where $k = \max_{f \in G} k_f$. Thus, one can compute $\overline{G}$ and split each expansion edge of $\overline{G}$ with $4n$ subdivision vertices (the maximum number of reflex corners that a face can have). The resulting graph $\overline{G}'$ has $O(n^2)$ vertices. Then one can apply the algorithm by Alam et al. on $\overline{G}'$ with $k_f = 4n$ for every face f of G. However, this would lead to a time complexity $O(n^{\frac{9}{2}})$. We conclude this section by observing that the number of crossings per edge is a critical parameter for the ortho-polygon representability of an embedded graph: Even two crossings per edge may give rise to a graph that cannot be represented (see Fig. 2). On the positive side, the following theorem can be proved by applying Lemma 3.

Theorem 2. *Every 1-plane graph admits an embedding-preserving OPVR.*

4   Bounds and Optimization for 1-Plane Graphs

Motivated by Theorem 2, in this section we study upper and lower bounds on the vertex complexity of 1-plane graphs. We present a result about partitioning the edges of a 3-connected 1-plane graph so that each partition set induces a plane graph and one of these plane graphs has maximum vertex degree six, which is a tight bound. This result may be of independent interest since it contributes to recent combinatorial studies about partitioning the edge set of 1-plane graphs into two plane subgraphs having special properties (see e.g. [1,10,23]). Next, we use this result to show an upper bound of 12 and a lower bound of 2 on the vertex complexity of 3-connected 1-plane graphs. Finally, we argue that the vertex complexity of OPVRs of 1-plane graphs strongly depends on their connectivity properties; namely, we show that if an n-vertex 1-plane graph G is 2-connected and it can be augmented to become 3-connected only at the expenses of loosing its 1-planarity, then the vertex complexity of any OPVR of G may be $\Omega(n)$. Also, for these graphs we show that a 1-planar embedding that guarantees constant vertex complexity can be computed in $O(n)$ time under the assumption that they do not have a certain type of crossing configuration.

We shall distinguish between the crossing configurations depicted in Fig. 4. Figure 4(a) is a *B-configuration* if the dotted edges are missing, and it is an *augmented B-configuration* otherwise. The crossing configurations of Figs. 4(b), (c), and (d) are a *kite*, a *W-configuration*, and a *T-configuration*, respectively [5,33]. Figure 4(e) depicts an *augmented T-configuration*. In the following we shall also refer to *crossing augmented 1-plane graphs* [7]. A 1-plane graph G is crossing augmented, when for each pair of crossing edges (u,v) and (w,z), the subgraph of G induced by $\{u,v,w,z\}$ is a K_4. We call *cycle edges* of (u,v) and (w,z) the four edges of the K_4 different from (u,v) and (w,z) (they form a 4-cycle). Note that a 1-plane graph can always be made crossing augmented in $O(n)$ time, by adding the missing cycle edges without introducing any new crossings [2,7,29].

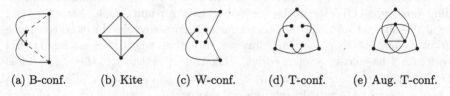

(a) B-conf.   (b) Kite   (c) W-conf.   (d) T-conf.   (e) Aug. T-conf.

Fig. 4. Crossing configurations of 1-plane graphs.

Edge Partitions. An *edge partition* of a 1-plane graph G is a coloring of its edges with one of two colors, *red* and *blue*, such that both the red graph G_R induced by the red edges and the blue graph G_B induced by the blue edges are plane.

Theorem 3. *Let G be a 3-connected 1-plane graph with n vertices. There is an edge partition of G such that the red graph has maximum vertex degree six and this bound is worst case optimal. Also, such an edge edge partition can be computed in $O(n)$ time.*

Proof Sketch: We assume that G is crossing augmented. The proof relies on claims that describe properties of the cycle edges of G which make it possible to construct the desired partition of the edges of G.

Claim 1. *There are no two cycle edges of G that cross each other.*

Claim 2. *Any edge of G is the cycle edge of at most two pairs of crossing edges.*

Let G_p be the plane graph obtained from G by removing an edge for each pair of crossing edges. We can arbitrarily choose what edges to remove, provided that we never remove a cycle edge. Claim 1 ensures that this choice is always feasible. Let G_p^+ be a plane graph obtained by edge-augmenting G_p so to become a plane triangulation. We apply a *Schnyder trees decomposition* to G_p^+, so to find an orientation of its internal edges such that each internal vertex has exactly three outgoing edges and the vertices of the outer face have no outgoing edge. Finally, we arbitrarily orient the edges of the outer face of G_p^+.

Claim 3. *Let (u,v) and (w,z) be two crossing edges of G. Then either $\{u,v\}$ or $\{w,z\}$ have both an outgoing edge in G_p^+, that is a cycle edge of (u,v) and (w,z).*

We use Claim 3 to partition the edge set of G as follows. For each pair of crossing edges (u,v) and (w,z) of G we color with the red color the edge connecting the pair, $\{u,v\}$ or $\{w,z\}$, for which Claim 3 holds. By this choice, each end-vertex of a red edge has one outgoing edge among the cycle edges of (u,v) and (w,z). Since every vertex is incident to at most three outgoing edges in G_p^+, and since each edge is the cycle edge of at most two pairs of crossing edges (Claim 2), by this procedure at most six edges for each vertex get the red color. The proof that this bound on the vertex degree of G_R is tight uses a graph constructed as follows: Start with a sufficiently large plane triangulation G_p, and insert an augmented T-configuration inside every face of G_p. The tightness of the bound can then be derived by a counting argument based on Euler's formula. The linear time complexity follows from the fact that G has $O(n)$ edges (see e.g. [29]) and that Schnyder trees can be constructed in $O(n)$ time [26].   □

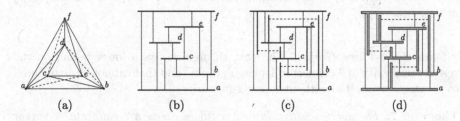

Fig. 5. (a) An edge partition of a 1-plane graph G; red (blue) edges are dashed (solid). (b) A strong BVR γ_B of G_B. (c) Insertion of the red edges into γ_B. (d) An OPVR of G.

Vertex Complexity Bounds for 3-Connected 1-Plane Graphs. Theorem 3 can be used to construct an OPVR of a 3-connected 1-plane graph whose vertex complexity does not depend on the input size. The idea for this construction is as follows. Let G_B and G_R be the plane graphs defined by the edge partition of Theorem 3; see e.g. Fig. 5(a). Under the assumption that G is crossing augmented, it can be proved that G_B is 2-connected, which implies that it admits a strong BVR γ_B (this can be computed in $O(n)$ time [31]); see e.g. Fig. 5(b). Assume that two vertices u and v are connected by a red edge and let $\gamma_B(u)$ and $\gamma_B(v)$ be the horizontal bars representing them. We attach a vertical bar to $\gamma_B(u)$ and a vertical bar to $\gamma_B(v)$ such that each vertical bar shares an end-vertex with the horizontal bar and the two vertical bars can see each other horizontally. This makes it possible to draw the horizontal red edge (u, v); see e.g. Fig. 5(c). Once all red edges have been added to γ_B, every vertex v is represented as a "rake"-shaped object consisting of one horizontal bar and at most six vertical bars (we have a vertical bar for each red edge incident to v and there are at most six such edges). This "rake"-shaped object can then be used as the skeleton of an orthogonal polygon that has two reflex corners per vertical bar; see e.g. Fig. 5(d).

Theorem 4. *Let G be a 3-connected 1-plane graph with n vertices. There exists an $O(n)$-time algorithm that computes an an embedding-preserving OPVR of G with vertex complexity at most 12, on an integer grid of size $O(n) \times O(n)$.*

Based on Theorem 4, we can significantly improve the time complexity of an algorithm that computes an optimal OPVR.

Theorem 5. *Let G be a 3-connected 1-plane graph with n vertices. There exists an $O(n^{\frac{7}{4}}\sqrt{\log n})$-time algorithm that computes an embedding-preserving optimal OPVR γ of G, on an integer grid of size $O(n) \times O(n)$. Also, γ has the minimum number of total reflex corners among all the embedding preserving optimal OPVRs of G.*

The following lower bound can be proved.

Theorem 6. *There is an infinite family $\mathcal{G}$ of 3-connected 1-plane graphs such that for any graph G of $\mathcal{G}$, any embedding preserving OPVR has vertex complexity at least two.*

2-Connected 1-Plane Graphs. The next theorem shows a lower bound on the vertex complexity of 2-connected 1-planar graphs (and that cannot be augmented to become 3-connected without losing 1-planarity).

Theorem 7. *For every positive integer n, there exists a 2-connected 1-planar graph G with $O(n)$ vertices such that, for every 1-planar embedding of G, any embedding preserving OPVR of G has vertex complexity $\Omega(n)$.*

Proof Sketch: We prove the claim for a fixed 1-planar embedding (the proof can be easily extended to all 1-planar embeddings of G). Consider the 1-plane

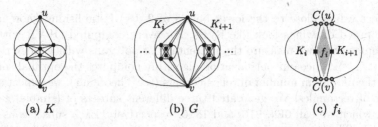

| (a) K | (b) G | (c) f_i |

Fig. 6. Illustration for the proof of Theorem 7.

graph K in Fig. 6(a). It has 2 vertices on its outer face, u and v, plus 6 inner vertices. We now construct G as follows. Attach $n + 1$ copies $K_1, \ldots, K_{n+1}$ of K all sharing u and v. The copies are attached in parallel without introducing any further crossing, as shown in Fig. 6(b). Also connect u and v with an edge on the outer face. The resulting graph G has $8(n + 1) - 2n = 6n + 8 = O(n)$ vertices. Also, G is 2-connected and 1-plane by construction, hence it admits an OPVR by Theorem 2. Consider now an embedding preserving OPVR of G and the corresponding orthogonal drawing Γ. Between any two consecutive copies K_i and K_{i+1} ($i = 1, \ldots, n$), there is a face f_i of G having two expansion vertices of $C(u)$ (the expansion cycle of u) and two expansion vertices of $C(v)$ on its boundary, together with two dummy vertices; see Fig. 6(c). Each dummy vertex forms one $\frac{\pi}{2}$ angle inside f_i. Each expansion vertex forms one $\frac{\pi}{2}$ angle inside f_i. Hence, there are at least six $\frac{\pi}{2}$ angles inside f_i. Since the original edges of f_i have no bends, by Property 1 the two expansion edges of f_i must form (at least) two $\frac{3\pi}{2}$ angles inside f_i. In Γ there are n of such faces requiring two angles of $\frac{3\pi}{2}$ each from an expansion edge. If every vertex of G is represented by a polygon with vertex complexity at most k, the edges of each expansion cycle form at most $4 + k$ angles of $\frac{3\pi}{2}$ inside their incident faces (that are not expansion cycles). At least ten of these angles are inside the outer face of Γ (Property 1), and hence it must be $(4 + k)2 - 10 \geq 2n$, that is $k \geq n + 1$.   □

The graphs used to prove Theorem 7 contain several W-configurations. For a contrast, we can show that the absence of W-configurations suffices to find a 1-planar embedding that admits an OPVR with constant vertex complexity.

Theorem 8. *Let G be a 2-connected 1-plane graph with n vertices and no W-configurations. A 1-planar OPVR of G with vertex complexity at most 22 on an integer grid of size $O(n) \times O(n)$ can be computed in $O(n)$ time.*

5   Experiments and Open Problems

We implemented the optimization algorithm of Theorem 1 using the GDToolkit library [12]. To evaluate the performance of the algorithm in practice, we tested it on a large set of 1-plane graphs, which always admit an OPVR (Theorem 2). In addition, we have the following two objectives: (*i*) Measure the vertex complexity of the computed OPVRs; in particular, for 3-connected 1-plane graphs we expect values close to the lower bound of 2. (*ii*) Establishing "how much" the computed drawings look like RVRs. For every computed OPVR with vertex complexity k, we measure the percentage of polygons with i reflex corners ($i \in [0, \ldots, k]$). Since our optimization algorithm computes the optimal solution having the minimum number of reflex corners (see Theorem 1), we expect a high number of rectangles. We generated three different subsets of (simple) 1-plane graphs, which we call GEN, BIC, and TRIC, respectively. Each subset consists of 170 graph. The number of vertices of each graph ranges from 20 to 100. The graphs in GEN are general 1-plane graphs, while those in BIC and in TRIC are 2-connected and 3-connected, respectively. All graphs are maximal (no further edges can be added in their embedding while preserving 1-planarity). The experiments confirmed both our expectations. The optimization algorithm took less than 15 s for all instances up to 60 vertices, and about 41 s on the largest instance with 100 vertices on a common laptop. The optimal solutions of all GEN graphs required vertex complexity 1, except two of them with vertex complexity 0. The average percentage of rectangles is around 90%, and never below 80% in any instance. Hence, most of the drawing looks like an RVR. The running times for BIC and TRIC reflect the behavior observed for GEN (with some more demanding large instances). For every graph of TRIC we found a drawing with vertex complexity either 1 or 2. Most of the BIC graphs required vertex complexity 2, some required vertex complexity 3, and only one graph required vertex complexity 4. The percentage of vertices drawn as rectangles is very high also for BIC and TRIC (around 80% for BIC and around 75% for TRIC).

The results in this paper naturally raise several interesting open problems. Among them: (1) Close the gap between the upper bound and the lower bound on the vertex complexity of OPVRs of 3-connected 1-plane graphs (see Theorems 5 and 6). (2) We find it interesting to study the problem of computing OPVRs that maximize the number of rectangular vertices, even at the expenses of sub-optimal vertex complexity. (3) Theorem 8 constructs 1-planar embeddings that guarantee constant vertex complexity. What 2-connected 1-plane graphs admit a 1-planar OPVR with constant vertex complexity?

References

1. Ackerman, E.: A note on 1-planar graphs. Discrete Appl. Math. **175**, 104–108 (2014)
2. Alam, M.J., Brandenburg, F.J., Kobourov, S.G.: Straight-line grid drawings of 3-connected 1-planar graphs. In: Wismath, S., Wolff, A. (eds.) GD 2013. LNCS, vol. 8242, pp. 83–94. Springer, Heidelberg (2013). doi:10.1007/978-3-319-03841-4_8
3. Alam, M.J., Kobourov, S.G., Mondal, D.: Orthogonal layout with optimal face complexity. In: Freivalds, R.M., Engels, G., Catania, B. (eds.) SOFSEM 2016. LNCS, vol. 9587, pp. 121–133. Springer, Heidelberg (2016). doi:10.1007/978-3-662-49192-8_10
4. Bannister, M.J., Cabello, S., Eppstein, D.: Parameterized complexity of 1-planarity. In: Dehne, F., Solis-Oba, R., Sack, J.-R. (eds.) WADS 2013. LNCS, vol. 8037, pp. 97–108. Springer, Heidelberg (2013). doi:10.1007/978-3-642-40104-6_9

5. Biedl, T.C., Liotta, G., Montecchiani, F.: On visibility representations of non-planar graphs. In: Fekete, S.P., Lubiw, A. (eds) SoCG 2016, LIPIcs, vol. 51, pp. 19:1–19: 16. Schloss Dagstuhl - Leibniz-Zentrum fuer Informatik (2016). http://www.dagstuhl.de/dagpub/978-3-95977-009-5

6. Brandenburg, F.J.: 1-Visibility representations of 1-planar graphs. J. Graph Algorithms Appl. **18**(3), 421–438 (2014)

7. Brandenburg, F.J.: On 4-map graphs, 1-planar graphs, their recognition problem. CoRR, abs/1509.03447 (2015). http://arxiv.org/abs/1509.03447

8. Cabello, S., Mohar, B.: Adding one edge to planar graphs makes crossing number and 1-planarity hard. SIAM J. Comput. **42**(5), 1803–1829 (2013)

9. Cornelsen, S., Karrenbauer, A.: Accelerated bend minimization. J. Graph Algorithms Appl. **16**(3), 635–650 (2012)

10. Czap, J., Hudák, D.: On drawings and decompositions of 1-planar graphs. Electr. J. Comb. **20**(2), P54 (2013)

11. Dean, A.M., Hutchinson, J.P.: Rectangle-visibility representations of bipartite graphs. Discrete Appl. Math. **75**(1), 9–25 (1997)

12. Di Battista, G., Didimo, W.: GDToolkit. In: Tamassia, R. (ed.) Handbook of Graph Drawing and Visualization, pp. 571–597. CRC Press, Boca Raton (2013)

13. Di Battista, G., Eades, P., Tamassia, R., Tollis, I.G.: Graph Drawing. Prentice-Hall, Upper Saddle River (1999)

14. Di Giacomo, E., Didimo, W., Evans, W.S., Liotta, G., Meijer, H., Montecchiani, F., Wismath, S.K.: Ortho-polygon visibility representations of embedded graphs. ArXiv e-prints, abs/1604.08797v2 (2016). http://arxiv.org/abs/1604.08797v2

15. Duchet, P., Hamidoune, Y., Las Vergnas, M., Meyniel, H.: Representing a planar graph by vertical lines joining different levels. Discrete Math. **46**(3), 319–321 (1983)

16. Eades, P., Hong, S.-H., Katoh, N., Liotta, G., Schweitzer, P., Suzuki, Y.: A linear time algorithm for testing maximal 1-planarity of graphs with a rotation system. Theor. Comput. Sci. **513**, 65–76 (2013)

17. Eades, P., Liotta, G.: Right angle crossing graphs and 1-planarity. Discrete Appl. Math. **161**(7–8), 961–969 (2013)

18. Evans, W.S., Kaufmann, M., Lenhart, W., Mchedlidze, T., Wismath, S.K.: Bar 1-visibility graphs vs. other nearly planar graphs. J. Graph Algorithms Appl. **18**(5), 721–739 (2014)

19. Garg, A., Tamassia, R.: A new minimum cost flow algorithm with applications to graph drawing. In: North, S. (ed.) GD 1996. LNCS, vol. 1190, pp. 201–216. Springer, Heidelberg (1997). doi:10.1007/3-540-62495-3_49

20. Hutchinson, J.P., Shermer, T.C., Vince, A.: On representations of some thickness-two graphs. Comput. Geom. **13**(3), 161–171 (1999)

21. Kant, G., Liotta, G., Tamassia, R., Tollis, I.G.: Area requirement of visibility representations of trees. Inf. Process. Lett. **62**(2), 81–88 (1997)

22. Korzhik, V.P., Mohar, B.: Minimal obstructions for 1-immersions and hardness of 1-planarity testing. J. Graph Theory **72**(1), 30–71 (2013)

23. Lenhart, W.J., Liotta, G., Montecchiani, F.: On partitioning the edges of 1-planar graphs. CoRR, abs/1511.07303 (2015). http://arxiv.org/abs/1511.07303

24. Otten, R.H.J.M., Van Wijk, J.G.: Graph representations in interactive layout design. In: IEEE ISCSS, pp. 914–918. IEEE (1978)

25. Rosenstiehl, P., Tarjan, R.E.: Rectilinear planar layouts and bipolar orientations of planar graphs. Discrete Comput. Geom. **1**, 343–353 (1986)

26. Schnyder, W.: Embedding planar graphs on the grid. In: Johnson, D.S. (ed.), SODA 1990, pp. 138–148. SIAM (1990)

27. Shermer, T.C.: On rectangle visibility graphs III. External visibility and complexity. In: Fiala, F., Kranakis, E., Sack, J.-R., (eds.) CCCG 1996, pp. 234–239. Carleton University Press (1996)

28. Streinu, I., Whitesides, S.: Rectangle visibility graphs: characterization, construction, and compaction. In: Alt, H., Habib, M. (eds.) STACS 2003. LNCS, vol. 2607, pp. 26–37. Springer, Heidelberg (2003). doi:10.1007/3-540-36494-3_4

29. Suzuki, Y.: Re-embeddings of maximum 1-planar graphs. SIAM J. Discrete Math. **24**(4), 1527–1540 (2010)

30. Tamassia, R.: On embedding a graph in the grid with the minimum number of bends. SIAM J. Comp. **16**(3), 421–444 (1987)

31. Tamassia, R., Tollis, I.G.: A unified approach to visibility representations of planar graphs. Discrete Comput. Geom. **1**(1), 321–341 (1986)

32. Thomassen, C.: Plane representations of graphs. In: Progress in Graph Theory, pp. 43–69. AP (1984)

33. Thomassen, C.: Rectilinear drawings of graphs. J. Graph Theory **12**(3), 335–341 (1988)

34. Wismath, S.K.: Characterizing bar line-of-sight graphs. In: Rourke, J.O. (ed), SoCG 1985, pp. 147–152. ACM (1985)

Obstructing Visibilities with One Obstacle

Steven Chaplick[1(✉)], Fabian Lipp[1], Ji-won Park[2], and Alexander Wolff[1]

[1] Lehrstuhl für Informatik I, Universität Würzburg, Würzburg, Germany
{steven.chaplick,fabian.lipp,alexander.wolff}@uni-wuerzburg.de
[2] KAIST, Daejeon, Korea
wldnjs1727@kaist.ac.kr

Abstract. Obstacle representations of graphs have been investigated quite intensely over the last few years. We focus on graphs that can be represented by a single obstacle. Given a (topologically open) non-self-intersecting polygon C and a finite set P of points in general position in the complement of C, the *visibility graph* $G_C(P)$ has a vertex for each point in P and an edge pq for any two points p and q in P that can *see* each other, that is, $\overline{pq} \cap C = \emptyset$. We draw $G_C(P)$ straight-line and call this a *visibility drawing*. Given a graph G, we want to compute an obstacle representation of G, that is, an obstacle C and a set of points P such that $G = G_C(P)$. The complexity of this problem is open, even when the points are exactly the vertices of a simple polygon and the obstacle is the complement of the polygon—the *simple-polygon visibility graph problem*.

There are two types of obstacles; *outside* obstacles lie in the unbounded component of the visibility drawing, whereas *inside* obstacles lie in the complement of the unbounded component. We show that the class of graphs with an inside-obstacle representation is incomparable with the class of graphs that have an outside-obstacle representation. We further show that any graph with at most seven vertices has an outside-obstacle representation, which does not hold for a specific graph with eight vertices. Finally, we show NP-hardness of the *outside-obstacle graph sandwich problem*: given graphs G and H on the same vertex set, is there a graph K such that $G \subseteq K \subseteq H$ and K has an outside-obstacle representation. Our proof also shows that the *simple-polygon visibility graph sandwich problem*, the *inside-obstacle graph sandwich problem*, and the *single-obstacle graph sandwich problem* are all NP-hard.

1 Introduction

Recognizing graphs that have a certain type of geometric representation is a well-established field of research dealing with, for example, interval graphs, unit disk graphs, coin graphs (which are exactly the planar graphs), and visibility graphs. In this paper, we are interested in visibilities of points in the presence of

The full version of this paper is available on arXiv [5]. Whenever we refer to the Appendix we mean the appendix of arXiV:1607.00278v2.

J.-w. Park acknowledges support by the NRF grant 2011-0030044 (SRC-GAIA) funded by the Korean government.

Y. Hu and M. Nöllenburg (Eds.): GD 2016, LNCS 9801, pp. 295–308, 2016.
DOI: 10.1007/978-3-319-50106-2_23

a single obstacle. Given a (topologically open) non-self-intersecting polygon C and a finite set P of points in general position in the complement of C, the *visibility graph* $G_C(P)$ has a vertex for each point in P and an edge pq for any two points p and q in P that can *see* each other, that is, $\overline{pq} \cap C = \emptyset$. Given a graph G, we want to compute a (single-) obstacle representation of G, that is, an obstacle C and a set of points P such that $G = G_C(P)$ (if such a representation exists). The complexity of this reconstruction problem is open, even for the case that the points are exactly the vertices of a simple polygon and the (outside) obstacle is the complement of the polygon. This special case is called the *simple-polygon visibility graph (reconstruction) problem*.

The *visibility drawing* is a straight-line drawing of the visibility graph. The visibility drawing allows us to differentiate two types of obstacles: *outside* obstacles lie in the unbounded component of the visibility drawing, whereas *inside* obstacles lie in the complement of the unbounded component.

If we drop the restriction to single obstacles, our problem can be seen as an optimization problem. For a graph G, let obs(G) be the smallest number of obstacles that suffices to represent G as a visibility graph. Analogously, let $\text{obs}_{\text{out}}(G)$ be the number of obstacles needed to represent G in the presence of an outside obstacle, and let $\text{obs}_{\text{in}}(G)$ be the number of obstacles needed to represent G in the absence of outside obstacles. Specifically, we say that G has an *outside-obstacle representation* if G can be represented by a single outside obstacle (e.g. Fig. 1), and G has an *inside-obstacle representation* if G can be represented by a single inside obstacle (e.g. Fig. 3b).

Previous Work. Not only have Alpert et al. [1] introduced the notion of the obstacle number of a graph, they also characterized the class of graphs that can be represented by a single simple obstacle, namely a convex polygon. They also asked many interesting questions, for example, given an integer o, is there a graph of obstacle number exactly o? If the previous question is true, given an integer $o > 1$, what is the smallest number of vertices of a graph with obstacle number o? Mukkamala et al. [13] showed the first question is true. For the second question, Alpert et al. [1] found a 12-vertex graph that needs two obstacles, namely $K_{5,7}^*$, where $K_{m,n}^*$ with $m \leq n$ is the complete bipartite graph minus a matching of size m. They also showed that for any $m \leq n$, obs($K_{m,n}^*$) ≤ 2. This result was improved by Pach and Sarıöz [14] who showed that the 10-vertex graph $K_{5,5}^*$ also needs two obstacles. More recently, Berman et al. [3] suggested some necessary conditions for a graph to have obstacle number 1 which they used to find a *planar* 10-vertex graph that cannot be represented by a single obstacle.

Alpert et al. [1] conjectured that every graph of obstacle number 1 has also outside-obstacle number 1. Berman et al. [3] further conjectured that every graph of obstacle number o has outside-obstacle number o. Alpert et al. [1] also showed that outerplanar graphs always have outside-obstacle representations and posed the question to bound the inside/convex obstacle number of outerplanar/planar graphs. Fulek et al. [7] partly answered this by showing that five convex obstacles are sufficient for outerplanar graphs—and that sometimes four are needed.

For the asymptotic bound on the obstacle number of a graph, it is obvious that any n-vertex graph has obstacle number $O(n^2)$. Balko et al. [2] showed that the obstacle number of an n-vertex graph is (at most) $O(n \log n)$. For the lower bound, improving on previous results [1,12,13], Dujmović and Morin [6] showed there are n-vertex graphs whose obstacle number is $\Omega(n/(\log \log n)^2)$.

Johnson and Sarıöz [10] investigated the special case where the visibility graph is required to be plane. They showed (by reduction from PLANARVERTEXCOVER) that in this case computing the obstacle number is NP-hard. By reduction to MAXDEG-3 PLANARVERTEXCOVER, they showed that the problem admits a polynomial-time approximation scheme and is fixed-parameter tractable. Koch et al. [11] also considered the plane case, restricted to outside obstacles. They gave a(n efficiently checkable) characterization of all biconnected graphs that admit a plane outside-obstacle representation.

A few years ago, Ghosh and Goswami [8] surveyed visibility graph problems, among them simple-polygon visibility graph problem. Open Problem 29 in their survey is the complexity of the recognition problem and Open Problem 33 is the complexity of the fore-mentioned reconstruction problem. Very recently, this question has been settled for an interesting variant of the problem where the points are not only the vertices of the graph but also the obstacles (which are closed in this case): Cardinal and Hoffmann [4] showed that recognizing point-visibility graphs is $\exists \mathbb{R}$-complete, that is, as hard as deciding the existence of a real solution to a system of polynomial inequalities (and hence, at least NP-hard).

The graph sandwich problem has been introduced by Golumbic et al. [9] as a generalization of the recognition problem. They set up the abstract problem formulation and gave efficient algorithms for some concrete graph properties—and hardness results for others.

Preliminaries. In this paper, we consider only finite simple graphs. Whenever we say cycles, we always mean simple cycles. Let G be a graph and let v, u be its vertices. The *circumference* of G, denoted by $\mathrm{circ}(G)$, is the length of its longest cycle. $v \sim u$ denotes that v and u are adjacent. We call v and u *twins* if $v \neq u$ and $N(v) \backslash \{u\} = N(u) \backslash \{v\}$. We say v is *exposed to the outside* if it is on the boundary of the unbounded component of the straight-line drawing of G given by the point set. All vertices are exposed to the outside in an *exposed outside-obstacle representation*. In all figures (of graphs), unless otherwise stated, edges are solid and non-edges are dashed.

Our Contribution. We have the following results. (Recall that a *co-bipartite* graph is the complement of a bipartite graph.)

- Every graph of circumference at most 6 has an outside-obstacle representation (Theorem 1).
- Every 7-vertex graph has an outside-obstacle representation (Theorem 2). Moreover, there is an 8-vertex co-bipartite graph that has no single-obstacle representation (Theorem 5).
- There is an 11-vertex co-bipartite graph with an inside-obstacle representation, but no outside-obstacle representation (Theorem 4). This resolves the above-mentioned open problems of Alpert et al. [1] and Berman et al. [3].

- The Outside-Obstacle Graph Sandwich Problem is NP-hard even for co-bipartite graphs. The same holds for the Simple-Polygon Visibility Graph Sandwich Problem. This does not solve, but sheds some light on a long-standing open problem: the recognition of visibility graphs of simple polygons. While little is known for the complexity of computing the obstacle number, the Single-Obstacle Graph Sandwich Problem is shown to be also NP-hard.

Remarks and Open Problems. The recognition of inside- and outside-obstacle graphs is currently open. We expect that testing either of these cases is NP-hard. Assuming that this is true, it would be interesting to show fixed-parameter tractability w.r.t. the number of vertices of the obstacle. We now know that $\mathrm{obs}_{\mathrm{in}}(G)$ and $\mathrm{obs}_{\mathrm{out}}(G)$ are usually different, but can we bound $\mathrm{obs}_{\mathrm{in}}(G)$ in terms of $\mathrm{obs}_{\mathrm{out}}(G)$? While we have shown that the trivial lower bound $\mathrm{obs}_{\mathrm{out}}(G) - 1$ is tight, an upper bound is only known for outerplanar graphs [1,7].

2   Graphs with Small Circumference

In this section we will describe how to construct an outside-obstacle representation for any graph whose circumference is at most 6. To prove this result we show that for every vertex v of a biconnected graph G with circumference at most 6, there is an exposed outside-obstacle representation of G with v on the convex hull of $V(G)$. Lemma 3 makes it easier to describe the outside-obstacle representation. We then apply Lemmas 1 and 2 to obtain an outside-obstacle representation of a graph.

We provide an 8-vertex graph of circumference 8 that requires at least two obstacles in the next section, so the only gap is the circumference-7 case. We conjecture that every graph of circumference 7 has an outside-obstacle representation. As a first step towards this conjecture, we show that every 7-vertex graph has an outside-obstacle representation by providing a list of point sets such that each 7-vertex graph can be represented by an outside obstacle when the vertices of the graph are mapped to a point set in our list.

Proofs of Lemmas 1, 2 and 3 are in Appendix A and brief ideas are sketched here.

Lemma 1. *Let G and H be graphs on different vertex sets. If $\mathrm{obs}_{\mathrm{out}}(G) = 1$ and $\mathrm{obs}_{\mathrm{out}}(H) = 1$, then $\mathrm{obs}_{\mathrm{out}}(G \cup H) = 1$.*

Proof (Sketch). Place two graphs far enough and merge outside obstacles.   □

Lemma 2. *Let G and H be graphs with exposed outside-obstacle representations. Let u be a vertex of G, and let v be a vertex of H. Assume that v lies on the convex hull of $V(H)$. If K is the graph obtained by identifying u and v, then K also has an exposed outside-obstacle representation.*

Proof (Sketch). Make the outside-obstacle representation of H small and narrow (with respect to v) enough to fit in some circular sector lying inside the obstacle centered at u in the outside-obstacle representation of G. Then replace the circular sector with above obstacle representation of H.   □

Lemma 3. *Let H be a graph, v be a vertex of H, A be the set of twins of v, and $G = H \setminus A$. If G that has an exposed outside-obstacle representation in which v lies on the convex hull of $V(G)$, then H has an exposed outside-obstacle representation in which all vertices in $A \cup \{v\}$ lie on the convex hull of $V(H)$.*

Proof (Sketch). Place twins close enough since their neighborhoods are same. □

The following observation helps to restrict the structure of biconnected graphs of given circumference where indices are taken modulo k.

Observation 1. *Let G be a graph of circumference k and let $C = v_1 v_2 \ldots v_k$ be a cycle. G doesn't contain a $v_i - v_{i+t}$ path P of length t' disjoint to $v_i C v_{i+t}$ where $0 < t < k$ and $t' > t$, since it would create $(k + t' - t)$-cycle. In particular, if $v \notin \{v_1, \ldots, v_k\}$ is adjacent to v_i, then v is neither adjacent to v_{i-1} nor v_{i+1}.*

Theorem 1. *If the circumference of a graph G is at most 6, then G has an outside-obstacle representation.*

Proof If G is disconnected, we give an outside-obstacle representation for each connected component and simply merge them by Lemma 1.

When G is connected, we decompose it into its biconnected components, i.e., the block decomposition tree of G. Starting in its root, we include representations of the children in turn using Lemma 2.

Let H be a biconnected component of G. It suffices to show that H satisfies the condition for Lemma 2: For each vertex v of H, H has an exposed outside-obstacle representation such that v is on the convex hull of $V(H)$.

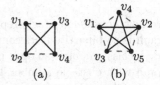

Fig. 1. Graphs of circumference 4 and 5 with outside-obstacle representations (Color figure online)

Case 1: $\text{circ}(H) = 3$

As H is biconnected, H is a triangle and trivially satisfies the condition.

Case 2: $\text{circ}(H) = 4$

Let $C = v_1 v_2 v_3 v_4 \subset H$ be a 4-cycle. If H contains exactly four vertices, there is an outside-obstacle representation; see Fig. 1a. Note that we can choose the (dashed blue) diagonals $v_1 v_3$ and $v_2 v_4$ to be edges or non-edges as desired. Otherwise, without loss of generality, there is a vertex $x \in H \setminus C$ with $x \sim v_1$. As H is biconnected, there is a path of length at least 2 from v_1 to another vertex of C containing x. Observation 1 implies that $x \not\sim v_2$, $x \sim v_3$, and $x \not\sim v_4$. Since we have another 4-cycle $C' = v_1 x v_3 v_4$, the same holds for v_2, implying $v_2 \not\sim v_4$. Hence x is a non-adjacent twin of v_2. It follows that any vertex in $H \setminus C$ is a non-adjacent twin of one of $v_1, \ldots, v_4$. Since the vertices in Fig. 1a are in convex position, we can embed H using Lemma 3.

Case 3: $\text{circ}(H) = 5$

Let $C = v_1 v_2 v_3 v_4 v_5 \subset H$ be a 5-cycle. If H contains exactly five vertices, see Fig. 1b for its outside-obstacle representation. Otherwise, without loss of generality, there is a vertex $x \in H \setminus C$ with $x \sim v_1$. Observation 1 implies $x \not\sim v_2, v_5$. As H is biconnected, there is either path $v_1 x v_3$ or $v_1 x v_4$. Without loss of generality, we assume $x \sim v_3$ and thus $x \not\sim v_4$. Then $v_2 \not\sim v_4, v_5$ since we have another 5-cycle $v_1 x v_3 v_4 v_5$ and can apply the same logic. Hence, x is a non-adjacent twin of v_2. As in the Case 2, we see that every vertex in $H \setminus C$ is a non-adjacent twin of one of $v_1, v_2, \ldots, v_5$ and we can embed H using Lemma 3.

Case 4: $\mathrm{circ}(H) = 6$ (We postpone this case to Appendix A.) □

Theorem 2. *Any graph with at most 7 vertices has an outside-obstacle representation.*

Proof (Sketch). By Theorem 1, it suffices to provide an outside-obstacle representation of each 7-vertex graph containing C_7. In Appendix A, we classify such graphs into 15 groups and give an outside-obstacle representation of each. □

3   Co-bipartite Graphs

We now consider obstacle representations of *co-bipartite* graphs. Recall that a graph is co-bipartite if its complement is bipartite. Using this seemingly simple graph class, we settle an open problem posed by Alpert et al. [1] who asked if each graph with obstacle number 1 has an outside-obstacle representation. Namely, we provide an 11-vertex graph B_{11} (see Fig. 3b) where not only is this not the case, but B_{11} in fact has an inside-obstacle representation where the obstacle is the simplest possible shape, i.e., a triangle.[1] We also provide a smallest graph with obstacle number 2; see the 8-vertex graph in Fig. 3c. This improves on the smallest previously known such graphs (e.g., the 10-vertex graphs of Pach and Sarıöz [14] and of Berman et al. [3]) and shows that Theorem 2 is tight.

Properties of Outside-Obstacle Representations. We build on the easy observation (see Observation 2 below) that in every outside-obstacle representation of a graph, for every clique Z, the convex hull $\mathrm{CH}(Z)$ of the point set of Z cannot be touched by the obstacle. In other words, the obstacle must occur outside of each such convex hull. Since we focus on co-bipartite graphs, this observation greatly restricts the ways one may realize an outside representation. Additionally, we will use this observation implicitly throughout this section whenever considering two cliques in a graph with an outside-obstacle representation.

Observation 2. *If G has an outside-obstacle representation (P, C), then for every clique $Z \subseteq V(G)$, the convex hull $\mathrm{CH}(Z)$ of the points corresponding to Z is disjoint from C, i.e., $C \cap \mathrm{CH}(Z) = \emptyset$.*

For a graph G containing two cliques $Z, Z' \subseteq V(G)$ and outside-obstacle representation, consider the convex hulls $\mathrm{CH}(Z)$ and $\mathrm{CH}(Z')$. We say that these convex hulls are k-*crossing* when $\mathrm{CH}(Z) \setminus \mathrm{CH}(Z')$ consists of $k+1$ disjoint regions.

---

[1] Note that for topologically closed obstacles, this obstacle could be a line segment.

Note that this condition is symmetric, i.e., when $\mathrm{CH}(Z) \setminus \mathrm{CH}(Z')$ consists of r disjoint regions so does $\mathrm{CH}(Z') \setminus \mathrm{CH}(Z)$. We refer to these disjoint regions of the difference as the *petals* of Z (Z' respectively).

We now introduce a special 6-vertex graph K_6^* which is used in the following technical lemma and our NP-hardness proof. This graph is the result of deleting a 3-edge matching from a 6-clique; see Fig. 3a.

Lemma 4. *Let G be a graph containing two cliques Z, Z'. For every outside-obstacle representation of G, the following properties hold.*

(a) *If $\mathrm{CH}(Z)$ and $\mathrm{CH}(Z')$ are t-crossing, then every vertex in Z has at least $t-1$ neighbors in Z' and vice versa. That is, if Z contains a vertex with only r neighbors in Z', then $\mathrm{CH}(Z)$ and $\mathrm{CH}(Z')$ are at most $(r+1)$-crossing.*

(b) *If G contains K_6^* (with missing edges $z_1 z_1'$, $z_2 z_2'$, $z_3 z_3'$; see Fig. 3a) as an induced subgraph, $\{z_1, z_2, z_3\} \subseteq Z$, and $\{z_1', z_2', z_3'\} \subseteq Z'$, then $\mathrm{CH}(\{z_1, z_2, z_3\})$ and $\mathrm{CH}(\{z_1', z_2', z_3'\})$ are at least 1-crossing. Furthermore, $\mathrm{CH}(Z)$ and $\mathrm{CH}(Z')$ are at least 1-crossing.*

(c) *If G contains a 4-cycle $z_1 z_1' z_2' z_2$ as an induced subgraph, $\{z_1, z_2\} \subseteq Z$, $\{z_1', z_2'\} \subseteq Z'$, $\mathrm{CH}(Z)$ and $\mathrm{CH}(Z')$ intersect, and z_1 and z_2 are contained in a petal Q^Z of Z, then z_1' and z_2' are contained in different petals of Z' which are both adjacent to Q^Z. This implies that, if $\mathrm{CH}(Z)$ and $\mathrm{CH}(Z')$ are 1-crossing, then either z_1 and z_2 or z_1' and z_2' are in different petals.*

Proof. (a) Suppose $\mathrm{CH}(Z)$ and $\mathrm{CH}(Z')$ are t-crossing for some $t \geq 2$. Note that $|Z|, |Z'| \geq t+1$ since the convex hull of each must contain at least $t+1$ points. For $A \in \{Z, Z'\}$, let $Q_0^A, \ldots, Q_t^A$ be the petals of $\mathrm{CH}(A)$ in clockwise order around $\mathrm{CH}(Z) \cap \mathrm{CH}(Z')$ where, for each $i \in \{0, \ldots, t\}$, Q_i^Z is between $Q_i^{Z'}$ and $Q_{i+1}^{Z'}$ and all indices are considered modulo $t+1$.

Consider a vertex $v \in Z$ ($v \in Z'$ follows symmetrically). If v is in $\mathrm{CH}(Z) \cap \mathrm{CH}(Z')$, then we are done since v sees every vertex in Z' and $|Z'| \geq t+1$. So, suppose $v \in Q_1^Z$. Consider the points $p_1 = Q_1^{Z'} \cap Q_0^Z$ and $p_2 = Q_2^{Z'} \cap Q_2^Z$. Define the subregion R (depicted as the grey region in Fig. 2a) of $\mathrm{CH}(Z) \cup \mathrm{CH}(Z')$ whose boundary, in clockwise order, is formed by $\overline{p_1 v}$, $\overline{v p_2}$, and the polygonal

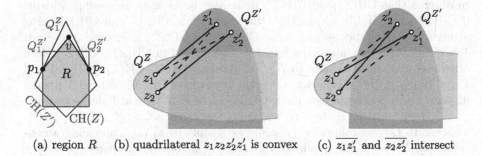

(a) region R (b) quadrilateral $z_1 z_2 z_2' z_1'$ is convex   (c) $\overline{z_1 z_1'}$ and $\overline{z_2 z_2'}$ intersect

Fig. 2. Aides for the proof of Lemma 4.

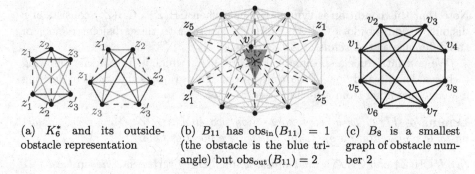

(a) K_6^* and its outside-obstacle representation

(b) B_{11} has $\mathrm{obs_{in}}(B_{11}) = 1$ (the obstacle is the blue triangle) but $\mathrm{obs_{out}}(B_{11}) = 2$

(c) B_8 is a smallest graph of obstacle number 2

Fig. 3. Three small graphs: K_6^*, B_{11} and B_8 (Color figure online)

chain from p_2 to p_1 along the boundary of $\mathrm{CH}(Z')$. Note that, for each $i \in \{0, 3, 4, \ldots, t\}$, $Q_i^{Z'} \subset R$ and R is convex, i.e., for every $u \in Q_i^{Z'}$, the line segment $\overline{vu}$ is contained in $\mathrm{CH}(Z) \cup \mathrm{CH}(Z')$. Thus, v has at least $t - 1$ neighbors in Z'.

(b) Consider the graph K_6^* as labeled in Fig. 3a. We first show that the convex hulls of $X = \{z_1, z_2, z_3\}$ and $Y = \{z_1', z_2', z_3'\}$ are at least 1-crossing.

Suppose that $\mathrm{CH}(X)$ and $\mathrm{CH}(Y)$ intersect but are 0-crossing. Since $|X| = |Y| = 3$, a vertex in $X \cup Y$ must be contained in $\mathrm{CH}(X) \cap \mathrm{CH}(Y)$. Hence, this vertex dominates $X \cup Y$, but K_6^* doesn't have such a vertex—a contradiction.

Now, suppose that $\mathrm{CH}(X)$ and $\mathrm{CH}(Y)$ are disjoint, and let $H = \mathrm{CH}(X \cup Y)$. Since $\mathrm{CH}(X)$ and $\mathrm{CH}(Y)$ are disjoint, the boundary ∂H of H contains at most two line segments that connect a vertex of X to a vertex of Y, i.e., at most two non-edges of K_6^* occur on ∂H. However, we will now see that every non-edge of K_6^* must occur on ∂H. Consider the line segment $\overline{z_1 z_1'}$ and suppose it is not on ∂H. This means that there are vertices u and v of $K_6^* \setminus \{z_1, z_1'\}$ where u and v occur on opposite sides of the line determined by $\overline{z_1 z_1'}$. However, since $z_1 z_1'$ is the only non-edge incident to either z_1 or z_1', the non-edge $z_1 z_1'$ is enclosed by $\overline{u z_1}$, $\overline{z_1 v}$, $\overline{v z_1'}$, $\overline{z_1' u}$, which provides a contradiction. Thus, every non-edge must occur on ∂H, which contradicts the fact that at most two line segments spanning between $\mathrm{CH}(X)$ and $\mathrm{CH}(Y)$ can occur on ∂H.

We now know that $\mathrm{CH}(X)$ and $\mathrm{CH}(Y)$ are at least 1-crossing. We use this to observe that $\mathrm{CH}(Z)$ and $\mathrm{CH}(Z')$ must also be at least 1-crossing. Clearly, if $\mathrm{CH}(Z)$ and $\mathrm{CH}(Z')$ are disjoint, this contradicts $\mathrm{CH}(X)$ and $\mathrm{CH}(Y)$ being at least 1-crossing. So, suppose that $\mathrm{CH}(Z)$ and $\mathrm{CH}(Z')$ intersect but are not 1-crossing. Note that no vertex v of K_6^* is contained in $\mathrm{CH}(Z) \cap \mathrm{CH}(Z')$ since otherwise v would dominate to K_6^*. In particular, $X \subseteq \mathrm{CH}(Z) \setminus \mathrm{CH}(Z')$ and $Y \subseteq \mathrm{CH}(Z') \setminus \mathrm{CH}(Z)$. However, we again would have $\mathrm{CH}(X)$ and $\mathrm{CH}(Y)$ being disjoint, i.e., a contradiction. Thus, $\mathrm{CH}(Z)$ and $\mathrm{CH}(Z')$ are at least 1-crossing.

(c) Suppose that z_1' and z_2' belong to the same petal $Q^{Z'}$. This petal is adjacent to Q^Z, as otherwise z_1 would be visible to z_2' (i.e., providing a contradiction). Now, if the quadrilateral $z_1 z_2 z_2' z_1'$ is convex, the non-edge $z_1 z_2'$ is not accessible from the outside (see Fig. 2b). If the quadrilateral $z_1 z_2 z_2' z_1'$ is non-convex, either a non-edge $z_1 z_2'$ or a non-edge $z_2 z_1'$ will not be accessible from the outside. Thus, $\overline{z_1 z_1'}$ and $\overline{z_2 z_2'}$ intersect since $\mathrm{CH}(\{z_1, z_2\})$ and $\mathrm{CH}(\{z_1', z_2'\})$ are disjoint. The edge $z_1 z_1'$ together with the boundary of $\mathrm{CH}(Z) \cup \mathrm{CH}(Z')$ split the plane into at most two bounded and one unbounded region. Then at least one of the non-edges $z_1 z_2'$ and $z_1' z_2$ lies inside the union of the bounded regions. This contradicts the fact that all non-edges should be accessible from the outside. For example, in Fig. 2c, the non-edge $z_1' z_2$ cannot intersect any outside obstacle. $\square$

Inside- vs. Outside-Obstacle Graphs. We now use Lemma 4 to show that there is an 11-vertex graph (see B_{11} in Fig. 3b) that has an inside-obstacle representation but no outside-obstacle representation. This resolves an open question of Alpert et al. [1]. We conjecture that, for any graph G with at most 10 vertices, $\mathrm{obs}_{\mathrm{in}}(G) = 1$ implies $\mathrm{obs}_{\mathrm{out}}(G) = 1$.

Theorem 3. *There is an 11-vertex graph (e.g., B_{11} in Fig. 3b) with inside-obstacle number 1, but outside-obstacle number 2.*

Proof. The 11-vertex co-bipartite graph B_{11} is constructed as follows. We start with K_{10} on the vertices $z_1, \ldots, z_5, z_1', \ldots, z_5'$. We then delete a 5-edge matching $\{z_i z_i' : i \in \{1, \ldots, 5\}\}$ from K_{10} to obtain K_{10}^*. Finally, we obtain B_{11} by adding a vertex v adjacent to $z_1, \ldots, z_5$. (Fig. 3b shows an inside-obstacle representation of B_{11} with a triangular obstacle.)

It remains to argue that B_{11} has no outside-obstacle representation. Note that B_{11} contains two cliques $Z = \{z_1, \ldots, z_5, v\}$ and $Z' = \{z_1', \ldots, z_5'\}$. Furthermore, the vertex $v \in Z$ has no neighbors in Z'. Thus, by Lemma 4(a), in any outside-obstacle representation, $\mathrm{CH}(Z)$ and $\mathrm{CH}(Z')$ are at most 1-crossing. Additionally, since each z_i has a non-neighbor in Z', no z_i is contained in $\mathrm{CH}(Z) \cap \mathrm{CH}(Z')$. In particular, since Z has only two petals, there are three z_i's, say z_1, z_2, z_3, that are contained in a single petal of Z. Now note that K_6^* is the subgraph of B_{11} induced by $\{z_1, z_2, z_3, z_1', z_2', z_3'\}$. Since z_1, z_2, z_3 are contained in a petal of Z, $\mathrm{CH}(\{z_1, z_2, z_3\})$ and $\mathrm{CH}(\{z_1', z_2', z_3'\})$ are disjoint, contradicting Lemma 4(b). Thus, B_{11} has outside-obstacle number 2. $\square$

Note that a graph with an inside-obstacle representation is either a clique or contains a cycle since an inside obstacle cannot (by definition) pierce the convex hull of the point set[2]. Thus, by Theorem 3 and this fact, we have the following.

Theorem 4. *The classes of inside-obstacle representable graphs and outside-obstacle representable graphs are incomparable.*

Obstacle Number 2. We present an 8-vertex graph (see B_8 in Fig. 3c) with obstacle number 2. To prove this result, we first apply Lemma 4 to show that B_8 has no outside-obstacle representation. In Lemma 5 (proven in Appendix B), we demonstrate that B_8 also has no inside-obstacle representation. In particular, these lemmas together with Theorem 2 provide the following theorem.

---

[2] In Appendix D, we show that $K_{2,3}$ is the smallest graph with a cycle and an outside-obstacle representation but no inside-obstacle representation.

Theorem 5. *The smallest graphs without a single-obstacle representation have eight vertices, e.g., the co-bipartite graph B_8 in Fig. 3c.*

Proof. The graph B_8 has 8 vertices $v_1, \ldots, v_8$. It has precisely the following set of non-edges: $v_1v_6, v_2v_5, v_3v_7, v_4v_5, v_4v_6, v_4v_7, v_8v_1, v_8v_2, v_8v_3$. Note that the subgraph induced by $\{v_1, v_2, v_3, v_5, v_6, v_7\}$ is a K_6^*. Further, note that $Z = \{v_1, v_2, v_3, v_4\}$ and $Z' = \{v_5, v_6, v_7, v_8\}$ are cliques.

Suppose (for a contradiction) B_8 has an outside-obstacle representation. By Lemma 4(b), $CH(Z)$ and $CH(Z')$ are at least 1-crossing. Additionally, since v_4 has only one neighbor in Z', we know that $CH(Z)$ and $CH(Z')$ are at most 2-crossing. We will consider these two cases separately. Let Q_0^Z, Q_1^Z, Q_2^Z be the petals of Z and $Q_0^{Z'}$, $Q_1^{Z'}$, $Q_2^{Z'}$ be the petals of Z' where the cyclic order of the petals around $CH(Z) \cap CH(Z')$ is $Q_2^{Z'}, Q_0^Z, Q_0^{Z'}, Q_1^Z, Q_1^{Z'}, Q_2^Z, Q_2^{Z'}$. Note that every vertex is contained in one of the petals.

Case 1: $CH(Z)$ and $CH(Z')$ are 2-crossing. Suppose $v_4 \in Q_0^Z$. Since v_8 is the only neighbor of v_4 in Z', we must have $v_8 \in Q_2^{Z'}$, and now the only vertex in Q_0^Z is v_4 and the only vertex in $Q_2^{Z'}$ is v_8. However, we now have $\{v_1, v_2, v_3\} \subset Q_1^Z \cup Q_2^Z$ and $\{v_5, v_6, v_7\} \subset Q_0^{Z'} \cup Q_1^{Z'}$, i.e., $CH(\{v_1, v_2, v_3\})$ and $CH(\{v_5, v_6, v_7\})$ are disjoint, contradicting Lemma 4(b).

Case 2: $CH(Z)$ and $CH(Z')$ are 1-crossing. Note that v_1, v_2, and v_3 cannot belong to the same petal (otherwise, we would contradict Lemma 4(b)). Similarly, v_5, v_6, and v_7 cannot belong to the same petal. Thus, without loss of generality, we have v_1 and v_2 in Q_0^Z, v_3 in Q_1^Z, v_5 and v_7 in $Q_0^{Z'}$, and v_6 in $Q_1^{Z'}$. When v_4 is in Q_0^Z and v_8 is in $Q_0^{Z'}$, the induced 4-cycle $v_4v_2v_7v_8$ contradicts Lemma 4(c). Similarly, when v_4 is in Q_0^Z and v_8 is in $Q_1^{Z'}$, we use the induced 4-cycle $v_4v_2v_6v_8$; when v_4 is in Q_1^Z and v_8 is in $Q_0^{Z'}$, we use the induced 4-cycle $v_4v_3v_5v_8$; and when v_4 is in Q_1^Z and v_8 is in $Q_1^{Z'}$, we use the induced 4-cycle $v_4v_3v_6v_8$.

It remains to show that B_8 has no inside-obstacle representation (formalized in Lemma 5 below). This is proven in Appendix B.   □

Lemma 5. *The graph B_8 in Fig. 3c has no inside-obstacle representation.*

4   NP-Hardness

In this section, we show that the single-obstacle, outside-obstacle, inside-obstacle graph sandwich problems as well as the simple-polygon visibility graph sandwich problem are all NP-hard. Note that the complexity of the obstacle graph sandwich problem yields an upper bound for the complexity of our (simpler) recognition problem.

Theorem 6. *The outside-obstacle graph sandwich problem is NP-hard. In other words, given two graphs G and H with the same vertex set and $G \subseteq H$, it is NP-hard to decide whether there is a graph K such that $G \subseteq K \subseteq H$ and $\mathrm{obs}_{\mathrm{out}}(K) = 1$. This holds even if G and H are co-bipartite.*

Proof. We reduce from MONOTONENOTALLEQUAL3SAT, which is NP-hard [15]. In this version of 3SAT, all literals are positive, and the task is to decide whether the given 3SAT formula φ admits a truth assignment such that in each clause at least one and at most two variables are true.

Given φ, we build a graph G_φ with edges, non-edges and "maybe"-edges such that φ is a yes-instance if and only if G_φ has a subgraph that has an outside-obstacle representation and contains all edges, no non-edges and an arbitrary subset of the maybe-edges. (In other words, the set of edges of G_φ yields G in the statement of the theorem, and the set of edges and maybe-edges yields H.) Let $\{v_1, \ldots, v_n\}$ be the set of variables, and let $\{C_1, \ldots, C_m\}$ be the set of clauses in φ. For $i = 1, \ldots, n$, let v_{ij} be the j-th occurrence of v_i in φ.

Now we can construct G_φ. For each variable, we introduce a *variable vertex* (of the same name). These n vertices form a clique. For each occurrence v_{ij} of a variable v_i in φ, we introduce an *occurrence vertex* (of the same name). These $3m$ vertices also form a clique. In order to restrict how the two cliques intersect, we add to G_φ a copy of K_6^* labeled as in Fig. 3a; vertices z_1, z_2, z_3 participate in the occurrence-vertex clique, whereas vertices z_1', z_2', z_3' participate in the variable-vertex clique. We add one more vertex u to the occurrence-vertex clique. The special vertex u is adjacent to z_3' and has non-edges to all other vertices in the variable-vertex clique. The edge set of G_φ depends on φ as follows. Each variable vertex v_i has

- an edge to any occurrence vertex v_{ij},
- a non-edge to any occurrence vertex $v_{k\ell}$ that represents an occurrence of a variable v_k that co-occurs with v_i in some clause of φ,
- a maybe-edge to any other occurrence vertex.

Next, we show how to use a feasible truth assignment of φ to lay out G_φ so that all its non-edges are accessible from the outside. We place the vertices on the boundary of two intersecting rectangles, one for each clique. Given these positions, we show that all non-edges intersect the outer face of the union of the edges. Finally, we bend the sides of the rectangles slightly into very flat circular arcs such that all of the previous (non-) visibilities remain and the vertices are in general position.

We take two axis-aligned rectangles R_1 and R_2 that intersect as a cross; see Fig. 4. Let X_1, X_2, X_3, X_4 be the corners of $R_1 \cap R_2$ in clockwise order, starting in the lower left corner. We place the variable vertices on the boundary of the "wide" rectangle R_1: the vertices $v_1, \ldots, v_p$ of the true variables are equally spaced from top to bottom on a segment on the left side, similarly the vertices $v_{p+1}, \ldots, v_n$ of the false variables go to a segment on the right side. (In Fig. 4(b), $p = 3$.) The two vertical segments are chosen such that they "see" four disjoint horizontal segments on the top and bottom edge of R_2; refer to Fig. 4(a) for the positions of the six segments in total.

In each clause, we sort the variables in increasing order of index. We place the occurrence vertices on the horizontal segments of R_2. For a true variable v_i (such as v_2 in Fig. 4(b)) the first occurrence vertex v_{i1} has two potential locations; the bottom location is where the ray from v_i through X_1 hits the bottom

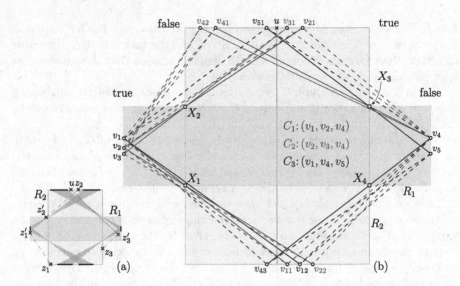

Fig. 4. NP-hardness: maybe-edges and the two cliques are not drawn.

right segment, the top location is where the ray from v_i trough X_2 hits the top right segment. We place v_{i1} to its bottom or top location depending on whether v_{i1} is the first or second occurrence of a true variable in its clause, respectively. (Remember that within each clause, at most two variables are true and at most two are false.) Occurrence vertices v_{i2} etc. go between the top or bottom locations of v_{i1} and $v_{i+1,1}$, again depending on whether they are the first or second occurrence of a true variable in their respective clauses. (E.g., in Fig. 4(b), v_{21} goes to the top, whereas v_{22} goes to the bottom.)

The special vertex u is placed in the center of the top edge of R_2; hence, it is not visible from any variable vertex; see Fig. 4(a). The vertices of K_6^* can be placed such that u sees only z_3', but neither z_1' nor z_2'; see Fig. 4(a).

By construction, all edges are inside $R_1 \cup R_2$. It remains to show that all non-edges (dashed in Fig. 4(b)) go through the complement of $R_1 \cup R_2$. This is due to the order of the variable vertices and the occurrence vertices along the boundary of $R_1 \cup R_2$ and due to the order of the variables in each clause. Suppose that a variable vertex v_i has a non-edge with occurrence vertex $v_{k\ell}$. This means that there is an occurrence v_{ij} of v_i in the same clause as $v_{k\ell}$. If v_i and v_k have different truth values, then v_i cannot see $v_{k\ell}$; refer to Fig. 4(a). So assume that both are true and that $i < k$. But then v_i lies above v_k on the left segment of R_1, and v_{ij} lies to the left of $v_{k\ell}$ on the bottom right segment of R_2. Hence, v_i cannot see $v_{k\ell}$.

It remains to show that an outside-obstacle representation of G_φ yields a feasible truth assignment for φ. By Lemmas 4(a) and (b), we know that the convex hulls of the two cliques are at least 1-crossing due to the presence of K_6^* and at most 2-crossing due to u. To see that these hulls are exactly 1-crossing, we suppose that G_φ has a 2-crossing drawing for a contradiction. Consider the subgraph H induced by u and the first clause $C_1 = \{v_i, v_j, v_k\}$, of φ i.e., $H = G[\{u, v_i, v_j, v_k, v_{i1}, v_{j1}, v_{k1}\}]$. Let Q_u be the petal containing u. Since the only neighbor of u in the variable-vertex clique is z_3', no other variable vertices belong to the petal opposite Q_u. Thus, two of $\{v_i, v_j, v_k\}$, say v_i and v_j, occur in one petal Q_1' adjacent to Q_u, and v_k occurs in the other petal Q_2' which is adjacent to Q_u. Notice that each of v_{i1}, v_{j1}, v_{k1} cannot belong to the petal opposite Q_1' since this would make it adjacent to both v_i and v_j. Similarly, no neither v_{i1} nor v_{j1} can occur in the petal opposite Q_2' since it would then be adjacent to v_k. Thus, v_{i1} and v_{j1} belong to the same petal and this petal is adjacent to Q_1'. However, this contradicts Lemma 4(c) since $\{v_i, v_j, v_{i1}, v_{j1}\}$ induces a 4-cycle.

Now, since the convex hulls are exactly 1-crossing, we have two groups (petals) of vertices in each of the two cliques. Without loss of generality, the variable-vertex clique is divided into a left and a right group, and the occurrence-vertex clique is divided into a top and a bottom group. We set those variables to true whose vertices lie on the left, the rest to false.

Now suppose that the three variables v_1, v_2, and v_3 of clause C_1 lie in the same group, say, on the left. Then two of their occurrence vertices (say v_{11} and v_{21}) lie in the same group, say, in the top group. Since $v_1 v_{21}$ and $v_2 v_{11}$ are non-edges, $v_1 v_{11} v_{21} v_2$ is an induced 4-cycle. Now Lemma 4(c), yields the desired contradiction. Hence, no three variable vertices in a clause can be in the same (left or right) group. Therefore, our truth assignment is indeed feasible. This completes the NP-hardness proof. $\qquad\square$

To show hardness for the simple-polygon visibility graph sandwich problem, we must make sure that any vertex of the obstacle is also a vertex of the graph. It suffices to add X_1, X_2, X_3, X_4 as vertices to G_φ that lie in both cliques.

Theorem 7. *The simple-polygon visibility graph sandwich problem is NP-hard. In other words, given two graphs G and H with the same vertex set and $G \subseteq H$, it is NP-hard to decide whether there is a graph K and a polygon Π such that $G \subseteq K \subseteq H$ and $K = G_\Pi(V(\Pi))$. This holds even if G and H are co-bipartite.*

We can also use the NP-hardness of the outside-obstacle sandwich problem to show NP-hardness for both the single-obstacle sandwich problem and the inside-obstacle sandwich problem. The idea is simply to combine a given graph G with a graph such as B_{11} which has outside-obstacle number greater than one, but inside-obstacle number one. The combined graph would then have inside-obstacle number one if and only if the graph G has outside-obstacle number one. The details of this are given in Appendix C.

References

1. Alpert, H., Koch, C., Laison, J.D.: Obstacle numbers of graphs. Discrete Comput. Geom. **44**(1), 223–244 (2009). http://dx.doi.org/10.1007/s00454-009-9233-8
2. Balko, M., Cibulka, J., Valtr, P.: Drawing graphs using a small number of obstacles. In: Di Giacomo, E., Lubiw, A. (eds.) GD 2015. LNCS, vol. 9411, pp. 360–372. Springer, Heidelberg (2015). doi:10.1007/978-3-319-27261-0_30

3. Berman, L.W., Chappell, G.G., Faudree, J.R., Gimbel, J., Hartman, C., Williams, G.I.: Graphs with obstacle number greater than one. Arxiv report arXiv.org/abs/1606.03782 (2016)
4. Cardinal, J., Hoffmann, U.: Recognition and complexity of point visibility graphs. In: Arge, L., Pach, J. (eds.) Proceedings of the 31st International Symposium Computational Geometry (SoCG 2015), LIPIcs, vol. 34, pp. 171–185. Schloss Dagstuhl - Leibniz-Zentrum für Informatik (2015)
5. Chaplick, S., Lipp, F., Park, J.w., Wolff, A.: Obstructing visibilities with one obstacle. Arxiv report arXiv.org/abs/1607.00278v2 (2016)
6. Dujmović, V., Morin, P.: On obstacle numbers. Electr. J. Combin. **33**(3), paper #P3.1, 7 p. (2015). arXiv.org/abs/1308.4321
7. Fulek, R., Saeedi, N., Sarıöz, D.: Convex obstacle numbers of outerplanar graphs and bipartite permutation graphs. In: Pach, J. (ed.) Thirty Essays on Geometric Graph Theory, pp. 249–261. Springer, New York (2013)
8. Ghosh, S.K., Goswami, P.P.: Unsolved problems in visibility graphs of points, segments and polygons. Arxiv report arXiv.org/abs/1012.5187v4 (2012)
9. Golumbic, M.C., Kaplan, H., Shamir, R.: Graph sandwich problems. J. Algorithms **19**(3), 449–473 (1995)
10. Johnson, M.P., Sarıöz, D.: Representing a planar straight-line graph using few obstacles. In: Proceedings of the 26th Canadian Conference Computational Geometry (CCCG 2014), pp. 95–99 (2014). http://www.cccg.ca/proceedings/2014/papers/paper14.pdf
11. Koch, A., Krug, M., Rutter, I.: Graphs with plane outside-obstacle representations. Arxiv report arXiv.org/abs/1306.2978 (2013)
12. Mukkamala, P., Pach, J., Pálvölgyi, D.: Lower bounds on the obstacle number of graphs. Electr. J. Combin. **19**(2), paper #P32, 8 p. (2012). http://www.combinatorics.org/ojs/index.php/eljc/article/view/v19i2p32
13. Mukkamala, P., Pach, J., Sarıöz, D.: Graphs with large obstacle numbers. In: Thilikos, D.M. (ed.) WG 2010. LNCS, vol. 6410, pp. 292–303. Springer, Heidelberg (2010). doi:10.1007/978-3-642-16926-7_27
14. Pach, J., Sarıöz, D.: On the structure of graphs with low obstacle number. Graphs Comb. **27**(3), 465–473 (2011). http://dx.doi.org/10.1007/s00373-011-1027-0
15. Schaefer, T.J.: The complexity of satisfiability problems. In: Proceedings of the 10th Annual ACM Symposium Theory Computing (STOC 1978), pp. 216–226 (1978). http://dx.doi.org/10.1145/800133.804350

Beyond Planarity

On the Size of Planarly Connected Crossing Graphs

Eyal Ackerman[1], Balázs Keszegh[2(✉)], and Mate Vizer[2]

[1] Department of Mathematics, Physics, and Computer Science,
University of Haifa at Oranim, 36006 Tivon, Israel
ackerman@sci.haifa.ac.il
[2] Alfréd Rényi Institute of Mathematics, Hungarian Academy of Sciences,
Budapest 1053, Hungary
keszegh@renyi.hu, vizermate@gmail.com

Abstract. We prove that if an n-vertex graph G can be drawn in the plane such that each pair of crossing edges is independent and there is a crossing-free edge that connects their endpoints, then G has $O(n)$ edges. Graphs that admit such drawings are related to quasi-planar graphs and to maximal 1-planar and fan-planar graphs.

Keywords: Planar graphs · Crossing edges · Crossing-free edge · Fan-planar graphs · 1-planar graphs

1 Introduction

Throughout this paper we consider graphs with no loops or parallel edges. A *topological graph* is a graph drawn in the plane with its vertices as distinct points and its edges as Jordan arcs that connect the corresponding points and do not contain any other vertex as an interior point. Every pair of edges in a topological graph has a finite number of intersection points, each of which is either a vertex that is common to both edges, or a crossing point at which one edge passes from one side of the other edge to its other side. A topological graph is *simple* if every pair of its edges intersect at most once. A *geometric* graph is a (simple) topological graph in which every edge is a straight-line segment. If the vertices of a geometric graph are in convex position, then the graph is a *convex geometric graph*.

Call a pair of independent[1] and crossing edges e and e' in a topological graph G *planarly connected* if there is a crossing-free edge in G that connects an endpoint of e and an endpoint of e'. A *planarly connected crossing* (PCC for short) topological graph is a topological graph in which every pair of independent crossing edges is planarly connected. An abstract graph is a PCC graph if it can be drawn as a topological PCC graph.

---

[1] Two edges are *independent* if they do not share a vertex. Note that in a simple topological graph two crossing edges must be independent.

© Springer International Publishing AG 2016
Y. Hu and M. Nöllenburg (Eds.): GD 2016, LNCS 9801, pp. 311–320, 2016.
DOI: 10.1007/978-3-319-50106-2_24

Our motivation for studying PCC graphs comes from two examples of topological graphs that satisfy this property: A graph is *k-planar* if it can be drawn as a topological graph in which each edge is crossed at most k times (we call such a topological graph *k-plane*). Suppose that G is an n-vertex 1-planar topological graph with the maximum possible number of edges (i.e., there is no n-vertex 1-planar graph with more edges than G). Now consider a drawing D of G as a 1-plane topological graph with the least number of crossings. Then it is easy to see that D is a simple topological graph. Moreover, D is a PCC topological graph. Indeed, if (u, v) and (w, z) are two independent edges that cross at a point x and are not planarly connected, then we can draw a crossing-free edge (u, w) that consists of the (perturbed) segments (u, x) and (w, x) of (u, v) and (w, z), respectively. This way we either increase the number of edges in the graph or we are able to replace a crossed edge with a crossing-free edge and get a 1-plane drawing of G with less crossings.

Another example for PCC topological graphs are certain drawings of *fan-planar* graphs. A graph is called *fan-planar* if it can be drawn as a simple topological graph such that for every edge e all the edges that cross e share a common endpoint on the same side of e. As before, it can be shown (see [11, Corollary 1]) that such an embedding of a maximum fan-planar graph with as many crossing-free edges as possible admits a PCC topological graph.

Both 1-plane topological graphs and fan-planar graphs are sparse, namely, their maximum number of edges is $4n - 8$ [14] and $5n - 10$ [11], respectively (where n denotes the number of vertices). Our main result shows that simple PCC topological graphs are always sparse.

Theorem 1. *Let G be an n-vertex topological graph such that for every two crossing edges e and e' it holds that e and e' are independent and there is a crossing-free edge that connects an endpoint of e and an endpoint of e'. Then G has at most cn edges, where c is an absolute constant.*

Note that by definition in a simple topological graph every pair of crossing edges must be independent, therefore, Theorem 1 holds for PCC simple topological graphs. We strongly believe that (not necessarily simple) PCC topological graphs also have linearly many edges, however, our proof currently falls short of showing that.

It follows from Theorem 1 that 1-plane and fan-planar graphs have linearly many edges, however, with a much weaker upper bound than the known ones. It would be interesting to improve our upper bound and to find the exact maximum size of a PCC (simple) topological graph. We show that this value is at least $9n - O(1)$ (see Sect. 3), which implies that not every PCC graph is a (maximum) 1-plane or fan-planar graph.

PCC graphs are also related to two other classes of topological graphs. Call a topological graph *k-quasi-plane* if it has no k pairwise crossing edges. According to a well-known and rather old conjecture (see e.g., [6,12]) *k-quasi-plane* graphs should have linearly many edges.

Conjecture 1. For any integer $k \geq 2$ there is a constant c_k such that every n-vertex k-quasi-plane graph has at most $c_k n$ edges.

It is easy to see that if G is a PCC simple topological graph, then G is 9-quasi-plane: Suppose for contradiction that G contains a set E' of 9 pairwise crossing edges and let V' be the set of their endpoints. Since G is a simple topological graph, no two edges in E' share an endpoint, therefore $|V'| = 18$. Let G' be the subgraph of G that is induced by V' and let E'' be the crossing-free edges of G'. Clearly (V', E'') is a plane graph. Moreover, all the edges in E' must lie in the same face f of this plane graph, since they are pairwise crossing. It follows that f is incident to every vertex in V' and therefore (V', E'') is an outerplanar graph. Thus, $|E''| \leq 2 \cdot 18 - 3 = 33$. On the other hand, since G' is also PCC and no two edges in E' share an endpoint, it follows that $|E''| \geq \binom{9}{2} = 36$, a contradiction.

Therefore, Conjecture 1, if true, would immediately imply Theorem 1 for simple topological graphs. However, this conjecture was only verified for $k = 3$ [4,5,13], for $k = 4$ [1], and (for any k) for convex geometric graphs [7]. For $k \geq 5$ the currently best upper bounds on the size of n-vertex k-quasi-plane graphs are $n(\log n)^{O(\log k)}$ by Fox and Pach [9,10], and $O_k(n \log n)$ for simple topological graphs by Suk and Walczak [16].

Another conjecture that implies Theorem 1 (also for topological graphs that are not necessarily simple) is related to *grids* in topological graphs. A k-*grid* in a topological graph is a pair of edge subsets E_1, E_2 such that $|E_1| = |E_2| = k$, and every edge in E_1 crosses every edge in E_2. Ackerman et al. [2] proved that every n-vertex topological graph that does not contain a k-grid with distinct vertices has at most $O_k(n \log^* n)$ edges and conjectured that this upper bound can be improved to $O_k(n)$. It is not hard to show, as before, that a PCC graph does not contain an 8-grid with distinct vertices. Therefore, this conjecture, if true, would also imply Theorem 1.

Outline. We prove Theorem 1 in the following section. In Sect. 3 we give a lower bound on the maximum size of a PCC simple topological graph, generalize the notion of planarly connected edges, and conclude with some open problems.

2   Proof of Theorem 1

Let $G = (V, E)$ be an n-vertex topological graph such that for every two crossing edges e and e' it holds that e and e' are independent and there is a crossing-free edge that connects an endpoint of e and an endpoint of e'. Denote by $E' \subseteq E$ the set of crossing-free (planar) edges in G, and by $E'' = E \setminus E'$ the set of crossed edges in G. Since $G' = (V, E')$ is a plane graph, we have $|E'| \leq 3n$, so it remains to prove that $|E''| = O(n)$.

Let $G'_1 = (V_1, E'_1), \ldots, G'_k = (V_k, E'_k)$ be the connected components of the graph G', and let $E''_{i,j} = \{(u, v) \in E'' \mid u \in V_i \text{ and } v \in V_j\}$.

Lemma 1. $|E''_{i,i}| \leq 96|V_i|$ for $1 \leq i \leq k$.

Proof. Assume without loss of generality that $i = 1$ and consider the graph G'_1. Let $f_1, \ldots, f_\ell$ be the faces of the plane graph G'_1. For a face f_j, let $V(f_j)$ be the vertices that are incident to f_j, and let $E''(f_j)$ be the edges in $E''_{1,1}$ that lie within f_j (thus, their endpoints are in $V(f_j)$). Denote by $|f_j|$ the size of f_j, that is, the length of the shortest closed walk that visits every edge on the boundary of f_j. Recall that in the Introduction we argued that a PCC simple topological graph is 9-quasi-plane. For the same arguments we have the following observation.

Observation 2. *There are no 9 pairwise crossing edges in* $E''(f_j)$.

Proposition 1. $|E''(f_j)| \leq 16|f_j|$, *for* $1 \leq j \leq \ell$.

Proof. Define first an auxiliary graph $\hat{G}_j$ as follows. When traveling along the boundary of f_j in clockwise direction, we meet every vertex in $V(f_j)$ at least once and possibly several times if the boundary of f_j is not a simple cycle. Let $v_1, v_2, \ldots, v_{|f_j|}$ be the list of vertices as they appear along the boundary of f_j, where a new instance of a vertex is introduced whenever a visited vertex is revisited. The edge set of $\hat{G}_j$ corresponds to $E''(f_j)$, however, we make sure to pick the "correct" instance of a vertex in $v_1, v_2, \ldots, v_{|f_j|}$ for a vertex in $V(f_j)$ that was visited more than once when traveling along the boundary of f_j (see Fig. 1 for an example).

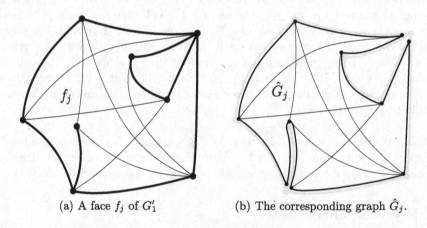

(a) A face f_j of G'_1 (b) The corresponding graph $\hat{G}_j$.

Fig. 1. Illustrations for the proof of Proposition 1.

Let $\hat{e}_1$ and $\hat{e}_2$ be a pair of crossing edges in $\hat{G}_j$ and let e_1 and e_2 be their corresponding edges in G. Clearly, e_1 and e_2 are crossing edges and therefore are independent and planarly connected. It follows from Observation 2 that $\hat{G}_j$ does not contain 9 pairwise crossing edges.

We now realize the underlying abstract graph of $\hat{G}_j$ as a convex geometric graph: The vertices $v_1, v_2, \ldots, v_{|f_j|}$ are the vertices of a convex polygon (in that order), and the edges of $\hat{G}_j$ are realized as straight-line segments. Suppose that two edges (v_{i_1}, v_{i_2}) and (v_{i_3}, v_{i_4}) cross in this realization. Assume without loss of generality that $i_1 < i_2$, $i_3 < i_4$ and $i_1 < i_3$. Since these edges are the chords of a convex polygon it must be that $i_1 < i_3 < i_2 < i_4$. It follows that (v_{i_1}, v_{i_2}) and (v_{i_3}, v_{i_4}) also cross in $\hat{G}_j$. Thus, the realization of $\hat{G}_j$ as a convex geometric graph does not contain 9 pairwise crossing edges. According to a result of Capoyleas and Pach [7], an n-vertex convex geometric graph with no $k+1$ pairwise crossing edges has at most $\binom{n}{2}$ edges if $n \leq 2k + 1$ and at most $2kn - \binom{2k+1}{2}$ edges if $n \geq 2k + 1$. Therefore, $|E''(f_j)| \leq 16|f_j|$. $\qquad\square$

We now return to proving that $|E''_{1,1}| = O(|V_1|)$. Using the fact that $\sum_{j=1}^{\ell} |f_j| = 2|E'_1| \leq 6|V_1|$, we have

$$|E''_{1,1}| = \sum_{j=1}^{\ell} E''(f_j) \leq \sum_{j=1}^{\ell} 16|f_j| \leq 96|V_1|,$$

which completes the proof of the lemma. $\qquad\square$

It remains to bound the number of edges in E'' between different connected components of G'. To this end, we introduce some more notations. For every $j \neq i$, let $V_{i,j}$ be the vertices of V_i that are connected to some vertex in V_j, i.e., $V_{i,j} = \{v_i \in V_i \mid (v_i, v_j) \in E'' \text{ for some } v_j \in V_j\}$. Let H be a simple (abstract) graph whose vertex set is $\{u_1, \ldots, u_k\}$ and whose edge set consists of the edges (u_i, u_j) such that $E''_{i,j} \neq \emptyset$.

Lemma 3. H is a planar graph.

Proof. For $1 \leq i \leq k$ identify u_i with one of the vertices of G'_i and let T_i be a spanning tree of G'_i. We draw every edge (u_i, u_j) of H as follows: Pick arbitrarily a pair $v_i \in V_i$ and $v_j \in V_j$ such that $(v_i, v_j) \in E''$. The edge (u_i, u_j) consists of the unique path in T_i from u_i to v_i, the edge (v_i, v_j) and the unique path in T_j from v_j to u_j. See Fig. 2 for an example. Note that in the drawing of H that is obtained this way all the crossing points are inherited from G, however, there are overlaps between edges. Still, each such (maximal) overlap contains an endpoint of an edge, and it is not hard to show that the edges in such a drawing can be slightly perturbed so that all the overlaps are removed and no new crossings are introduced (see [3, Lemma 2.4]). We denote such a drawing of H by H'.

The important observation is that if two edges in H' cross, then they must share an endpoint. Indeed, suppose for contradiction that (u_a, u_b) and (u_c, u_d) are two independent and crossing edges. Then it follows that G contains two independent and crossing edges (v_a, v_b) and (v_c, v_d), such that $v_a \in V_a$, $v_b \in V_b$, $v_c \in V_c$ and $v_d \in V_d$. Since these two edges are planarly connected, there should be a crossing-free edge that connects a vertex in $\{v_a, v_b\}$ with a vertex in $\{v_c, v_d\}$. However, this is impossible since these four vertices belong to distinct connected components of G'.

Finally, a graph that can be drawn so that each crossing is between two edges that share a common vertex is planar: this follows from the strong Hanani-Tutte Theorem (see, e.g., [8,15,18]). $\qquad\square$

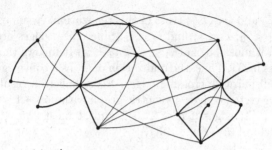

(a) G' has three connected components.

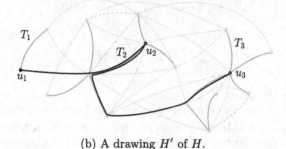

(b) A drawing H' of H.

Fig. 2. Illustrations for the proof of Lemma 3.

Lemma 4. $|E''_{i,j}| \leq 8(|V_{i,j}| + |V_{j,i}|)$ for every $1 \leq i < j \leq k$.

Proof. Since G'_i and G'_j are planar graphs, we can properly color their vertices with four colors. Denote the colors by $1, 2, 3, 4$, and let $V^c_{i,j}$ (resp., $V^c_{j,i}$) be the vertices of color c in $V_{i,j}$ (resp., $V_{j,i}$). We claim that the number of edges in $E''_{i,j}$ that connect a vertex from $V^c_{i,j}$ and a vertex from $V^{c'}_{j,i}$ is at most $2(|V^c_{i,j}| + |V^{c'}_{j,i}|)$ for every $c, c' \in \{1, 2, 3, 4\}$. Indeed, denote the graph that consists of these edges by G^* and consider its drawing as inherited from G. It is not hard to see that G^* is a planar graph: Suppose that two edges in G^* cross and denote them by (u, v) and (x, y) such that $u, x \in V^c_{i,j}$ and $v, y \in V^{c'}_{j,i}$. Since u and x are both of color c, there is no crossing-free edge in G'_i that connects them. Similarly, there is no crossing-free edge in G'_j that connects v and y. Since there are also no crossing-free edges in $E''_{i,j}$, it follows that (u, v) and (x, y) are not independent, a contradiction.

Therefore, G^* is a plane graph. Because G^* is also bipartite, its number of edges is at most twice its number of vertices. Thus,

$$|E''_{i,j}| \leq 2 \sum_{1 \leq c \leq 4} \sum_{1 \leq c' \leq 4} (|V^c_{i,j}| + |V^{c'}_{j,i}|) = 8(|V_{i,j}| + |V_{j,i}|),$$

and the lemma follows. □

Lemma 5. $\sum_{j \neq i} |V_{i,j}| \leq 3(|V_i| + 4 \deg_H(u_i))$ for every $1 \leq i \leq k$.

Proof. We use again ideas from the proofs of Lemmas 3 and 4. Assume without loss of generality that $i = 1$ and consider the graph G'_1. Since G'_1 is a planar graph, we can properly color its vertices with four colors. Denote the colors by $1, 2, 3, 4$, and let V_1^c (resp., $V_{1,j}^c$) be the vertices of color c in V_1 (resp., $V_{1,j}$). Clearly, $\sum_{j=2}^k |V_{1,j}| = \sum_{c=1}^4 \sum_{j=2}^k |V_{1,j}^c|$. Therefore it is enough to consider $\sum_{j=2}^k |V_{1,j}^c|$ for a fixed color c.

Recall that in the proof of Lemma 3, for $1 \le i \le k$, we have identified u_i with one of the vertices of G_i' and denoted by T_i a spanning tree of G_i'. We define a graph H^c whose vertex set consists of V_1^c and the vertices u_j that are adjacent to u_1 in H. For each such vertex u_j and every vertex $v_1 \in V_{1,j}^c$ pick arbitrarily an edge (v_1, v_j) such that $v_j \in V_j$ (such an edge exists by the definition of $V_{1,j}$), and draw an edge (v_1, u_j) as follows: (v_1, u_j) consists of the edge (v_1, v_j) in G and the unique path in T_j from v_j to u_j.

Observe that H^c is a simple graph (i.e., it has no parallel edges or loops). Moreover, in the drawing of H^c that is obtained as above, all the crossing points are inherited from G, however, there are overlaps between edges. Still, each such (maximal) overlap contains an endpoint of an edge, and thus, as in the proof of Lemma 3, the edges of H^c can be slightly perturbed so that all the overlaps are removed and no new crossings are introduced.

Consider such a drawing of H^c and observe that if two edges cross in this drawing, then they must share an endpoint. Indeed, suppose for contradiction that (v_1, u_a) and (v_1', u_b) are two independent and crossing edges. Then G contains two independent and crossing edges (v_1, v_a) and (v_1', v_b), such that $v_1, v_1' \in V_1$, $v_a \in V_a$, and $v_b \in V_b$. Since these two edges are planarly connected, there should be a crossing-free edge that connects a vertex in $\{v_1, v_a\}$ with a vertex in $\{v_1', v_b\}$. However, this is impossible because there is no crossing-free edge between two vertices from different connected components of G' and there is also no crossing-free edge (v_1, v_1') since both v_1 and v_1' are of color c.

This implies that H^c is a planar graph. Observe that $\sum_{j=2}^k |V_{1,j}^c|$ is precisely the number of edges in H^c. Thus, $\sum_{j=2}^k |V_{1,j}^c| \le 3|V(H^c)| = 3(|V_1^c| + \deg_H(u_1))$, and it follows that $\sum_{j=2}^k |V_{1,j}| = \sum_{c=1}^4 \sum_{j=2}^k |V_{1,j}^c| \le 3|V_1| + 12 \deg_H(u_1)$. $\square$

Recall that it remains to show that $|E''| = O(n)$:

$$|E''| = \sum_{1 \le i \le k} |E_{i,i}''| + \sum_{1 \le i < j \le k} |E_{i,j}''|$$

$$\le 96n + 8 \sum_{1 \le i < j \le k} (|V_{i,j}| + |V_{j,i}|)$$

$$= 96n + 8 \sum_{1 \le i \le k} \sum_{j \ne i} |V_{i,j}|$$

$$\le 96n + 24 \sum_{1 \le i \le k} (|V_i| + 4 \deg_H(u_i))$$

$$\le 96n + 24n + 96 \cdot 2|E(H)| \le 120n + 192 \cdot 3n = 696n.$$

Note that in the last inequality we used the fact that H is a planar graph. We conclude that $|E| = |E'| + |E''| \le 699n$. Theorem 1 is proved.

3   Discussion

Recall that we leave open the question of whether Theorem 1 holds for PCC topological graphs in which every pair of crossing edges shares a vertex or is planarly connected.

It would also be interesting to find the maximum size of an n-vertex PCC simple topological graph. The proof of Theorem 1 shows that this quantity is at most $699n$, but we believe that a linear bound with a much smaller multiplicative constant holds. Figure 3 describes a construction of an n-vertex PCC simple topological graph with $9n - O(1)$ edges. This construction was given by Géza Tóth [17], and it improves a construction of ours with $6.6n - O(1)$ edges that appeared in an earlier version of this paper. It goes as follows: place $n - 6$ points on the y-axis, say at $(0, i)$ for $i = 0, 1, \ldots, n - 7$; for every $i = 0, \ldots, n - 8$ add a straight-line edge connecting $(0, i)$ and $(0, i + 1)$ (these edges will be crossing-free); for every $i = 0, \ldots, n - 9$ add an edge connecting $(0, i)$ and $(0, i + 2)$ that goes slightly to the left of the y-axis; for every $i = 0, \ldots, n - 10$ add an edge connecting $(0, i)$ and $(0, i + 3)$ that goes slightly to the right of the y-axis; add three points with the same x coordinate to the left (resp., right) of the y-axis and connect each of them by straight-line edges to each of the points on the y-axis; connect every pair of points to the left (resp., right) of the y-axis by a crossing-free edge. One can easily verify that the resulting graph is indeed a PCC simple topological graph and has $9n - O(1)$ edges.

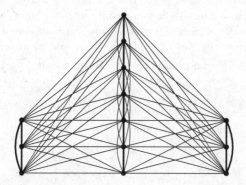

Fig. 3. A construction of a topological PCC graph with $9n - O(1)$ edges.

The notion of planarly connected edges can be generalized as follows. For an integer $k \geq 0$, we say that two crossing edges e and e' in a topological graph G are k-*planarly connected* if there is a path of at most k *crossing-free* edges in G that connects an endpoint of e with an endpoint of e'. Call a graph k-*planarly connected crossing* (k-PCC for short) graph if it can be drawn as a topological graph in which every pair of crossing edges is k-planarly connected. Thus, PCC graphs are 1-PCC graphs.

For $k = 0$, graphs that can be drawn as topological graphs in which every pair of crossing edges share a vertex are actually planar graphs, as noted in the proof of Lemma 3. For $k \geq 2$ we can no longer claim that a k-PCC graph is sparse. Indeed, it is easy to see that K_n is a 2-PCC graph: simply pick a vertex v and draw it with all of its neighbors as a crossing-free star. Now every remaining edge can be drawn such that we get a simple topological graph in which for any two crossing edges there is a path (through v) of two crossing-free edges that connects their endpoints.

Note that if G is a k-PCC graph and G' is a subgraph of G, then this does not imply that G' is also a k-PCC graph. For example, it is not hard to see that for any k there is a (sparse) graph that is not k-PCC: simply replace every edge of K_5 (or any non-planar graph) with a path of length $k + 1$. Call the resulting graph G' and observe that any drawing of it must contain two independent and crossing edges such that there is no path of length at most k between their endpoints. On the other hand, if $k \geq 2$ then clearly G' is a subgraph of a k-PCC graph (K_n).

We conclude with a few interesting questions one can ask about the notion of planarly connected crossings: Is it possible to construct for any n and k a graph with quadratically many edges which is not k-PCC? Can we recognize (k-)PCC graphs efficiently? Given that a graph is a (k-)PCC graph, is it possible to find efficiently such an embedding?

Acknowledgments. We thank Géza Tóth for his permission to include his construction for a lower bound on the size of a PCC graph in this paper. We also thank an anonymous referee for pointing out an error in an earlier version of this paper.

Most of this work was done during a visit of the first author to the Rényi Institute that was partially supported by the National Research, Development and Innovation Office – NKFIH under the grant PD 108406 and by the ERC Advanced Research Grant no. 267165 (DISCONV). The second author was supported by the National Research, Development and Innovation Office – NKFIH under the grant PD 108406 and K 116769 and by the János Bolyai Research Scholarship of the Hungarian Academy of Sciences. The third author was supported by Development and Innovation Office – NKFIH under the grant SNN 116095.

References

1. Ackerman, E.: On the maximum number of edges in topological graphs with no four pairwise crossing edges. Discrete Computat. Geom. **41**(3), 365–375 (2009). http://dx.doi.org/10.1007/s00454-009-9143-9
2. Ackerman, E., Fox, J., Pach, J., Suk, A.: On grids in topological graphs. Comput. Geom. **47**(7), 710–723 (2014). http://dx.doi.org/10.1016/j.comgeo.2014.02.003
3. Ackerman, E., Fulek, R., Tóth, C.D.: Graphs that admit polyline drawings with few crossing angles. SIAM J. Discrete Math. **26**(1), 305–320 (2012). http://dx.doi.org/10.1137/100819564
4. Ackerman, E., Tardos, G.: On the maximum number of edges in quasi-planar graphs. J. Comb. Theory Ser. A **114**(3), 563–571 (2007). http://dx.doi.org/10.1016/j.jcta.2006.08.002

5. Agarwal, P.K., Aronov, B., Pach, J., Pollack, R., Sharir, M.: Quasi-planar graphs have a linear number of edges. Combinatorica **17**(1), 1–9 (1997). http://dx.doi.org/10.1007/BF01196127
6. Brass, P., Moser, W.O.J., Pach, J.: Research Problems in Discrete Geometry. Springer, New York (2005)
7. Capoyleas, V., Pach, J.: A Turán-type theorem on chords of a convex polygon. J. Comb. Theory Ser. B **56**(1), 9–15 (1992). http://dx.doi.org/10.1016/0095-8956(92)90003-G
8. Chojnacki, C.: Über wesentlich unplättbare Kurven im dreidimensionalen Raume. Fundamenta Mathematicae **23**(1), 135–142 (1934)
9. Fox, J., Pach, J.: Coloring K_k-free intersection graphs of geometric objects in the plane. Eur. J. Comb. **33**(5), 853–866 (2012). http://dx.doi.org/10.1016/j.ejc.2011.09.021
10. Fox, J., Pach, J.: Applications of a new separator theorem for string graphs. Comb. Probab. Comput. **23**(1), 66–74 (2014). http://dx.doi.org/10.1017/S0963548313000412
11. Kaufmann, M., Ueckerdt, T.: The density of fan-planar graphs. CoRR abs/1403.6184 (2014). http://arxiv.org/abs/1403.6184
12. Pach, J.: Notes on geometric graph theory. In: Goodman, J., Pollack, R., Steiger, W. (eds.) Discrete and Computational Geometry: Papers from DIMACS special year, DIMACS series, vol. 6, pp. 273–285. AMS, Providence (1991)
13. Pach, J., Radoičić, R., Tóth, G.: Relaxing planarity for topological graphs. In: Gőri, E., Katona, G.O., Lovász, L. (eds.) More Graphs, Sets and Numbers, Bolyai Society Mathematical Studies, vol. 15, pp. 285–300. Springer, Heidelberg (2006)
14. Pach, J., Tóth, G.: Graphs drawn with few crossings per edge. Combinatorica **17**(3), 427–439 (1997). http://dx.doi.org/10.1007/BF01215922
15. Pelsmajer, M.J., Schaefer, M., Štefankovič, D.: Removing even crossings. J. Comb. Theory Ser. B **97**(4), 489–500 (2007). http://dx.doi.org/10.1016/j.jctb.2006.08.001
16. Suk, A., Walczak, B.: New bounds on the maximum number of edges in k-quasi-planar graphs. Comput. Geom. **50**, 24–33 (2015). http://dx.doi.org/10.1016/j.comgeo.2015.06.001
17. Tóth, G.: Private communication (2015)
18. Tutte, W.: Toward a theory of crossing numbers. J. Comb. Theory **8**(1), 45–53 (1970). http://www.sciencedirect.com/science/article/pii/S0021980070800072

Re-embedding a 1-Plane Graph into a Straight-Line Drawing in Linear Time

Seok-Hee Hong[1(✉)] and Hiroshi Nagamochi[2]

[1] University of Sydney, Sydney, Australia
seokhee.hong@sydney.edu.au
[2] Kyoto University, Kyoto, Japan
nag@amp.i.kyoto-u.ac.jp

Abstract. Thomassen characterized some 1-plane embedding as the forbidden configuration such that a given 1-plane embedding of a graph is drawable in straight-lines if and only if it does not contain the configuration [C. Thomassen, Rectilinear drawings of graphs, J. Graph Theory, 10(3), 335–341, 1988].

In this paper, we characterize some 1-plane embedding as the forbidden configuration such that a given 1-plane embedding of a graph can be re-embedded into a straight-line drawable 1-plane embedding of the same graph if and only if it does not contain the configuration. Re-embedding of a 1-plane embedding preserves the same set of pairs of crossing edges. We give a linear-time algorithm for finding a straight-line drawable 1-plane re-embedding or the forbidden configuration.

1 Introduction

Since the 1930s, a number of researchers have investigated *planar* graphs. In particular, a beautiful and classical result, known as *Fáry's Theorem*, asserts that every plane graph admits a *straight-line drawing* [5]. Indeed, a straight-line drawing is the most popular drawing convention in Graph Drawing.

More recently, researchers have investigated *1-planar graphs* (i.e., graphs that can be embedded in the plane with at most one crossing per edge), introduced by Ringel [13]. Subsequently, the structure of 1-planar graphs has been investigated [4,12]. In particular, Pach and Toth [12] proved that a 1-planar graph with n vertices has at most $4n - 8$ edges, which is a tight upper bound. Unfortunately, testing the 1-planarity of a graph is NP-complete [6,11], however linear-time algorithms are available for special subclasses of 1-planar graphs [1,3,7].

Thomassen [14] proved that every 1-plane graph (i.e., a 1-planar graph embedded with a given *1-plane embedding*) admits a straight-line drawing if and only if it does not contain any of two special 1-plane graphs, called *B-configuration* or *W-configuration*, see Fig. 1.

Research supported by ARC Future Fellowship and ARC Discovery Project DP160104148. This is an extended abstract. For a full version with omitted proofs, see [9].

Y. Hu and M. Nöllenburg (Eds.): GD 2016, LNCS 9801, pp. 321–334, 2016.
DOI: 10.1007/978-3-319-50106-2_25

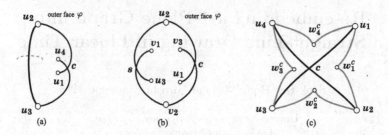

Fig. 1. (a) B-configuration with three edges u_1u_2, u_2u_3 and u_3u_4 and one crossing c made by an edge pair $\{u_1u_2, u_3u_4\}$, where edge u_2u_3 may have a crossing when the configuration is part of a 1-plane embedding; (b) W-configuration with four edges u_1u_2, u_2u_3, v_1v_2 and v_2v_3 and two crossings c and s made by edge pairs $\{u_1u_2, v_2v_3\}$ and $\{u_2u_3, v_1v_2\}$, where possibly $u_1 = v_1$ and $u_3 = v_3$; (c) Augmenting a crossing $c \in \chi$ made by edges u_1u_3 and u_2u_4 with a new cycle $Q_c = (u_1, w_1^c, u_2, w_2^c, u_3, w_3^c, u_4, w_4^c)$ depicted by gray lines.

Recently, Hong et al. [8] gave an alternative constructive proof, with a linear-time testing algorithm and a drawing algorithm. They also showed that some 1-planar graphs need an exponential area with straight-line drawing.

We call a 1-plane embedding *straight-line drawable* (*SLD* for short) if it admits a straight-line drawing, i.e., it does not contain a B- or W-configuration by Thomassen [14]. In this paper, we investigate a problem of "re-embedding" a given non-SLD 1-plane embedding γ into an SLD 1-plane embedding γ'. For a given 1-plane embedding γ of a graph G, we call another 1-plane embedding γ' of G a *cross-preserving embedding* of γ if exactly the same set of edge pairs make the same crossings in γ'.

More specifically, we first characterize the *forbidden configuration* of 1-plane embeddings that cannot admit an SLD cross-preserving 1-plane embedding. Based on the characterization, we present a linear-time algorithm that either detects the forbidden configuration in γ or computes an SLD cross-preserving 1-plane embedding γ'.

Formally, the main problem considered in this paper is defined as follows.

Re-embedding a 1-Plane Graph into a Straight-line Drawing
Input: A 1-planar graph G and a 1-plane embedding γ of G.
Output: Test whether γ admits an SLD cross-preserving 1-plane embedding γ', and construct such an embedding γ' if one exists, or report the forbidden configuration.

To design a linear-time implementation of our algorithm in this paper, we introduce a *rooted-forest representation of non-intersecting cycles* and an efficient procedure of flipping subgraphs in a plane graph. Since these data structure and procedure can be easily implemented, it has advantage over the complicated decomposition of biconnected graphs into triconnected components [10] or the SPQR tree [2].

2   Plane Embeddings and Inclusion Forests

Let U be a set of n elements, and let $\mathcal{S}$ be a family of subsets $S \subseteq U$. We say that two subsets $S, S' \subseteq U$ are *intersecting* if none of $S \cap S'$, $S - S'$ and $S' - S$ is empty. We call $\mathcal{S}$ a *laminar* if no two subsets in $\mathcal{S}$ are intersecting. For a laminar $\mathcal{S}$, the *inclusion-forest* of $\mathcal{S}$ is defined to be a forest $\mathcal{I} = (\mathcal{S}, \mathcal{E})$ of a disjoint union of rooted trees such that (i) the sets in $\mathcal{S}$ are regarded as the vertices of $\mathcal{I}$, and (ii) a set S is an ancestor of a set S' in $\mathcal{I}$ if and only if $S' \subseteq S$.

Lemma 1. *For a cyclic sequence $(u_1, u_2, \ldots, u_\delta)$ of $\delta \geq 2$ elements, define an interval (i, j) to be the set of elements u_k with $i \leq k \leq j$ if $i \leq j$ and $(i, j) = (i, \delta) \cup (1, j)$ if $i > j$. Let $\mathcal{S}$ be a set of intervals. A pair of two intersecting intervals in $\mathcal{S}$ (when $\mathcal{S}$ is not a laminar) or the inclusion-forest of $\mathcal{S}$ (when $\mathcal{S}$ is a laminar) can be obtained in $O(\delta + |\mathcal{S}|)$ time.*

Throughout the paper, a graph $G = (V, E)$ stands for a simple undirected graph. The set of vertices and the set of edges of a graph G are denoted by $V(G)$ and $E(G)$, respectively. For a vertex v, let $E(v)$ be the set of edges incident to v, $N(v)$ be the set of neighbors of v, and $\deg(v)$ denote the degree $|N(v)|$ of v. A simple path with end vertices u and v is called a u, v-*path*. For a subset $X \subseteq V$, let $G - X$ denote the graph obtained from G by removing the vertices in X together with the edges in $\cup_{v \in X} E(v)$.

A *drawing* D of a graph G is a geometric representation of the graph in the plane, such that each vertex of G is mapped to a point in the plane, and each edge of G is drawn as a curve. A drawing D of a graph $G = (V, E)$ is called *planar* if there is no edge crossing. A planar drawing D of a graph G divides the plane into several connected regions, called *faces*, where a face enclosed by a closed walk of the graph is called an *inner face* and the face not enclosed by any closed walk is called the *outer face*.

A planar drawing D *induces* a plane embedding γ of G, which is defined to be a pair (ρ, φ) of the *rotation system* (i.e., the circular ordering of edges for each vertex) ρ, and the outer face φ whose facial cycle C_φ gives the outer boundary of D. Let $\gamma = (\rho, \varphi)$ be a plane embedding of a graph $G = (V, E)$. We denote by $F(\gamma)$ the set of faces in γ, and by C_f the facial cycle determined by a face $f \in F$, where we call a subpath of C_f a *boundary path* of f. For a simple cycle C of G, the plane is divided by C in two regions, one containing only inner faces and the other containing the outer area, where we say that the former is *enclosed* by C or the *interior* of C, while the latter is called the *exterior* of C. We denote by $F_{in}(C)$ the set of inner faces in the interior of C, by $E_{in}(C)$ the set of edges in $E(C_f)$ with $f \in F_{in}(C)$, and by $V_{in}(C)$ the set of end-vertices of edges in $E_{in}(C)$. Analogously define $F_{ex}(C)$, $E_{ex}(C)$ and $V_{ex}(C)$ in the exterior of C. Note that $E(C) = E_{in}(C) \cap E_{ex}(C)$ and $V(C) = V_{in}(C) \cap V_{ex}(C)$.

For a subgraph H of G, we define the embedding $\gamma|_H$ of γ induced by H to be a sub-embedding of γ obtained by removing the vertices/edges not in H, keeping the same rotation system around each of the remaining vertices/crossings and the same outer face.

2.1 Inclusion Forests of Inclusive Set of Cycles

In this and next subsections, let (G, γ) stand for a plane embedding of $\gamma = (\rho, \varphi)$ of a biconnected simple graph $G = (V, E)$ with $n = |V| \geq 3$.

Let C be a simple cycle in G. We define the *direction* of C to be an ordered pair (u, v) with $uv \in E(C)$ such that the inner faces in $F_{in}(C)$ appear on the right hand side when we traverse C in the order that we start u and next visit v.

For simplicity, we say that two simple cycles C and C' are *intersecting* if $F_{in}(C)$ and $F_{in}(C')$ are intersecting.

Let $\mathcal{C}$ be a set of simple cycles in G. We call $\mathcal{C}$ *inclusive* if no two cycles in $\mathcal{C}$ are intersecting, i.e., $\{F_{in}(C) \mid C \in \mathcal{C}\}$ is a laminar. When $\mathcal{C}$ is inclusive, the *inclusion-forest* of $\mathcal{C}$ is defined to be a forest $\mathcal{I} = (\mathcal{C}, \mathcal{E})$ of a disjoint union of rooted trees such that:

(i) the cycles in $\mathcal{C}$ are regarded as the vertices of $\mathcal{I}$, and
(ii) a cycle C is an ancestor of a cycle C' in $\mathcal{I}$ if and only if $F_{in}(C') \subseteq F_{in}(C)$.

Let $\mathcal{I}(\mathcal{C})$ denote the inclusion-forest of $\mathcal{C}$. For a vertex subset $X \subseteq V$, let $\mathcal{C}(X)$ denote the set of cycles $C \in \mathcal{C}$ such that $x \in V(C)$ for some vertex $x \in X$, where we denote $\mathcal{C}(\{v\})$ by $\mathcal{C}(v)$ for short.

Lemma 2. *For (G, γ), let $\mathcal{C}$ be a set of simple cycles of G. Then any of the following tasks can be executed in $O(n + \sum_{C \in \mathcal{C}} |E(C)|)$ time.*

(i) *Decision of the directions of all cycles in $\mathcal{C}$;*
(ii) *Detection of a pair of two intersecting cycles in $\mathcal{C}$ when $\mathcal{C}$ is not inclusive, and construction of the inclusion-forests $\mathcal{I}(\mathcal{C}(v))$ for all vertices $v \in V$ when $\mathcal{C}$ is inclusive; and*
(iii) *Construction of the inclusion-forest $\mathcal{I}(\mathcal{C})$ when $\mathcal{C}$ is inclusive.*

2.2 Flipping Spindles

A simple cycle C of G is called a *spindle* (or a u, v-*spindle*) of γ if there are two vertices $u, v \in V(C)$ such that no vertex in $V(C) - \{u, v\}$ is adjacent to any vertex in the exterior of C, where we call vertices u and v the *junctions* of C. Note that each of the two subpaths of C between u and v is a boundary path of some face in $F(\gamma)$.

Given (G, γ), we denote the rotation system around a vertex $v \in V$ by $\rho_\gamma(v)$. For a spindle C in γ, let $J(C)$ denote the set of the two junctions of C.

Flipping a u, v-spindle C means to modify the rotation system of vertices in $V_{in}(C)$ as follows:

(i) For each vertex $w \in V_{in}(C) - J(C)$, reverse the cyclic order of $\rho_\gamma(w)$; and
(ii) For each vertex $u \in J(C)$, reverse the order of subsequence of $\rho_\gamma(u)$ that consists of vertices $N(u) \cap V_{in}(C)$.

Every two distinct spindles C and C' in γ are non-intersecting, and they always satisfy one of $E_{\text{in}}(C) \cap E_{\text{in}}(C') = \emptyset$, $E_{\text{in}}(C) \subseteq E_{\text{in}}(C')$, and $E_{\text{in}}(C') \subseteq E_{\text{in}}(C)$. Let $\mathcal{C}$ be a set of spindles in γ, which is always inclusive, and let $\mathcal{I}(\mathcal{C})$ denote the inclusion-forest of $\mathcal{C}$.

When we modify the current embedding γ by flipping each spindle in $\mathcal{C}$, the resulting embedding $\gamma_{\mathcal{C}}$ is the same, independent from the ordering of the flipping operation to the spindles, since for two spindles C and C' which share a common junction vertex $u \in J(C) \cap J(C')$, the sets $N(u) \cap V_{\text{in}}(C)$ and $N(u) \cap V_{\text{in}}(C')$ do not intersect, i.e., they are disjoint or one is contained in the other.

Define the *depth* of a vertex $v \in V$ in $\mathcal{I}$ to be the number of spindles $C \in \mathcal{C}$ such that $v \in V_{\text{in}}(C) - J(C)$, and denote by $\text{p}(v)$ the parity of depth of vertex v, i.e., $\text{p}(v) = 1$ if the depth is odd and $\text{p}(v) = -1$ otherwise.

For a vertex $v \in V$, let $\mathcal{C}[v]$ denote the set of spindles $C \in \mathcal{C}$ such that $v \in J(C)$, and let $\gamma_{\mathcal{C}[v]}$ be the embedding obtained from γ by flipping all spindles in $\mathcal{C}[v]$. Let $\text{rev}\langle\sigma\rangle$ mean the reverse of a sequence σ. Then we see that $\rho_{\gamma_{\mathcal{C}}}(v) = \rho_{\gamma_{\mathcal{C}[v]}}(v)$ if $\text{p}(v) = 1$; and $\rho_{\gamma_{\mathcal{C}}}(v) = \text{rev}\langle\rho_{\gamma_{\mathcal{C}[v]}}(v)\rangle$ otherwise. To obtain the embedding $\gamma_{\mathcal{C}}$ from the current embedding γ by flipping each spindle in $\mathcal{C}$, it suffices to show how to compute each of $\text{p}(v)$ and $\rho_{\gamma_{\mathcal{C}[v]}}(v)$ for all vertices $v \in V$.

Lemma 3. *Given (G, γ), let $\mathcal{C}$ be a set of spindles of γ. Then any of the following tasks can be executed in $O(n + \sum_{C \in \mathcal{C}} |E(C)|)$ time.*

(i) *Decision of parity $\text{p}(v)$ of all vertices $v \in V$; and*
(ii) *Computation of $\rho_{\gamma_{\mathcal{C}[v]}}(v)$ for all vertices $v \in V$.*

3   Re-embedding 1-Plane Graph and Forbidden Configuration

A drawing D of a graph $G = (V, E)$ is called a *1-planar drawing* if each edge has at most one crossing. A 1-planar drawing D of graph G induces a *1-plane embedding* γ of G, which is defined to be a tuple (χ, ρ, φ) of the *crossing system* χ of E, the *rotation system* ρ of V, and the *outer face* φ of D. The *planarization* $\mathcal{G}(G, \gamma)$ of a 1-plane embedding γ of graph G is the plane embedding obtained from γ by regarding crossings also as graph vertices, called *crossing-vertices*. The set of vertices in $\mathcal{G}(G, \gamma)$ is given by $V \cup \chi$. For a notational convenience, we refer to a subgraph/face of $\mathcal{G}(G, \gamma)$ as a subgraph/face in γ.

Let $\gamma = (\chi, \rho, \varphi)$ be a 1-plane embedding of graph G. We call another 1-plane embedding $\gamma' = (\chi', \rho', \varphi')$ of graph G a *cross-preserving* 1-plane embedding of γ when the same set of edge pairs makes crossings, i.e., $\chi = \chi'$. In other words, the planarization $\mathcal{G}(G, \gamma')$ is another plane embedding of $\mathcal{G}(G, \gamma)$ such that the alternating order of edges incident to each crossing-vertex $c \in \chi$ is preserved.

To eliminate the additional constraint on the rotation system on each crossing-vertex $c \in \chi$, we introduce "circular instances." We call an instance (G, γ) of 1-plane embedding *circular* when for each crossing $c \in \chi$, the four end-vertices of the two crossing edges $u_1 u_3$ and $u_2 u_4$ that create c (where u_1, u_2, u_3 and u_4 appear in the clockwise order around c) are contained in a

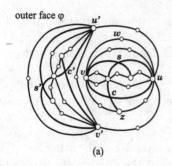 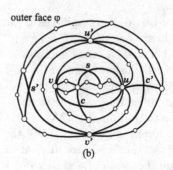

         (a)   (b)

Fig. 2. Circular instances (G, γ) with a cut-vertex u of $\mathcal{G}$, where the crossing edges are depicted by slightly thicker lines: (a) hard B-cycles $C = (u, c, v, s)$ and $C' = (u', c', v', s')$, (b) hard B-cycle $C = (u, c, v, s)$ and a nega-cycle $C' = (u', c', v', s')$ whose reversal is a hard B-cycle, where vertices $u, v, u', v' \in V$ and crossings $c, s, c', s' \in \chi$.

cycle $Q_c = (u_1, w_1^c, u_2, w_2^c, u_3, w_3^c, u_4, w_4^c)$ of eight crossing-free edges for some vertices w_i^c, $i = 1, 2, 3, 4$ of degree 2, as shown in Fig. 1(c). By definition, c and each w_i^c not necessarily appear along the same facial cycle in the planarization $\mathcal{G}(G, \gamma)$. For example, path (v, w, u) is part of such a cycle Q_s for the crossing s in the circular instance in Fig. 2(a), but c and w are not on the same facial cycle in the planarization.

    A given instance can be easily converted into a circular instance by augmenting the end-vertices of each pair of crossing edges as follows. In the plane graph, $\mathcal{G}(G, \gamma)$, for each crossing-vertex $c \in \chi$ and its neighbors u_1, u_2, u_3 and u_4 that appear in the clockwise order around c, we add a new vertex w_i^c, $i = 1, 2, 3, 4$ and eight new edges $u_i w_i^c$ and $w_i^c u_{i+1}$, $i = 1, 2, 3, 4$ (where u_5 means u_1) to form a cycle Q_c of length 8 whose interior contains no other vertex than c.

    Let H be the resulting graph augmented from G, and let Γ be the resulting 1-plane embedding of H augmented from γ. Note that $|V(H)| \leq |V(G)| + 4|\chi|$ holds. We easily see that if γ admits an SLD cross-preserving embedding γ' then Γ admits an SLD cross-preserving embedding Γ'. This is because a straight-line drawing $D_{\gamma'}$ of γ' can be changed into a straight-line drawing $D_{\Gamma'}$ of some cross-preserving embedding Γ' of Γ by placing the newly introduced vertices w_i^c within the region sufficiently close to the position of c. We here see that cycle Q_c can be drawn by straight-line segments without intersecting with other straight-line segments in $D_{\gamma'}$.

    Note that the instance (G, γ') remains circular for any cross-preserving embedding γ' of γ. In the rest of paper, let (G, γ) stand for a circular instance $(G = (V, E), \gamma = (\chi, \rho, \varphi))$ with $n \geq 3$ vertices and let $\mathcal{G}$ denote its planarization $\mathcal{G}(G, \gamma)$. Figure 2 shows examples of circular instances (G, γ), where the vertex-connectivity of $\mathcal{G}$ is 1.

    As an important property of a circular instance, the subgraph $G_{(0)}$ with crossing-free edges is a spanning subgraph of G and the four end-vertices of any two crossing edges are contained in the same block of the graph $G_{(0)}$. The biconnectivity is necessary to detect certain types of cycles by applying Lemma 2.

3.1 Candidate Cycles, B/W Cycle, Posi/Nega Cycle, Hard/Soft Cycle

For a circular instance (G, γ), finding a cross-preserving embedding of γ is effectively equivalent to finding another plane embedding of G so that all the current B- and W-configurations are eliminated and no new B- or W-configurations are introduced. To detect the cycles that can be the boundary of a B- or W-configuration in changing the plane embedding of G, we categorize cycles containing crossing vertices in G.

A *candidate posi-cycle* (resp., *candidate nega-cycle*) in G is defined to be a cycle $C = (u, c, v)$ or $C = (u, c, v, s)$ in G with $u, v \in V$ and $c, s \in \chi$ such that the interior (resp., exterior) of C does not contain a crossing-free edge $uv \in E$ and any other crossing vertex c' adjacent to both u and v.

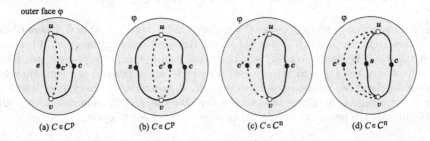

Fig. 3. Candidate posi- and nega-cycles $C = (u, c, v)$ and $C = (u, c, v, s)$ in G, where white circles represent vertices in V while black ones represent crossings in χ: (a) candidate posi-cycle of length 3, (b) candidate posi-cycle of length 4, (c) candidate nega-cycle of length 3, and (d) candidate nega-cycle of length 4.

Figure 3(a)–(b) and (c)–(d) illustrate candidate posi-cycles and candidate nega-cycles, respectively. Let C^P and C^n be the sets of candidate posi-cycles and candidate nega-cycles, respectively. By definition we see that the set $C^P \cup C^n \cup \{C_f \mid f \in F(\gamma)\}$ is inclusive, and hence $|C^P \cup C^n \cup \{C_f \mid f \in F(\gamma)\}| = O(n)$.

A candidate posi-cycle C with $C = (u, c, v)$ (resp., $C = (u, c, v, s)$) is called a *B-cycle* if

(a)-(B): the exterior of C contains no vertices in $V - \{u, v\}$ adjacent to c (resp., contains exactly one vertex in $V - \{u, v\}$ adjacent to c or s).

Note that $uv \in E$ when $C = (u, c, v)$ is a B-cycle, as shown in Fig. 4(a). Figure 4(b) and (d) illustrate the other types of B-cycles.

A candidate posi-cycle $C = (u, c, v, s)$ is called a *W-cycle* if

(a)-(W): the exterior of C contains no vertices in $V - \{u, v\}$ adjacent to c or s.

Figure 4(c) and (e) illustrate W-cycles.

Let C_W (resp., C_B) be the set of W-cycles (resp., B-cycles) in γ. Clearly a W-cycle (resp., B-cycle) gives rise to a W-configuration (resp., B-configuration). Conversely, by choosing a W-configuration (resp., B-configuration) so that the interior is minimal, we obtain a W-cycle (resp., B-cycle). Hence we observe that the current embedding γ admits a straight-line drawing if and only if $C_W = C_B = \emptyset$.

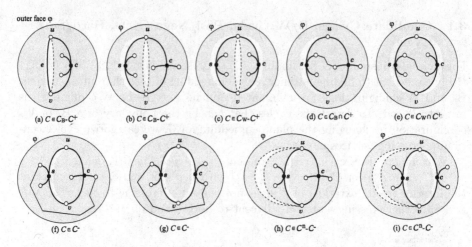

Fig. 4. Illustration of types of cycles $C = (u, c, v)$ and $C = (u, c, v, s)$ in $\mathcal{G}$, where white circles represent vertices in V while black ones represent crossings in χ: (a) B-cycle of length 3, which is always soft, (b) soft B-cycle of length 4, (c) soft W-cycle, (d) hard B-cycle of length 4, (e) hard W-cycle, (f) nega-cycle whose reversal is a hard B-cycle, (g) nega-cycle whose reversal is a hard W-cycle, (h) candidate nega-cycle of length 4 that is not a nega-cycle whose reversal is a hard B-cycle, and (i) candidate nega-cycle of length 4 that is not a nega-cycle whose reversal is a hard W-cycle.

A W- or B-cycle C is called *hard* if
(b): length of C is 4, and the interior of $C = (u, c, v, s)$ contains no inner face f whose facial cycle C_f contains both vertices u and v, i.e., some path connects c and s without passing through u or v.

On the other hand, a W- or B-cycle $C = (u, c, v, s)$ of length 4 that does not satisfy condition (b) or a B-cycle of length 3 is called *soft*. We also call a hard B- or W-cycle a *posi-cycle*.

Figure 4(d) and (e) illustrate a hard B-cycle and a hard W-cycles, respectively, whereas Fig. 4(a) and (b) (resp., (c)) illustrate soft B-cycles (resp., a soft W-cycle).

A cycle $C = (u, c, v, s)$ is called a *nega-cycle* if it becomes a posi-cycle when an inner face in the interior of C is chosen as the outer face. In other words, a nega-cycle is a candidate nega-cycle $C = (u, c, v, s)$ of length 4 that satisfies the following conditions (a') and (b'), where (a') (resp., (b')) is obtained from the above conditions (a)-(B) and (a)-(W) (resp., (b)) by exchanging the roles of "interior" and "exterior":
(a'): the interior of C contains at most one vertex in $V - \{u, v\}$ adjacent to c or s; and
(b'): the exterior of C contains no face f whose facial cycle C_f contains both vertices u and v.

Figure 4(f) and (g) illustrate nega-cycles, whereas Fig. 4(h) and (i) illustrate candidate nega-cycles that are not nega-cycles.

Let $\mathcal{C}^+$ (resp., $\mathcal{C}^-$) denote the set of posi-cycles (resp., nega-cycles) in γ. By definition, it holds that $\mathcal{C}^+ \subseteq \mathcal{C}_\mathrm{W} \cup \mathcal{C}_\mathrm{B} \subseteq \mathcal{C}^\mathrm{p}$ and $\mathcal{C}^- \subseteq \mathcal{C}^\mathrm{n}$.

3.2   Forbidden Cycle Pairs

We define a forbidden configuration that characterizes 1-plane embeddings, which cannot be re-embedded into SLD ones. A *forbidden cycle pair* is defined to be a pair $\{C, C'\}$ of a posi-cycle $C = (u, c, v, s)$ and a posi- or nega-cycle $C' = (u', c', v', s')$ in $\mathcal{G}$ with $u, v, u', v' \in V$ and $c, s, c', s' \in \chi$ to which $\mathcal{G}$ has a u, u'-path P_1 and a v, v'-path P_2 such that:

(i) when $C' \in \mathcal{C}^+$, paths P_1 and P_2 are in the exterior of C and C', i.e., $V(P_1) - \{u, u'\}, V(P_2) - \{v, v'\} \subseteq V_\mathrm{ex}(C) \cap V_\mathrm{ex}(C')$, where C and C' cannot have any common inner face; and

(ii) when $C' \in \mathcal{C}^-$, paths P_1 and P_2 are in the exterior of C and the interior of C', i.e., $V(P_1) - \{u, u'\}, V(P_2) - \{v, v'\} \subseteq V_\mathrm{ex}(C) \cap V_\mathrm{in}(C')$, where C is enclosed by C'.

In (i) and (ii), P_1 and P_2 are not necessary disjoint, and possibly one of them consists of a single vertex, i.e., $u = u'$ or $v = v'$.

The pair of cycles C and C' in Fig. 5(a) (resp., Fig. 5(b)) is a forbidden cycle pair, because there is a pair of a u, u'-path $P_1 = (u, x, z, y, u')$ and a v, v'-path $P_2 = (v, x', z, y', v')$ that satisfy the above conditions (i) (resp., (ii)). Note that the pair of cycles C and C' in Fig. 2(a)–(b) is not forbidden cycle pair, because there are no such paths.

Our main result of this paper is as follows.

Theorem 1. *A circular instance (G, γ) admits an SLD cross-preserving embedding if and only if it has no forbidden cycle pair. Finding an SLD cross-preserving embedding of γ or a forbidden cycle pair in $\mathcal{G}$ can be computed in linear time.*

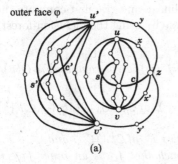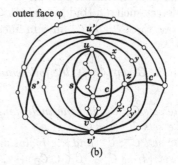

Fig. 5. Illustration of circular instances (G, γ) with a cut-vertex z of $\mathcal{G}$, where the crossing edges are depicted by slightly thicker lines: (a) forbidden cycle pair with hard B-cycles $C = (u, c, v, s)$ and $C' = (u', c', v', s')$ (b) forbidden cycle pair with a hard B-cycle $C = (u, c, v, s)$ and a nega-cycle $C' = (u', c', v', s')$ whose reversal is a hard B-cycle, where vertices $u, v, u', v' \in V$ and crossings $c, s, c', s' \in \chi$.

Proof of necessity: The necessity of the theorem follows from the next lemma.

For a cycle $C = (u, c, v, s) \in C^+$ (resp., C^-) with $u, v \in V$ and $c, s \in \chi$ in $\mathcal{G}$, we call a vertex $z \in V$ an *in-factor* of C if the exterior of $C \in C^+$ (resp., the interior of $C \in C^-$) has a z, u-path $P_{z,u}$ and a z, v-path $P_{z,v}$, i.e., $V(P_{z,u} - \{u\}) \cup V(P_{z,v} - \{v\})$ is in $V_{ex}(C)$ (resp., $V_{in}(C)$). Paths $P_{z,u}$ and $P_{z,v}$ are not necessarily disjoint.

Lemma 4. *Given $\mathcal{G} = \mathcal{G}(G, \gamma)$, let γ' be a cross-preserving embedding of γ. Then:*

(i) *Let $z \in V$ be an in-factor of a cycle $C \in C^+ \cup C^-$ in $\mathcal{G}$. Then cycle C is a posi-cycle (resp., a nega-cycle) in $\mathcal{G}(G, \gamma')$ if and only if z is in the exterior (resp., interior) of C in γ';*

(ii) *For a forbidden cycle pair $\{C, C'\}$, one of C and C' is a posi-cycle in $\mathcal{G}(G, \gamma')$ (hence any cross-preserving embedding of γ contains a B- or W-configuration and (G, γ) admits no SLD cross-preserving embedding).*

Proof of sufficiency: In the rest of paper, we prove the sufficiency of Theorem 1 by designing a linear-time algorithm that constructs an SLD cross-preserving embedding of an instance without a forbidden cycle pair.

4   Biconnected Case

In this section, (G, γ) stands for a circular instance such that the vertex-connectivity of the plane graph $\mathcal{G}$ is at least 2. In a biconnected graph $\mathcal{G}$, any two posi-cycles $C = (u, c, v, s)$, $C' = (u', c', v', s') \in C^+$ with $u, v, u', v' \in V$ give a forbidden cycle pair if they do not share an inner face, because there is a pair of u, u'-path and v, v'-path in the exterior of C and C'. Analogously any pair of a posi-cycle C and a nega-cycle C' such that C' encloses C is also a forbidden cycle pair in a biconnected graph $\mathcal{G}$.

To detect such a forbidden pair in $\mathcal{G}$ in linear time, we first compute the sets C_p, C_n, C_W, C_B, C^+ and C^- in γ in linear time by using the inclusion-forest from Lemma 2.

Lemma 5. *Given (G, γ), the following in (i)–(iv) can be computed in $O(n)$ time.*

(i) *The sets C_p, C_n and the inclusion-forest $\mathcal{I}$ of $C_p \cup C_n \cup \{C_f \mid f \in F(\gamma)\}$;*

(ii) *The sets C_W and C_B;*

(iii) *The sets C^+, C^- and the inclusion-forest $\mathcal{I}^*$ of $C^+ \cup C^-$; and*

(iv) *A set $\{f_C \mid C \in (C_W \cup C_B) - C^+\}$ such that f_C is an inner face in the interior of a soft B- or W-cycle C with $V(C_f) \supseteq V(C)$.*

Given (G, γ), a face $f \in F(\gamma)$ is called *admissible* if all posi-cycles enclose f but no nega-cycle encloses f. Let $A(\gamma)$ denote the set of all admissible faces in $F(\gamma)$.

Lemma 6. *Given (G, γ), it holds $A(\gamma) \neq \emptyset$ if and only if no forbidden cycle pair exists in γ. A forbidden cycle pair, if one exists, and $A(\gamma)$ can be obtained in $O(n)$ time.*

By the lemma, if (G, γ) has no forbidden cycle pair, i.e., $A(\gamma) \neq \emptyset$, then any new embedding obtained from γ by changing the outer face with a face in $A(\gamma)$ is a cross-preserving embedding of γ which has no hard B- or W-cycle.

4.1   Eliminating Soft B- and W-cycles

Suppose that we are given a circular instance (G, γ) such that G is biconnected and $\mathcal{C}^+ = \emptyset$. We now show how to eliminate all soft B- and W-cycles in G in linear time using the inclusion-forest from Lemma 2 and the spindles from Lemma 3.

Lemma 7. *Given (G, γ) with $\mathcal{C}^+ = \emptyset$, there exists an SLD cross-preserving embedding $\gamma' = (\chi, \rho', \varphi')$ of γ such that $V(C_{\varphi'}) \supseteq V(C_\varphi)$ for the facial cycle C_φ (resp., $C_{\varphi'}$) of the outer face φ (resp., φ'), which can be constructed in $O(n)$ time.*

Given an instance (G, γ) with a biconnected graph G, we can test whether it has either a forbidden cycle pair or an admissible face by Lemmas 5 and 6. In the former, it cannot have an SLD cross-preserving embedding by Lemma 4. In the latter, we can eliminate all hard B- and W-cycles by choosing an admissible face as a new outer face, and then eliminate all soft B- and W-cycles by a flipping procedure based on Lemma 7. All the above can be done in linear time.

To treat the case where the vertex-connectivity of G is 1 in the next section, we now characterize 1-plane embeddings that can have an SLD cross-preserving embedding such that a specified vertex appears along the outer boundary. For a vertex $z \in V$ in a graph G, we call a 1-plane embedding γ of G *z-exposed* if vertex z appears along the outer boundary of γ. We call (G, γ) *z-feasible* if it admits a z-exposed SLD cross-preserving embedding γ' of γ.

Lemma 8. *Given (G, γ) such that $A(\gamma) \neq \emptyset$, let z be a vertex in V. Then:*

(i) *The following conditions are equivalent:*
  (a) *γ admits no z-exposed SLD cross-preserving embedding;*
  (b) *$A(\gamma)$ contains no face f with $z \in V(C_f)$; and*
  (c) *G has a posi- or nega-cycle C to which z is an in-factor;*
(ii) *A z-exposed SLD cross-preserving embedding or a posi- or nega-cycle C to which z is an in-factor can be computed in $O(n)$ time.*

5   One-Connected Case

In this section, we prove the sufficiency of Theorem 1 by designing a linear-time algorithm claimed in the theorem. Given a circular instance (G, γ), where G may be disconnected, obviously we only need to test each connected component of

$\mathcal{G}$ separately to find a forbidden cycle pair. Thus we first consider a circular instance (G, γ) such that the vertex-connectivity of $\mathcal{G}$ is 1; i.e., $\mathcal{G}$ is connected and has some cut-vertices.

A block B of $\mathcal{G}$ is a maximal biconnected subgraph of $\mathcal{G}$. For a biconnected graph $\mathcal{G}$, we already know how to find a forbidden cycle pair or an SLD cross-preserving embedding from the previous section. For a trivial block B with $|V(B)| = 2$, there is nothing to do. If some block B of $\mathcal{G}$ with $|V(B)| \geq 3$ contains a forbidden cycle pair, then (G, γ) cannot admit any SLD cross-preserving embedding by Lemma 4.

We now observe that $\mathcal{G}$ may contain a forbidden cycle pair even if no single block of $\mathcal{G}$ has a forbidden cycle pair.

Lemma 9. *For a circular instance (G, γ) such that the vertex-connectivity of $\mathcal{G}$ is 1, let B_1 and B_2 be blocks of $\mathcal{G}$ and let $P_{1,2}$ be a z_1, z_2-path of $\mathcal{G}$ with the minimum number of edges, where $V(B_i) \cap V(P_{1,2}) = \{z_i\}$ for each $i = 1, 2$. If $\gamma|_{B_i}$ has a posi- or nega-cycle C_i to which z_i is an in-factor for each $i = 1, 2$, then $\{C_1, C_2\}$ is a forbidden cycle pair in $\mathcal{G}$.*

For a linear-time implementation, we do not apply the lemma for all pairs of blocks in $\mathcal{B}$. A block of $\mathcal{G}$ is called a *leaf block* if it contains only one cut-vertex of $\mathcal{G}$, where we denote the cut-vertex in a leaf block B by v_B. Without directly searching for a forbidden cycle pair in $\mathcal{G}$, we use the next lemma to reduce a given embedding by repeatedly removing leaf blocks.

Lemma 10. *For a circular instance (G, γ) such that the vertex-connectivity of $\mathcal{G} = \mathcal{G}(G, \gamma)$ is 1 and a leaf block B of $\mathcal{G}$ such that $\gamma|_B$ is v_B-feasible, let $H = G - (V(B) - \{v_B\})$ be the graph obtained by removing the vertices in $V(B) - \{v_B\}$. Then*

(i) *The instance $(H, \gamma|_H)$ is circular; and*
(ii) *If $(H, \gamma|_H)$ admits an SLD cross-preserving embedding γ_H^*, then an SLD cross-preserving embedding γ^* of γ can be obtained by placing a v_B-exposed SLD cross-preserving embedding γ_B^* of $\gamma|_B$ within a space next to the cut-vertex v_B in γ_H^*.*

Given a circular instance (G, γ) such that $\mathcal{G} = \mathcal{G}(G, \gamma)$ is connected, an algorithm **Algorithm Re-Embed-1-Plane** for Theorem 1 is designed by the following three steps.

The first step tests whether $\mathcal{G}$ has a block B such that $\gamma|_B$ has a forbidden cycle pair, based on Lemma 8. If one exists, the algorithm outputs a forbidden cycle pair and halts.

After the first step, no block has a forbidden cycle pair. In the current circular instance (G, γ), one of the following holds:

(i) the number of blocks in $\mathcal{G}$ is at least two and there is at most one leaf block B such that $\gamma|_B$ is not v_B-feasible;
(ii) $\mathcal{G}$ has two leaf blocks B and B' such that $\gamma|_B$ is not v_B-feasible and $\gamma|_{B'}$ is not $v_{B'}$-feasible; and
(iii) the number of blocks in $\mathcal{G}$ is at most one.

In (ii), v_B is an in-factor of a cycle C in $\gamma|_B$ and $v_{B'}$ is an in-factor of a cycle C' in $\gamma|_{B'}$ by Lemma 8, and we obtain a forbidden cycle pair $\{C, C'\}$ by Lemma 9. Otherwise if (i) holds, then we can remove all leaf blocks B such that $\gamma|_B$ is not v_B-feasible by Lemma 10. The second step keeps removing all leaf blocks B such that $\gamma|_B$ is not v_B-feasible until (ii) or (iii) holds to the resulting embedding. If (i) occurs, then the algorithm outputs a forbidden cycle pair and halts.

When all the blocks of $\mathcal{G}$ can be removed successfully, say in an order of $B^1, B^2, \ldots, B^m$, the third step constructs an embedding with no B- or W-cycles by starting with such an SLD embedding of B^m and by adding an SLD embedding of B^i to the current embedding in the order of $i = m-1, m-2, \ldots, 1$. By Lemma 10, this results in an SLD cross-preserving embedding of the input instance (G, γ).

Note that we can obtain an SLD cross-preserving embedding $\gamma^*_{H^1}$ of γ in the third step when the first and second step did not find any forbidden cycle pair. Thus the algorithm finds either an SLD cross-preserving embedding of γ or a forbidden cycle pair. This proves the sufficiency of Theorem 1.

By the time complexity result from Lemma 8, we see that the algorithm can be implemented in linear time.

References

1. Auer, C., Bachmaier, C., Brandenburg, F.J., Gleißner, A., Hanauer, K., Neuwirth, D., Reislhuber, J.: Outer 1-planar graphs. Algorithmica **74**(4), 1293–1320 (2016)
2. Di Battista, G., Tamassia, R.: On-line planarity testing. SIAM J. Comput. **25**(5), 956–997 (1996)
3. Eades, P., Hong, S.-H., Katoh, N., Liotta, G., Schweitzer, P., Suzuki, Y.: A linear time algorithm for testing maximal 1-planarity of graphs with a rotation system. Theor. Comput. Sci. **513**, 65–76 (2013)
4. Fabrici, I., Madaras, T.: The structure of 1-planar graphs. Discrete Math. **307**(7–8), 854–865 (2007)
5. Fáry, I.: On straight line representations of planar Graphs. Acta Sci. Math. Szeged **11**, 229–233 (1948)
6. Grigoriev, A., Bodlaender, H.: Algorithms for graphs embeddable with few crossings per edge. Algorithmica **49**(1), 1–11 (2007)
7. Hong, S.-H., Eades, P., Katoh, N., Liotta, G., Schweitzer, P., Suzuki, Y.: A linear-time algorithm for testing outer-1-planarity. Algorithmica **72**(4), 1033–1054 (2015)
8. Hong, S.-H., Eades, P., Liotta, G., Poon, S.-H.: Fáry's theorem for 1-planar graphs. In: Gudmundsson, J., Mestre, J., Viglas, T. (eds.) COCOON 2012. LNCS, vol. 7434, pp. 335–346. Springer, Heidelberg (2012). doi:10.1007/978-3-642-32241-9_29
9. Hong, S.-H., Nagamochi, H.: Re-embedding a 1-plane graph into a straight-line drawing in linear time. Technical report TR 2016–002, Department of Applied Mathematics and Physics, Kyoto University (2016)
10. Hopcroft, J.E., Tarjan, R.E.: Dividing a graph into triconnected components. SIAM J. Comput. **2**, 135–158 (1973)
11. Korzhik, V.P., Mohar, B.: Minimal obstructions for 1-immersions and Hardness of 1-planarity testing. J. Graph Theory **72**(1), 30–71 (2013)

12. Pach, J., Toth, G.: Graphs drawn with few crossings per edge. Combinatorica **17**(3), 427–439 (1997)
13. Ringel, G.: Ein Sechsfarbenproblem auf der Kugel. Abh. Math. Semin. Univ. Hamb. **29**, 107–117 (1965)
14. Thomassen, C.: Rectilinear drawings of graphs. J. Graph Theory **10**(3), 335–341 (1988)

1-Bend RAC Drawings of 1-Planar Graphs

Walter Didimo[1], Giuseppe Liotta[1], Saeed Mehrabi[2],
and Fabrizio Montecchiani[1]([envelope])

[1] Dipartimento di Ingegneria, Università degli Studi di Perugia, Perugia, Italy
{walter.didimo,giuseppe.liotta,fabrizio.montecchiani}@unipg.it
[2] David R. Cheriton School of Computer Science,
University of Waterloo, Waterloo, Canada
smehrabi@uwaterloo.ca

Abstract. A graph is 1-*planar* if it has a drawing where each edge
is crossed at most once. A drawing is *RAC (Right Angle Crossing)* if
the edges cross only at right angles. The relationships between 1-planar
graphs and RAC drawings have been partially studied in the literature.
It is known that there are both 1-planar graphs that are not straight-line
RAC drawable and graphs that have a straight-line RAC drawing but
that are not 1-planar [22]. Also, straight-line RAC drawings always exist
for *IC-planar graphs* [9], a subclass of 1-planar graphs. One of the main
questions still open is whether every 1-planar graph has a RAC drawing
with at most one bend per edge. We positively answer this question.

1 Introduction

An emerging research line in Graph Drawing studies families of non-planar
graphs that can be drawn so that crossing edges verify some desired properties. This topic is informally recognized as "beyond planarity". Different
types of properties give rise to different families of beyond planar graphs.
Among them, particular attention has been devoted to 1-*planar graphs* (see,
e.g., [1,2,7–9,17,23,24,29,31,35]) and to *RAC (Right Angle Crossing) graphs*
(see, e.g., [4,6,13–15,18–21,27,30]). A graph is 1-planar if it has a drawing where
each edge is crossed at most once, while it is RAC if it has a polyline drawing
where the edges cross only at right angles. From an application point of view,
the study of these two families is motivated by several cognitive experiments,
suggesting that the readability of a layout is negatively correlated to the number of crossings [33,34,38] and that user task performances are not affected too
much if edges cross at large angles [25,26,28]. Also, users often prefer straight-
line drawings or layouts whose edges have few bends [32], and several algorithms
optimize this aesthetic criterion [11]. Note that, every graph admits a polyline
RAC drawing with at most three bends per edge [18].

For the reasons above, it is interesting to study what graphs can be drawn
with at most one crossing per edge, right angle crossings, and few bends per

Research supported in part by the MIUR project AMANDA "Algorithmics for
MAssive and Networked DAta", prot. 2012C4E3KT_001.

edge at the same time. We recall that n-vertex 1-planar graphs have at most $4n - 8$ edges [31] and that straight-line 1-planar drawings have at most $4n - 9$ edges [17]. Also, straight-line RAC graphs have at most $4n - 10$ edges [18], while RAC drawings with at most one bend per edge or two bends per edge, have at most $6.5n - 13$ and $74.2n$ edges, respectively [5]. These results immediately imply that there are 1-planar graphs not admitting 1-planar drawings with straight-line edges and 1-planar graphs not admitting straight-line drawings with right angle crossings. Also, there exist straight-line RAC drawable graphs that are not 1-planar [22]. In this scenario, one of the main questions still open is whether every 1-plane graph admits a RAC drawing with at most one bend per edge. This paper positively answers this question, by proving the following result.

Theorem 1. *Let G be an n-vertex 1-planar graph. Then G admits a 1-planar RAC drawing Γ with at most one bend per edge. Also, if a 1-planar embedding of G is given as part of the input, Γ can be computed in $O(n)$ time.*

We remark that a characterization of the 1-planar graphs that can be drawn with straight-line edges was given by Thomassen in 1988 [37]. The characterization is described in terms of the existence of a 1-planar embedding that does not contain two primitive forbidden configurations. This result immediately implies that every 1-planar graph admits a 1-planar drawing with at most one bend per edge (which is not necessarily RAC); it is sufficient to subdivide each crossing edge of any given 1-planar embedding with a dummy vertex, so to remove any possible forbidden configuration. Dummy vertices will correspond to bends in the final drawing. Moreover, Alam et al. [2], proved that every 3-connected 1-plane graph can be drawn with straight-line edges, except for at most one edge that may require one bend. We also remark that straight-line RAC drawings always exist for *IC-planar graphs* [9], a subclass of 1-planar graphs.

Some proofs and technicalities are omitted and can be found in [16].

2   Preliminaries

We assume familiarity with basic terminology of graph drawing [11]. In the following we only consider *simple* drawings of graphs, i.e., drawings where two edges have at most one point in common (which is either a common endpoint or a common interior point where the two edges properly cross each other). A *k-bend drawing* of a graph is a drawing where each edge is represented as a polyline with at most $k > 0$ bends. A graph G is *planar* if it admits a planar (i.e., crossing-free) drawing. Such a drawing subdivides the plane into topologically connected regions, called *faces*. The infinite region is the *outer face*. The number of vertices encountered in the closed walk along the boundary of a face f is the *degree* of f. If G is not 2-connected a vertex may be encountered more than once, thus contributing with more than one unit to the degree of f. A *planar embedding* of G is an equivalence class of planar drawings of G having the same set of faces. A *plane graph* is a planar graph with a given planar embedding.

The concept of planar embedding can be extended to non-planar drawings. Given a non-planar drawing Γ, interpret every crossing as a vertex. The resulting planarized drawing has a planar embedding. An *embedding* of a (non-planar) graph G is an equivalence class of drawings whose planarized versions have the same planar embedding. A *1-plane* graph is a 1-planar graph with a given *1-planar embedding*, i.e., an embedding where each edge is crossed at most once. Each face of a 1-planar embedding is composed of both vertices and/or crossings, and its degree is the number of vertices or crossings encountered in the closed walk along its boundary. A *kite* K is a 1-plane graph isomorphic to K_4 with an embedding such that all the vertices are on the boundary of the outer face, the four edges on the boundary are crossing-free, and the remaining two edges cross each other. Given a 1-plane graph G and a kite $K = \{a, b, c, d\}$, such that $K \subseteq G$, we say that K is *empty* if it does not contain any vertex of G inside the 4-cycle $\{a, b, c, d\}$ (it contains only the crossing point). A pair of crossing edges of G *forms an empty kite* if their four end-vertices induce an empty kite. A 1-plane graph G, possibly containing parallel edges, is *triangulated* if each face is a triangle, formed by either three vertices or by one crossing and two vertices. Clearly, a triangulated 1-plane graph is 2-connected. The next observation follows from the definition of a triangulated 1-plane graph.

Observation 1. *Let G be a triangulated 1-plane graph. Every pair of crossing edges of G forms an empty kite, except for at most one pair of crossing edges if their crossing point is on the outer face of G.*

3   1-Bend RAC Drawings of 1-Planar Graphs

To prove Theorem 1 we give an algorithm that takes as input a simple 1-plane graph G with n vertices (see, e.g., Fig. 1(a)), and computes a 1-bend 1-planar RAC drawing Γ of G in $O(n)$ time. We assume that G is connected, as otherwise we can draw independently each connected component. The high-level idea is as follows. Augment G and modify its embedding to get a triangulated 1-plane graph, possibly containing parallel edges. Execute a suitable decomposition of the triangulated 1-plane graph and apply a recursive technique that computes a 1-bend 1-planar RAC drawing. Remove dummy vertices and edges.

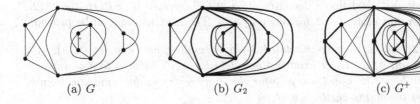

(a) G (b) G_2 (c) G^+

Fig. 1. Illustration for the augmentation step.

Augmentation. The first step of the algorithm transforms G into a triangulated 1-plane graph G^+ by adding edges and vertices. The 1-planar embedding of G^+ may be different from that of G for the common part. Let (a, c) and (b, d) be two edges of G that cross in a point p. Let $\{a, b, c, d\}$ be the circular order of the vertices around p. For each such pair of crossing edges, we add an edge (a, b), and draw[1] it such that it follows the curves (a, p) and (p, b). Similarly, we draw the three edges (b, c), (c, d) and (d, a). This operation ensures that each pair of crossing edges forms an empty kite. Also, this operation does not introduce edge crossings but it may create parallel edges. We denote by G_1 the resulting (multi)graph. For each pair of parallel edges e and e' of G_1, such that $e \in G$ and $e' \in G_1$, we remove e from G_1. This immediately implies that no parallel edge is crossed in G_1. We then remove one edge for each pair of parallel edges e_1 and e_2 such that the curve $e_1 \cup e_2$ does not contain any vertex in its interior. We let G_2 be the resulting graph, which can be easily computed in $O(n)$ time, since G has $O(n)$ crossings (see, e.g., [36]). Figure 1(b) shows the graph G_2 obtained from the graph G of Fig. 1(a). We remark that a similar operation has been used by Alam et al. [2] in order to compute a straight-line drawable 1-planar embedding of a 3-connected 1-planar graph. However, only 3-connected graphs are considered by Alam et al., and in this case the augmented graph does not contain parallel edges [2]. We do not have any restriction on the connectivity of G, which poses additional issues in the construction and in the drawing of a suitable 1-planar embedding. We transform G_2 into a triangulated 1-plane graph. Note that a face of degree two consists of two parallel edges, thus only the outer face of G_2 can have degree two. In this case, each of the two parallel edges is part of an empty kite. Thus, we remove one of these two edges to make the degree of the outer face equal to three (it will be formed by two vertices and one crossing). Let f be an inner face of G_2 that is not a triangle. Such a face contains no crossings on its boundary, since each crossing is shared by exactly four triangular faces by the empty kite property. We add an *extra vertex* v_f inside f and connect it to all vertices (with multiplicity) on the boundary of f. Figure 1(c) shows the graph G^+ obtained from the graph G_2 in Fig. 1(b), extra vertices are drawn as squares. Since G_2 has $O(n)$ faces, G^+ has $O(n)$ vertices and edges, and it is computed in $O(n)$ time. The next lemma follows from the above discussion.

Lemma 1. *Graph* G^+ *is a triangulated 1-plane (multi)graph.*

Decomposition. We define a decomposition of G^+ inspired by $SPQR$-trees [12], but simpler and more direct for our purposes. The next lemma can be proved.

Lemma 2. *Let* G *be a triangulated 1-plane (multi)graph and let* $\{u, v\}$ *be a separation pair of* G. *There exist two parallel edges* e, e' *incident to* u *and* v *such that* $\{u, v\}$ *is not a separation pair for the graph obtained by removing from* G *all vertices inside the cycle* $\{e, u, e', v\}$.

---

[1] For ease of description, here we are interpreting an embedding as a drawing.

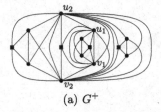 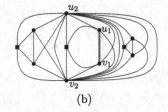 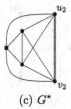

(a) G^+ (b)   (c) G^*

Fig. 2. Illustration for the decomposition step. Thick edges are thicker (and red). (Color figure online)

By Lemma 2, for each separation pair $\{u, v\}$ of G^+, there exist $k > 1$ parallel edges $\{e_1, \ldots, e_k\}$ between u and v, such that the cycle $\{u, e_1, v, e_k\}$ encloses all other copies in its interior. We call the *inner graph* of (u, v) the subgraph G_{uv} of G^+ whose outer face is $\{u, e_1, v, e_k\}$, and an *inner component* of (u, v) each subgraph C_{uv}^i of G_{uv} whose outer face is $\{u, e_i, v, e_{i+1}\}$, for $i = 1, \ldots, k-1$. Let G_{uv} be an inner graph of G^+ that does not contain any inner graph as a subgraph. Replace G_{uv} with an edge between u and v, called *thick edge*; the resulting graph is still a triangulated 1-plane graph. Iterate this procedure until there are no more inner graphs to be replaced. This is done in $O(n)$ time and results in a simple triangulated 1-plane graph G^*, which is 3-connected by Lemma 2. Figure 2(c) shows the graph G^* obtained from the graph G^+ in Fig. 2(a), through the intermediate step in Fig. 2(b). The next lemma follows.

Lemma 3. *Graph G^* is a simple 3-connected triangulated 1-plane graph.*

Drawing. The overview of the drawing algorithm is as follows. Start with a 1-bend 1-planar RAC drawing of G^*, and then recursively replace thick edges with a 1-bend 1-planar RAC drawing of the corresponding inner graphs. Deleting the edges and vertices added by the augmentation step we get a 1-bend 1-planar RAC drawing of G. To compute a 1-bend 1-planar RAC drawing of G^*, first remove from G^* all pairs of crossing edges and denote by H^* the resulting plane graph (see Fig. 3(a)). Note that thick edges are never crossed by construction, and all faces of H^* have either degree 3 or degree 4. We can prove the following.

Lemma 4. *Graph H^* is 3-connected.*

Compute a planar straight-line drawing γ^* of H^* where all faces are strictly convex and the outer face is a prescribed polygon P; this can be done by applying the linear-time algorithm by Chiba et al. [10] (see Fig. 3(b)). If the outer face of H^* has degree four, we let P be a trapezoid, else P is a triangle. Since all faces are either triangles or quadrangles, we can avoid three collinear vertices by slight perturbations (which cannot cause a face to become non convex). To reinsert the crossing edges, we distinguish between the inner faces and the outer face of H^*. Two crossing edges can be easily reinserted in an inner face, just drawing one of the two with no bend and the other with one bend, such that they cross at right angles (see, e.g., [3] and Fig. 3(c)). To reinsert two crossing edges e_1, e_2 in

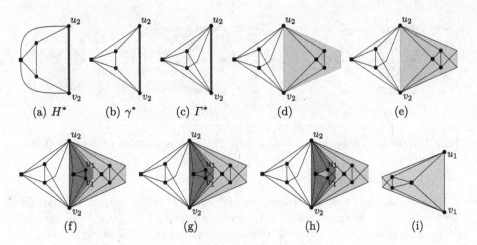

Fig. 3. Illustration for the drawing step. (Color figure online)

the outer face of H^* so that they form a right angle, we can draw e_1 and e_2 with one bend each. Namely, P is a trapezoid by construction. Assume that the minor base m and the greater base M of P are aligned with the horizontal axis. The first segment of e_1 is such that its rightmost endpoint p_1 coincides with the rightmost endpoint of m, and its leftmost endpoint q_1 is b units above the leftmost endpoint of m, where b is equal to the length of m. The second segment of e_1 has q_1 as rightmost endpoint, and its leftmost endpoint r_1 coincides with the leftmost endpoint of M. Edge e_2 is drawn symmetrically.

Consider now a thick edge (u, v) of G^* and its inner graph G_{uv}. Recall that G_{uv} consists of $k - 1 \geq 1$ inner components $C_1, \ldots, C_{k-1}$. Each C_i $(i = 1, \ldots, k-1)$ has two parallel edges e_i, e_{i+1} as outer face. Also, analogously to G^*, $C_i^- = C_i \setminus \{e_{i+1}\}$ is a simple 3-connected triangulated 1-plane graph (it is a subgraph of G^+ and all its inner graphs have been replaced by thick edges). Remove all crossing edges of C_{k-1}^- and let H_{k-1}^- be the resulting 3-connected plane graph. Compute a planar straight-line drawing γ_{k-1} of H_{k-1}^- such that all faces are strictly convex polygons and the outer face is a prescribed polygon P. If the outer face of H_{k-1}^- has degree three, P is a triangle whose side with corners u and v has length equal to the length of the thick edge (u, v) in Γ^*, and its height is small enough so that the thick edge (u, v) can be replaced with P without introducing crossings. If the outer face of H_{k-1}^- has degree four, P is a trapezoid such that its greater base has u and v as corners and the same length as the thick edge (u, v) in Γ^*. The height of P is such that the thick edge (u, v) can be replaced with P without introducing crossings. Also, the minor base of P is sufficiently short so that the pair of crossing edges on the outer face of H_{k-1}^- can be reinserted without introducing crossings in Γ^*, as described for H^* (see Fig. 3(d)). By the same argument used for H^*, all pairs of crossing edges can be reinserted so as to form right angle crossings and have at most one bend each (see Fig. 3(e)). If $k - 1 > 1$, we iterate this procedure and compute a drawing Γ_i^-

for each C_i^-, for $i = k-2, \ldots, 1$. The polygon representing the outer face of each Γ_i can be suitably chosen so to fit inside the face containing edge e_{i+1} of drawing of Γ_{i+1}. The union of all such drawings is a 1-bend 1-planar RAC drawing Γ_{uv} of G_{uv} (see Figs. 3(f) and 3(g)), with the exception of some parallel edges. Namely, the parallel edges $e_1, \ldots, e_k$ are represented by overlapping segments between u and v, and for our needs all of them but one can be removed from the drawing.

Repeat this procedure for each thick edge of G^*, and recursively apply the same technique for each inner graph of G^*; see Figs. 3(h) and 3(i) for a complete illustration. The resulting drawing Γ is a 1-bend 1-planar RAC drawing of G^+ (except for some parallel edges). Removing dummy vertices and edges, we get the desired drawing of G. In terms of time complexity, each planar straight-line drawing with (strictly) convex faces is computed in linear time in the size of the input graph [10], and in linear time we can reinsert the crossing edges. Thus the whole procedure takes $O(n)$ time. This concludes the proof of Theorem 1.

4   Conclusions and Open Problems

We proved that every 1-planar graph admits a 1-planar RAC drawing with at most one bend per edge. The proof is constructive and based on a drawing algorithm, which may produce 1-bend 1-planar RAC drawings with exponential area: Is this area requirement necessary for some 1-planar graphs? Also, our algorithm may change the embedding of the input graph: Are there 1-planar embeddings that are not realizable as 1-bend RAC drawings? Characterizing straight-line 1-planar RAC drawable graphs is also an interesting problem.

References

1. Ackerman, E.: A note on 1-planar graphs. Discrete Appl. Math. **175**, 104–108 (2014). http://dx.doi.org/10.1016/j.dam.2014.05.025
2. Alam, M.J., Brandenburg, F.J., Kobourov, S.G.: Straight-line grid drawings of 3-connected 1-planar graphs. In: Wismath, S., Wolff, A. (eds.) GD 2013. LNCS, vol. 8242, pp. 83–94. Springer, Heidelberg (2013). doi:10.1007/978-3-319-03841-4_8
3. Angelini, P., Cittadini, L., Didimo, W., Frati, F., Battista, G.D., Kaufmann, M., Symvonis, A.: On the perspectives opened by right angle crossing drawings. J. Graph Algorithms Appl. **15**(1), 53–78 (2011)
4. Argyriou, E.N., Bekos, M.A., Symvonis, A.: Maximizing the total resolution of graphs. Comput. J. **56**(7), 887–900 (2013)
5. Arikushi, K., Fulek, R., Keszegh, B., Morić, F., Tóth, C.D.: Graphs that admit right angle crossing drawings. Comput. Geom. **45**(4), 169–177 (2012)
6. Bekos, M.A., van Dijk, T.C., Kindermann, P., Wolff, A.: Simultaneous drawing of planar graphs with right-angle crossings and few bends. J. Graph Algorithms Appl. **20**(1), 133–158 (2016). http://dx.doi.org/10.7155/jgaa.00388
7. Biedl, T.C., Liotta, G., Montecchiani, F.: On visibility representations of non-planar graphs. In: Fekete, S.P., Lubiw, A. (eds.) SoCG 2016. LIPIcs, vol. 51, pp. 19:1–19:16. Schloss Dagstuhl - Leibniz-Zentrum fuer Informatik (2016). http://www.dagstuhl.de/dagpub/978-3-95977-009-5

8. Brandenburg, F.J.: 1-visibility representations of 1-planar graphs. J. Graph Algorithms Appl. **18**(3), 421–438 (2014)
9. Brandenburg, F.J., Didimo, W., Evans, W.S., Kindermann, P., Liotta, G., Montecchiani, F.: Recognizing and drawing IC-planar graphs. Theor. Comput. Sci. **636**, 1–16 (2016). http://dx.doi.org/10.1016/j.tcs.2016.04.026
10. Chiba, N., Yamanouchi, T., Nishizeki, T.: Linear algorithms for convex drawings of planar graphs. In: Progress in Graph Theory, pp. 153–173 (1984)
11. Di Battista, G., Eades, P., Tamassia, R., Tollis, I.G.: Graph Drawing. Prentice Hall, Upper Saddle River (1999)
12. Di Battista, G., Tamassia, R.: On-line planarity testing. SIAM J. Comput. **25**(5), 956–997 (1996)
13. Di Giacomo, E., Didimo, W., Eades, P., Liotta, G.: 2-layer right angle crossing drawings. Algorithmica **68**(4), 954–997 (2014)
14. Di Giacomo, E., Didimo, W., Grilli, L., Liotta, G., Romeo, S.A.: Heuristics for the maximum 2-layer RAC subgraph problem. Comput. J. **58**(5), 1085–1098 (2015)
15. Di Giacomo, E., Didimo, W., Liotta, G., Montecchiani, F.: Area requirement of graph drawings with few crossings per edge. Comput. Geom. **46**(8), 909–916 (2013)
16. Didimo, W., Liotta, G., Mehrabi, S., Montecchiani, F.: 1-Bend RAC Drawings of 1-Planar Graphs. ArXiv e-prints abs/1608.08418 (2016). http://arxiv.org/abs/1608.08418v1
17. Didimo, W.: Density of straight-line 1-planar graph drawings. Inf. Process. Lett. **113**(7), 236–240 (2013)
18. Didimo, W., Eades, P., Liotta, G.: Drawing graphs with right angle crossings. Theor. Comput. Sci. **412**(39), 5156–5166 (2011)
19. Didimo, W., Liotta, G.: The crossing angle resolution in graph drawing. In: Pach, J. (ed.) Thirty Essays on Geometric Graph Theory. Springer, New York (2012)
20. Didimo, W., Liotta, G., Romeo, S.A.: Topology-driven force-directed algorithms. In: Brandes, U., Cornelsen, S. (eds.) GD 2010. LNCS, vol. 6502, pp. 165–176. Springer, Heidelberg (2011). doi:10.1007/978-3-642-18469-7_15
21. Didimo, W., Liotta, G., Romeo, S.A.: A graph drawing application to web site traffic analysis. J. Graph Algorithms Appl. **15**(2), 229–251 (2011)
22. Eades, P., Liotta, G.: Right angle crossing graphs and 1-planarity. Discrete Appl. Math. **161**(7–8), 961–969 (2013)
23. Grigoriev, A., Bodlaender, H.L.: Algorithms for graphs embeddable with few crossings per edge. Algorithmica **49**(1), 1–11 (2007)
24. Hong, S.-H., Eades, P., Liotta, G., Poon, S.-H.: Fáry's theorem for 1-planar graphs. In: Gudmundsson, J., Mestre, J., Viglas, T. (eds.) COCOON 2012. LNCS, vol. 7434, pp. 335–346. Springer, Heidelberg (2012). doi:10.1007/978-3-642-32241-9_29
25. Huang, W.: Using eye tracking to investigate graph layout effects. In: APVIS 2007, pp. 97–100 (2007)
26. Huang, W., Eades, P., Hong, S.: Larger crossing angles make graphs easier to read. J. Vis. Lang. Comput. **25**(4), 452–465 (2014)
27. Huang, W., Eades, P., Hong, S., Lin, C.: Improving force-directed graph drawings by making compromises between aesthetics. In: VL/HCC, pp. 176–183. IEEE (2010)
28. Huang, W., Hong, S.H., Eades, P.: Effects of crossing angles. In: PacificVis 2008, pp. 41–46. IEEE (2008)
29. Korzhik, V.P., Mohar, B.: Minimal obstructions for 1-immersions and hardness of 1-planarity testing. J. Graph Theory **72**(1), 30–71 (2013)

30. Nguyen, Q., Eades, P., Hong, S.-H., Huang, W.: Large crossing angles in circular layouts. In: Brandes, U., Cornelsen, S. (eds.) GD 2010. LNCS, vol. 6502, pp. 397–399. Springer, Heidelberg (2011). doi:10.1007/978-3-642-18469-7_40

31. Pach, J., Tóth, G.: Graphs drawn with few crossings per edge. Combinatorica 17(3), 427–439 (1997)

32. Purchase, H.: Which aesthetic has the greatest effect on human understanding? In: DiBattista, G. (ed.) GD 1997. LNCS, vol. 1353, pp. 248–261. Springer, Heidelberg (1997). doi:10.1007/3-540-63938-1_67

33. Purchase, H.C.: Effective information visualisation: a study of graph drawing aesthetics and algorithms. Interact. Comput. 13(2), 147–162 (2000)

34. Purchase, H.C., Carrington, D.A., Allder, J.: Empirical evaluation of aesthetics-based graph layout. Empirical Softw. Eng. 7(3), 233–255 (2002)

35. Ringel, G.: Ein Sechsfarbenproblem auf der Kugel. Abh. Math. Semin. Univ. Hambg. 29(1–2), 107–117 (1965)

36. Suzuki, Y.: Re-embeddings of maximum 1-planar graphs. SIAM J. Discrete Math. 24(4), 1527–1540 (2010)

37. Thomassen, C.: Rectilinear drawings of graphs. J. Graph Theory 12(3), 335–341 (1988)

38. Ware, C., Purchase, H.C., Colpoys, L., McGill, M.: Cognitive measurements of graph aesthetics. Inf. Vis. 1(2), 103–110 (2002)

On the Density of Non-simple 3-Planar Graphs

Michael A. Bekos[1]([✉]), Michael Kaufmann[1], and Chrysanthi N. Raftopoulou[2]

[1] Institut für Informatik, Universität Tübingen, Tubingen, Germany
{bekos,mk}@informatik.uni-tuebingen.de
[2] School of Applied Mathematics and Physical Sciences, NTUA, Athens, Greece
crisraft@mail.ntua.gr

Abstract. A *k-planar graph* is a graph that can be drawn in the plane such that every edge is crossed at most k times. For $k \leq 4$, Pach and Tóth [20] proved a bound of $(k+3)(n-2)$ on the total number of edges of a k-planar graph, which is tight for $k = 1, 2$. For $k = 3$, the bound of $6n - 12$ has been improved to $\frac{11}{2}n - 11$ in [19] and has been shown to be optimal up to an additive constant for simple graphs. In this paper, we prove that the bound of $\frac{11}{2}n - 11$ edges also holds for non-simple 3-planar graphs that admit drawings in which non-homotopic parallel edges and self-loops are allowed. Based on this result, a characterization of *optimal 3-planar graphs* (that is, 3-planar graphs with n vertices and exactly $\frac{11}{2}n - 11$ edges) might be possible, as to the best of our knowledge the densest known simple 3-planar is not known to be optimal.

1 Introduction

Planar graphs play an important role in graph drawing and visualization, as the avoidance of crossings and occlusions is central objective in almost all applications [10,18]. The theory of planar graphs [15] could be very nicely applied and used for developing great layout algorithms [13,22,23] based on the planarity concepts. Unfortunately, real-world graphs are usually not planar despite of their sparsity. With this background, an initiative has formed in recent years to develop a suitable theory for *nearly planar graphs*, that is, graphs with various restrictions on their crossings, such as limitations on the number of crossings per edge (e.g., k-planar graphs [21]), avoidance of local crossing configurations (e.g., quasi planar graphs [2], fan-crossing free graphs [9], fan-planar graphs [17]) or restrictions on the crossing angles (e.g., RAC graphs [11], LAC graphs [12]). For precise definitions, we refer to the literature mentioned above.

The most prominent is clearly the concept of k-planar graphs, namely graphs that allow drawings in the plane such that each edge is crossed at most k times by other edges. The simplest case $k = 1$, i.e., 1-planar graphs [21], has been subject of intensive research in the past and it is quite well understood, see e.g. [4,6–8,14,20]. For $k \geq 2$, the picture is much less clear. Only few papers on special cases appeared, see e.g., [3,16].

This work has been supported by DFG grant Ka812/17-1.

Y. Hu and M. Nöllenburg (Eds.): GD 2016, LNCS 9801, pp. 344–356, 2016.
DOI: 10.1007/978-3-319-50106-2_27

Pach and Tóth's paper [20] stands out and contributed a lot to the understanding of nearly planar graphs. The paper considers the number of edges in simple k-planar graphs for general k. Note the well-known bound of $3n - 6$ edges for planar graphs deducible from Euler's formula. For small $k = 1, 2, 3$ and 4, bounds of $4n - 8$, $5n - 10$, $6n - 12$ and $7n - 14$ respectively, are proven which are tight for $k = 1$ and $k = 2$. This sequence seems to suggest a bound of $O(kn)$ for general k, but Pach and Tóth also gave an upper bound of $4.1208\sqrt{k}n$. Unfortunately, this bound is still quite large even for medium k (for $k = 9$, it gives $12.36n$). Meanwhile for $k = 3$ and $k = 4$, the bounds above have been improved to $5.5n - 11$ and $6n - 12$ in [19] and [1], respectively. In this paper, we prove that the bound on the number of edges for $k = 3$ also holds for non-simple 3-planar graphs that do not contain homotopic parallel edges and homotopic self-loops. Our extension required substantially different approaches and relies more on geometric techniques than the more combinatorial ones given in [19] and [1]. We believe that it might also be central for the characterization of *optimal* 3-planar graphs (that is, 3-planar graphs with n vertices and exactly $\frac{11}{2}n - 11$ edges), since the densest known simple 3-planar graph has only $\frac{11n}{2} - 15$ edges and does not reach the known bound.

The remaining of this paper is structured as follows: Some definitions and preliminaries are given in Sect. 2. In Sects. 3 and 4, we give significant insights in structural properties of 3-planar graphs in order to prove that 3-planar graphs on n vertices cannot have more than $\frac{11}{2}n - 11$ edges. We conclude in Sect. 5 with open problems.

2   Preliminaries

A *drawing* of a graph G is a representation of G in the plane, where the vertices of G are represented by distinct points and its edges by Jordan curves joining the corresponding pairs of points, so that: (i) no edge passes through a vertex different from its endpoints, (ii) no edge crosses itself and (iii) no two edges meet tangentially. In the case where G has multi-edges, we will further assume that both the bounded and the unbounded closed regions defined by any pair of self-loops or parallel edges of G contain at least one vertex of G in their interior. Hence, the drawing of G has no *homotopic* edges. In the following when referring to 3-planar graphs we will mean that non-homotopic edges are allowed in the corresponding drawings. We call such graphs *non-simple*.

Following standard naming conventions, we refer to a 3-planar graph with n vertices and maximum possible number of edges as *optimal 3-planar*. Let H be an optimal 3-planar graph on n vertices together with a corresponding 3-planar drawing $\Gamma(H)$. Let also H_p be a subgraph of H with the largest number of edges, such that in the drawing of H_p (that is inherited from $\Gamma(H)$) no two edges cross each other. We call H_p a *maximal planar substructure* of H. Among all possible optimal 3-planar graphs on n vertices, let $G = (V, E)$ be the one with the following two properties: (a) its maximal planar substructure, say $G_p = (V, E_p)$, has maximum number of edges among all possible planar substructures of all optimal 3-planar graphs, (b) the number of crossings in the drawing of G is minimized over all optimal 3-planar graphs subject to (a). We refer to G as *crossing-minimal optimal* 3-*planar graph*.

With slight abuse of notation, let $G - G_p$ be obtained from G by removing only the edges of G_p and let e be an edge of $G - G_p$. Since G_p is maximal, edge e must cross at least one edge of G_p. We refer to the part of e between an endpoint of e and the nearest crossing with an edge of G_p as *stick*. The parts of e between two consecutive crossings with G_p are called *middle parts*. Clearly, e consists of exactly 2 sticks and 0, 1, or 2 middle parts. A stick of e lies completely in a face of G_p and crosses at most two other edges of $G - G_p$ and an edge of this particular face. A stick of e is called *short*, if there is a walk along the face boundary from the endpoint of the stick to the nearest crossing point with G_p, which contains only one other vertex of the face boundary. Otherwise, the stick of e is called *long*; see Fig. 1a. A middle part of e also lies in a face of G_p. We say that e *passes through* a face of G_p, if there exists a middle part of e that completely lies in the interior of this particular face. We refer to a middle part of an edge that crosses consecutive edges of a face of G_p as *short middle part*. Otherwise, we call it *far middle part*.

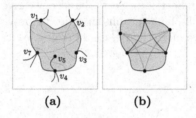

Fig. 1. (a) Illustration of a non-simple face $\{v_1, v_2, \ldots, v_7\}$; v_6 is identified with v_4. The sticks from v_1 and v_2 are short, while the one from v_7 is long. All other edge segments are middle-parts. (b) The case, where two triangles of type $(3, 0, 0)$ are associated to the same triangle.

Let $\mathcal{F}_s = \{v_1, v_2, \ldots, v_s\}$ be a face of G_p with $s \geq 3$. The order of the vertices (and subsequently the order of the edges) of $\mathcal{F}_s$ is determined by a walk around the boundary of $\mathcal{F}_s$ in clockwise direction. Since $\mathcal{F}_s$ is not necessarily simple, a vertex (or an edge, respectively) may appear more than once in this order; see Fig. 1a. We say that $\mathcal{F}_s$ is of type $(\tau_1, \tau_2, \ldots, \tau_s)$ if for each $i = 1, 2, \ldots, s$ vertex v_i is incident to τ_i sticks of $\mathcal{F}_s$ that lie between (v_{i-1}, v_i) and (v_i, v_{i+1})[1].

Lemma 1 (Pach and Tóth [20]). *A triangular face of G_p contains at most 3 sticks.*

Proof. Consider a triangular face $\mathcal{T}$ of G_p of type (τ_1, τ_2, τ_3). Clearly, $\tau_1, \tau_2, \tau_3 \leq 3$, as otherwise an edge of G_p has more than three crossings. Since a stick of $\mathcal{T}$ cannot cross more than two other sticks of $\mathcal{T}$, it follows that $\tau_1 + \tau_2 + \tau_3 \leq 3$. $\square$

---

[1] In the remainder of the paper, all indices are subject to $(mod\ s) + 1$.

3   The Density of Non-simple 3-Planar Graphs

Let $G = (V, E)$ be a crossing-minimal optimal 3-planar graph with n vertices drawn in the plane. Let also $G_p = (V, E_p)$ be the maximal planar substructure of G. In this section, we will prove that G cannot have more than $\frac{11n}{2} - 11$ edges, assuming that G_p is fully triangulated, i.e., $|E_p| = 3n - 6$. This assumption will be proved in Sect. 4. Next, we prove that the number of triangular faces of G_p with exactly 3 sticks cannot be larger than those with at most 2 sticks.

Lemma 2. *We can uniquely associate each triangular face of G_p with 3 sticks to a neighboring triangular face of G_p with at most 2 sticks.*

Proof. Let $\mathcal{T} = \{v_1, v_2, v_3\}$ be a triangular face of G_p. By Lemma 1, we have to consider three types for $\mathcal{T}$: $(3,0,0)$, $(2,1,0)$ and $(1,1,1)$.

- $\mathcal{T}$ *is of type* $(3,0,0)$: Since v_1 is incident to 3 sticks of $\mathcal{T}$, edge (v_2, v_3) is crossed three times. Let $\mathcal{T}'$ be the triangular face of G_p neighboring $\mathcal{T}$ along (v_2, v_3). We have to consider two cases: (a) one of the sticks of $\mathcal{T}$ ends at a corner of $\mathcal{T}'$, and (b) none of the sticks of $\mathcal{T}$ ends at a corner of $\mathcal{T}'$. In Case (a), the two remaining sticks of $\mathcal{T}$ might use the same or different sides of $\mathcal{T}'$ to exit it. In both subcases, it is not difficult to see that $\mathcal{T}'$ can have at most two sticks. In Case (b), we again have to consider two subcases, depending on whether all sticks of $\mathcal{T}$ use the same side of $\mathcal{T}'$ to pass through it or two different ones. In the former case, it is not difficult to see that $\mathcal{T}'$ cannot have any stick, while in the later $\mathcal{T}'$ can have at most one stick. In all aforementioned cases, we associate $\mathcal{T}$ with $\mathcal{T}'$.
- $\mathcal{T}$ *is of type* $(2,1,0)$: Since v_2 is incident to one stick of $\mathcal{T}$, edge (v_1, v_3) is crossed at least once. We associate $\mathcal{T}$ with the triangular face $\mathcal{T}'$ of G_p neighboring $\mathcal{T}$ along (v_1, v_3). Since the stick of $\mathcal{T}$ that is incident to v_2 has three crossings in $\mathcal{T}$, $\mathcal{T}'$ has no sticks emanating from v_1 or v_3. In particular, $\mathcal{T}'$ can have at most one additional stick emanating from its third vertex.
- $\mathcal{T}$ *is of type* $(1,1,1)$: This actually cannot occur. Indeed, if $\mathcal{T}$ is of type $(1,1,1)$, then all sticks of $\mathcal{T}$ have already three crossings each. Hence, the three triangular faces adjacent to $\mathcal{T}$ define a 6-gon in G_p, which contains only six interior edges. So, we can easily remove them and replace them with 8 interior edges (see, e.g., Fig. 1b), contradicting thus the optimality of G.

Note that our analysis also holds for non-simple triangular faces. We now show that the assignment is unique. This holds for triangular faces of type $(2,1,0)$, since a triangular face that is associated with one of type $(2,1,0)$ cannot contain two sides each with two crossings, which implies that it cannot be associated with another triangular face with three sticks. This leaves only the case that two $(3,0,0)$ triangles are associated with the same triangle $\mathcal{T}'$ (see, e.g., the triangle with the gray-colored edges in Fig. 1b). In this case, there exists another triangular face (bottommost in Fig. 1b), which has exactly two sticks because of 3-planarity. In addition, this face cannot be associated with some other triangular face. Hence, one of the two type-$(3,0,0)$ triangular faces associated with $\mathcal{T}'$ can be assigned to this triangular face instead resolving the conflict.   □

We are now ready to prove the main theorem of this section.

Theorem 1. *A 3-planar graph of n vertices has at most $\frac{11}{2}n - 11$ edges, which is a tight bound.*

Proof. Let t_i be the number of triangular faces of G_p with exactly i sticks, $0 \leq i \leq 3$. The argument starts by counting the number of triangular faces of G_p with exactly 3 sticks. From Lemma 2, we conclude that the number t_3 of triangular faces of G_p with exactly 3 sticks is at most as large as the number of triangular faces of G_p with 0, 1 or 2 sticks. Hence $t_3 \leq t_0 + t_1 + t_2$. We conclude that $t_3 \leq t_p/2$, where t_p denotes the number of triangular faces in G_p, since $t_0 + t_1 + t_2 + t_3 = t_p$. Note that by Euler's formula $t_p = 2n - 4$. Hence, $t_3 \leq n - 2$. Thus, we have: $|E| - |E_p| = (t_1 + 2t_2 + 3t_3)/2 = (t_1 + t_2 + t_3) + (t_3 - t_1)/2 = (t_p - t_0) + (t_3 - t_1)/2 \leq t_p + t_3/2 \leq 5t_p/4$. So, the total number of edges of G is at most: $|E| \leq |E_p| + 5t_p/4 \leq 3n - 6 + 5(2n - 4)/4 = 11n/2 - 11$. In [5] we prove that our bound is tight by a construction similar to the one of Pach et al. [19]. $\square$

4 The Density of the Planar Substructure

Let $G = (V, E)$ be a crossing-minimal optimal 3-planar graph with n vertices drawn in the plane. Let also $G_p = (V, E_p)$ be the maximal planar substructure of G. In this section, we will prove that G_p is fully triangulated, i.e., $|E_p| = 3n - 6$ (see Theorem 2). To do so, we will explore several structural properties of G_p (see Lemmas 3–13), assuming that G_p has at least one non-triangular face, say $\mathcal{F}_s = \{v_1, v_2, \ldots, v_s\}$ with $s \geq 4$. In the first observations, we do not require that G_p is connected. This is proved in Lemma 6. Recall that in general $\mathcal{F}_s$ is not necessarily simple, which means that a vertex may appear more than once along $\mathcal{F}_s$. Our goal is to contradict either the *optimality* of G (that is, the fact that G contains the maximum number of edges among all 3-planar graphs with n vertices) or the *maximality* of G_p (that is, the fact that G_p has the maximum number of edges among all planar substructures of all optimal 3-planar graphs with n vertices) or the *crossing minimality* of G (that is, the fact that G has the minimum number of crossings subject to the size of the planar substructure).

Lemma 3. *Let $\mathcal{F}_s = \{v_1, v_2, \ldots, v_s\}$, $s \geq 4$ be a non-triangular face of G_p. Then, each stick of $\mathcal{F}_s$ is crossed at least once within $\mathcal{F}_s$.*

Proof (Sketch). Assume to the contrary that there exists a stick of $\mathcal{F}_s$ that is not crossed within $\mathcal{F}_s$. W.l.o.g. let (v_1, v_1') be the edge containing this stick and assume that (v_1, v_1') emanates from vertex v_1 and leads to vertex v_1' by crossing the edge (v_i, v_{i+1}) of $\mathcal{F}_s$. We initially prove that $i + 1 = s$. Next, we show that there exist two edges e_1 and e_2 which cross (v_i, v_{i+1}) and are not sticks emanating from v_1. The desired contradiction follows from the observation that we can remove edges e_1, e_2 and (v_1, v_1') from G and replace them with the chord (v_1, v_{s-1}) and two additional edges that are both sticks either at v_1 or at v_s. In this way, a new graph is obtained, whose maximal planar substructure has more edges than G_p, which contradicts the maximality of G_p. The detailed proof is given in [5]. $\square$

Lemma 4. *Let* $\mathcal{F}_s = \{v_1, v_2, \ldots, v_s\}$, $s \geq 4$ *be a non-triangular face of* G_p. *Then, each middle part of* $\mathcal{F}_s$ *is short, i.e., it crosses consecutive edges of* $\mathcal{F}_s$.

Proof. (Sketch). For a proof by contradiction, assume that (u, u') is an edge that defines a middle part of $\mathcal{F}_s$ which crosses two non-consecutive edges of $\mathcal{F}_s$, say w.l.o.g. (v_1, v_2) and (v_i, v_{i+1}), where $i \neq 2$ and $i + 1 \neq s$. We distinguish two main cases. Either (u, u') is not involved in crossings in the interior of $\mathcal{F}_s$ or (u, u') is crossed by an edge, say e, within $\mathcal{F}_s$. In both cases, it is possible to lead to a contradiction to the maximality of G_p; refer to [5] for more details. □

Lemma 5. *Let* $\mathcal{F}_s = \{v_1, v_2, \ldots, v_s\}$, $s \geq 4$ *be a non-triangular face of* G_p. *Then, each stick of* $\mathcal{F}_s$ *is short.*

Proof. Assume for a contradiction that there exists a far stick. Let w.l.o.g. (v_1, v_1') be the edge containing this stick and assume that (v_1, v_1') emanates from vertex v_1 and leads to vertex v_1' by crossing the edge (v_i, v_{i+1}) of $\mathcal{F}_s$, where $i \neq 2$ and $i + 1 \neq s$. If we can replace (v_1, v_1') either with chord (v_1, v_i) or with chord (v_1, v_{i+1}), then the maximal planar substructure of the derived graph would have more edges than G_p; contradicting the maximality of G_p. Thus, there exist two edges, say e_1 and e_2, that cross (v_i, v_{i+1}) to the left and to the right of (v_1, v_1'), respectively; see Fig. 2a. By Lemma 3, edge (v_1, v_1') is crossed by at least one other edge, say e, inside $\mathcal{F}_s$. Note that by 3-planarity edge (v_1, v_1') might also be crossed by a second edge, say e', inside $\mathcal{F}_s$. Suppose first, that (v_1, v_1') has a single crossing inside $\mathcal{F}_s$. To cope with this case, we propose two alternatives: (a) replace e_1 with chord (v_1, v_{i+1}) and make vertex v_{i+1} an endpoint of e, or (b) replace e_2 with chord (v_1, v_i) and make vertex v_i an endpoint of both e; see Figs. 2b and c, respectively. Since e and (v_i, v_{i+1}) are not homotopic, it follows that at least one of the two alternatives can be applied, contradicting the maximality of G_p.

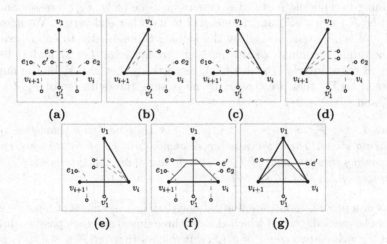

Fig. 2. Different configurations used in the proof of Lemma 5.

Consider now the case where (v_1, v_1') has two crossings inside $\mathcal{F}_s$, with edges e and e'. Similarly to the previous case, we propose two alternatives: (a) replace e_1 with chord (v_1, v_{i+1}) and make vertex v_{i+1} an endpoint of both e and e', or (b) replace e_2 with chord (v_1, v_i) and make vertex v_i an endpoint of both e and e'; see Figs. 2d and e, respectively. Note that in both alternatives the maximal planar substructure of the derived graph has more edges than G_p, contradicting the maximality of G_p. Since e and e' are not homotopic, it follows that one of the two alternatives is always applicable, as long as, e and e' are not simultaneously sticks from v_i and v_{i+1}, respectively; see Fig. 2f. In this scenario, both alternatives would lead to a situation, where (v_i, v_{i+1}) has two homotopic copies. To cope with this case, we observe that e, e' and (v_1, v_1') are three mutually crossing edges inside $\mathcal{F}_s$. We proceed by removing from G edges e_1 and e_2, which we replace by (v_1, v_i) and (v_1, v_{i+1}); see Fig. 2g. In the derived graph the maximal planar substructure contains more edges than G_p (in particular, edges (v_1, v_i) and (v_1, v_{i+1})), contradicting its maximality.   □

Lemma 6. *The planar substructure G_p of a crossing-minimal optimal 3-planar graph G is connected.*

Proof. Assume to the contrary that the maximum planar substructure G_p of G is not connected and let G_p' be a connected component of G_p. Since G is connected, there is an edge of $G - G_p$ that bridges G_p' with $G_p - G_p'$. By definition, this edge is either a stick or a passing through edge for the common face of G_p' and $G - G_p'$. In both cases, it has to be short (by Lemmas 4 and 5); a contradiction.   □

In the next two lemmas, we consider the case where a non-triangular face $\mathcal{F}_s = \{v_1, v_2, \ldots, v_s\}$, $s \geq 4$ of G_p has no sticks. Let $br(\mathcal{F}_s)$ and $\overline{br}(\mathcal{F}_s)$ be the set of bridges and non-bridges of $\mathcal{F}_s$, respectively (in Fig. 1a, edge (v_4, v_5) is a bridge). In the absence of sticks, a passing through edge of $\mathcal{F}_s$ *originates* from one of its end-vertices, crosses an edge of $\overline{br}(\mathcal{F}_s)$ to *enter* $\mathcal{F}_s$, passes through $\mathcal{F}_s$ (possibly by defining two middle parts, if it crosses an edge of $br(\mathcal{F}_s)$), crosses another edge of $\overline{br}(\mathcal{F}_s)$ to *exit* $\mathcal{F}_s$ and *terminates* to its other end-vertex. We *associate* the edge of $\overline{br}(\mathcal{F}_s)$ that is used by the passing through edge to enter (exit) $\mathcal{F}_s$ with the origin (terminal) of this passing through edge. Let $\overline{s_b}$ and s_b be the number of edges in $\overline{br}(\mathcal{F}_s)$ and $br(\mathcal{F}_s)$, respectively. Let also $\widehat{s_b}$ be the number of edges of $\overline{br}(\mathcal{F}_s)$ that are crossed by no passing through edge of $\mathcal{F}_s$. Clearly, $\widehat{s_b} \leq \overline{s_b}$ and $s = \overline{s_b} + 2s_b$.

Lemma 7. *Let $\mathcal{F}_s = \{v_1, v_2, \ldots, v_s\}$, $s \geq 4$ be a non-triangular face of G_p that has no sticks. Then, the number $\widehat{s_b}$ of non-bridges of $\mathcal{F}_s$ that are crossed by no passing through edge of $\mathcal{F}_s$ is strictly less than half the number $\overline{s_b}$ of of non-bridges of $\mathcal{F}_s$, that is, $\widehat{s_b} < \frac{\overline{s_b}}{2}$.*

Proof. For a proof by contradiction assume that $\widehat{s_b} \geq \frac{\overline{s_b}}{2}$. Since at most $\frac{\overline{s_b}}{2}$ edges of $\mathcal{F}_s$ can be crossed (each of which at most three times) and each passing through edge of $\mathcal{F}_s$ crosses two edges of $\overline{br}(\mathcal{F}_s)$, it follows that $|pt(\mathcal{F}_s)| \leq \lfloor \frac{3\overline{s_b}}{4} \rfloor$, where $pt(\mathcal{F}_s)$ denotes the set of passing through edges of $\mathcal{F}_s$. To obtain a contradiction, we remove from G all edges that pass through $\mathcal{F}_s$ and we introduce $2s - 6$ edges $\{(v_1, v_i) : 2 < i < s\} \cup \{(v_i, v_i + 2) : 2 \leq i \leq s - 2\}$ that lie completely in the interior of $\mathcal{F}_s$. This simple operation will lead to a larger graph (and therefore to a contradiction to the optimality of G) or to a graph of the same size but with larger planar substructure (and therefore to a contradiction to the maximality of G_p) as long as $s > 4$. For $s = 4$, we need a different argument. By Lemma 4, we may assume that all three passing through edges of $\mathcal{F}_s$ cross two consecutive edges of $\mathcal{F}_s$, say w.l.o.g. (v_1, v_2) and (v_2, v_3). This implies that chord (v_1, v_3) can be safely added to G; a contradiction to the optimality of G. $\square$

Lemma 8. *Let $\mathcal{F}_s = \{v_1, v_2, \ldots, v_s\}$, $s \geq 4$ be a non-triangular face of G_p. Then, $\mathcal{F}_s$ has at least one stick.*

Proof (Sketch). For a proof by contradiction, assume that $\mathcal{F}_s$ has no sticks. By Lemma 7, it follows that there exist at least two incident edges of $\overline{br}(\mathcal{F}_s)$ that are crossed by passing through edges of $\mathcal{F}_s$, say w.l.o.g. (v_s, v_1) and (v_1, v_2). Note that these two edges are not bridges of $\mathcal{F}_s$. If $s + \widehat{s_b} + 2s_b \geq 6$, then as in the proof of Lemma 7, it is possible to construct a graph that is larger than G or of equal size as G but with larger planar substructure. The same holds when $s + \widehat{s_b} + 2s_b = 5$ (that is, $s = 5$ and $\widehat{s_b} = s_b = 0$ or $s = 4$, $\widehat{s_b} = 1$ and $s_b = 0$). Both cases, contradict either the optimality of G or the maximality of G_p. The case where $s + \widehat{s_b} + 2s_b = 4$ is slightly more involved; refer to [5]. $\square$

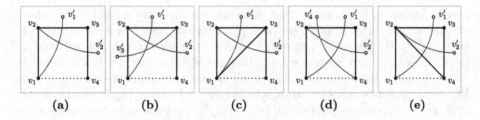

Fig. 3. Different configurations used in Lemma 9.

By Lemma 5, all sticks of $\mathcal{F}_s$ are short. A stick (v_i, v_i') of $\mathcal{F}_s$ is called *right*, if it crosses edge (v_{i+1}, v_{i+2}) of $\mathcal{F}_s$. Otherwise, stick (v_i, v_i') is called *left*. Two sticks are called *opposite*, if one is left and the other one is right.

Lemma 9. *Let $\mathcal{F}_s = \{v_1, v_2, \ldots, v_s\}$, $s \geq 4$ be a non-triangular face of G_p. Then, $\mathcal{F}_s$ has not three mutually crossing sticks.*

Proof. Suppose to the contrary that there exist three mutually crossing sticks of $\mathcal{F}_s$ and let e_i, for $i = 1, 2, 3$ be the edges containing these sticks. W.l.o.g. we assume that at least two of them are right sticks, say e_1 and e_2. Let $e_1 = (v_1, v_1')$. Then, $e_2 = (v_2, v_2')$; see Fig. 3a. Since e_1, e_2 and e_3 mutually cross, e_3 can only contain a left stick. By Lemma 5 its endpoint on $\mathcal{F}_s$ is v_3 or v_4. The first case is illustrated in Fig. 3b. Observe that (v_1, v_2) of $\mathcal{F}_s$ is only crossed by e_3. Indeed, if there was another edge crossing (v_1, v_2), then it would also cross e_1 or e_2, both of which have three crossings. Hence, e_3 can be replaced with (v_1, v_3); see Fig. 3c. The maximal planar substructure of the derived graph would have more edges than G_p, contradicting the maximality of G_p. The case where v_4 is the endpoint of e_3 on $\mathcal{F}_s$ is illustrated in Fig. 3e. Suppose that there exists an edge crossing (v_2, v_3) of $\mathcal{F}_s$ to the left of e_3. This edge should also cross e_2 or e_3, which is not possible since both edges have three crossings. So, we can replace e_3 with chord (v_2, v_4) as in Fig. 3e, contradicting the maximality of G_p.   □

Lemma 10. *Let $\mathcal{F}_s = \{v_1, v_2, \ldots, v_s\}$, $s \geq 4$ be a non-triangular face of G_p. Then, each stick of $\mathcal{F}_s$ is crossed exactly once within $\mathcal{F}_s$.*

Proof (Sketch). The detailed proof is given in [5]. By Lemma 3, each stick of $\mathcal{F}_s$ is crossed at least once within $\mathcal{F}_s$. So, the proof is given by contradiction either to the optimality of G or to the maximality of G_p, assuming the existence of a stick of $\mathcal{F}_s$ that is crossed twice within $\mathcal{F}_s$, say by edges e_1 and e_2. Note that by 3-planarity a stick of $\mathcal{F}_s$ cannot be further crossed within $\mathcal{F}_s$. First, we prove that e_1 and e_2 do not cross each other. Then, we show that e_1 and e_2 cannot be simultaneously passing through $\mathcal{F}_s$. The desired contradiction is obtained by considering two main cases: Either e_1 passes through $\mathcal{F}_s$ (and therefore, e_2 is a stick of $\mathcal{F}_s$) or both e_1 and e_2 are sticks of $\mathcal{F}_s$.   □

Lemma 11. *Let $\mathcal{F}_s = \{v_1, v_2, \ldots, v_s\}$, $s \geq 4$ be a non-triangular face of G_p. Then, there are no crossings between sticks and middle parts of $\mathcal{F}_s$.*

Proof. Assume to the contrary that there exists a stick, say of edge (v_1, v_1') that emanates from vertex v_1 of $\mathcal{F}_s$ (towards v_1'), which is crossed by a middle part of (u, u') of $\mathcal{F}_s$. By Lemma 10, this stick cannot have another crossing within $\mathcal{F}_s$. By Lemma 5, we can assume w.l.o.g. that (v_1, v_1') is a right stick, i.e., (v_1, v_1') crosses (v_2, v_3). By Lemma 4, edge (u, u') crosses two consecutive edges of $\mathcal{F}_s$. We distinguish two cases based on whether (v_1, v_1') crosses (v_s, v_1) and (v_1, v_2) of $\mathcal{F}_s$ or (v_1, v_1') crosses (v_1, v_2) and (v_2, v_3) of $\mathcal{F}_s$; see Figs. 4a and c respectively.

In the first case, we can assume w.l.o.g. that u is the vertex associated with (v_1, v_2), while u' is the one associated with (v_s, v_1). Hence, there exists an edge, say f_1, that crosses (v_1, v_2) to the right of (u, u'), as otherwise we could replace (u, u') with stick (v_2, u') and reduce the total number of crossings by one, contradicting the crossing minimality of G. Edge f_1 passes through $\mathcal{F}_s$ and also crosses

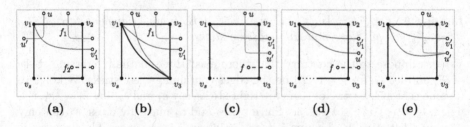

Fig. 4. Different configurations used in Lemma 11.

edge (v_2, v_3) above (v_1, v_1'). Similarly, there exists an edge f_2 that crosses (v_2, v_3) below (v_1, v_1'), as otherwise replacing (v_1, v_1') with chord (v_1, v_3) would contradict the maximality of G_p. We proceed by removing edges (u, u') and f_2 from G and by replacing them with (v_3, u) and chord (v_1, v_3); see Fig. 4b. The maximal planar substructure of the derived graph is larger than G_p; a contradiction.

In the second case, we assume that u is associated with (v_1, v_2) and u' with (v_2, v_3); see Fig. 4c. In this scenario, there exists an edge, say f, that crosses (v_2, v_3) below (v_1, v_1'), as otherwise we could replace (v_1, v_1') with chord (v_1, v_3), contradicting the maximality of G_p. If (v_1, u') does not belong to G, then we remove (u, u') from G and replace it with stick (v_1, u'); see Fig. 4d. In this way, the derived graph has fewer crossings than G; a contradiction. Note that (v_1, v_1') and (v_1, u') cannot be homotopic (if $v_1' = u'$), as otherwise edge (v_1, v_1') and (u, u') would not cross in the initial configuration. Hence, edge (v_1, u') already exists in G. In this case, f is identified with (v_1, u'); see Fig. 4e. But, in this case f is an uncrossed stick of $\mathcal{F}_s$, contradicting Lemma 3. $\qquad\square$

Lemma 12. *Let $\mathcal{F}_s = \{v_1, v_2, \ldots, v_s\}$, $s \geq 4$ be a non-triangular face of G_p. Then, any stick of $\mathcal{F}_s$ is only crossed by some opposite stick of $\mathcal{F}_s$.*

Proof. By Lemma 5, each stick of $\mathcal{F}_s$ is short. By Lemma 10, each stick of $\mathcal{F}_s$ is crossed exactly once within $\mathcal{F}_s$ and this crossing is not with a middle part due to Lemma 11. For a proof by contradiction, consider two crossing sticks that are not opposite and assume w.l.o.g. that the first stick emanates from vertex v_1 (towards vertex v_1') and crosses edge (v_2, v_3), while the second stick emanates from vertex v_2 (towards vertex v_2') and crosses edge (v_3, v_4); see Fig. 5a.

If we can replace (v_1, v_1') with the chord (v_1, v_3), then the maximal planar substructure of the derived graph would have more edges than G_p; contradicting the maximality of G_p. Thus, there exists an edge, say e, that crosses (v_2, v_3) below (v_1, v_1'). By Lemma 11, edge e is passing through $\mathcal{F}_s$. Symmetrically, we can prove that there exists an edge, say e', which crosses (v_3, v_4) right next to v_4, that is, e' defines the closest crossing point to v_4 along (v_3, v_4). Note that e' can be either a passing through edge or a stick of $\mathcal{F}_s$. We proceed by removing from G edges e' and (v_1, v_1') and by replacing them by the chord (v_2, v_4) and edge (v_4, v_1'); see Fig. 5b. The maximal planar substructure of the derived graph has more edges than G_p (in the presence of edge (v_2, v_4)), a contradiction. $\qquad\square$

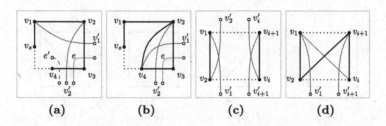

(a)   (b)   (c)   (d)

Fig. 5. Different configurations used in (a)–(b) Lemma 12 and (c)–(d) Lemma 13.

Lemma 13. *Let $\mathcal{F}_s = \{v_1, v_2, \ldots, v_s\}$, $s \geq 4$ be a non-triangular face of G_p. Then, $\mathcal{F}_s$ has exactly two sticks.*

Proof. By Lemmas 8 and 12 there exists at least one pair of opposite crossing sticks. To prove the uniqueness, assume that $\mathcal{F}_s$ has two pairs of crossing opposite sticks, say (v_1, v_1'), (v_2, v_2') and (v_i, v_i'), (v_{i+1}, v_{i+1}'), $2 < i < s$; see Fig. 5c. We remove edges (v_2, v_2') and (v_i, v_i') and replace them by (v_1, v_i) and (v_2, v_{i+1}); see Fig. 5d. By Lemmas 4 and 5, the newly introduced edges cannot be involved in crossings. The maximal planar substructure of the derived graph has more edges than G_p (in the presence of (v_1, v_i) or (v_2, v_{i+1})); a contradiction. □

We are ready to state the main theorem of this section.

Theorem 2. *The planar substructure G_p of a crossing-minimal optimal 3-planar graph G is fully triangulated.*

Proof. For a proof by contradiction, assume that G_p has a non-triangular face $\mathcal{F}_s = \{v_1, v_2, \ldots, v_s\}$, $s \geq 4$. By Lemmas 10, 12 and 13, face $\mathcal{F}_s$ has exactly two opposite sticks, that cross each other. Assume w.l.o.g. that these two sticks emanate from v_1 and v_2 (towards v_1' and v_2') and exit $\mathcal{F}_s$ by crossing (v_2, v_3) and (v_1, v_s), respectively; recall that by Lemma 5 all sticks are short; see Fig. 6a.

If we can replace (v_1, v_1') with the chord (v_1, v_3), then the maximal planar substructure of the derived graph would have more edges than G_p; contradicting the maximality of G_p. Thus, there exists an edge, say e, that crosses (v_2, v_3) below (v_1, v_1'). By Lemma 13, edge e is passing through $\mathcal{F}_s$. We consider two cases: (a) edge (v_2, v_3) is only crossed by e and (v_1, v_1'), (b) there is a third edge, say e', that crosses (v_2, v_3) (which by Lemma 13 is also passing through $\mathcal{F}_s$).

In Case (a), we can remove from G edges e and (v_1, v_1'), and replace them by (v_1, v_3) and the edge from v_2 to the endpoint of e that is below (v_3, v_4); see Fig. 6b. In Case (b), there has to be a (passing through) edge, say e'', surrounding v_4 (see Fig. 6c), as otherwise we could replace e' with a stick emanating from v_4 towards the endpoint of e' that is to the right of (v_2, v_3), which contradicts Lemma 13. We proceed by removing from G edges e'' and (v_1, v_1') and by replacing them by (v_2, v_4) and the edge from v_2 to the endpoint of e'' that is associated with (v_3, v_4); see Fig. 6d. The maximal planar substructure of the derived graph has more edges than G_p (in the presence of (v_1, v_2) in Case (a) and (v_2, v_4) in

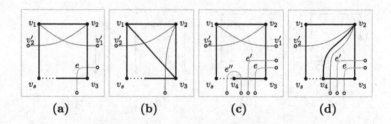

Fig. 6. Different configurations used in Theorem 2.

Case (b)), which contradicts the maximality of G_p. Since G_p is connected, there cannot exist a face consisting of only two vertices.   □

5   Discussion and Conclusion

This paper establishes a tight upper bound on the number of edges of non-simple 3-planar graphs containing no homotopic parallel edges or self-loops. Our work is towards a complete characterization of all optimal such graphs. In addition, we believe that our technique can be used to achieve better bounds for larger values of k. We demonstrate it for the case where $k = 4$, where the known bound for simple graphs is due to Ackerman [1].

If we could prove that a crossing-minimal optimal 4-planar graph $G = (V, E)$ has always a fully triangulated planar substructure $G_p = (V, E_p)$ (as we proved in Theorem 2 for the corresponding 3-planar ones), then it is not difficult to prove a tight bound on the number of edges for 4-planar graphs. Similar to Lemma 1, we can argue that no triangle of G_p has more than 4 sticks. Then, we associate each triangle of G_p with 4 sticks to a neighboring triangle with at most 2 sticks. This would imply $t_4 \leq t_1 + t_2$, where t_i denotes the number of triangles of G_p with exactly i sticks. So, we would have $|E| - |E_p| = (4t_4 + 3t_3 + 2t_2 + t_1)/2 \leq 3(t_4 + t_3 + t_2 + t_1)/2 = 3(2n - 4)/2 = 3n - 6$. Hence, the number of edges of a 4-planar graph G is at most $6n - 12$. We conclude with some open questions.

- A nice consequence of our work would be the complete characterization of optimal 3-planar graphs, as exactly those graphs that admit drawings where the set of crossing-free edges form hexagonal faces which contain 8 additional edges each
- We also believe that for simple 3-planar graphs (i.e., where even non-homotopic parallel edges are not allowed) the corresponding bound is $5.5n - 15$.
- We conjecture that the maximum number of edges of 5- and 6-planar graphs are $\frac{19}{3}n - O(1)$ and $7n - 14$, respectively.
- More generally, is there a closed function on k which describes the maximum number of edges of a k-planar graph for $k > 3$? Recall the general upper bound of $4.1208\sqrt{k}n$ by Pach and Tóth [20].

Acknowledgment. We thank E. Ackerman for bringing to our attention [1] and [19].

References

1. Ackerman, E.: On topological graphs with at most four crossings per edge. CoRR abs/1509.01932 (2015)
2. Agarwal, P.K., Aronov, B., Pach, J., Pollack, R., Sharir, M.: Quasi-planar graphs have a linear number of edges. Combinatorica **17**(1), 1–9 (1997)
3. Auer, C., Brandenburg, F.J., Gleißner, A., Hanauer, K.: On sparse maximal 2-planar graphs. In: Didimo, W., Patrignani, M. (eds.) GD 2012. LNCS, vol. 7704, pp. 555–556. Springer, Heidelberg (2013). doi:10.1007/978-3-642-36763-2_50

4. Bekos, M.A., Bruckdorfer, T., Kaufmann, M., Raftopoulou, C.: 1-planar graphs have constant book thickness. In: Bansal, N., Finocchi, I. (eds.) ESA 2015. LNCS, vol. 9294, pp. 130–141. Springer, Heidelberg (2015). doi:10.1007/978-3-662-48350-3_12

5. Bekos, M.A., Kaufmann, M., Raftopoulou, C.N.: On the density of 3-planar graphs. CoRR abs/1602.04995v3 (2016)

6. Borodin, O.V.: A new proof of the 6 color theorem. J. Graph Theory 19(4), 507–521 (1995)

7. Brandenburg, F.J.: 1-visibility representations of 1-planar graphs. J. Graph Algorithms Appl. 18(3), 421–438 (2014)

8. Brandenburg, F.J., Eppstein, D., Gleißner, A., Goodrich, M.T., Hanauer, K., Reislhuber, J.: On the density of maximal 1-planar graphs. In: Didimo, W., Patrignani, M. (eds.) GD 2012. LNCS, vol. 7704, pp. 327–338. Springer, Heidelberg (2013). doi:10.1007/978-3-642-36763-2_29

9. Cheong, O., Har-Peled, S., Kim, H., Kim, H.-S.: On the number of edges of fan-crossing free graphs. In: Cai, L., Cheng, S.-W., Lam, T.-W. (eds.) ISAAC 2013. LNCS, vol. 8283, pp. 163–173. Springer, Heidelberg (2013). doi:10.1007/978-3-642-45030-3_16

10. Di Battista, G., Eades, P., Tamassia, R., Tollis, I.G.: Graph Drawing: Algorithms for the Visualization of Graphs. Prentice-Hall, Upper Saddle River (1999)

11. Didimo, W., Eades, P., Liotta, G.: Drawing graphs with right angle crossings. Theoret. Comput. Sci. 412(39), 5156–5166 (2011)

12. Dujmovic, V., Gudmundsson, J., Morin, P., Wolle, T.: Notes on large angle crossing graphs. Chicago J. Theor. Comput. Sci. 4, 1–14 (2011)

13. de Fraysseix, H., Pach, J., Pollack, R.: How to draw a planar graph on a grid. Combinatorica 10(1), 41–51 (1990)

14. Grigoriev, A., Bodlaender, H.L.: Algorithms for graphs embeddable with few crossings per edge. Algorithmica 49(1), 1–11 (2007)

15. Harary, F.: Graph Theory. Addison-Wesley, Boston (1991)

16. Hong, S.-H., Nagamochi, H.: Testing full outer-2-planarity in linear time. In: Mayr, E.W. (ed.) WG 2015. LNCS, vol. 9224, pp. 406–421. Springer, Heidelberg (2016). doi:10.1007/978-3-662-53174-7_29

17. Kaufmann, M., Ueckerdt, T.: The density of fan-planar graphs. CoRR abs/1403.6184 (2014)

18. Kaufmann, M., Wagner, D. (eds.): Drawing Graphs, Methods and Models. LNCS, vol. 2025. Springer, Heidelberg (2001)

19. Pach, J., Radoicic, R., Tardos, G., Tóth, G.: Improving the crossing lemma by finding more crossings in sparse graphs. Discrete Comput. Geom. 36(4), 527–552 (2006)

20. Pach, J., Tóth, G.: Graphs drawn with few crossings per edge. Combinatorica 17(3), 427–439 (1997)

21. Ringel, G.: Ein sechsfarbenproblem auf der kugel. Abhandlungen aus dem Mathematischen Seminar der Universität Hamburg (in German) 29, 107–117 (1965)

22. Tamassia, R.: On embedding a graph in the grid with the minimum number of bends. SIAM J. Comput. 16(3), 421–444 (1987)

23. Tutte, W.T.: How to draw a graph. Proc. London Math. Soc. 3(13), 743–767 (1963)

A Note on the Practicality of Maximal Planar Subgraph Algorithms

Markus Chimani[1], Karsten Klein[2], and Tilo Wiedera[1(✉)]

[1] University Osnabrück, Osnabrück, Germany
{markus.chimani,tilo.wiedera}@uni-osnabrueck.de
[2] Uni Konstanz, Konstanz, Germany
karsten.klein@uni-konstanz.de

Abstract. Given a graph G, the NP-hard *Maximum Planar Subgraph* problem (*MPS*) asks for a planar subgraph of G with the maximum number of edges. There are several heuristic, approximative, and exact algorithms to tackle the problem, but—to the best of our knowledge—they have never been compared competitively in practice.

We report on an exploratory study on the relative merits of the diverse approaches, focusing on practical runtime, solution quality, and implementation complexity. Surprisingly, a seemingly only theoretically strong approximation forms the building block of the strongest choice.

1   Introduction

We consider the problem of finding a large planar subgraph in a given non-planar graph $G = (V, E)$; $n := |V|$, $m := |E|$. We distinguish between algorithms that find a *large*, *maximal*, or *maximum* such graph: while the latter (MPS) is one with largest edge cardinality and NP-hard to find [18], a subgraph is inclusionwise maximal if it cannot be enlarged by adding any further edge of G. Sometimes, the inverse question—the *skewness* of G—is asked: find the smallest number $skew(G)$ of edges to remove, such that the remaining graph is planar.

The problem is a natural non-trivial graph property, and the probably best known non-planarity measure next to crossing number. This already may be reason enough to investigate its computation. Moreover, MPS/skewness arises at the core of several other applications: E.g., the practically strongest heuristic to draw G with few crossings—the *planarization method* [2,7][1]—starts with a large planar subgraph, and then adds the other edges into it.

Recognizing graphs of small skewness also plays a crucial role in parameterized complexity: Many problems become easier when considering a planar graph; e.g., maximum flow can be computed in $\mathcal{O}(n \log n)$ time, the Steiner tree problem allows a PTAS, the maximum cut can be found in polynomial time, etc. It hence can be a good idea to (in a nutshell) remove a couple of edges to obtain a planar graph, solve the problem on this subgraph, and then consider suitable

---

[1] In contrast to this meaning, the MPS problem itself is also sometimes called *planarization*. We refrain from the latter use to avoid confusion.

© Springer International Publishing AG 2016
Y. Hu and M. Nöllenburg (Eds.): GD 2016, LNCS 9801, pp. 357–364, 2016.
DOI: 10.1007/978-3-319-50106-2_28

modifications to the solution to accommodate for the previously ignored edges. E.g., we can compute a maximum flow in time $\mathcal{O}(skew(G)^3 \cdot n \log n)$ [13].

While solving MPS is NP-hard, there are diverse polynomial-time approaches to compute a large or maximal planar subgraph, ranging from very trivial to sophisticated. By Euler's formula we know that already a spanning tree gives a 1/3-approximation for MPS. Hence all reasonable algorithms achieve this ratio. Only the *cactus algorithms* (see below) are known to exhibit better ratios. We will also consider an exact MPS algorithm based on integer linear programs (ILPs).

All algorithms considered in this paper are known (for quite some time, in fact), and are theory-wise well understood both in terms of worst case solution quality and running time. To our knowledge, however, they have never been practically compared. In this paper we are in particular interested in the following quality measures, and their interplay:

- What is the practical difference in terms of running time?
- What is the practical difference in solution quality (i.e., subgraph density)?
- What is the implementation effort of the various approaches?

Overall, understanding these different quality measures as a multi-criteria setting, we can argue for each of the considered algorithms that it is pareto-optimal. We are in particular interested in studying a somewhat "blurred" notion of pareto-optimality: We want to investigate, e.g., in which situations the additional sophistication of algorithms gives "significant enough" improvements.[2]

Also the measure of "implementation complexity" is surprisingly hard to concisely define, and even in the field of software-engineers there is no prevailing notion; "lines of code" are, for example, very unrelated to the intricacies of algorithm implementation. We will hence only argue in terms of direct comparisons between pairs of algorithms, based on our experience when implementing them.[3]

As we will see in the next section, there are certain building blocks all algorithms require, e.g., a graph data structure and (except for C, see below) an efficient planarity test. When discussing implementation complexity, it seems safe to assume that a programmer will already start off with some kind of graph library for her basic datastructure needs.[4] In the context of the ILP-based approach, we assume that the programmer uses one of the various freely available (or commercial) frameworks. Writing a competitive branch-and-cut framework from ground up would require a staggering amount of knowledge, experience, time, and finesse. The ILP method is simply not an option if the programmer may not use a preexisting framework.

---

[2] Clearly, there is a systematic weakness, as it may be highly application-dependent what one considers "significant enough". We hence cannot universally answer this question, but aim to give a guideline with which one can come to her own conclusions.

[3] This measure is still intrinsically subjective (although we feel that the situation is quite clear in most cases), and opinions may vary depending on the implementor's knowledge, experience, etc. We discuss these issues when they become relevant.

[4] Many freely available graph libraries contain linear-time planarity tests. They usually lack sophisticated algorithms for finding large planar subgraphs.

In the following section, we discuss our considered algorithms and their implementation complexity. In Sect. 3, we present our experimental study. We first consider the pure problem of obtaining a planar subgraph. Thereafter, we investigate the algorithm choices when solving MPS as a subproblem in a typical graph drawing setting—the planarization heuristic.

2 Algorithms

Naïve Approach (Nï). The algorithmically simplest way to find a maximal planar subgraph is to start with the empty graph and to insert each edge (in random order) unless the planarity test fails. Given an $\mathcal{O}(n)$ time planarity test (we use the algorithm by Boyer and Myrvold [3], which is also supposed to be among the practically fastest), this approach requires $\mathcal{O}(nm)$ overall time.[5]

In our study, we consider a trivial multi-start variant that picks the best solution after several runs of the algorithm, each with a different randomized order. The obvious benefit of this approach is the fact that it is trivial to understand and implement—once one has any planarity test as a black box.

Augmented Planarity Test (BM, BM+). Planarity tests can be modified to allow the construction of large planar subgraphs. We will briefly sketch these modifications in the context of the above mentioned $\mathcal{O}(n)$ planarity test by Boyer and Myrvold [3]: In the original test, we start with a DFS tree and build the original graph bottom-up; we incrementally add a vertex together with its DFS edges to its children and the backedges from its decendents. The test fails if at least one backedge cannot be embedded.

We can obtain a large (though in general not maximal) planar subgraph by ignoring whether some backedges have not been embedded, and continuing with the algorithm (BM). If we require maximality, we can use Nï as a prostprocessing to grow the obtained graph further (BM+). While this voids the linear runtime, it will be faster than the pure naïve approach. Given an implementation of the planarity testing algorithm, the required modifications are relatively simple per se—however, they are potentially hard to get right as the implementor needs to understand side effects within the complex planarity testing implementation.

Alternatively, Hsu [14] showed how to overcome the lack of maximality directly within the planarity testing algorithm [19] (which is essentially equivalent to [3]), retaining linear runtime. While this approach is the most promising in terms of running time, it would require the most demanding implementation of all approaches discussed in this paper (including the next subsection)—it means to implement a full planarity testing algorithm plus intricate additional procedures. We know of no implementation of this algorithm.

---

[5] We *could* speed-up the process in practice by starting with a spanning tree plus three edges. However, there are instances where the initial inclusion of a spanning tree prohibits the optimal solution and restricts one to approximation ratios $\leq 2/3$ [8].

Cactus Algorithm (C, C+). The only non-trivial approximation ratios are achieved by two cactus-based algorithms [4]. Thereby, one starts with the disconnected vertices of G. To obtain a ratio of $7/18$ (C), we iteratively add triangles connecting formerly disconnected components. This process leaves a forest F of tree-like structures made out of triangles—*cactusses*. Finally, we make F connected by adding arbitrary edges of E between disconnected components. Since this subgraph will not be maximal in general, we can use Nï to grow it further (C+).

From the implementation point of view, this algorithm is very trivial and—unless one requires maximality—does not even involve any planarity test. While a bit more complex than the naïve approach, it does not require modifications to complex and potentially hard-to-understand planarity testing code like BM.

For the best approximation ratio of $4/9$ one seeks not a maximal but a maximum cactus forest. However, this approach is of mostly theoretical interest as it requires non-trivial polynomial time matroid algorithms.

ILP Approach (ILP). Finally, we use an integer linear program (ILP) to solve MPS exactly in reasonable (but formally non-polynomial) time, see [15]. With binary variables for each edge, specifying whether it is in the solution, we have

$$\max\left\{\sum_{e\in E} x_e \;\middle|\; \sum_{e\in K} x_e \le |K|-1 \text{ for all Kuratowski subdivisions } K \subseteq G\right\}.$$

Kuratowski's theorem [17] states that a graph is planar if and only if it does not contain a K_5 or a $K_{3,3}$ as a subdivision—*Kuratowski* subdivisions. Hence we guarantee a planar solution by requiring to remove at least one edge in each such subgraph K. While the set of these constraints is exponential in size, we can separate them heuristically within a branch-and-cut framework, see [15]: after each LP relaxation, we round the fractional solution and try to identify a Kuratowski subdivision that leads to a violated constraint.

This separation in fact constitutes the central implementation effort. Typical planarity testing algorithms initially only answer *yes* or *no*. In the latter case, however, all known linear-time algorithms can be extended to extract a *witness* of non-planarity in the form of a Kuratowski subdivision in $\mathcal{O}(n)$ time. If the available implementation does not support this additional query, it can be simulated using $\mathcal{O}(n)$ calls to the planarity testing algorithm, by incrementally removing edges whenever the graph stays non-planar after the removal. Both methods result in a straight-forward implementation (assuming some familiarity with ILP frameworks), but an additional tuning step to decide, e.g., on rounding thresholds, is necessary. The overall complexity is probably somewhere in-between C and BM. In our study, we decided to use the effective extraction scheme described in [10] which gives several Kuratowski subdivions via a single call. We propose, however, to use this feature only if it is already available in the library: its implementation effort would otherwise be comparable to a full planarity test, and in particular for harder instances its benefit is not significant.

3   Experiments

For an exploratory study we conducted experiments on several benchmark sets. We summarize the results as follows—observe the inversion between F1 and F2.

F1. C+ yields the best solutions. Choosing a "well-growable" initial subgraph—in our case a good cactus—is practically important. The better solution of BM is a weak starting point for BM+; even Ni gives clearly better solutions.

F2. BM gives better solutions than C; both are the by far fastest approaches. Thus, if runtime is more crucial than maximality, we suggest to use BM.

F3. ILP only works for small graphs. Expander graphs (they are sparse but well-connected) seem to be among the hardest instances for the approach.

F4. Larger planar subgraphs lead to fewer crossings for the planarization method. However, this is much less pronounced with modern insertion methods.

Setup and Instances. All considered algorithms are implemented in C++ (g++ 5.3.1, 64bit, -03) as part of OGDF [5], the ILP is solved via CPLEX 12.6. We use an Intel Xeon E5-2430 v2, 2.50 GHz running Debian 8; algorithms use singles cores out of twelve, with a memory limit of 4 GB per process.

We use the non-planar graphs of the established benchmark sets NORTH [12] (423 instances), ROME [11] (8249), and STEINLIB [16] (586), all of which include real-world instances. In our plots, we group instances according to $|V|$ rounded to the nearest multiple of 10; for ROME we only consider graph with ≥ 25 vertices.

Additionally, we consider two artificial sets: BAAL [1] are scale-free graphs, and REGULAR [20] (implemented as part of the OGDF) are random regular graphs; they are *expander graphs* w.h.p. [folklore]. Both sets contain 20 instances for each combination of $|V| \in \{10^2, 10^3, 10^4\}$ and $|E|/|V| \in \{2, 3, 5, 10, 20\}$.

Evaluation. Our results confirm the need for heuristic approaches, as ILP solves less than 25% of the larger graphs of the (comparably simple) ROME set within 10 min. Even deploying strong preprocessing [6] (+PP) and doubling the computation time does not help significantly, cf. Fig. 1(d). Already 30-vertex graphs with density 3, generated like REGULAR, cannot be solved within 48 hours ($\rightarrow$F3).

We measure solution quality by the *density* (edges per vertices) of the computed planar subgraph. Independently of the benchmark set, C+ always achieves the best solutions, cf. Fig. 1(a), (b) (table) ($\rightarrow$F1). We know instances where Ni is only a 1/3 approximation whereas the worst ratio known for BM+ is 2/3 [8]. Surprisingly, Ni yields distinctly better solutions than BM+ in practice ($\rightarrow$F1).

On STEINLIB, BAAL, and REGULAR, both C and BM behave similar w.r.t. solution quality. For ROME and NORTH, however, BM yields solutions that are 20–30% better, respectively ($\rightarrow$F2). This discrepancy seems to be due the fact that the found subgraphs are generally very sparse for both algorithms on BAAL and REGULAR (average density of 1.1 and 1.2, respectively, for the largest graphs).

Both C and BM are extremly (and similarly) fast; Fig. 1(c) (table) ($\rightarrow$F2). For BM+ and C+, the Ni-postprocessing dominates the running time: Ni is worst, followed by C+ and BM+—a larger initial solution leads to fewer trys for growing.

	density, relative to best										runtime [s]									
	STEINLIB				BAAL			REGULAR			STEINLIB				BAAL			REGULAR		
	B-E	I*	S†	V†	2	3	4	2	3	4	B-E	I*	S	V	2	3	4	2	3	4
BM	.86	.85	.82	.84	.72	.64	.66	.82	.84	.95	.06	.07	.01	.11	.00	.02	1	.00	.02	2
BM+	.90	.90	.88	.86	.85	.74	.73	.89	.89	.97	47.34	8.85	2.70	90.97	.06	10.57	143	.03	4.60	236
C	.84	.67	.81	.87	.60	.68	.75	.74	.88	.95	.04	.04	.01	.07	.00	.01	0	.00	.01	1
C+	1.0	1.0	1.0	1.0	1.0	1.0	1.0	1.0	1.0	1.0	9.19	16.22	3.13	16.35	.06	12.32	152	.04	5.34	217
NÏ	.92	.98	.91	.91	.96	.94	.92	.92	.91	.97	49.38	28.59	2.43	95.08	.05	7.92	239	.04	4.85	252

†S (= constr. sparse): PUC, SP;   V (= VLSI): ALUE, ALUT, LIN, TAQ, DIW, DMXA, GAP, MSM, 1R, 2R

(a) ROME, solution quality,
C = 1.02[1.01, 1.06]

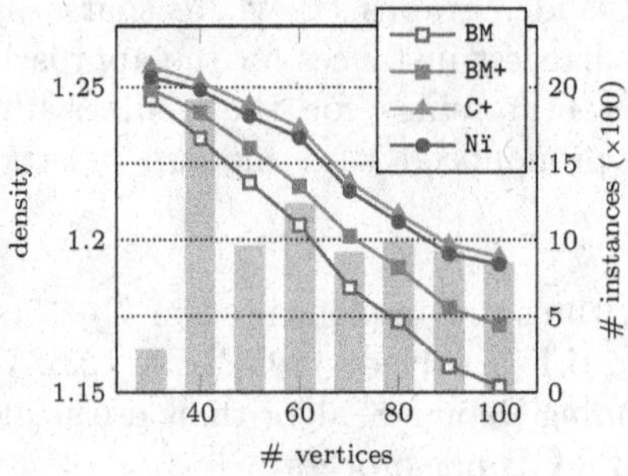

(b) NORTH, solution quality relative to ILP
(over instances solved by ILP),
C = 0.76[0.64, 0.85]

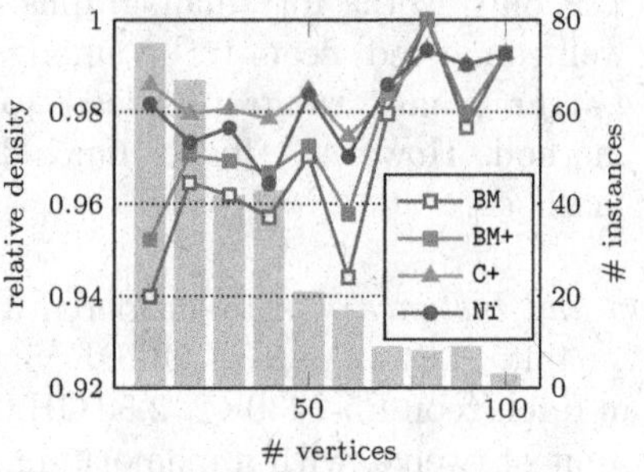

(c) ROME, running time,
BM ≈ 0, C ≈ 0

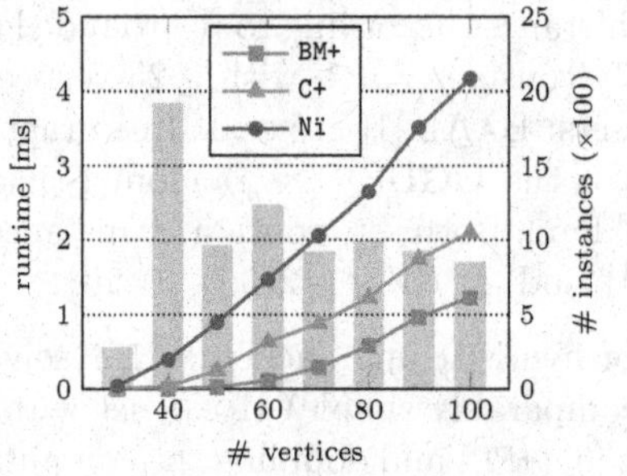

(d) ROME, success-rate of ILP

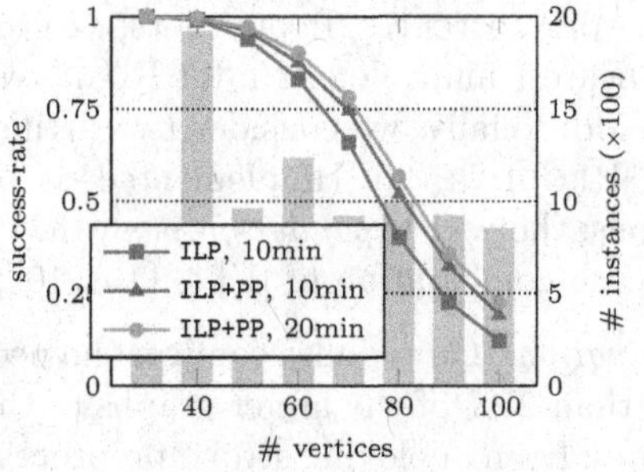

(e) ROME, simple planarization,
C = 102[61, 173]

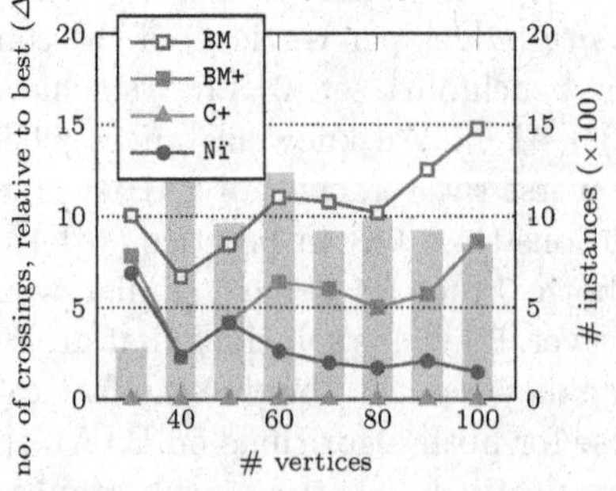

(f) ROME, state-of-the-art planarization

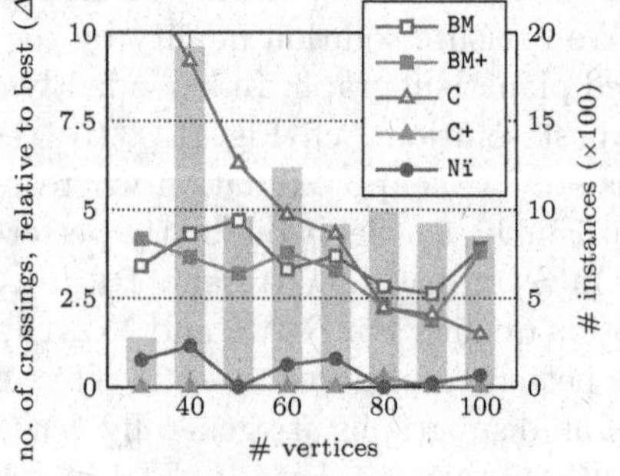

Fig. 1. We may omit algorithms whose values are unsuitable for a plot; instead we give their average[min, max] in the caption.

Nonetheless, we observe that the (weaker) solution of C allows for significantly more *successful* growing steps that BM ($\rightarrow$F1).

Finally, we investigate the importance of the subgraph selection for the planarization method, cf. Fig. 1(e), (f). For the simplest insertion algorithms (iterative edge insertions, fixed embedding, no postprocessing, [2]), a strong subgraph method (C+) is important; C leads to very bad solutions. For state-of-the-art insertion routines (simultaneous edge insertions, variable embedding, strong postprocessing, [7,9]) the subgraph selection is less important; even C is feasible.

References

1. **BarabasiAlbertGenerator** of the java universal network/graph framework. http:// jung.sourceforge.net
2. Batini, C., Talamo, M., Tamassia, R.: Computer aided layout of entity relationship diagrams. J. Syst. Softw. **4**(2–3), 163–173 (1984)
3. Boyer, J.M., Myrvold, W.J.: On the cutting edge: simplified $\mathcal{O}(n)$ planarity by edge addition. J. Graph Algorithms Appl. **8**(2), 241–273 (2004)
4. Călinescu, G., Fernandes, C., Finkler, U., Karloff, H.: A better approximation algorithm for finding planar subgraphs. J. Algorithms **27**, 269–302 (1998)
5. Chimani, M., Gutwenger, C., Jünger, M., Klau, G.W., Klein, K., Mutzel, P.: The open graph drawing framework (OGDF). In: Tamassia, R. (ed.) Handbook of Graph Drawing and Visualization, Chap. 17, pp. 543–569. CRC Press (2014). http://www.ogdf.net
6. Chimani, M., Gutwenger, C.: Non-planar core reduction of graphs. Discrete Math. **309**(7), 1838–1855 (2009)
7. Chimani, M., Gutwenger, C.: Advances in the planarization method: effective multiple edge insertions. J. Graph Algorithms Appl. **16**(3), 729–757 (2012)
8. Chimani, M., Hedtke, I., Wiedera, T.: Limits of greedy approximation algorithms for the maximum planar subgraph problem. In: Mäkinen, V., Puglisi, S.J., Salmela, L. (eds.) IWOCA 2016. LNCS, vol. 9843, pp. 334–346. Springer, Heidelberg (2016). doi:10.1007/978-3-319-44543-4_26
9. Chimani, M., Hliněný, P.: Inserting multiple edges into a planar graph. CoRR abs/1509.07952 (2015). http://arxiv.org/abs/1509.07952
10. Chimani, M., Mutzel, P., Schmidt, J.M.: Efficient extraction of multiple kuratowski subdivisions. In: Hong, S.-H., Nishizeki, T., Quan, W. (eds.) GD 2007. LNCS, vol. 4875, pp. 159–170. Springer, Heidelberg (2008). doi:10.1007/978-3-540-77537-9_17
11. Di Battista, G., Garg, A., Liotta, G., Tamassia, R., Tassinari, E., Vargiu, F.: An experimental comparison of four graph drawing algorithms. Comput. Geom. **7**(56), 303–325 (1997)
12. Di Battista, G., Garg, A., Liotta, G., Parise, A., Tassinari, R., Tassinari, E., Vargiu, F., Vismara, L.: Drawing directed acyclic graphs: an experimental study. Int. J. Comput. Geom. Appl. **10**(06), 623–648 (2000)
13. Hochstein, J.M., Weihe, K.: Maximum s-t-flow with k crossings in $\mathcal{O}(k^3 n \log n)$. In: Proceedings of the SODA 2007, pp. 843–847. ACM-SIAM (2007)
14. Hsu, W.-L.: A linear time algorithm for finding a maximal planar subgraph based on PC-trees. In: Wang, L. (ed.) COCOON 2005. LNCS, vol. 3595, pp. 787–797. Springer, Heidelberg (2005). doi:10.1007/11533719_80
15. Jünger, M., Mutzel, P.: Maximum planar subgraphs and nice embeddings: practical layout tools. Algorithmica **16**(1), 33–59 (1996)

16. Koch, T., Martin, A., Voß, S.: SteinLib: An updated library on steiner tree problems in graphs. Technical report ZIB-Report 00–37, Konrad-Zuse-Zentrum für Informationstechnik Berlin (2000). http://elib.zib.de/steinlib
17. Kuratowski, C.: Sur le problme des courbes gauches en topologie. Fundamenta Mathematicae **15**(1), 271–283 (1930)
18. Liu, P.C., Geldmacher, R.C.: On the deletion of nonplanar edges of a graph. In: Proceedings of the 10th Southeastern Conference on Combinatorics, Graph Theory and Computing, pp. 727–738 (1979). Congress. Numer., XXIII-XXIV, Utilitas Math., Winnipeg, Man
19. Shih, W., Hsu, W.: A new planarity test. Theor. Comput. Sci. **223**(1–2), 179–191 (1999)
20. Steger, A., Wormald, N.C.: Generating random regular graphs quickly. Comb. Probab. Comput. **8**(4), 377–396 (1999)

Crossing Minimization and Crossing Numbers

Crossing Minimization in Storyline Visualization

Martin Gronemann[1], Michael Jünger[1], Frauke Liers[2],
and Francesco Mambelli[1(✉)]

[1] Department of Computer Science, University of Cologne, Cologne, Germany
{gronemann,mjuenger,mambelli}@informatik.uni-koeln.de
[2] Department of Mathematics, University of Erlangen-Nürnberg, Erlangen, Germany
frauke.liers@math.uni-erlangen.de

Abstract. A storyline visualization is a layout that represents the temporal dynamics of social interactions along time by the convergence of chronological lines. Among the criteria oriented at improving aesthetics and legibility of a representation of this type, a small number of line crossings is the hardest to achieve. We model the crossing minimization in the storyline visualization problem as a multi-layer crossing minimization problem with tree constraints. Our algorithm can compute a layout with the minimum number of crossings of the chronological lines. Computational results demonstrate that it can solve instances with more than 100 interactions and with more than 100 chronological lines to optimality.

1 Introduction

Visualizing time-varying relationships between entities using converging and diverging curves on a timeline has received a considerable amount of interest recently. The ability to display interactions among entities, while at the same time being able to put these in a chronological context has found applications beyond its initial purpose which coined its name. Munroe [26] introduced the *storyline visualization* as hand-drawn illustrations in xkcd's "Movie Narrative Charts", where lines represent the characters of various popular movies and the scenes are ordered chronologically and represented by bundling the lines of the corresponding characters. This concept has been used to visualize various spatiotemporal data, like communities in time-varying graphs [25,33], software projects [28], topic analysis [9], etc.

However, hand crafted or semi-automated methods are limited in their applicability in a world of ever growing datasets. In order to obtain a storyline visualization automatically, Tanahashi and Ma [32] discuss various aspects of a well-designed storyline visualization and present an evolutionary algorithm that incorporates these in its objective function. They identify three important criteria that one usually wants to optimize: line crossings, whose number should be small, line wiggles, that should be avoided by drawing every chronological line as straight as possible, and space efficiency. Based on these aspects, Liu et al. [24] describe a technique which further improves the layout and runs significantly faster compared to the evolutionary algorithm in [32]. Being able to

© Springer International Publishing AG 2016
Y. Hu and M. Nöllenburg (Eds.): GD 2016, LNCS 9801, pp. 367–381, 2016.
DOI: 10.1007/978-3-319-50106-2_29

create storyline visualizations of bigger instances, Tanahashi et al. [31] take this one step further and show how to create storyline visualizations from streaming data.

In this paper, we study the crossing minimization problem in storyline visualization from a combinatorial optimization point of view. While most approaches tackle this problem with heuristics, Kostitsyna et al. [22] recently shed some light onto its combinatorial properties. Besides noting that the decision problem is NP-complete (by reduction from bipartite crossing number), they provide a lower bound for the number of crossings in a restricted variant of the problem and show that the general problem is fixed-parameter tractable in the number of characters. But a straightforward implementation of the algorithm is impractical, even for a small number of characters.

However, the problem is also similar to a few already well-studied problems in graph-drawing. It may seem that the problem is related to a special case of metro-line crossing minimization, in particular, the so called two-sided models in which metro-lines run only from left to right [5,12]. However, all metro-line crossing minimization problems have in common that they are defined on a rail network whose embedding is fixed due to its geographical context. This difference makes a straightforward transformation difficult.

As already observed by Kostitsyna et al. [22], storyline crossing minimization has a strong relationship to *multi-layer crossing minimization (MLCM)*. Here each node of the graph is assigned to one of the layers (parallel straight lines) in such a way that each edge connects two vertices on consecutive layers. The aim is to find an ordering of the nodes on every layer such that the total number of edge crossings is minimized. Although the corresponding planarity testing problem is linear-time solvable [19], the MLCM problem itself remains NP-hard even when restricted to two layers [13]. This led to the development of various heuristics, but also to exact approaches [6,8,16,18].

In order to exploit existing techniques for solving MLCM instances, a straightforward transformation can be sketched as follows. We represent the characters as paths in an MLCM instance, in which the layers mark important points in time, e.g., a new bundle has to be created. Of course, a bundle of lines (paths) requires the corresponding vertices to be consecutive on the layer, a constraint which is problematic in the general MLCM setting.

But we can borrow ideas from another crossing minimization problem type, the so called *tanglegrams*. The general tanglegram problem consists of two trees and a set of edges connecting the leaves of one tree with the leaves of the other, i.e., the leaves and the connecting edges form a bipartite graph. The objective is essentially to perform a bipartite (or two-layer) crossing minimization with the additional constraint that leaves of the same subtree appear consecutively on the layers. However, when consulting the literature on tanglegrams, attention must be paid to the details. Some definitions require the trees to be binary, while others restrict the edge set to be a perfect matching, or both [7,11,27]. Since the focus of the paper is not on tanglegrams, we restrict ourselves to the general case. Here two works are of interest, Baumann et al. [4] describe an ILP-based approach, whereas Wotzlaw et al. [34] employ a SAT-formulation. However, not only the techniques differ, in [4] only two layers are considered, whereas the SAT approach in [34] works on multiple layers but requires that the tree constraints are k-ary with $k > 1$ fixed.

The related problem of testing level planarity under tree constraints is discussed by Angelini et al. [1]. They show that if edges are restricted to run between consecutive layers, then the problem can be solved in quadratic time, whereas if this restriction does not hold, the problem is NP-complete.

In this paper we solely focus on the crossing minimization problem in storyline visualization. Therefore, we neglect other design aspects and restrict ourselves to the combinatorial problem, i.e., determining an ordering of the lines such that the number of crossings is minimum. We model this problem as a special variant of the MLCM problem under tree constraints and provide an ILP formulation for it. Computational results show that we are able to solve instances of moderate size to optimality within a few seconds. Moreover, we provide solutions for storyline instances from the literature, some of which have been solved to optimality for the first time. These are of particular value, since they offer a reference when comparing the crossing minimization performance of heuristics.

2 Modelling Storyline Visualization as Multi-layer Crossing Minimization with Tree Constraints

We begin with a formal definition of the multi-layer crossing minimization problem with tree constraints (MLCM-TC). The input for MLCM-TC consists of a graph $G = (V, E, \mathcal{T})$, where the set of the nodes $V = \bigcup_{r=1}^{p} V_r$ is partitioned into p different layers. $E = \bigcup_{r=1}^{p-1} E_r$ is the set of the edges such that $E_r \subseteq V_r \times V_{r+1}$ for every $r \in \{1, 2, \ldots, p-1\}$, i.e., each edge of E_r has one end in V_r and the other in V_{r+1}. $\mathcal{T} = \{T_r \mid r = 1, 2, \ldots, p\}$ is a family of rooted trees with at least one internal node (root node), whose leaves are exactly the nodes of V_r. In the following, whenever we consider a graph, we implicitly assume that it is of this type, which is known in the literature as "(proper) $\mathcal{T}$-level graph" [1].

Given an instance $G = (V, E, \mathcal{T})$ of MLCM-TC, the task is to determine, for each layer $r \in \{1, 2, \ldots, p\}$, permutations $\pi_r = \langle v_1, v_2, \ldots, v_{|V_r|} \rangle$ of the nodes in V_r such that for each internal node τ of T_r, all t leaves in the subtree rooted at τ are adjacent in π_r, i.e., they form a sub-permutation $\langle v_i, v_{i+1}, \ldots, v_{i+t-1} \rangle$ for some $i \in \{1, 2, \ldots, |V_r| - t + 1\}$.

An easy reduction of the NP-hard MLCM problem to the MLCM-TC problem (add a trivial tree with the root as the only internal node to each layer) shows that MLCM-TC is NP-hard. This justifies the usage of integer programming techniques in the next section. Now we give a formal description of the storyline visualization problem in order to support our hypothesis that MLCM-TC captures its core when the criteria "line wiggle avoidance" and "space efficiency" are neglected in favour of crossing minimization. A story consists of a set of characters $C = \{c_1, c_2, \ldots, c_n\}$ and a set of scenes $S \subseteq 2^C$. For each scene $s \in S$, b_s and

e_s are the points in time when s begins and ends, respectively. The time intervals $[b_{s_1}, e_{s_1}]$ and $[b_{s_2}, e_{s_2}]$ of two distinct scenes s_1 and s_2 may have a non-empty intersection, but if they do, we require $s_1 \cap s_2 = \emptyset$.

The storyline visualization problem requires depicting each character $c \in C$ by a curve in the Euclidean plane that is strictly monotone on the time axis that we arbitrarily fix to the horizontal x-axis. The curve begins at the x-coordinate $x_c^b = \min\{b_s \mid c \in s\}$ and ends at $x_c^e = \max\{e_s \mid c \in s\}$. We call the interval $[x_c^b, x_c^e]$ the lifespan of character c.

The curves must be such that for every scene $s = \{c_{\sigma_1}, c_{\sigma_2}, \ldots, c_{\sigma_k}\} \in S$ the k corresponding curves in the interval $[b_s, e_s]$ are horizontal parallel lines that are equally spaced with vertical distance 1. Furthermore, the curves of all $c \notin s$ are restricted to y-coordinates that have an absolute difference of at least 2 to the y-coordinates of the curves $c_{\sigma_i} \in s$ in the interval $[b_s, e_s]$, and to all curves for characters that are not members of any scene that intersects with $[b_s, e_s]$. An example is given in Fig. 1.

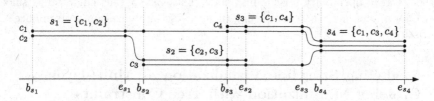

Fig. 1. An example of a story with four scenes and four characters, where characters c_3 and c_4 enter late, character c_2 leaves early, and the time intervals $[b_{s_2}, e_{s_2}]$ and $[b_{s_3}, e_{s_3}]$ have a non-empty intersection.

Given a story $(C, S, \{[b_s, e_s] \mid s \in S\})$, we construct an MLCM-TC instance $G = (V, E, \mathcal{T})$ as follows:

1. Sort the points in time $\{b_s \mid s \in S\} \cup \{e_s \mid s \in S\}$ in non-decreasing order, and let $\langle t_1, t_2, \ldots, t_p \rangle$ be the sorted sequence.
2. Associate a layer V_r with each t_r ($r \in \{1, 2, \ldots, p\}$), create a node $v_{c,r}$ for each character c for which t_r is within its lifespan, i.e., for which $t_r \in [x_c^b, x_c^e]$, and let $V_r = \{v_{c,r} \mid t_r \in [x_c^b, x_c^e]\}$.
3. Let $V = \bigcup_{r=1}^{p} V_r$.
4. Let $E = \big\{\{v_{c,r}, v_{c,r+1}\}$ for all $c \in C$ such that $t_r, t_{r+1} \in [x_c^b, x_c^e]\big\}$.
5. For each layer V_r create a tree T_r as follows:
    i. For each scene $s = \{c_{\sigma_1}, c_{\sigma_2}, \ldots, c_{\sigma_k}\}$ such that $t_r \in [b_s, e_s]$ create an internal tree node $v_{s,r}$ and tree edges $\{v_{s,r}, v_{c_{\sigma_i},r}\}$ for all $i \in \{1, 2, \ldots, k\}$.
    ii. Unless the above results in a rooted tree with all nodes in V_r as leaves, create a tree root ρ_r and tree edges connecting ρ_r to all previously created internal tree nodes of T_r and to all character nodes in V_r that are not joined to a previously added internal tree node.

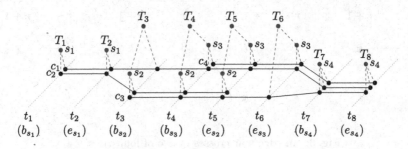

Fig. 2. The MLCM-TC instance of the story of Fig. 1.

In Fig. 2, we demonstrate the construction for our example instance from Fig. 1.

Notice that the trees in $\mathcal{T}$ are all of height up to 2, which means that storyline visualization instances yield a special subclass of MLCM-TC instances. By construction, an optimal solution of this MLCM-TC instance induces a storyline visualization with the minimum number of crossings, and, conversely, any instance of this special MLCM-TC subclass with trees of height up to 2 is the result of the given transformation for some story. Thus, both problems are equivalent. As MLCM can be reduced to this special subclass, NP-hardness is maintained.

3   Integer Linear Programming Formulation

We present an integer linear programming (ILP) formulation of MLCM-TC. ILP formulations have already been introduced for the general MLCM problem [16,18] as well as for MLCM-TC, when restricted to the special case of two layers only [4]. Both models use quadratic ordering formulations. In this section, we will extend these formulations to an ILP model for MLCM-TC.

To this end, let $G = (V, E, \mathcal{T})$ be an instance of MLCM-TC, as described in Sect. 2. For every layer $r \in \{1, 2, \ldots, p\}$, let $V_r^{(2)} = \{(i, j) \in V_r \times V_r : i < j\}$ be the set of all the ordered pairs of nodes on the considered layer with the first index smaller than the second. As the total number of edge crossings is the sum of all crossings in adjacent layers r and $r + 1$, summed up for all $r \in \{1, 2, \ldots, p-1\}$, let us consider the problem for a pair of adjacent layers r and $r + 1$, with $r \in \{1, 2, \ldots, p-1\}$.

A permutation of the nodes in V_r is characterized by variables $x_{ij}^r \in \{0, 1\}$ associated with the pairs $(i, j) \in V_r^{(2)}$ as follows:

$$x_{ij}^r = 1 \quad \text{if and only if } i \text{ is placed above } j \text{ on layer } r.$$

Then a pair of edges $(i, k), (j, \ell) \in E_r$ crosses if and only if

i is placed above j on layer r and ℓ is placed above k on layer $r + 1$

or

j is placed above i on layer r and k is placed above ℓ on layer $r + 1$,

see Fig. 3.

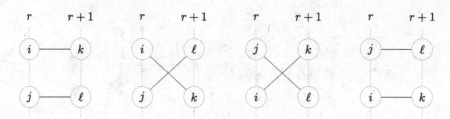

Fig. 3. An edge pair crosses in two of four cases.

Therefore, if $\{x_{ij}^r \mid (i,j) \in V_r^{(2)}\}$ and $\{x_{k\ell}^{r+1} \mid (k,\ell) \in V_{r+1}^{(2)}\}$ describe node permutations on layers r and $r+1$, respectively, we have

$$c_{ijk\ell} := x_{ij}^r(1 - x_{k\ell}^{r+1}) + (1 - x_{ij}^r)x_{k\ell}^{r+1} \in \{0,1\}$$

and $c_{ijk\ell} = 1$ if and only if the edges (i,k) and (j,ℓ) cross.

It is well known (see, e.g., [14]) that $\{x_{ij}^r \in \{0,1\} \mid (i,j) \in V_r^{(2)}\}$ characterizes a node permutation on V_r if and only if the *transitivity conditions*

$$0 \le x_{hi}^r + x_{ij}^r - x_{hj}^r \le 1 \qquad (h < i < j)$$

are satisfied for all $r \in \{1, 2, \ldots, p\}$.

It remains to model the tree conditions implied by the elements of $\mathcal{T}$. Given a layer $r \in \{1, 2, \ldots, p\}$ and two nodes i and j in V_r, we denote by $P(i,j)$ the lowest common ancestor of i and j in T_r. Let $V_r^{(3)} = \{(h,i,j) \in V_r \times V_r \times V_r : h < i < j\}$. For every $r \in \{1, 2, \ldots, p\}$ and every triple $(h,i,j) \in V_r^{(3)}$, we impose the *tree constraints*

$$\begin{aligned} x_{hj}^r = x_{ij}^r \quad & \text{if } P(h,i) \ne P(P(h,i),j), \\ x_{hi}^r = x_{hj}^r \quad & \text{if } P(i,j) \ne P(h,P(i,j)). \end{aligned}$$

The first equation forbids the placement of j between h and i in case j does not belong to the smallest subtree containing h and i. Similarly, the second equation forbids the placement of h between i and j in case h is not contained in the smallest subtree of i and j.

Putting it all together, we obtain the following model for MLCM-TC based on a combination of [18] for MLCM and [4] for the special case of MLCM-TC for two layers:

$$\text{minimize} \quad \sum_{r=1}^{p-1} \sum_{\substack{(i,j) \in V_r^{(2)},\ (k,\ell) \in V_{r+1}^{(2)} \\ (i,k),(j,\ell) \in E_r}} x_{ij}^r(1 - x_{k\ell}^{r+1}) + (1 - x_{ij}^r)x_{k\ell}^{r+1}$$

subject to

$$
\begin{aligned}
0 \le x_{hi}^r + x_{ij}^r - x_{hj}^r \le 1 \quad &\text{for all } r \in \{1, 2, \ldots, p\} \text{ and } (h, i, j) \in V_r^{(3)} \\
x_{hj}^r = x_{ij}^r \quad &\text{for all } r \in \{1, 2, \ldots, p\} \text{ and } (h, i, j) \in V_r^{(3)} \\
&\text{if } P(h, i) \ne P(P(h, i), j) \\
x_{hi}^r = x_{hj}^r \quad &\text{for all } r \in \{1, 2, \ldots, p\} \text{ and } (h, i, j) \in V_r^{(3)} \\
&\text{if } P(i, j) \ne P(h, P(i, j)) \\
x_{ij}^r \in \{0, 1\} \quad &\text{for all } r \in \{1, 2, \ldots, p\} \text{ and } (i, j) \in V_r^{(2)}.
\end{aligned}
$$

This is a quadratic 0–1-programming problem with linear constraints, namely, the transitivity conditions and the tree conditions. (Without the tree conditions, the problem is also called a *quadratic linear ordering problem*.)

When we temporarily ignore the transitivity conditions and the tree conditions, the remaining problem is known as *quadratic 0–1-optimization* of the form

$$
\begin{aligned}
\text{minimize } & z^T Q z + q^T z \\
\text{s.t. } & z \in \{0, 1\}^N
\end{aligned}
$$

for an upper triangular matrix $Q \in \mathbb{Z}^{N \times N}$ and a vector $q \in \mathbb{Z}^N$. A well known construction of Hammer [15], see also [2, 10, 23], results in an equivalent formulation as a maximum cut problem on a graph $G_{mc} = (V_{mc}, E_{mc})$ with $N+1$ nodes, all but one are identified with the z_i, $i \in \{1, 2, \ldots, N\}$. Let us call the additional node z_0, so $V_{mc} = \{z_0, z_1, \ldots, z_N\}$. The undirected edges (z_i, z_j), $1 \le i < j \le N$, correspond to the nonzero entries of the matrix Q, and there are additional N edges (z_0, z_i) for $1 \le i \le N$, giving the edge set E_{mc}. The edge weights $w_e = w_{ij}$, $0 \le i < j \le N$, are easily computed from Q and q. For $W \subseteq V_{mc}$ the edge set $\delta(W) = \{(i, j) \in E_{mc} \mid i \in W, j \in V_{mc} \setminus W\}$ is called a *cut* in G_{mc}. Then the resulting *maximum cut problem* has the form

$$
\max\{w(\delta(W)) \mid W \subseteq V_{mc}\}.
$$

By introducing variables $y_e \in \{0, 1\}$ for each $e \in E_{mc}$, the maximum cut problem can be formulated as

$$
\text{maximize} \quad \sum_{e \in E_{mc}} w_e y_e
$$

subject to

$$
\begin{aligned}
\sum_{e \in F} y_e - \sum_{e \in C \setminus F} y_e \le |F| - 1 \quad &\text{for all cycles } C \subseteq E_{mc} \text{ and all } F \subseteq C, |F| \text{ odd} \\
y_e \in \{0, 1\} \quad &\text{for all } e \in E_{mc},
\end{aligned}
$$

see [3]. The constraints are called *odd cycle constraints*.

Applying this transformation is the key to our algorithm: The edges $e \in E_{mc}$ not incident to z_0 correspond to edge pairs $(i, k), (j, \ell) \in E_r$, $r \in \{1, 2, \ldots, p-1\}$. The edges $e \in E_{mc}$ that are incident to z_0 correspond to our variables x_{ij}^r for $r \in \{1, 2, \ldots, p\}$, $i < j$. In view of the latter property, we can formulate MLCM-TC as a maximum cut problem with the additional transitivity and tree constraints, and we can solve it using a branch and cut approach for the maximum cut problem like in [2] that additionally enforces these extra constraints.

4   Implementation

The implementation used to determine the minimum number of crossings in a storyline visualization consists of two main phases, a preprocessing phase and a branch and cut phase. During the preprocessing, we first reduce the number of layers of the problem (if possible), by identifying two consecutive layers r and $r + 1$ in case the corresponding trees T_r and T_{r+1} are identical and every node in V_r and V_{r+1} is an end of one edge of E_r (e.g., layers 4 and 5 of Fig. 2 can be identified). Then, a variant of the barycenter heuristic proposed by Sugiyama et al. [29], in which the presence of the trees on layers is taken into account, is executed in order to obtain an initial feasible solution that defines the indexing within the layers: In this heuristic, the nodes of the trees are sorted according to their barycenters. The barycenter of a given leaf t is computed by assigning to each edge, that has t as end, the relative position of the other end as weight. The barycenter of each internal node τ is the mean of the barycenters of all the leaves of the subtree rooted at τ.

During the creation of the maximum cut graph induced by the heuristic solution, we exploit the fact that the tree constraints force many variables to assume the same value, so that we can identify them. Moreover, this procedure reduces also the number of constraints consistently after all variables have been replaced by their representatives: On the one hand, the tree constraints are not needed in the formulation anymore; on the other hand, some transitivity constraints become deactivated or redundant. It is important to point out that, during this first phase, the problem is initialized without constraints and they are added according to need during the subsequent branch and cut phase.

The branch and cut phase is realized in C++ using ABACUS [20] and CPLEX [17]. The initial relaxation consists just of the objective function together with lower bounds 0 and upper bounds 1 for the variables. Odd cycle constraints and transitivity constraints are generated via separation, the former with the same strategy as described in [2], the latter by complete enumeration.

5   Computational Results

Our test-bed consists of:

- three movie instances [30], namely "Inception", the original trilogy of "Star Wars" and "The Matrix";
- three book instances from the Stanford GraphBase database [21], namely "Adventures of Huckleberry Finn", "Anna Karenina" and "Les Misérables".

These instances have been converted to MLCM-TC by using the procedure described in Sect. 2. In the conversion of the book instances, a slight change is required: Since these instances do not report time intervals, but just the list of the characters involved in each scene of each chapter, a layer has been created for each of these scenes, instead of for each beginning and ending time point.

The three movie instances have been generated using the raw data set from [30] in order to compare them with results in the literature. We obtained "Inception", "Star Wars" and "The Matrix" following the principles described in Sect. 2. However, after having solved them, we realized that the number of crossings given by our algorithm for "Inception" was 35, while it was 24 in [31] and 23 in [24]. After a careful study of the layouts provided in [24,31,32], we noticed that the storylines of "Inception" and "The Matrix" in [24,31,32] differ from the raw data set provided by [30], and therefore are not comparable with our instances.

In order to make a comparison possible, "Inception" required three major modifications. This modified instance is called "Inception-sf" and is generated by incorporating the following changes that are based on a careful study of the layouts provided in [24,31,32]. The storyline for the character "Mal" is allowed to take shortcuts, i.e., in long periods of absence it is drawn as a thin curve that may cross other storylines without accounting for these crossings (see Fig. 12 in [24]). Moreover, the grouping at the end of the movie does not correspond to the last scene in the data set. To keep our layout comparable, we enforced in our new instance the same grouping at the end. The third discrepancy is the number of characters. In the data from [30] there are ten characters listed in the corresponding file, whereas the layouts from the literature [24,31,32] contain only eight storylines, in which "Arch" and "Asian" are missing. A major modification was also necessary in "The Matrix", where the storylines for the characters "Brown", "Smith" and "Jones" are allowed to take shortcuts as well. We call it "The Matrix-sf".

Since the instances "Anna Karenina" and "Les Misérables" are very big, we have split them into chapters and sequences of chapters. The resulting test-bed is made of eight chapters, seven pairs of chapters, six triples of chapters and five quadruples of chapters from "Anna Karenina", and five chapters, four pairs of chapters and three triples of chapters from "Les Misérables", plus the entire "Adventures of Huckleberry Finn", "Inception-sf", "Inception", "Star Wars", "The Matrix-sf", and "The Matrix".

To the best of our knowledge, this is the first time in which ILP techniques are applied to storyline visualizations. Thus comparisons of computational results are not possible. Runs were performed on one node of the HPC Cluster of the Computer Science Department of the University of Cologne. The node used consists of two Intel E5-2690v2 CPUs with ten cores each and 128GB RAM.

While the book instances generated from the Stanford GraphBase database are introduced here for the first time, the literature provides crossing counts for the three movie instances ("Inception", "Star Wars", and "The Matrix"). Table 1 shows a comparison of the minimum number of crossings (OPT) from our approach with the numbers of crossings obtained by the streaming-oriented approach from Tanahashi et al. [31] (THM), the Storyflow approach from Liu et al. [24] (LIU), and the evolutionary algorithm from Tanahashi and Ma [32] (TM). Crossing counts for THM, LIU and TM are taken from Table 3 in [31]. We can confirm that the best solution reported by Liu et al. [24] for the movie "Incep-

Table 1. Comparison of the solution of the movies.

	OPT	THM [31]	LIU [24]	TM [32]
Inception-sf	**23**	24	**23**	99
Star Wars	**39**	**41**	48	51
The Matrix-sf	**10**	22	**14**	43

tion" is optimal. For "Star Wars" the approach from Tanahashi et al. [31] comes very close to the optimal solution, even though the instance is the biggest and has the highest crossing count. One may conclude that the heuristics in [24,31] deliver solutions with a good crossing count, especially when considering the fact that they do not optimize the crossing count alone.

In Table 2, we report the information about the solution of the considered instances: The number of layers (p), of nodes ($|V|$), of edges ($|E|$), the minimum

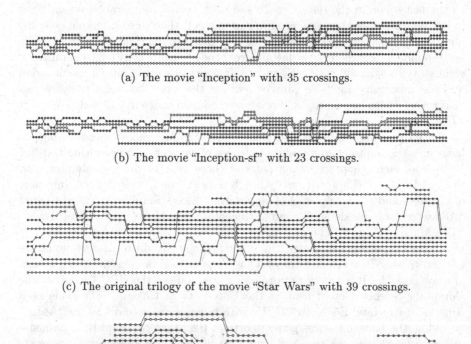

(a) The movie "Inception" with 35 crossings.

(b) The movie "Inception-sf" with 23 crossings.

(c) The original trilogy of the movie "Star Wars" with 39 crossings.

(d) The movie "The Matrix" with 12 crossings.

Fig. 4. Storyline visualizations with minimum number of crossings of the three movies from [30].

Table 2. Information about the solution of the considered instances.

	p	$|V|$	$|E|$	cr	n_{var}	n_{oddc}	n_{trans}	n_{sub}	n_{LPs}	Time
anna1	58	409	368	20	1944	2684	60	31	344	13.03
anna2	58	525	489	12	3689	2665	1	1	126	0.88
anna3	48	265	219	0	951	0	0	1	1	0.01
anna4	49	364	334	20	2116	2231	48	13	159	4.86
anna5	71	615	565	17	3821	3182	60	3	197	2.60
anna6	56	522	495	31	3586	4368	49	3	150	3.89
anna7	62	467	420	9	2525	2278	82	17	191	7.88
anna8	28	192	175	6	1036	850	1	1	45	0.15
anna1–2	117	1454	1397	57	16433	18284	89	5	545	196.24
anna2–3	108	1461	1394	28	18763	16849	29	3	469	48.96
anna3–4	100	1015	951	34	8473	8516	45	3	328	12.66
anna4–5	126	1808	1748	78	23742	26129	181	3	814	306.32
anna5–6	129	1760	1697	76	19967	23155	252	3	656	281.26
anna6–7	120	1445	1385	79	14464	32396	671	5	3008	1387.57
anna7–8	90	905	850	32	7248	8711	265	3	365	19.16
anna1–3	166	2948	2865	[100, 199]	52072	61743	631	1	1155	t.l.
anna2–4	158	2637	2557	78	40789	46600	351	3	2042	1284.03
anna3–5	174	3100	3012	[115, 224]	51814	60646	366	7	1391	t.l.
anna4–6	178	3115	3044	[124, 298]	50106	207148	232	3	1697	t.l.
anna5–7	191	3742	3656	[144, 361]	69156	77742	653	1	1216	t.l.
anna6–8	146	2205	2140	[117, 200]	28767	45396	864	3	2052	t.l.
anna1–4	216	4627	4534	[115, 339]	98525	100149	251	1	1252	t.l.
anna2–5	232	5366	5266	[102, 350]	116249	111255	261	1	1001	t.l.
anna3–6	226	5262	5168	[122, 424]	119573	121148	180	1	1345	t.l.
anna4–7	240	5467	5375	[117, 504]	119974	123020	238	1	1166	t.l.
anna5–8	217	4624	4534	[123, 470]	93832	97792	377	1	1088	t.l.
huck	107	1059	985	42	7942	11024	357	29	1098	111.31
jean1	95	502	462	10	1777	1265	49	3	167	0.90
jean2	59	226	212	6	461	385	0	1	44	0.08
jean3	99	873	838	13	6559	3407	801	7	360	6.31
jean4	76	909	876	42	9219	10116	177	3	335	22.22
jean5	73	491	471	17	2608	2412	4	3	138	1.52
jean1–2	154	1102	1055	20	5823	4172	111	3	226	3.93
jean2–3	159	1808	1767	33	18882	14128	1512	3	732	48.90
jean3–4	176	3249	3208	[115, 232]	57746	66222	482	1	1698	t.l.
jean4–5	149	1943	1907	96	24584	32573	619	3	1037	1012.44
jean1–3	254	2853	2780	53	27720	20886	1991	3	1177	143.34
jean2–4	235	4182	4135	[130, 302]	75150	81236	429	1	1928	t.l.
jean3–5	248	4429	4386	[101, 372]	79208	83279	503	1	1529	t.l.
Inception-sf	137	798	787	23	1401	1756	7	3	108	0.89
Inception	139	925	915	35	1784	2376	6	3	130	2.02
Star Wars	100	940	926	39	2132	2441	8	1	168	0.99
The Matrix-sf	82	678	660	10	1343	1219	18	3	125	0.72
The Matrix	82	683	669	12	1388	1328	45	3	98	0.77

number of crossings (cr) in boldface or a pair [lower bound, best known number of crossings], the number of variables (n_{var}), of odd cycle constraints added during the separation (n_{oddc}), of transitivity constraints added during the separation (n_{trans}), of subproblems in the branch and cut tree (n_{sub}), of linear programming relaxations solved (n_{LPs}), and the runtime expressed in seconds (Time) where "t.l." means that the run was aborted due to the time limit of one hour, in which cases the cr column contains an interval. While 29 of the 42 instances have been solved to optimality, for the remaining 13 instances the best lower bound for the number of crossings differs from the best solution found at timeout termination.

When we analyze the behaviour of our algorithm, we have to distinguish between movie and book instances: Since the original instances from [30] allow more than one scene per layer, the trees on the layers of the movie instances restrict consistently the possible permutations of the corresponding nodes and consequently reduce the number of variables. On the other hand, this is not the case for the book instances, where only one scene per layer occurs. We can observe that MLCM-TC for movies tends to be much easier in comparison to a book instance with similar numbers of layers, nodes, and edges.

The difficulty of a book instance is mainly influenced by the combination of two parameters: the number of layers p and the number of nodes $|V|$. If the number of nodes is fixed, the higher the number of layers is, the easier the solution is, since the distribution of the nodes on more layers reduces the number of variables of the problem. On the other hand, if the number of layers is fixed, the difficulty increases with the number of nodes.

The hardest instance we have been able to solve to optimality is "anna2–4", where 2 637 nodes are distributed on only 158 layers which results in 40 789

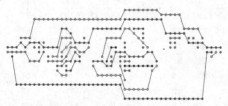

(a) The third chapter of the book "Anna Karenina" with 0 crossings.

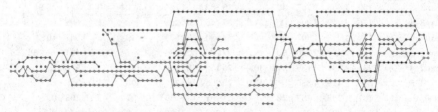

(b) The first chapter of the book "Les Misérables" with 10 crossings.

Fig. 5. Storyline visualizations of two chapters from "Anna Karenina" and "Les Misérables" [21].

variables. The biggest solved instance in terms of number of layers is "jean1–3" with 254 layers but only 2 853 nodes, which results in 27 720 variables.

We present crossing minimal storyline visualizations of the three movie instances in Fig. 4 and the two book instances in Fig. 5.

6   Conclusion

In this work we have tackled the crossing minimization problem in storyline visualization via an ILP formulation. Despite being an NP-hard problem, computational results show that with our approach one can handle instances of medium size within a reasonable time frame. However, our approach is of purely combinatorial nature, thus, extending it to automatically generate storyline visualizations such that other design criteria are taken into account is not straightforward.

Acknowledgments. The authors are grateful to Käte Zimmer who made her MLCM code, developed in the context of her Master's thesis [35], available to us. Her code served as the basis for our experimental MLCM-TC implementation. Our work is supported by the EU grant FP7-PEOPLE-2012-ITN - Marie-Curie Action "Initial Training Networks" no. 316647 "Mixed-Integer Nonlinear Optimization" (MINO).

References

1. Angelini, P., Da Lozzo, G., Di Battista, G., Frati, F., Roselli, V.: The importance of being proper: (in clustered-level planarity and T-level planarity). Theoret. Comput. Sci. **571**, 1–9 (2015)
2. Barahona, F., Jünger, M., Reinelt, G.: Experiments in quadratic 0–1 programming. Math. Program. **44**, 127–137 (1989)
3. Barahona, F., Mahjoub, A.R.: On the cut polytope. Math. Program. **36**(2), 157–173 (1986)
4. Baumann, F., Buchheim, C., Liers, F.: Exact bipartite crossing minimization under tree constraints. In: Festa, P. (ed.) SEA 2010. LNCS, vol. 6049, pp. 118–128. Springer, Heidelberg (2010). doi:10.1007/978-3-642-13193-6_11
5. Bekos, M.A., Kaufmann, M., Potika, K., Symvonis, A.: Line crossing minimization on metro maps. In: Hong, S.-H., Nishizeki, T., Quan, W. (eds.) GD 2007. LNCS, vol. 4875, pp. 231–242. Springer, Heidelberg (2008). doi:10.1007/978-3-540-77537-9_24
6. Buchheim, C., Wiegele, A., Zheng, L.: Exact algorithms for the quadratic linear ordering problem. INFORMS J. Comput. **22**(1), 168–177 (2010)
7. Buchin, K., Buchin, M., Byrka, J., Nöllenburg, M., Okamoto, Y., Silveira, R.I., Wolff, A.: Drawing (Complete) binary tanglegrams. Algorithmica **62**(1), 309–332 (2012)
8. Chimani, M., Hungerländer, P., Jünger, M., Mutzel, P.: An SDP approach to multi-level crossing minimization. In: Müller-Hannemann, M., Werneck, R. (eds.) Proceedings of the 13th Workshop on Algorithm Engineering and Experiments, ALENEX 2011, pp. 116–126. Society for Industrial and Applied Mathematics (2011)

9. Cui, W., Liu, S., Tan, L., Shi, C., Song, Y., Gao, Z.J., Tong, X., Qu, H.: TextFlow: towards better understanding of evolving topics in text. IEEE Trans. Vis. Comput. Graph. **17**(12), 2412–2421 (2011)

10. De Simone, C.: The cut polytope and the Boolean quadric polytope. Discrete Math. **79**(1), 71–75 (1990)

11. Fernau, H., Kaufmann, M., Poths, M.: Comparing trees via crossing minimization. J. Comput. Syst. Sci. **76**(7), 593–608 (2010)

12. Fink, M., Pupyrev, S.: Metro-line crossing minimization: hardness, approximations, and tractable cases. In: Wismath, S., Wolff, A. (eds.) GD 2013. LNCS, vol. 8242, pp. 328–339. Springer, Heidelberg (2013). doi:10.1007/978-3-319-03841-4_29

13. Garey, M.R., Johnson, D.S.: Crossing number is NP-complete. SIAM J. Algebr. Discrete Methods **4**(3), 312–316 (1983)

14. Grötschel, M., Jünger, M., Reinelt, G.: Facets of the linear ordering polytope. Math. Program. **33**(1), 43–60 (1985)

15. Hammer, P.: Some network flow problems solved with pseudo-Boolean programming. Oper. Res. **13**, 388–399 (1965)

16. Healy, P., Kuusik, A.: Algorithms for multi-level graph planarity testing and layout. Theoret. Comput. Sci. **320**(2–3), 331–344 (2004)

17. IBM: IBM ILOG CPLEX Optimization Studio 12.6 (2014). http://www-01.ibm.com/software/commerce/optimization/cplex-optimizer/

18. Jünger, M., Lee, E.K., Mutzel, P., Odenthal, T.: A polyhedral approach to the multi-layer crossing minimization problem. In: Di Battista, G. (ed.) GD 1997. LNCS, vol. 1353, pp. 13–24. Springer, Heidelberg (1997). doi:10.1007/3-540-63938-1_46

19. Jünger, M., Leipert, S., Mutzel, P.: Level planarity testing in linear time. In: Whitesides, S.H. (ed.) GD 1998. LNCS, vol. 1547, pp. 224–237. Springer, Heidelberg (1998). doi:10.1007/3-540-37623-2_17

20. Jünger, M., Thienel, S.: The ABACUS system for branch-and-cut-and-price algorithms in integer programming and combinatorial optimization. Softw. Pract. Exp. **30**, 1325–1352 (2000)

21. Knuth, D.E.: The Stanford GraphBase source (1993). ftp://ftp.cs.stanford.edu/pub/sgb/sgb.tar.gz

22. Kostitsyna, I., Nöllenburg, M., Polishchuk, V., Schulz, A., Strash, D.: On minimizing crossings in storyline visualizations. In: Di Giacomo, E., Lubiw, A. (eds.) GD 2015. LNCS, vol. 9411, pp. 192–198. Springer, Heidelberg (2015). doi:10.1007/978-3-319-27261-0_16

23. Liers, F., Jünger, M., Reinelt, G., Rinaldi, G.: Computing exact ground states of hard ising spin glass problems by branch-and-cut. In: Hartmann, A.K., Rieger, H. (eds.) New Optimization Algorithms in Physics, pp. 47–69. Wiley-VCH, Weinheim (2004)

24. Liu, S., Wu, Y., Wei, E., Liu, M., Liu, Y.: StoryFlow: tracking the evolution of stories. IEEE Trans. Vis. Comput. Graph. **19**(12), 2436–2445 (2013)

25. Muelder, C.W., Crnovrsanin, T., Sallaberry, A., Ma, K.L.: Egocentric storylines for visual analysis of large dynamic graphs. In: Proceedings of the 2013 IEEE International Conference on Big Data, pp. 56–62 (2013)

26. Munroe, R.: xkcd #657: Movie Narrative Charts (2009). http://xkcd.com/657/

27. Nöllenburg, M., Völker, M., Wolff, A., Holten, D.: Drawing binary tanglegrams: an experimental evaluation. In: Finocchi, I., Hershberger, J. (eds.) Proceedings of the 11th Workshop on Algorithm Engineering and Experiments, ALENEX 2009, pp. 106–119. Society for Industrial and Applied Mathematics (2009)

28. Ogawa, M., Ma, K.L.: Software evolution storylines. In: Telea, A.C. (ed.) Proceedings of the 5th International Symposium on Software Visualization, SOFTVIS 2010, pp. 35–42. ACM (2010)
29. Sugiyama, K., Tagawa, S., Toda, M.: Methods for visual understanding of hierarchical system structures. IEEE Trans. Syst. Man Cybern. **11**(2), 109–125 (1981)
30. Tanahashi, Y.: Movie data set (2013). http://vis.cs.ucdavis.edu/~tanahashi/data_downloads/storyline_visualizations/story_data.tar
31. Tanahashi, Y., Hsueh, C.H., Ma, K.L.: An efficient framework for generating storyline visualizations from streaming data. IEEE Trans. Vis. Comput. Graph. **21**(6), 730–742 (2015)
32. Tanahashi, Y., Ma, K.L.: Design considerations for optimizing storyline visualizations. IEEE Trans. Vis. Comput. Graph. **18**(12), 2679–2688 (2012)
33. Vehlow, C., Beck, F., Auwärter, P., Weiskopf, D.: Visualizing the evolution of communities in dynamic graphs. Comput. Graph. Forum **34**(1), 277–288 (2015)
34. Wotzlaw, A., Speckenmeyer, E., Porschen, S.: Generalized k-ary tanglegrams on level graphs: a satisfiability-based approach and its evaluation. Discrete Appl. Math. **160**(16–17), 2349–2363 (2012)
35. Zimmer, K.: Ein Branch-and-Cut-Algorithmus für Mehrschichten-Kreuzungsminimierung. Master's thesis, Institut für Informatik, Universität zu Köln (2013)

Block Crossings in Storyline Visualizations

Thomas C. van Dijk[1], Martin Fink[2], Norbert Fischer[1], Fabian Lipp[1]([⊠]),
Peter Markfelder[1], Alexander Ravsky[3], Subhash Suri[2], and Alexander Wolff[1]

[1] Lehrstuhl für Informatik I, Universität Würzburg, Würzburg, Germany
fabian.lipp@uni-wuerzburg.de
[2] University of California, Santa Barbara, USA
{fink,suri}@cs.ucsb.edu
[3] Pidstryhach Institute for Applied Problems of Mechanics and Mathematics,
National Academy of Sciences of Ukraine, Lviv, Ukraine
oravsky@mail.ru
http://www1.informatik.uni-wuerzburg.de

Abstract. Storyline visualizations help visualize encounters of the characters in a story over time. Each character is represented by an x-monotone curve that goes from left to right. A meeting is represented by having the characters that participate in the meeting run close together for some time. In order to keep the visual complexity low, rather than just minimizing pairwise crossings of curves, we propose to count *block crossings*, that is, pairs of intersecting bundles of lines.

Our main results are as follows. We show that minimizing the number of block crossings is NP-hard, and we develop, for meetings of bounded size, a constant-factor approximation. We also present two fixed-parameter algorithms and, for meetings of size 2, a greedy heuristic that we evaluate experimentally.

1 Introduction

A storyline visualization is a convenient abstraction for visualizing the complex narrative of interactions among people, objects, or concepts. The motivation comes from the setting of a movie, novel, or play where the narrative develops as a sequence of interconnected scenes, each involving a subset of characters. See Fig. 1 for an example.

The storyline abstraction of characters and events occurring over time can be used as a metaphor for visualizing other situations, from physical events involving groups of people meeting in corporate organizations, political leaders managing global affairs, and groups of scholars collaborating on research to abstract co-occurrences of "topics" such as a global event being covered on the front pages of multiple leading news outlets, or different organizations turning their attention to a common cause.

A storyline visualization maps a set of characters of a story to a set of curves in the plane and a sequence of meetings between the characters to regions in the

---

The full version of this paper is available on arXiv [3]. Whenever we refer to the *Appendix* we mean the appendix of arXiv:1609.00321v1.

© Springer International Publishing AG 2016
Y. Hu and M. Nöllenburg (Eds.): GD 2016, LNCS 9801, pp. 382–398, 2016.
DOI: 10.1007/978-3-319-50106-2_30

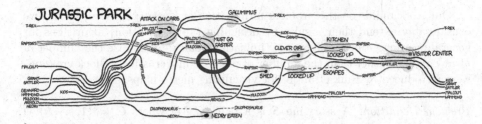

Fig. 1. Storyline visualization for *Jurassic Park* by xkcd [11] with a block crossing (highlighted by a bold green ellipse). (Color figure online)

plane where the corresponding curves come close to each other. The current form of storyline visualizations seems to have been invented by Munroe [11] (compare Fig. 1), who used it to visualize, in a compact way, which subsets of characters meet over the course of a movie. Each character is shown as an x-monotone curve. Meetings occur at certain times from left to right. A meeting corresponds to a point in time where the characters that meet are next to each other with only small gaps between them. Munroe highlights meetings by underlaying them with a gray shaded region, while we use a vertical line for that purpose. Hence, a storyline visualization can be seen as a drawing of a hypergraph whose vertices are represented by the curves and whose edges come in at specific points in time.

A natural objective for the quality of a storyline visualization is to minimize unnecessary "crossings" among the character lines. The number of crossings alone, however, is a poor measure: two blocks of "locally parallel" lines crossing each other are far less distracting than an equal number of crossings randomly scattered throughout the drawing. Therefore, instead of pairwise crossings, we focus on minimizing the number of *block crossings*, where each block crossing involves two arbitrarily large sets of parallel lines forming a crossbar, with no other line in the crossing area; see Fig. 1 for an example.

Previous Work. Kim et al. [6] used storylines to visualize genealogical data; meetings correspond to marriages and special techniques are used to indicate child–parent relationships. Tanahashi and Ma [12] computed storyline visualizations automatically and showed how to adjust the geometry of individual lines to improve the aesthetics of their visualizations. Muelder et al. [10] visualized clustered, dynamic graphs as storylines, summarizing the behavior of the local network surrounding user-selected foci.

Only recently a more theoretical and principled study was initiated by Kostitsyna et al. [8], who considered the problem of minimizing pairwise (not *block*) crossings in storylines. They proved that the problem is NP-hard in general, and showed that it is fixed-parameter tractable with respect to the (total) number of characters. For the special case of 2-character meetings without repetitions, they developed a lower bound on the number of crossings, as well as as an upper bound of $O(k \log k)$ when the meeting graph—whose edges describe the pairwise meetings of characters—is a tree.

Our work builds on the problem formulation of Kostitsyna et al. [8] but we considerably extend their results by designing (approximation) algorithms for general meetings—for a different optimization goal: we minimize the number of *block crossing* rather than the number of pairwise line crossings. Block crossings were introduced by Fink et al. [5] for visualizing metro maps.

Problem Definition. A storyline S is a pair (C, M) where $C = \{1, \ldots, k\}$ is a set of *characters* and $M = [m_1, m_2, \ldots, m_n]$ with $m_i \subseteq C$ and $|m_i| \geq 2$ for $i = 1, 2, \ldots, n$ is a sequence of *meetings* of at least two characters. We call any set $g \subseteq C$ of characters that has at least one meeting, a *group*. We define the *group hypergraph* $\mathcal{H} = (C, \Gamma)$ whose vertices are the characters and whose hyperedges are the groups that are involved in at least one meeting. The group hypergraph does not include the temporal aspect of the storyline—it models only the graph-theoretical structure of groups participating in the storyline meetings; it can be built by lexicographically sorting the meetings in M in $O(nk \log n)$ time.

Note that we do not encode the exact times of the meetings: In a given visualization, at any time t, there is a unique vertical order π of the characters. Without changing π by crossings, we can increase or decrease vertical gaps between lines. If a group g forms a contiguous interval in π^t, then we can bring g's lines within a short distance δ_{group} without any crossing, and also make sure that all other lines are at a larger distance of at least δ_{sep}. Since any group must be supported at a time just before its meeting starts, computing an output drawing consists mainly of changing the permutation of characters over time so that during a meeting its group is supported by the current permutation. We therefore focus on changing the permutation by crossings over time, and only have to be concerned about the order of meetings; the final drawing can be obtained by a simple post-processing from this discrete set of permutations.

If $\{\pi_1, \pi_2, \ldots, \pi_k\} = \{1, 2, \ldots, k\}$, then $\langle \pi_1, \pi_2, \ldots, \pi_k \rangle$ is a permutation of length k of C. For $a \leq b < c$, a *block crossing* (a, b, c) on the permutation $\pi = \langle 1, \ldots, k \rangle$ is the exchange of two consecutive blocks $\langle a, \ldots, b \rangle$ and $\langle b+1, \ldots, c \rangle$; see Fig. 2. A meeting m *fits* a permutation π (or a permutation π *supports* a meeting m) if the characters participating in m form an interval in π. In other words, there is a permutation of m that is part of π. If we apply a sequence B of block crossings to a permutation π in the given order, we denote the resulting permutation by $B(\pi)$.

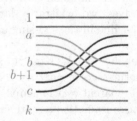

Fig. 2. Block crossing (a, b, c)

Problem 1 (Storyline Block Crossing Minimization (SBCM)). Given a storyline instance (C, M) find a solution consisting of a start permutation π^0 of C and a sequence B of (possibly empty) sequences of block crossings $B_1, B_2, \ldots, B_n$ such that the total number of block crossings is minimized and $\pi^1 = B_1(\pi^0)$ supports m_1, $\pi^2 = B_2(\pi^1)$ supports m_2, etc.

We also consider d-SBCM, a special case of SBCM where meetings involve groups of size at most d, for an arbitrary constant d. E.g., 2-SBCM allows only 2-character meetings, a setting that was also studied by Kostitsyna et al. [8].

Our Results. We observe that a storyline has a crossing-free visualization if and only if its group hypergraph is an interval hypergraph. A hypergraph can be tested for the interval property in $O(n^2)$ time, where n is the number of hyperedges. We show that 2-SBCM is NP-hard (see Sect. 3) and that SBCM is fixed-parameter tractable with respect to k (Sect. 4). The latter can be modified to handle pairwise crossings, where its runtime improves on Kostitsyna et al. [8].

We present a greedy algorithm for 2-SBCM that runs in $O(k^3 n)$ time for k characters. We do some preliminary experiments where we compare greedy solutions to optimal solutions; see Sect. 5. One of our main results is a constant-factor approximation algorithm for d-SBCM for the case that d is bounded and that meetings cannot be repeated; see Sect. 6. Our algorithm is based on a solution for the following NP-complete hypergraph problem, which may be of independent interest. Given a hypergraph $\mathcal{H}$, we want to delete the minimum number of hyperedges so that the remainder is an interval hypergraph. We develop a $(d+1)$-approximation algorithm, where d is the maximum size of a hyperedge in $\mathcal{H}$; see Sect. 7. Finally, we list some open problems in Appendix H.

2 Preliminaries

First, we consider the special case where every meeting consists of two characters. For these restricted instances, every meeting can be realized from any permutation by a single block crossing. This raises the question whether there is also an *optimal* solution that fulfills this condition. The answer is negative—if we may prescribe the start permutation; see Appendix A for details.

Observation 2. *Given an instance of 2-SBCM, there is a solution with at most one block crossing before each of the meetings. In particular, there is a solution with at most n block crossings in total.*

Detecting Crossing-Free Storylines. If a storyline admits a crossing-free visualization, then the vertical permutation of the character lines remains the same over time, and all meetings involve groups that form contiguous subsets in that permutation. (The visualization can be obtained by placing characters along a vertical line in the correct permutation and for each meeting bringing its lines together for the duration of the meeting and then separating them apart again.) In other words, a single permutation supports each group of $\mathcal{H} = (C, \Gamma)$. This holds if and only if $\mathcal{H}$ is an *interval hypergraph*. This is the case if there exists a permutation $\pi = \langle v_1, \dots, v_k \rangle$ of C such that each hyperedge $e \in \Gamma$ corresponds to a contiguous block of characters in this permutation. As an anonymous reviewer pointed out, this is equivalent to the hypergraph having path support [1]. An interval hypergraph can be visualized by placing all of its vertices on a line, and drawing each of its hyperedges as an interval that includes all vertices of e

and no vertex of $V \setminus e$. Checking whether a k-vertex hypergraph is an interval hypergraph takes $O(k^2)$ time [13]. Recall that we can build $\mathcal{H}$ in $O(nk \log n)$ time.

Theorem 3. *Given the group hypergraph $\mathcal{H}$ of an instance of SBCM with k characters, we can check in $O(k^2)$ time whether a crossing-free solution exists.*

For 2-SBCM we only need to check (in $O(k)$ time) whether $\mathcal{H}$ is a collection of vertex-disjoint paths; this is dominated by the time ($O(n)$) for building $\mathcal{H}$.

3   NP-Completeness of SBCM

In this section we prove that SBCM is NP-complete. This is known for BCM. But SBCM is not simply a generalization of BCM because in SBCM we can choose an arbitrary start permutation. Therefore, the idea of our hardness proof is to force a certain start permutation by adding some characters and meetings. We reduce from SORTING BY TRANSPOSITIONS (SBT), which has also been used to show the hardness of BCM [5]. In SBT, the problem is to decide whether there is a sequence of transpositions (which are equivalent to block crossings) of length at most k that transforms a given permutation π to the identity. SBT was recently shown NP-hard by Bulteau et al. [2].

We show hardness for 2-SBCM, which also implies that SBCM is NP-hard. It is easy to see that SBCM is in NP: Obviously, the maximum number of block crossings needed for any number of characters and meetings is bounded by a polynomial in k and n. Therefore also the size of the solutions is bounded by a polynomial. To test the feasibility of a solution efficiently, we simply test whether the permutations between the block crossings support the meetings in the right order from left to right. We will use the following obvious fact.

Observation 4. *If permutation π needs c block crossings to be sorted, any permutation containing π as subsequence needs at least c block crossings to be sorted.*

Theorem 5. *2-SBCM is NP-complete.*

Proof. It remains to show the NP-hardness. We reduce from SBT. Given an instance of SBT, that is, a permutation π of $\{1, \ldots, k\}$, we show how to use a hypothetical, efficient algorithm for 2-SBCM to determine the minimum number of transpositions (i.e., block crossings) that transforms π to the identity $\iota = \langle 1, 2, \ldots, k \rangle$. Note that π can be sorted by at most k block crossings. So k is an upper bound for an optimal solution of instance π of SBT.

We extend the set of characters $\{1, 2, \ldots, k\}$ to $C = \{1, \ldots, k, c_1, c_2, \ldots, c_{2k}\}$. Correspondingly, we extend $\pi = \langle \pi_1, \pi_2, \ldots, \pi_k \rangle$ to $\pi' = \langle c_1, \ldots, c_{2k}, \pi_1, \ldots, \pi_k \rangle$ and ι to $\iota' = \langle c_1, c_2, \ldots, c_{2k}, 1, 2, \ldots, k \rangle$. Let $M_{\pi'}$ and $M_{\iota'}$ be the sequences of meetings of all neighboring pairs in π' and ι', respectively. Let M_1 and M_2 be the concatenations of $k + 1$ copies of $M_{\pi'}$ and $M_{\iota'}$, respectively. By repeating we get $M_1 = M_{\pi'}^{k+1}$ and $M_2 = M_{\iota'}^{k+1}$. This yields the instance $\mathcal{S} = (C, M)$ of 2-SBCM, where M is the concatenation of M_1 and M_2; see Fig. 3.

We show that the number of block crossings needed for the 2-SBCM instance $\mathcal{S}$ equals the number of block crossings to solve instance π of SBT.

First, let B be a shortest sequence of block crossings to sort π. Then, (π', B) is a feasible solution for $\mathcal{S}$. The start permutation π' supports all meetings in M_1 without any block crossing. Using B, the lines are sorted to ι', and this permutation supports all meetings in M_2 without any further block crossings; see Fig. 3. Hence, the number of block crossings in any solution of π is an upper bound for the minimum number of block crossings needed for $\mathcal{S}$.

For the other direction, let (π^*, B^*) be an optimal solution for $\mathcal{S}$. Any solution of 2-SBCM gives rise to a symmetric solution that is obtained by reversing the order of the characters. Without loss of generality, we assume that π' (rather than the reverse permutation π'^R) occurs somewhere in M_1.

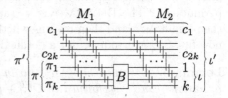

Fig. 3. Solution for the 2-SBCM instance $\mathcal{S}$ corresponding to a solution B of instance π of SBT. The box B represents the block crossings.

Next, we show that the start permutation π' occurs somewhere in M_1 and that ι' occurs somewhere in M_2. If there is a sequence $M_{\pi'}$ of meetings between which there is no block crossing, the permutation at this position can only be the start permutation π' or its reverse. For a contradiction, assume that π' does not occur during M_1 in the layout induced by (π^*, B^*). Then there is no such sequence without any block crossing in it. As this sequence is repeated $k + 1$ times, the solution would need at least $k + 1$ block crossings. This contradicts our upper bound, which is k. Analogously, we can show that the permutation ι' or its reverse occurs in M_2.

We now want to show that the unreversed version of ι' occurs in M_2. For a contradiction, assume the opposite. We forget about the lines $1, \ldots, k$ and only consider the sequence $\pi'' = \langle c_1, \ldots, c_{2k} \rangle$ in π' which is reversed to $\iota'' = \langle c_{2k}, \ldots, c_1 \rangle$ in ι'^R. Eriksson et al. [4] showed that we need $\lceil (l + 1)/2 \rceil$ block crossings to reverse a permutation of l elements. This implies that we need $k + 1$ block crossings to transform π'' to ι''. As π' and ι'^R contain these sequences as subsequences, Observation 4 implies that the transformation from π' to ι'^R also needs at least $k + 1$ block crossings. As the optimal solution uses at most k block crossings, we know that we cannot reach ι'^R and thus the sequence of permutations contains π' and ι'.

The sequence of block crossings that transforms π' to ι' yields a sequence B of block crossings of the same length that transforms π to ι. This shows that the length of a solution for $\mathcal{S}$ is an upper bound for the length of an optimal solution of the corresponding SBT instance π. Thus, the two are equal.   □

Hardness Without Repetitions. With arbitrarily large meetings, SBCM is hard even without repeating meetings. We can emulate a repeated sequence of 2-character meetings by gradually increasing group sizes; see Appendix B.

4   Exact Algorithms

We present two exact algorithms. Conceptually, both build up a sequence of block crossings while keeping track of how many meetings have already been accomplished. The first uses polynomial space; the second improves the runtime at the cost of exponential space.

We start with a data structure that keeps track of permutations, block crossings and meetings. It is initialized with a given permutation and has two operations. The CHECK operation returns whether a given meeting fits the current permutation. The BLOCKMOVE operation performs a given block crossing on the permutation and then returns whether the most-recently CHECKed meeting now fits. See Appendix C for a detailed description.

Lemma 6. *A sequence of arbitrarily interleaved* BLOCKMOVE *and* CHECK *operations can be performed in* $O(\beta + \mu)$ *time, where* β *is the number of block crossings and* μ *is sum of cardinalities of the meetings given to* CHECK. *Space usage is* $O(k)$.

A block crossing can be represented by indices (a, b, c) with $1 \leq a \leq b < c \leq k$; hence, there are $\frac{k^3 - k}{6}$ distinct block crossings on a permutation of length k.

Now we provide an output-sensitive algorithm for SBCM whose runtime depends on the number of block crossings required by the optimum.

Theorem 7. *An instance of SBCM can be solved in* $O(k! \cdot (\frac{k^3 - k}{6})^{\beta} \cdot (\beta + \mu))$ *time and* $O(\beta k)$ *working space if a solution with* β *block crossings exists, where* $\mu = \sum_{i \in M} |m_i|$.

Proof. Consider a branching algorithm that starts from a permutation of the characters and keeps trying all possible block crossings. This has branching factor $\frac{k^3 - k}{6}$ and we can enumerate the children of a node in constant time each by enumerating triples (a, b, c). While applying block crossings, the algorithm keeps track of how many meetings fit this sequence of permutations using the data structure from Lemma 6. We use depth-first iterative-deepening search [7] from all possible start permutations until we find a sequence of permutations that fulfills all meetings. Correctness follows from the iterative deepening: we want an (unweighted) shortest sequence of block crossings. The runtime and space bounds follow from the standard analysis of iterative-deepening search, observing that a node uses $O(k)$ space and it takes $O(\beta + \mu)$ time in total to evaluate a path from root to leaf.   □

We have that μ is $O(kn)$ since there are n meetings and each consists of at most k characters. At the cost of exponential space, we can improve the runtime and get rid of the dependence on β, showing the problem to be fixed parameter linear for k. We note that the following algorithm can easily be adapted to handle pairwise crossings rather than block crossings; in this case the runtime improves upon the original result of Kostisyna et al. [8] by a factor of $k!$.

Theorem 8. *An instance of SBCM can be solved in $O(k! \cdot k^3 \cdot n)$ time and $O(k! \cdot k \cdot n)$ space.*

Proof. Let $f(\pi, \ell)$ be the optimal number of block crossings in a solution to the given instance when restricted to the first ℓ meetings and to have π as its final permutation. Note that by definition the solution for the actual instance is given by $\min_{\pi^*} f(\pi^*, n)$, where the minimum ranges over all possible permutations. As a base case, $f(\pi, 0) = 0$ for all π, since the empty set of meetings is supported by any permutation. Let π and π' be permutations that are one block crossing apart and let $0 \leq \ell \leq \ell'$. If the meetings $\{m_{\ell+1}, \ldots, m_{\ell'}\}$ fit π', then $f(\pi', \ell') \leq f(\pi, \ell) + 1$: if we can support the first ℓ meetings and end on π, then with one additional block crossing we can support the first ℓ' meetings and end with π'.

We now model this as a graph. Let G be an unweighted directed graph on nodes (π, ℓ) and call a node *start node* if $\ell = 0$. There is an arc from (π, ℓ) to (π', ℓ') if and only if π and π' are one block crossing apart, $\ell \leq \ell'$, and the meetings $\{m_{\ell+1}, \ldots, m_{\ell'}\}$ fit π'. Note that we allow $\ell = \ell'$ since we may need to allow block crossings that do not immediately achieve an additional meeting (cf. Proposition 18), so G is not acyclic. In the constructed graph, $f(\pi, \ell)$ equals the graph distance from the node (π, ℓ) to the closest start node. Call a path to a start node that realizes this distance *optimal*.

In G, consider any path $[(\pi_1, \ell_1), (\pi_2, \ell_2), (\pi_3, \ell_3)]$ with $\ell_3 > \ell_2$. If meeting $\ell_2 + 1$ fits π_2, then $[(\pi_1, \ell_1), (\pi_2, \ell_2 + 1), (\pi_3, \ell_3)]$ is also a path. Repeating this transformation shows that for all π, the node (π, n) has an optimal path in which every arc maximally increases ℓ. Let G' be the graph where we drop all arcs from G that do not maximally increase ℓ. Note that G' still contains a path that corresponds to the global optimum.

The graph G' has $O(k! \cdot n)$ nodes and each node has outdegree $O(k^3)$. Then a breadth-first search from all start nodes to any node (π^*, n) achieves the claimed time and space bounds, assuming we can enumerate the outgoing arcs of a node in constant time each.

For a given node (π, ℓ) we can enumerate all possible block crossings in constant time each, as before. In G', we also need to know the maximum ℓ' such that all meetings $\ell + 1$ up to ℓ' fit π'. Note that ℓ' only depends on ℓ and π'. We precompute a table $M(\pi, \ell)$ that gives this value. Computing $M(\pi, \ell)$ for given π and all ℓ takes a total of $O(kn)$ time: first compute for every m_i whether it fits π, then compute the implied 'forward pointers' using a linear scan. So using $O(k! \cdot k \cdot n)$ preprocessing time and $O(k! \cdot n)$ space, we have an efficient implementation of the breadth-first search. The theorem follows. □

5   SBCM with Meetings of Two Characters

A Greedy Algorithm. To quickly draw good storyline visualizations for 2-SBCM, we develop an $O(kn)$-time greedy algorithm. Given an instance $\mathcal{S} = (C, M)$, we reserve a list $B = []$ that the algorithm will use to store the block crossings. The algorithm starts with an arbitrary permutation π^0 of C. In every step the algorithm removes all meetings from the beginning of M that fit the current permutation π^i of the algorithm. Subsequently, the algorithm picks a block crossing b such that the resulting permutation $\pi^{i+1} = b(\pi^i)$ supports the maximum number of meetings from the beginning of M. Then b is appended to the list B. This process repeats until M is empty. The algorithm returns (π^0, B).

Note that there are at most $O(k^3)$ possible block crossings. Thus to find the appropriate block crossings, the algorithm could simply check all of them. Many of those, however, will result in permutations that do not even support the next meeting, which would be a bad choice. Hence, our algorithm considers only *relevant* block crossings, i.e., block crossings yielding a permutation that supports the next meeting. Let $\{c, c'\}$ be the next meeting in M. If x and y are the positions of c and c' in the current permutation, i.e., $\pi^i_x = c$ and $\pi^i_y = c'$ (without loss of generality, assume $x < y$), the relevant block crossings are:

$$\{(z, x, y-1) \colon 1 \le z \le x\} \cup \{(x, z, y) \colon x \le z < y\} \cup \{(x+1, y-1, z) \colon y \le z \le k\}.$$

So the number of relevant block crossings in each step is $k + 1$. Let n_i be the maximum number of meetings at the beginning of M we can achieve by one of these block crossings. We use the data structure in Lemma 6 and check for each relevant block crossing how many meetings can be done with this permutation. Hence, we can identify a block crossing achieving the maximum number in $O(kn_i)$ time since we have to check $k + 1$ paths containing up to n_i meetings each. Clearly, the numbers of meetings n_i in each iteration of the algorithm sum up to n and therefore the algorithm runs in $O(kn)$ total time.

The way we described the greedy algorithm, it starts with an arbitrary permutation. Instead, we could start with a permutation that supports the maximum number of meetings before the first block crossing needs to be done. In other words, we want to find a maximal prefix M' of M such that (C, M') can be represented without any block crossings. We can find M' in $O(kn)$ time: we start with an empty graph and add the meetings successively. In each step we check whether the graph is still a collection of paths, which can be done in $O(k)$ time. It is easy to construct a permutation that supports all meetings in M'. While this is a sensible heuristic, we do not prove that this reduces the total number of block crossings. Indeed, we experimentally observe that while the heuristic is generally good, this is not always the case; see Fig. 4 for an example that uses the heuristic start permutation.

Note that the greedy algorithm yields optimal solutions for special cases of 2-SBCM. The proof for the following theorem can be found in Appendix D.

Theorem 9. *For $k = 3$, the greedy algorithm produces optimal solutions.*

Experimental Evaluation. In this section, we report on some preliminary experimental results. We only consider 2-SBCM. We generated random instances as follows. Given n and k, we generate n pairs of characters as meetings, uniformly at random using rejection sampling to ensure that consecutive meetings are different. (Repeated meetings are not sensible.)

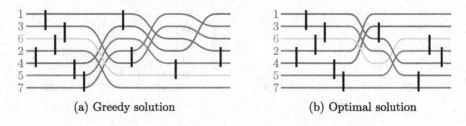

(a) Greedy solution   (b) Optimal solution

Fig. 4. The greedy algorithm is not optimal.

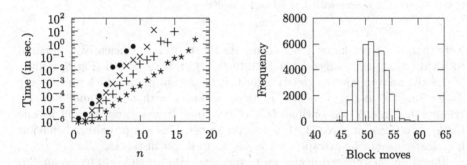

Fig. 5. Left: Runtime of the exact algorithm of Theorem 7 on random instances with $k = 4(\star), 5(+), 6(\times), 7(\bullet)$. Each data point is the average of 50 random instances. Right: Histogram of the number of block crossings used by the greedy algorithm for all $k!$ different start permutations, on a single random instance with $n = 100$ and $k = 8$.

First, we consider the exact algorithm of Theorem 7. As expected, its runtime depends heavily on k (Fig. 5, left). Perhaps unexpectedly, we observe exponential runtime in n. This is actually a property of our random instances, in which β tends to increase linearly with n. Note that this does not invalidate the algorithm since we may be interested in instances for which β is indeed small.

Since the exact algorithm is feasible only for rather small instances, we now shift our focus to the greedy algorithm. Recall that it starts with an arbitrary permutation and proceeds greedily. The histogram in Fig. 5 (right) shows the number of block crossings used by the greedy algorithm depending on the start permutation, for a single random instance: this bell curve is typical. We see that there are "rare" start permutations that do strictly better than almost all others. Indeed, for the reported instance, a random start permutation does 7.2 block crossings worse in expectation than the best possible start permutation.

We call the best possible result of the greedy algorithm over all start permutations BestGreedy, which we calculate by brute force. Let RandomGreedy start with a permutation chosen uniformly at random, and let Heuristic-Greedy start with the heuristic start permutation that we have described above. The histogram in Fig. 6 (left) shows how many more block crossings HeuristicGreedy uses than BestGreedy on random instances. This distribution is heaviest near zero, but there are instances where performance is poor.

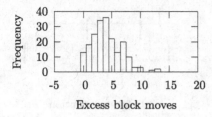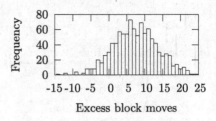

Fig. 6. Left: histogram of HEURISTICGREEDY minus BESTGREEDY, 200 instances with with $k = 7$ and $n = 100$. Right: histogram of RANDOMGREEDY minus HEURISTIC-GREEDY, 1000 instances with $k = 30$ and $n = 200$.

Note that we do not know how to compute BESTGREEDY efficiently. Compared to RANDOMGREEDY, we see that HEURISTICGREEDY fares well (Fig. 6, right).

Lastly, we compare the greedy algorithm to the optimum, which we can only do for small k and n. On 1000 random instances with $k = 5$ and $n = 12$, HEURISTICGREEDY was optimal 56% of the time. It was sometimes off by one (38%), two (5%), or three (1%), but never worse. This is a promising behavior, but clearly cannot be extrapolated verbatim to larger instances.

Based on these experiments, we recommend HEURISTICGREEDY as an efficient, reasonable heuristic.

6 Approximation Algorithm

We now develop a constant-factor approximation algorithm for d-SBCM where d is a constant. We initially assume that each group meeting occurs exactly once, but later show how to extend our results to the setting where the same group can meet a bounded number of times.

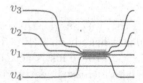

Fig. 7. Meeting $\{v_1, v_2, v_3, v_4\}$

Overview. Our approximation algorithm has the following three main steps.

1. Reduce the input group hypergraph $\mathcal{H} = (C, \Gamma)$ to an *interval hypergraph* $\mathcal{H}_f = (C, \Gamma \setminus \Gamma_p)$ by deleting a subset $\Gamma_p \subseteq \Gamma$ of the edges of $\mathcal{H}$.
2. Choose a permutation π^0 of the characters that supports all groups of this interval hypergraph $\mathcal{H}_f$. Thus, π^0 is the order of characters at the beginning of the timeline.
3. Incrementally create support for each deleted meeting of Γ_p in order of increasing time, as follows. Suppose that $g \in \Gamma_p$ is the group meeting to support. Keep one of the character lines involved in this meeting fixed and bring, for the duration of the meeting, the remaining (at most $d - 1$) lines close to it. Then retract those lines to their original position in π^0; see Fig. 7.

Step 2 is straightforward: Sect. 2 shows how to find a permutation supporting all the groups for an interval hypergraph. In Step 3, we introduce at most $2(d-1)$ block crossings for each meeting $g \in \Gamma_p$ not initially supported. The main technical parts of the algorithm are Step 1 and an analysis to charge at most a constant number of block crossings in Step 3 to a block crossing in the optimal visualization. Step 1 requires solving a hypergraph problem; this is technically the most challenging part, and consumes the entire Sect. 7.

Bounds and Analysis. We call Γ_p *paid* edges, and the remainder $\Gamma_f = \Gamma \setminus \Gamma_p$ *free* edges. Intuitively, free edges can be realized without block crossings because $\mathcal{H}_f$ is an interval hypergraph, while the edges of Γ_p must be charged to block crossings of the optimal drawing. We initialize the drawing by placing the characters in the vertical order π^0, which supports all the groups in Γ_f. Now we consider the paid edges in left-to-right order. Suppose that the next meeting involves a group $g' \in \Gamma_p$. We have $|g'| \leq d$. We arbitrarily fix one of its characters, leaving its line intact, and bring the remaining $(d-1)$ lines in its vicinity to realize the meeting. This creates at most $(d-1)$ block crossings, one per line. When the meeting is over, we again use up to $(d-1)$ block crossings to revert the lines back to their original position prescribed by π^0; see Fig. 7.

We do this for each paid hyperedge, giving rise to at most $2(d-1)|\Gamma_p|$ block crossings. We now prove that this bound is within a constant factor of optimal. We first establish a lower bound on the optimal number of block crossings *assuming* that π^0 is the optimal start permutation.

Lemma 10. *Let π be a permutation of the characters, let Γ_f be the groups supported by π, and let $\Gamma_p = \Gamma \setminus \Gamma_f$. Any storyline visualization that uses π as the start permutation has at least $4|\Gamma_p|/(3d^2)$ block crossings.*

Proof. Let $g \in \Gamma_p$. Since g is not supported by π, the optimal drawing does not contain the characters of g as a contiguous block initially. However, in order to support this meeting, these characters must eventually become contiguous before the meeting starts. The order changes only through (block) crossings; we bound the number of groups that can become supported after each block crossing.

After a block crossing, at most three pairs of lines that were not neighbors before can become neighbors in the permutation: after the blocks $C_1, C_2 \subseteq C$ cross, there is one position in the permutation where a line of C_1 is next to a line of C_2, and two positions with a line of C_1 (C_2, respectively) and a line of $C \setminus (C_1 \cup C_2)$. Any group that was not supported, but is supported after the block crossing, must contain one of these pairs. We can describe each such group in the new permutation by specifying the new pair and the numbers d_1 and d_2 of characters of the group above and below the new pair in the permutation. Since the group size is at most d, we have $d_1 + d_2 \leq d$. The product $d_1(d - d_1)$ achieves its maximum value for $d_1 = d_2 = d/2$, and so there are at most $d^2/4$ possible groups for each new pair. Thus, the total number of newly supported groups after a block crossing is at most $3d^2/4$, which shows that the optimal number of block crossings is at least $4|\Gamma_p|/(3d^2)$, completing the proof. $\qquad\square$

We now bound the loss of optimality caused by not knowing the initial permutation used by the optimal solution. The key idea here is to use a constant-factor approximation for the problem of deleting the minimum number of hyper-edges from $\mathcal{H}$ so that it becomes an interval hypergraph (INTERVAL HYPER-GRAPH EDGE DELETION). We prove the following theorem in Sect. 7.

Theorem 11. *We can find a* $(d + 1)$-*approximation for* INTERVAL HYPER-GRAPH EDGE DELETION *on group hypergraphs with* n *meetings of rank* d *in* $O(n^2)$ *time.*

Let Γ_{OPT} be the set of paid edges in the optimal solution, and Γ_p the set of paid edges in our algorithm. By Theorem 11, we have $|\Gamma_p| \leq (d+1)|\Gamma_{\text{OPT}}|$. Let ALG and OPT be the numbers of block crossings for our algorithm and the optimal solution, respectively. By Lemma 10, we have OPT $\geq 4|\Gamma_{\text{OPT}}|/(3d^2)$, which gives $|\Gamma_{\text{OPT}}| \leq 3d^2/4 \cdot$ OPT. On the other hand, we have ALG $\leq 2(d-1)|\Gamma_p| \leq 2(d-1)(d+1)|\Gamma_{\text{OPT}}|$. Combining the two inequalities, we get ALG $\leq 3(d^2 - 1)d^2/2 \cdot$ OPT, which establishes our main result.

Theorem 12. *d-SBCM admits a* $(3(d^2 - 1)d^2/2)$-*approximation algorithm.*

Remark. We assumed that each group meets only once, but we can extend the result if each group can meet c times, for constant c. Our algorithm then yields a $(c \cdot 3(d^2-1)d^2/2)$-factor approximation; each repetition of a meeting may trigger a constant number of block crossings not present in the optimal solution.

Runtime Analysis. We have to consider the permutation (of length k) of characters before and after each of the n meetings, as well as after each of the $O(n)$ block crossings. This results in $O(kn)$ time for the last part of the algorithm, but this is dominated by the time $(O(n^2))$ needed for finding Γ_p and for determining the start permutation.

We can improve the running time to $O(kn)$ by a slight modification: using the approximation algorithm for INTERVAL HYPERGRAPH EDGE DELETION is only necessary for sparse instances. If $\mathcal{H}$ has sufficiently many edges, any start permutation will yield a good approximation. Since no meeting involves more than d characters, no start permutation can support more than dk meetings. If $n \geq 2dk$, then even the optimal solution must therefore remove at least half of the edges. Hence, taking an arbitrary start permutation yields an approximation factor of at most $2 < d + 1$.

We now change the algorithm to use an arbitrary start permutation if $n \geq 2dk$ and only use the approximation for INTERVAL HYPERGRAPH EDGE DELETION otherwise, i.e., especially only if there are $O(k)$ edges. Hence, for sparse instances we have $O(n^2) = O(k^2)$, and for dense instances, the $O(n^2)$ runtime is not necessary. We get the following improved result. (The runtime is worst-case optimal since the output complexity is of the same order.)

Theorem 13. *d-SBCM admits an* $O(kn)$-*time* $(3(d^2 - 1)d^2/2)$-*approximation algorithm.*

Using some special properties of the 2-character case, we can improve the approximation factor for 2-SBCM from 18 to 12; see Appendix E.

7 Interval Hypergraph Edge Deletion

We now describe the main missing piece from our approximation algorithm: how to approximate the minimum number of edges whose deletion reduces a hypergraph to an interval hypergraph, i.e., how to solve the following problem.

Problem 14 (INTERVAL HYPERGRAPH EDGE DELETION). Given a hypergraph $\mathcal{H} = (V, E)$ find a smallest set $E_p \subseteq E$ such that $\mathcal{H}_f = (V, E \setminus E_p)$ is an interval hypergraph.

Note that a graph contains a Hamiltonian path if and only if one can remove all but $n - 1$ edges so that only vertex-disjoint paths (here, a single path) remain; hence, our problem is hard even for graphs.

Theorem 15. INTERVAL HYPERGRAPH EDGE DELETION *is NP-hard.*

We now present a $(d + 1)$-approximation algorithm for rank-d hypergraphs, in which each hyperedge has at most d vertices. In this section we give all main ideas. Detailed proofs can be found in Appendix F; they are mostly not too hard to obtain, but require the distinction of many cases.

For our algorithm, we use the following characterization: A hypergraph is an interval hypergraph if and only if it contains none of the hypergraphs shown in Fig. 8 as a subhypergraph [9,13]. Due to the bounded rank, the families of F_k and M_k are finite with F_{d-2} and M_{d-1} as largest members. Cycles are the only arbitrarily large forbidden subhypergraphs in our setting. Let $\mathcal{F} = \{O_1, O_2, F_1, \ldots, F_{d-2}, M_1, \ldots, M_{d-1}, C_3, \ldots, C_{d+1}\}$. A hypergraph is $\mathcal{F}$-free if it does not contain any hypergraph of $\mathcal{F}$ as a subhypergraph. Note that a cycle in a hypergraph consists of hyperedges $e_1, \ldots, e_k$ so that there are vertices $v_1, \ldots, v_k$ with $v_i \in e_{i-1} \cap e_i$ for $2 \leq i \leq k$ (and $v_1 \in e_1 \cap e_k$) and no edge e_i contains a vertex of $v_1, \ldots, v_k$ except for v_i and v_{i+1}.

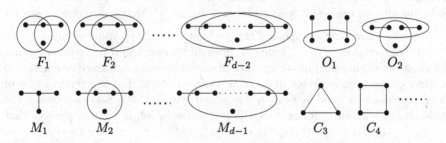

Fig. 8. Forbidden subhypergraphs for interval hypergraphs (edges represent pairwise hyperedges, circles/ellipses show hyperedges of higher cardinality).

Our algorithm consists of two steps. First, we search for subhypergraphs contained in $\mathcal{F}$, and remove all edges involved in these hypergraphs. In the second step, we break remaining (longer) cycles by removing some more hyperedges after carefully analyzing the structure of connected components. Subhypergraphs in $\mathcal{F}$ consist of at most $d + 1$ hyperedges. A given optimal solution must remove at least one of the hyperedges; removing all of them instead yields a factor of at most $d + 1$. The second step will not negatively affect this approximation factor.

Intuitively, allowing long cycles, but forbidding subhypergraphs of $\mathcal{F}$, results in a generalization of interval hypergraphs where the vertices may be placed on a cycle instead of a vertical line. This is not exactly true, but we will see that the connected components after the first step have a structure similar to this, which will help us find a set of edges whose removal destroys all remaining long cycles.

Lemma 22 (Appendix F) shows that any vertex is contained in at most three hyperedges of a cycle, where the case of three hyperedges with a common vertex occurs only if a hyperedge is contained in the union of its two neighbors in the cycle. Assume that e_1, e_2, and e_3 are consecutive edges of a cycle C. If all three edges are present in an interval representation, we know that we will first encounter vertices that are only contained in e_1, then vertices that are in $(e_1 \cap e_2) \setminus e_3$, then vertices in $e_1 \cap e_2 \cap e_3$, followed by vertices of $(e_2 \cap e_3) \setminus e_1$, and vertices of $e_3 \setminus (e_1 \cup e_2)$. Some of the sets (except for pairwise intersections) may be empty. We do not know the order of vertices within one set, but we know the relative order of any pair of vertices of different sets. By generalizing this to the whole cycle, we get a cyclic order—describing the local order in a possible interval representation—of sets defined by containment in 1, 2, or 3 hyperedges. We call these sets *cycle-sets* and their cyclic order the *cycle-order* of C.

We can analyze how an edge $e \notin C$ relates to the order of cycle-sets; e can contain a cycle-set completely, can be disjoint from it, or can contain only part of its vertices. We call a consecutive sequence of cycle-sets contained in edge e—potentially starting and ending with cycle-sets partially contained in e—an *interval* of e on C. The following lemma shows that every edge forms only a single interval on a given cycle.

Lemma 16. *If a hyperedge $e \in E$ intersects two cycle-sets of a cycle C, then e fully contains all cycle-sets lying in between in one of the two directions along C.*

We now know that by opening the cycle at a single position within a cycle-set not contained in e, $C + e$ forms an interval hypergraph. Edge e adds further information: If only part of the vertices of a cycle-set are contained in e and also vertices of the next cycle-set in one direction, we know that the vertices of e in the first cycle-set should be next to the second cycle-set. We use this to refine the cycle-sets to a cyclic order of *cells*, the *cell order* (a cell is a set of vertices that should be contiguous in the cyclic order). Initially, the cells are the cycle-sets. In each step we refine the cell-order by inserting an edge containing vertices of more than one cell, possibly splitting two cells into two subcells each. The following lemma shows that during this process of refinements, as an invariant each remaining edge forms a single interval on the cell order.

Lemma 17. *If a hyperedge $e \in E$ intersects two cells, then e fully contains all cells lying in between in one of the two directions along the cyclic order.*

After refining cells as long as possible, each edge of the connected component that we did not insert lies completely within a single cell. Several edges can lie within the same cell, forming a hypergraph that imposes restrictions on the order of vertices within the cell. However, the cell contains fewer than d vertices. Hence, this small hypergraph cannot contain any cycles, since we removed all short cycles, and must be an interval hypergraph.

With this cell-structure, it is not too hard to show that the following strategy to make the connected component an interval hypergraph is optimal (see Lemmas 24, 25 and 26 in Appendix F): For each pair of adjacent cells we determine the number of edges containing both cells, select the pair minimizing that number, and remove all edges containing both. The cell order then yields an order of the connected component's vertices that supports all remaining edges. Since this last step of the algorithm is done optimally, we do not further change the approximation ratio, which, overall, is $d + 1$, because we never remove more than $d + 1$ edges for at least one edge that the optimal solution removes.

Runtime. Our algorithm can be implemented to run in $O(m^2)$ time for m hyperedges. We give the main ideas here and present details in Appendix G. When searching for forbidden subhypergraphs, we first remove all cycles of length $k \leq d$ using a modified breadth-first search in $O(m^2)$ time. The remaining types of forbidden subhypergraphs each contain an edge that contains all but one (O_2 and F_k), two (M_k), or three (O_1) vertices of the subhypergraph. We always start searching from such an edge and use that all short cycles have already been removed. In the second phase, we determine the connected components and initialize the cell order for each of them, in $O(n + m)$ time. Stepwise refinement requires $O(m^2)$ time. Counting hyperedges between adjacent cells, determining optimal splitting points, and finding the final order can all be done in linear time.

Theorem 11 *We can find a $(d + 1)$-approximation for* INTERVAL HYPERGRAPH EDGE DELETION *on hypergraphs with m hyperedges of rank d in $O(m^2)$ time.*

References

1. Buchin, K., van Kreveld, M.J., Meijer, H., Speckmann, B., Verbeek, K.: On planar supports for hypergraphs. J. Graph Algorithms Appl. **15**(4), 533–549 (2011)
2. Bulteau, L., Fertin, G., Rusu, I.: Sorting by transpositions is difficult. SIAM J. Discrete Math. **26**(3), 1148–1180 (2012)
3. van Dijk, T.C., Fink, M., Fischer, N., Lipp, F., Markfelder, P., Ravsky, A., Suri, S., Wolff, A.: Block crossings in storyline visualizations (2016). http://arxiv.org/abs/1609.00321
4. Eriksson, H., Eriksson, K., Karlander, J., Svensson, L., Wästlund, J.: Sorting a bridge hand. Discrete Math. **241**(1), 289–300 (2001)

5. Fink, M., Pupyrev, S., Wolff, A.: Ordering metro lines by block crossings. J. Graph Algorithms Appl. **19**(1), 111–153 (2015)
6. Kim, N.W., Card, S.K., Heer, J.: Tracing genealogical data with timenets. In: Proceedings of the International Conference Advanced Visual Interfaces (AVI 2010), pp. 241–248 (2010)
7. Korf, R.E.: Depth-first iterative-deepening: an optimal admissible tree search. Artif. Intell. **27**(1), 97–109 (1985)
8. Kostitsyna, I., Nöllenburg, M., Polishchuk, V., Schulz, A., Strash, D.: On minimizing crossings in storyline visualizations. In: Di Giacomo, E., Lubiw, A. (eds.) GD 2015. LNCS, vol. 9411, pp. 192–198. Springer, Heidelberg (2015). doi:10.1007/978-3-319-27261-0_16
9. Moore, J.I.: Interval hypergraphs and D-interval hypergraphs. Discrete Math. **17**(2), 173–179 (1977)
10. Muelder, C., Crnovrsanin, T., Sallaberry, A., Ma, K.: Egocentric storylines for visual analysis of large dynamic graphs. In: Proceedings of the IEEE International Conference Big Data, pp. 56–62 (2013)
11. Munroe, R.: Movie narrative charts. https://xkcd.com/657/
12. Tanahashi, Y., Ma, K.: Design considerations for optimizing storyline visualizations. IEEE Trans. Vis. Comput. Graph. **18**(12), 2679–2688 (2012)
13. Trotter, W.T., Moore, J.I.: Characterization problems for graphs, partially ordered sets, lattices, and families of sets. Discrete Math. **16**(4), 361–381 (1976)

The Bundled Crossing Number

Md. Jawaherul Alam[1], Martin Fink[2], and Sergey Pupyrev[3,4(✉)]

[1] Department of Computer Science, University of California, Irvine, USA
jawaherul@gmail.com
[2] Department of Computer Science, University of California, Santa Barbara, USA
fink@cs.ucsb.edu
[3] Department of Computer Science, University of Arizona, Tucson, USA
spupyrev@gmail.com
[4] Institute of Mathematics and Computer Science, Ural Federal University,
Yekaterinburg, Russia

Abstract. We study the algorithmic aspect of edge bundling. A bundled crossing in a drawing of a graph is a group of crossings between two sets of parallel edges. The bundled crossing number is the minimum number of bundled crossings that group all crossings in a drawing of the graph.

We show that the bundled crossing number is closely related to the orientable genus of the graph. If multiple crossings and self-intersections of edges are allowed, the two values are identical; otherwise, the bundled crossing number can be higher than the genus.

We then investigate the problem of minimizing the number of bundled crossings. For circular graph layouts with a fixed order of vertices, we present a constant-factor approximation algorithm. When the circular order is not prescribed, we get a $\frac{6c}{c-2}$-approximation for a graph with n vertices having at least cn edges for $c > 2$. For general graph layouts, we develop an algorithm with an approximation factor of $\frac{6c}{c-3}$ for graphs with at least cn edges for $c > 3$.

1 Introduction

For many real-world networks with substantial numbers of links between objects, traditional graph drawing algorithms produce visually cluttered and confusing drawings. Reducing the number of edge crossings is one way to improve the quality of the drawings. However, minimizing the number of crossings is very difficult [6,8], and a large number of crossings is sometimes unavoidable. Another way to alleviate this problem is to employ the edge bundling technique in which some edge segments running close to each other are collapsed into bundles to reduce the clutter [9,13,17,20–22,26]. While these methods produce simplified drawings of graphs and significantly reduce visual clutter, they are typically heuristics and provide no guarantee on the quality of the result.

We study the algorithmic aspect of edge bundling, which is listed as one of the open questions in a recent survey on crossing minimization by Schaefer [27]. Our goal is to formalize the underlying geometric problem and design efficient algorithms with provable theoretical guarantees. In our model, *pairwise* edge

© Springer International Publishing AG 2016
Y. Hu and M. Nöllenburg (Eds.): GD 2016, LNCS 9801, pp. 399–412, 2016.
DOI: 10.1007/978-3-319-50106-2_31

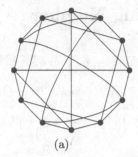 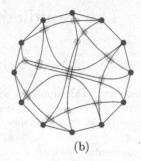

(a)   (b)

Fig. 1. Circular layout of the Chvátal graph: (a) 28 pairwise edge crossings, (b) 13 bundled crossings.

crossings are merged into bundles of crossings, reducing the number of *bundled crossings*, where a bundled crossing is the intersection of two groups of edges; see Fig. 1. We consider both the general setting, where multiple crossings and self-intersections of the edges are allowed, and the more natural restricted setting in which only simple drawings are allowed.

1.1 Our Contribution

We first prove that in the most general setting (when a pair of edges is allowed to cross multiple times and an edge may be crossed by itself or by an incident edge) the bundled crossing number coincides with the orientable genus of the graph (Sect. 2); thus, computing it exactly is NP-hard [30]. In the more natural setting restricted to simple drawings—without double- and self-crossings—, the bundled crossing number of some graphs is strictly greater than the genus.

Next, we consider the *circular bundled crossing number* (Sect. 3), that is, the minimum number of bundled crossings that can be achieved in a circular graph layout. For a fixed circular order of vertices, we present a 16-approximation algorithm and a fixed-parameter algorithm with respect to the number of bundled crossings. For circular layouts without a given vertex order, we develop an algorithm with the approximation factor $\frac{6c}{c-2}$ for graphs with n vertices having at least cn edges for $c > 2$.

In Sect. 4, we study the *bundled crossing number* for general drawings. The algorithm for circular layouts can also be applied for this setting; we show that it guarantees the approximation factor $\frac{6c}{c-3}$ for graphs with at least cn edges for $c > 3$. We then suggest an alternative algorithm that produces fewer bundled crossings for graphs with a large planar subgraph.

Finally, by extending our analysis for circular layouts, we resolve one of the open problems stated by Fink et al. [16] for an ordering problem of paths on a graph arising in visualizing metro maps (Sect. 5).

1.2 Related Work

Edge Crossings. Crossing minimization is a rich topic in graph drawing [6] but still poorly understood from the algorithmic point of view. The best currently known algorithm implies an $\mathcal{O}(n^{9/10})$-approximation for the minimum crossing number on graphs having bounded maximum degree [8]. In contrast, the problem is NP-hard even for cubic graphs and a hardness of constant-factor approximation is known [7]. Minimizing crossings in circular layouts is also NP-hard, and several heuristics have been proposed [4,18]. For graphs with $m \geq 4n$, an $\mathcal{O}(\log^2 n)$-approximation algorithm exists [29]. Our algorithm guarantees an $\mathcal{O}(1)$-approximation for bundled crossings under that condition.

Bundled crossings are closely related to the model of *degenerate crossings* in which multiple edge crossings at the same point in the plane are counted as a single crossing if all pairs of edges passing through the point intersect. An unrestricted variant, called the *genus crossing number* ($\mathrm{gcr}(G)$), allows for self-crossings of edges and multiple crossings between pairs of edges. Mohar showed that the genus crossing number equals the *non-orientable genus* of a graph [24]; thus, $\mathrm{gcr}(G) = \mathcal{O}(m)$. This is similar to our result that the bundled crossing number in this unrestricted setting equals the *orientable genus* of the graph. If self-crossings are not allowed, then we obtain the *degenerate crossing number* ($\mathrm{dcr}(G)$) [25,28]. It was conjectured by Mohar [24] that the genus crossing number always equals the degenerate crossing number; Schaefer and Štefankovič show that $\mathrm{dcr}(G) \leq 6 \cdot \mathrm{gcr}(G) = \mathcal{O}(m)$. A further restriction of the problem forbids multiple crossings between a pair of edges. The corresponding *simple degenerate crossing number* is $\Omega(m^3/n^2)$ for graphs with $m \geq 4n$ edges [1]. Thus multiple crossings between pairs of edges are significant for the corresponding value of the crossing number. Notice the difference to the bundled crossing number, which is always $\mathcal{O}(m)$, even when no self- and multiple crossings are allowed.

Recently, Fink et al. [15] introduced the bundled crossing number. However, they only study the bundled crossing number of a given embedding and show that determining the number is NP-hard. They also present a heuristic that in some cases, e.g., in circular layouts, yields a constant-factor approximation. In contrast, we study the variable-embedding setting: minimize the bundled crossing number over all embeddings of a graph, which is posed as an open problem in [15].

Edge Bundling. Improving the quality of layouts via edge bundling is related to the idea of confluent drawings, when a non-planar graph is presented in a planar way by merging groups of edges [11,12]. The first discussion of bundled edges in the graph drawing literature appeared in [18], where the authors improve circular layouts by routing edges either on the outer or on the inner face of a circle. The hierarchical approach by Holten [20] bundles the edges based on an additional tree structure, and the method is also applied for circular layouts. Similar to [12,18,20], we study circular graph layouts. Edge bundling methods for general graph layouts are suggested in [9,13,17,21,22]. While these methods create an overview drawing, they allow the edges within a bundle to cross and overlap each other arbitrarily, making individual edges hard to follow. The issue is addressed in [5,26], where the edges within a bundle are drawn parallel, as lines in metro maps. To the best of our knowledge, none of the above works on edge bundling provides a guarantee on the quality of the result, though they can be applied in conjunction with our algorithms to provide a better visualization.

Metro Maps. Crossing minimization has also been studied in the context of visualizing metro maps. There, a planar graph (the metro network) and a set of paths in the graph (metro lines) are given. The goal is to order the paths along the edges of the graph so as to minimize the number of crossings. Fink, Pupyrev, and Wolff [16] suggest to merge single line crossings into crossings of blocks of lines minimizing the number of *block crossings* in the map. They devise approximation algorithms for several classes of simple underlying networks (paths, upward trees) and an asymptotically worst-case optimal algorithm for general networks. While we use some ideas of [16] (Sect. 3.1), bundled crossings are more general, since the edges are not restricted to be routed along a specified planar graph. Furthermore, we resolve an open question stated in [16].

2   Bundled Crossings and Graph Genus

Let $G = (V, E)$ with $n = |V|$ and $m = |E|$ be a graph drawn in the plane (with crossings). A *bundled crossing* is a subset C of the crossings so that the following conditions hold:

(i) Every crossing in C belongs to edges $e_1 \in E_1$ and $e_2 \in E_2$, for two subsets $E_1, E_2 \subseteq E$ (E_1 and E_2 are the *bundles* of the bundled crossing), and C contains a crossing of each edge pair e_1, e_2, for $e_1 \in E_1$ and $e_2 \in E_2$.

(ii) One can find a pseudodisk D—a closed polygonal region crossing every edge at most twice—that separates C (in its interior) from all remaining crossings of the embedding. No edge $e \notin E_1 \cup E_2$ intersects D. The requirement ensures that the bundled crossing is visually separated from the rest of the drawing.

The bundled crossing number of a drawing is the minimum number of bundled crossings into which the crossings can be partitioned (with disjoint pseudodisks). The *bundled crossing number* $\mathrm{bc}(G)$ of G is the minimum number of bundled crossings in a drawing of G. For a circular layout, we denote the *circular bundled crossing number* by $\mathrm{bc}^\circ(G)$. If the circular order π of vertices is prescribed, we speak of the *fixed circular bundled crossing number*, $\mathrm{bc}^\circ(G, \pi)$. Clearly, $\mathrm{bc}(G) \leq \mathrm{bc}^\circ(G) \leq \mathrm{bc}^\circ(G, \pi)$.

We now discuss the relation of the bundled crossing number to the orientable genus of the graph. More specifically, consider the unrestricted drawing style for graphs in which double crossings of edges are allowed, as well as self intersections and crossings of adjacent edges. Let $\mathrm{bc}'(G)$ be the minimum number of bundled crossings achievable for G in this unrestricted drawing style. We show that $\mathrm{bc}'(G)$ equals the graph genus.

Theorem 1. *For every graph G with genus $\mathrm{g}(G)$, it holds that $\mathrm{bc}'(G) = \mathrm{g}(G)$.*

Proof. It is easy to show that $\mathrm{g}(G) \leq \mathrm{bc}'(G)$. We take a drawing of G with the minimum number of bundled crossings, $\mathrm{bc}'(G)$, on the sphere. Then, for every bundled crossing, we add a handle to the sphere, where we route one of the bundles through the handle and one on top of it. This way we get a crossing-free drawing of G on a surface of genus $\mathrm{bc}'(G)$.

For the other direction, assume that we have a crossing-free drawing of G on a surface of genus $g = \mathrm{g}(G)$. It is known that such a drawing can be modeled using the representation of a genus-g surface by a *fundamental polygon* with $4g$ sides in the plane [23]. More precisely, the sides of the polygon are numbered in circular order $a_1, b_1, a'_1, b'_1, \ldots, a_g, b_g, a'_g, b'_g$; for $1 \le k \le g$, the pairs (a_k, a'_k) and (b_k, b'_k) of sides are identified in opposite direction, meaning that an edge leaving side a_k appears on the corresponding position of edge a'_k; see Fig. 2 for an example showing K_6 drawn in a fundamental square that models a drawing on the torus. Directly transforming a drawing on the surface into the fundamental polygon can lead to vertices appearing multiple times on the polygon's boundary; however, small movements of the vertices on the surface fix this. Thus, we assume that all vertices lie in the interior of the fundamental polygon, and all edges leave the polygon only in the relative interior of a side of the polygon; especially, every point of an edge appears at most twice on the boundary of the fundamental polygon. (There can be parts of edges connecting two points on different sides of the polygon without directly touching a vertex as in Fig. 2).

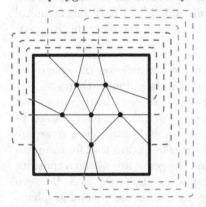

Fig. 2. K_6 drawn in a fundamental square modeling a torus.

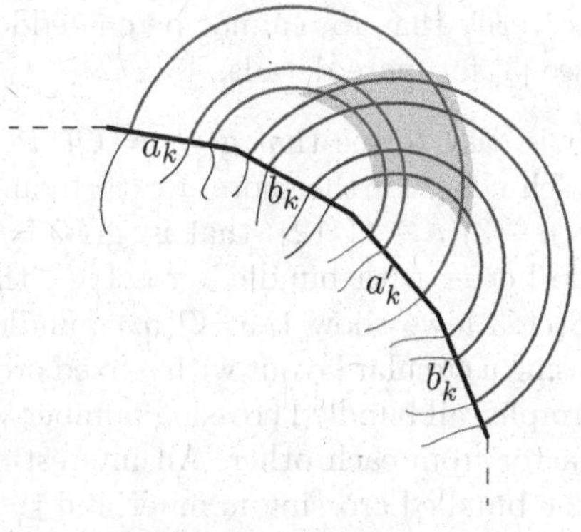

Fig. 3. A single bundled crossing outside the fundamental polygon.

Given such a crossing-free representation of the drawing of G via the fundamental polygon, we create a new drawing of G in the plane by connecting parts of the edges outside of the fundamental polygon. For every $1 \le k \le g$, we connect identified points of edges on $a_k, a'_k, b_k,$ and b'_k as shown in Fig. 3. It is easy to see that for every k, only one bundled crossing is necessary; furthermore, all g tuples of four consecutive sides are independent. Hence, we get a drawing with g bundled crossing, which proves that $\mathrm{bc}'(G) \le \mathrm{g}(G)$.   □

When creating a drawing as in the second part of the above proof, it may happen that we introduce (i) double crossings of edges, (ii) crossings between adjacent edges, or (iii) self intersections of an edge. Certainly, a drawing avoiding such configurations—that is, a *simple* drawing—is preferred. From now on, we only consider simple drawings. Let $\mathrm{bc}(G)$ denote the minimum number of bundled crossings achievable with a simple drawing of G. It turns out that insisting on a simple drawing sometimes makes additional bundled crossing necessary.

Lemma 1. *For every graph $G = (V, E)$, $\mathrm{bc}(G) \geq \mathrm{g}(G)$, and there are graphs G for which $\mathrm{bc}(G) > \mathrm{g}(G)$.*

Proof. Since we only restrict the allowed drawings, we clearly have $\mathrm{bc}(G) \geq \mathrm{bc}'(G) = \mathrm{g}(G)$ and the first claim follows.

For the second part of the lemma, consider the complete graph on six vertices, K_6, with genus $\mathrm{g}(K_6) = 1$; there is a crossing-free drawing of K_6 on the torus. Every realization of K_6 with only one bundled crossing leads to a drawing on the torus. Consider such a drawing in the fundamental polygon model of the torus; in this case, the fundamental polygon can be seen as an axis-aligned square where edges can go to the upper, lower, left, and right side of the square. If two edges incident to the same vertex v leave the square to adjacent sides, the edges cross in the bundled crossing, which is forbidden. Furthermore, no part of an edge can enter and leave the square on adjacent sides since this would result in a forbidden self-intersection. Given these constraints, it is not hard but technical to verify that K_6 cannot be embedded on the torus and, therefore, $\mathrm{bc}(K_6) > 1$; see [3] for more details. □

It is easy to see that $\mathrm{g}(G) = \mathcal{O}(m)$ by introducing a handle on the sphere for each edge. Furthermore, for the complete graph K_n, it is known that $\mathrm{g}(K_n) = \lceil (n-3)(n-4)/12 \rceil$, that is, $\mathrm{g}(G) = \Theta(m)$ for some graphs. Clearly, we cannot do better with bundled crossings, that is, $\mathrm{bc}(G) = \Omega(m)$ for some graphs. In Sect. 3.1 we show that $\mathcal{O}(m)$ bundled crossings always suffice, even if we are using a circular layout with a fixed order of vertices. This means that for complete graphs, all bundled crossing number variants and the genus are within a constant factor from each other. An interesting question is how large the ratio between the bundled crossing number and the graph genus can get for general graphs.

It is known that $\Omega(m^3/n^2)$ single crossings are necessary for graphs with n vertices and $m \geq 4n$ edges [2]. For dense graphs with $m = \Theta(n^2)$ edges, $\Theta(m^2)$ crossings are required, while the bundled crossing number is $\mathcal{O}(m)$. Therefore, using edge bundles can significantly reduce visual complexity of a drawing.

3 Circular Layouts

Now we consider circular graph layouts. Let $G = (V, E)$ be a graph and let $\pi = [v_1, \ldots, v_n]$ be a permutation of its vertices. The goal is to draw G in such a way that the vertices are placed on the boundary of a disk in the circular order prescribed by π, all edges are drawn inside the circle, and the number of bundled crossings, $\mathrm{bc}^\circ(G, \pi)$, is minimized. We start with a scenario when π is predefined.

3.1 Circular Layouts with Fixed Order

Since in our model adjacent edges are not allowed to cross and the circular order of the vertices is fixed, the order of outgoing edges for every vertex is unique for π.

Hence, we may assume that G is a matching. Note that in this case the circular layout can be seen as a *weak pseudoline arrangement*, that is, an arrangement of pseudolines in which not every pair of pseudolines has to cross [10].

Assume that edges e_1 and e_2 are *parallel*, that is, they do not have to cross, and they start and end as immediate neighbors. Clearly, in any simple drawing, e_1 and e_2 do not cross and they are crossed by exactly the same set of other edges; otherwise we would have a forbidden double crossing. Therefore, we can remove e_2 from the instance, find a drawing for the remaining graph, and then reintroduce e_2 without an additional bundled crossing. To this end, we route e_2 parallel to e_1 and let it participate in e_1's bundled crossings in the same bundle as e_1. Thus, we may assume that (i) the input contains no parallel pairs of edges. Additionally, we assume that (ii) every edge of the input graph has to be crossed by an edge (which can be checked by looking at the given circular order); otherwise, such an edge is removed from the input and later reinserted without crossings. In the following we assume that the input satisfied both conditions (i) and (ii) and such a graph is called *simplified*.

Next we develop an approximation algorithm for $\mathrm{bc}^\circ(G, \pi)$ by showing how to find a solution with only a linear number of bundled crossings, and proving that every feasible solution, even an optimum one, must have a linear number of bundled crossings. We start with the lower bound.

Lemma 2. *Let* $G = (V, E)$ *be a simplified graph with fixed circular vertex order* π. *Then,* $\mathrm{bc}^\circ(G, \pi) \geq m/16$.

Proof. Assume we are given a circular drawing of G with the minimum number of bundled crossings. Such a drawing is a weak pseudoline arrangement. Let H be the embedded planar graph that we get by planarizing the drawing, that is, by replacing each crossing by a crossing vertex and adding the cycle $(v_1, v_2, \ldots, v_n)$. We consider the faces of H. Some faces are bounded by original edges and an additional edge stemming from the cycle. Next we lower bound the number of triangles in the pseudoline arrangement and, hence, the triangular faces in H.

Assume that we follow some edge in the drawing and analyze the faces at one of its sides. If all faces were quadrilaterals, then the edge would be completely parallel to a neighboring edge, which is not possible in a simplified instance. Hence, on both sides of the edge we find at least one face that is either a triangle or a k-gon with $k \geq 5$. For $k \geq 3$, let f_k be the number of faces in the drawing of H of degree k. Since we see at least $2m$ sides of such faces and every side only once, we have $2m \leq 3f_3 + \sum_{k \geq 5} k f_k$. Fink et al. [15] show that $f_3 = 4 + \sum_{k \geq 5}(k-4)f_k$. Hence, $2m \leq 3f_3 + \sum_{k \geq 5}(k-4)f_k + 4\sum_{k \geq 5} f_k \leq 3f_3 + (f_3 - 4) + 4(f_3 - 4) \leq 8f_3$, which implies $f_3 \geq m/4$. Note that the bound is tight; see [3].

To complete the proof of the lemma, we use a result of Fink et al. [15], who show that the crossings in a fixed drawing can be partitioned into no less than $f_3/4$ bundled crossings. Since every drawing has at least m/4 triangles, $\mathrm{bc}^\circ(G, \pi) \geq m/16$. □

Note that, as Fink et al. [15] point out, there exist circular drawings whose crossings can be partitioned into no less than $\Theta(m^2)$ bundled crossings. However,

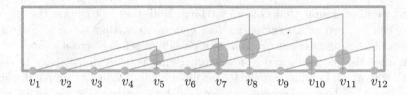

Fig. 4. Finding a circular layout with $m - 1$ bundled crossings (gray shaded).

we can choose the drawing as long as we follow the cyclic order, π, of vertices. We use this freedom and show how to construct a solution with $\mathcal{O}(m)$ bundled crossings.

Lemma 3. *Let $G = (V, E)$ be a graph with a fixed circular vertex order. We can find a circular layout with at most $m - 1$ bundled crossings in $\mathcal{O}(m^2)$ time.*

Proof. Recall that we may assume that the input graph is a matching. Since only the circular order of the vertices matters for the combinatorial embedding, we transform the circle into a rectangle with $v_1, \ldots, v_n$ placed on the lower side from left to right; see Fig. 4. We produce a drawing in which every edge $e = (v_i, v_j)$ with $i < j$ consists of two straight-line segments.[1] The first segment leaves v_i with a slope α; when the segment is above v_j it is followed by a vertical segment connecting down to v_j. Since there are only two slopes, every crossing is between a vertical segment and a segment of slope α. It is easy to see that two edges (v_i, v_j) and $(v_{i'}, v_{j'})$ cross if the endvertices are interleaved, that is, if $i < i' < j < j'$ or $i' < i < j' < j$. In that case, the edges have to cross in any possible embedding and we do not introduce additional crossings.

Finally, we create a single bundled crossing for each edge e consisting of all crossings of e's vertical segment. It is easy to see that this yields a feasible partitioning of all crossings into bundled crossings. Since the edge ending at vertex v_n will not have any crossing on its vertical segment, the number of bundled crossings is at most $m - 1$. The drawing is created in $\mathcal{O}(m)$ time but the time needed to produce a combinatorial embedding depends on the number of crossings; it is bounded by $\mathcal{O}(m^2)$. $\square$

The upper bound of $m - 1$ is tight: a matching in which every edge crosses every other edge requires that many bundled crossings. Combining the algorithm and the lower bound of Lemma 2, we get the following result.

Theorem 2. *For a graph G with a fixed circular vertex order, we can find a 16-approximation for the fixed circular bundled crossing number in $\mathcal{O}(m^2)$ time.*

Fixed-Parameter Tractability. We now show that deciding whether a solution with at most k bundled crossings exists is fixed-parameter tractable with respect to k. The crucial instruments for achieving this are the graph simplification and

---

[1] We thank an anonymous reviewer for suggesting this simplified proof.

the lower bound of Lemma 2. If after the simplification, G has more than $16k$ edges, we know that $\mathrm{bc}^\circ(G, \pi) > 16k/16 = k$ and we can reject the instance. Otherwise, if at most k edges remain, we can afford to solve the problem exhaustively.

Theorem 3. *Let $G = (V, E)$ be a graph with a fixed circular vertex order π. Deciding whether $\mathrm{bc}^\circ(G, \pi) \leq k$ is fixed-parameter tractable with respect to k with a running time of $\mathcal{O}(2^{0.657k^2} k^{128k^2} + m)$.*

Proof. We simplify the graph in $\mathcal{O}(m)$ time. Afterwards, we check every combination of circular order, combinatorial embedding, and partitioning of the crossings into up to k sets. If any such combination yields a feasible partitioning into bundled crossings, we accept the instance; otherwise, we reject it.

There are at most $\binom{16k}{2} \leq 128k^2$ pairs of edges that need to cross. Hence, there are up to k^{128k^2} ways to partition the crossings into up to k sets. Since every pair of edges crosses at most once, the circular embedding can be extended to a pseudoline arrangement (in which every pair crosses exactly once). Felsner and Valtr proved [14] that there are at most $2^{0.657k^2}$ arrangements of k pseudolines, and Yamanaka et al. [31] presented a method that iterates over all pseudoline arrangements using $\mathcal{O}(k^2)$ total space and $\mathcal{O}(1)$ time per arrangement. For each pseudoline arrangement, we can check whether an embedding with the prescribed circular order occurs as a part in $\mathcal{O}(k^3)$ time; within the same time bound, we can check whether a given partitioning of the crossings yields feasible bundled crossings. In total this takes $\mathcal{O}(2^{0.657k^2} k^{128k^2} + m)$ time. □

3.2 Circular Layouts with Free Order

We now study the variant of the problem in which the circular order of the vertices is not known. How can one find a suitable order? A possible approach would be finding an order that optimizes some aesthetic criteria (e.g., the total length of the edges [18] or the number of pairwise crossings [4]) and then applying the algorithm of Lemma 3. Next we analyze such an approach.

In Sect. 2, we have already seen that $\mathrm{bc}(G) \geq \mathrm{g}(G)$. We can use this for getting a lower bound for the bundled crossing number.

Lemma 4. *For every graph $G = (V, E)$ with n vertices and m edges, $\mathrm{bc}(G) \geq (m - (3n - 6))/6$ and $\mathrm{bc}^\circ(G) \geq (m - (2n - 3))/6$.*

Proof. Assume we have a crossing-free drawing of graph G on a surface of genus $\mathrm{g} = \mathrm{g}(G)$. The relation between vertices, edges, and faces is described by the Euler formula $n - m + f = 2 - 2\,\mathrm{g}$. Combining this with $2m \geq 3f$, we get that $\mathrm{bc}(G) \geq \mathrm{g}(G) \geq (m - (3n - 6))/6$.

Now consider a circular drawing with the minimum number $k = \mathrm{bc}^\circ(G)$ of bundled crossings. All n vertices lie on the outer face. Hence, we can add $n - 3$ edges triangulating the outer face without introducing new crossings. We get a new graph G' with $m' = m + n - 3$ edges and a (non-circular) drawing of G' with k bundled crossings. Hence, $k \geq (m' - (3n - 6))/6 = (m - (2n - 3))/6$. □

For dense graphs with more than $2n$ edges, we can get a constant-factor approximation using the upper bound of $m - 1$ with an arbitrary order.

Theorem 4. *Let $G = (V, E)$ be a graph with $m \geq cn$ for some $c > 2$. There is an $\mathcal{O}(n^2)$-time algorithm that computes a solution for the circular bundled crossing number with an approximation factor of $\frac{6c}{c-2}$.*

Proof. Using the algorithm of Lemma 3, we find a solution with at most $m - 1$ bundled crossings. By Lemma 4, $(m - (2n - 3))/6$ crossings are required. Then the approximation factor is $\frac{m-1}{(m-(2n-3))/6} = 6\left(1 + \frac{2n-4}{m-2n+3}\right) \leq 6\left(1 + \frac{2n}{m-2n}\right) \leq 6\left(1 + \frac{2}{c-2}\right) = \frac{6c}{c-2}$, which is constant for every $c > 2$ and $n \geq 1$. $\square$

For constructing a constant-factor approximation algorithm for sparse graphs with $m \leq 2n$ one would need better bounds. We next suggest a possible direction by improving our algorithm for some input graphs. The idea is to save some crossings by first drawing an outerplanar subgraph of G.

Lemma 5. *Let $G = (V, E)$ be a graph and $G^* = (V, E^*)$ be a subgraph of G having $m^* = |E^*|$ edges that is outerplanar with respect to a vertex order π. Then $\mathrm{bc}^\circ(G) \leq \mathrm{bc}^\circ(G, \pi) \leq 2(m - m^*)$ and we can find such a solution in $\mathcal{O}(m^2)$ time.*

Proof. The algorithm is similar to the one used in Lemma 3 in which every edge consists of two segments. This time we initialize the embedding by adding the edges of E^* without crossings, each with a segment of slope α. Next, we add the remaining edges from left to right ordered by their first vertex. When adding edge $e = (v_i, v_j)$ with $i < j$, we route the edge with two vertical segments and a middle segment of slope α. We start upward from v_i so that the first segment crosses all edges present at $x = x(v_i)$ that have to cross e, but no other edge. We start the middle segment with slope α there and complete with a vertical segment at $x = x(v_j)$. It is easy to see that any edge of E^* whose vertical segment could intersect e must start left of v_i. However, our routing of e places the possible crossing on a vertical segment of e. Hence, all vertical segments of edges of E^* are crossing-free. Creating a bundled crossing for each vertical segment of the edges of $E - E^*$ results, therefore, in at most $2(m - m^*)$ bundled crossings. $\square$

This bound is asymptotically tight; see [3].

4   General Drawings

We now consider general (non-circular) drawings. Note that Lemma 4 provides a lower bound for the bundled crossing number, and Lemma 3 gives an algorithm that can be applied for general drawings. Combining the lower and the upper bounds, we get the following result for dense graphs.

Theorem 5. *Let $G = (V, E)$ be a graph with $m \geq cn$ for some $c > 3$. There is an $\mathcal{O}(n^2)$-time algorithm that computes a solution for the bundled crossing number with an approximation factor of $\frac{6c}{c-3}$.*

Proof. By Lemma 4, $\mathrm{bc}(G) \geq (m-(3n-6))/6$, and by Lemma 3, $\mathrm{bc}(G) \leq m-1$. Then the approximation factor of the algorithm of Lemma 3 is $\frac{m-1}{(m-(3n-6))/6} =$
$6\left(1 + \frac{3n-7}{m-3n+6}\right) \leq 6\left(1 + \frac{3n}{m-3n}\right) \leq 6\left(1 + \frac{3}{c-3}\right) = \frac{6c}{c-3}$. $\qquad\square$

Can we improve the algorithm for general drawings? Next we develop an alternative upper bound based on a planar subgraph $G^* = (V, E^*)$ of G, which produces fewer bundled crossings if $m^* = |E^*| > 3m/4$.

Lemma 6. *Let $G = (V, E)$ be a graph, let $G^* = (V, E^*)$ be its planar subgraph, and let $m^* = |E^*|$. Then, $\mathrm{bc}(G) \leq 4(m - m^*)$.*

Proof. We start with a topological book embedding of G^*, that is, a planar embedding with all vertices on the x-axis and the edges composed of circular arcs whose center is on the x-axis. Giordano et al. [19] show how to construct such an embedding with at most two circular arcs per edge and all edges being x-monotone (that is, edges with two circular arcs cannot change the direction). We add the edges of $E' = E \setminus E^*$ to get a non-planar topological book embedding (with up to two circular arcs per edge) and keep the drawing simple, that is, free of self-intersections, double crossings, and crossings of adjacent edges. Then we split the drawing at the spine and interpret each half as a circular layout with fixed order. Using the algorithm of Lemma 5, we get an embedding with at most $2(m - m^*)$ crossings for each side and $4(m - m^*)$ crossings in total.

It remains to show how to add an edge $e = (u, v) \in E'$. Consider all planar edges incident to u and v. If we can add e as a single circular arc above or below the spine without crossing any of these edges, we do so. Otherwise, two edges e_1 adjacent to u and e_2 adjacent to v exist (see Fig. 5), and e must be inserted using two circular arcs. We consider all these obstructing two-bend edges incident to u and v and insert e by placing its bend next to the rightmost bend of an edge incident to u (see Fig. 6), avoiding all intersections with planar edges. Bends of the edges incident to u are ordered by their endvertex so that they do not cross.

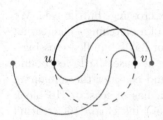

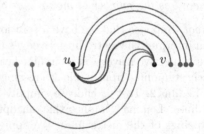

Fig. 5. Adding edge $e = (u, v)$ requires two circular arcs.

Fig. 6. Inserting edge $e = (u, v)$ with two circular arcs.

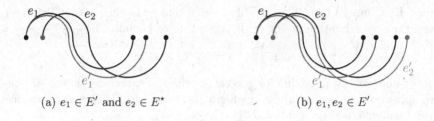

(a) $e_1 \in E'$ and $e_2 \in E^\star$ （b) $e_1, e_2 \in E'$

Fig. 7. Double crossings of edges are not possible

It is easy to see that there are no self-intersections and no crossings of adjacent edges. There are also no double crossings: Otherwise, let e_1 and e_2 be a pair of edges that cross both above and below the spine. Assume that $e_1 \in E', e_2 \in E^\star$. Since e_1 consists of two segments, there must be adjacent planar edges that caused e_1's shape. We find such an edge e_1' that crosses with the planar edge e_2, a contradiction; see Fig. 7a. If $e_1, e_2 \in E'$, we find a planar edge e_2' causing the two-arc shape of e_2, such that e_1' and e_2' cross, another contradiction; see Fig. 7b. □

5 Block Crossings in Metro Maps

Our analysis has an interesting application for block crossings in metro maps [16]. The block crossing minimization problem (BCM) asks to order simple paths (metro lines) along the edges of a plane graph (underlying metro network) so as to minimize the total number of block crossings. Fink et al. [16] present a method that uses two block crossings per line on a tree network, and ask whether a (constant-factor) approximation is possible. With the help of the lower bound of Lemma 2, we affirmatively answer the question. We provide a sketch of the proof; see [3] for details.

Theorem 6. *There is an $\mathcal{O}(\ell^2)$-time 32-approximation algorithm for BCM, where ℓ is the number of metro lines and the underlying network is a tree.*

Proof. Suppose that we have a solution with k block crossings on the tree. We can interpret the metro lines as edges in the drawing of a matching—connecting the respective leaves—in a circular layout. This layout has k bundled crossings, each stemming from a block crossing. Hence, we could use the lower bound of Lemma 2. To this end, we simplify the instance and consider the remaining m lines. Lemma 2 implies that an optimum solution has at least $m/16$ block crossings of the metro lines. We apply the method of Fink et al. [16] creating $2m$ block crossings in $\mathcal{O}(m^2)$ time, and reinsert the simplified lines. □

6 Conclusion

We have considered the bundled crossing number problem and devised upper and lower bounds for general as well as circular layouts with and without fixed circular vertex order. We have also shown the relation of the bundled crossing number to the orientable graph genus and resolved an open problem for block crossings of metro lines on trees. The setting of bundled crossings still has several interesting questions to offer. It seems very likely that the circular bundled crossing number problem is NP-hard, but a proof is missing. Furthermore, an approximation or a fixed-parameter algorithm for the version with free circular vertex order is desirable. Both questions are also interesting for general graph layouts.

References

1. Ackerman, E., Pinchasi, R.: On the degenerate crossing number. Discret. Comput. Geom. **49**(3), 695–702 (2013)
2. Ajtai, M., Chvátal, V., Newborn, M.M., Szemerédi, E.: Crossing-free subgraphs. N.-Holl. Math. Stud. **60**, 9–12 (1982)
3. Alam, M.J., Fink, M., Pupyrev, S.: The bundled crossing number. CoRR, cs.CG/1608.08161 (2016)
4. Baur, M., Brandes, U.: Crossing reduction in circular layouts. In: Hromkovič, J., Nagl, M., Westfechtel, B. (eds.) WG 2004. LNCS, vol. 3353, pp. 332–343. Springer, Heidelberg (2004). doi:10.1007/978-3-540-30559-0_28
5. Bouts, Q.W., Speckmann, B.: Clustered edge routing. In: PacificVis 2015, pp. 55–62 (2015)
6. Buchheim, C., Chimani, M., Gutwenger, C., Jünger, M., Mutzel, P.: Crossings and planarization. In: Handbook of Graph Drawing and Visualization. CRC Press (2013)
7. Cabello, S.: Hardness of approximation for crossing number. Discret. Comput. Geom. **49**(2), 348–358 (2013)
8. Chuzhoy, J.: An algorithm for the graph crossing number problem. In: STOC 2011, pp. 303–312 (2011)
9. Cui, W., Zhou, H., Qu, H., Wong, P.C., Li, X.: Geometry-based edge clustering for graph visualization. TVCG **14**(6), 1277–1284 (2008)
10. Fraysseix, H., Mendez, P.O.: Stretching of Jordan arc contact systems. In: Liotta, G. (ed.) GD 2003. LNCS, vol. 2912, pp. 71–85. Springer, Heidelberg (2004). doi:10.1007/978-3-540-24595-7_7
11. Dickerson, M., Eppstein, D., Goodrich, M.T., Meng, J.Y.: Confluent drawings: visualizing non-planar diagrams in a planar way. JGAA **9**(1), 31–52 (2005)
12. Eppstein, D., Holten, D., Löffler, M., Nöllenburg, M., Speckmann, B., Verbeek, K.: Strict confluent drawing. In: Wismath, S., Wolff, A. (eds.) GD 2013. LNCS, vol. 8242, pp. 352–363. Springer, Heidelberg (2013). doi:10.1007/978-3-319-03841-4_31
13. Ersoy, O., Hurter, C., Paulovich, F.V., Cantareiro, G., Telea, A.: Skeleton-based edge bundling for graph visualization. TVCG **17**(12), 2364–2373 (2011)
14. Felsner, S., Valtr, P.: Coding and counting arrangements of pseudolines. Discret. Comput. Geom. **46**(3), 405–416 (2011)
15. Fink, M., Hershberger, J., Suri, S., Verbeek, K.: Bundled crossings in embedded graphs. In: Kranakis, E., Navarro, G., Chávez, E. (eds.) LATIN 2016. LNCS, vol. 9644, pp. 454–468. Springer, Heidelberg (2016). doi:10.1007/978-3-662-49529-2_34
16. Fink, M., Pupyrev, S., Wolff, A.: Ordering metro lines by block crossings. JGAA **19**(1), 111–153 (2015)

17. Gansner, E. Hu, Y., North, S., Scheidegger, C.: Multilevel agglomerative edge bundling for visualizing large graphs. In: PacificVis 2011, pp. 187–194. IEEE (2011)
18. Gansner, E.R., Koren, Y.: Improved circular layouts. In: Kaufmann, M., Wagner, D. (eds.) GD 2006. LNCS, vol. 4372, pp. 386–398. Springer, Heidelberg (2007). doi:10.1007/978-3-540-70904-6_37
19. Giordano, F., Liotta, G., Mchedlidze, T., Symvonis, A., Whitesides, S.: Computing upward topological book embeddings of upward planar digraphs. J. Discret. Algorithms 30, 45–69 (2015)
20. Holten, D.: Hierarchical edge bundles: visualization of adjacency relations in hierarchical data. TVCG 12(5), 741–748 (2006)
21. Holten, D., van Wijk, J.J.: Force-directed edge bundling for graph visualization. Comput. Graph. Forum 28(3), 983–990 (2009)
22. Lambert, A., Bourqui, R., Auber, D.: Winding roads: routing edges into bundles. Comput. Graph. Forum 29(3), 853–862 (2010)
23. Lazarus, F., Pocchiola, M., Vegter, G., Verroust, A.: Computing a canonical polygonal schema of an orientable triangulated surface. In: SoCG 2001, pp. 80–89. ACM (2001)
24. Mohar, B.: The genus crossing number. ARS Math. Contempo. 2(2), 157–162 (2009)
25. Pach, J., Tóth, G.: Degenerate crossing numbers. Discret. Comput. Geom. 41(3), 376–384 (2009)
26. Pupyrev, S., Nachmanson, L., Bereg, S., Holroyd, A.E.: Edge routing with ordered bundles. Comput. Geom. 52, 18–33 (2016)
27. Schaefer, M.: The graph crossing number and its variants: a survey. Electron. J. Comb. Dyn. Surv. 21 (2013)
28. Schaefer, M., Štefankovič, D.: The degenerate crossing number and higher-genus embeddings. In: Di Giacomo, E., Lubiw, A. (eds.) GD 2015. LNCS, vol. 9411, pp. 63–74. Springer, Heidelberg (2015). doi:10.1007/978-3-319-27261-0_6
29. Shahrokhi, F., Sýkora, O., Székely, L.A., Vrt'o, I.: Book embeddings and crossing numbers. In: Mayr, E.W., Schmidt, G., Tinhofer, G. (eds.) WG 1994. LNCS, vol. 903, pp. 256–268. Springer, Heidelberg (1995). doi:10.1007/3-540-59071-4_53
30. Thomassen, C.: The graph genus problem is NP-complete. J. Algorithms 10(4), 568–576 (1989)
31. Yamanaka, K., Nakano, S., Matsui, Y., Uehara, R., Nakada, K.: Efficient enumeration of all ladder lotteries and its application. Theor. Comput. Sci. 411(16–18), 1714–1722 (2010)

Approximating the Rectilinear Crossing Number

Jacob Fox[1], János Pach[2], and Andrew Suk[3($\boxtimes$)]

[1] Stanford University, Stanford, CA, USA
jacobfox@stanford.edu
[2] EPFL, Lausanne, Switzerland and Courant Institute, Newyork, NY, USA
pach@cims.nyu.edu
[3] University of Illinois at Chicago, Chicago, IL, USA
suk@uic.edu

Abstract. A *straight-line* drawing of a graph G is a mapping which assigns to each vertex a point in the plane and to each edge a straight-line segment connecting the corresponding two points. The *rectilinear crossing number* of a graph G, $\mathrm{cr}(G)$, is the minimum number of pairs of crossing edges in any straight-line drawing of G. Determining or estimating $\overline{\mathrm{cr}}(G)$ appears to be a difficult problem, and deciding if $\overline{\mathrm{cr}}(G) \leq k$ is known to be NP-hard. In fact, the asymptotic behavior of $\overline{\mathrm{cr}}(K_n)$ is still unknown.

In this paper, we present a deterministic $n^{2+o(1)}$-time algorithm that finds a straight-line drawing of any n-vertex graph G with $\overline{\mathrm{cr}}(G) + o(n^4)$ pairs of crossing edges. Together with the well-known Crossing Lemma due to Ajtai et al. and Leighton, this result implies that for any dense n-vertex graph G, one can efficiently find a straight-line drawing of G with $(1 + o(1))\overline{\mathrm{cr}}(G)$ pairs of crossing edges.

1 Introduction

A *drawing* of a graph G is a mapping f that assigns to each vertex a distinct point in the plane and to each edge uv a continuous arc connecting $f(u)$ and $f(v)$, not passing through the image of any other vertex. Two edges in a drawing *cross* if their interiors have a point in common. The *crossing number* of G, denoted by $\mathrm{cr}(G)$, is the minimum number of pairs of crossing edges in any drawing of G. Hence, $\mathrm{cr}(G) = 0$ if and only if G is planar. Determining or estimating the crossing number of a graph is one of the oldest problems in graph theory, with over 700 papers written on the subject. We refrain here from attempting to give an overview of the long history of crossing numbers and their applications in discrete and computational geometry, and refer the reader to the survey articles by Pach and Tóth [31], Schaefer [33], and the extensive bibliography maintained by Vrt'o [40].

J. Fox—Supported by a Packard Fellowship, by NSF CAREER award DMS 1352121, and by an Alfred P. Sloan Fellowship.
J. Pach—Supported by a Hungarian Science Foundation NKFI grant, by Swiss National Science Foundation Grants 200021-165977 and 200020-162884.
A. Suk—Supported by NSF grant DMS-1500153.

© Springer International Publishing AG 2016
Y. Hu and M. Nöllenburg (Eds.): GD 2016, LNCS 9801, pp. 413–426, 2016.
DOI: 10.1007/978-3-319-50106-2_32

In the present paper, we focus on *straight-line drawings* of a graph G, that is, drawings of G where the edges are represented by straight-line segments. We will assume that in all such drawings, no three vertices are collinear, and no point lies in the interior of three distinct edges. The *rectilinear crossing number* of G, denoted by $\overline{cr}(G)$, is the minimum number of pairs of crossing edges in any straight-line drawing of G. Clearly $cr(G) \le \overline{cr}(G)$, and a theorem of Fáry [18] states that $\overline{cr}(G) = 0$ when G is planar. On the other hand, it was shown by Bienstock and Dean [9] that there are graphs with crossing number four, whose rectilinear crossing numbers are arbitrarily large.

Determining the rectilinear crossing number of a graph appears to be a difficult problem. In fact, the asymptotic value of $\overline{cr}(K_n)$ is still unknown. The exact values for $\overline{cr}(K_n)$ are known for $n \le 27$ and $n = 30$, and for large n, the current best known bounds are

$$0.379972 \binom{n}{4} < \overline{cr}(K_n) < 0.380473 \binom{n}{4},$$

due to Ábrego et al. [1] and Fabila-Monroy and López [17] respectively. For more details on $\overline{cr}(K_n)$, including its striking connection to Sylvester's four-point problem [37, 38], see [2, 34].

From an algorithmic point of view, computing $\overline{cr}(G)$ is known to be NP-hard [8]. More precisely, it is known to be $\exists \mathbb{R}$-complete, that is, complete for the existential theory of the reals (see [32, 33]). On the other hand, many researchers have designed polynomial time algorithms for approximating crossing numbers of *sparse* graphs. In particular, a seminal result of Hopcroft and Tarjan [25] is that there is a linear time algorithm for testing planarity of a graph. Kawarabayashi and Reed [26] generalized their result and established a linear time algorithm that decides whether $cr(G) \le k$ when k is fixed. Leighton and Rao [28] obtained an efficient algorithm that finds a drawing of any bounded-degree n vertex graph G with at most $O(\log^4 n)(n + cr(G))$ pairs of crossing edges. This was later improved by Even, Guha, and Schieber [16] to $O(\log^3 n)(n + cr(G))$, and further improved by Arora, Rao, and Vazirani [6] to $O(\log^2 n)(n + cr(G))$. For more results on computing $cr(G)$ for bounded degree graphs, see [14].

For *dense* graphs G, very little is known about $\overline{cr}(G)$, and as mentioned above, not even the asymptotic value of $\overline{cr}(K_n)$. Our main result is the following.

Theorem 1. *There is a deterministic $n^{2+o(1)}$-time algorithm for constructing a straight-line drawing of any n-vertex graph G in the plane with*

$$\overline{cr}(G) + O(n^4/(\log \log n)^\delta)$$

crossing pairs of edges, where $\delta > 0$ is an absolute constant.

A classic result of Ajtai et al. [5] and Leighton [27], known as the Crossing Lemma, implies that the rectilinear crossing number of any n-vertex graph with e edges is at least $\frac{e^3}{64n^2} - 4n$. Hence all n-vertex graphs G with $\Omega(n^2)$ edges satisfy $\overline{cr}(G) \ge \Omega(n^4)$. This implies the following.

Corollary 1. *There is a deterministic $n^{2+o(1)}$-time algorithm for constructing a straight-line drawing of any n-vertex graph G with $|E(G)| \geq \varepsilon n^2$, where $\varepsilon > 0$ is fixed, such that the drawing has at most $(1+o(1))\overline{cr}(G)$ crossing pairs of edges.*

A sequence $(G_n : n = 1, 2, \ldots)$ of graphs with $|V(G_n)| = n$ is called *quasi-random with density p* (where $0 < p < 1$) if, for all subsets $X, Y \subset V(G_n)$, $e_{G_n}(X, Y) = p|X||Y| + o(n^2)$. An important result of Chung, Graham, and Wilson [12] shows that being quasi-random with density p is equivalent to many other properties almost surely satisfied by the random graph $G(n, p)$. Studying properties of quasi-random graphs has been an important research direction with numerous applications. In Sect. 5, we prove the following result.

Theorem 2. *Fix $0 < p < 1$ and let $(G_n : n = 1, 2, \ldots)$ be a sequence of graphs that is quasi-random with density p. Then*

$$\overline{cr}(G_n) = (1 + o(1))p^2 \cdot \overline{cr}(K_n).$$

More generally, we show any two edge-weighted graphs which are close in cut-distance have rectilinear crossing numbers which are close (see Lemma 3). For results on crossing numbers of random graphs, consult [36].

Organization. In the next section, we collect several geometric results on planar point sets and give an exponential time algorithm for computing the rectilinear crossing number of a (small) graph. In Sect. 3, we show that if two graphs are close in cut-distance, then their rectilinear crossing numbers are approximately the same. In Sect. 4, we prove Theorem 1. Finally in Sect. 5, we prove Theorem 2.

We omit floor and ceiling signs whenever they are not crucial. All logarithms are base 2.

2   Order Types and Same-Type Transversals

Let $V = (v_1, \ldots, v_n)$ be an n-element point sequence in $\mathbb{R}^2$ in general position, that is, no three members of V are collinear. The *order type* of V is the mapping $\chi : \binom{V}{3} \to \{+1, -1\}$ (positive orientation, negative orientation), assigning each triple of V its orientation. By setting $v_i = (x_i, y_i) \in \mathbb{R}^2$, for $i_1 < i_2 < i_3$,

$$\chi(\{v_{i_1}, v_{i_2}, v_{i_3}\}) = \text{sgn} \det \begin{pmatrix} 1 & 1 & 1 \\ x_{i_1} & x_{i_2} & x_{i_3} \\ y_{i_1} & y_{i_2} & y_{i_3} \end{pmatrix}.$$

Therefore, two n-element point sequences $V = (v_1, \ldots, v_n)$ and $U = (u_1, \ldots, u_n)$ have the same order type if they are "combinatorially equivalent". By lexicographically ranking each triple (i_1, i_2, i_3), where $1 \leq i_1 < i_2 < i_3 \leq n$, we can describe each order type χ with the vector $(\chi_1, \chi_2, \ldots) \in \{-1, +1\}^{\binom{n}{3}}$, such that $\chi_j = +1$ if and only if $\chi(\{v_{i_1}, v_{i_2}, v_{i_3}\}) > 0$ and $\text{Rank}(i_1, i_2, i_3) = j$. We will call vectors $\chi^* \in \{-1, +1\}^{\binom{n}{3}}$ *abstract order types*, and we say that an abstract order type χ^* is *realizable* if there is a point set V in the plane whose order type realizes χ^*. The concept of order types was introduced by Goodman and Pollack [23] and has played a crucial role in gathering knowledge about crossing numbers. See [22,23] for more background on order types.

Given k disjoint subsets $V_1, \ldots, V_k \subset V$, a *transversal* of $(V_1, \ldots, V_k)$ is any k-element sequence $(v_1, \ldots, v_k)$ such that $v_i \in V_i$ for all i. We say that the k-tuple of parts $(V_1, \ldots, V_k)$ has *same-type* transversals if all of its transversals have the same order type. One of the key ingredients in the proof of Theorem 1 is the following regularity lemma for same-type transversals established by the authors in [20]. A partition on a finite set V is called *equitable* if any two parts differ in size by at most one.

Theorem 3. *There is an absolute constant C such that the following holds. For each $0 < \varepsilon < 1$ and for any finite point set V in $\mathbb{R}^2$, there is an equitable partition $V = V_1 \cup V_2 \cup \cdots \cup V_K$, with $1/\varepsilon < K < \varepsilon^{-C}$, such that all but at most $\varepsilon \binom{K}{4}$ quadruples of parts $\{V_{i_1}, V_{i_2}, V_{i_3}, V_{i_4}\}$ have same-type transversals.*

For small graphs $G = (V, E)$ with $|V(G)| = K$, we can compute $\overline{cr}(G)$ as follows. We generate $\binom{K}{3}$ polynomials $f_1, f_2, \ldots, f_{\binom{K}{3}} \in \mathbb{R}[x_1, \ldots, x_K, y_1, \ldots, y_K]$, where for $1 \leq i_1 < i_2 < i_3 \leq K$ and $\mathrm{Rank}(i_1, i_2, i_3) = j$, we have

$$f_j = \det \begin{pmatrix} 1 & 1 & 1 \\ x_{i_1} & x_{i_2} & x_{i_3} \\ y_{i_1} & y_{i_2} & y_{i_3} \end{pmatrix}.$$

Fix an abstract order type $\chi^* \in \{+1, -1\}^{\binom{K}{3}}$, and let $j_1, \ldots, j_r$ be the indices for which $\chi^*_{j_\ell} = +1$, and let $j'_1, \ldots, j'_s$ be the indices for which $\chi^*_{j'_\ell} = -1$. In order to decide if χ^* is realizable, we need to see if there are real solutions to the polynomial system

$$f_{j_1} > 0, \ldots, f_{j_r} > 0 \qquad f_{j'_1} < 0, \ldots, f_{j'_s} < 0.$$

This is a special case of the satisfiability problem in the existential theory of the reals (see [10]). By an algorithm of Basu, Pollack, and Roy [7], we can decide if the polynomial system above has real solutions in $2^{O(K \log K)}$ time. Moreover, if there are solutions, the algorithm will output a solution $(x_1, \ldots, x_K, y_1, \ldots, y_K)$, where each coordinate uses at most $2^{O(K \log K)}$ bits. Hence if χ^* is realizable, we obtain a point set $V = \{v_1, \ldots, v_K\}$ in the plane that realizes χ^*, and each point has at most $2^{O(K \log K)}$ bits.

If we do obtain such a point set V, we then compute the minimum number of pairs of crossing edges over all straight-line drawings of G which uses V as its vertex set. This can be done in $2^{O(K \log K)}$ time. By repeating the procedure above over all $2^{\binom{K}{3}}$ abstract order types χ^*, we have the following.

Lemma 1. *Given a graph G on K vertices, we can find a straight-line drawing of G with $\overline{cr}(G)$ pairs of crossing edges in $2^{O(K^3)}$ time.*

See [3,4] for an alternative heuristic method for computing $\overline{cr}(G)$.

3   Cut-Distance and the Frieze–Kannan Regularity Lemma

An *edge weighted graph* $G = (V, E)$ is a graph with weights $w_G(uv) \in [0, 1]$ associated with each edge $uv \in E(G)$. For convenience, set $w_G(uv) = 0$ if $uv \notin E(G)$. For $S, T \subset V(G)$, we define

$$e_G(S, T) = \sum_{u \in S, v \in T} w_G(uv).$$

Note that if the sets S and T have a nonempty intersection, the weights of the edges running in $S \cap T$ are counted twice. Let G and G' be two edge weighted labeled graphs with the same vertex set $V = \{v_1, \ldots, v_n\}$. The *cut-distance* between G and G' is defined as

$$d(G, G') = \max_{S, T \subset V} |e_G(S, T) - e_{G'}(S, T)|.$$

Hence, the cut-distance between two labeled graphs measures how different the two graphs are when considering the size of various cuts. This concept has played a crucial role in the work of Frieze and Kannan [21] on efficient approximation algorithms for dense graphs. See [11] and the book [29] for more results on cut-distance.

We generalize the concept of crossing numbers to edge weighted graphs as follows. Let $\mathcal{D}$ be a straight-line drawing of G in the plane, and let $X_{\mathcal{D}} \subset \binom{E(G)}{2}$ denote the set of pairs of crossing edges in the drawing. The *rectilinear crossing number* of the *edge-weighted graph* G is defined as

$$\overline{\mathrm{cr}}(G) = \min_{\mathcal{D}} \sum_{(uv, st) \in X_{\mathcal{D}}} w_G(uv) \cdot w_G(st),$$

where the minimum is taken over all straight-line drawings of G. Thus for any unweighted graph $G = (V, E)$, we can assign weights $w_G(uv) = 1$ for $uv \in E(G)$ and $w_G(uv) = 0$ for $uv \notin E(G)$ so that the definition of $\overline{\mathrm{cr}}(G)$ remains consistent. By copying the proof of Lemma 1 almost verbatim, we have the following lemma.

Lemma 2. *Let G be an edge weighted graph on K vertices, where the weight of each edge uses at most B bits. Then we can find a straight-line drawing of G with $\overline{\mathrm{cr}}(G)$ weighted edge crossings in $2^{O(K^3)} B^2$ time.*

Another key ingredient used in the proof of Theorem 1 is a variant of Szemerédi's regularity lemma developed by Frieze and Kannan. Szemerédi's regularity lemma [39] is one of the most powerful tools in modern combinatorics and gives a rough structural characterization of all graphs. According to the lemma, for every $\varepsilon > 0$ there is $K = K(\varepsilon)$ such that every graph has an equitable vertex partition into at most K parts such that all but at most an ε fraction of the pairs of parts behave "regularly".[1] The dependence of K on $1/\varepsilon$ is notoriously strong.

---

[1] For a pair (V_i, V_j) of vertex subsets, the density $d(V_i, V_j)$ is defined as $\frac{e_G(V_i, V_j)}{|V_i||V_j|}$. The pair (V_i, V_j) is called ε-regular if for all $V_i' \subset V_i$ and $V_j' \subset V_j$ with $|V_i'| \geq \varepsilon |V_i|$ and $|V_j'| \geq \varepsilon |V_j|$, we have $|d(V_i', V_j') - d(V_i, V_j)| \leq \varepsilon$.

It follows from the proof that $K(\varepsilon)$ may be taken to be an exponential tower of twos of height $\varepsilon^{-O(1)}$. Gowers [24] used a probabilistic construction to show that such an enormous bound is indeed necessary. This is quite unfortunate, because in algorithmic applications of the regularity lemma this parameter typically has a negative impact on the efficiency. Consult [13], [35], [19] for other proofs that improve on various aspects of the result.

Frieze and Kannan [21] developed a weaker notion of regularity which is sufficient for certain algorithmic applications, and for which the dependence on the approximation parameter ε is much better. Let $\varepsilon > 0$ and let $G = (V, E)$ be a graph on n vertices. An equitable partition $\mathcal{P} : V = V_1 \cup \cdots \cup V_K$ is said to be ε-*Frieze-Kannan-regular* if for all subsets $S, T \subset V(G)$, we have

$$\left| e_G(S, T) - \sum_{1 \leq i, j \leq K} e_G(V_i, V_j) \frac{|S \cap V_i||T \cap V_j|}{|V_i||V_j|} \right| < \varepsilon n^2.$$

Frieze and Kannan [21] showed that for any $\varepsilon > 0$, every graph $G = (V, E)$ has an ε-Frieze-Kannan-regular partition with K parts, where $1/\varepsilon < K < 2^{O(\varepsilon^{-2})}$. Moreover, such a partition can be found in randomized $O(n^2)$-time. For the algorithm we present in the next section, we will use the following more recent algorithmic version due to Dellamonica et al.

Theorem 4 [15]. *There is an absolute constant c such that the following holds. For each $\varepsilon > 0$ and for any graph $G = (V, E)$ on n vertices, there is a deterministic algorithm which finds an ε-Frieze-Kannan-regular partition on V with at most $2^{\varepsilon^{-c}}$ parts, and runs in $2^{2^{\varepsilon^{-c}}} n^2$-time.*

Given an n-vertex graph $G = (V, E)$, let $\mathcal{P} : V = V_1 \cup \cdots \cup V_K$ be an ε-Frieze-Kannan-regular partition obtained from Theorem 4. We now define two edge-weighted graphs $G/\mathcal{P}$ and $G_{\mathcal{P}}$ as follows. Let $G/\mathcal{P}$ be the edge-weighted graph on the vertex set $\{1, \ldots, K\}$ and with edge weights

$$w_{G/\mathcal{P}}(ij) = \frac{e_G(V_i, V_j)}{(n/K)^2} \qquad 1 \leq i \neq j \leq K.$$

Let $G_{\mathcal{P}}$ be an edge-weighted graph with vertex set $V = V(G)$, and with edge weights

$$w_{G_{\mathcal{P}}}(uv) = \begin{cases} \frac{e_G(V_i, V_j)}{(n/K)^2} & \text{if } u \in V_i, v \in V_j, 1 \leq i \neq j \leq K; \\ 0 & \text{if } u, v \in V_i, 1 \leq i \leq K. \end{cases}$$

Thus, the Frieze–Kannan regularity lemma says that $d(G, G_{\mathcal{P}}) < \varepsilon n^2$, which implies that $G/\mathcal{P}$ is a small graph that gives a good approximation of G. We now prove the following lemmas which establish a relationship between $\overline{\mathrm{cr}}(G), \overline{\mathrm{cr}}(G_{\mathcal{P}})$, and $\overline{\mathrm{cr}}(G/\mathcal{P})$.

Lemma 3. *Let $\varepsilon \in (0, 1/2)$ and let G and G' be two n-vertex edge-weighted graphs on the same vertex set V. If $d(G, G') < \varepsilon n^2$, then we have*

$$|\overline{\mathrm{cr}}(G) - \overline{\mathrm{cr}}(G')| \leq \varepsilon^{\frac{1}{4C}} n^4,$$

where C is an absolute constant from Theorem 3.

Proof: Consider a straight-line drawing $\mathcal{D}$ of $G = (V, E)$ in the plane such that if $X_{\mathcal{D}} \subset \binom{E}{2}$ denotes the set of pairs of crossing edges in $\mathcal{D}$, we have

$$\overline{\mathrm{cr}}(G) = \sum_{(e_1, e_2) \in X_{\mathcal{D}}} w_G(e_1) w_G(e_2). \tag{1}$$

With slight abuse of notation, let V be the point set in the plane representing the vertices of G in the drawing $\mathcal{D}$. We can assume that V is in general position. With approximation parameter $\varepsilon^{1/(4C)}$, we apply Theorem 3 to the point set V and obtain an equitable partition $V = V_1 \cup \cdots \cup V_K$, where $K \leq \varepsilon^{-1/4}$, such that all but at most $\varepsilon^{1/(4C)} \binom{K}{4}$ quadruples of parts $(V_{i_1}, V_{i_2}, V_{i_3}, V_{i_4})$ have same-type transversals. Let $T \subset \binom{[K]}{4}$ be the set of quadruples (i_1, i_2, i_3, i_4) such that $(V_{i_1}, V_{i_2}, V_{i_3}, V_{i_4})$ has same type transversal and every such transversal is in convex position. Then for each such quadruple, we can order the elements $(i_1, i_2, i_3, i_4) \in T$ so that every segment with one endpoint in V_{i_1} and the other in V_{i_2} crosses every segment with one endpoint in V_{i_3} and the other in V_{i_4}. Therefore, we have

$$\overline{\mathrm{cr}}(G) \geq \sum_{(i_1, i_2, i_3, i_4) \in T} e_G(V_{i_1}, V_{i_2}) e_G(V_{i_3}, V_{i_4}). \tag{2}$$

On the other hand, let us consider the drawing $\mathcal{D}'$ of G' on the same point set $V = V_1 \cup \cdots \cup V_K$. We say that the quadruple $(v_1, v_2, v_3, v_4) \in \binom{V}{4}$ is *bad* if two members lie in a single part V_j, or if all four members lie in distinct parts $V_{i_1}, V_{i_2}, V_{i_3}, V_{i_4}$ such that $(V_{i_1}, V_{i_2}, V_{i_3}, V_{i_4})$ does not have same-type transversals. By Theorem 3, we have at most

$$K \binom{\lceil n/K \rceil}{2} \binom{n}{2} + \varepsilon^{\frac{1}{4C}} \binom{K}{4} \left\lceil \frac{n}{K} \right\rceil^4 \leq \frac{n^4}{4K} + Kn^2 + \varepsilon^{\frac{1}{4C}} \binom{n}{4} \leq 2\varepsilon^{\frac{1}{4C}} \binom{n}{4},$$

bad quadruples. Since each edge has weight at most one, we have

$$\overline{\mathrm{cr}}(G') \leq \sum_{(i_1, i_2, i_3, i_4) \in T} e_{G'}(V_{i_1}, V_{i_2}) e_{G'}(V_{i_3}, V_{i_4}) + 2\varepsilon^{\frac{1}{4C}} \binom{n}{4}.$$

Since $d(G, G') < \varepsilon n^2$, and by (2), we have

$$\overline{\mathrm{cr}}(G') \leq \sum_{(i_1, i_2, i_3, i_4) \in T} e_{G'}(V_{i_1}, V_{i_2}) e_{G'}(V_{i_3}, V_{i_4}) + 2\varepsilon^{\frac{1}{4C}} \binom{n}{4}$$

$$\leq \sum_{(i_1, i_2, i_3, i_4) \in T} (e_G(V_{i_1}, V_{i_2}) + \varepsilon n^2)(e_G(V_{i_3}, V_{i_4}) + \varepsilon n^2) + 2\varepsilon^{\frac{1}{4C}} \binom{n}{4}$$

$$\leq \overline{\mathrm{cr}}(G) + \frac{\varepsilon^{1/2} n^4}{2} + \varepsilon \frac{n^4}{4!} + 2\varepsilon^{\frac{1}{4C}} \binom{n}{4}$$

$$\leq \overline{\mathrm{cr}}(G) + \varepsilon^{\frac{1}{4C}} n^4.$$

The last inequality follows from the fact that C is a sufficiently large constant. A symmetric argument also shows that $\overline{\mathrm{cr}}(G) \leq \overline{\mathrm{cr}}(G') + \varepsilon^{\frac{1}{4C}} n^4$, and the statement follows. □

Let G be an edge-weighted graph on the vertex set $V = \{1, \ldots, K\}$ with weights $w_G(i, j)$. The blow-up $G[m]$ of G is the edge-weighted graph obtained from G by replacing each vertex i by an independent set U_i of order m, and each edge between U_i and U_j has weight $w_G(i, j)$ for $i \neq j$.

Lemma 4. *Let G and $G[m]$ be described as above. Then*

$$0 \leq \overline{\mathrm{cr}}(G[m]) - m^4 \overline{\mathrm{cr}}(G) \leq K^3 m^4.$$

Proof: We start by proving the second inequality first. Fix a drawing $\mathcal{D}$ of G such that if X denotes the set of pairs of crossing edges in $\mathcal{D}$, we have

$$\sum_{(e_1, e_2) \in X} w_G(e_1) w_G(e_2) = \overline{\mathrm{cr}}(G).$$

Let V be the point set in the plane representing the vertices of G in the drawing. We can assume that V is in general position. We draw the blow-up graph $G[m]$ as follows. For each point $v \in V$ in the plane, we choose a very small δ and add $m - 1$ points in the disk centered at v with radius δ. These points will represent U_v. By choosing δ sufficiently small, every quadruple of parts $(U_{i_1}, U_{i_2}, U_{i_3}, U_{i_4})$ will have same-type transversals. Moreover, we can do this so that the resulting point set is in general position. Finally if $uv \in E(G)$, we draw all edges between the point sets U_u and U_v. Let X_m denote the set of pairs of crossing edges in our drawing of $G[m]$.

Set $U = U_1 \cup \cdots \cup U_K$. We say that the quadruple (u_1, u_2, u_3, u_4) of points in U is bad if two of its members lie in a single part U_i. Hence the number of bad quadruples in U is at most $K\binom{m}{2}\binom{Km}{2}$. Since each edge has weight at most one, we have

$$\overline{\mathrm{cr}}(G[m]) \le \sum_{(e_1,e_2)\in X_m} w_{G[m]}(e_1)w_{G[m]}(e_2)$$

$$\le m^4\overline{\mathrm{cr}}(G) + K\binom{m}{2}\binom{Km}{2}$$

$$\le m^4\overline{\mathrm{cr}}(G) + K^3 m^4.$$

On the other hand, now consider a drawing $\mathcal{D}'$ of $G[m]$ such that if X' denotes the set of pairs of crossing edges in $\mathcal{D}'$, we have

$$\overline{\mathrm{cr}}(G[m]) = \sum_{(e_1,e_2)\in X'} w_{G[m]}(e_1)w_{G[m]}(e_2).$$

Let $V(G[m]) = U_1 \cup \cdots \cup U_K$. By selecting one point from each U_i, we obtain a drawing of G which has at least $\overline{\mathrm{cr}}(G)$ weighted pairs of crossing edges. Summing over all of these m^K distinct drawings of G, each weighted crossing appears m^{K-4} times. Therefore,

$$\overline{\mathrm{cr}}(G[m]) \ge \overline{\mathrm{cr}}(G) \cdot m^K / m^{K-4} = m^4\overline{\mathrm{cr}}(G).$$

This completes the proof.   □

4   Proof of Theorem 1

The Algorithm. Input: Let G be a graph with vertex set $V = \{v_1, v_2, \ldots, v_n\}$.

1. Set $\varepsilon = (\log\log n)^{\frac{-1}{2c}}$, where c is defined in Theorem 4. We apply Theorem 4 to G with approximation parameter ε, and obtain an equitable partition $\mathcal{P}$: $V = V_1 \cup \cdots \cup V_K$ on our vertex set with the desired properties such that $1/\varepsilon < K < 2^{\varepsilon^{-c}} = 2^{\sqrt{\log\log n}}$. This can be done deterministically in $n^{2+o(1)}$- time using the algorithm of Dellamonica et al. [15].

2. Let $G/\mathcal{P}$ be the edge-weighted graph on the vertex set $\{1, \ldots, K\}$ with edge weights $w_{G/\mathcal{P}}(ij) = \frac{e_G(V_i,V_j)}{|V_i||V_j|}$. Using Lemma 2, we can find a drawing of $G/\mathcal{P}$ with $\overline{\mathrm{cr}}(G/\mathcal{P})$ weighted pairs of crossing edges. Let $U = \{u_1, \ldots, u_K\}$ be the point set for such a drawing where each point uses at most $2^{O(K\log K)}$ bits. This can be done in $2^{O(K^3)} = n^{o(1)}$ time.

3. We draw $G = (V, E)$ as follows. Let L be the set of lines spanned by U, and let δ be the minimum positive distance[2] between the points in U and the lines in L. Note that δ uses at most $2^{O(K\log K)}$ bits since the line spanned by any two points in U will have the form $y = m_0 x + b_0$, where m_0 and b_0 uses at most $2^{O(K\log K)}$ bits. Therefore, the distance between a point in U and the line $y = m_0 x + b_0$ will use at most $2^{O(K\log K)}$ bits. Set $D(i, \delta/10)$ to be the disk centered at u_i with radius $\delta/10$. We place the points of V_i in

---

[2] The distance between a point and a line in the plane is the length of the line segment which joins the point to the line and is perpendicular to the line.

$D(i, \delta/10)$ so that the point set $V_1 \cup \cdots \cup V_K$ is in general position, and each point uses at most $2^{O(K \log K)} < O(n)$ bits. Notice that every quadruple of parts $(V_{i_1}, V_{i_2}, V_{i_3}, V_{i_4})$ has same-type transversals. We then draw all edges of G on this point set. This can be done in $O(n^2)$ time.

4. Return: the drawing of G.

The total running time for the algorithm above is $n^{2+o(1)}$.

Let $\mathcal{D}$ be the drawing of $G = (V, E)$ obtained from the algorithm above, where $V = \{v_1, \ldots, v_n\} \subset \mathbb{R}^2$, and let X denote the set of pairs of crossing edges in $\mathcal{D}$. We say that the quadruple of points $\{v_{i_1}, v_{i_2}, v_{i_3}, v_{i_4}\}$ in V is bad if two of its members lie in a single disk $D(j, \delta/10)$. Hence there are at most $n^4/(2K)$ bad quadruples. Therefore

$$|X| \leq \left(\frac{n}{K}\right)^4 \overline{\mathrm{cr}}(G/\mathcal{P}) + \frac{n^4}{2K}. \tag{3}$$

Just as above, let $G_{\mathcal{P}}$ be the edge weighted graph with vertex set V (same as G), with edge weights $w_{G_{\mathcal{P}}}(uv) = e_G(V_i, V_j)/(n/K)^2$, if $u \in V_i$, $v \in V_j$ and $i \neq j$, and $w_{G_{\mathcal{P}}}(uv) = 0$ otherwise. Since $G_{\mathcal{P}}$ is an (n/K)-blow-up of $G/\mathcal{P}$, by the proof of Lemma 4, we have

$$\left(\frac{n}{K}\right)^4 \overline{\mathrm{cr}}(G/\mathcal{P}) \leq \overline{\mathrm{cr}}(G_{\mathcal{P}}). \tag{4}$$

Since Theorem 4 implies that the cut-distance between G and $G_{\mathcal{P}}$ satisfies $d(G, G_{\mathcal{P}}) < \varepsilon n^2$, Lemma 3 implies that

$$\overline{\mathrm{cr}}(G_{\mathcal{P}}) \leq \overline{\mathrm{cr}}(G) + \varepsilon^{\frac{1}{4C}} n^4. \tag{5}$$

Putting together (3), (4), and (5), shows that

$$|X| < \overline{\mathrm{cr}}(G) + O\left(\frac{n^4}{(\log \log n)^\delta}\right),$$

where δ is an absolute constant. This completes the proof of Theorem 1.

5   The Rectilinear Crossing Number of Quasi-random Graphs

Proof of Theorem 2. Let $\mathcal{D}$ be a straight-line drawing of G_n in the plane with exactly $\overline{\mathrm{cr}}(G_n)$ edge crossings, and let $V = \{v_1, v_2, \ldots, v_n\}$ be the point set in the plane that represents the vertices of G_n in the drawing. Without loss of generality, we can assume no three members of V are collinear, no two members of V share the same x-coordinate, and V is ordered by increasing x-coordinate.

Set $\varepsilon = n^{-1/2C}$, where C is defined in Theorem 3. By Theorem 3, there is an equitable partition $V = V_1 \cup \cdots \cup V_K$ into K parts, where $K \leq \varepsilon^{-C} = \sqrt{n}$, such that all but at most $\varepsilon\binom{K}{4}$ quadruples $(V_{i_1}, V_{i_2}, V_{i_3}, V_{i_4})$ of parts have same-type transversals. Let $Q \subset \binom{V}{4}$ be the set of quadruples in V that are in convex position. We say that a quadruple $(v_{i_1}, v_{i_2}, v_{i_3}, v_{i_4}) \in Q$ is *bad* if two of its members lie in a single part V_j, or if they lie in distinct parts of $(V_{j_1}, V_{j_2}, V_{j_3}, V_{j_4})$ the does not have same-type transversals. Hence the number of bad quadruples in Q is at most

$$K\binom{n/K}{2}\binom{n}{2} + \varepsilon\binom{K}{4}\left\lceil \frac{n}{K}\right\rceil^4 \le \frac{n^4}{4K} + \varepsilon\binom{n}{4}.$$

Let T denote the number of quadruples of parts $(V_{i_1}, V_{i_2}, V_{i_3}, V_{i_4})$, where each such quadruple $(V_{i_1}, V_{i_2}, V_{i_3}, V_{i_4})$ has same-type transversals and each such transversal is in convex position. Then we have

$$T \cdot \left(\frac{n}{K}\right)^4 \ge |Q| - \left(\frac{n^4}{4K} + \varepsilon\binom{n}{4}\right) \ge \overline{\mathrm{cr}}(K_n) - \left(\frac{n^4}{4K} + \varepsilon\binom{n}{4}\right).$$

Since

$$\overline{\mathrm{cr}}(G_n) \ge T\left(p\lfloor n/K\rfloor^2 - o(n^2)\right)^2 = Tp^2\lfloor n/K\rfloor^4 - o(n^4),$$

this implies

$$\overline{\mathrm{cr}}(G_n) \ge p^2\overline{\mathrm{cr}}(K_n) - p^2\left(\frac{n^4}{4K} + \varepsilon\binom{n}{4}\right) - o(n^4) = p^2\overline{\mathrm{cr}}(K_n) - o(n^4).$$

On the other hand, drawing G_n in the plane by placing its vertices on a point set that minimizes the number of quadruples in convex position, one can follow the arguments above to show that

$$\overline{\mathrm{cr}}(G_n) \le p^2\overline{\mathrm{cr}}(K_n) + o(n^4).$$

This completes the proof of Theorem 2.   □

6   Concluding Remarks

Pach et al. [30] introduced the following alternative notion of crossing number. For any positive integer $k \ge 1$, the *geometric k-planar crossing number* of G, denoted by $\overline{\mathrm{cr}}_k$, is the minimum number of crossings between edges of the same color over all k-edge-colorings of G and all straight-line drawings of G. By following the proof of Theorem 1 almost verbatim, we have the following theorem. We note that one needs to slightly modify the proof of Lemma 3, by coloring the edges of G' between parts V_{i_1} and V_{i_2} (V_{i_3} and V_{i_4}), so that the number of edges in color i between the two parts in G' is roughly the same as the number of edges in color i between the two parts in G.

Theorem 5. *Let $k \ge 1$ be a fixed constant. Given any n-vertex graph G, there is a deterministic $n^{2+o(1)}$-time algorithm that finds a straight-line drawing of G in the plane, and a k-coloring of the edges in G, such that the number of monochromatic pairs of crossing edges in the drawing is at most $\overline{\mathrm{cr}}_k(G) + O(n^4/(\log\log n)^\delta)$, where δ is an absolute constant.*

We suspect that Theorem 1 also holds for other crossing number variants.

Let us also remark that the rectilinear crossing number is a testable parameter, which means that there is a constant time randomized algorithm for approximating the rectilinear crossing number. More precisely, for each $\epsilon > 0$ there is $t = t(\epsilon) > 0$ such that the following holds. If G is a graph on n vertices, by sampling a random induced subgraph H of G on t vertices, we can approximate with probability of success at least .99 the rectilinear crossing number of G with error at most ϵn^4. We do this by noting that the random sample H is, with probability at least .99, close in cut-distance to G (see the Lovász book [29] for details). By Lemma 3, if they are close in cut-distance, we get that $\overline{cr}(G)$ is within ϵn^4 of $\overline{cr}(H)\frac{n^4}{t^4}$.

References

1. Ábrego, B.M., Cetina, M., Fernández-Merchant, S., Leaños, J., Salazar, G.: On ($\leq k$)-edges, crossings, and halving lines of geometric drawings of K_n. Discret. Comput. Geom. **48**, 192–215 (2012)
2. Ábrego, B., Fernández-Merchant, S., Salazar, G.: The rectilinear crossing number of K_n: closing in (or are we?). Thirty Essays on Geometric Graph Theory, pp. 5–18. Springer, Heidelberg (2012)
3. Aichholzer, O., Aurenhammer, F., Krasser, H.: Enumerating order types for small point sets with applications. Order **19**, 265–281 (2002)
4. Aichholzer, O., Krasser, H.: Abstract order type extension and new results on the rectilinear crossing number. Comput. Geom. **36**, 2–15 (2007). Special Issue on the 21st European Workshop on Computational Geometry
5. Ajtai, M., Chvátal, V., Newborn, M., Szemerédi, E.: Crossing-free subgraphs. In: Theory and practice of combinatorics, North Holland Mathematics Studies, vol. 60, pp. 9–12 (1982)
6. Arora, S., Rao, S., Vazirani, U.: Expander flows, geometric embeddings and graph partitioning. J. ACM **56**, 37 (2009)
7. Basu, S., Pollack, R., Roy, M.-F.: On the combinatorial and algebraic complexity of quantifier elimination. J. ACM **43**, 1002–1045 (1996)
8. Bienstock, D.: Some provably hard crossing number problems. Discret. Comput. Geom. **6**, 443–459 (1991)
9. Bienstock, D., Dean, N.: Bounds for rectilinear crossing numbers. J. Graph Theory **17**, 333–348 (1993)
10. Björner, A., Las Vergnas, M., Sturmfels, B., White, N., Ziegler, G.: Oriented Matroids. Cambridge University Press, Cambridge (1993)
11. Borgs, C., Chayes, J.T., Lovász, L., Sós, V.T., Vesztergombi, K.: Convergent sequences of dense graphs I: subgraph frequencies, metric properties and testing. Adv. Math. **219**, 1801–1851 (2008)
12. Chung, F.R.K., Graham, R.L., Wilson, R.M.: Quasi-random graphs. Combinatorica **9**, 345–362 (1989)
13. Conlon, D., Fox, J.: Bounds for graph regularity and removal lemmas. Geom. Funct. Anal. **22**, 1191–1256 (2012)
14. Chuzhoy, J.: An algorithm for the graph crossing number problem. In: Proceedings of the 43rd annual ACM Symposium on Theory of Computing, pp. 303–312 (2011)

15. Dellamonica, D., Kalyanasundaram, S., Martin, D., Rödl, V., Shapira, A.: An optimal algorithm for finding Frieze-Kannan regular partitions. Comb. Probab. Comput. **24**, 407–437 (2015)
16. Even, G., Guha, S., Schieber, B.: Improved approximations of crossings in graph drawings and VLSI layout areas. SIAM J. Comput. **32**, 231–252 (2002)
17. Faila-Monroy, R., López, J.: Computational search of small point sets with small rectilinear crossing number. J. Graph Algorithms Appl. **18**, 393–399 (2014)
18. Fáry, I.: On straight line representation of planar graphs. Acta Univ. Szeged. Sect. Math. **11**, 229–233 (1948)
19. Fox, J., Lovász, L.M.: A tight lower bound for Szemerédi's regularity lemma (preprint)
20. Fox, J., Pach, J., Suk, A.: Density, regularity theorems for semi-algebraic hypergraphs. In: Proceedings of the 26th Annual ACM-SIAM Symposium on Discrete Algorithms (SODA 2115). San Diego, pp. 1517–1530 (2015)
21. Frieze, A., Kannan, R.: Quick approximation to matrices and applications. Combinatorica **19**, 175–220 (1999)
22. Goodman, J.E., Pollack, R.: Allowable sequences and order types in discrete and computational geometry. In: Pach, J. (ed.) New Trends in Discrete and Computational Geometry. Algorithms and Combinatorics, vol. 10, pp. 103–134. Springer, Heidelberg (1993)
23. Goodman, J.E., Pollack, R.: Multidimensional sorting. SIAM J. Comput. **12**, 484–507 (1983)
24. Gowers, W.T.: Lower bounds of tower type for Szemerédi's uniformity lemma. Geom. Funct. Anal. **7**, 322–337 (1997)
25. Hopcroft, J., Tarjan, R.: Efficient planarity testing. J. ACM **21**, 549–568 (1974)
26. Kawarabayashi, K., Reed, B.: Computing crossing number in linear time. In: Proceedings of the Thirty-Ninth Annual ACM Symposium on Theory of Computing (STOC 2007), pp. 382–390. ACM, New York (2007)
27. Leighton, F.T.: Complexity Issues in VLSI. Foundations of Computing Series. MIT Press, Cambridge (1983)
28. Leighton, F.T., Rao, S.: Multicommodity max-flow min-cut theorems and their use in designing approximation algorithms. J. ACM **46**, 787–832 (1999)
29. Lovász, L.: Large Networks and Graph Limits. American Mathematical Society Colloquium Publications, vol. 60. American Mathematical Society, Providence (2012)
30. Pach, J., Székely, L., Tóth, C.D., Tóth, G.: Note on k-planar crossing numbers (manuscript)
31. Pach, J., Tóth, G.: Thirteen problems on crossing numbers. Geombinatorics **9**, 195–207 (2000)
32. Schaefer, M.: Complexity of some geometric and topological problems. In: Eppstein, D., Gansner, E.R. (eds.) GD 2009. LNCS, vol. 5849, pp. 334–344. Springer, Heidelberg (2010). doi:10.1007/978-3-642-11805-0_32
33. Schaefer, M.: The graph crossing number, its variants: a survey, Electron. J. Combin. DS21, 100 (2014)
34. Scheinerman, E., Wilf, H.S.: The rectilinear crossing number of a complete graph and Sylvester's "four point" problem of geometric probability. Am. Math. Mon. **101**, 939–943 (1994)
35. Moshkovitz, G., Shapira, A.: A short proof of Gowers' lower bound for the regularity lemma (preprint)
36. Spencer, J., Tóth, G.: Crossing numbers of random graphs. Random Struct. Algorithms **21**, 347–358 (2002)

37. Sylvester, J.J.: Question 1491. The Educational Times, London (1864)
38. Sylvester, J.J.: Rep. Br. Assoc. **35**, 8–9 (1865)
39. Szemerédi, E.: Regular partitions of graphs. In: Problèmes combinatoires et théorie des graphes (Colloq. Internat. CNRS, Univ. Orsay), vol. 260, pp. 399–401. CNRS, Paris (1978)
40. Vrt'o, I.: Crossing numbers bibliography. www.ifi.savba.sk/~imrich

The Crossing Number of the Cone of a Graph

Carlos A. Alfaro[1]([⊠]), Alan Arroyo[2], Marek Derňár[3], and Bojan Mohar[4]

[1] Banco de México, Mexico City, Mexico
carlos.alfaro@banxico.org.mx
[2] Department of Combinatorics and Optimization, University of Waterloo,
Waterloo, Canada
amarroyo@uwaterloo.ca
[3] Faculty of Informatics, Masaryk University, Brno, Czech Republic
m.dernar@gmail.com
[4] Department of Mathematics, Simon Fraser University, Burnaby, Canada
mohar@sfu.ca

Abstract. Motivated by a problem asked by Richter and by the long standing Harary-Hill conjecture, we study the relation between the crossing number of a graph G and the crossing number of its cone CG, the graph obtained from G by adding a new vertex adjacent to all the vertices in G. Simple examples show that the difference $cr(CG) - cr(G)$ can be arbitrarily large for any fixed $k = cr(G)$. In this work, we are interested in finding the smallest possible difference, that is, for each non-negative integer k, find the smallest $f(k)$ for which there exists a graph with crossing number at least k and cone with crossing number $f(k)$. For small values of k, we give exact values of $f(k)$ when the problem is restricted to simple graphs, and show that $f(k) = k + \Theta(\sqrt{k})$ when multiple edges are allowed.

1 Introduction

Little is known on the relation between the crossing number and the chromatic number. In this sense Albertson's conjecture (see [1]), that if $\chi(G) \geq r$, then $cr(G) \geq cr(K_r)$, has taken a great interest. Albertson's conjecture has been proved [1,3,14] for $r \leq 16$. It is related to Hajós' Conjecture that every r-chromatic graph contains a subdivision of K_r. If G contains a subdivision of K_r, then $cr(G) \geq cr(K_r)$. Thus Albertson's conjecture is weaker than Hajós' conjecture, however Hajós' conjecture is false for any $r \geq 7$ [6].

The *cone* of a graph G is the graph CG obtained from G by adding an *apex*, a new vertex that is adjacent to each vertex in G. Many properties of a graph are automatically transferred to its cone. For example, if G is r-coloring-critical,

C.A. Alfaro—Supported by SNI and CONACYT grant 166059.
A. Arroyo—Suported by CONACYT.
B. Mohar—Supported in part by an NSERC Discovery Grant (Canada), by the Canada Research Chairs program, and by a Research Grant of ARRS (Slovenia). On leave from: IMFM, Department of Mathematics, Ljubljana, Slovenia.

Y. Hu and M. Nöllenburg (Eds.): GD 2016, LNCS 9801, pp. 427–438, 2016.
DOI: 10.1007/978-3-319-50106-2_33

then CG is $(r+1)$-coloring-critical. During the Crossing Numbers Workshop in 2013, in an attempt to understand Alberston's conjecture, Richter proposed the following problem: Given an integer $n \geq 5$ and a graph G with crossing number at least $cr(K_n)$, does it follow that the crossing number of its cone CG is at least $cr(K_{n+1})$? There are examples where these two values can differ arbitrarily (for instance, if G is the disjoint union of K_4's and K_5's). What is less clear is how close these values can be.

The answer to Richter's question is positive for the first interesting case when $n = 5$: Kuratowski's theorem implies that the cone of any graph with crossing number at least $cr(K_5) = 1$ contains a subdivision of CK_5 or $CK_{3,3}$, and each of these graphs has crossing number at least $cr(K_6) = 3$. Unfortunately, the answer is negative for the next case, as the graph in Fig. 1 shows. This graph has crossing number 3, and a cone with crossing number at most 6, and this is less than $cr(K_7) = 9$. This motivated us to investigate the following question.

Problem 1. For each $k \geq 0$, find the smallest integer $f(k)$ for which there is a graph G with crossing number at least k and its cone has $cr(CG) = f(k)$.

Fig. 1. A counterexample to Richter's question when $n = 6$.

Note that $f(k)$ can also be defined as the largest integer such that every graph with $cr(G) \geq k$, has $cr(CG) \geq f(k)$. An upper bound to the function $f(k)$ is obtained from the graph in Fig. 1, by changing the multiplicity of each edge to r. Any drawing of the new graph has at least $3r^2$ crossings, and its cone has crossing number $3r^2 + 3r$. This shows that $f(k) \leq k + \sqrt{3k}$. Our main result shows that this is close to be best possible.

Theorem 2. *Let G be a graph with $cr(G) \geq k$. Then $cr(CG) \geq k + \sqrt{k/2}$.*

Thus we have the following:

Corollary 3. *For multigraphs we have $f(k) = k + \Theta(\sqrt{k})$.*

The paper is organized as follows. Page drawings, a concept intimately related to drawings of the cone of a graph, are defined in Sect. 2 and used throughout the subsequent sections. Although, there seems to be a connection between 1-page drawings and drawings of the cone, their exact relationship is much more subtle. Our proofs are instructive in this manner and provide further understanding of these concepts.

The proof of our main result, Theorem 2 is provided in Sect. 3. In Sect. 4, we restrict Problem 1 to the case of simple graphs. To distinguish between these two problems we use $f_s(k)$ instead of $f(k)$. Along this paper, a graph is allowed to have multiple edges but no loops; when our graphs have no multiple edges, then we refer them as simple graphs. We find the smallest values of f_s by showing that $f_s(1) = 3$, $f_s(2) = 5$, $f_s(3) = 6$, $f_s(4) = 8$ and $f_s(5) = 10$. These initial values may suggest that $f_s(k) \geq 2k$. However, in Sect. 5 we show that

$$f_s(k) = k + o(k),$$

and provide additional justification for a more specific conjecture that

$$f_s(k) = k + \sqrt{2}\, k^{3/4}(1 + o(1)).$$

2   Page Drawings

In this section we describe a perspective provided from considering *page drawings* of graphs, a concept that has been studied in its own and has interesting applications. The relation between 1- and 2-page drawings has shown to be handy as it is used in the proofs of Theorems 2 and 7. A more detailed discussion on the relevant aspects of this section can be found in [2,5,13].

For an integer $k \geq 1$, a k-*page book* consists of k half planes sharing their boundary line ℓ (spine). A k-*page-drawing* is a drawing of a graph in which vertices are placed in the spine of a k-page book, and each edge arc is contained in one page. A convenient way to visualize a k-page drawing is by means of the *circular model*. In this model each page is represented by a unit 2-dimensional disk, so that the vertices are arranged identically on each disk boundary and each edge is drawn entirely in exactly one disk. In this work we are only interested in 1 and 2-page drawings, and, to be more precise, in the following problem.

Problem 4. Given a 1-page drawing of a graph G with k crossings, find an upper bound on the number of crossings of an optimal 2-page drawing of G while having the order of vertices of G on the spine unchanged.

In other words, if the drawing of G in the plane is such that all the vertices are incident to the outer-face (which is equivalent to having a 1-page drawing), what is the most efficient way to redraw some edges in the outer-face to reduce the number of crossings? For this purpose, we define the *circle graph* C_D of any 1-page drawing D of G as the graph whose vertices are the edges of G, and any two elements are adjacent if they cross in D. Note that C_D depends only on the cyclic order of the vertices of G in the spine.

A related problem was previously formulated by Kainen in [11], where he studied the *outerplanar crossing number* of a graph as the minimum number of crossings in any drawing of G so that all its vertices are incident to the same face. Clearly, the crossing number of CG is at most the outer-planar crossing number of G. Although, Kainen was interested in finding an n-vertex graph that has the largest difference between its crossing number and its outer-planar crossing number, for us it will be useful to consider drawings in which the vertices are incident to the same face.

Turning a 1-page drawing into a 2-page drawing is equivalent to finding a bipartition $(X, V(C_D) \setminus X)$ of the vertices of C_D, each part representing the set of edges of G drawn in one of the pages. Minimizing the number of crossings in the obtained 2-page drawing of G is equivalent to maximize the number edges in C_D between X and $V(C_D) \setminus X$. This last problem is known as the *max-cut problem*, and if the considered graph C_D has m edges, then, a well-known result of Erdős [7] states that its maximum edge-cut has size more than $m/2$. Improvements to this general bound are known (see [8,9] and a more recent survey [4]). For our purpose the following bound of Edwards will be useful.

Lemma 5 (Edwards [8,9]). *Suppose that G is a graph of order n with $m \geq 1$ edges. Then G contains a bipartite subgraph with at least $\frac{1}{2}m + \sqrt{\frac{1}{8}m + \frac{1}{64}} - \frac{1}{8} > \frac{1}{2}m$ edges.*

In our context, this result translates to the following observation that we will use.

Corollary 6. *Let D be a 1-page drawing of a graph G with $k \geq 1$ crossings. Then some edges of G can be redrawn in a new page, obtaining a 2-page drawing with at most $\frac{1}{2}k - \sqrt{\frac{1}{8}k + \frac{1}{64}} + \frac{1}{8}$ crossings. Such a drawing can be found in time $O(|E(G)| + k)$.*

The proof of Corollary 6 will be provided in the full version.

3   Lower Bound on the Crossing Number of the Cone

This section contains the proof of our main result.

Proof (of Theorem 2). Let $\widehat{D}$ be an optimal drawing of the cone CG of G with apex a, and suppose $\widehat{D}$ has less than $k + \sqrt{k/2}$ crossings. We consider $D = \widehat{D}|_G$, the drawing of G induced by $\widehat{D}$. If we let t to be the number of crossings in D, then we have

$$k \leq t < k + \sqrt{k/2}. \tag{1}$$

For each vertex $v \in V(G) \cup \{a\}$, let s_v be the number of crossings in $\widehat{D}$ involving edges incident with v. Using that $cr(\widehat{D}) < k + \sqrt{k/2}$ and the left-hand side inequality in (1), we obtain that $s_a < \sqrt{k/2}$.

Consider $x_1, \ldots, x_{s_a}$, the crossings involving edges incident with a. Since $\widehat{D}$ is optimal, each of these crossings is between an edge incident to a and an edge in G. Let $e_1, \ldots, e_{s_a}$ be the list of edges in G (we allow repetitions) so that x_i is the crossing between e_i and an edge incident with a. We subdivide each edge e_i in D using two points close to the crossing x_i, and we remove the edge segment σ_i joining these new two vertices, in order to obtain a drawing D_0 of a graph G_0 with t crossings (see Fig. 2).

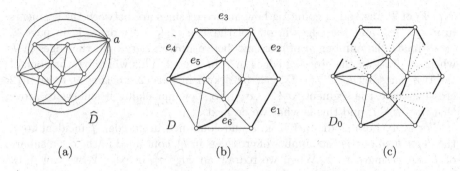

Fig. 2. A drawing where the crossed edges are cut.

The obtained drawing D_0 has all its vertices incident to the face of D_0 containing the point corresponding to the apex vertex a of CG in $\widehat{D}$. For simplicity, we may assume that this is the unbounded face of D_0. It follows that there exists a simple closed curve ℓ in the closure of this face, containing all the vertices of G_0. Thus, D_0 gives rise to a 1-page drawing of G_0 with spine ℓ.

Now construct a new drawing of G as follows:

1. Start with the 1-page drawing D_0. Partition the edges of G_0 according to Corollary 6, and draw the edges of one part in page 2 outside ℓ.
2. Reinsert edge segments $\sigma_1, \ldots, \sigma_{s_a}$ as they where drawn in D, to obtain a drawing D_1 (of a subdivision) of G. These segments do not cross each other, but they may cross some of the edges of G_0 that we placed in page 2 in step 1.

Now we estimate the number of crossings in D_1. According to Corollary 6, after step 1 we obtain a 2-page drawing D_0 with less than $t/2 - \sqrt{t/8} + 1/8$ crossings. After step 2 we gain some new crossings between the added segments $\sigma_1, \ldots, \sigma_{s_a}$ and the edges of G_0 drawn on page 2 in step 1.

Claim. The number of new crossings between $\sigma_1, \ldots, \sigma_{s_a}$ and the edges drawn on page 2 in step 1 is at most $(k - 1)/2$.

Proof. We may assume that, for each $v \in V(G)$, $s_v < \sqrt{k/2}$. Otherwise, by removing v and all the edges incident to v, we obtain a drawing of $CG - v$ containing a subdrawing of G, in which v is represented as the apex, and this drawing has less than k crossings, a contradiction.

Let $e \in E(G)$ be an edge having ends $u, v \in V(G)$. Suppose that $ay_1, \ldots, ay_{r_e}$ are the edges incident to a that cross e in $\widehat{D}$. We may assume that, for every i, j with $1 \leq i < j \leq r_e$, when we traverse e from u to v, the crossing $x_i = e \cap ay_i$ precedes the crossing $x_j = e \cap ay_j$. It is convenient to let $x_0 = u$ and $x_{r_e+1} = v$.

The edges of G_0 included in $D[e]$ are the segments of $D[e] - \{\sigma_1, \ldots, \sigma_{s_a}\}$. We enumerate these edges as $\tau_0^e, \ldots, \tau_{r_e}^e$, so that τ_i^e is included in the $x_i x_{i+1}$-arc of $D[e]$. Note that τ_1^e is incident to u, while τ_{r_e} is incident to v.

Let $T = \{\tau_i^e : e \in E(G) \text{ and } 0 \leq i \leq r_e\}$ be the set of edges of G_0. In Step 1, when we apply Corollary 2.2 to the edges in D_0, we obtain a partition

$T_1 \cup T_2$ of T. Instead of counting how many crossings are between the segments in $\sigma_1, \ldots, \sigma_{s_a}$ and the edges in one of the T_i's when we redraw T_i in page 2, we estimate the number m of crossings between $\sigma_1, \ldots, \sigma_{s_a}$ and the edges in T when we draw all the crossing edges in T in page 2. This will show that one of the two parts, either T_1 or T_2, can be drawn in page 2 creating at most $m/2$ crossings with the segments $\sigma_1, \ldots, \sigma_{s_a}$. To show our claim, it suffices to prove that $m \leq k - 1$, and this is what we do next.

For every point p distinct from a and contained in an edge f incident to a, the *depth* $h(p)$ of p is the number of crossings in $\widehat{D}$, contained in the open subarc of f connecting a to p. When we redraw an edge τ_i^e in page 2, we can draw it so that it crosses at most $h(x_i) + h(x_{i+1})$ segments in $\sigma_1, \ldots, \sigma_{s_a}$. Such new drawing of τ_i^e is obtained from letting the segment of τ_i^e near to x_i follow the same dual path in D that x_i follows to reach a via ay_i. Likewise the new end of τ_i^e near x_{i+1} is defined. The new τ_i^e is obtained from connecting the two *end segments* of τ_i^e inside the face of D containing a.

Let $X(a)$ be the set of crossings involving edges incident to a. For every $x \in X(a)$, there are precisely two elements in T, so that when they are redrawn in page 2, one of its end segments mimics the arc between x and a inside the edge including x and a. Each $v \in V(G)$ is incident to at most s_v edges crossing in D_0. Then, for every $v \in V(G)$, there are are most s_v edges in T, so that when we redraw them in page 2, one of their ends mimics the dual path followed by the edge $\widehat{D}[xa]$. These two observations together imply that

$$m \leq \sum_{x \in X(a)} 2h(x) + \sum_{v \in V} h(v)s_v$$

$$< 2\sum_{v \in V} (1 + 2 + \ldots + (h(v) - 1)) + \sqrt{k/2}\sum_{v \in V} h(v)$$

$$\leq \sum_{v \in V} h(v)^2 + (\sqrt{k/2})s_a \leq \left(\sum_{v \in V} h(v)\right)^2 + k/2$$

$$= s_a^2 + k/2 < k.$$

Because m is an integer less than k, $m \leq k - 1$ as desired.   □

At the end, we obtained a drawing D_1 of (a subdivision of) G with less than $t/2 - \sqrt{t/8} + 1/8 + (k-1)/2$ crossings. Using (1) it follows that

$$cr(D_1) < \frac{1}{2}(k + \sqrt{k/2}) - \sqrt{t/8} + 1/8 + k/2 - 1/2 = k + \sqrt{k/8} - \sqrt{t/8} - 3/8 < k,$$

contradicting the fact that $cr(D_1) \geq cr(G) \geq k$.   □

4   Exact Values of the Crossing Number of the Cone for Simple Graphs

In this section, we investigate the minimum crossing number of a cone, with the restriction of only considering simple graphs. We are interested in finding the smallest integer $f_s(k)$ for which there is a simple graph with crossing number at least k, whose cone has crossing number $f_s(k)$. On one hand, we describe below a family of simple graphs that shows that $f_s(k) \leq 2k$. Our best general lower bound is obtained from Theorem 2. The main result in this section, Theorem 7, help us to obtain exact values on $f_s(k)$ for cases when k is small.

Theorem 7. *Let G be a simple graph with crossing number k. Then*

(1) if $k \geq 2$, then $cr(CG) \geq k + 3$;
(2) if $k \geq 4$, then $cr(CG) \geq k + 4$; and
(3) if $k \geq 5$, then $cr(CG) \geq k + 5$.

Before proving Theorem 7, we describe a family of examples that is used to find an upper bound for $f_s(k)$, that is exact for the values $k = 3, 4, 5$. Given an integer $k \geq 3$, the graph F_k (Fig. 3) is obtained from two disjoint cycles $C_1 = x_0 \ldots x_{k-1} x_0$ and $C_2 = y_0 \ldots y_{2k-1} y_0$ by adding, for each $i = 0, \ldots, k-1$, the edges $x_i y_{2i-2}$, $x_i y_{2i-1}$, $x_i y_{2i}$, $x_i y_{2i+1}$ (where the indices of the vertices y_j are taken modulo $2k$). It is not hard to see that F_k has crossing number k: a drawing with k crossings is shown in Fig. 3. To show that $cr(F_k) \geq k$, for $i \in \{0, \ldots, k-1\}$, consider L^i, the K_4 induced by the vertices in $\{x_i, x_{i+1}, y_{2i}, y_{2i+1}\}$. Every L^i is a subgraph of a K_5 subdivision of F_k, thus, in an optimal drawing of F_k, at least one of the edges in L^i is crossed. This only guarantees that $cr(F_k) \geq k/2$, as two edges from distinct L^i's might be crossed. However, if an edge from L^i crosses an edge e_j from some other L^j, then $F_k - e_j$ has a K_5 subdivision including L^i, exhibiting a new crossing in some edge in L^i. Therefore, every L^i either has a crossing not involving an edge in another L^j, or there are least two crossings involving edges in L^i. This shows that $cr(F_k) \geq k$.

The graph shown in Fig. 4 has crossing number 2, and its cone has crossing number 5. This shows that $f_s(2) \leq 5$. On the other hand, F_3, F_4, and F_5 serve as examples to show that $f_s(k) \leq 2k$ for $k = 3, 4, 5$. These bounds are tight for $2 \leq k \leq 5$ by Theorem 7.

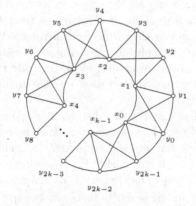

Fig. 3. The graph F_k.

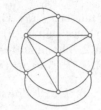

Fig. 4. A graph with crossing number 2 whose cone has crossing number 5.

Proof (Proof of Theorem 7). Suppose G is a graph with $cr(G) = k$. Let $\widehat{D}$ be an optimal drawing of the cone CG, D its restriction to G, and F_a be the face of D containing the apex a. The vertices of G incident to F_a are the *planar neighbors* of a.

Assume that $k \geq 2$, and suppose $\widehat{D}$ has exactly $k + t$ crossings. Theorem 2 guarantees that $t \geq 1$. Since each edge from a to a non-planar-neighbor introduce at least one crossing, the apex a has either 0, 1, 2, 3 or 4 non-planar neighbors (if a has more than 4 non-planar neighbours, then any of the items in Theorem 7 is satisfied).

We start by assuming that a has no non-planar neighbors. In this case, D is a 1-page drawing of G. Corollary 6 implies that we can obtain a new drawing of G with less than $(k + t)/2$ crossings. Thus $(k + t)/2 > cr(G) = k$, which implies that $t \geq k + 1$. In any case of the theorem, this implies the conclusion, thus we may now assume that a has at least one and at most t non-planar neighbors.

(1) Let us now assume that $k \geq 2$ and $t \leq 2$.

Suppose a has exactly one non-planar neighbor u. Then $cr(D)$ has at most $k + 1$ crossings. At least one edge incident to u is crossed in D, otherwise, all the crossed edges have ends in F_a, and using Corollary 6, we obtain a drawing of G with less than $(k + 1)/2$ crossings, contradicting that $cr(G) = k$. If at least two crossings in D involve edges incident to u, or if D has k crossings, then by redrawing u in F_a, and adding all the edges to its neighbors without creating any crossings, we obtain a drawing of G with less than k crossings. Therefore D has $k + 1$ crossings, and exactly k of them involve edges not incident to u. Again, we apply Corollary 6 to obtain a drawing of G with at most $\frac{1}{2}(k-1) < k$ crossings (this time we are more careful by setting our two pages in such way that the edge not incident to u that crosses an edge incident to u is redrawn in the page contained in F_a).

Finally, suppose a has exactly two distinct non-planar neighbors u and v. Then, $\widehat{D}$ has $k + 2$ crossings; D has k crossings, and the edges au, av are crossed exactly once. Notice that any crossed edge in D is incident to either u or v; otherwise, we can redraw such edge inside F_a, obtaining a drawing of G with less than k crossings. Redraw v in $\widehat{D}[a]$ (where $\widehat{D}[a]$ denotes the point representing a in $\widehat{D}$); draw the edge uv (if it exists in G) as the arc $\widehat{D}[au]$, and draw the edges from v to its neighbors distinct from u, inside F_a without creating new crossings. Since every crossing in D involves an edge incident with v, we obtain a drawing of G with at most one crossing, a contradiction.

(2) Now, suppose that $k \geq 4$ and that $t = 3$.

The case when the apex a has only one non-planar neighbor u is similar to the above. If at least three crossings in D involve edges incident with u, then by redrawing u and the edges incident to u in F_a, we obtain a drawing with less than k crossings, a contradiction. Thus, at most two crossings involve edges incident to u. We redraw the remaining crossed edges according to Corollary 6 (if there is an edge e that crosses an edge incident to u, in order to remove an extra crossing, we may choose this new drawing so that e is redrawn in the page contained in F_a). If 2 crossings involve edges incident to u, then the obtained drawing has at most $\frac{k}{2} + 1$ crossings, where the $+1$ comes from the fact that e was drawn in the page contained in F_a. If at most one of the edges at u is crossed, then the new drawing has at most $(k+1)/2$ crossings. In any case, since $k \geq 4$, the new drawing has less than k crossings, a contradiction.

Let us now consider the case when the apex has two non-planar neighbors u and v. In this case, the drawing D has either k or $k + 1$ crossings, and one of $\{au, av\}$, say au, is crossed only once. Let L be the set of crossed edges in D that are not incident to u or v. Suppose there are at least two crossings involving only edges in L. Then, either there are two edges in L that do not cross, or L has an edge e that crosses two other edges in L. In the former case, we redraw such pair of edges in F_a; in the latter case, we redraw e in F_a. Any of these modifications yield a drawing with less than k crossings. Thus, we may assume that at most one crossing in D involves two edges not incident to u or v. Redraw v in $\widehat{D}[a]$; draw the edge vu (if such edge exists in G) as $\widehat{D}[au]$; and the remaining edges from v to its neighbors distinct from u without creating new crossings. The new drawing of G has at most two crossings: possibly one in $\widehat{D}[av]$ and another between edges in L, a contradiction.

Finally suppose that the apex a has three non-planar neighbors u, v, w. In this case D has precisely k crossings, and the edges au, av, aw are crossed exactly once. Observe that any crossed edge in D is incident to one of $\{u, v, w\}$, otherwise we can redraw such edge in F_a, obtaining a drawing of G with less than k crossings.

Let H be the graph induced by $\{u, v, w\}$. If, for $x \in \{u, v, w\}$, $d_H(x)$ denotes the degree of x in H, then at most $d_H(x)$ crossings involve edges at x. Otherwise, by redrawing x in $\widehat{D}[a]$; drawing the edges from x to its neighbors in H by using the respective edges from a; and, by drawing the remaining edges at x in F_a without creating new crossings, we obtain a drawing of G with less than k crossings. So for each vertex $x \in \{u, v, w\}$, there are at most two crossings involving edges at x. Hence D has at most three crossings, a contradiction.

(3) The proof will be included in the full version of the paper.   □

5   Asymptotics for Simple Graphs

Lastly, we try to understand the behaviour of $f_s(k)$ when k is large. The important part is the increase of the crossing number after adding the apex, thus we define

$$\phi_s(k) = f_s(k) - k.$$

We have proved that $\phi(k) = f(k) - k \geq \frac{1}{2}k^{1/2}$. The term $k^{1/2}$ is asymptotically tight in the case when we allow multiple edges. However, it is unclear how large $\phi_s(k)$ is. This question is treated next.

Theorem 8. $\phi_s(k) = O(k^{3/4})$.

Proof. Let us consider a positive integer k and let n be the smallest integer such that $cr(K_n) \geq k$. Then $G = K_n$ has a crossing number at least k and its cone is K_{n+1}.

To find an upper bound for $cr(K_{n+1})$ in terms of $cr(K_n)$, start with a drawing of K_n with $cr(K_n)$ crossings. Then clone a vertex, that is, place a new vertex very close to an original vertex, and draw the new edges along the original edges. Each edge incident to the new vertex cross $O(n^2)$ edges, thus the obtained drawing has $cr(K_n) + O(n^3)$ crossings. Therefore

$$\phi_s(k) \leq cr(K_{n+1}) - cr(K_n) \leq O(n^3).$$

It is known [12] that

$$\frac{3}{10}\binom{n}{4} \leq cr(K_n) \leq \frac{3}{8}\binom{n}{4}.$$

(The constant $3/10$ in the lower bound has been recently improved to 0.32025, see [12] for more information.) Then $\phi_s(k) = O(n^3) = O(k^{3/4})$.   □

The *Harary-Hill Conjecture* [10] states that

$$cr(K_n) = \begin{cases} \frac{1}{64}n(n-2)^2(n-4), & n \text{ is even}; \\ \frac{1}{64}(n-1)^2(n-3)^2, & n \text{ is odd}. \end{cases}$$

Proposition 9. *If the Harary-Hill conjecture holds, then*

$$\phi_s(k) \leq \sqrt{2}\,k^{3/4}(1 + o(1)).$$

Proof. As in the proof of Theorem 8, but with a slight twist for added precision, we take n such that $cr(K_{n-1}) < k \leq cr(K_n)$. We also take n_1 such that for $k_1 = k - cr(K_{n-1})$ we have $cr(K_{n_1-1}) < k_1 \leq cr(K_{n_1})$. Let $G = K_{n-1} \cup K_{n_1}$. Then $cr(G) = cr(K_{n-1}) + cr(K_{n_1}) \geq k$ and $cr(CG) = cr(K_n) + cr(K_{n_1+1})$. Therefore,

$$\phi_s(k) \leq cr(K_n) + cr(K_{n_1+1}) - cr(K_{n-1}) - cr(K_{n_1})$$
$$\leq cr(K_n) - cr(K_{n-1}) + cr(K_{n_1+1}) - cr(K_{n_1-1}).$$

By inserting the values for the crossing number from the Harary-Hill Conjecture, we obtain (the calculation given is for odd n and odd n_1, it is similar when n or n_1 is even):

$$cr(K_n) - cr(K_{n-1}) = \frac{1}{64}((n-1)^2(n-3)^2 - (n-1)(n-3)^2(n-5)) = \frac{1}{16}n^3(1+o(1))$$

and

$$cr(K_{n_1+1}) - cr(K_{n_1-1}) = \tfrac{1}{64}((n_1+1)(n_1-1)^2(n_1-3)$$
$$-(n_1-1)(n_1-3)^2(n_1-5))$$
$$= \tfrac{1}{8}n_1^3(1+o(1)).$$

Noticing that $k = \tfrac{1}{64}n^4(1+o(1))$ and $k_1 = \tfrac{1}{64}n_1^4(1+o(1)) = O(n^3)$ because $k_1 \le cr(K_n) - cr(K_{n-1})$, we conclude that $n_1^3 = O(n^{9/4}) = o(k^{3/4})$ and henceforth

$$\phi_s(k) \le \tfrac{1}{16}n^3(1+o(1)) + \tfrac{1}{8}n_1^3(1+o(1)) = \sqrt{2}\,k^{3/4}(1+o(1)).$$

$\square$

The above proof works even under a weaker hypothesis that $cr(K_n) = \alpha n^4 + \beta n^3(1+o(1))$, where α and β are constants. This would imply that $\phi_s(k) = O(k^{3/4})$. Our conjecture is that (9) gives the precise asymptotics.

Conjecture 10. $\phi_s(k) = \sqrt{2}\,k^{3/4}(1+o(1))$.

A reviewer noted that this asymptotic is matched when the graph we are considering is dense.

Remark 11. Let G be a graph with n vertices, m edges, $cr(G) = k$ and such that $m \ge 4n$. If $m = \Omega(n^2)$, then $cr(CG) \ge k + \Omega(k^{3/4})$.

The details will be provided in the full version.

Summary

To put the results of this paper into context, let us overview some of the motivation and some of directions for future work. The starting point of this paper was an attempt to understand Albertson's conjecture. The results of the paper (and their proofs) show that the crossing number behavior when adding an apex vertex is intimately related to 1-page drawings, but the exact relationship is quite subtle. There is some evidence that the minimal increase of the crossing number when an apex is added should be achieved with very dense graphs, close to the complete graphs. Our Conjecture 10 entails this problem. Although very dense graphs have fewer vertices than sparser graphs with the same crossing number and thus need fewer connections to be made from the apex to their vertices, their near optimal drawings are far from 1-page drawings and therefore more crossings are needed. The full understanding of this antinomy would shed new light on the Harary-Hill conjecture.

Finally, it is worth pointing out that neither exact nor approximation algorithm is known for computing the crossing number of graphs of bounded treewidth. Adding an apex to a graph increases the tree-width of the graph by 1, thus understanding the crossing number of the cone is an important special case that would need to be understood before devising an algorithm for general graphs of bounded tree-width.

References

1. Albertson, M.O., Cranston, D.W., Fox, J.: Crossings, colorings and cliques. Electron. J. Comb. **16**, #R45 (2009)
2. Bannister, M.J., Eppstein, D.: Crossing minimization for 1-page and 2-page drawings of graphs with bounded treewidth. In: Duncan, C., Symvonis, A. (eds.) GD 2014. LNCS, vol. 8871, pp. 210–221. Springer, Heidelberg (2014). doi:10.1007/978-3-662-45803-7_18
3. Barát, J., Tóth, G.: Towards the Albertson conjecture. Electron. J. Comb. **17**, #R73 (2010)
4. Bollobás, B., Scott, A.D.: Better bounds for Max Cut. In: Contemporary Combinatorics. Bolyai Soc. Math. Stud. vol. 10, pp. 185–246. János Bolyai Math. Soc., Budapest (2002)
5. Buchheim, C., Zheng, L.: Fixed linear crossing minimization by reduction to the maximum cut problem. In: Computing and Combinatorics, pp. 507–516 (2006)
6. Catlin, P.A.: Hajos' graph-coloring conjecture: variations and counterexamples. J. Comb. Theory Ser. B **26**, 268–274 (1979)
7. Erdős, P.: Gráfok páros körüljárású résgráfjairól (On bipartite subgraphs of graphs in Hungarian). Mat. Lapok **18**, 283–288 (1967)
8. Edwards, C.S.: Some extremal properties of bipartite subgraphs. Can. J. Math. **25**, 475–485 (1973)
9. Edwards, C.S.: An improved lower bound for the number of edges in a largest bipartite subgraph. In: Recent Advances in Graph Theory, pp. 167–181 (1975)
10. Harary, F., Hill, A.: On the number of crossings in a complete graph. Proc. Edinb. Math. Soc. **13**, 333–338 (1963)
11. Kainen, P.C.: The book thickness of a graph II. Congr. Numerantium **71**, 127–132 (1990)
12. De Klerk, E., Pasechnik, D.V., Schrijver, A.: Reduction of symmetric semidefinite programs using the regular ∗-representation. Math. Program. **109**, 613–624 (2007)
13. de Klerk, E., Pasechnik, D., Salazar, G.: Improved lower bounds on book crossing numbers of complete graphs. SIAM J. Discret. Math. **27**, 619–633 (2013)
14. Oporowski, B., Zhao, D.: Coloring graphs with crossings. Discret. Math. **309**, 2948–2951 (2009)

Topological Graph Theory

Topological Drawings
of Complete Bipartite Graphs

Jean Cardinal[1(✉)] and Stefan Felsner[2]

[1] Université Libre de Bruxelles (ULB), Brussels, Belgium
jcardin@ulb.ac.be
[2] Technische Universität (TU) Berlin, Berlin, Germany
felsner@math.tu-berlin.de

Abstract. Topological drawings are natural representations of graphs in the plane, where vertices are represented by points, and edges by curves connecting the points. We consider a natural class of simple topological drawings of *complete bipartite* graphs, in which we require that one side of the vertex set bipartition lies on the outer boundary of the drawing. We investigate the combinatorics of such drawings. For this purpose, we define combinatorial encodings of the drawings by enumerating the distinct drawings of subgraphs isomorphic to $K_{2,2}$ and $K_{3,2}$, and investigate the constraints they must satisfy. We prove in particular that for complete bipartite graphs of the form $K_{2,n}$ and $K_{3,n}$, such an encoding corresponds to a drawing if and only if it obeys consistency conditions on triples and quadruples. In the general case of $K_{k,n}$ with $k \geq 2$, we completely characterize and enumerate drawings in which the order of the edges around each vertex is the same for vertices on the same side of the bipartition.

1 Introduction

We consider *topological graph drawings*, which are drawings of simple undirected graphs where vertices are represented by points in the plane, and edges are represented by simple curves that connect the corresponding points. We typically restrict those drawings to satisfy some natural nondegeneracy conditions. In particular, we consider *simple* drawings, in which every pair of edges intersect at most once. A common vertex counts as an intersection.

While being perhaps the most natural and the most used representations of graphs, simple drawings are far from being understood from the combinatorial point of view. For the smallest number of edge crossings in a simple topological drawing of K_n [1,2,8] or of $K_{k,n}$ [4,12] there are long standing conjectures but the actual minimum remains unknown.

In order to cope with the inherent complexity of the drawings, it is useful to consider combinatorial abstractions. Those abstractions are discrete structures encoding some features of a drawing. One such abstraction, introduced by Kratochvíl, Lubiw, and Nešetřil, is called *abstract topological graphs* (AT-graph) [9]. An AT-graph consists of a graph (V, E) together with a set $\mathcal{X} \subseteq \binom{E}{2}$.

© Springer International Publishing AG 2016
Y. Hu and M. Nöllenburg (Eds.): GD 2016, LNCS 9801, pp. 441–453, 2016.
DOI: 10.1007/978-3-319-50106-2_34

A topological drawing is said to *realize* an AT-graph if the pairs of edges that cross are exactly those in $\mathcal{X}$. Another abstraction of a topological drawing is called the *rotation system*. The rotation system associates a circular permutation with every vertex v, which in a realization must correspond to the order in which the neighbors of v are connected to v. Natural realizability problems are: given an AT-graph or a rotation system, is it realizable as a topological drawing? The realizability problem for AT-graphs is known to be NP-complete [10].

For simple topological drawings of complete graphs, the two abstractions are actually equivalent [11]. It is possible to reconstruct the set of crossing pairs of edges by looking at the rotation system, and vice-versa. Kynčl recently proved the remarkable result that a complete AT-graph (an AT-graph for which the underlying graph is complete) can be realized as a simple topological drawing of K_n if and only if all the AT-subgraphs on at most 6 vertices are realizable [5,6]. This directly yields a polynomial-time algorithm for the realizability problem. While this provides a key insight on topological drawings of complete graphs, similar realizability problems already appear much more difficult when they involve complete *bipartite* graphs. In that case, knowing the rotation system is not sufficient for reconstructing the intersecting pairs of edges.

We propose a fine-grained analysis of simple topological drawings of complete bipartite graphs. In order to make the analysis more tractable, we introduce a natural restriction on the drawings, by requiring that one side of the vertex set bipartition lies on a circle at infinity. This gives rise to meaningful, yet complex enough, combinatorial structures.

Definitions. We wish to draw the complete bipartite graph $K_{k,n}$ in the plane in such a way that:

1. vertices are represented by points,
2. edges are continuous curves that connect those points, and do not contain any other vertices than their two endpoints,
3. no more than two edges intersect in one point,
4. edges pairwise intersect at most once; in particular, edges incident to the same vertex intersect only at this vertex,
5. the k vertices of one side of the bipartition lie on the outer boundary of the drawing.

Properties 1–4 are the usual requirements for *simple topological drawings* also known as *good drawings*. As we will see, property 5 leads to drawings with interesting combinatorial structures. Throughout this paper, the term *drawing* always refers to drawings satisfying the above properties.

The set of vertices of a bipartite graph $K_{k,n}$ will be denoted by $P \cup V$, where P and V are the two sides of the bipartition, with $|P| = k$ and $|V| = n$. When we consider a given drawing, we will use the word "vertex" and "edge" to denote both the vertex or edge of the graph, and their representation as points and curves. Without loss of generality, we can assume that the k outer vertices $p_1, \ldots, p_k$ lie in clockwise order on the boundary of a disk that contains all

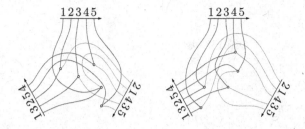

Fig. 1. Two drawings of $K_{3,5}$ satisfying the constraints. In both drawings the rotation system is $(12345, 21435, 13254)$.

the edges, or on the line at infinity. The vertices of V are labeled $1, \ldots, n$. An example of such a drawing is given in Fig. 1.

The *rotation system* of the drawing is a sequence of k permutations on n elements associated with the vertices of P in clockwise order. For each vertex of P, its permutation encodes the (say) counterclockwise order in which the n vertices of V are connected to it. Due to our last constraint on the drawings, the rotations of the k vertices of P around each vertex of V are fixed and identical, they reflect the clockwise order of $p_1, \ldots, p_k$ on the boundary.

Unlike for complete graphs, the rotation system of a drawing of a complete bipartite graph does not completely determine which pairs of edges are intersecting. This is exemplified with the two drawings in Fig. 1.

Results. The paper is organized as follows. In Sect. 2, we consider drawings with a *uniform* rotation system, in which the k permutations of the vertices of P are all equal to the identity. In this case, we can state a general structure theorem that allows us to completely characterize and count drawings of arbitrary bipartite graphs $K_{k,n}$.

In Sect. 3, we consider drawings of $K_{2,n}$ with arbitrary rotation systems. We consider a natural combinatorial encoding of such drawings, and state two necessary consistency conditions involving triples and quadruples of points in V. We show that these conditions are also sufficient, yielding a polynomial-time algorithm for checking consistency of a drawing.

In Sect. 4, we consider drawings of $K_{3,n}$ and study a complete classification of all drawings of $K_{3,3}$. This directly gives a necessary consistency condition on triples of vertices in V. We also provide an additional necessary condition on *quadruples*. A proof that the consistency conditions on triples and quadruples are sufficient for drawings of $K_{3,n}$ can be found in the long version of the paper [3].

2   Drawings with Uniform Rotation System

We first consider the case where k is arbitrary but the rotation system is uniform, that is, the permutation around each of the k vertices p_i is the same. Without

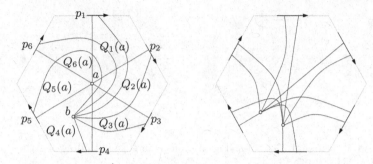

Fig. 2. Having placed b in $Q_4(a)$ the crossing pairs of edges and the order of crossings on each edge is prescribed. In particular $a \in Q_4(b)$. On the right a symmetric drawing of the pair.

loss of generality we assume that this permutation is the identity permutation on $[n]$.

In a given drawing, each of the n vertices of V splits the plane into k regions $Q_1, Q_2, \ldots, Q_k$, where each Q_i is bounded by the edges from v to p_i and p_{i+1}, with the understanding that $p_{k+1} = p_1$. We denote by $Q_i(v)$ the ith region defined by vertex v and further on call these regions *quadrants*. We let type$(a, b) = i$, for $a, b \in V$ and $i \in [k]$, whenever $a \in Q_i(b)$. This implies that $b \in Q_i(a)$, see Fig. 2. Indeed if $a < b$ and $j \neq i + 1$, then edge $p_{i+1}b$ has to intersect all the edges $p_j a$, while edge $p_j b$ has to avoid $p_{i+1}b$ until they meet in b. It follows that none of the edges $p_j b$ can intersect $p_{i+1}a$. This shows that $a \in Q_i(b)$.

Observation 1 (Symmetry).
For all a, b in uniform rotation systems: type$(a, b) =$ type(b, a).

For the case $k = 2$, we have exactly two types of pairs, that we will denote by A and B. The two types are illustrated on Fig. 3.

The drawings of $K_{2,n}$ with uniform rotations can be viewed as *colored pseudo-line arrangements*, where each pseudoline is split into two segments of distinct colors, and no crossing is monochromatic. This is illustrated on Fig. 4. The pseudoline of a vertex $v \in V$ is denoted by $\ell(v)$. The left (red) and right (blue) parts of this pseudoline are denoted by $\ell_L(v)$ and $\ell_R(v)$. Now having type$(a, b) =$ type$(b, a) = A$ means that b lies above $\ell(a)$ and a lies above $\ell(b)$. While having type$(a, b) =$ type$(b, a) = B$ means that b lies below $\ell(a)$ and a lies below $\ell(b)$.

The Triple Rule.

Lemma 1 (Triple rule).
For uniform rotation systems and three vertices $a, b, c \in V$ with $a < b < c$

$$\text{type}(a, c) \in \{\text{type}(a, b), \text{type}(b, c)\}.$$

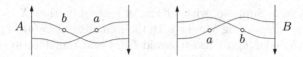

Fig. 3. The two types of pairs for drawings of $K_{2,n}$ with uniform rotation systems. (Color figure online)

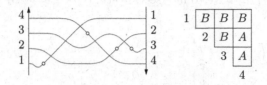

Fig. 4. Drawing $K_{2,4}$ as a colored pseudoline arrangement. The type of each pair is given in the table on the right. (Color figure online)

Proof. **Case** $k = 2$. If type$(a,b) \neq$ type(b,c) there is nothing to show since there are only two types. Without loss of generality, suppose that type$(a,b) =$ type$(b,c) = B$. This situation is illustrated in the left part of Fig. 5. The pseudo-line $\ell(c)$ must cross $\ell(b)$ on $\ell_R(b)$, otherwise we would have type$(b,c) = A$. Hence the point c is on the right of this intersection. Pseudoline $\ell(a)$ must cross $\ell(b)$ on $\ell_L(b)$, and a is left of this intersection. It follows that $\ell(a)$ and $\ell(c)$ cross on $\ell_R(a)$ and $\ell_L(c)$, i.e., type$(a,c) = B$.

Case $k > 2$. For the general case assume that type$(a,b) = i$ and type$(a,c) = j$. If $i = j$ there is nothing to show. Now suppose $i \neq j$. From $c \in Q_j(a)$ it follows that $p_{j+1}a$ and p_jc are disjoint. Edges p_jb and p_jc only share the endpoint p_j, hence c has to be in the region delimited by p_jb and $p_{j+1}a$, see the right part of Fig. 5. This region is contained in $Q_j(b)$, hence type$(b,c) = j$. □

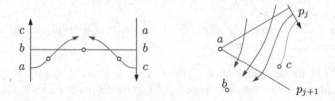

Fig. 5. Illustrations for the $k = 2$ case of Lemma 1 (left), and the $k > 2$ case of Lemma 1 (right).

The Quadruple Rule

Lemma 2. *For four vertices* $a,b,c,d \in V$ *with* $a < b < c < d$ *and* $X \in \{A,B\}$: *if* type$(a,c) =$ type$(b,c) =$ type$(b,d) = X$ *then* type$(a,d) = X$.

Proof. **Case** $k = 2$. Suppose, without loss of generality, that $X = B$. Consider the pseudolines representing b and c with their crossing at $\ell_R(b) \cap \ell_L(c)$. Coming from the left the edge $\ell_L(d)$ has to avoid $\ell_L(c)$ and therefore intersects $\ell_R(b)$. On $\ell_R(b)$ the crossing with $\ell_L(c)$ is left of the crossing with $\ell_L(d)$, see Fig. 6. Symmetrically from the right the edge $\ell_R(a)$ has to intersect $\ell_L(c)$ and this intersection is left of $\ell_R(b) \cap \ell_L(c)$. To reach the crossings with $\ell_L(c)$ and $\ell_R(b)$ edges $\ell_R(a)$ and $\ell_L(d)$ have to intersect, hence, type$(a, d) = B$.

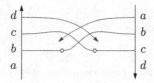

Fig. 6. Illustration for the $k > 2$ case of Lemma 1.

Case $k > 2$. In the general case, we let $X = i$, and consider the pseudoline arrangement defined by the two successive vertices p_i and p_{i+1} of P defining the quadrants Q_i. Proving that type$(a, d) = i$, that is, that $a \in Q_i(d)$, can be done as above for $k = 2$ on the drawing of $K_{2,n}$ induced by $\{p_i, p_{i+1}\}$ and V. $\qquad\square$

Decomposability and Counting

We can now state a general structure theorem for all drawings of $K_{k,n}$ with uniform rotation systems.

Theorem 1. *Given a type for each pair of vertices in V, there exists a drawing realizing those types if and only if:*

1. *there exists $s \in \{2, \ldots, n\}$ and $X \in [k]$ such that type$(a, b) = X$ for all pairs a, b with $a < s$ and $b \geq s$,*
2. *the same holds recursively when the interval $[1, n]$ is replaced by any of the two intervals $[1, s - 1]$ and $[s, n]$.*

Proof. ($\Rightarrow$) Let us first show that if there exists a drawing, then the types must satisfy the above structure. We proceed by induction on n. Pick the smallest $s \in \{2, \ldots, n\}$ such that type$(1, b) =$ type$(1, s)$ for all $b \geq s$. Set $X :=$ type$(1, s)$. We claim that type$(a, b) = X$ for all a, b such that $1 \leq a < s \leq b \leq n$. For $a = 1$ this is just the condition on s. Now let $1 < a$.

First suppose that type$(1, a) \neq X$. We can apply the triple rule on $1, a, b$. Since type$(1, b) \in \{$type$(1, a),$ type$(a, b)\}$, we must have that type$(a, b) = X$.

Now suppose that type$(1, a) = X$. We have type$(1, s - 1) = Y \neq X$ by definition. As in the previous case we obtain type$(s - 1, b) = X$ from the triple rule for $1, s - 1, b$. Applying the triple rule on $1, a, s - 1$ yields type$(a, s - 1) = Y$.

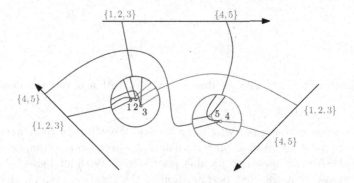

Fig. 7. Illustration of the recursive structure of the drawings in the uniform case.

Now apply the quadruple rule on $1, a, s-1, b$. We know that $\text{type}(1, s-1) = \text{type}(a, s-1) = Y$, and by definition $\text{type}(1, b) = X$. Hence we must have that $\text{type}(a, b) \neq Y$.

Finally, apply the triple rule on $a, s-1, b$. We know that $\text{type}(a, s-1) = Y$, $\text{type}(s-1, b) = X$. Since $\text{type}(a, b) \neq Y$, we must have $\text{type}(a, b) = X$. This yields the claim.

($\Leftarrow$) Now given the recursive structure, it is not difficult to construct a drawing. Consider the two subintervals as a single vertex, then recursively blow up these two vertices. (See Fig. 7 for an illustration). $\qquad\square$

The recursive structure yields a corollary on the number of distinct drawings.

Corollary 1 (Counting drawings with uniform rotation systems). *For every pair of integers $k, n > 0$ denote by $T(k, n)$ the number of simple topological drawings of the complete bipartite graph isomorphic to $K_{k,n}$ with uniform rotation systems. Then*

$$T(n + 1, k + 1) = \sum_{j=0}^{n} \binom{n+j}{2j} C_j \, k^j$$

where C_j is the jth Catalan number.

3 Drawings with $k = 2$

In this section we deal with drawings with $k = 2$ and arbitrary rotation system. We now have three types of pairs, that we call N, A, and B, as illustrated on Fig. 8. The type N (for noncrossing) is new, and is forced whenever the pair corresponds to an inversion in the two permutations.

Recall that a drawing of $K_{2,n}$, in which no pair is of type N, can be seen as a colored pseudoline arrangement as defined previously. Similarly, a drawing of $K_{2,n}$ in which some pairs are of type N can be seen as an arrangement of colored

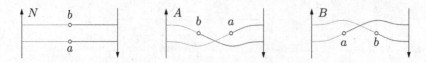

Fig. 8. The three types of pairs for drawings of $K_{2,n}$ with arbitrary rotation systems.

monotone curves crossing pairwise *at most* once. We will refer to arrangement of monotone curves that cross *at most* once as *quasi-pseudoline arrangements*. The pairs of type N correspond to parallel pseudolines. Without loss of generality, we can suppose that the first permutation in the rotation system, that is, the order of the pseudolines on the left side, is the identity. We denote by π the permutation on the right side.

Note that every permutation π is feasible in the sense that there is a drawing of $K_{2,n}$ such that the rotations are (id, π). To realize this, take the point set $\{(i, \pi(i)) : i \in [n]\}$ and consider horizontal and rays starting from each of these points to the left and upward respectively.

Triples. For $a, b, c \in V$, with $a < b < c$, we are interested in the triples of types $(\mathrm{type}(a,b), \mathrm{type}(a,c), \mathrm{type}(b,c))$ that are possible in a topological drawing of $K_{2,n}$, i.e., all possible topological drawings of $K_{2,3}$. Such triples will be called *legal*. We like to display triples in little tables, e.g., the triple $\mathrm{type}(a,b) = X$,

$$
\begin{array}{c|c|c}
a & X & Y \\
\cline{1-3}
& b & Z \\
\cline{2-3}
& & c
\end{array}
$$

$\mathrm{type}(a,c) = Y$, and $\mathrm{type}(b,c) = Z$ is represented as

Lemma 3 (Decomposable Triples[1]). *A triple with $Y \in \{X, Z\}$ is always legal. There are 15 triples of this kind.*

Lemma 4.

$$
\begin{array}{c|c|c} a & N & A \\ \cline{1-3} & b & B \\ \cline{2-3} & & c \end{array}
\qquad
\begin{array}{c|c|c} a & A & B \\ \cline{1-3} & b & N \\ \cline{2-3} & & c \end{array}
$$

There are exactly two non-decomposable legal triples: *and*   .

The proofs of the two lemmas can be found in [3]. With the two lemmas we have classified all 17 legal triples.

Observation 2 (Triple Rule). *Any three vertices of V in a drawing of $K_{2,n}$ must induce one of the 17 legal triples of types.*

Quadruples. We aim at a characterization of collections of types that correspond to drawings. Already in the case of uniform rotations we had to add Lemma 2, a condition for quadruples. In the general case the situation is more complex than in the uniform case, see Fig. 9.

---

[1] These triples of this lemma are decomposable in the sense of Theorem 1.

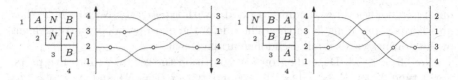

Fig. 9. The quadruple rule from Lemma 2 does not hold in the presence of N types.

Reviewing the proof of Lemma 2 we see that in the case discussed there, where given B types are intended to enforce $\text{type}(a, d) = B$, we need that in π element a is before b, this is equivalent to $\text{type}(a, b) \neq N$. Symmetrically, three A types enforce $\text{type}(a, d) = A$ when d is the last in π, i.e., if $\text{type}(c, d) \neq N$.

Lemma 5. *Consider four vertices $a, b, c, d \in V$ such that $a < b < c < d$.*
If $\text{type}(a, b) \neq N$ and $\text{type}(a, c) = \text{type}(b, c) = \text{type}(b, d) = B$ then $\text{type}(a, d) = B$.
If $\text{type}(c, d) \neq N$ and $\text{type}(a, c) = \text{type}(b, c) = \text{type}(b, d) = A$ then $\text{type}(a, d) = A$.

Consistency. With the next theorem we show that consistency on triples and quadruples is enough to grant the existence of a drawing.

Theorem 2 (Consistency of drawings for $k = 2$). *Given a type for each pair of vertices in V, there exists a drawing realizing those types if and only if all triples are legal and the quadruple rule (Lemma 5) is satisfied.*

The proof of this result is based on the characterization of *local sequences* in pseudoline arrangements. Given an arrangement of n pseudolines, the local sequences are the permutations α_i of $[n] \setminus \{i\}$, $i \in [n]$, representing the order in which the ith pseudoline intersects the $n - 1$ others.

Lemma 6 (Theorem 6.17 in [7]). *The set $\{\alpha_i\}_{i \in [n]}$ is the set of local sequences of an arrangement of n pseudolines if and only if*

$$ij \in \text{inv}(\alpha_k) \Leftrightarrow ik \in \text{inv}(\alpha_j) \Leftrightarrow jk \in \text{inv}(\alpha_i),$$

for all triples i, j, k, where $\text{inv}(\alpha)$ is the set of inversions of the permutation α.

Proof (Theorem 2). The necessity of the condition is implied by Observation 2 and Lemma 5.

We proceed by giving an algorithm for constructing an appropriate drawing. From the proof of Lemma 4, we know that having legal triples implies that the sets of inversion pairs and its complement, the set of non-inversion pairs, are both transitive. Hence, there is a well defined permutation π representing the rotation at p_2.

We aim at defining the local sequences α_i that allow an application of Lemma 6. This will yield a pseudoline arrangement. A drawing of $K_{2,n}$, however, will only correspond to a quasi-pseudoline arrangement. Therefore, we first construct a quasi-pseudoline arrangement T for the pair $(\overline{\pi}, \text{id})$, i.e., only the quasi-pseudolines corresponding to i and j with $\text{type}(i,j) = N$ cross in T. The idea is that appending T on the right side of the quasi-pseudoline arrangement of the drawing yields a full pseudoline arrangement.

Now fix $i \in [n]$. Depending on i we partition the set $[n] \setminus i$ into five parts. For a type X let $X_<(i) = \{j : j < i \text{ and } \text{type}(j,i) = X\}$ and $X_>(i) = \{j : j > i \text{ and } \text{type}(i,j) = X\}$, the five relevant parts are $A_<(i)$, $A_>(i)$, $B_<(i)$, $B_>(i)$, and $N(i) = N_<(i) \cup N_>(i)$. The pseudoline ℓ_i has three parts. The edge incident to p_1 (the red edge) is crossed by pseudolines ℓ_j with $j \in A_>(i) \cup B_<(i)$. The edge incident to p_2 (the blue edge) is crossed by pseudolines ℓ_j with $j \in A_<(i) \cup B_>(i)$. The part of ℓ_i belonging to T is crossed by pseudolines ℓ_j with $j \in N(i)$. The order of the crossings in the third part, i.e., the order of crossings with pseudolines ℓ_j with $j \in N(i)$, is prescribed by T.

Regarding the order of the crossings on the second part we know that the lines for $j \in A_<(i)$ have to cross ℓ_i from left to right in order of decreasing indices and the lines for $j \in B_>(i)$ have to cross ℓ_i from left to right in order of increasing indices, see Fig. 10. If $j \in A_<(i)$ and $j' \in B_>(i)$, then consistency of triples implies that $\text{type}(j,j') \in \{A, B\}$. If $\text{type}(j,j') = A$, then on ℓ_i the crossing of j' has to be left of the crossing of j. If $\text{type}(j,j') = B$, then on ℓ_i the crossing of j has to be left of the crossing of j'.

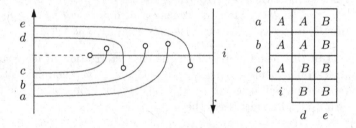

Fig. 10. Crossings on the edge $i\,p_2$.

The described conditions yield a "left–to–right" relation $\to_i$ such that for all $x, y \in A_<(i) \cup B_>(i)$ one of $x \to_i y$ and $y \to_i x$ holds. We have to show that $\to_i$ is acyclic. Since $\to_i$ is a tournament it is enough to show that $\to_i$ is transitive.

Suppose there is a cycle $x \to_i y \to_i z \to_i x$. If $x, y < i$ and $z > i$, then $\text{type}(x,i) = \text{type}(y,i) = A$, moreover, from $x \to_i y$ we get $y < x$ and from $y \to_i z \to_i x$ we get $\text{type}(x,z) = A$, and $\text{type}(y,z) = B$. Since $\text{type}(i,z) = B \neq N$ this is a violation of the second quadruple rule of Lemma 5.

If $x < i$ and $y, z > i$, then we have $\text{type}(i,y) = \text{type}(i,z) = B$. From this together with $y \to_i z$ we obtain $y < z$, and $z \to_i x \to_i y$ yields $\text{type}(x,y) = B$, and $\text{type}(x,z) = A$. This is a violation of the first quadruple rule of Lemma 5.

Adding the corresponding arguments for the order of crossings on the first part of line ℓ_i we conclude that the permutation α_i is uniquely determined by the given types and the choice of T.

The consistency condition on triples of local sequences needed for the application of Lemma 6 is trivially satisfied because legal triples of types correspond to drawings of $K_{2,3}$ and each such drawing together with the extensions of the lines in T consists of three pairwise crossing pseudolines. □

Since the two rules we enforced only involve at most four vertices of V, we immediately get the following corollary.

Corollary 2. *Consistency on all 4-tuples of V is sufficient and necessary for drawings of $K_{2,n}$, yielding an $O(n^4)$ time algorithm for checking consistency of an assignment of types.*

4   Drawings with $k = 3$

At the beginning of the previous section we have seen that any pair of rotations is feasible for drawings of $K_{2,n}$. This is not true in the case of $k > 2$. For $k = 4$ the system of rotations $([1, 2], [2, 1], [1, 2], [2, 1])$ is easily seen to be infeasible. In the case $k = 3$ it is less obvious that infeasible systems of rotations exist. We will show later (Proposition 1) that $([1, 2, 3, 4], [4, 2, 1, 3], [2, 4, 3, 1])$ is infeasible.

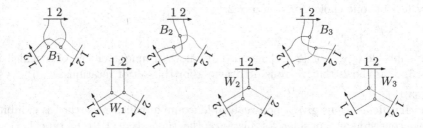

Fig. 11. The six types of drawings of $K_{3,2}$. (Color figure online)

Pairs. We again start by looking at the types for pairs, i.e., at all possible drawings of $K_{3,2}$. We already know that if the rotation system is uniform $(\mathrm{id}_2, \mathrm{id}_2, \mathrm{id}_2)$, then there are three types of drawings. The other three options $(\mathrm{id}_2, \overline{\mathrm{id}_2}, \overline{\mathrm{id}_2})$, $(\mathrm{id}_2, \overline{\mathrm{id}_2}, \mathrm{id}_2)$, and $(\mathrm{id}_2, \mathrm{id}_2, \overline{\mathrm{id}_2})$, each have a unique drawing. Figure 11 shows the six possible types and associates them to the symbols B_α, and W_α, for $\alpha = 1, 2, 3$.

The three edges emanating from a vertex $i \in [n]$ partition the drawing area into three regions. Define $Q_\alpha(i)$ as the region bounded by the two edges $i\,p_{\alpha+1}$ and $i\,p_{\alpha-1}$ not containing p_α When the types have been prescribed for all pairs of vertices we know which vertices are located in which region of i. Conversely, if we know the α for which $j \in Q_\alpha(i)$, then only one B-type and one W-type remain eligible for the pair (i, j).

Triples. We now classify the triples, i.e., drawings of $K_{3,3}$. It turns out that there are 92 types. A complete description can be found in [3].

Now suppose that rotations $(\mathrm{id}, \pi_2, \pi_3)$ are prescribed. We want to decide whether there is a corresponding drawing. The first step would be to determine the type of the drawing for each pair of vertices. For all non-uniform pairs $\mathrm{type}(i,j) \in \{W_1, W_2, W_3\}$ is uniquely given by the system.

The type of the remaining pairs is B_α for some α. Beforehand each $\alpha \in \{1,2,3\}$ is possible but of course the types of every triple must also correspond to a drawing, i.e., the types of each triple must be among the 89 drawable types of the classification. This may force the types of additional pairs.

Before giving a larger example we show that by looking at triples we can deduce that not all choices $(\mathrm{id}, \pi_2, \pi_3)$ of prescribed rotations are feasible, i.e., there are choices that have no corresponding drawing.

Proposition 1. *The system* $(\mathrm{id}_4, [4,2,1,3], [2,4,3,1])$ *is an infeasible set of rotations.*

Proof. The table of types for the given permutations is shown on the right. Consider the subtable $\begin{array}{|c|c|}\hline W_1 & W_3 \\\hline B_\alpha & \\\hline\end{array}$ corresponding to $\{1,2,3\}$. From the classification of triples it follows that the only feasible one choice for α is $\alpha = 2$.

1	W_1	W_3	W_1
2	B_α	W_2	
3	W_1		
4			

The subtable $\begin{array}{|c|c|}\hline B_\alpha & W_2 \\\hline W_1 & \\\hline\end{array}$ of $\{2,3,4\}$ again only allows a unique choice of α which is $\alpha = 3$. This shows that there is no drawing for this set of rotations. $\square$

Quadruples. Let us give a non-realizable example which nevertheless exhibits triple consistency. Consider for instance the types $\mathrm{type}(1,2) = \mathrm{type}(1,4) = \mathrm{type}(3,4) = B_1$ and $\mathrm{type}(1,3) = \mathrm{type}(2,3) = \mathrm{type}(2,4) = B_2$. Every triple is decomposable, i.e., we have consistency on triples, however, the full table is not decomposable. Since all the rotations/permutations are the identity, ie.e., the system is uniform we know from Theorem 1 that there is no corresponding drawing.

The need for a condition on quadruples is not restricted to tables of uniform systems. The table on the right is consistent on all triples, still it is not realizable. This can be shown by looking at the table corresponding to the green-blue $K_{2,n}$ subgraph, which reveals a bad quadruple. Note that for the table of the green-blue $K_{2,n}$ the elements have to be sorted according to $\pi_3 = (2,1,3,4)$.

1	W_1	B_1	B_1
2	B_1	B_2	
3	B_2		
4			

Let T be an assignment of types, e.g., in form of a table. From T we know the corresponding system (π_1, π_2, π_3) of rotations.

Definition 1. T *is* consistent on quadruples *if for any four vertices* a, b, c, d *and* $i \in \{1, 2, 3\}$ *the assignment of types from* A, B, N *induced by the restriction of* π_{i-1} *and* π_{i+1} *to* a, b, c, d *satisfies the condition from Lemma 5.*

Note that checking the condition requires sorting a, b, c, d according to π_{i-1}.

4.1   The Consistency Theorem

Theorem 3 (Consistency of drawings for $k = 3$, see [3]). *Given a type for each pair of vertices in V, there exists a drawing realizing those types if and only if all triples and quadruples are consistent.*

Acknowledgments. This work was started at the *Workshop on Order Types, Rotation Systems, and Good Drawings* in Strobl 2015. We thank the organizers and participants for fruitful discussions, in particular Pedro Ramos who suggested to look at complete bipartite graphs. We thank the anonymous referees for their insightful remarks.

S. Felsner also acknowledges support from DFG grant Fe 340/11-1.

References

1. Ábrego, B.M., Aichholzer, O., Fernández-Merchant, S., Ramos, P., Salazar, G.: The 2-page crossing number of K_n. Discret. Comput. Geom. **49**(4), 747–777 (2013)
2. Blažek, J., Koman, M.: A minimal problem concerning complete plane graphs. In: Fiedler, M. (ed.) Theory of Graphs and Its Applications, pp. 113–117. Czech. Acad. of Sci. (1964)
3. Cardinal, J., Felsner, S.: Topological drawings of complete bipartite graphs (full version). arXiv:1608.08324 [cs.CG]
4. Christian, R., Richter, R.B., Salazar, G.: Zarankiewicz's conjecture is finite for each fixed m. J. Comb. Theory Ser. B **103**(2), 237–247 (2013)
5. Kynčl, J.: Simple realizability of complete abstract topological graphs in P. Discret. Comput. Geom. **45**(3), 383–399 (2011)
6. Kynčl, J.: Simple realizability of complete abstract topological graphs simplified. In: Di Giacomo, E., Lubiw, A. (eds.) GD 2015. LNCS, vol. 9411, pp. 309–320. Springer, Heidelberg (2015). doi:10.1007/978-3-319-27261-0_26
7. Felsner, S.: Geometric Graphs and Arrangements. Advanced Lectures in Mathematics. Vieweg Verlag, Berlin (2004)
8. Harary, F., Hill, A.: On the number of crossings in a complete graph. Proc. Edinburgh Math. Soc. **13**, 333–338 (1963)
9. Kratochvíl, J., Lubiw, A., Nešetřil, J.: Noncrossing subgraphs in topological layouts. SIAM J. Discrete Math. **4**(2), 223–244 (1991)
10. Kratochvíl, J., Matoušek, J.: NP-hardness results for intersection graphs. Commentationes Math. Univ. Carol. **30**, 761–773 (1989)
11. Pach, J., Tóth, G.: How many ways can one draw a graph? Combinatorica **26**(5), 559–576 (2006)
12. Zarankiewicz, K.: On a problem of P Turán concerning graphs. Fundamenta Mathematicae **41**, 137–145 (1954)

A Direct Proof of the Strong Hanani–Tutte Theorem on the Projective Plane

Éric Colin de Verdière[1], Vojtěch Kaluža[2(✉)], Pavel Paták[3],
Zuzana Patáková[3], and Martin Tancer[2]

[1] CNRS and Département d'informatique, École normale supérieure, Paris, France
Eric.Colin.De.Verdiere@ens.fr
[2] Department of Applied Mathematics, Charles University,
Prague, Czech Republic
{kaluza,tancer}@kam.mff.cuni.cz
[3] Einstein Institute of Mathematics, The Hebrew University of Jerusalem,
Jerusalem, Israel
{patak,zuzka}@kam.mff.cuni.cz

Abstract. We reprove the strong Hanani–Tutte theorem on the projective plane. In contrast to the previous proof by Pelsmajer, Schaefer and Stasi, our method is constructive and does not rely on the characterization of forbidden minors, which gives hope to extend it to other surfaces. Moreover, our approach can be used to provide an efficient algorithm turning a Hanani–Tutte drawing on the projective plane into an embedding.

Keywords: Graph drawing · Graph embedding · Hanani–Tutte theorem · Projective plane · Topological graph theory

1 Introduction

A drawing of a graph on a surface is a *Hanani–Tutte drawing* (shortly an *HT-drawing*) if no two vertex-disjoint edges cross an odd number of times. We call vertex-disjoint edges *independent*.

Pelsmajer, Schaefer and Stasi [14] proved the following theorem via consideration of the forbidden minors for the projective plane.

Theorem 1 (Strong Hanani–Tutte for the projective plane, [14]). *A graph G can be embedded into the projective plane if and only if it admits an HT-drawing on the projective plane.*[1]

The project was partially supported by the Czech-French collaboration project EMBEDS (CZ: 7AMB15FR003, FR: 33936TF). É.C.V. was partially supported by the French ANR Blanc project ANR- 12-BS02-005 (RDAM). V.K. was partially supported by the project GAUK 926416. V.K. and M.T. were partially supported by the project GAČR 16-01602Y. P.P. was supported by the ERC Advanced grant no. 320924. Z.P. was partially supported by Israel Science Foundation grant ISF-768/12.

[1] Of course, the "only if" part is trivial.

© Springer International Publishing AG 2016
Y. Hu and M. Nöllenburg (Eds.): GD 2016, LNCS 9801, pp. 454–467, 2016.
DOI: 10.1007/978-3-319-50106-2_35

Our main result is a constructive proof of Theorem 1. The need for a constructive proof is motivated by the strong Hanani–Tutte conjecture, which states that an analogous result is valid on an arbitrary (closed) surface. This conjecture is known to be valid only on the sphere (plane) and on the projective plane. The approach via forbidden minors is relatively simple on the projective plane; however, this approach does not seem applicable to other surfaces, because there is no reasonable characterization of forbidden minors for them. (Already for the torus or the Klein bottle, the exact list is not known.)

On the other hand, our approach reveals a number of difficulties that have to be overcome in order to obtain a constructive proof. If the conjecture is true, our approach may serve as a basis for its proof on a general surface. If the conjecture is not true, then our approach may perhaps help to reveal appropriate structure needed for a construction of a counterexample.

The Hanani–Tutte Theorem on the Plane and Related Results. Let us now briefly describe the history of the problem; for complete history and relevant results we refer to a nice survey by Schaefer [17]. Following the work of Hanani [2], Tutte [19] made a remarkable observation now known as the (strong) Hanani–Tutte theorem: a graph is planar if and only if it admits an HT-drawing in the plane. The theorem has also a parallel history in algebraic topology, where it follows from the ideas of van Kampen, Flores, Shapiro and Wu [11,20,21].

It is a natural question whether the strong Hanani–Tutte theorem can be extended to graphs on other surfaces; as we already said before, it has been confirmed only for the projective plane [14] so far. On general surfaces, only the weak version [1,16] of the theorem is known to be true: if a graph is drawn on a surface so that every pair of edges crosses an even number of times[2], then the graph can be embedded into the surface while preserving the cyclic order of the edges at all vertices. Note that in the strong version we require that only independent edges cross even number of times, while in the weak version this condition has to hold for all pairs of edges.

We remark that other variants of the Hanani–Tutte theorem generalizing the notion of embedding in the plane have also been considered. For instance, the strong Hanani–Tutte theorem was proved for partially embedded graphs [18] and both weak and strong Hanani–Tutte theorem were proved also for 2-clustered graphs [6].

The strong Hanani–Tutte theorem is important from the algorithmic point of view, since it implies the Trémaux crossing theorem, which is used to prove de Fraysseix-Rosenstiehl's planarity criterion [4]. This criterion has been used to justify the linear time planarity algorithms including the Hopcroft-Tarjan [8] and the Left-Right [3] algorithms. For more details we again refer to [17].

One of the reasons why the strong Hanani–Tutte theorem is so important is that it turns planarity question into a system of linear equations. For general surfaces, the question whether there exists a Hanani–Tutte drawing of G leads to a system of quadratic equations [11] over $\mathbb{Z}_2$. If the strong Hanani–Tutte theorem

---

[2] Including 0 times.

is true for the surface, any solution to the system then serves as a certificate that G is embeddable. Moreover, if the proof of the Hanani–Tutte theorem is constructive, it gives a recipe how to turn the solution into an actual embedding. Unfortunately, solving systems of quadratic equations is NP-complete.

For completeness we mention that for each surface there exists a polynomial time algorithm that decides whether a graph can be embedded into that surface [9,12]; however, the hidden constant depends exponentially on the genus.

The original proofs of the strong Hanani–Tutte theorem in the plane used Kuratowski's theorem [10], and therefore are non-constructive. In 2007, Pelsmajer, Schaefer and Štefankovič [15] published a constructive proof. They showed a sequence of moves that change an HT-drawing into an embedding.

A key step in their proof is their Theorem 2.1. We say that an edge is *even* if it crosses every other edge an even number of times (including the adjacent edges).

Theorem 2 (Theorem 2.1 of [15]). *If D is a drawing of a graph G in the plane, and E_0 is the set of even edges in D, then G can be drawn in the plane so that no edge in E_0 is involved in an intersection and there are no new pairs of edges that intersect an odd number of times.*

Unfortunately, an analogous result is simply not true on other surfaces, as is shown in [16]. In particular, this is an obstacle for a constructive proof of Theorem 1. The key step of our approach is to provide a suitable replacement of Theorem 2 on the projective plane. This is provided by Theorem 7 in Sect. 3.

The version of the paper identical to the present one can be found on arXiv: http://arxiv.org/abs/1608.07855v1. We refer to the full version of this paper in the subsequent submission on arXiv, which contains many details missing in this extended abstract.

2   Hanani–Tutte Drawings

In this section, we consider Hanani–Tutte drawings of graphs on the sphere and on the projective plane. We use the standard notation from graph theory. Namely, if G is a graph, then $V(G)$ and $E(G)$ denote the set of vertices and the set of edges of G, respectively. Given a vertex v or an edge e, by $G - v$ or $G - e$ we denote the graph obtained from G by removing v or e, respectively.

Regarding drawings of graphs, first, let us recall a few standard definitions considered on an arbitrary surface. We put the standard general position assumptions on the drawings. That is, we consider only drawings of graphs on a surface such that no edge contains a vertex in its interior and every pair of edges meets only in a finite number of points, where they *cross* transversally. However, we allow three or more edges meeting in a single point.[3]

Let D be a drawing of a graph G on a surface S. Given two distinct edges e and f of G by $\mathrm{cr}(e, f) = \mathrm{cr}_D(e, f)$ we denote the number of crossings between e

---

[3] We do not mind them because we study pairwise interactions of edges only.

and f in D modulo 2. We say that an edge e of G is *even* if $\mathrm{cr}(e, f) = 0$ for any $f \in E(G)$ distinct from e. We emphasize that we consider the crossing number as an element of $\mathbb{Z}_2$ and all computations throughout the paper involving it are done in $\mathbb{Z}_2$.

HT-Drawings on $\mathbb{R}P^2$. It is convenient for us to set up some conventions for working with the HT-drawings on the (real) projective plane, $\mathbb{R}P^2$. There are various ways to represent $\mathbb{R}P^2$. Our convention will be the following: we consider the sphere S^2 and a disk (2-ball) B in it. We remove the interior of B and identify the opposite points on the boundary ∂B. This way, we obtain a representation of $\mathbb{R}P^2$. Let γ be the curve coming from ∂B after the identification. We call this curve a *crosscap*. It is a homologically (homotopically) non-trivial simple cycle (loop) in $\mathbb{R}P^2$, and conversely, any homologically (homotopically) nontrivial simple cycle (loop) may serve as a crosscap up to a self-homeomorphism of $\mathbb{R}P^2$. In drawings, we use the symbol $\otimes$ for the crosscap coming from the removal of the disk 'inside' this symbol.

Given an HT-drawing of a graph on $\mathbb{R}P^2$, it can be slightly shifted so that it meets the crosscap in a finite number of points and only transversally, still keeping the property that we have an HT-drawing. Therefore, we may add to our conventions that this is the case for our HT-drawings on $\mathbb{R}P^2$.

Now, we consider a map $\lambda \colon E(G) \to \mathbb{Z}_2$. For an edge e, we let $\lambda(e)$ be the number of crossings of e and the crosscap γ modulo 2. We emphasize that λ depends on the choice of the crosscap.

Given a (graph-theoretic) cycle Z in G, we can distinguish whether Z is drawn as a homologically nontrivial cycle by checking the value $\lambda(Z) := \sum \lambda(e) \in \mathbb{Z}_2$ where the sum is over all edges of Z. The cycle Z is homologically nontrivial if and only if $\lambda(Z) = 1$. In particular, it follows that $\lambda(Z)$ does not depend on the choice of the crosscap.

Projective HT-Drawings on S^2. Let D be an HT-drawing of a graph G on $\mathbb{R}P^2$. It is not hard to deduce a drawing D' of the same graph on S^2 such that every pair (e, f) of *independent* edges satisfies $\mathrm{cr}(e, f) = \lambda(e)\lambda(f)$. Indeed, it is sufficient to 'undo' the crosscap, glue back the disk B and then let the edges intersect on B. See the two leftmost pictures below.

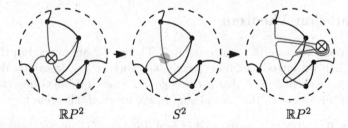

This motivates the following definition.

Definition 3. *Let D be a drawing of a graph G on S^2 and $\lambda \colon E(G) \to \mathbb{Z}_2$ be a function. Then the pair (D, λ) is a* projective HT-drawing *of G on S^2 if $\mathrm{cr}(e, f) = \lambda(e)\lambda(f)$ for any pair of independent edges e and f of G.*

It turns out that a projective HT-drawing on S^2 can also be transformed to an HT-drawing on $\mathbb{R}P^2$.

Proposition 4. *A graph G admits a projective HT-drawing on S^2 (with respect to some function $\lambda \colon E(G) \to \mathbb{Z}_2$) if and only if it admits an HT-drawing on $\mathbb{R}P^2$.*

The full proof of the missing implication is not too difficult and it is given in the full version of the paper (see Corollary 5). The core of the proof can be deduced from the two rightmost pictures above.

The main strength of Proposition 4 relies in the fact that in projective HT-drawings on S^2 we can ignore the actual geometric position of the crosscap and work in S^2 instead, which is simpler. This is especially helpful when we need to merge two drawings.

In order to distinguish the usual HT-drawings on S^2 from the projective HT-drawings, we will sometimes refer to the former as to the *ordinary* HT-drawings on S^2.

Nontrivial Walks. Let (D, λ) be a projective HT-drawing of a graph G and ω be a walk in G. We define $\lambda(\omega) := \sum_{e \in E(\omega)} \lambda(e)$ where $E(\omega)$ is the multiset of edges appearing in ω. Equivalently, it is sufficient to consider only the edges appearing an odd number of times in ω, because $2\lambda(e) = 0$ for any edge e. We say that ω is *trivial* if $\lambda(\omega) = 0$ and *nontrivial* otherwise. We often use this terminology in special cases when ω is an edge, a path, or a cycle.

Now let us consider a subgraph P of G such that every cycle in P is trivial. Then P essentially behaves as a planar subgraph of G, which we make more precise by the following lemma. For its proof, see Lemma 8 in the full version of the paper.

Lemma 5. *Let (D, λ) be a projective HT-drawing of G on S^2 and let P be a subgraph of G such that every cycle in P is trivial. Then there is a projective HT-drawing (D', λ') of G on S^2 such that $\lambda'(e) = 0$ for any edge e of $E(P)$.*

3   Separation Theorem

In this section, we state the replacement of Theorem 2 announced in the introduction. First we introduce some terminology; as we see from the definition below, a simple cycle Z such that every edge of Z is even splits G into two parts. This fact is analogous to the crucial step in the proof of Theorem 2.

Definition 6. *Let G be a graph and D be a drawing of G on S^2. Let us assume that Z is a cycle of G such that every edge of Z is even and it is drawn as a simple cycle in D. Let S^+ and S^- be the two components of $S^2 \setminus D(Z)$. We call a vertex $v \in V(G) \setminus V(Z)$ an* inside vertex *if it belongs to S^+ and an* outside *vertex otherwise. Given an edge $e = uv \in E(G) \setminus E(Z)$, we say that e is an* inside *edge if either u is an inside vertex or if $u \in V(Z)$ and $D(e)$ points locally to S^+ next to $D(u)$. Analogously we define an* outside *edge.[4] We let V^+ and E^+ be the sets of the inside vertices and the inside edges, respectively. Analogously, we define V^- and E^-. We also define the graphs $G^{+0} := (V^+ \cup V(Z), E^+ \cup E(Z))$ and $G^{-0} := (V^- \cup V(Z), E^- \cup E(Z))$.*

Now, we may formulate our main technical tool—the separation theorem for projective HT-drawings.

Theorem 7. *Let (D, λ) be a projective HT-drawing of a 2-connected graph G on S^2 and Z a cycle of G that is simple in D and such that every edge of Z is even. Moreover, we assume that every edge e of Z is trivial, that is, $\lambda(e) = 0$. Then there is a projective HT-drawing (D', λ') of G on S^2 satisfying the following properties.*

- *The drawings D and D' coincide on Z;*
- *the cycle Z is free of crossings and all of its edges are trivial in D';*
- *$D'(G^{+0})$ is contained in $S^+ \cup D'(Z)$;*
- *$D'(G^{-0})$ is contained in $S^- \cup D'(Z)$; and*
- *either all edges of G^{+0} or all edges of G^{-0} are trivial (according to λ'); that is, at least one of the drawings $D'(G^{+0})$ or $D'(G^{-0})$ is an ordinary HT-drawing on S^2.*

In the remainder of this section, we describe the main ingredients of the proof of Theorem 7 and we also derive this theorem from the ingredients. We will often encounter the setting when G, (D, λ) and Z satisfy the assumptions of Theorem 7. Therefore, we say that G, (D, λ) and Z satisfy the *separation assumptions* if (1) G is a 2-connected graph; (2) (D, λ) is a projective HT-drawing of G; (3) Z is a cycle in G drawn as a simple cycle in D; (4) every edge of Z is even in D and trivial.

Arrow Graph. From now on, let us fix G, (D, λ) and Z satisfying the separation assumptions. This also fixes the distinction between the outside and the inside.

Definition 8. *A* bridge *B of G (with respect to Z) is a subgraph of G that is either an edge not in Z but with both endpoints in Z (and its endpoints also belong to B), or a connected component of $G - V(Z)$ together with all edges (and their endpoints in Z) with one endpoint in that component and the other endpoint in Z.[5]*

We say that B is an inside bridge *if it is a subgraph of G^{+0}, and an* outside bridge *if it is a subgraph of G^{-0} (every bridge is thus either an inside bridge or an outside bridge).*

A walk *ω in G is a* proper walk *if no vertex in ω belongs to $V(Z)$, except possibly its endpoints, and no edge of ω belongs to $E(Z)$. In particular, each proper walk belongs to a single bridge.*

---

[4] It turns out that every edge $e \in E(G) \setminus E(Z)$ is either an outside edge or an inside edge, because every edge of Z is even.

[5] This is a standard definition; see, e.g., Mohar and Thomassen [13, p. 7].

Since we assume that G is 2-connected, every inside bridge contains at least two vertices of Z. The bridges induce partitions of $E(G) \setminus E(Z)$ and of $V(G) \setminus V(Z)$.

We want to record which pairs of vertices on $V(Z)$ are connected with a nontrivial and proper walk inside or outside. For this purpose, we create two new graphs A^+ and A^-, possibly with loops but without multiple edges. In order to distinguish these graphs from G, we draw their edges with double arrows and we call these graphs an *inside arrow graph* and an *outside arrow graph*, respectively. The edges of these graphs are called the *inside/outside arrows*. We set $V(A^+) = V(A^-) = V(Z)$.

Now we describe the *arrows*, that is, $E(A^+)$ and $E(A^-)$. Let u and v be two vertices of $V(Z)$, not necessarily distinct. By W_{uv}^+ we denote the set of all proper nontrivial walks in G^{+0} with endpoints u and v. We have an *inside arrow* connecting u and v in $E(A^+)$ if and only if W_{uv}^+ is nonempty. In order to distinguish the edges of G from the arrows, we denote an arrow by $\overline{uv} = \overline{vu}$. An arrow which is a loop at a vertex v is denoted by $\overline{vv}$. (This convention will allow us to work with arrows $\overline{uv}$ without a distinction whether $u = v$ or $u \neq v$.) Analogously, we define the set W_{uv}^- and the *outside arrows*. Below, we provide an example of an unusual HT-drawing of K_5 on $\mathbb{R}P^2$, the corresponding projective HT-drawing on S^2 and the corresponding arrow graphs.

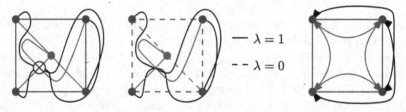

Given an inside arrow $\overline{uv}$ and an inside bridge B, we say that B *induces* $\overline{uv}$ if there is a walk in B which belongs to W_{uv}^+. An inside bridge B is *nontrivial* if it induces at least one arrow. Given two inside arrows $\overline{uv}$ and $\overline{xy}$, we say that $\overline{uv}$ and $\overline{xy}$ *are induced by different bridges* if there are two different inside bridges B and B' such that B induces $\overline{uv}$ and B' induces $\overline{xy}$. As usual, we define analogous notions for the outside as well.

Possible Configurations of Arrows. Now, we utilize the arrow graph to show that certain configurations of arrows are not possible.

Lemma 9.

(a) *Every inside arrow shares a vertex with every outside arrow.*

(b) *Let $\overline{ab}$ and $\overline{xy}$ be two arrows induced by different inside bridges of G^{+0}. If the two arrows do not share an endpoint, their endpoints have to interleave along Z.*

(c) *There are no three vertices a, b, c on Z, an inside bridge B^+, and an outside bridge B^- such that B^+ induces the arrows $\overline{ab}$ and $\overline{ac}$ (and no other arrows) and B^- induces the arrows $\overline{ab}$ and $\overline{bc}$ (and no other arrows).*

For proof, see Lemmas 12, 13 and 14 in the full version of the paper.

Lemma 9 is, of course, also valid if we swap the inside and the outside. Schematically, the forbidden configurations from Lemma 9 are drawn in the picture below. The cyclic order in (a) may be arbitrary whereas it is important in (b) that the arrows there do not interleave. Different dashing of lines in (b) correspond to arrows induced by different inside bridges. The arrows of the same colour in (c) are induced by the same bridge.

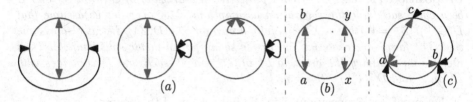

Now we describe important configurations that may occur.

Definition 10. *We say that G forms*

(a) an inside fan *if there is a vertex common to all inside arrows. (The arrows may come from various inside bridges.)*

(b) an inside square *if it contains four vertices a, b, c and d ordered in this cyclic order along Z and the inside arrows are precisely $\overline{ab}$, $\overline{bc}$, $\overline{cd}$ and $\overline{ad}$. In addition, we require that the inside graph G^{+0} has only one nontrivial inside bridge.*

(c) an inside split triangle *if there exist three vertices a, b and c such that the arrows of G are $\overline{ab}$, $\overline{ac}$ and $\overline{bc}$. In addition, we require that every nontrivial inside bridge induces either the two arrows $\overline{ab}$ and $\overline{ac}$, or just a single arrow.*

We have analogous definitions for an outside fan, outside square *and* outside split triangle.

More precisely the notions in Definition 10 depend on G, (D, λ) and Z satisfying the separation assumptions.

The picture below shows schematic drawings of the configurations of arrows from Definition 10. Different dashing of lines correspond to different inside bridges. The loop in the right drawing in (a) is an inside loop (drawn outside due to lack of space). The drawing in (c) is only one instance of an inside split triangle.

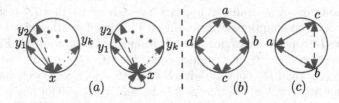

A relatively direct case analysis, using Lemma 9, reveals the following fact.

Proposition 11. *Let (D, λ) be a projective HT-drawing on S^2 of a graph G and let Z be a cycle in G satisfying the separation assumptions. Then G forms an (inside or outside) fan, square, or split triangle.*

For proof, see Proposition 16 in the full version of the paper. On the other hand, any configuration from Definition 10 can be redrawn without using the crosscap:

Proposition 12. *Let (D, λ) be a projective HT-drawing of G^{+0} on S^2 and Z be a cycle satisfying the separation assumptions. Moreover, let us assume that $D(G^{+0}) \cap S^- = \emptyset$ (that is, G^{+0} is fully drawn on $S^+ \cup D(Z)$). Let us also assume that G^{+0} forms an inside fan, an inside square or an inside split triangle. Then there is an ordinary HT-drawing D' of G^{+0} on S^2 such that D coincides with D' on Z and $D'(G^{+0}) \cap S^- = \emptyset$.*

For proof, see Proposition 17 in the full version of the paper.

Now we are missing only one tool to finish the proof of Theorem 7. This tool is the "redrawing procedure" of Pelsmajer, Schaefer and Štefankovič [15]. More concretely, we need the following variant of Theorem 2. (Note that the theorem below is not in the setting of projective HT-drawings. However, the notions used in the statement are still well defined according to Definition 6.)

Theorem 13. *Let D be a drawing of a graph G on the sphere S^2. Let Z be a cycle in G such that every edge of Z is even and Z is drawn as a simple cycle. Then there is a drawing D'' of G such that*

- *D'' coincides with D on Z;*
- *$D''(G^{+0})$ belongs to $S^+ \cup D(Z)$ and $D''(G^{-0})$ belongs to $S^- \cup D(Z)$;*
- *whenever (e, f) is a pair of edges such that both e and f are inside edges or both e and f are outside edges, then $\mathrm{cr}_{D''}(e, f) = \mathrm{cr}_D(e, f)$.*

It is easy to check that the proof of Theorem 2 in [15] proves Theorem 13 as well. Additionally, we note that an alternative proof of Theorem 2 in [7, Lemma 3] can also be extended to yield Theorem 13. For completeness, we provide its proof in Sect. 8 of the full version of the paper.

Finally, we prove Theorem 7, assuming the validity of the aforementioned auxiliary results.

Proof sketch (of Theorem 7). First, we use Theorem 13 to G and D to obtain a drawing D'' keeping in mind that all edges of Z are even. By Proposition 11, G forms one of the redrawable configurations from Definition 10 on one of the sides. Without loss of generality, it appears inside. It means that D'' restricted to G^{+0} satisfies the assumptions of Proposition 12. Therefore, there is an ordinary HT-drawing D^+ of G^{+0} satisfying the conclusions of Proposition 12. Finally, we let D' be the drawing of G on S^2 which coincides with D^+ on G^{+0} and with D'' on G^{-0}. Both D'' and D^+ coincide with D on Z; therefore, D' is well defined. We set λ' so that $\lambda'(e) := \lambda(e)$ for an edge $e \in E^-$ and $\lambda'(e) := 0$ for any other edge. Now, it is easy to verify that (D', λ') is the required projective HT-drawing. The picture below provides an example of the drawings in the proof. □

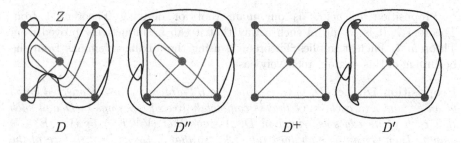

D D'' D^{+} D'

4   Proof of the Strong Hanani–Tutte Theorem on $\mathbb{R}P^2$

In this section we sketch a proof of Theorem 1 from Theorem 7 and the auxiliary results from the previous section.

Given a graph G that admits an HT-drawing on the projective plane, we need to show that G is actually projective-planar. By Proposition 4, we may assume that G admits a projective HT-drawing (D, λ) on S^2. We head for using Theorem 7. For this, we need that G is 2-connected and contains a suitable trivial cycle Z that may be redrawn so that it satisfies the assumptions of Theorem 7. Therefore, we start with auxiliary claims that will bring us to this setting. Many of them are similar to auxiliary steps in [15] (sometimes they are almost identical, adapted to a new setting). The proofs are at the beginning of Sect. 4 of the full version of the paper.

Before we state the next lemma, we recall the well known fact that any graph admits a (unique) decomposition into *blocks of 2-connectivity* [5, Chap. 3]. Here, we also allow the case that G is disconnected.

Lemma 14. *If G admits a projective HT-drawing on S^2, then at most one block of 2-connectivity in G is non-planar. Moreover, if all blocks are planar, G is planar as well.*

Observation 15. *Let (D, λ) be a drawing of a 2-connected graph. If D does not contain any trivial cycle, then G is planar.*

Lemma 16. *Let (D, λ) be a projective HT-drawing on S^2 of a graph G and let Z be a cycle in G. Then G can be redrawn only by local changes next to the vertices of Z to a projective HT-drawing D' on S^2 so that λ remains unchanged and $cr_{D'}(e, f) = \lambda(e)\lambda(f)$, for any pair $(e, f) \in E(Z) \times E(G)$ of distinct (not necessarily independent) edges. In particular, if $\lambda(e) = 0$ for every edge e of Z, then every edge of Z becomes even in D'.*

Once we know that the edges of a cycle can be made even we also need to know that such a cycle can be made simple.

Lemma 17. *Let (D, λ) be a projective HT-drawing on S^2 of a graph G and let Z be a cycle in G such that each of its edges is even. Then G can be redrawn so that Z becomes a simple cycle, its edges remain even and the resulting drawing is still a projective HT-drawing (with λ unchanged).*

Proposition 18 below is our main tool for deriving Theorem 1 from Theorem 7. It is set up in such a way that it can be inductively proved from Theorem 7. Then it implies Theorem 1, using the auxiliary lemmas from the beginning of this section, relatively easily.

Proposition 18. *Let (D, λ) be a projective HT-drawing of a 2-connected graph G on S^2 and Z a cycle in G that is completely free of crossings in D and such that each of its edges is trivial in D. Assume that (V^+, E^+) or (V^-, E^-) is empty. Then G can be embedded into $\mathbb{R}P^2$ so that Z bounds a disk face of the resulting embedding. If, in addition, D is an ordinary HT-drawing on S^2, then G can be embedded into S^2 so that Z bounds a face of the resulting embedding.*[6]

First we prove Theorem 1 assuming the validity of Proposition 18. Then, we sketch a proof of the proposition. See Proposition 24 in the full version of the paper for the complete proof.

Proof (of Theorem 1). We prove the result by induction on the number of vertices of G. We can trivially assume that G has at least three vertices.

If G has at least two blocks of 2-connectivity, G can be written as $G_1 \cup G_2$, where $G_1 \cap G_2$ is a minimal cut of G, and therefore, has at most one vertex. By Lemma 14, we may assume that G_1 is planar and G_2 non-planar. By induction, there exists an embedding D_2 of G_2 into $\mathbb{R}P^2$. So G_1 is planar, G_2 is embeddable in $\mathbb{R}P^2$, and $G_1 \cap G_2$ has at most one vertex. From these two embeddings, we easily derive an embedding of $G = G_1 \cup G_2$ in $\mathbb{R}P^2$.

We are left with the case when G is 2-connected. By Observation 15, we may assume that there is at least one trivial cycle Z in (D, λ). We can also make each of its edges trivial by Lemma 5 and even by Lemma 16. In addition, we make Z simple using Lemma 17. Hence G, Z and the current projective HT-drawing satisfy the separation assumptions.

Then we use Z to redraw G as follows. At first, we apply Theorem 7 to get a projective HT-drawing (D', λ') that separates G^{+0} and G^{-0}. We define $D^+ := D'(G^{+0})$ and $D^- := D'(G^{-0})$—without loss of generality, D^- is an ordinary HT-drawing on S^2, while D^+ is a projective HT-drawing on S^2.

Next, we apply Proposition 18 above to D^+ and D^- separately. Thus, we get embeddings of G^{+0} and G^{-0}—one of them in S^2, the other one in $\mathbb{R}P^2$. In addition, Z bounds a face in both of them; hence, we can easily glue them to get an embedding of the whole graph G into $\mathbb{R}P^2$.   □

Proof sketch (of Proposition 18). The proof proceeds by induction on the number of edges of G. The base case is when G is a cycle.

Without loss of generality, we assume that (V^-, E^-) is empty. That is, $G = G^{+0}$. If (V^+, E^+) is also empty, G consists only of Z and such a graph can easily be embedded into the plane or projective plane as required. Therefore, we assume that (V^+, E^+) is nonempty.

---

[6] We need to consider the case of ordinary HT-drawings in this proposition for a well working induction.

We find a path γ in $(V(G^{+0}), E(G^{+0}) \setminus E(Z))$ connecting two points x and y lying on Z. We may choose x, y so that $x \neq y$ since G is 2-connected.

Case 1: There exists a trivial γ. We provide only a very brief sketch for this case. First, we achieve, without redrawing Z, that γ is drawn as a simple path and every edge of γ is even and trivial. This can be done by steps similar to those in the proof of Theorem 1. Since γ is inside Z, now it splits the interior of Z into two disks. Once we carefully identify the two arcs of Z determined by the endpoints of γ, we get a cycle (in a different graph) separating the two disks. This way, we achieve essentially the same situation as in the proof of Theorem 1 and we can resolve it using Theorem 7.

Case 2: All choices of γ are nontrivial. Now, we need to resolve the case when all possible choices of γ are nontrivial. Let A^{+0} be the graph obtained from the inside arrow graph A^+ by adding the edges of Z (in particular, Z is a subgraph of A^{+0}). We aim to show that A^{+0} admits an embedding in $\mathbb{R}P^2$ such that Z bounds a disk face. As soon as we show this, we aim to replace the embedding of each arrow of A^{+0} by an embedding of the inside bridges inducing this arrow (if there are more such bridges, we embed them in parallel). The key fact that makes it possible is that each inside bridge meets Z in exactly two points and induces a single arrow and no loop. (Here, we leave this fact without a proof.) We also need to check that each of the bridges, together with Z, admits an embedding. This follows from Proposition 12 for inside fans and from Case 1 of this proof.

It remains to sketch why A^{+0} admits the required embedding. We know that any two disjoint arrows interleave using Lemma 9(b). Let us consider two concentric closed disks E_1 and E_2 such that E_1 belongs to the interior of E_2. Let us draw Z to the boundary of E_2. Let a be the number of arrows of A^+ and let us consider $2a$ points on the boundary of E_1 forming the vertices of a regular $2a$-gon. These points will be marked by ordered pairs xy where $\overline{xy}$ is an inside arrow. We mark the points so that the cyclic order of the points respects the cyclic order on Z in the first coordinate (the pairs with the same first coordinate are consecutive). However, for a fixed x, the pairs $xy_1, \ldots, xy_k$ corresponding to all arrows emanating from x are ordered in the reverted order when compared with the order of $y_1, \ldots, y_k$ on Z.

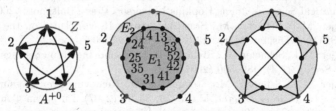

It is not hard to check that the points marked xy and yx are precisely the opposite points. Now, we get the required drawing in the following way. For any arrow $\overline{xy}$ we connect x with the point marked xy and y with yx. We can do all the connections simultaneously for all arrows without introducing any crossing since we have respected the cyclic order in the first coordinate. We remove the interior of E_1 and identify the opposite points on its boundary. This way we introduce a crosscap. Finally, we glue another disk along its boundary to Z and we get the required drawing on $\mathbb{R}P^2$.   □

Acknowledgment. We would like to thank Alfredo Hubard for fruitful discussions and valuable comments.

References

1. Cairns, G., Nikolayevsky, Y.: Bounds for generalized thrackles. Discret. Comput. Geom. **23**(2), 191–206 (2000)
2. Chojnacki, C.: Über wesentlich unplättbare Kurven im dreidimensionalen Raume. Fundamenta Mathematicae **23**(1), 135–142 (1934)
3. de Fraysseix, H., Ossona de Mendez, P.: Trémaux trees and planarity. Eur. J. Comb. **33**(3), 279–293 (2012)
4. de Fraysseix, H., Rosenstiehl, P.: A characterization of planar graphs by Trémaux orders. Combinatorica **5**(2), 127–135 (1985)
5. Diestel, R.: Graph Theory. Graduate Texts in Mathematics, vol. 173, 4th edn. Springer, Heidelberg (2010)
6. Fulek, R., Kynčl, J., Malinović, I., Pálvölgyi, D.: Clustered planarity testing revisited. Electron. J. Comb. **22**(4), P4–P24 (2015)
7. Fulek, R., Pelsmajer, M.J., Schaefer, M., Štefankovič, D.: Adjacent crossings do matter. J. Graph Algorithms Appl. **16**(3), 759–782 (2012)
8. Hopcroft, J., Tarjan, R.: Efficient planarity testing. J. ACM **21**(4), 549–568 (1974)
9. Kawarabayashi, K., Mohar, B., Reed, B.: A simpler linear time algorithm for embedding graphs into an arbitrary surface and the genus of graphs of bounded tree-width. In: 49th Annual IEEE Symposium on Foundations of Computer Science, 2008, pp. 771–780, October 2008
10. Kuratowski, C.: Sur le problème des courbes gauches en Topologie. Fundamenta Mathematicae **15**(1), 271–283 (1930)
11. Levow, R.B.: On Tutte's algebraic approach to the theory of crossing numbers. In: Proceedings of the Third Southeastern Conference on Combinatorics, Graph Theory, and Computing, pp. 314–315. Florida Atlantic Univ., Boca Raton (1972)
12. Mohar, B.: A linear time algorithm for embedding graphs in an arbitrary surface. SIAM J. Discret. Math. **12**(1), 6–26 (1999)
13. Mohar, B., Thomassen, C.: Graphs on Surfaces. Johns Hopkins Studies in the Mathematical Sciences. Johns Hopkins University Press, Baltimore (2001)
14. Pelsmajer, M.J., Schaefer, M., Stasi, D.: Strong Hanani-Tutte on the projective plane. SIAM J. Discret. Math. **23**(3), 1317–1323 (2009)
15. Pelsmajer, M.J., Schaefer, M., Štefankovič, D.: Removing even crossings. J. Comb. Theory Ser. B **97**(4), 489–500 (2007)
16. Pelsmajer, M.J., Schaefer, M., Štefankovič, D.: Removing even crossings on surfaces. Electron. Notes Discret. Math. **29**, 85–90 (2007). European Conference on Combinatorics, Graph Theory and Applications
17. Schaefer, M.: Hanani-Tutte and related results. In: Bárány, I., Böröczky, K.J., Tóth, G.F., Pach, J. (eds.) Geometry – Intuitive. Discrete, and Convex: A Tribute to László Fejes Tóth, pp. 259–299. Springer, Heidelberg (2013)

18. Schaefer, M.: Toward a theory of planarity: Hanani-Tutte and planarity variants. J. Graph Algorithms Appl. **17**(4), 367–440 (2013)
19. Tutte, W.T.: Toward a theory of crossing numbers. J. Comb. Theory **8**(1), 45–53 (1970)
20. van Kampen, E.R.: Komplexe in euklidischen Räumen. Abhandlungen aus dem Mathematischen Seminar der Universität Hamburg **9**(1), 72–78 (1933)
21. Wu, W.: On the planar imbedding of linear graphs I. J. Syst. Sci. Math. Sci. **5**(4), 290–302 (1985)

Hanani-Tutte for Radial Planarity II

Radoslav Fulek[1], Michael Pelsmajer[2], and Marcus Schaefer[3(✉)]

[1] IST Austria, Am Campus 1, 3400 Klosterneuburg, Austria
radoslav.fulek@gmail.com
[2] Illinois Institute of Technology, Chicago, IL 60616, USA
pelsmajer@iit.edu
[3] DePaul University, Chicago, IL 60604, USA
mschaefer@cs.depaul.edu

Abstract. A drawing of a graph G is *radial* if the vertices of G are placed on concentric circles $C_1, \ldots, C_k$ with common center c, and edges are drawn *radially*: every edge intersects every circle centered at c at most once. G is *radial planar* if it has a radial embedding, that is, a crossing-free radial drawing. If the vertices of G are ordered or partitioned into ordered levels (as they are for leveled graphs), we require that the assignment of vertices to circles corresponds to the given ordering or leveling. A pair of edges e and f in a graph is *independent* if e and f do not share a vertex.

We show that a graph G is radial planar if G has a radial drawing in which every two independent edges cross an even number of times; the radial embedding has the same leveling as the radial drawing. In other words, we establish the *strong* Hanani-Tutte theorem for radial planarity. This characterization yields a very simple algorithm for radial planarity testing.

1 Introduction

This paper continues work begun in "Hanani-Tutte for Radial Planarity" [16] by the same authors; to make the current paper self-contained we repeat some of the background and terminological exposition from the previous paper.

In a leveled graph every vertex is assigned a level in $\{1, \ldots, k\}$. A radial drawing visualizes the leveling of the graph by placing the vertices on concentric circles corresponding to the levels of G. Edges must be drawn as *radial* curves: for any circle concentric with the others, a radial curve intersects the circle at most once. A leveled graph is *radial planar* if it admits a radial embedding, that is, a radial drawing without crossings. The concept of *radial planarity* generalizes *level planarity* [8] in which levels are parallel lines instead of concentric circles (radially-drawn edges are replaced with *x-monotone edges*).

A version of the paper containing all proofs is available on arxiv.

R. Fulek—The research leading to these results has received funding from the People Programme (Marie Curie Actions) of the European Union's Seventh Framework Programme (FP7/2007-2013) under REA grant agreement no [291734].

© Springer International Publishing AG 2016
Y. Hu and M. Nöllenburg (Eds.): GD 2016, LNCS 9801, pp. 468–481, 2016.
DOI: 10.1007/978-3-319-50106-2_36

We previously established the weak Hanani-Tutte theorem for radial planarity: a leveled graph G is radial planar if it has a radial drawing (respecting the leveling) in which every two edges cross an even number of times [16, Theorem 1]. Our main result is the following strengthening of the weak Hanani-Tutte theorem for radial planarity, also generalizing the strong Hanani-Tutte theorem for level-planarity [15]:

Theorem 1. *If a leveled graph has a radial drawing in which every two independent edges cross an even number of times, then it has a radial embedding.*

The weak variant of a Hanani-Tutte theorem makes the stronger assumption that *every* two edges cross an even number of times. Often, weak variants lead to stronger conclusions. For example, it is known that if a graph can be drawn in a surface so that every two edges cross evenly, then the graph has an embedding on that surface with the same rotation system, i.e. the cyclic order of ends at each vertex remains the same [7,20]. This, in a way, is a disadvantage, since it implies that the original drawing is already an embedding, so that weak Hanani-Tutte theorems do not help in finding embeddings. On the other hand, strong Hanani-Tutte theorems are often algorithmic. Theorem 1 yields a very simple algorithm for radial planarity testing (described in Sect. 5) which is based on solving a system of linear equations over $\mathbb{Z}/2\mathbb{Z}$, see also [22, Sect. 1.4]. Our algorithm runs in time $O(n^{2\omega})$, where $n = |V(G)|$ and $O(n^{\omega})$ is the complexity of multiplication of two square $n \times n$ matrices. Since our linear system is sparse, it is also possible to use Wiedemann's randomized algorithm [24], with expected running time $O(n^4 \log^2 n)$ in our case.

Radial planarity was first studied by Bachmaier et al. [3]. Radial and other layered layouts are a popular visualization tool (see [4, Sect. 11], [9]); early examples of radial graph layouts can be found in the literature on sociometry [19]. Bachmaier et al. [3] showed that radial planarity can be tested, and an embedding can be found, in linear time. Their algorithm is based on a variant of PQ-trees [6] and is rather intricate. It generalizes an earlier linear-time algorithm for level-planarity testing by Jünger and Leipert [18]. Angelini et al. [2] devised recently a conceptually simpler algorithm for radial planarity with running time $O(n^4)$ (quadratic if the leveling is proper, that is, edges occur between consecutive levels only), by reducing the problem to a tractable case of Simultaneous PQ-ordering [5].

We prove Theorem 1 by ruling out the existence of a minimal counter-example. By the weak Hanani-Tutte theorem [16, Theorem 1] a minimal counter-example must contain a pair of independent edges crossing an odd number of times. Thus, [16, Theorem 1] serves as the base case in our argument (mirroring the development for level-planarity). In place of Theorem 1 we establish a stronger version, Theorem 4, which we discuss in Sect. 4.

We refer the reader to [16,21–23] for more background on the family of Hanani-Tutte theorems, but suffice it to say that strong variants are still rather rare, so we consider the current result as important evidence that Hanani-Tutte is a viable route to graph-drawing questions.

2 Terminology

For the purposes of this paper, graphs may have multiple edges, but no loops. An *ordered* graph $G = (V, E)$ is a graph whose vertices are equipped with a total order $v_1 < v_2 < \cdots < v_n$. We can think of an ordered graph as a special case of a *leveled* graph, in which every vertex of G is assigned a *level*, a number in $\{1, \ldots, k\}$ for some k. The leveling of the vertices induces a weak ordering of the vertices.

For convenience we represent radial drawings as drawings on a (standing) cylinder. Intuitively, imagine placing a cylindrically-shaped mirror in the center of a radial drawing as described in the introduction.[1] The *cylinder* is $\mathcal{C} = I \times \mathbb{S}^1$, where I is the unit interval $[0, 1]$ and $\mathbb{S}^1$ is the unit circle. Thus, a *point* on the cylinder is a pair (i, s), where $i \in I$ and $s \in \mathbb{S}^1$. The *projection* to I, or $\mathbb{S}^1$ maps $(i, s) \in \mathcal{C}$ to i, or s. We denote the projection of a point or a subset α of $I \times \mathbb{S}^1$ to I by $I(\alpha)$ and to $\mathbb{S}^1$ by $\mathbb{S}^1(\alpha)$. We define a relation to represent relative heights on the standing cylinder model: for $x, y \in \mathcal{C}$, let $x \leq y$ iff $I(x) \leq I(y)$. This allows us to write $\min I(X)$ and $\max I(X)$ as simply $\min X$ and $\max X$ (for $X \subset \mathcal{C}$), write $\min X \leq x$ to assert that $\min I(X) \leq I(x)$ (for $x \in \mathcal{C}$), etc.

The *winding number* of a closed curve on a cylinder is the number of times the projection to $\mathbb{S}^1$ of the curve winds around $\mathbb{S}^1$, i.e., the number of times the projection passes through an arbitrary point of $\mathbb{S}^1$ in the counterclockwise sense minus the number of times the projection passes through the point in the clockwise sense. A closed curve (or a closed walk in a graph) on a cylinder is *essential*[2] if its winding number is odd. A graph drawn on the cylinder is *essential* if it contains an essential cycle.

A *radial drawing* $\mathcal{D}(G)$ of an ordered graph G is a drawing of G on the cylinder in which every edge is *radial*, that is, its projection to I is injective (it does not "turn back"), and for every pair of vertices $u < v$ we have $I(u) < I(v)$. In a radial drawing, an *upper (lower) edge* at v is an edge incident to v for which $\min e = v$ ($\max e = v$). A vertex v is a *sink (source)*, if v has no upper (lower) edges. To avoid unnecessary complications, we may assume that $I(G)$ is contained in the interior of I. We think of $\mathcal{D}$ as a function from G (treated as a topological space) to $\mathcal{C}$. Thus, $\mathcal{D}(G')$, for $G' \subseteq G$, is a restriction of G to G'.

The *rotation* at a vertex in a drawing (on any surface) of a graph is the cyclic, clockwise order of the ends of edges incident to the vertex in the drawing. The *rotation system* is the set of rotations at all the vertices in the drawing. For radial drawings, we define the *upper (lower)* rotation at a vertex v to be the linear order of the ends of the upper (lower) edges in the rotation at v, starting with the direction corresponding to the clockwise orientation of $\mathbb{S}^1$. The rotation at a vertex in a radial drawing is completely determined by its upper and lower rotation. The *rotation system* of a radial drawing is the set of the upper and lower rotations at all the vertices in the drawing.

---

[1] Search for "cylindrical mirror anamorphoses" on the web for many cool pictures of this transformation.

[2] Note that we define an essential curve slightly differently than usual.

For any closed, possibly self-intersecting, curve in the plane (or cylinder), we can *two-color* the complement of the curve so that connected regions each get one color and crossing the curve switches colors. A point can appear more than once on the curve, and in such cases the color switches if the closed curve is crossed an odd number of times. For example, a plane graph (embedding) can have a face bounded by a closed curve that uses an edge e twice (once in each direction); crossing e means crossing the boundary walk twice, so the two-coloring will have the same color on both sides of e. If the closed curve is non-essential, the region incident to $1 \times \mathbb{S}^1$ and all other regions of the same color form the *exterior* (which includes the region incident to $0 \times \mathbb{S}^1$); the remaining regions form the *interior* of the curve.

Each pair of edges in a graph drawing crosses *evenly* or *oddly* according to the parity of the number of crossings between the two edges. A drawing of a graph is *even* if every pair of edges cross evenly. A drawing of a graph is *independently even* if every two independent edges in the drawing cross an even number of times; two edges that share an endpoint may cross oddly or evenly.

For any (non-degenerate) continuous deformation of a drawing of G, the parity of the number of crossings between pairs of independent edges changes only when an edge passes through a vertex. We call this event an *edge-vertex switch*. When an edge e passes through a vertex v, the crossing-parity changes between e and every edge incident to v.

3   Weak Hanani-Tutte for Radial Drawings

Let us first recall the weak variant of the result that we want to prove.

Theorem 2 (Fulek, Pelsmajer, Schaefer [16, Theorem 1]). *If a leveled graph has a radial drawing in which every two edges cross an even number of times, then it has a radial embedding with the same rotation system and leveling.*

We need a stronger version of this result that also keeps track of the parity of winding numbers.

Theorem 3 (Fulek, Pelsmajer, Schaefer [16, Theorem 2]). *If an ordered graph G has an even radial drawing, then it has a radial embedding with the same ordering, the same rotation system, and so that the winding number parity of every cycle remains the same.*

Theorem 2 follows from Theorem 3 using the construction from [15, Sect. 4.2] that was used to reduce level-planarity to x-monotonicity.

3.1   Working with Radial Embeddings and Even Drawings

Given a connected graph G with a rotation system, one can define a *facial walk* purely combinatorially by following the edges according to the rotation system (see, for example, [17, Sect. 3.2.6]), by traversing consecutive edges at each vertex, in clockwise order.

We need some terminology for embeddings of an ordered graph G with $v_1 < v_2 < \ldots < v_n$. The *maximum (minimum)* of a facial walk W in the radial drawing of G is the maximum (minimum) v that lies on W. A *local maximum (local minimum)* of a facial walk W is a vertex v on W so that is larger (smaller) than both the previous and subsequent vertices on W. (A vertex v might appear more than once on W; the previous definition implicitly refers to one such appearance.)

Let $e = uv$ and $e' = vw$ be two consecutive edges on the facial walk W of a face f in an embedding $\mathcal{E}(G)$ of G. We call (e, v, e') a *wedge* at v in f, and we can identify it with a small neighborhood of v in the interior of f. Intuitively, we think of wedges as being the corners of a face. Given an even drawing $\mathcal{D}(G)$ of G with the same rotation system as in $\mathcal{E}(G)$, we identify the wedge (e, v, e') with a small neighborhood of v on the left side of W. A point in the complement of W in $\mathcal{D}(G)$ is in the *interior (exterior)* of W if it receives the same (opposite) color as a wedge of W when we two-color the complement of W. Note that in an even drawing, all the wedges of W have the same color, and hence, the definition is consistent.

For general (not necessarily facial) walks W, we call $(e, v, e', -)$ and $(e, v, e', +)$ a *wedge* at v in an oriented walk W, and identify it with a small neighborhood of v on left and right, respectively, side of W. A wedge of an essential walk W in a radial drawing is *above* W if its color in the two-coloring of the complement of W is the same as the color of the region incident to $1 \times \mathbb{S}^1$, Otherwise, we say the edge is *below* W; in that case, it has the same color as the region incident to $0 \times \mathbb{S}^1$.

At each sink (source) v, the wedge that contains the region directly above (below) v is called a *concave wedge*. A facial walk in an even radial drawing is an *upper (lower) facial walk* if it contains $1 \times \mathbb{S}^1$ ($0 \times \mathbb{S}^1$) in its interior. An *outer facial walk* is an upper or lower facial walk; other facial walks are *inner facial walks*. If a radial embedding of G has only one outer facial walk (and one outer face) then it also has an x-monotone embedding, using the technique described in Sect. 2.

Let C denote a cycle in an even radial drawing. Let e, f and e', f', respectively, be edges incident to the maximum v and minimum u of C. Let $<_v$ be the lower rotation at v and let $<_u$ be the upper rotation at u. Suppose that $e <_v f$ and suppose that e, e', f', f (we allow $e = e'$ or $f = f'$) appear in this order along C. Then C is essential if and only if $f' <_u e'$. See Fig. 1

Lemma 1. *The cycle C is essential if and only if the two paths connecting its extreme vertices do so in inverse order.*

4   Strong Hanani-Tutte for Radial Drawings

Theorem 3 preserves the parity of the winding number of cycles in even radial drawings of ordered graphs, but there are examples showing that we cannot hope to do this when the drawings are only independently even. We will make do with a somewhat weaker property:

Given an ordered graph G with a radial drawings D_1 and D_2, we say that D_2 *is supported by* D_1 if for every essential cycle C_2 in D_2, there is an essential cycle C_1 in D_1 such that $[\min C_1, \max C_1] \subseteq [\min C_2, \max C_2]$.

A radial drawing of an ordered connected graph is *weakly essential* if every essential cycle in the drawing passes through v_1 or v_n (the first or the last vertex). With this definition we can state the strengthened version of our main result which we need for the proof.

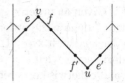

Fig. 1. Essential cycle (ends of path in inverse order at u and v).

Theorem 4. *Let G be an ordered graph. Suppose that G has an independently even radial drawing. Then G has a radial embedding. Moreover, (i) if the given drawing of G is weakly essential then G has an x-monotone embedding; and (ii) the new embedding is supported by the original drawing.*

Theorem 1 follows from Theorem 4 by the construction from [15, Sect. 4.2].

4.1 Working with Independently Even Radial Drawings

Call an edge drawn on C *bounded* if its points lie between its endpoints (with respect to the I-coordinate); that is, $u < p < v$ for every point p in the interior of edge uv. Call a drawing of G *bounded* if all edges are bounded.

Lemma 2. *If a graph has an (independently) even bounded drawing, then it has an (independently) even radial drawing with the same rotation system.*

We often make use of the following fact.

Lemma 3. *Let P be a path and let C be an essential cycle, vertex disjoint from P, in an independently even radial drawing of a graph. Then $I(P)$ does not contain $I(C)$.*

If two consecutive edges in the rotation at a vertex v cross oddly, we can make them cross evenly by a local redrawing: we "flip" the order of the two edges in the rotation at v, adding a crossing, which makes them cross evenly. For x-monotonicity it is known [15] that if the edges incident to a vertex cannot be made to cross evenly using edge flips, then there must be a connected component H of $G \setminus \{v, w\}$ satisfying $v \leq \min H < \max H \leq w$ or a multi-edge vw, but both cases can be dealt with. An application of the weak Hanani-Tutte theorem for x-monotonicity completes the proof. We want to use the same approach for radial drawings, and we already know that the weak Hanani-Tutte theorem holds for radial drawings. However, for radial drawings there may be a vertex v whose incident edges cannot be made to cross evenly using flips, but no obstacle like H exists. However, a closer look reveals that this can only happen when the vertex v is either the first or the last vertex of the ordered graph. The next lemma helps us deal with this case.

Given an ordered graph G with vertices $v_1 < \ldots < v_n$ without the edge $v_1 v_n$, let G' denote the ordered graph obtained by removing v_1, v_n, and replacing each edge $v v_i$ with $i \in \{1, n\}$ by a "pendant edge" vu where u is a new vertex of degree one, with u is placed in the order before v_2 if $i = 1$ and after v_{n-1} if $i = n$ (and otherwise ordered arbitrarily).

Lemma 4. *If G is a connected ordered graph with an (independently) even radial drawing $\mathcal{D}(G)$, then G' has an (independently) even radial drawing $\mathcal{D}'(G')$ such that $\mathcal{D}'(G \setminus \{v_1, v_n\}) = \mathcal{D}(G \setminus \{v_1, v_n\})$.*

Using Lemma 4 we can establish part (i) of Theorem 4.

Lemma 5. *Suppose that G has an independently even radial drawing that is weakly essential. Then G has an x-monotone embedding.*

4.2   Components of a Minimal Counterexample

We establish various properties of a minimal counter-example G—first with respect to vertices, then edges—to Theorem 4 given by an independently even radial drawing $\mathcal{D}(G)$.

Lemma 6. *G is connected.*

Let v be a vertex and suppose that B is a connected component of $G \setminus v$ with $\min B > v$. By Lemma 6, there exists at least one edge from v to a vertex in B.

Lemma 7. *Let v be a vertex and B be a connected component of $G \setminus v$ with $\min B > v$. Then either $|V(B)| = 1$ (and the vertex of B has just one neighbor, v) or B is essential.*

Lemma 8. *Let v be a vertex and B be a connected component of $G \setminus v$ with $\min B > v$. If B is essential, then $v = v_1$.*

Lemma 9. *Suppose that $v, w \in V$ and B is a connected component of $V \setminus \{v, w\}$ with $v < \min B$, $\max B < w$, and there is at least one edge from B to v and at least one edge from B to w. Then B is essential.*

4.3   Proof of Theorem 4

Let G be a minimal counter-example to the theorem given by an independently even radial drawing $\mathcal{D}(G)$. We already established part (i) of the theorem in Lemma 5. We also know, by Lemma 6, that G is connected.

If the drawing is even, then Theorem 3 gives us a radial embedding of G, and part (ii) of the theorem is satisfied since Theorem 3 maintains the parity of winding numbers of cycles. So there must be two adjacent edges crossing oddly.

Recall that flipping a pair of consecutive edges in an upper or lower rotation at a vertex changes the parity of crossing between the edges.

First, we repeatedly flip pairs of consecutive edges that cross oddly in the upper or lower rotation at a vertex until none remain. Let e, f be an odd pair of minimum distance in the—without loss of generality—upper rotation of a vertex v. Then let g be any edge in the upper rotation between e and f, which must cross both evenly.

Lemma 10. *Suppose that in $\mathcal{D}(G)$ there exist three paths P, Q and Q' starting at v such that $\{e, f, g\}$ is the set of first edges on those paths, with $\min P < v = \min Q = \min Q'$ and $V(P) \cap V(Q) = V(P) \cap V(Q') = \{v\}$. Then it cannot be that both $\max Q > \max P$ and $\max Q' > \max P$.*

The proof of Theorem 4 splits into two cases.

Case 1: Assume that $v \neq v_1, v_n$. Let G_e, G_f and G_g be the (not necessarily distinct) components of $G \setminus \{v\}$ containing an endpoint of e, f, g respectively. We are interested in showing that one or more of these intersects $\{u \in V : u < v\}$. For each of G_e, G_f and G_g, if it does not intersect that region, then by Lemma 7, it must be either (i) a single vertex or (ii) essential. However, case (ii) is impossible, since then $v_1 = v$ by Lemma 8 (contradiction). Suppose that two of them, let's say G_e and G_f are

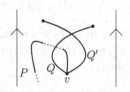

Fig. 2. P, Q, and Q' from Lemma 10.

single vertices w_e and w_f, respectively, with $w_e > w_f$. Then remove w_f and the edge vw_f from G and apply the minimality of G. Thus, we obtain a radial embedding of $G \setminus w_f$ supported by $\mathcal{D}(G \setminus w_f)$. We re-insert vw_f into the obtained embedding of $G \setminus w_f$ without crossings by drawing the edge vw_f alongside the longer edge vw_e. Thus, we can assume that at most one of G_e, G_f, G_g is of type (i) and none is of type (ii), which means that at least one of them must intersect $\{u \in V : u < v\}$.

Let P be a path from v through e, f, or g, which ends in the region $I < v$, chosen so as to minimize $\max P$, see Fig. 2. Let w_P be its vertex of max I-coordinate. Choose a minimal such P, so that every vertex except its last is in the region $I \geq v$. Let P_1 be the initial portion of P, from v to w_P, and let P_2 be the later portion of P, from w_P to the region $I < v$.

Let H be the subgraph induced by $\{u \in V : v < u < w_P\}$. Let H_2 be the component of H that intersects P_2 (empty if P_2 has just one edge) and let H_e, H_f, H_g be the (not necessarily distinct) components of H incident to e, f, g, respectively (empty if the upper endpoint of that edge is in the region $I > w_P$). By the choice of P, H_2 is disjoint (and distinct) from each of H_e, H_f, H_g and there is no edge from $H_e \cup H_f \cup H_g$ to the region $I < v$.

Suppose that H_e is non-empty and that v is the only vertex adjacent to H_e. By Lemmas 7 and 8 H_e is a single vertex. Then its only incident edge is e and we proceed as follows.

Remove e and its upper endpoint "v_e" and apply the minimality of G. Thus, we obtain a radial embedding $\mathcal{E}(G \setminus v_e)$ of $G \setminus v_e$ that is supported by $\mathcal{D}(G \setminus v_e)$. If there is an edge in $G \setminus v_e$ from v to w' with $w' \geq w_P$, simply draw e alongside that edge in the obtained embedding. So let's assume that there is no such edge.

Then P_1 contains at least one vertex in the region $v < I < w_P$. Let H_1 be the component of H that intersects P_1 and let H'_1 be the subgraph induced by $V(H_1) \cup \{v\}$. By the choice of P, H_1 is not incident to an edge intersecting the region $I < v$. The radial embedding $\mathcal{E}(H'_1)$ is non-essential because $I(H'_1) \subseteq I(P_2)$, by Lemma 3. Let H''_1 denote the union of H'_1 with all its incident edges (if any) intersecting the region $I > w_P$. We can draw e alongside the boundary of the lower outer face of H''_1 in $\mathcal{E}(H''_1)$ so that it is bounded. Hence, we can apply Lemma 2 to re-embed e without crossings (contradiction).

Thus, if H_e is non-empty then it must be adjacent to vertices other than v, which means vertices in either region $I < v$ or $I \geq w_P$, where the former is ruled out due to the choice of P. By similar arguments, H_f and H_g have neighbors in $I \geq w_P$ unless they are empty. Hence, we have the following.

Lemma 11. *Every non-empty subgraph H_e, H_f and H_g is non-essential and adjacent to a vertex in $I \geq w_P$. (If H_e is empty, then $\max e \geq w_P$, and similarly for H_f and H_g.)*

Without loss of generality we suppose that P goes through e, since otherwise we can redraw near v to flip the relative order of the ends of e, f, g at v so that P ends at the leftmost edge, renaming it e, renaming the middle one g, and the right one f—then flipping if needed we recover the earlier crossing parities between each pair of edges (e, f), (e, g), (f, g).

Consider minimal paths P_f and P_g from v, with first edge f and g, respectively, that end in $\{u : u < v \text{ or } u \geq w_P\}$. (These must exist because if H_f does not have neighbors in $I \geq w_P$ then H_f must be empty, which means that f is an edge from v to $I \geq w_P$ and thus we can let P_f be f with its endpoints. Likewise for H_g to get P_g.)

If neither P_e nor P_f intersects P_1, then neither intersects P and both end in $I > w_P$, which contradicts Lemma 10. Thus, we may assume that there exists a path from v through f or g to P_1 which lies in the region $v \leq I \leq w_P$. Hence, there exists a cycle through e, v, f or e, v, g which lies in the region $v \leq I \leq w_P$.

We choose C so as to minimize $\max C$ and to be essential if possible. Let w be the vertex with $w = \max C$. Without loss of generality, we may assume that $f \in E(C)$. Let B_e be the component of $\{u : v < u < w\}$ that is incident to e, and let B'_e be the graph formed from the union of B_e and all incident edges (including e) with their endpoints. Define B_f, B'_f, B_g, B'_g similarly, but in the case that the upper endpoint of g is not in the region $v < I < w$, let $B_g = \emptyset$ and let B'_g be just g with its endpoints.

By the choice of C we have $B_e \cap B_f = B_e \cap B_g = B_f \cap B_g = \emptyset$. None of B_e, B_f and B_g is joined by an edge with a vertex $u < v$ by the choice of P.

First consider the case that C is not essential. Since g crosses every edge of C evenly and g is between e and f near v—which is in the interior of C— the other endpoint of g must be in the interior of C or on C. In the former case, every vertex of B_g must be in the interior of C because C and B_g are disjoint so their edges cross evenly. For the same reason B_g cannot be adjacent to any vertices in the region $I > w$, so $V(B'_g) \setminus V(B_g) \subseteq \{v, w\}$. By Lemma 11, B'_g

includes w and $w = w_P$ and H_e is non-essential, but then B_g is essential by Lemma 9, a contradiction since $B_g \subseteq H_g$. Hence the upper endpoint of g is in C. By the choice of C, it must be w; i.e., $g = vw$. Let C' be the cycle formed from g and either of the paths from v to w in C. since $\max C' = \max C$, when we chose C, we could have chosen C' instead. Then C' must be non-essential. Thus, the preceding argument all applies with C' replacing C, implying that $e = vw$ or $f = vw$. The multi-edge vw gives either a direct contradiction if you want to think of it that way, or else remove one, apply induction, and redraw it alongside its parallel edge.

Hence, C is essential, and by the choice of C and P, and by Lemma 3 applied to C and P_2 we obtain that $w_P = w$. Thus, $B_e = H_e$, $B_f = H_f$ and $B_g = H_g$. Suppose that $g = vw$. The union $C \cup g$ contains two cycles through g one of which is non-essential in $\mathcal{D}(G)$, since otherwise C would not be essential by a simple parity argument. By applying the argument in the previous paragraph to the non-essential cycle, it follows that $B_e = e = vw$ or $B_f = f = vw$. Then we can remove one of the multi-edges vw and obtain a contradiction with the choice of G. Hence, we assume that $g \neq vw$. By Lemma 11, $\mathcal{D}(B_g')$ intersects the region $I \geq w = w_P$. Furthermore, by Lemmas 9 and 11, $V(B_g') \not\subseteq V(B) \cup \{\{v, w\}$, so $\mathcal{D}(B_g')$ intersects the region $I > w$. By applying Lemmas 9 and 11 to B_e likewise, the drawing $\mathcal{D}(B_e')$ also intersects the region $I > w$, unless $e = vw$. In what follows we show that the former cannot happen. Then by symmetry the same applies to f, and hence, we obtain a multi-edge vw contradicting the minimality of G, which completes the proof. Let Q' be a shortest path in B_g' starting at v with g and ending in the region $I > w$. By assuming that $\mathcal{D}(B_e')$ intersects the region $I > w$, we may let $Q \subseteq B_e' \setminus v$ be a shortest path in $B_e' \setminus v$ starting with e' and ending in the region $I > w$.

We modify the drawing of e and f near v so that they switch positions in the rotation at v; then g crosses both e and f oddly, e and f cross evenly, and the edge g is between e and f. Furthermore, we modify drawings of edges of C near the vertices of C so that every pair of edges of C cross each other evenly; this will not affect the upper rotation at v. We correct the lower rotation at w so that the first edge on P_2, let's say g', is between the edges $e' \in B_e'$ and $f' \in B_f'$ on C. Since C is essential, the edge g' crosses C evenly since P_2 begins and ends below C. Since G is independently even, g' crosses both e' and f' either oddly or evenly. In the next paragraph, we show that the edge g' crosses e' oddly.

Let C_e be a cycle consisting of Q', the part of C between v and w through B_e', and a new edge edge from w to the upper endpoint of Q'. For convenience, we can make the new edge drawn radially, such that it crosses its incident edge in Q' evenly (it cannot cross any other edge of C'). Then every two edges in C_e cross evenly except for the pair g, e. Consider the two-coloring of the complement of C_e in C. Because the two paths in C_e from v to w cross each other oddly, the color immediately to the right (left) of e at v will be the same as the color to immediately to the left (right) of e' at w. By Lemma 1 applied to the essential cycle C, f and g are to the right (left) of e at v if and only if f' and g' are to the left (right) of e' at w. Therefore, the upper wedge at v between e and g

will have the same color as the lower wedge between e' and g' at w; the latter implies that the end of g' near w will have have that color as well. The entire region $I < v$ must have the opposite color as the upper wedge at v, and P_2 has an endpoint in this region. Thus, the two ends of P_2 have different colors. Then P_2 must cross C_e oddly. Since the drawing of G is independently even, it must be that g' crosses e' oddly (Fig. 3).

As shown earlier, if g' crosses e' oddly then g' also crosses f' oddly. Let $Q \subseteq B'_e \setminus v$ be a shortest path in $B'_e \setminus v$ starting with w, e' and ending in the region $I > w_P$. Then we can apply Lemma 10 upside down: the role of "v" in Lemma 10 is played by w, the role of "P" by Q, "Q" is P_2 and "Q'" is the part of C between v and w through B'_f (contradiction). Indeed, P_2 is internally disjoint from both Q and C by the choice of P, and Q is internally disjoint from B_f.

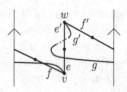

Fig. 3.   Applying Lemma 10 upside down.

Case 2: Assume that $v = v_1$ or $v = v_n$. We can assume that G does not contain edge $v_1 v_n$, since otherwise $\mathcal{D}(G)$ is weakly essential by Lemma 3, and we are done by Lemma 5. We can also suppose that only pairs of edges at v_1 or v_n cross an odd number of times. Otherwise, we end up in the previous case. Turn G into a graph G' with pendant edges as described in the paragraph preceding Lemma 4; then Lemma 4 implies that G' has an even drawing, so by Theorem 3 it has a radial embedding. We can redraw pendant edges and identify endpoints to obtain a radial embedding of G, but we need to do this carefully to satisfy part (ii) of the theorem if G' is essential: We "expose" (see Fig. 4) the maximum vertex of the lower face boundary of G' so that it remains on the lower face boundary of G (and likewise for the the minimum vertex of the upper face boundary). Any essential cycle C in G not present in G' passes through v_1 or v_n. In order to satisfy part (ii) we need an essential cycle C' in the embedding of G' for which $[\min C', \max C'] \subseteq [\min C, \max C]$. However, a lower or upper facial walk of G' contains such a cycle.

5   Algorithm

Theorem 1 reduces radial planarity testing to a system of linear equations over $\mathbb{Z}/2\mathbb{Z}$. For planarity testing, systems like this were first constructed by Wu and Tutte [22, Sect. 1.4.2].

Unlike in the case of x-monotone drawings, two drawings of an edge e with end vertices fixed cannot necessarily be obtained one from another by a continuous deformation during which we keep the drawing of e radial: up to a continuous deformation, two radial drawings of an edge differ by a certain number of (Dehn) twists. We perform a twist of $e = uv$, $u < v$ very close to v, i.e., the twist is carried out by removing a small portion P_e of e such that we have $w \notin I(P_e)$, for all vertices w, and reconnecting the severed pieces of e by a curve intersecting every edge e', s.t. $I(P_e) \subset I(e')$, exactly once. Observe that with respect to the parity of crossings between edges performing a twist of e close to v equals performing an edge-vertex switch of e with all the vertices $w < v$ (including those for which $w < u$). Hence, the orientation of the twist does not matter, and any twist of e keeping e radial can be simulated by a twist of e very close to v and a set of edge-vertex switches of e with certain vertices w, for which $u < w < v$.

By the previous paragraph a linear system for testing radial planarity can be constructed as follows. The system has a variable $x_{e,v}$ for every edge-vertex switch (e, v) such that $v \in I(e)$, and a variable x_e for every edge twist. Given an arbitrary radial drawing of G we denote by $\mathrm{cro}(e, f)$ the parity of the number of crossings between e and f.

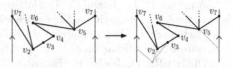

Fig. 4. Exposing v_6 on the lower outer face.

In the linear system, for each pair of independent edges $(e, f) = (uv, wz)$, where $u < v$, $w < z$, $u < w$, and $w < v$, we require

$$\mathrm{cr}(e, f) \equiv \begin{cases} x_{e,w} + x_{e,z} + x_f \bmod 2 \text{ if } z < v, \text{ and} \\ x_{e,w} + x_{f,v} + x_e \bmod 2 \text{ if } z > v. \end{cases}$$

Then G is radial planar if and only if this linear system has a solution.

6   Open Questions

We conjecture that Theorem 3, and—in light of [14, Sect. 2]—its algorithmic consequences, extend to *bounded drawings* [13] on a cylinder, defined as follows. We are given a pair (G, γ) of a graph G and a map $\gamma : V \to \mathbb{N}$, and consider cylindrical drawings of G in which (i) $u < v$ whenever $\gamma(u) < \gamma(v)$ for $u, v \in V$, and (ii) $\gamma(u) \le \gamma(w) \le \gamma(v)$, where $uv \in E$ and $\gamma(u) \le \gamma(v)$, whenever $I(w) \in I(uv)$. By Lemma 2, radial planarity is the special case in which γ is injective.

In the plane, such a result is already known: a weak Hanani-Tutte variant for bounded embeddings in the plane [12]. (A more general result was proved by M. Skopenkov in a different context [23, Theorem 1.5].) This together with a result showing that edges can be made x-monotone [13, Lemma 1] shows that the corresponding planarity variant coincides with strip planarity [1]. We do not know whether projections of edges to I can be made injective in bounded embeddings on the cylinder, though we conjecture that this is the case.

If the previous conjecture holds, bounded embeddings on the cylinder can be treated as clustered planar embeddings [10,11] where all the clusters are pairwise nested. The complexity status of this special case of c-planarity is open to the best of our knowledge. The counter-examples in [14, Sect. 6,8], a Hanani-Tutte theorem for this setting would be the most general direct extension of the Hanani-Tutte theorem to clustered planar drawings that we can hope for.

References

1. Angelini, P., Lozzo, G., Battista, G., Frati, F.: Strip planarity testing. In: Wismath, S., Wolff, A. (eds.) GD 2013. LNCS, vol. 8242, pp. 37–48. Springer, Heidelberg (2013). doi:10.1007/978-3-319-03841-4_4
2. Angelini, P., Da Lozzo, G., Di Battista, G., Frati, F., Patrignani, M., Rutter, I.: Beyond level planarity. arXiv preprint arXiv:1510.08274 (2015)
3. Bachmaier, C., Brandenburg, F.J., Forster, M.: Radial level planarity testing and embedding in linear time. J. Graph Algorithms Appl. 9, 53–97 (2005)
4. Di Battista, G., Eades, P., Tamassia, R., Tollis, I.G.: Graph Drawing: Algorithms for the Visualization of Graphs, 1st edn. Prentice Hall PTR, Upper Saddle River (1998)
5. Bläsius, T., Rutter, I.: Simultaneous PQ-ordering with applications to constrained embedding problems. ACM Trans. Algorithms 12(2), 16 (2016)
6. Booth, K.S., Lueker, G.S.: Testing for the consecutive ones property, interval graphs, and graph planarity using PQ-tree algorithms. J. Comput. Syst. Sci. 13(3), 335–379 (1976)
7. Cairns, G., Nikolayevsky, Y.: Bounds for generalized thrackles. Discrete Comput. Geom. 23(2), 191–206 (2000)
8. Di Battista, G., Nardelli, E.: Hierarchies and planarity theory. IEEE Trans. Syst. Man Cybernet. 18(6), 1035–1046 (1989)
9. Di Giacomo, E., Didimo, W., Liotta, G.: Spine and radial drawings, Chap. 8, pp. 247–284. Discrete Mathematics and Its Applications. Chapman and Hall/CRC (2013)
10. Feng, Q.-W., Cohen, R.F., Eades, P.: How to draw a planar clustered graph. In: Ding-Zhu, D., Li, M. (eds.) Computing and Combinatorics. LNCS, vol. 959, pp. 21–30. Springer, Berlin Heidelberg (1995)
11. Feng, Q.-W., Cohen, R.F., Eades, P.: Planarity for clustered graphs. In: Spirakis, P. (ed.) ESA 1995. LNCS, vol. 979, pp. 213–226. Springer, Heidelberg (1995). doi:10.1007/3-540-60313-1_145
12. Fulek, R.: Towards the Hanani-Tutte theorem for clustered graphs. In: Graph-Theoretic Concepts in Computer Science - 40th International Workshop, WG 2014, Nouan-le-Fuzelier, France, 25–27 June 2014. Revised Selected Papers, pp. 176–188 (2014). arXiv:1410.3022v2
13. Fulek, R.: Bounded embeddings of graphs in the plane. In: Combinatorial Algorithms - 27th International Workshop, IWOCA 2016, Helsinki, Finland, 17–19 August 2016, Proceedings, pp. 31–42 (2016)
14. Fulek, R., Kynčl, J., Malinovic, I., Pálvölgyi, D.: Clustered planarity testing revisited. Electron. J. Combinatorics 22 (2015)
15. Fulek, R., Pelsmajer, M., Schaefer, M., Štefankovič, D.: Hanani-Tutte, monotone drawings, and level-planarity. In: Pach, J. (ed.) Thirty Essays on Geometric Graph Theory, pp. 263–287. Springer, New York (2013)
16. Fulek, R., Pelsmajer, M.J., Schaefer, M.: Hanani-Tutte for radial planarity. In: Graph Drawing and Network Visualization - 23rd International Symposium, GD 2015, Los Angeles, CA, USA, 24–26 September 2015, Revised Selected Papers, pp. 99–110 (2015)
17. Gross, J.L., Tucker, T.W.: Topological Graph Theory. Dover Publications Inc., Mineola (2001). Reprint of the 1987 original
18. Jünger, M., Leipert, S.: Level planar embedding in linear time. J. Graph Algorithms Appl. 6(1), 72–81 (2002)

19. Northway, M.L.: A method for depicting social relationships obtained by sociometric testing. Sociometry **3**(2), 144–150 (1940)
20. Pelsmajer, M.J., Schaefer, M., Štefankovič, D.: Removing even crossings. J. Combin. Theory Ser. B **97**(4), 489–500 (2007)
21. Schaefer, M.: Toward a theory of planarity: Hanani-Tutte and planarity variants. J. Graph Algortihms Appl. **17**(4), 367–440 (2013)
22. Schaefer, M.: Hanani-Tutte and related results. In: Bárány, I., Böröczky, K.J., Fejes Tóth, G., Pach (eds.) Geometry-Intuitive, Discrete, and Convex–A Tribute to László Fejes Tóth, vol. 24. Bolyai Society Mathematical Studies (2014)
23. Skopenkov, M.: On approximability by embeddings of cycles in the plane. Topology Appl. **134**(1), 1–22 (2003)
24. Wiedemann, D.H.: Solving sparse linear equations over finite fields. IEEE Trans. Inform. Theory **32**(1), 54–62 (1986)

Beyond Level Planarity

Patrizio Angelini[1], Giordano Da Lozzo[2], Giuseppe Di Battista[2],
Fabrizio Frati[2(✉)], Maurizio Patrignani[2], and Ignaz Rutter[3]

[1] Tübingen University, Tübingen, Germany
angelini@informatik.uni-tuebingen.de
[2] Roma Tre University, Rome, Italy
{dalozzo,gdb,frati,patrigna}@dia.uniroma3.it
[3] Karlsruhe Institute of Technology, Karlsruhe, Germany
rutter@kit.edu

Abstract. In this paper we settle the computational complexity of two
open problems related to the extension of the notion of level planarity
to surfaces different from the plane. Namely, we show that the problems
of testing the existence of a level embedding of a level graph on the surface of the rolling cylinder or on the surface of the torus, respectively
known by the name of CYCLIC LEVEL PLANARITY and TORUS LEVEL
PLANARITY, are polynomial-time solvable.

Moreover, we show a complexity dichotomy for testing the SIMULTA-
NEOUS LEVEL PLANARITY of a set of level graphs, with respect to both
the number of level graphs and the number of levels.

1 Introduction and Overview

The study of level drawings of level graphs has spanned a long time; the seminal
paper by Sugiyama *et al.* on this subject [23] dates back to 1981, well before graph
drawing was recognized as a distinguished research area. This is motivated by
the fact that level graphs naturally model hierarchically organized data sets and
level drawings are a very intuitive way to represent such graphs.

Formally, a *level graph* (V, E, γ) is a directed graph (V, E) together with a
function $\gamma : V \to \{1, \ldots, k\}$, with $1 \leq k \leq |V|$. The set $V_i = \{v \in V : \gamma(v) = i\}$ is
the i-th *level* of (V, E, γ). A level graph (V, E, γ) is *proper* if for each $(u, v) \in E$,
either $\gamma(u) = \gamma(v) - 1$, or $\gamma(u) = k$ and $\gamma(v) = 1$. Let $l_1, \ldots, l_k$ be k horizontal
straight lines on the plane ordered in this way with respect to the y-axis. A *level
drawing* of (V, E, γ) maps each vertex $v \in V_i$ to a point on l_i and each edge
$(u, v) \in E$ to a curve monotonically increasing in the y-direction from u to v.
Note that a level graph (V, E, γ) containing an edge $(u, v) \in E$ with $\gamma(u) > \gamma(v)$
does not admit any level drawing. A level graph is *level planar* if it admits a
level embedding, i.e., a level drawing with no crossing; see Fig. 1(a). The LEVEL
PLANARITY problem asks to test whether a given level graph is level planar.

Research was partially supported by DFG grant Ka812/17-1, by MIUR project
AMANDA, prot. 2012C4E3KT_001, and by DFG grant WA 654/21-1.

Y. Hu and M. Nöllenburg (Eds.): GD 2016, LNCS 9801, pp. 482–495, 2016.
DOI: 10.1007/978-3-319-50106-2_37

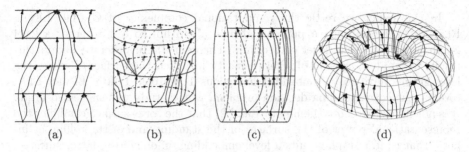

Fig. 1. Level embeddings (a) on the plane, (b) on the standing cylinder, (c) on the rolling cylinder, and (d) on the torus.

Problem LEVEL PLANARITY has been studied for decades [12,15,18,20,22], starting from a characterization of the single-source level planar graphs [12] and culminating in a linear-time algorithm for general level graphs [20]. A characterization of level planarity in terms of "minimal" forbidden subgraphs is still missing [13,17]. The problem has also been studied to take into account a clustering of the vertices (CLUSTERED LEVEL PLANARITY [2,14]) or consecutivity constraints for the vertex orderings on the levels ($\mathcal{T}$-LEVEL PLANARITY [2,24]).

Differently from the standard notion of planarity, the concept of level planarity does not immediately extend to representations of level graphs on surfaces different from the plane[1]. When considering the surface $\mathbb{O}$ of a sphere, level drawings are usually defined as follows: The vertices have to be placed on the k circles given by the intersection of $\mathbb{O}$ with k parallel planes, and each edge is a curve on $\mathbb{O}$ that is monotone in the direction orthogonal to these planes. The notion of level planarity in this setting goes by the name of RADIAL LEVEL PLANARITY and is known to be decidable in linear time [4]. This setting is equivalent to the one in which the level graph is embedded on the "standing cylinder": Here, the vertices have to be placed on the circles defined by the intersection of the cylinder surface $\mathbb{S}$ with planes parallel to the cylinder bases, and the edges are curves on $\mathbb{S}$ monotone with respect to the cylinder axis; see [3,4,10] and Fig. 1(b).

Problem LEVEL PLANARITY has been also studied on the surface $\mathbb{R}$ of a "rolling cylinder"; see [3,5,6,10] and Fig. 1(c). In this setting, k straight lines $l_1, \ldots, l_k$ parallel to the cylinder axis lie on $\mathbb{R}$, where $l_1, \ldots, l_k$ are seen in this clockwise order from a point p on one of the cylinder bases, the vertices of level V_i have to be placed on l_i, for $i = 1, \ldots, k$, and each edge (u, v) is a curve λ lying on $\mathbb{R}$ and flowing monotonically in clockwise direction from u to v as seen from p. Within this setting, the problem takes the name of CYCLIC LEVEL PLANARITY [6]. Note that a level graph (V, E, γ) may now admit a level embedding even if it contains edges (u, v) with $\gamma(u) > \gamma(v)$. Contrary to the other mentioned settings, the complexity of testing CYCLIC LEVEL PLANARITY is still unknown, and a polynomial (in fact, linear) time algorithm has been presented only for *strongly connected graphs* [5], which are level graphs such that for each pair of vertices there exists a directed cycle through them.

---

[1] We consider connected orientable surfaces; the *genus* of a surface is the maximum number of cuttings along non-intersecting closed curves that do not disconnect it.

In this paper we settle the computational complexity of CYCLIC LEVEL PLANARITY by showing a polynomial-time algorithm to test whether a level graph admits a cyclic level embedding (Theorem 3). In order to obtain this result, we study a version of level planarity in which the surface $\mathbb{T}$ where the level graphs have to be embedded has genus 1; we call TORUS LEVEL PLANARITY the corresponding decision problem, whose study was suggested in [6]. It is not difficult to note (Lemmata 1 and 2) that the torus surface combines the representational power of the surfaces of the standing and of the rolling cylinder – that is, if a graph admits a level embedding on one of the latter surfaces, then it also admits a level embedding on the torus surface. Furthermore, both RADIAL LEVEL PLANARITY and CYCLIC LEVEL PLANARITY (and hence LEVEL PLANARITY) reduce in linear time to TORUS LEVEL PLANARITY.

The main result of the paper (Theorem 2) is a quadratic-time algorithm for proper instances of TORUS LEVEL PLANARITY and a quartic-time algorithm for general instances. Our solution is based on a linear-time reduction (Observation 1 and Lemmata 3–6) that, starting from any proper instance of TORUS LEVEL PLANARITY, produces an equivalent instance of the SIMULTANEOUS PQ-ORDERING problem [8] that can be solved in quadratic time (Theorem 1).

Motivated by the growing interest in simultaneous embeddings of multiple planar graphs, which allow to display several relationships on the same set of entities in a unified representation, we define a new notion of level planarity in which multiple level graphs are considered and the goal is to find a simultaneous level embedding of them. The problem SIMULTANEOUS EMBEDDING (see the seminal paper [11] and a recent survey [7]) takes as input k planar graphs $(V, E_1), \ldots, (V, E_k)$ and asks whether they admit planar drawings mapping each vertex to the same point of the plane. We introduce the problem SIMULTANEOUS LEVEL PLANARITY, which asks whether k level graphs $(V, E_1, \gamma), \ldots, (V, E_k, \gamma)$ admit level embeddings mapping each vertex to the same point along the corresponding level. As an instance of SIMULTANEOUS LEVEL PLANARITY for two graphs on two levels is equivalent to one of CYCLIC LEVEL PLANARITY on two levels (Theorem 5), we can solve SIMULTANEOUS LEVEL PLANARITY in polynomial time in this case. This positive result cannot be extended (unless P=NP), as the problem becomes NP-complete even for two graphs on three levels and for three graphs on two levels (Theorem 4). Altogether, this establishes a tight border of tractability for SIMULTANEOUS LEVEL PLANARITY.

Complete proofs can be found in the full version of the paper [1].

2 Preliminaries

A *tree* T is a connected acyclic graph. The degree-1 vertices of T are the *leaves* of T, denoted by $\mathcal{L}(T)$, while the remaining vertices are the *internal* vertices.

A digraph $G = (V, E)$ without directed cycles is a *directed acyclic graph* (*DAG*). An edge $(u, v) \in E$ directed from u to v is an *arc*; vertex u is a *parent* of v and v is a *child* of u. A vertex is a *source* (*sink*) if it has no parents (children).

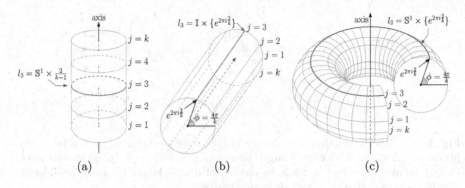

Fig. 2. Levels on (a) $\mathbb{S}$, on (b) $\mathbb{R}$, and on (c) $\mathbb{T}$, respectively.

Embeddings on Levels. An *embedding* of a graph on a surface $\mathbb{Q}$ is a mapping Γ of each vertex v to a distinct point on $\mathbb{Q}$ and of each edge $e = (u, v)$ to a simple Jordan curve on $\mathbb{Q}$ connecting u and v, such that no two curves cross except at a common endpoint. Let $\mathbb{I}$ and $\mathbb{S}^1$ denote the unit interval and the boundary of the unit disk, respectively. We define the surface $\mathbb{S}$ of the standing cylinder, $\mathbb{R}$ of the rolling cylinder, and $\mathbb{T}$ of the torus as $\mathbb{S} = \mathbb{S}^1 \times \mathbb{I}$, as $\mathbb{R} = \mathbb{I} \times \mathbb{S}^1$, and as $\mathbb{T} = \mathbb{S}^1 \times \mathbb{S}^1$, respectively. The *j-th level* of surfaces $\mathbb{S}$, $\mathbb{R}$, and $\mathbb{T}$ with k levels is defined as $l_j = \mathbb{S}^1 \times \{\frac{j-1}{k-1}\}$, $l_j = \mathbb{I} \times \{e^{2\pi i \frac{j-1}{k}}\}$, and $l_j = \mathbb{S}^1 \times \{e^{2\pi i \frac{j-1}{k}}\}$, respectively; see Fig. 2. An edge (x, y) on $\mathbb{S}$, on $\mathbb{R}$, or on $\mathbb{T}$ is *monotone* if it intersects the levels $\gamma(x), \gamma(x) + 1, \ldots, \gamma(y)$, where $k + 1 = 1$, exactly once and does not intersect any of the other levels.

Problems RADIAL, CYCLIC, and TORUS LEVEL PLANARITY take as input a level graph $G = (V, E, \gamma)$ and ask to find an embedding Γ of G on $\mathbb{S}$, on $\mathbb{R}$, and on $\mathbb{T}$, respectively, in which each vertex $v \in V$ lies on $l_{\gamma(v)}$ and each edge $(u, v) \in E$ is monotone. Embedding Γ is called a *radial*, a *cyclic*, and a *torus level embedding* of G, respectively. A level graph admitting a radial, cyclic, or torus level embedding is called *radial*, *cyclic*, or *torus level planar*, respectively.

Lemmata 1 and 2 show that the torus surface combines the power of representation of the standing and of the rolling cylinder. To strengthen this fact, we present a level graph in Fig. 3a that is neither radial nor cyclic level planar, yet it is torus level planar; note that the underlying (non-level) graph is also planar.

Lemma 1. *Every radial level planar graph is also torus level planar. Further,* RADIAL LEVEL PLANARITY *reduces in linear time to* TORUS LEVEL PLANARITY.

Lemma 2. *Every cyclic level planar graph is also torus level planar. Further,* CYCLIC LEVEL PLANARITY *reduces in linear time to* TORUS LEVEL PLANARITY.

Orderings and PQ-Trees. Let A be a finite set. We call *linear ordering* any permutation of A. When considering the first and the last elements of the permutation as consecutive, we talk about *circular orderings*. Let $\mathcal{O}$ be a circular

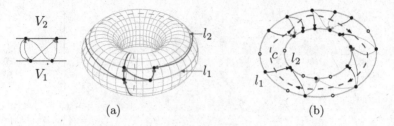

Fig. 3. (a) A level graph that is neither cyclic nor radial level planar, yet it is torus level planar. (b) A radial level embedding Γ of a level graph $(V_1 \cup V_2, E, \gamma)$ on two levels. Colors are used for edges incident to vertices of degree larger than one to illustrate that the edge ordering on E in Γ is v-consecutive.

ordering on A and let $\mathcal{O}'$ be the circular ordering on $A' \subseteq A$ obtained by restricting $\mathcal{O}$ to the elements of A'. Then $\mathcal{O}'$ is a *suborder* of $\mathcal{O}$ and $\mathcal{O}$ is an *extension* of $\mathcal{O}'$. Let A and S be finite sets, let $\mathcal{O}' = s_1, s_2, \ldots, s_{|S|}$ be a circular ordering on S, let $\phi : S \to A$ be an injective map, and let $A' \subseteq A$ be the image of S under ϕ; then $\phi(\mathcal{O}')$ denotes the circular ordering $\phi(s_1), \phi(s_2), \ldots, \phi(s_{|S|})$. We also say that a circular ordering $\mathcal{O}'$ on S is a *suborder* of a circular ordering $\mathcal{O}$ on A (and $\mathcal{O}$ is an *extension* of $\mathcal{O}'$) if $\phi(\mathcal{O}')$ is a suborder of $\mathcal{O}$.

An *unrooted PQ-tree* T is a tree whose leaves are the elements of a ground set A. PQ-tree T can be used to represent all and only the circular orderings $\mathcal{O}(T)$ on A satisfying a given set of *consecutivity constraints* on A, each of which specifies that a subset of the elements of A has to appear consecutively in all the represented circular orderings on A. The internal nodes of T are either *P-nodes* or *Q-nodes*. The orderings in $\mathcal{O}(T)$ are all and only the circular orderings on the leaves of T obtained by arbitrarily ordering the neighbours of each P-node and by arbitrarily selecting for each Q-node either a given circular ordering on its neighbours or its reverse ordering. Note that possibly $\mathcal{O}(T) = \emptyset$, if T is the empty tree, or $\mathcal{O}(T)$ represents all possible circular orderings on A, if T is a star centered at a P-node. In the latter case, T is the *universal* PQ-tree on A.

We illustrate three linear-time operations on PQ-trees (see [9,16,19]). Let T and T' be PQ-trees on A and let $X \subseteq A$: The *reduction of T by X* builds a new PQ-tree on A representing the circular orderings in $\mathcal{O}(T)$ in which the elements of X are consecutive. The *projection of T to X, denoted by $T|_X$*, builds a new PQ-tree on X representing the circular orderings on X that are suborders of circular orderings in $\mathcal{O}(T)$. The *intersection of T and T', denoted by $T \cap T'$*, builds a new PQ-tree on A representing the circular orderings in $\mathcal{O}(T) \cap \mathcal{O}(T')$.

Simultaneous PQ-Ordering. Let $D = (N, Z)$ be a DAG with vertex set $N = \{T_1, \ldots, T_k\}$, where T_i is a PQ-tree, such that each arc $(T_i, T_j; \phi) \in Z$ consists of a source T_i, of a target T_j, and of an injective map $\phi : \mathcal{L}(T_j) \to \mathcal{L}(T_i)$ from the leaves of T_j to the leaves of T_i. Given an arc $a = (T_i, T_j; \phi) \in Z$ and circular orderings $\mathcal{O}_i \in \mathcal{O}(T_i)$ and $\mathcal{O}_j \in \mathcal{O}(T_j)$, we say that arc a is *satisfied* by $(\mathcal{O}_i, \mathcal{O}_j)$ if $\mathcal{O}_i$ extends $\phi(\mathcal{O}_j)$. The SIMULTANEOUS PQ-ORDERING problem asks to find circular orderings $\mathcal{O}_1 \in \mathcal{O}(T_1), \ldots, \mathcal{O}_k \in \mathcal{O}(T_k)$ on $\mathcal{L}(T_1), \ldots, \mathcal{L}(T_k)$, respectively, such that each arc $(T_i, T_j; \phi) \in Z$ is satisfied by $(\mathcal{O}_i, \mathcal{O}_j)$.

Let $(T_i, T_j; \phi)$ be an arc in Z. An internal node μ_i of T_i is *fixed by* an internal node μ_j of T_j (and μ_j *fixes* μ_i *in* T_i) if there exist leaves $x, y, z \in \mathcal{L}(T_j)$ and $\phi(x), \phi(y), \phi(z) \in \mathcal{L}(T_i)$ such that (i) removing μ_j from T_j makes x, y, and z pairwise disconnected in T_j, and (ii) removing μ_i from T_i makes $\phi(x)$, $\phi(y)$, and $\phi(z)$ pairwise disconnected in T_i. Note that by (i) the three paths connecting μ_j with x, y, and z in T_j share no node other than μ_j, while by (ii) those connecting μ_i with $\phi(x)$, $\phi(y)$, and $\phi(z)$ in T_i share no node other than μ_i. Since any ordering $\mathcal{O}_j$ determines a circular ordering around μ_j of the paths connecting it with x, y, and z in T_j, any ordering $\mathcal{O}_i$ extending $\phi(\mathcal{O}_j)$ determines the same circular ordering around μ_i of the paths connecting it with $\phi(x)$, $\phi(y)$, and $\phi(z)$ in T_i; this is why we say that μ_i is fixed by μ_j.

Theorem 1 below will be a key ingredient in the algorithms of the next section. However, in order to exploit it, we need to consider *normalized* instances of SIMULTANEOUS PQ-ORDERING, namely instances $D = (N, Z)$ such that, for each arc $(T_i, T_j; \phi) \in Z$ and for each internal node $\mu_j \in T_j$, tree T_i contains exactly one node μ_i that is fixed by μ_j. This property can be guaranteed by an operation, called *normalization* [8], defined as follows. Consider each arc $(T_i, T_j; \phi) \in Z$ and replace T_j with $T_i|_{\phi(\mathcal{L}(T_j))} \cap T_j$ in D, that is, replace tree T_j with its intersection with the projection of its parent T_i to the set of leaves of T_i obtained by applying mapping ϕ to the leaves $\mathcal{L}(T_j)$ of T_j.

Consider a normalized instance $D = (N, Z)$. Let μ be a P-node of a PQ-tree T with parents $T_1, \ldots, T_r$ and let $\mu_i \in T_i$ be the unique node in T_i, with $1 \leq i \leq r$, fixed by μ. The *fixedness* $fixed(\mu)$ of μ is defined as $fixed(\mu) = \omega + \sum_{i=1}^{r}(fixed(\mu_i) - 1)$, where ω is the number of children of T fixing μ. A P-node μ is *k-fixed* if $fixed(\mu) \leq k$. Also, instance D is *k-fixed* if all the P-nodes of any PQ-tree $T \in N$ are k-fixed.

Theorem 1 (Bläsius and Rutter [8], Theorems 3.2 and 3.3). *2-fixed instances of* SIMULTANEOUS PQ-ORDERING *can be tested in quadratic time.*

3 Torus Level Planarity

In this section we provide a polynomial-time testing and embedding algorithm for TORUS LEVEL PLANARITY that is based on the following simple observation.

Observation 1. *A proper level graph* $G = (\bigcup_{i=1}^{k} V_i, E, \gamma)$ *is torus level planar if and only if there exist circular orderings* $\mathcal{O}_1, \ldots, \mathcal{O}_k$ *on* $V_1, \ldots, V_k$ *such that, for each* $1 \leq i \leq k$ *with* $k + 1 = 1$, *there exists a radial level embedding of the level graph* $(V_i \cup V_{i+1}, (V_i \times V_{i+1}) \cap E, \gamma)$ *in which the circular orderings on* V_i *along* l_i *and on* V_{i+1} *along* l_{i+1} *are* $\mathcal{O}_i$ *and* $\mathcal{O}_{i+1}$, *respectively.*

In view of Observation 1 we focus on a level graph $G = (V_1 \cup V_2, E, \gamma)$ on two levels l_1 and l_2. Denote by V_1^+ and by V_2^- the subsets of V_1 and of V_2 that are incident to edges in E, respectively. Let Γ be a radial level embedding of G. Consider a closed curve c separating levels l_1 and l_2 and intersecting all the edges in E exactly once. The *edge ordering on E in Γ* is the circular ordering in which the edges in E intersect c according to a clockwise orientation of c on the surface $\mathbb{S}$ of the standing cylinder; see Fig. 3b. Further, let $\mathcal{O}$ be a circular ordering on the edge set E. Ordering $\mathcal{O}$ is *vertex-consecutive (v-consecutive)* if all the edges incident to each vertex in $V_1 \cup V_2$ are consecutive in $\mathcal{O}$.

Let $\mathcal{O}$ be a v-consecutive ordering on E. We define orderings $\mathcal{O}_1^+$ on V_1^+ and $\mathcal{O}_2^-$ on V_2^- *induced by* $\mathcal{O}$, as follows. Consider the edges in E one by one as they appear in $\mathcal{O}$. Append the end-vertex in V_1^+ of the currently considered edge to a list $\mathcal{L}_1^+$. Since $\mathcal{O}$ is v-consecutive, the occurrences of the same vertex appear consecutively in $\mathcal{L}_1^+$, regarding such a list as circular. Hence, $\mathcal{L}_1^+$ can be turned into a circular ordering $\mathcal{O}_1^+$ on V_1^+ by removing repetitions of the same vertex. Circular ordering $\mathcal{O}_2^-$ can be constructed analogously. We have the following.

Lemma 3. *Let $\mathcal{O}$ be a circular ordering on E and $(\mathcal{O}_1, \mathcal{O}_2)$ be a pair of circular orderings on V_1 and V_2. There exists a radial level embedding of G whose edge ordering is $\mathcal{O}$ and such that the circular orderings on V_1 and V_2 along l_1 and l_2 are $\mathcal{O}_1$ and $\mathcal{O}_2$, respectively, if and only if $\mathcal{O}$ is v-consecutive, and $\mathcal{O}_1$ and $\mathcal{O}_2$ extend the orderings $\mathcal{O}_1^+$ and $\mathcal{O}_2^-$ on V_1^+ and V_2^- induced by $\mathcal{O}$, respectively.*

Proof. The necessity is trivial. For the sufficiency, assume that $\mathcal{O}$ is v-consecutive and that $\mathcal{O}_1$ and $\mathcal{O}_2$ extend the orderings $\mathcal{O}_1^+$ and $\mathcal{O}_2^-$ on V_1^+ and V_2^- induced by $\mathcal{O}$, respectively. We construct a radial level embedding Γ of G with the desired properties, as follows. Let Γ^* be a radial level embedding consisting of $|E|$ non-crossing curves, each connecting a distinct point on l_1 and a distinct point on l_2. We associate each curve with a distinct edge in E, so that the edge ordering of Γ^* is $\mathcal{O}$. Note that, since $\mathcal{O}$ is v-consecutive, all the occurrences of the same vertex of V_1^+ and of V_2^- appear consecutively along l_1 and l_2, respectively. Hence, we can transform Γ^* into a radial level embedding Γ' of $G' = (V_1^+ \cup V_2^-, E, \gamma)$, by continuously deforming the curves in Γ^* incident to occurrences of the same vertex in V_1^+ (in V_2^-) so that their end-points on l_1 (on l_2) coincide. Since the circular orderings on V_1^+ and on V_2^- along l_1 and l_2 are $\mathcal{O}_1^+$ and $\mathcal{O}_2^-$, respectively, we can construct Γ by inserting the isolated vertices in $V_1 \setminus V_1^+$ and $V_2 \setminus V_2^-$ at suitable points along l_1 and l_2, so that the circular orderings on V_1 and on V_2 along l_1 and l_2 are $\mathcal{O}_1$ and $\mathcal{O}_2$, respectively. □

We construct an instance $I(G)$ of SIMULTANEOUS PQ-ORDERING starting from a level graph $G = (V_1 \cup V_2, E, \gamma)$ on two levels as follows; refer to Fig. 4, where $I(G)$ corresponds to the subinstance $I(G_{i,i+1})$ in the dashed box. We define the *level trees* T_1 and T_2 as the universal PQ-trees on V_1 and V_2, respectively. Also, we define the *layer tree* $T_{1,2}$ as the PQ-tree on E representing exactly the edge orderings for which a radial level embedding of G exists, which are the v-consecutive orderings on E, by Lemma 3. The PQ-tree $T_{1,2}$ can be constructed in $O(|G|)$ time [9, 19]. We define the *consistency trees* T_1^+ and T_2^- as the universal PQ-trees on V_1^+ and V_2^-, respectively. Instance $I(G)$ contains $T_1, T_2, T_{1,2}, T_1^+$, and T_2^-, together with the arcs (T_1, T_1^+, ι), (T_2, T_2^-, ι), $(T_{1,2}, T_1^+, \phi_1^+)$, and

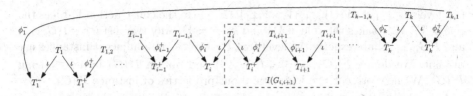

Fig. 4. Instance $I^*(G)$ of SIMULTANEOUS PQ-ORDERING for a level graph $G = (V, E, \gamma)$. Instance $I(G_{i,i+1})$ corresponding to the level graph $(V_i \cup V_{i+1}, (V_i \times V_{i+1}) \cap E, \gamma)$ induced by levels i and $i+1$ of G is enclosed in a dashed box.

$(T_{1,2}, T_2^-, \phi_2^-)$, where ι denotes the identity map and ϕ_1^+ (ϕ_2^-) assigns to each vertex in V_1^+ (in V_2^-) an incident edge in E. We have the following.

Lemma 4. *Level graph G admits a radial level embedding in which the circular ordering on V_1 along l_1 is $\mathcal{O}_1$ and the circular ordering on V_2 along l_2 is $\mathcal{O}_2$ if and only if instance $I(G)$ of* SIMULTANEOUS PQ-ORDERING *admits a solution in which the circular ordering on $\mathcal{L}(T_1)$ is $\mathcal{O}_1$ and the one on $\mathcal{L}(T_2)$ is $\mathcal{O}_2$.*

Proof. We prove the necessity. Let Γ be a radial level embedding of G. We construct an ordering on the leaves of each tree in $I(G)$ as follows. Let $\mathcal{O}_1$, $\mathcal{O}_2$, $\mathcal{O}_1^+$, and $\mathcal{O}_2^-$ be the circular orderings on V_1 along l_1, on V_2 along l_2, on V_1^+ along l_1, and on V_2^- along l_2 in Γ, respectively. Let $\mathcal{O}$ be the edge ordering on E in Γ. Note that $\mathcal{O} \in \mathcal{O}(T_{1,2})$ since $\mathcal{O}$ is v-consecutive by Lemma 3. The remaining trees are universal, hence $\mathcal{O}_1 \in \mathcal{O}(T_1)$, $\mathcal{O}_2 \in \mathcal{O}(T_2)$, $\mathcal{O}_1^+ \in \mathcal{O}(T_1^+)$, and $\mathcal{O}_2^- \in \mathcal{O}(T_2^-)$.

We prove that all arcs of $I(G)$ are satisfied. Arc (T_1, T_1^+, ι) is satisfied if and only if $\mathcal{O}_1$ extends $\mathcal{O}_1^+$. This is the case since ι is the identity map, since $V_1^+ \subseteq V_1$, and since $\mathcal{O}_1$ and $\mathcal{O}_1^+$ are the circular orderings on V_1 and V_1^+ along l_1. Analogously, arc (T_2, T_2^-, ι) is satisfied. Arc $(T_{1,2}, T_1^+, \phi_1^+)$ is satisfied if and only if $\mathcal{O}$ extends $\mathcal{O}_1^+$. This is due to the fact that ϕ_1^+ assigns to each vertex in V_1^+ an incident edge in E and to the fact that, by Lemma 3, ordering $\mathcal{O}$ is v-consecutive and $\mathcal{O}_1^+$ is induced by $\mathcal{O}$. Analogously, arc $(T_{1,2}, T_2^-, \phi_2^-)$ is satisfied.

We prove the sufficiency. Suppose that $I(G)$ is a positive instance of SIMULTANEOUS PQ-ORDERING, that is, there exist orderings $\mathcal{O}_1$, $\mathcal{O}_2$, $\mathcal{O}_1^+$, $\mathcal{O}_2^-$, and $\mathcal{O}$ of the leaves of the trees T_1, T_2, T_1^+, T_2^-, and $T_{1,2}$, respectively, satisfying all arcs of $I(G)$. Since ι is the identity map and since arcs (T_1, T_1^+, ι) and (T_2, T_2^-, ι) are satisfied, we have that $\mathcal{O}_1^+$ and $\mathcal{O}_2^-$ are restrictions of $\mathcal{O}_1$ and $\mathcal{O}_2$ to V_1^+ and V_2^-, respectively. Also, since $(T_{1,2}, T_1^+, \phi_1^+)$ and $(T_{1,2}, T_2^-, \phi_2^-)$ are satisfied, we have that $\mathcal{O}$ extends both $\mathcal{O}_1^+$ and $\mathcal{O}_2^-$. By the construction of $T_{1,2}$, ordering $\mathcal{O}$ is v-consecutive. By Lemma 3, a radial level embedding Γ of G exists in which the circular ordering on V_i along l_i is $\mathcal{O}_i$, for $i = 1, 2$. $\qquad\square$

We now show how to construct an instance $I^*(G)$ of SIMULTANEOUS PQ-ORDERING from a proper level graph $G = (\bigcup_{i=1}^k V_i, E, \gamma)$ on k levels; refer to Fig. 4. For each $i = 1, \ldots, k$, let $I(G_{i,i+1})$ be the instance of SIMULTANEOUS PQ-ORDERING constructed as described above starting from the level graph on two levels $G_{i,i+1} = (V_i \cup V_{i+1}, (V_i \times V_{i+1}) \cap E, \gamma)$ (in the construction V_i takes the role of V_1, V_{i+1} takes the role of V_2, and $k + 1 = 1$). Any two instances $I(G_{i-1,i})$ and $I(G_{i,i+1})$ share exactly the level tree T_i, whereas non-adjacent instances are disjoint. We define $I^{\cup}(G) = \bigcup_{i=1}^{k} I(G_{i,i+1})$ and obtain $I^*(G)$ by normalizing $I^{\cup}(G)$. We now present two lemmata about properties of instance $I^*(G)$.

Lemma 5. *$I^*(G)$ is 2-fixed, has $O(|G|)$ size, and can be built in $O(|G|)$ time.*

Proof. Every PQ-tree T in $I^{\cup}(G)$ is either a source with exactly two children or a sink with exactly two parents, and the normalization of $I^{\cup}(G)$ to obtain $I^*(G)$ does not alter this property. Thus every P-node in a PQ-tree T in $I^*(G)$ is at most 2-fixed. In fact, recall that for a P-node μ of a PQ-tree T with parents $T_1, \ldots, T_r$, we have that $fixed(\mu) = \omega + \sum_{i=1}^{r}(fixed(\mu_i) - 1)$, where ω is the number of children of T fixing μ, and $\mu_i \in T_i$ is the unique node in T_i, with $1 \le i \le r$, fixed by μ. Hence, if T is a source PQ-tree, it holds $\omega = 2$ and $r = 0$; whereas, if T is a sink PQ-tree, it holds $\omega = 0$, $r = 2$, and $fixed(\mu_i) = 2$ for each parent T_i of T. Therefore $I^*(G)$ is 2-fixed.

Since every internal node of a PQ-tree in $I^*(G)$ has degree greater than 2, to prove the bound on $|I^*(G)|$ it suffices to show that the total number of leaves of all PQ-trees in $I^*(G)$ is in $O(|G|)$. Since $\mathcal{L}(T_i) = V_i$ and $\mathcal{L}(T_i^-), \mathcal{L}(T_i^+) \subseteq V_i$, the number of leaves of all level and consistency trees is at most $3 \sum_{i=1}^{k} |V_i| \in O(|G|)$. Also, since $\mathcal{L}(T_{i,i+1}) = (V_i \times V_{i+1}) \cap E$, the number of leaves of all layer trees is at most $\sum_{i=1}^{k} |(V_i \times V_{i+1}) \cap E| \in O(|G|)$. Thus $|I^*(G)| \in O(|G|)$.

We have already observed that each layer tree $T_{i,i+1}$ can be constructed in $O(|G_{i,i+1}|)$ time; level and consistency trees are stars, hence they can be constructed in linear time in the number of their leaves. Finally, the normalization of each arc $(T_i, T_j; \phi)$ can be performed in $O(|T_i| + |T_j|)$ time [8]. Hence, the $O(|G|)$ time bound follows.   □

Lemma 6. *Level graph G admits a torus level embedding if and only if $I^*(G)$ is a positive instance of* SIMULTANEOUS PQ-ORDERING.

Proof. Suppose that G admits a torus level embedding Γ. For $i = 1, \ldots, k$, let $\mathcal{O}_i$ be the circular ordering on V_i along l_i. By Observation 1, embedding Γ determines a radial level embedding $\Gamma_{i,i+1}$ of $G_{i,i+1}$. By Lemma 4, for $i = 1, \ldots, k$, there exists a solution for the instance $I(G_{i,i+1})$ of SIMULTANEOUS PQ-ORDERING in which the circular ordering on $\mathcal{L}(T_i)$ ($\mathcal{L}(T_{i+1})$) is $\mathcal{O}_i$ (resp. $\mathcal{O}_{i+1}$). Since the circular ordering on $\mathcal{L}(T_i)$ is $\mathcal{O}_i$ both in $I(G_{i-1,i})$ and $I(G_{i,i+1})$ and since each arc of $I^*(G)$ is satisfied as it belongs to exactly one instance $I(G_{i,i+1})$, which is a positive instance of SIMULTANEOUS PQ-ORDERING, it follows that the circular orderings deriving from instances $I(G_{i,i+1})$ define a solution for $I^*(G)$.

Suppose that $I^*(G)$ admits a solution. Let $\mathcal{O}_1, \ldots, \mathcal{O}_k$ be the circular orderings on the leaves of the level trees $T_1, \ldots, T_k$ in this solution. By Lemma 4, for each $i = 1, \ldots, k$ with $k + 1 = 1$, there exists a radial level embedding of level graph $G_{i,i+1}$ in which the circular orderings on V_i along l_i and V_{i+1} along l_{i+1} are $\mathcal{O}_i$ and $\mathcal{O}_{i+1}$, respectively. By Observation 1, G is torus level planar.   □

We thus get the main result of this paper.

Theorem 2. TORUS LEVEL PLANARITY *can be tested in quadratic (quartic) time for proper (non-proper) instances.*

Proof. Consider any instance G of TORUS LEVEL PLANARITY. Assume first that G is proper. By Lemmata 5 and 6, a 2-fixed instance $I^*(G)$ of SIMULTANEOUS PQ-ORDERING equivalent to G can be constructed in linear time with $|I^*(G)| \in O(|G|)$. By Theorem 1 instance $I^*(G)$ can be tested in quadratic time.

If G is not proper, then subdivide every edge (u, v) that spans $h > 2$ levels with $h-2$ vertices, assigned to levels $\gamma(u)+1, \gamma(u)+2, \ldots, \gamma(v)-1$. This increases the size of the graph at most quadratically, and the time bound follows.   □

Theorem 2 and Lemma 2 imply the following result.

Theorem 3. CYCLIC LEVEL PLANARITY *can be solved in quadratic (quartic) time for proper (non-proper) instances.*

Our techniques allow us to solve a more general problem, that we call TORUS $\mathcal{T}$-LEVEL PLANARITY, in which a level graph $G = (\bigcup_{i=1}^{k} V_i, E, \gamma)$ is given together with a set of PQ-trees $\mathcal{T} = \{\overline{T}_1, \ldots, \overline{T}_k\}$ such that $\mathcal{L}(\overline{T}_i) = V_i$, where each tree $\overline{T}_i$ encodes consecutivity constraints on the ordering on V_i along l_i. The goal is then to test the existence of a level embedding of G on $\mathbb{T}$ in which the circular ordering on V_i along l_i belongs to $\mathcal{O}(\overline{T}_i)$. This problem has been studied in the plane [2,24] under the name of $\mathcal{T}$-LEVEL PLANARITY; it is NP-hard in general and polynomial-time solvable for proper instances. While the former result implies the NP-hardness of TORUS $\mathcal{T}$-LEVEL PLANARITY, the techniques of this paper show that TORUS $\mathcal{T}$-LEVEL PLANARITY can be solved in polynomial time for proper instances. Namely, in the construction of instance $I^*(G)$ of SIMULTANEOUS PQ-ORDERING, it suffices to replace level tree T_i with PQ-tree $\overline{T}_i$. Analogous considerations allow us to extend this result to RADIAL $\mathcal{T}$-LEVEL PLANARITY and CYCLIC $\mathcal{T}$-LEVEL PLANARITY.

4   Simultaneous Level Planarity

In this section we prove that SIMULTANEOUS LEVEL PLANARITY is NP-complete for two graphs on three levels and for three graphs on two levels, while it is polynomial-time solvable for two graphs on two levels.

Both NP-hardness proofs rely on a reduction from the NP-complete problem BETWEENNESS [21], which asks for a ground set S and a set X of ordered triplets of S, with $|S| = n$ and $|X| = k$, whether a linear order $\prec$ of S exists such that, for any $(\alpha, \beta, \gamma) \in X$, it holds true that $\alpha \prec \beta \prec \gamma$ or that $\gamma \prec \beta \prec \alpha$. Both proofs exploit the following gadgets.

The *ordering gadget* is a pair $\langle G_1, G_2 \rangle$ of level graphs on levels l_1 and l_2, where l_1 contains nk vertices $u_{i,j}$, with $i = 1, \ldots, k$ and $j = 1, \ldots, n$, and l_2 contains $n(k-1)$ vertices $v_{i,j}$, with $i = 1, \ldots, k-1$ and $j = 1, \ldots, n$. For $i = 1, \ldots, k-1$ and $j = 1, \ldots, n$, G_1 contains edge $(u_{i,j}, v_{i,j})$ and G_2 contains edge $(u_{i+1,j}, v_{i,j})$.

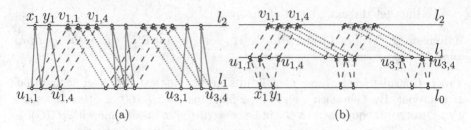

Fig. 5. Instances (a) $\langle G_1, G_2, G_3 \rangle$ and (b) $\langle G_1, G_2 \rangle$ corresponding to an instance of BETWEENNESS with $X = \{(u_{1,1}, u_{1,2}, u_{1,4}), (u_{1,2}, u_{1,3}, u_{1,4}), (u_{1,1}, u_{1,3}, u_{1,4})\}$.

See G_1 and G_2 in Fig. 5(a). Consider any simultaneous level embedding Γ of $\langle G_1, G_2 \rangle$ and assume, w.l.o.g. up to a renaming, that $u_{1,1}, \dots, u_{1,n}$ appear in this left-to-right order along l_1.

Lemma 7. *For every $i = 1, \dots, k$, vertices $u_{i,1}, \dots, u_{i,n}$ appear in this left-to-right order along l_1 in Γ; also, for every $i = 1, \dots, k-1$, vertices $v_{i,1}, \dots, v_{i,n}$ appear in this left-to-right order along l_2 in Γ.*

The *triplet gadget* is a path $T = (w_1, \dots, w_5)$ on two levels, where w_1, w_3, and w_5 belong to a level l_i and w_2 and w_4 belong to a level $l_j \neq l_i$. See G_3 in Fig. 5(a), with $i = 1$ and $j = 2$. We have the following.

Lemma 8. *In every level embedding of T, vertex w_3 is between w_1 and w_5 along l_i.*

We are now ready to prove the claimed NP-completeness results.

Theorem 4. SIMULTANEOUS LEVEL PLANARITY *is NP-complete even for three graphs on two levels and for two graphs on three levels.*

Proof. Both problems clearly are in $\mathcal{NP}$. We prove the $\mathcal{NP}$-hardness only for three graphs on two levels (see Fig. 5(a)), as the other proof is analogous (see Fig. 5(b)). We construct an instance $\langle G_1(V, E_1, \gamma), G_2(V, E_2, \gamma), G_3(V, E_3, \gamma) \rangle$ of SIMULTANEOUS LEVEL PLANARITY from an instance $(S = \{u_{1,1}, \dots, u_{1,n}\}, X = \{(u_{1,a_i}, u_{1,b_i}, u_{1,c_i}) : i = 1, \dots, k\})$ of BETWEENNESS as follows: Pair $\langle G_1, G_2 \rangle$ contains an ordering gadget on levels l_1 and l_2, where the vertices $u_{1,1}, \dots, u_{1,n}$ of G_1 are (in bijection with) the elements of S. Graph G_3 contains k triplet gadgets $T_i(u_{i,a_i}, x_i, u_{i,b_i}, y_i, u_{i,c_i})$, for $i = 1, \dots, k$. Vertices $x_1, y_1, \dots, x_k, y_k$ are all distinct and are on l_2. Clearly, the construction can be carried out in linear time. We prove the equivalence of the two instances.

($\Longrightarrow$) Suppose that a simultaneous level embedding Γ of $\langle G_1, G_2, G_3 \rangle$ exists. We claim that the left-to-right order of $u_{1,1}, \dots, u_{1,n}$ along l_1 satisfies the betweenness constraints in X. Suppose, for a contradiction, that there exists a triplet $(u_{1,a_i}, u_{1,b_i}, u_{1,c_i}) \in X$ with u_{1,b_i} not between u_{1,a_i} and u_{1,c_i} along l_1. By Lemma 7, u_{i,b_i} is not between u_{i,a_i} and u_{i,c_i}. By Lemma 8, we have that $T_i(u_{i,a_i}, x_i, u_{i,b_i}, y_i, u_{i,c_i})$ is not planar in Γ, a contradiction.

($\Longleftarrow$) Suppose that (S, X) is a positive instance of BETWEENNESS, and assume, w.l.o.g. up to a renaming, that $\prec := u_{1,1}, \ldots, u_{1,n}$ is a solution for (S, X). Construct a straight-line simultaneous level planar drawing of $\langle G_1, G_2, G_3 \rangle$ with: (i) vertices $u_{1,1}, \ldots, u_{1,n}, \ldots, u_{k,1}, \ldots, u_{k,n}$ in this left-to-right order along l_1, (ii) vertices $v_{1,1}, \ldots, v_{1,n}, \ldots, v_{k-1,1}, \ldots, v_{k-1,n}$ in this left-to-right order along l_2, (iii) vertices x_i and y_i to the left of vertices x_{i+1} and y_{i+1}, for $i = 1, \ldots, k-1$, and (iv) vertex x_i to the left of vertex y_i if and only if $u_{1,a_i} \prec u_{1,c_i}$.

Properties (i) and (ii) guarantee that, for any two edges $(u_{i,j}, v_{i,j})$ and $(u_{i',j'}, v_{i',j'})$, vertex $u_{i,j}$ is to the left of $u_{i',j'}$ along l_1 if and only if $v_{i,j}$ is to the left of $v_{i',j'}$ along l_2, which implies the planarity of G_1 in Γ. The planarity of G_2 in Γ is proved analogously. Properties (i) and (iii) imply that no two paths T_i and T_j cross each other, while Property (iv) guarantees that each path T_i is planar. Hence, the drawing of G_3 in Γ is planar.   □

The graphs in Theorem 4 can be made connected, by adding vertices and edges, at the expense of using one additional level. Also, the theorem holds true even if the simultaneous embedding is *geometric* or *with fixed edges* (see [7,11] for definitions).

In contrast to the NP-hardness results, a reduction to a proper instance of CYCLIC LEVEL PLANARITY allows us to decide in polynomial time instances composed of two graphs on two levels. Namely, the edges of a graph are directed from l_1 to l_2, while those of the other graph are directed from l_2 to l_1.

Theorem 5. SIMULTANEOUS LEVEL PLANARITY *is quadratic-time solvable for two graphs on two levels.*

5   Conclusions and Open Problems

In this paper we have settled the computational complexity of two of the main open problems in the research topic of level planarity by showing that the CYCLIC LEVEL PLANARITY and the TORUS LEVEL PLANARITY problems are polynomial-time solvable. Our algorithms run in quartic time in the graph size; it is hence an interesting challenge to design new techniques to improve this time bound. We also introduced a notion of simultaneous level planarity for level graphs and we established a complexity dichotomy for this problem.

An intriguing research direction is the one of extending the concept of level planarity to surfaces with genus larger than one. However, there seems to be more than 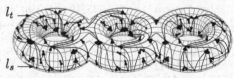 one meaningful way to arrange k levels on a high-genus surface. A reasonable choice would be the one shown in the figure, in which the levels are arranged in different sequences between two distinguished levels l_s and l_t (and edges only connect vertices on two levels in the same sequence). RADIAL LEVEL PLANARITY and TORUS LEVEL PLANARITY can be regarded as special cases of this setting (with only one and two paths of levels between l_s and l_t, respectively).

References

1. Angelini, P., Da Lozzo, G., Di Battista, G., Frati, F., Patrignani, M., Rutter, I.: Beyond level planarity. CoRR abs/1510.08274v3 (2015). http://arxiv.org/abs/1510.08274v3

2. Angelini, P., Da Lozzo, G., Di Battista, G., Frati, F., Roselli, V.: The importance of being proper: (in clustered-level planarity and T-level planarity). Theor. Comput. Sci. **571**, 1–9 (2015). http://dx.doi.org/10.1016/j.tcs.2014.12.019

3. Auer, C., Bachmaier, C., Brandenburg, F.J., Gleißner, A.: Classification of planar upward embedding. In: Kreveld, M., Speckmann, B. (eds.) GD 2011. LNCS, vol. 7034, pp. 415–426. Springer, Heidelberg (2012). doi:10.1007/978-3-642-25878-7_39

4. Bachmaier, C., Brandenburg, F., Forster, M.: Radial level planarity testing and embedding in linear time. J. Graph Algorithms Appl. **9**(1), 53–97 (2005). http://jgaa.info/accepted/2005/BachmaierBrandenburgForster2005.9.1.pdf

5. Bachmaier, C., Brunner, W.: Linear time planarity testing and embedding of strongly connected cyclic level graphs. In: Halperin, D., Mehlhorn, K. (eds.) ESA 2008. LNCS, vol. 5193, pp. 136–147. Springer, Heidelberg (2008). doi:10.1007/978-3-540-87744-8_12

6. Bachmaier, C., Brunner, W., König, C.: Cyclic level planarity testing and embedding. In: Hong, S.-H., Nishizeki, T., Quan, W. (eds.) GD 2007. LNCS, vol. 4875, pp. 50–61. Springer, Heidelberg (2008). doi:10.1007/978-3-540-77537-9_8

7. Bläsius, T., Kobourov, S.G., Rutter, I.: Simultaneous embeddings of planar graphs (Chap. 11). In: Tamassia, R. (ed.) Handbook of Graph Drawing and Visualization. Discrete Mathematics and Its Applications, pp. 349–382. Chapman and Hall/CRC, Boca Raton (2013)

8. Bläsius, T., Rutter, I.: Simultaneous PQ-ordering with applications to constrained embedding problems. ACM Trans. Algorithms **12**(2), 16 (2016). http://doi.acm.org/10.1145/2738054

9. Booth, K.S., Lueker, G.S.: Testing for the consecutive ones property, interval graphs, and graph planarity using PQ-tree algorithms. J. Comput. Syst. Sci. **13**(3), 335–379 (1976). http://dx.doi.org/10.1016/S0022-0000(76)80045-1

10. Brandenburg, F.: Upward planar drawings on the standing and the rolling cylinders. Comput. Geom. **47**(1), 25–41 (2014). http://dx.doi.org/10.1016/j.comgeo.2013.08.003

11. Braß, P., Cenek, E., Duncan, C.A., Efrat, A., Erten, C., Ismailescu, D., Kobourov, S.G., Lubiw, A., Mitchell, J.S.B.: On simultaneous planar graph embeddings. Comput. Geom. **36**(2), 117–130 (2007). http://dx.doi.org/10.1016/j.comgeo.2006.05.006

12. Di Battista, G., Nardelli, E.: Hierarchies and planarity theory. IEEE Trans. Syst. Man Cybern. **18**(6), 1035–1046 (1988). http://dx.doi.org/10.1109/21.23105

13. Estrella-Balderrama, A., Fowler, J.J., Kobourov, S.G.: On the characterization of level planar trees by minimal patterns. In: Eppstein, D., Gansner, E.R. (eds.) GD 2009. LNCS, vol. 5849, pp. 69–80. Springer, Heidelberg (2010). doi:10.1007/978-3-642-11805-0_9

14. Forster, M., Bachmaier, C.: Clustered level planarity. In: Emde Boas, P., Pokorný, J., Bieliková, M., Štuller, J. (eds.) SOFSEM 2004. LNCS, vol. 2932, pp. 218–228. Springer, Heidelberg (2004). doi:10.1007/978-3-540-24618-3_18

15. Fulek, R., Pelsmajer, M.J., Schaefer, M., Štefankovič, D.: Hanani-Tutte, monotone drawings, and level-planarity. In: Pach, J. (ed.) Thirty Essays on Geometric Graph Theory, pp. 263–287. Springer, New York (2013). http://dx.doi.org/10.1007/978-1-4614-0110-0_14

16. Haeupler, B., Raju Jampani, K., Lubiw, A.: Testing simultaneous planarity when the common graph is 2-connected. J. Graph Algorithms Appl. **17**(3), 147–171 (2013). http://dx.doi.org/10.7155/jgaa.00289

17. Healy, P., Kuusik, A., Leipert, S.: A characterization of level planar graphs. Discrete Math. **280**(1–3), 51–63 (2004). http://dx.doi.org/10.1016/j.disc.2003.02.001

18. Heath, L.S., Pemmaraju, S.V.: Recognizing leveled-planar dags in linear time. In: Brandenburg, F.J. (ed.) GD 1995. LNCS, vol. 1027, pp. 300–311. Springer, Heidelberg (1996). doi:10.1007/BFb0021813

19. Hsu, W., McConnell, R.M.: PQ trees, PC trees, and planar graphs. In: Mehta, D.P., Sahni, S. (eds.) Handbook of Data Structures and Applications. Chapman and Hall/CRC, Boca Raton (2004). http://dx.doi.org/10.1201/9781420035179

20. Jünger, M., Leipert, S., Mutzel, P.: Level planarity testing in linear time. In: Whitesides, S.H. (ed.) GD 1998. LNCS, vol. 1547, pp. 224–237. Springer, Heidelberg (1998). doi:10.1007/3-540-37623-2_17

21. Opatrny, J.: Total ordering problem. SIAM J. Comput. **8**(1), 111–114 (1979)

22. Randerath, B., Speckenmeyer, E., Boros, E., Hammer, P.L., Kogan, A., Makino, K., Simeone, B., Cepek, O.: A satisfiability formulation of problems on level graphs. Electron. Notes Discrete Math. **9**, 269–277 (2001). http://dx.doi.org/10.1016/S1571-0653(04)00327-0

23. Sugiyama, K., Tagawa, S., Toda, M.: Methods for visual understanding of hierarchical system structures. IEEE Trans. Syst. Man Cybern. **11**(2), 109–125 (1981). http://dx.doi.org/10.1109/TSMC.1981.4308636

24. Wotzlaw, A., Speckenmeyer, E., Porschen, S.: Generalized k-ary tanglegrams on level graphs: a satisfiability-based approach and its evaluation. Discrete Appl. Math. **160**(16–17), 2349–2363 (2012). http://dx.doi.org/10.1016/j.dam.2012.05.028

Special Graph Embeddings

Track Layout Is Hard

Michael J. Bannister[1], William E. Devanny[2($\boxtimes$)], Vida Dujmović[3],
David Eppstein[2], and David R. Wood[4]

[1] Department of Mathematics and Computer Science,
Santa Clara University, Santa Clara, CA, USA
mbannister@fastmail.fm
[2] Department of Computer Science,
University of California, Irvine, Irvine, CA, USA
{wdevanny,eppstein}@uci.edu
[3] School of Computer Science and Electrical Engineering,
University of Ottawa, Ottawa, Canada
vida.dujmovic@uottawa.ca
[4] School of Mathematical Sciences, Monash University, Melbourne, Australia
david.wood@monash.edu

Abstract. We show that testing whether a given graph has a 3-track
layout is hard, by characterizing the bipartite 3-track graphs in terms
of leveled planarity. Additionally, we investigate the parameterized complexity of track layouts, showing that past methods used for book layouts do not work to parameterize the problem by treewidth or almost-
tree number but that the problem is (non-uniformly) fixed-parameter
tractable for tree-depth. We also provide several natural classes of bipartite planar graphs, including the bipartite outerplanar graphs, squaregraphs, and dual graphs of arrangements of monotone curves, that always
have 3-track layouts.

1 Introduction

A *k-track layout* of a graph is a partition of the vertices into k ordered independent sets called *tracks*, and a partition of the edges into non-crossing subsets that
connect pairs of tracks. The *track-number* of a graph is the minimum k for which
it has a k-track layout. Track layouts are connected with the existence of low-
volume three-dimensional graph drawings: a graph has a three-dimensional drawing in an $O(1) \times O(1) \times O(n)$ grid if and only if it has track-number $O(1)$ [1,2].

Already in 2004, Dujmović et al. [3] asked whether it is computationally feasible to construct optimal track layouts. A graph has track-number 2 if and only
if it is a forest of caterpillars [3]. So we can efficiently recognize and construct
optimal track layouts for track-number 2 graphs. In this paper we show that the

Michael J. Bannister and David Eppstein were supported in part by NSF grant
CCF-1228639. William E. Devanny was supported by an NSF Graduate Research
Fellowship under grant DGE-1321846. Vida Dujmović was supported by NSERC.
David Wood was supported by the Australian Research Council.

© Springer International Publishing AG 2016
Y. Hu and M. Nöllenburg (Eds.): GD 2016, LNCS 9801, pp. 499–510, 2016.
DOI: 10.1007/978-3-319-50106-2_38

answer to the general question is negative: even recognizing the graphs with 3-track layouts is NP-complete. Our proof is based on the known NP-completeness of level planarity [4], and uses a new characterization of the bipartite graphs with 3-track layouts as being exactly the leveled planar graphs, undirected graphs that can be given a Sugiyama-style layered graph drawing with no crossings and no dummy vertices.

Additionally, we show that known methods of obtaining fixed-parameter tractable algorithms for other types of planar embedding, based on Courcelle's theorem for treewidth [5], or on kernelization of the 2-core for k-almost-trees [6], do not generalize to track number. However, for any fixed bound on the tree-depth of an input graph, the track number can be obtained in linear time.

We also provide several natural classes of bipartite planar graphs, including the bipartite outerplanar graphs, squaregraphs, and dual graphs of arrangements of monotone curves, that always have 3-track layouts.

2   Definitions

A *track layout* of a graph is a partition of its vertices into sequences, called *tracks*, such that the vertices in each sequence form an independent set and the edges between each pair of tracks form a non-crossing set. This means that there do not exist edges uv and $u'v'$ such that u is before u' in one track, but v is after v' in another track; such a pair of edges is said to form a *crossing*. (This ordering constraint on endpoints of pairs of edges connecting two tracks is the same as the constraint on the left-to-right ordering within levels on the endpoints of two edges connecting the same two levels of a layered drawing.)

The *track-number* of a graph G is the minimum number of tracks in a track layout of G; this is finite, since the layout in which each vertex forms its own track is always non-crossing. The set of edges between two tracks form a forest of caterpillars (a forest in which the non-leaf vertices of each component induce a path); in particular, the graphs with track-number 1 are the independent sets, and the graphs with track-number 2 are the forests of caterpillars [7].

A *tree-decomposition* of a graph G is given by a tree T whose nodes index a collection $(B_x \subseteq V(G) : x \in V(T))$ of sets of vertices in G called *bags*, such that:

– For every edge vw of G, some bag B_x contains both v and w, and
– For every vertex v of G, the set $\{x \in V(T) : v \in B_x\}$ induces a non-empty (connected) subtree of T.

The *width* of a tree-decomposition is $\max_x |B_x| - 1$, and the *treewidth* of a graph G is the minimum width of any tree decomposition of G. Treewidth was introduced (with a different but equivalent definition) by Halin [8] and tree decompositions were introduced by Robertson and Seymour [9].

A *layering* of a graph is a partition of the vertices into a sequence of disjoint subsets (called *layers*) such that each edge connects vertices in the same layer or consecutive layers. One way, but not the only way, to obtain a layering is the *breadth first layering* in which we partition the vertices by their distances from a fixed starting vertex, using breadth-first search [10,11].

The class of *leveled planar graphs* was introduced in 1992 by Heath and Rosenberg [4] in their study of queue layouts of graphs. A leveled planar drawing of a graph is a planar drawing in which the vertices are placed on a collection of parallel lines, and each edge must connect vertices in two consecutive parallel lines. Another equivalent way to state this is that this kind of drawing is a Sugiyama-style layered drawing [12] that achieves perfect quality according to two of the most important quality measures for the drawing, the number of edge crossings [13] and the number of dummy vertices [14].

3   Track Layouts and Leveled Planarity

We begin by demonstrating an equivalence between leveled planarity and bipartite 3-track layout.

Lemma 1 (implicit in [15]). *Every leveled planar graph has a 3-track layout.*

Proof. Assign the vertices of the graph to tracks according to the number of their level in the layered drawing, modulo 3, as shown in Fig. 1. Within each track, order the vertices within each level contiguously, and order the levels by their positions in the layered drawing. Two edges that connect the same pair of levels cannot cross because of the chosen vertex ordering within the levels, and two edges that connect different pairs of levels but are mapped to the same pair of tracks cannot cross because of the ordering of the levels within the tracks. □

Lemma 1 can be interpreted as 'wrapping' a layered drawing on to 3 tracks; see [3] for a more general wrapping lemma. As Fig. 1 shows, a 3-track layout can also be interpreted geometrically, as a planar drawing in which the tracks are represented as three rays from the origin; it follows from this interpretation that 3-track graphs have universal point sets of size $O(n)$, consisting of n points on each ray. However, for more than three tracks, a similar embedding of the tracks as rays in the plane would not lead to a planar drawing, because

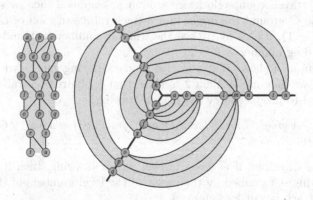

Fig. 1. Converting a layered drawing to a 3-track layout

there is no requirement that edges of the graph connect only consecutive rays. Indeed, all graphs (for example, arbitrarily large complete graphs) have 4-track subdivisions [16].

Define an *arc* of an undirected graph G to be a directed edge formed by orienting one of the edges of G. For a graph G with a 3-track layout, define a function δ from arcs to ± 1 as follows: if an arc uv goes from track i to track $i+1$ (mod 3) (that is, if it is oriented clockwise in the planar embedding described above), let $\delta(uv) = +1$; otherwise (if it is oriented counterclockwise), let $\delta(uv) = -1$. For an oriented cycle C, we define (by abuse of notation) $\delta(C) = \sum_{uv \in C} \delta(uv)$.

Lemma 2. *Let C be a cycle embedded in a 3-track layout. Cyclically orient the edges of C. If C is even then $\delta(C) = 0$. If C is odd then $|\delta(C)| = 3$. (Here $|x|$ is the absolute value of x.)*

Proof. We proceed by induction on $|C| := |V(C)|$. If $|C| = 3$, then C has one vertex on each track and $\delta(C) \in \{3, -3\}$. If $|C| = 4$, then C has two edges with $\delta = +1$ and two edges with $\delta = -1$, implying $\delta(C) = 0$. Now assume that $|C| \geq 5$. Use the 3-track layout to embed C in the plane as described above, but with straight edges instead of the curved edges shown in the figure. As a planar polygon, C has at least two ears, triangles formed by two of its edges that are empty of other vertices of C (which may be found as the leaf edges in the tree formed as the dual graph of a triangulation of C). If one ear has the same sign of δ for both of the edges that form it, these edges must connect pairs of vertices that are the innermost on their tracks. Therefore, two such ears with same-sign edges could only exist if C is a triangle. For any longer cycle, let uvw be an ear for which $\delta(uv) = -\delta(vw)$; thus edges uv and vw both connect the same two tracks, and (by the assumption that triangle uvw is empty) u and w are consecutive in their track. By deleting v and merging uw into a single vertex, we construct a cycle C' with $|C'| = |C| - 2$, and a 3-track layout of C' with $\delta(C') = \delta(C)$. The result follows by induction. ☐

The previous lemma can be restated in terms of winding number. The *winding number* of a closed curve C in the plane around a given point x is the number of times that C travels counterclockwise around x. Lemma 2 then says that for an oriented cycle C around the origin in a 3-track representation of C with three rays (as in Fig. 1), if C is even then the winding number is 0, and if C is odd then the winding number is 1.

While Lemma 1 shows that a leveled planar drawing can be wrapped on to three tracks, we now use Lemma 2 to show that a bipartite 3-track layout can be unwrapped to produce a leveled planar drawing.

Theorem 1. *A graph G has a leveled planar drawing if and only if G is bipartite and has a 3-track layout.*

Proof. In one direction, if G has a leveled planar drawing, then it is bipartite (with a coloring determined by the parity of the level numbers of the drawing) and has a 3-track layout by Lemma 1.

In the other direction, suppose that G is bipartite and has a 3-track layout. We may assume without loss of generality that G is connected, for otherwise we can draw each connected component of G separately; let T be a spanning tree of G, and let v be an arbitrary vertex of G. Assign v to level zero of a layered drawing, and assign each other vertex w to the level given by the sum of the numbers $\delta(xy)$ for the edges xy of the oriented path from v to w in T. (Some of these level numbers may be negative.) By construction, the endpoints of each edge of T are assigned to consecutive levels, and by applying Lemma 2 to the oriented cycle formed by a non-tree edge together with the tree path connecting its endpoints, the same can be shown to be true of each edge of $G - E(T)$.

Within each level of the drawing, the vertices all come from the same track, determined by the value of the level modulo 3. Assign the vertices to positions in left-to-right order on this level according to their ordering within this track. Then no two consecutive levels of the drawing can have crossing edges, because such a crossing would also be a crossing in the track layout. Therefore, this assignment of vertices to levels and to positions within these levels gives a leveled planar drawing of G. □

Theorem 2. *Testing whether a given graph has a k-track layout for any constant $k \geq 3$ is NP-complete.*

Proof. For $k = 3$ this follows from Theorem 1 and from the known NP-completeness of level planarity, proven by Heath and Rosenberg [4]. For $k > 3$ this follows by adding $k - 3$ additional vertices, adjacent to all other vertices, to a hard instance of the 3-track layout problem. □

4 Parameterized Complexity

A fixed-parameter tractable problem is also *strongly uniform fixed-parameter tractable*. A problem is *uniformly fixed-parameter tractable* if there is an algorithm that solves it in polynomial time for any value of the parameter, but we cannot compute the dependence on the parameter. Lastly a problem is *non-uniformly fixed-parameter tractable* if there is a collection of algorithms such that for each possible value of the parameter one of the algorithms solves the problem in polynomial time.

4.1 Treewidth

We sketch an argument as to why it is not possible to use Courcelle's Theorem (or any automata methods based on tree decompositions) to produce a fixed-parameter tractable algorithm for leveled planarity with respect to treewidth. Consider the family of graphs depicted in Fig. 2. These graphs have bounded treewidth (in fact pathwidth at most 12) and are leveled planar precisely when $p = q$. However, since p and q are unbounded it is necessary to carry more than a finite amount of state between bags in a treewidth decomposition when parsing the decomposition. Thus, the decompositions corresponding to leveled planar graphs cannot be recognized by automata or methods using automata such as

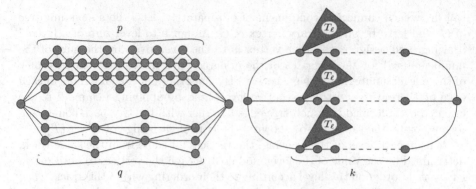

Fig. 2. A family of graphs with bounded treewidth demonstrating that the family of leveled planar graphs is not finite state.

Fig. 3. A family of 2-almost trees for which the standard kernelization cannot decide leveled planarity. The subgraphs T_ℓ are complete binary trees of depth ℓ.

Courcelle's Theorem. This intuitive argument is made formal below using the Myhill-Nerode Theorem for tree automata below.

Following Downey and Fellows [17], we define a *t-boundaried graph* to be a graph G with t designated *boundary* vertices labeled $1, 2, \ldots, t$. Given two t-boundaried graphs G_1 and G_2 we define their *gluing* $G_1 \oplus G_2$ by identifying each boundary vertex of G_1 with the boundary vertex of G_2 having the same label.

An *n-ary t-boundaried operator* $\otimes$ consists of a t-boundaried graph $G_\otimes = (V_\otimes, E_\otimes)$ and injections $f_i : \{1, \ldots, t\} \to V_\otimes$ for $1 \leq i \leq n$. Then for t-boundaried graphs $G_1, \ldots, G_n$ we define the t-boundaried graph $G_1 \otimes \cdots \otimes G_n$ by gluing each G_i to $G_\otimes$ after applying f_i to the boundary labels of $G_\otimes$. After the gluing the labels of G_i are forgotten.

It can be shown that there exists a *standard set* of t-boundaried operators on t-boundaried graphs that can be used to generate the set of all graphs of treewidth t. Furthermore, it is possible to convert (in linear time) a tree decomposition of width t into a parse tree that uses these standard operators; see Theorem 12.7.1 in [17]. Define $\mathcal{U}_t^{small}$ to be the *small universe* of t-boundaried graphs obtained by parse trees, using these standard operators. Given a family of graphs F, we define the equivalence relation $\sim_F$ on $\mathcal{U}_t^{small}$, such that $G_1 \sim_F G_2$ if and only if for all $H \in \mathcal{U}_t^{small}$, we have $G_1 \oplus H \in F \Leftrightarrow G_2 \oplus H \in F$.

A family of graphs F is said to be *t-finite state* if the family of parse trees for graphs in $F_t = F \cap \mathcal{U}_t^{small}$ is finite state. Equivalently, such a family of parse trees may be recognized by a finite tree automaton. We can now state the analog of the Myhill–Nerode Theorem (characterizing recognizability of sets of strings by finite state machines) for treewidth t graphs in place of strings and finite tree automata in place of finite state machines.

Lemma 3 (Theorem 12.7.2 of [17]). *Let F be a family of graphs. Then F is t-finite state if and only if $\sim_F$ has finite index over $\mathcal{U}_t^{small}$.*

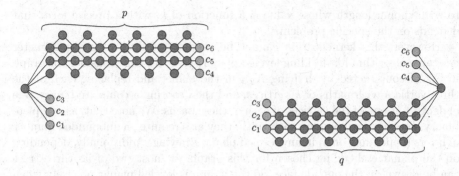

Fig. 4. The 6-boundaried graphs U_p (left) and L_q (right) from the proof of Theorem 3.

As we now show, leveled planarity is not t-finite state when t is sufficiently large.

Theorem 3. *For all $t \geq 6$, the families of leveled planar graphs and of 3-track graphs are not t-finite state.*

Proof. Let F be the family of leveled planar graphs. It suffices to prove the theorem in the case when $t = 6$. Consider the 6-boundaried graphs U_p and L_q shown in Fig. 4, and observe that $U_p \oplus L_q$ is leveled planar if and only if $p = q$. So $U_p \sim_F U_\ell$ if and only if $p = \ell$, which implies that $\sim_{F_6}$ does not have finite index and that in turn F is not 6-finite state by Lemma 3.   □

Theorem 3 implies that (when $t \geq 6$) the parse trees of leveled planar graphs with treewidth t are not recognizable by tree automata. Therefore automata-based methods such as Courcelle's Theorem cannot be used to show leveled planarity to be fixed-parameter tractable with respect to treewidth. In particular, leveled planarity cannot be expressed using the forms of monadic second-order graph logic to which Courcelle's Theorem applies.

4.2 Almost-Trees

The *cyclomatic number* (also called *circuit rank*) of a graph is defined to be $r = m - n + c$ where m is the number of edges, n is the number of vertices, and c is the number of connected components in the graph. We say that a graph G is a k-*almost tree* if every biconnected component of G has cyclomatic number at most k. The problems of 1-page and 2-page crossing minimization and testing 1-planarity were shown to be fixed-parameter tractable with respect to the k-almost tree parameter, via the kernelization method [6,18].

In these previous papers, the "standard kernelization" used for a k-almost tree G is constructed by first iteratively removing degree one vertices until no more remain, leaving what is called the *2-core* of G. The 2-core consists of vertices of degree greater than two and paths of degree two vertices connecting these high degree vertices. The paths of degree two vertices are then shortened to a maximum length whose value is a function of k, with a precise form that depends on the specific problem.

However, this kernelization cannot be used to produce a fixed-parameter tractable algorithm for deciding leveled planarity. To see this, consider the graph in Fig. 3, constructed by drawing $K_{2,3}$ in the plane, and replacing each of the three vertices with paths of k vertices, and then rooting a complete binary tree of depth ℓ at one of the vertices of each of these paths. We note that, as complete binary trees have unbounded pathwidth, they also require an unbounded number of layers (depending on ℓ) in any leveled planar drawing. Additionally, depending on the planar embedding chosen for this graph, at most two of its three trees can be drawn on the outside face. So this graph is leveled planar precisely when ℓ is small enough for the remaining tree T_ℓ to be drawn within one of the two bounded faces of the drawing, i.e., the leveled planarity of the graph depends on the relationship between k and ℓ. Since this relationship is not preserved in the kernelization it can not be used to produce a fixed-parameter tractable algorithm for leveled planarity.

4.3   Tree-Depth

The *tree-depth* of a graph G is the minimum height of a forest of rooted trees on the same vertex set as G such that edges in G only go from ancestors to descendants in the forest. It is bounded by pathwidth, and therefore by track-number: track-number$(G) \leq$ pathwidth$(G) + 1 \leq$ tree-depth(G); see [1,19].

Theorem 4. *Computing the track-number of a graph G is non-uniformly fixed-parameter linear in the tree-depth of G.*

Proof. Track-number and layered pathwidth are both monotone (cannot increase) under taking induced subgraphs. The graphs with tree-depth bounded by a constant are well-quasi-ordered under taking induced subgraphs and so for any fixed bound on tree-depth and either track-number or layered pathwidth there exist only a finite number of forbidden induced subgraphs [19]. Since the track-number and pathwidth are both bounded by the tree-depth, the same is true for any fixed bound on tree-depth, regardless of track-number or layered pathwidth.

Because induced subgraph testing is linear time for graphs with tree-depth bounded by a fixed number d, we can for each $t \leq d$ test if the graph has any of the forbidden induced subgraphs to track-number t each in linear time [19].   □

However, this argument does not tell us how to find the set of forbidden induced subgraphs, nor what the dependence of the time bound on the tree-depth is. It would be of interest to replace this existence proof with a more constructive algorithm.

5   Special Classes of Graphs

We consider here particular graph families such as the outerplanar graphs, and prove that these families are leveled planar. Our results are based on

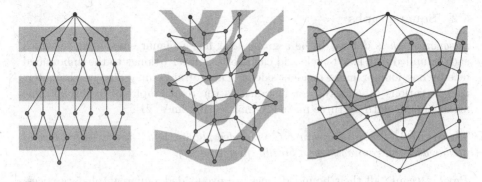

Fig. 5. Examples of graphs with planar breadth-first layerings (start vertex shown in red, and layering in yellow): left, a bipartite outerplanar graph (Theorem 5); center, a squaregraph (Theorem 6); and right, the dual graph of an arrangement of doubly-unbounded monotonic curves (Theorem 7). (Color figure online)

breadth-first layerings; we define a layering of a graph to be *planar* if there exists a non-crossing layered drawing of the graph in which the layers of the drawing are the same as the layers of the layering.

5.1   Bipartite Outerplanar Graphs

Theorem 5 (implicit in [15]). *Every bipartite outerplanar graph is leveled planar and 3-track. Every breadth first layering of such a graph G gives a leveled planar drawing.*

Proof. Let v be the starting vertex of a breadth first layering. Then for each face cycle C of the outerplanar embedding of G, there must be a unique nearest neighbor in C to v. For, if v were nearest to distinct vertices u and w in C, then by bipartiteness these two vertices must be non-adjacent in C. In this case, the graph formed by C together with the shortest paths from v to u and w would contain a subdivision of $K_{2,3}$ (with u and w as the degree three vertices, two paths between them in C, and one more path between them through the shortest path tree rooted at v), an impossibility for an outerplanar graph. For the same reason, the distances in v from this nearest neighbor or pair of nearest neighbors must increase monotonically in both directions around C until reaching a unique farthest neighbor, because in the same way any non-monotonicity could be used to construct a subdivision of $K_{2,3}$.

Thus, each face cycle of G has a planar breadth first layering. The result follows from the fact that in a plane graph with an assignment of levels to the vertices, there is a planar drawing consistent with this level assignment and with the given embedding of the graph, if and only if every face cycle of the given graph has a planar drawing consistent with the level assignment [20].   □

5.2   Squaregraphs

A *squaregraph* is defined to be a graph that has a planar embedding in which each bounded face is a 4-cycle and each vertex either belongs to the unbounded face or has four or more incident edges. These graphs may also be character-ized in various other ways, for instance as the dual graphs of hyperbolic line arrangements with no three mutually-intersecting lines [21].

Theorem 6. *Every squaregraph G is leveled planar, and 3-track, with a leveled planar drawing coming from a breadth first layering.*

Proof. Because all their bounded faces are even-sided, squaregraphs are neces-sarily bipartite, so every choice of a starting vertex gives a valid breadth first layering. Bandelt et al. [21, Lemma 12.2] prove that, for every choice of a starting vertex, we can add extra edges to the squaregraph to form a plane multigraph in which the added edges link each layer into a cycle, and in which these cycles are all nested within each other.

Now, choose the starting vertex v to be a vertex of the outer face. Then each cycle added in this augmentation of G contains an edge that separates v from the unbounded face of the augmented graph. If we remove each such edge from the augmented graph, we break each cycle into a path in a consistent way, such that the path ordering within each layer matches the given planar embedding of G.

□

5.3   Dual Graphs of Monotone Curves

Theorem 7. *Let A be a collection of finitely many x-monotone curves in the plane, each of whose projection onto the x-axis covers the entire axis, such that any two curves intersect at finitely many crossing points. Then the dual graph of the arrangement of the curves in A is leveled planar and 3-track.*

Proof. Each vertex of the dual graph corresponds to a connected component of the complement of $\bigcup A$; we call this the *region* of the vertex. We may assign each vertex to a layer according to the number of curves in A that pass above it; this is a breadth first layering starting from the vertex corresponding to the topmost (unbounded upward) connected component. Because a single curve separates adjacent regions, vertices in adjacent regions will be assigned to consecutive regions. No two vertices in the same layer have regions that project to overlapping subsets of the x-axis, so we may order the vertices within each layer according to the left-to-right ordering of these projections. This ordering is compatible with the planar embedding of the dual graph given by placing a representative point within each region and connecting each two adjacent regions by a curve crossing their shared boundary.

□

See Fig. 5 for examples of the graphs shown to have planar layerings by these theorems. Figure 6 gives another example, demonstrating that Theorem 7 cannot be generalized to monotone curves whose projections do not cover the

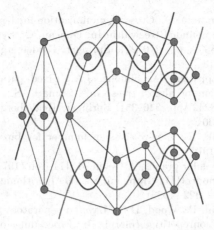

Fig. 6. An arrangement of monotone curves whose dual graph has no planar layering

entire axis: it gives a family of monotone curves, all ending within the outer face of their arrangement, such that the dual graph of the arrangement is not leveled planar. The dual graph is made of multiple $K_{2,3}$ subgraphs, each of which must have the 2-vertex side of its bipartition drawn on two layers with the 3-vertex side of its bipartition in a single layer between them; thus, up to top-bottom reflection, there is only a single layering for this graph that could possibly be planar. However, this layering forced by the planarity of the individual $K_{2,3}$ subgraphs is not planar globally, because it forces one of the two arms of the graph (upper and lower right) to collide with the "armpit" where the other arm meets the body of the graph (left). The graph is drawn without crossings in the figure, but in a way that does not respect any layering of the graph. The dual used in Fig. 6 is non-standard; there is a vertex in the outer face for each pair of consecutive curve endpoints on the outer face. The dual shown can be made as a subgraph of the more standard dual with one vertex on the outer face by adding additional curves. This example is also a series-parallel graph, and shows that Theorem 5 cannot be generalized to series-parallel, treewidth-2, or 2-outerplanar graphs: none of these classes of graphs is leveled planar.

References

1. Dujmović, V., Morin, P., Wood, D.R.: Layout of graphs with bounded tree-width. SIAM J. Comput. **34**, 553–579 (2005)
2. Dujmović, V., Wood, D.R.: Three-dimensional grid drawings with sub-quadratic volume. In: Pach, J., ed.: Towards a Theory of Geometric Graphs, vol. 342, pp. 55–66. Contemporary Mathematics - American Mathematical Society (2004)
3. Dujmović, V., Pór, A., Wood, D.R.: Track layouts of graphs. Discrete Math. Theor. Comput. Sci. **6**, 497–521 (2004)
4. Heath, L.S., Rosenberg, A.L.: Laying out graphs using queues. SIAM J. Comput. **21**, 927–958 (1992)

5. Bannister, M.J., Eppstein, D.: Crossing minimization for 1-page and 2-page drawings of graphs with bounded treewidth. In: Duncan, C., Symvonis, A. (eds.) GD 2014. LNCS, vol. 8871, pp. 210–221. Springer, Heidelberg (2014). doi:10.1007/978-3-662-45803-7_18

6. Bannister, M.J., Eppstein, D., Simons, J.A.: Fixed parameter tractability of crossing minimization of almost-trees. In: Wismath, S., Wolff, A. (eds.) GD 2013. LNCS, vol. 8242, pp. 340–351. Springer, Heidelberg (2013). doi:10.1007/978-3-319-03841-4_30

7. Harary, F., Schwenk, A.: A new crossing number for bipartite graphs. Utilitas Math. 1, 203–209 (1972)

8. Halin, R.: S-functions for graphs. J. Geom. 8, 171–186 (1976)

9. Robertson, N., Seymour, P.D.: Graph minors. II. Algorithmic aspects of tree-width. J. Algorithms 7, 309–322 (1986)

10. Dujmović, V., Morin, P., Wood, D.R.: Layered separators for queue layouts, 3D graph drawing and nonrepetitive coloring. In: Proceedings of 54th Symposium on Foundations of Computer Science (FOCS 2013), pp. 280–289. IEEE Computer Society (2013)

11. Dujmović, V., Morin, P., Wood, D.R.: Layered separators in minor-closed graph classes with applications. Electronic preprint arXiv:1306.1595 (2013)

12. Sugiyama, K., Tagawa, S., Toda, M.: Methods for visual understanding of hierarchical system structures. IEEE Trans. Syst. Man Cybern. SMC–11, 109–125 (1981)

13. Eades, P., Wormald, N.C.: Edge crossings in drawings of bipartite graphs. Algorithmica 11, 379–403 (1994)

14. Healy, P., Nikolov, N.S.: How to layer a directed acyclic graph. In: Mutzel, P., Jünger, M., Leipert, S. (eds.) GD 2001. LNCS, vol. 2265, pp. 16–30. Springer, Heidelberg (2002). doi:10.1007/3-540-45848-4_2

15. Felsner, S., Liotta, G., Wismath, S.K.: Straight-line drawings on restricted integer grids in two and three dimensions. J. Graph Algorithms Appl. 7, 363–398 (2003)

16. Dujmović, V., Wood, D.R.: Stacks, queues and tracks: layouts of graph subdivisions. Discrete Math. Theor. Comput. Sci. 7, 155–201 (2005)

17. Downey, R.G., Fellows, M.R.: Fundamentals of Parameterized Complexity. Texts in Computer Science. Springer, Berlin (2013)

18. Bannister, M.J., Cabello, S., Eppstein, D.: Parameterized complexity of 1-planarity. In: Dehne, F., Solis-Oba, R., Sack, J.-R. (eds.) WADS 2013. LNCS, vol. 8037, pp. 97–108. Springer, Heidelberg (2013). doi:10.1007/978-3-642-40104-6_9

19. Nešetřil, J., de Mendez, P.O.: Sparsity (Graphs, Structures, and Algorithms). Algorithms and Combinatorics, vol. 28. Springer, Heidelberg (2012)

20. Abel, Z., Demaine, E.D., Demaine, M.L., Eppstein, D., Lubiw, A., Uehara, R.: Flat foldings of plane graphs with prescribed angles and edge lengths. In: Duncan, C., Symvonis, A. (eds.) GD 2014. LNCS, vol. 8871, pp. 272–283. Springer, Heidelberg (2014). doi:10.1007/978-3-662-45803-7_23

21. Bandelt, H.J., Chepoi, V., Eppstein, D.: Combinatorics and geometry of finite and infinite squaregraphs. SIAM J. Discrete Math. 24, 1399–1440 (2010)

Stack and Queue Layouts via Layered Separators

Vida Dujmović[1] and Fabrizio Frati[2(✉)]

[1] School of Computer Science and Electrical Engineering, University of Ottawa,
Ottawa, Canada
vida.dujmovic@uottawa.ca
[2] Dipartimento di Ingegneria, Roma Tre University, Rome, Italy
frati@dia.uniroma3.it

Abstract. It is known that every proper minor-closed class of graphs
has bounded stack-number (a.k.a. book thickness and page number).
While this includes notable graph families such as planar graphs and
graphs of bounded genus, many other graph families are not closed under
taking minors. For fixed g and k, we show that every n-vertex graph that
can be embedded on a surface of genus g with at most k crossings per edge
has stack-number $\mathcal{O}(\log n)$; this includes k-planar graphs. The previously
best known bound for the stack-number of these families was $\mathcal{O}(\sqrt{n})$,
except in the case of 1-planar graphs. Analogous results are proved for
map graphs that can be embedded on a surface of fixed genus. None of
these families is closed under taking minors. The main ingredient in the
proof of these results is a construction proving that n-vertex graphs that
admit constant layered separators have $\mathcal{O}(\log n)$ stack-number.

1 Introduction

A *stack layout* of a graph G consists of a total order σ of $V(G)$ and a partition
of $E(G)$ into sets (called *stacks*) such that no two edges in the same stack *cross*;
that is, there are no edges vw and xy in a single stack with $v <_\sigma x <_\sigma w <_\sigma y$.
The minimum number of stacks in a stack layout of G is the *stack-number* of
G. Stack layouts, first defined by Ollmann [22], are ubiquitous structures with a
variety of applications (see [17] for a survey). A stack layout is also called a *book
embedding* and stack-number is also called *book thickness* and *page number*. The
stack-number is known to be bounded for planar graphs [24], bounded genus
graphs [20] and, most generally, all proper minor-closed graph families [4,5].

The purpose of this note is to bring the study of the stack-number beyond the
proper minor-closed graph families. *Layered separators* are a key tool for proving
our results. They have already led to progress on long-standing open problems
related to 3D graph drawings [11,15] and nonrepetitive graph colourings [13].
A *layering* $\{V_0, \ldots, V_p\}$ of a graph G is a partition of $V(G)$ into *layers* V_i such
that, for each $e \in E(G)$, there is an i such that the endpoints of e are both in

The research of Vida Dujmović was partially supported by NSERC, and Ontario
Ministry of Research and Innovation. The research of Fabrizio Frati was partially
supported by MIUR Project "AMANDA" under PRIN 2012C4E3KT.

Y. Hu and M. Nöllenburg (Eds.): GD 2016, LNCS 9801, pp. 511–518, 2016.
DOI: 10.1007/978-3-319-50106-2_39

V_i or one in V_i and one in V_{i+1}. A graph G has a *layered ℓ-separator* for a fixed layering $\{V_0, \ldots, V_p\}$ if, for every subgraph G' of G, there exists a set $S \subseteq V(G')$ with at most ℓ vertices in each layer (i.e., $V_i \cap S \le \ell$, for $i = 0, \ldots, p$) such that each connected component of $G' - S$ has at most $|V(G')|/2$ vertices. Our main technical contribution is the following theorem.

Theorem 1. *Every n-vertex graph that has a layered ℓ-separator has stack-number at most $5\ell \cdot \log_2 n$.*

We discuss the implications of Theorem 1 for two well-known non-minor-closed classes of graphs. A graph is (g, k)-*planar* if it can be drawn on a surface of Euler genus at most g with at most k crossings per edge. Then $(0,0)$-planar graphs are *planar graphs*, whose stack-number is at most 4 [24]. Further, $(0, k)$-planar graphs are k-*planar graphs* [23]; Bekos *et al.* [3] have recently proved that 1-planar graphs have bounded stack-number (see Alam *et al.* [1] for an improved constant). The family of (g, k)-planar graphs is not closed under taking minors[1] even for $g = 0$, $k = 1$; thus the result of Blankenship and Oporowski [4,5], stating that proper minor-closed graph families have bounded stack-number, does not apply to (g, k)-planar graphs. Dujmović *et al.* [12] showed that (g, k)-planar graphs have layered $(4g + 6)(k + 1)$-separators[2]. This and our Theorem 1 imply the following corollary. For all $g \ge 0$ and $k \ge 2$, the previously best known bound was $\mathcal{O}(\sqrt{n})$, following from the $\mathcal{O}(\sqrt{m})$ bound for m-edge graphs [21].

Corollary 1. *For any fixed g and k, every n-vertex (g, k)-planar graph has stack-number $\mathcal{O}(\log n)$.*

A (g, d)-*map graph* G is defined as follows. Embed a graph H on a surface of Euler genus g and label some of its faces as "nations" so that any vertex of H is incident to at most d nations; then the vertices of G are the faces of H labeled as nations and the edges of G connect nations that share a vertex of H. The $(0, d)$-map graphs are the well-known d-*map graphs* [6–9,18]. The $(g, 3)$-map graphs are the graphs of Euler genus at most g [8], thus they are closed under taking minors. However, for every $g \ge 0$ and $d \ge 4$, the (g, d)-map graphs are not closed under taking minors [12], thus the result of Blankenship and Oporowski [4,5] does not apply to them. The (g, d)-map graphs have layered $(2g+3)(2d+1)$-separators [12]. This and our Theorem 1 imply the following corollary. For all $g \ge 0$ and $d \ge 4$, the best previously known bound was $\mathcal{O}(\sqrt{n})$ [21].

Corollary 2. *For any fixed g and d, every n-vertex (g, d)-map graph has stack-number $\mathcal{O}(\log n)$.*

---

[1] The $n \times n \times 2$ grid graph is a well-known example of 1-planar graph with an arbitrarily large complete graph minor. Indeed, contracting the i-th row in the front $n \times n$ grid with the i-th column in the back $n \times n$ grid, for $1 \le i \le n$, gives a K_n minor.

[2] More precisely, Dujmović *et al.* [12] proved that (g, k)-planar graphs have *layered treewidth* at most $(4g + 6)(k + 1)$ and (g, d)-map graphs have layered treewidth at most $(2g+3)(2d+1)$. Just as the graphs of treewidth t have (classical) separators of size $t - 1$, so do the graphs of layered treewidth ℓ have layered ℓ-separators [15,16].

A "dual" concept to that of stack layouts are queue layouts. A *queue layout* of a graph G consists of a total order σ of $V(G)$ and a partition of $E(G)$ into sets (called *queues*), such that no two edges in the same queue *nest*; that is, there are no edges vw and xy in a single queue with $v <_\sigma x <_\sigma y <_\sigma w$. If $v <_\sigma x <_\sigma y <_\sigma w$ we say that xy *nests inside* vw. The minimum number of queues in a queue layout of G is called the *queue-number* of G. Queue layouts, like stack layouts, have been extensively studied. In particular, it is a long standing open problem to determine if planar graphs have bounded queue-number. Logarithmic upper bounds have been obtained via layered separators [2,11]. In particular, a result similar to Theorem 1 is known for the queue-number: Every n-vertex graph that has layered ℓ-separators has queue-number $\mathcal{O}(\ell \log n)$ [11]; this bound was refined to $3\ell \cdot \log_3(2n+1) - 1$ by Bannister *et al.* [2]. These results were established via a connection with the *track-number* of a graph [14]. Together with the fact that planar graphs have layered 2-separators [13,19], these results imply an $\mathcal{O}(\log n)$ bound for the queue-number of planar graphs, improving on a earlier result by Di Battista *et al.* [10]. The polylog bound on the queue-number of planar graphs extends to all proper minor-closed families of graphs [15,16]. Our approach to prove Theorem 1 also gives a new proof of the following result (without using track layouts). We include it for completeness.

Theorem 2. *Every n-vertex graph that has a layered ℓ-separator has queue-number at most $3\ell \cdot \log_2 n$.*

2   Proofs of Theorems 1 and 2

Let G be a graph and $L = \{V_0, \ldots, V_p\}$ be a layering of G such that G admits a layered ℓ-separator for layering L. Each edge of G is either an *intra-layer* edge, that is, an edge between two vertices in a set V_i, or an *inter-layer* edge, that is, an edge between a vertex in a set V_i and a vertex in a set V_{i+1}.

A total order on a set of vertices $R \subseteq V(G)$ is a *vertex ordering* of R. The stack layout construction computes a vertex ordering σ^s of $V(G)$ satisfying the *layer-by-layer* invariant, which is defined as follows: For $0 \le i < p$, the vertices in V_i precede the vertices in V_{i+1} in σ^s. Analogously, the queue layout construction computes a vertex ordering σ^q of $V(G)$ satisfying the layer-by-layer invariant.

Let S be a layered ℓ-separator for G with respect to L. Let $G_1, \ldots, G_k$ be the graphs induced by the vertices in the connected components of $G - S$ (the vertices of S do not belong to any graph G_j). These graphs are labeled $G_1, \ldots, G_k$ arbitrarily. Recall that, by the definition of a layered ℓ-separator for G, we have $|V(G_j)| \le n/2$, for each $1 \le j \le k$. Let $S_i = S \cap V_i$ and let ρ_i be an arbitrary vertex ordering of S_i, for $i = 0, \ldots, p$.

Both the stack and the queue layout constructions recursively construct vertex orderings of $V(G_j)$ satisfying the layer-by-layer invariant, for $j = 1, \ldots, k$. Let σ_j^s be the vertex ordering of $V(G_j)$ computed by the stack layout construction; we also denote by $\sigma_{j,i}^s$ the restriction of σ_j^s to the vertices in layer V_i. Note that $\sigma_j^s = \sigma_{j,1}^s, \sigma_{j,2}^s, \ldots, \sigma_{j,p}^s$ by the layer-by-layer invariant. Vertex orderings σ_j^q and $\sigma_{j,i}^q$ are defined analogously for the queue layout construction.

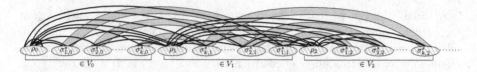

Fig. 1. Illustration for the stack layout construction. Edges incident to vertices in S are black and thick. Edges in graphs $G_1, \ldots, G_k$ are represented by gray regions.

We now show how to combine the recursively constructed vertex orderings to obtain a vertex ordering of $V(G)$. The way this combination is performed differs for the stack layout construction and the queue layout construction.

Stack Layout Construction. Vertex ordering σ^s is defined as (refer to Fig. 1)

$$\rho_0, \sigma^s_{1,0}, \sigma^s_{2,0}, \ldots, \sigma^s_{k-1,0}, \sigma^s_{k,0}, \rho_1, \sigma^s_{k,1}, \sigma^s_{k-1,1}, \ldots, \sigma^s_{2,1}, \sigma^s_{1,1},$$
$$\rho_2, \sigma^s_{1,2}, \sigma^s_{2,2}, \ldots, \sigma^s_{k-1,2}, \sigma^s_{k,2}, \rho_3, \sigma^s_{k,3}, \sigma^s_{k-1,3}, \ldots, \sigma^s_{2,3}, \sigma^s_{1,3}, \ldots.$$

The vertex ordering σ^s satisfies the layer-by-layer invariant, given that vertex ordering σ^s_j does, for $j = 1, \ldots, k$. Then Theorem 1 is implied by the following.

Lemma 1. *G has a stack layout with $5\ell \cdot \log_2 n$ stacks with vertex ordering σ^s.*

Proof: We use distinct sets of stacks for the intra- and the inter-layer edges.

Stacks for the intra-layer edges. We assign each intra-layer edge uv with $u \in S$ or $v \in S$ to one of ℓ stacks $P_1, \ldots, P_\ell$ as follows. Since uv is an intra-layer edge, $\{u, v\} \subseteq V_i$, for some $0 \le i \le p$. Assume w.l.o.g. that $u <_{\sigma^s} v$. Then $u \in S$ and let it be x-th vertex in ρ_i (recall that ρ_i contains at most ℓ vertices). Assign uv to P_x. The only intra-layer edges that are not yet assigned to stacks belong to graphs $G_1, \ldots, G_k$. The assignment of these edges to stacks is the one computed recursively; however, we use the same set of stacks to assign the edges of all graphs $G_1, \ldots, G_k$.

We now prove that no two intra-layer edges in the same stack cross. Let e and e' be two intra-layer edges of G and let both the endpoints of e be in V_i and both the endpoints of e' be in $V_{i'}$. Assume w.l.o.g. that $i \le i'$. If $i < i'$, then, since σ^s satisfies the layer-by-layer invariant, the endpoints of e precede those of e' in σ^s, hence e and e' do not cross. Suppose now that $i = i'$. If e and e' are in some stack P_x for $x \in \{1, \ldots, \ell\}$, then they are both incident to the x-th vertex in ρ_i, thus they do not cross. If e and e' are in some stack different from $P_1, \ldots, P_\ell$, then $e \in E(G_j)$ and $e' \in E(G_{j'})$, for some $j, j' \in \{1, \ldots, k\}$. If $j = j'$, then e and e' do not cross by induction. Otherwise, both the endpoints of e precede both the endpoints of e' or vice versa, since the vertices in $\sigma^s_{\min\{j,j'\},i}$ precede those in $\sigma^s_{\max\{j,j'\},i}$ in σ^s or vice versa, depending on whether i is even or odd; hence e and e' do not cross.

We now bound the number of stacks we use for the intra-layer edges of G; we claim that this number is at most $\ell \cdot \log_2 n$. The proof is by induction on n; the base case $n = 1$ is trivial. For any subgraph H of G, let $p_1(H)$ be the number of stacks we use for the intra-layer edges of H, and let $p_1(n') = \max_H\{p_1(H)\}$ over all subgraphs H of G with n' vertices. As proved above, $p_1(G) \le \ell + \max\{p_1(G_1), \dots, p_1(G_k)\}$. Since each graph G_j has at most $n/2$ vertices, we get that $p_1(G) \le \ell + p_1(n/2)$. By induction $p_1(G) \le \ell + \ell \cdot \log_2(n/2) = \ell \cdot \log_2 n$.

Stacks for the inter-layer edges. We use distinct sets of stacks for the *even inter-layer edges* – connecting vertices on layers V_i and V_{i+1} with i even – and for the *odd inter-layer edges* – connecting vertices on layers V_i and V_{i+1} with i odd. We only describe how to assign the even inter-layer edges to $2\ell \cdot \log_2 n$ stacks so that no two edges in the same stack cross; the assignment for the odd inter-layer edges is analogous.

We assign each even inter-layer edge uv with $u \in S$ or $v \in S$ to one of 2ℓ stacks $P'_1, \dots, P'_{2\ell}$ as follows. Since uv is an inter-layer edge, u and v respectively belong to layers V_i and V_{i+1}, for some $0 \le i \le p - 1$. If $u \in S$, then u is the x-th vertex in ρ_i, for some $1 \le x \le \ell$; assign edge uv to P'_x. If $u \notin S$, then $v \in S$ is the y-th vertex in ρ_{i+1}, for some $1 \le y \le \ell$; assign edge uv to $P'_{\ell+y}$. The only even inter-layer edges that are not yet assigned to stacks belong to graphs $G_1, \dots, G_k$. The assignment of these edges to stacks is the one computed recursively; however, we use the same set of stacks to assign the edges of all graphs $G_1, \dots, G_k$.

We prove that no two even inter-layer edges in the same stack cross. Let e and e' be two even inter-layer edges of G. Let V_i and V_{i+1} be the layers containing the endpoints of e. Let $V_{i'}$ and $V_{i'+1}$ be the layers containing the endpoints of e'. Assume w.l.o.g. that $i \le i'$. If $i < i'$, then $i + 1 < i'$, given that both i and i' are even. Then, since σ^s satisfies the layer-by-layer invariant, both the endpoints of e precede both the endpoints of e', thus e and e' do not cross. Suppose now that $i = i'$. If e and e' are in some stack P'_h for $h \in \{1, \dots, 2\ell\}$, then e and e' are both incident either to the h-th vertex of ρ_i or to the $(h - \ell)$-th vertex of ρ_{i+1}, hence they do not cross. If e and e' are in some stack different from $P'_1, \dots, P'_{2\ell}$, then $e \in E(G_j)$ and $e' \in E(G_{j'})$, for $j, j' \in \{1, \dots, k\}$. If $j = j'$, then e and e' do not cross by induction. Otherwise, $j \ne j'$ and then e nests inside e' or vice versa, since the vertices in $\sigma^s_{\min\{j,j'\},i}$ precede those in $\sigma^s_{\max\{j,j'\},i}$ and the vertices in $\sigma^s_{\max\{j,j'\},i+1}$ precede those in $\sigma^s_{\min\{j,j'\},i+1}$ in σ^s; hence e and e' do not cross.

We now bound the number of stacks we use for the even inter-layer edges of G; we claim that this number is at most $2\ell \cdot \log_2 n$. The proof is by induction on n; the base case $n = 1$ is trivial. For any subgraph H of G, let $p_2(H)$ be the number of stacks we use for the even inter-layer edges of H, and let $p_2(n') = \max_H\{p_2(H)\}$ over all subgraphs H of G with n' vertices. As proved above, $p_2(G) \le 2\ell + \max\{p_2(G_1), \dots, p_2(G_k)\}$. Since each graph G_j has at most $n/2$ vertices, we get that $p_2(G) \le 2\ell + p_2(n/2)$. By induction $p_2(G) \le 2\ell + 2\ell \cdot \log_2(n/2) = 2\ell \cdot \log_2 n$.

The described stack layout uses $\ell \cdot \log_2 n$ stacks for the intra-layer edges, $2\ell \cdot \log_2 n$ stacks for the even inter-layer edges, and $2\ell \cdot \log_2 n$ stacks for the odd inter-layer edges, thus $5\ell \cdot \log_2 n$ stacks in total. This concludes the proof.   □

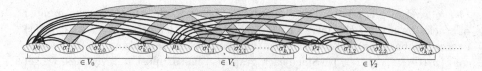

Fig. 2. Illustration for the queue layout construction.

Queue Layout Construction. Vertex ordering σ^q is defined as (refer to Fig. 2)
$$\rho_0, \sigma_{1,0}^q, \sigma_{2,0}^q, \ldots, \sigma_{k,0}^q, \rho_1, \sigma_{1,1}^q, \sigma_{2,1}^q, \ldots, \sigma_{k,1}^q, \ldots, \rho_p, \sigma_{1,p}^q, \sigma_{2,p}^q, \ldots, \sigma_{k,p}^q.$$

The vertex ordering σ^q satisfies the layer-by-layer invariant, given that vertex ordering σ_j^q does, for $j = 1, \ldots, k$. Then Theorem 2 is implied by the following.

Lemma 2. *G has a queue layout with $3\ell \cdot \log_2 n$ queues with vertex ordering σ^q.*

Proof: We use distinct sets of queues for the intra- and the inter-layer edges.

Queues for the intra-layer edges. We assign each intra-layer edge uv with $u \in S$ or $v \in S$ to one of ℓ queues $Q_1, \ldots, Q_\ell$ as follows. Since uv is an intra-layer edge, $\{u, v\} \subseteq V_i$, for some $0 \le i \le p$. Assume w.l.o.g. that $u <_{\sigma^q} v$. Then $u \in S$ and let it be the x-th vertex of ρ_i. Assign uv to Q_x. The only intra-layer edges that are not yet assigned to queues belong to graphs $G_1, \ldots, G_k$. The assignment of these edges to queues is the one computed recursively; however, we use the same set of queues to assign the edges of all graphs $G_1, \ldots, G_k$.

The proof that no two intra-layer edges in the same queue nest is the same as the proof no two intra-layer edges in the same stack cross in Lemma 1 (with the word "nest" replacing "cross" and with σ^q replacing σ^s). The proof that the number of queues we use for the intra-layer edges is at most $\ell \cdot \log_2 n$ is also the same as the proof that the number of stacks we use for the intra-layer edges is at most $\ell \cdot \log_2 n$ in Lemma 1.

Queues for the inter-layer edges. We assign each inter-layer edge uv with $u \in S$ or $v \in S$ to one of 2ℓ queues $Q_1', \ldots, Q_{2\ell}'$ as follows. Since uv is an inter-layer edge, u and v respectively belong to layers V_i and V_{i+1}, for some $0 \le i \le p - 1$. If $u \in S$, then u is the x-th vertex in ρ_i, for some $1 \le x \le \ell$; assign edge uv to Q_x'. If $u \notin S$, then $v \in S$ is the y-th vertex in ρ_{i+1}, for some $1 \le y \le \ell$; assign edge uv to $Q_{\ell+y}'$. The only inter-layer edges that are not yet assigned to queues belong to graphs $G_1, \ldots, G_k$. The assignment of these edges to queues is the one computed recursively; however, we use the same set of queues to assign the edges of all graphs $G_1, \ldots, G_k$.

We prove that no two inter-layer edges e and e' in the same queue nest. Let V_i and V_{i+1} be the layers containing the endpoints of e. Let $V_{i'}$ and $V_{i'+1}$ be the layers containing the endpoints of e'. Assume w.l.o.g. that $i \le i'$. If $i < i'$, then both endpoints of e precede the endpoint of e' in $V_{i'+1}$ (hence e' is not nested inside e) and both endpoints of e' follow the endpoint of e in V_i (hence e is not nested inside e'), since σ^q satisfies the layer-by-layer invariant; thus e and e' do not nest. Suppose now that $i = i'$. If e and e' are in some queue Q_h' for

$h \in \{1, \ldots, 2\ell\}$, then e and e' are both incident either to the h-th vertex of ρ_i or to the $(h-\ell)$-th vertex of ρ_{i+1}, hence they do not nest. If e and e' are in some queue different from $Q'_1, \ldots, Q'_{2\ell}$, then $e \in E(G_j)$ and $e' \in E(G_{j'})$, for $j, j' \in \{1, \ldots, k\}$. If $j = j'$, then e and e' do not nest by induction. Otherwise, $j \neq j'$ and then the endpoints of e alternate with those of e' in σ^q, since the vertices in $\sigma^q_{\min\{j,j'\},i}$ precede those in $\sigma^q_{\max\{j,j'\},i}$ and the vertices in $\sigma^q_{\min\{j,j'\},i+1}$ precede those in $\sigma^q_{\max\{j,j'\},i+1}$ in σ^q; hence e and e' do not nest.

We now bound the number of queues we use for the inter-layer edges of G; we claim that this number is at most $2\ell \cdot \log_2 n$. The proof is by induction on n; the base case $n = 1$ is trivial. For any subgraph H of G, let $q(H)$ be the number of queues we use for the inter-layer edges of H, and let $q(n') = \max_H\{q(H)\}$ over all subgraphs H of G with n' vertices. As proved above, $q(G) \leq 2\ell + \max\{q(G_1), \ldots, q(G_k)\}$. Since each graph G_j has at most $n/2$ vertices, we get that $q(G) \leq 2\ell + q(n/2)$. By induction $q(G) \leq 2\ell + 2\ell \cdot \log_2(n/2) = 2\ell \cdot \log_2 n$.

Thus, the described queue layout uses $\ell \cdot \log_2 n$ queues for the intra-layer edges and $2\ell \cdot \log_2 n$ queues for the inter-layer edges, thus $3\ell \cdot \log_2 n$ queues in total. This concludes the proof. $\qquad\square$

Acknowledgments. The authors wish to thank David R. Wood for stimulating discussions and comments on the preliminary version of this article.

References

1. Alam, M.J., Brandenburg, F.J., Kobourov, S.G.: On the book thickness of 1-planar graphs (2015). http://arxiv.org/abs/1510.05891
2. Bannister, M.J., Devanny, W.E., Dujmović, V., Eppstein, D., Wood, D.R.: Track layouts, layered path decompositions, and leveled planarity (2015). http://arxiv.org/abs/1506.09145
3. Bekos, M.A., Bruckdorfer, T., Kaufmann, M., Raftopoulou, C.: 1-planar graphs have constant book thickness. In: Bansal, N., Finocchi, I. (eds.) ESA 2015. LNCS, vol. 9294, pp. 130–141. Springer, Heidelberg (2015). doi:10.1007/978-3-662-48350-3_12
4. Blankenship, R.: Book embeddings of graphs. Ph.D. thesis, Department of Mathematics, Louisiana State University, USA (2003)
5. Blankenship, R., Oporowski, B.: Book embeddings of graphs and minor-closed classes. In: 32nd Southeastern International Conference on Combinatorics, Graph Theory and Computing. Department of Mathematics, Louisiana State University (2001)
6. Chen, Z.Z.: Approximation algorithms for independent sets in map graphs. J. Algorithms **41**(1), 20–40 (2001)
7. Chen, Z.Z.: New bounds on the edge number of a k-map graph. J. Graph Theory **55**(4), 267–290 (2007)
8. Chen, Z.Z., Grigni, M., Papadimitriou, C.H.: Map graphs. J. ACM **49**(2), 127–138 (2002)
9. Demaine, E.D., Fomin, F.V., Hajiaghayi, M., Thilikos, D.M.: Fixed-parameter algorithms for (k, r)-center in planar graphs and map graphs. ACM Trans. Algorithms **1**(1), 33–47 (2005)

10. Di Battista, G., Frati, F., Pach, J.: On the queue number of planar graphs. SIAM J. Comput. **42**(6), 2243–2285 (2013)

11. Dujmović, V.: Graph layouts via layered separators. J. Comb. Theory Ser. B **110**, 79–89 (2015)

12. Dujmović, V., Eppstein, D., Wood, D.R.: Genus, treewidth, and local crossing number. In: Di Giacomo, E., Lubiw, A. (eds.) GD 2015. LNCS, vol. 9411, pp. 87–98. Springer, Heidelberg (2015). doi:10.1007/978-3-319-27261-0_8

13. Dujmović, V., Joret, G., Frati, F., Wood, D.R.: Nonrepetitive colourings of planar graphs with O(log n) colours. Electron. J. Comb. **20**(1), P51 (2013)

14. Dujmović, V., Morin, P., Wood, D.R.: Layout of graphs with bounded tree-width. SIAM J. Comput. **34**(3), 553–579 (2005)

15. Dujmović, V., Morin, P., Wood, D.R.: Layered separators for queue layouts, 3D graph drawing and nonrepetitive coloring. In: 54th Annual IEEE Symposium on Foundations of Computer Science, pp. 280–289. IEEE Computer Society (2013)

16. Dujmović, V., Morin, P., Wood, D.R.: Layered separators in minor-closed families with applications (2014). http://arxiv.org/abs/1306.1595

17. Dujmović, V., Wood, D.R.: On linear layouts of graphs. Discrete Math. Theor. Comput. Sci. **6**(2), 339–358 (2004)

18. Fomin, F.V., Lokshtanov, D., Saurabh, S.: Bidimensionality and geometric graphs. In: Rabani, Y. (ed.) 23rd Annual ACM-SIAM Symposium on Discrete Algorithms, pp. 1563–1575. SIAM (2012)

19. Lipton, R.J., Tarjan, R.E.: A separator theorem for planar graphs. SIAM J. Appl. Math. **36**(2), 177–189 (1979)

20. Malitz, S.M.: Genus g graphs have pagenumber $O(\sqrt{g})$. J. Algorithms **17**(1), 85–109 (1994)

21. Malitz, S.M.: Graphs with E edges have pagenumber $O(\sqrt{E})$. J. Algorithms **17**(1), 71–84 (1994)

22. Ollmann, L.T.: On the book thicknesses of various graphs. In: Hoffman, F., Levow, R.B., Thomas, R.S.D. (eds.) 4th Southeastern Conference on Combinatorics, Graph Theory, and Computing, vol. VIII, p. 459. Congressus Numerantium (1973)

23. Pach, J., Tóth, G.: Graphs drawn with few crossings per edge. Combinatorica **17**(3), 427–439 (1997)

24. Yannakakis, M.: Embedding planar graphs in four pages. J. Comput. Sys. Sci. **38**(1), 36–67 (1989)

Gabriel Triangulations and Angle-Monotone Graphs: Local Routing and Recognition

Nicolas Bonichon[1], Prosenjit Bose[2], Paz Carmi[3], Irina Kostitsyna[4], Anna Lubiw[5(✉)], and Sander Verdonschot[6]

[1] LaBRI, University of Bordeaux, Bordeaux, France
bonichon@labri.fr
[2] School of Computer Science, Carleton University, Ottawa, Canada
jit@scs.carleton.ca
[3] Department of Computer Science, Ben-Gurion University of the Negev, Beersheba, Israel
carmip@cs.bgu.ac.il
[4] Université Libre de Bruxelles, Brussels, Belgium
irina.kostitsyna@ulb.ac.be
[5] David R. Cheriton School of Computer Science, University of Waterloo, Waterloo, Canada
alubiw@uwaterloo.ca
[6] School of Electrical Engineering and Computer Science, University of Ottawa, Ottawa, Canada
sander@cg.scs.carleton.ca

Abstract. A geometric graph is *angle-monotone* if every pair of vertices has a path between them that—after some rotation—is x- and y-monotone. Angle-monotone graphs are $\sqrt{2}$-spanners and they are increasing-chord graphs. Dehkordi, Frati, and Gudmundsson introduced angle-monotone graphs in 2014 and proved that Gabriel triangulations are angle-monotone graphs. We give a polynomial time algorithm to recognize angle-monotone geometric graphs. We prove that every point set has a plane geometric graph that is *generalized angle-monotone*—specifically, we prove that the half-θ_6-graph is generalized angle-monotone. We give a local routing algorithm for Gabriel triangulations that finds a path from any vertex s to any vertex t whose length is within $1 + \sqrt{2}$ times the Euclidean distance from s to t. Finally, we prove some lower bounds and limits on local routing algorithms on Gabriel triangulations.

1 Introduction

A geometric graph has vertices that are points in the plane, and edges that are drawn as straight-line segments, with the weight of an edge being its Euclidean length. A geometric graph need not be planar. Geometric graphs that have relatively short paths are relevant in many applications for routing and network design, and have been a subject of intense research. A main scenario is that we

© Springer International Publishing AG 2016
Y. Hu and M. Nöllenburg (Eds.): GD 2016, LNCS 9801, pp. 519–531, 2016.
DOI: 10.1007/978-3-319-50106-2_40

are given a point set and must construct a sparse geometric graph on that point set with good shortest path properties.

If the shortest path between every pair of points has length at most t times the Euclidean distance between the points, then the geometric graph is called a *t-spanner*, and the minimum such t is called the *spanning ratio*. Since their introduction by Paul Chew in 1986 [10], spanners have been heavily studied [18].

Besides the existence of short paths, another issue is *routing*—how to find short paths in a geometric graph. One goal is to find paths using *local routing* where the path is found one vertex at a time using only local information about the neighbours of the current vertex plus the coordinates of the destination. A main example of such a method is *greedy routing*: from the current vertex u take any edge to a vertex v that is closer (in Euclidean distance) to the destination than u is. The geometric graphs for which greedy routing succeeds in finding a path are called *greedy drawings*. These have received considerable attention because of their potential ability to replace routing tables for network routing, and because of the noted conjecture of Papadimitriou and Ratajczak [19] (proved in [5,16]) that every 3-connected planar graph has a greedy drawing. One drawback is that a path found by greedy routing may be very long compared to the Euclidean distance between the endpoints. Of course this is inevitable if the geometric graph has large spanning ratio.

When a geometric graph is a t-spanner, we can ideally hope for a local routing algorithm that finds a path whose length is at most k times the Euclidean distance between the endpoints, for some k, where, of necessity, $k \geq t$. The maximum ratio, k, of path length to Euclidean distance is called the *routing ratio*. For example, the Delaunay triangulation, which is a t-spanner for $t \leq 1.998$ [21], permits local routing with routing ratio $k \leq 5.90$ [7]. It is an open question whether the spanning ratio and routing ratio are equal, though there is a provable gap for L_1-Delaunay triangulations [7] and TD-Delaunay triangulations [9].

Other "good" Paths. Recently, a number of other notions of "good" paths in geometric graphs have been investigated. Alamdari et al. [2] introduced *self-approaching* graphs, where any two vertices s and t are joined by a *self-approaching path*—a path such that a point moving continuously along the path from s to any intermediate destination r on the path always gets closer to r in Euclidean distance. In an *increasing-chord* graph, this property also holds for the reverse path from t to s. The self-approaching path property is stronger than the greedy path property in two ways: it applies to every intermediate destination r, and it requires that continuous motion (not just the vertices) along the path to r always gets closer to r. The significance of the stronger property is that self-approaching and increasing-chord graphs have bounded spanning ratios of 5.333 [15] and 2.094 [20], respectively. An important characterization is that a path is self-approaching if and only if at each point on the path, there is a 90° wedge that contains the rest of the path [15].

Angelini et al. [4] introduced *monotone drawings*, where any two vertices s and t are joined by a path that is monotone in some direction. This is a natural desirable property, but not enough to guarantee a bounded spanning ratio.

Angle-Monotone Paths. In this paper we explore properties of another class of geometric graphs with good path properties. These are the *angle-monotone graphs* which were first introduced by Dehkordi, Frati, and Gudmundsson [12] as a tool to investigate increasing-chord graphs. (We note that Dehkordi et al. [12] did not give a name to their graph class.)

A polygonal path with vertices $v_0, v_1, \ldots, v_n$ is *β-monotone* for some angle β if the vector of every edge (v_i, v_{i+1}) lies in the closed $90°$ wedge between $\beta - 45°$ and $\beta + 45°$. (In the terminology of Dehkordi et al. [12] this is a *θ-path*.) In particular, an *x-y-monotone* path (where x and y coordinates are both non-decreasing along the path) is a β-monotone path for $\beta = 45°$ (measured from the positive x-axis). A path is *angle-monotone* if there is some angle β for which it is β-monotone. To visualize this, note that a path is angle-monotone if and only if it can be rotated to be x-y-monotone. An angle-monotone path is a special case of a self-approaching path where the wedges containing the rest of the path all have the same orientation. See Fig. 1. This implies that an angle-monotone path is also angle-monotone when traversed in the other direction, and thus, has the increasing-chord property. Observe that angle-monotone paths have spanning ratio $\sqrt{2}$—this is because x-y-monotone paths do.

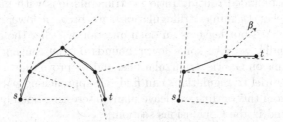

Fig. 1. The difference between a self-approaching st path (left) with $90°$ wedges each containing the rest of the path, and an angle-monotone path (right) where the $90°$ wedges all have the same orientation β.

A geometric graph is *angle-monotone* if for every pair of vertices u, v, there is an angle-monotone path from u to v. Note that the angle β may be different for different pairs u, v. Dehkhori et al. [12] introduced angle-monotone graphs, and proved that they include the class of Gabriel triangulations (triangulations with no obtuse angle). Their main goal was to prove that any set of n points in the plane has a planar increasing-chord graph with $O(n)$ Steiner points and $O(n)$ edges. Given their result that Gabriel graphs are increasing chord, this follows from a result of Bern et al. [6] that any point set can be augmented with $O(n)$ points to a point set whose Delaunay triangulation is Gabriel.

The notion of angle-monotone graphs can be generalized to wedges of angle γ different from $90°$. (A precise definition is given below.) We call these *angle-monotone graphs with width γ*, or *generalized angle-monotone graphs*. For $\gamma < 180°$, they still have bounded spanning ratios.

Results. The main themes we explore are: Which geometric graphs are angle-monotone? Can we create a sparse (generalized) angle-monotone graph on any given point set? Do angle-monotone graphs permit local routing?

Our first main result is a polynomial time algorithm to test if a geometric graph is angle-monotone. This is significant because it is not known whether increasing chord graphs can be recognized in polynomial time (or whether the problem is NP-hard). Our algorithm extends to generalized angle-monotone graphs for any width $\gamma < 180°$.

Our next result is that for any point set in the plane, there is a plane geometric graph on that point set that is angle-monotone with width 120°. In particular, we prove that the *half-θ_6-graph* has this property. Width 90° cannot always be achieved because it would imply spanning ratio $\sqrt{2}$ which is known to be impossible for some point sets, as discussed below under Further Background.

The rest of the paper is about local routing algorithms, where we concentrate on a subclass of angle-monotone graphs, namely the Gabriel triangulations. We give a local routing algorithm for Gabriel triangulations that achieves routing ratio $1+\sqrt{2} \approx 2.41$. This is better than the best known routing ratio for Delaunay triangulations of 5.90 [7]. Also, our algorithm is simpler. The algorithm succeeds, i.e. finds a path to the destination, for any triangulation, and we prove that the algorithm has a bounded routing ratio for triangulations with maximum angle less than 120°. For Delaunay triangulations, we prove a lower bound on the routing ratio of 5.07, but leave as an open question whether the algorithm ever does worse. Finally, we give some lower bounds on the routing ratio of local routing algorithms on Gabriel triangulations, and we prove that no local routing algorithm on Gabriel triangulations can find self-approaching paths.

As is clear from this outline, we leave many interesting open questions, some of which are listed in the Conclusions section.

Further Background. The standard Delaunay triangulation is not self-approaching in general [2], and therefore not angle-monotone.

The *Gabriel graph* of point set P is a graph in which for every edge (u, v) the circle with diameter uv contains no points of P. A Gabriel graph that is a triangulation is called a *Gabriel triangulation*. Any Gabriel triangulation is a Delaunay triangulation. Observe that a triangulation is Gabriel if and only if it has no obtuse angles. Not every point set has a Gabriel triangulation, e.g. three points forming an obtuse triangle.

There are several results on constructing self-approaching/increasing-chord graphs on a given set of points. Alamdari et al. [2] constructed an increasing chord network of linear size using Steiner points, and Dehkordi et al. [12] improved this to a plane network. It is an open question whether every point set admits a plane increasing-chord graph without adding Steiner points. However, for the more restrictive case of angle-monotone graphs, the answer is no: any angle-monotone graph has spanning ratio $\sqrt{2}$ but there is a point set (specifically, the vertices of a regular 23-gon) for which any planar geometric graph has spanning ratio at least 1.4308 [13]. An earlier example was given by Mulzer [17].

Preliminaries and Definitions. A polygonal path with vertices $v_0, v_1, \ldots, v_n$ is β-*monotone with width* γ for some angles β and γ with $\gamma < 180°$ if the vector of every edge (v_i, v_{i+1}) lies in the closed wedge of angle γ between $\beta - \frac{\gamma}{2}$ and $\beta + \frac{\gamma}{2}$. When we have no need to specify β, we say that the path is *angle-monotone with width* γ, or "generalized angle-monotone". A path that is generalized angle-monotone is a *generalized self-approaching path* [1] and thus has bounded spanning ratio depending on γ [1]. But in fact, we can do better:

Observation 1 *[proof in long version]. The spanning ratio of an angle-monotone path with width* $\gamma < 180°$ *is at most* $1/\cos\frac{\gamma}{2}$.

A geometric graph is *angle-monotone with width* γ if for every pair of vertices u, v, there is an angle-monotone path with width γ from u to v. When we have no need to specify γ, we say that the graph is "generalized angle-monotone".

Note that in an angle-monotone path (with width $90°$) the distances from v_0 to later vertices form an increasing sequence. Furthermore, any β-monotone path from u to v lies in a rectangle with u and v at opposite corners and with two sides at angles $\beta \pm 45°$, and the union of such rectangles over all $\beta \in [0, 360°)$ forms the disc with diameter uv. (See Figure in long version.) This implies:

Lemma 1. *Any angle-monotone path from* u *to* v *lies inside the disc with diameter* uv.

2   Recognizing Angle-Monotone Graphs

In this section we give an $O(nm^2)$ time algorithm to test if a geometric graph with n vertices and m edges is angle-monotone. The idea is to look for angle-monotone paths from a node s to all other nodes, and then repeat over all choices of s. For a given source vertex s, the algorithm explores nodes u in non-decreasing order of their distance from s. At each vertex u we store information to capture all the possible angles β for which there is a β-monotone path from s to u. We show how to propagate this information along an edge from u to v.

We begin with some notation. We will measure angles counterclockwise from the positive x-axis, modulo $360°$. To any ordered pair u, v of vertices (points) of our geometric graph we associate the vector $v - u$ and we denote its angle by $\alpha(u, v)$. If S is a set of angles that lie within a wedge of angle less than $180°$, then we define the *minimum* of S to be the most clockwise angle, and the *maximum* of S to be the most counter-clockwise angle. More formally, α is the *minimum* of S if for any other $\beta \in S$, $\beta - \alpha \in [0, 180°)$, and similarly for *maximum*.

Although there may be exponentially many angle-monotone paths from s to u, each such path has two extreme edges. More precisely, if P is an angle-monotone path from s to u, then the angles, $\alpha(e), e \in P$, lie in a $90°$ wedge, and so this set has a minimum and maximum that differ by at most $90°$. We will store a list of all such min-max pairs with vertex u. Each pair defines a wedge of at most $90°$. Since each pair is defined by two edges, there are at most $O(m^2)$ such pairs (though we will show below that we only need to store $O(m)$ of them).

The algorithm starts off by looking at every edge (s, u) and adding the pair $(\alpha(s, u), \alpha(s, u))$ to u's list. Then the algorithm explores vertices $u \neq s$ in non-decreasing order of their distance from s. To explore vertex u, consider each edge (u, v) and each pair $(\alpha(e), \alpha(f))$ stored with u, and update the list of pairs for vertex v as follows. If $\alpha(u, v)$ is within $90°$ of $\alpha(e)$ and within $90°$ of $\alpha(f)$ then add to v's list the pair $(\min\{\alpha(u, v), \alpha(e)\}, \max\{\alpha(u, v), \alpha(f)\})$.

If ever the algorithm tries to explore a vertex that has no pairs stored with it, then halt—the graph is not angle-monotone. To justify correctness we prove:

Lemma 2. *When the algorithm has explored all the vertices closer to s than v, then there exists an angle-monotone path from s to v with extreme edges e and f if and only if the pair $(\alpha(e), \alpha(f))$ is in v's list.*

Proof. The proof is by induction on the distance from s to v.

For the "only if" direction, let P be an angle-monotone path from s to v with extreme edges e and f, and let u be the penultimate vertex of P. The subpath of P from s to u is an angle-monotone path. Suppose its extreme edges are e' and f' where $e = e'$ or $f = f'$ or both. Now, u is closer to s so by induction the pair $(\alpha(e'), \alpha(f'))$ is in u's list. Because P is angle-monotone, $\alpha(u, v)$ is within $90°$ of $\alpha(e')$ and $\alpha(f')$. Thus the update step applies. During the update step we add the angle $\alpha(u, v)$ to the pair $(\alpha(e'), \alpha(f'))$, which gives the pair $(\alpha(e), \alpha(f))$. Thus we add the pair $(\alpha(e), \alpha(f))$ to v's list.

For the "if" direction, suppose that the pair $(\alpha(e), \alpha(f))$ is in v's list. This pair was added to v's list because of an update from some vertex u closer to s applied to some pair $(\alpha(e'), \alpha(f'))$ in u's list. By induction, there exists an angle-monotone path P from s to u with extreme edges e' and f', and because the update is only performed when $\alpha(u, v)$ is within $90°$ degrees of $\alpha(e')$ and $\alpha(f')$ therefore the edge (u, v) can be added to P to produce an angle-monotone path with extreme edges e and f. □

To improve the efficiency of the algorithm we observe that it is redundant to store at a vertex v a pair whose wedge contains the wedge of another pair. Therefore, we only need to store $O(m)$ pairs at each vertex, at most one pair whose first element is $\alpha(e)$ for each edge e. We can simply keep with each vertex v a vector indexed by edges e, in which we store the minimal pair $(\alpha(e), \alpha(f))$ (if any) associated with v so far. Finally, observe that during the course of the algorithm, each edge (u, v) is handled once in an update step. With the refinement just mentioned, handling an edge costs $O(m)$. Therefore the algorithm runs in time $O(m^2)$ for a single choice of s, and in time $O(nm^2)$ overall.

The algorithm can be generalized to recognize angle-monotone graphs of width γ for fixed $\gamma < 180°$. It is no longer legitimate to explore vertices in order of distance from s, since a generalized angle-monotone path will not necessarily respect this ordering. However, we can run the algorithm in phases, where phase i captures all the angle-monotone paths of width γ that start at s and have at most i edges. Since no angle-monotone path can repeat a vertex, there are at most $n - 1$ edges in any angle-monotone path. Thus we need $n - 1$ phases. In each phase, for each directed edge (u, v) we update each pair $(\alpha(e), \alpha(f))$ stored at u as follows. If $\alpha(u, v)$ is within γ of $\alpha(e)$ and within γ of $\alpha(f)$ then add to

v's list the pair $(\min\{\alpha(u,v),\alpha(e)\}, \max\{\alpha(u,v),\alpha(f)\})$. In this way, each of the $n-1$ phases takes time $O(m^2)$, so the total run-time of the algorithm over all choices of s becomes $O(n^2m^2)$.

3   A Class of Generalized Angle-Monotone Graphs

In this section we show that every point set in the plane has a plane geometric graph that is angle-monotone with width $120°$. In particular, we will prove that the *half-θ_6-graph* has this property. As noted in the Introduction, there are point sets for which no plane graph is angle-monotone with width $90°$. It is an open question to narrow this gap and find the minimum angle γ for which every point set has a plane graph that is angle-monotone with width γ (and thus spanning ratio $1/\cos\frac{\gamma}{2}$).

We first define the half-θ_6-graph. For each point $u \in P$, partition the plane into $60°$ cones with apex u, with each cone defined by two rays at consecutive multiples of $60°$ from the positive x-axis. Label the cones C_0, C_1, C_2, C_3, C_4, and C_5 in clockwise order around u, starting from the cone containing the positive y-axis. See Fig. 2(a).

For two vertices u and v the *canonical triangle* T_{uv} is the triangle bounded by: the cone of u that contains v; and the line through v perpendicular to the bisector of that cone. See Fig. 2(b). Notice that if v is in an even cone of u, then u is in an odd cone of v. We build the half-θ_6-graph as follows. For each vertex u and each even $i = 0, 2, 4$, add the edge uv provided that v is in the C_i cone of u and T_{uv} is empty. We call v the *C_i-neighbour of u*. For simplicity, we assume that no two points lie on a line parallel to a cone boundary, guaranteeing that each vertex connects to exactly one vertex in each even cone. Hence the graph has at most $3n$ edges in total. The half-θ_6-graph is a type of Delaunay triangulation where the empty region is an equilateral triangle in a fixed orientation as opposed to a disk [8]. It can be computed in $O(n\log n)$ time [18].

To prove angle-monotonicity properties of the half-θ_6-graph, we use an idea like the one used by Angelini [3]. His goal was to show that every abstract triangulation has an embedding that is monotone, i.e. angle-monotone with

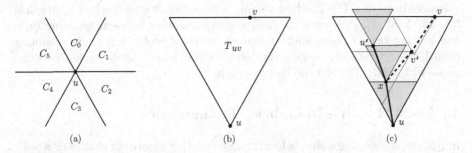

Fig. 2. (a) 6 cones originating from point u, (b) Canonical triangle T_{uv}, (c) path σ_u (solid) with its empty canonical triangles shaded, path σ_v (dashed) and their common vertex x.

width 180°. (The same result was obtained in [14] with a different proof.) Angelini did this by showing that the Schnyder drawing of any triangulation is monotone, and in fact, upon careful reading, his proof shows that any Schnyder drawing is angle-monotone with the smaller width 120°. Schnyder drawings are a special case of half-θ_6-graphs [8] so it is not surprising that Angelini's proof idea extends to the half-θ_6-graph in general.

Theorem 1. *The half-θ_6-graph is angle-monotone with width 120°.*

Proof. We must prove that for any points u and v, there is an angle-monotone path from u to v of width 120°. Assume without loss of generality that v is in the C_0 cone of u. See Fig. 2(b).

Our path from u to v will be the union of two paths, each of which is angle-monotone of width 60°. We begin by constructing a path σ_u from u in which each vertex is joined to its C_0 neighbour. This is a β-monotone path of width 60° for $\beta = 90°$. If the path contains v we are done, so assume otherwise. Let u' be the last vertex of the path that lies in T_{uv}. Note that v cannot lie in the C_0 cone of u'. Let S be the subpath of σ_u from u to u', together with the C_0 cone of u'. Then S separates T_{uv} into two parts. Suppose that v lies in the right-hand part (the other case is symmetric). See Fig. 2(c).

Next, construct a path σ_v from v in which each vertex is joined to its C_4 neighbour. This is a β-monotone path of width 60° for $\beta = 210°$.

We now claim that σ_u and σ_v have a common vertex x. Then as our final path from u to v we take the portion of σ_u from u to x followed by the portion of σ_v backwards from x to v. Since the reverse of σ_v is β-monotone with width 60° for $\beta = 30°$, the final path is β-monotone with width 120° for $\beta = 60°$.

It remains to prove that x exists. Let v' be the last vertex of σ_v that lies strictly to the right of S. Let u'' be the last vertex of σ_u that lies below v'. We claim that u'' is the C_4 neighbour of v', and thus that u'' provides our vertex x. Let T be the empty canonical triangle from u'' to its C_0-neighbour (or the empty C_0 cone of u'' in case u'' has no C_0-neighbour). First note that u'' is in the C_4 cone of v'—otherwise v' would be in T. Next note that $T_{v'u''}$ is empty—otherwise v would have a C_4-neighbour that is in T or is to the right of S.   □

Theorem 1 implies that the spanning ratio of the half-θ_6-graph is 2, which was already known [11]. The best routing ratio achievable for the half-θ_6-graph is $5/\sqrt{3} \approx 2.887$ [9]. (This was the first proved separation between spanning ratio and routing ratio.) Since angle-monotone paths of width 120° have spanning ratio 2, this implies that no local routing algorithm can compute angle-monotone paths with width 120° on the half-θ_6-graph.

4 Local Routing in Gabriel Triangulations

In this section we give a simple local "angle" routing algorithm that finds a path from s to t in any triangulation. Like previous algorithms, the path walks only along edges of triangles that intersect the line segment st. The novelty is that the next edge of the path is chosen based on angles relative to the vector st.

The details of the algorithm are in Sect. 4.1. In Sect. 4.2 we prove that the algorithm has routing ratio $1 + \sqrt{2}$ on Gabriel graphs, and discuss its behaviour on Delaunay triangulations. In Sect. 4.3 we give lower bounds on the routing and competitive ratios of local routing algorithms on Gabriel graphs.

4.1 Local Angle Routing

Our algorithm is simple to describe: Suppose we want a route from s to t in a triangulation. Orient st horizontally, t to the right. Suppose we have reached vertex p. Consider the last (rightmost) triangle that is incident to p and intersects the line segment st. The triangle has two edges incident to p. Of these two edges, take the one that has the minimum angle to the horizontal ray from p to the right. See Fig. 3. Pseudo-code can be found below in Algorithm 1. Note that in the pseudo-code, the angle test is equivalently replaced by two tests, identifying steps of type A and B for easier case analysis. For an example of a path computed by the algorithm, see Fig. 4. Observe that the algorithm always succeeds in finding a route from s to t because it always advances rightward in the sequence of triangles that intersect line segment st.

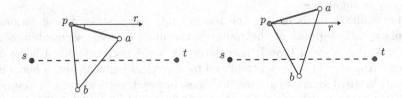

Fig. 3. Local routing from s to t. At vertex p, with pab being the rightmost triangle incident to p that intersects line segment st, we route from p to a because the (unsigned) angle apr is less than angle bpr. A step of type A is shown on the left and a step of type B on the right.

Algorithm 1. Local angle routing

```
1  p ← s
2  while p ≠ t do
3      Let T = pab be the rightmost triangle containing p that intersects segment
          st, with p and a on the same side of line st.
4      if a is closer to line st than p then              /* step of type A */
5          p ← a
6      else                                               /* step of type B */
7          if |slope(pa)| ≤ |slope(pb)| then
8              p ← a
9          else
10             p ← b
```

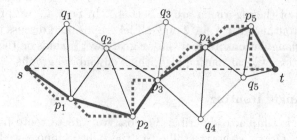

Fig. 4. Example of route computed by Algorithm 1 (heavy blue path). In dotted red, a longer route obtained by replacing each segment of the route by the most extreme angle. Both routes are within $(1 + \sqrt{2})$ of $||st||$. (Color figure online)

4.2 Analysis of the Algorithm

In this section we will prove that the above algorithm has routing ratio exactly $1 + \sqrt{2}$ on Gabriel triangulations, which have maximum angle at most 90°. In the last part of the section we generalize the analysis to triangulations with a larger maximum angle, and we show that the routing ratio is at least 5.07 on Delaunay triangulations.

The intuition for bounding the routing ratio on Gabriel triangulations is to replace each segment of the route by the most extreme segment possible. See Fig. 4. Any step of type B is replaced by a 45° segment plus a horizontal segment. Any step of type A is replaced by a vertical segment plus a horizontal segment. Vertical segments are the bad ones, but each vertical must be preceded by 45° segments, which means that instead of travelling 1 unit horizontally (the optimum route) we have travelled $\sqrt{2}$ along a 45° segment plus 1 vertically, giving us the $1 + \sqrt{2}$ ratio. We now give a more formal proof.

For each edge $e = (p_i, p_{i+1})$ of the path, let $d_x(e) = ||x(p_i) - x(q_{p+1})||$ and $d_y(e) = ||y(p_i) - y(p_{i+1})||$. Let A (resp. B) be the set of edges of the path where the algorithm makes a step of type A (resp. type B). (Context will distinguish edge sets from steps.) Let $x_B = \sum_{e \in B} d_x(e)$ and $x_A = \sum_{e \in A} d_x(e)$.

Lemma 3. *On any Gabriel triangulation the path computed by Algorithm 1 is x-increasing.*

Proof. Let us show that each step is x-increasing. Consider a step from p, with a and b as defined in Algorithm 1. Assume without loss of generality that p and a are above line st and b is below. Since T is the last triangle incident to p that intersects st, the clockwise ordering of T is pab. Refer to Fig. 3.

If the algorithm takes a step of type B then a is above p (in y coordinate) and b is below p. Since $\angle bpa \le 90°$, thus $x(a)$ and $x(b)$ are greater than $x(p)$. If the algorithm takes a step of type A then since b is below st and a is above st and $\angle bap \le 90°$, thus $x(a)$ is greater than $x(p)$.   □

Theorem 2. *On any Gabriel triangulation, Algorithm 1 has a routing ratio of $1 + \sqrt{2}$ and this bound is tight.*

Proof. We first bound $\sum_{e \in B} ||e||$. Observe that each edge in B forms an angle with the horizontal line through p that is at most $45°$. Thus $\sum_{e \in B} d_y(e) \leq x_B$ and $\sum_{e \in B} ||e|| \leq \sqrt{2} x_B$.

We next bound $\sum_{e \in A} ||e||$. Observe that edges in A move us closer to the line st, and must be balanced by previous steps (of type B) that moved us farther from the line st. This implies that $\sum_{e \in A} d_y(e) \leq \sum_{e \in B} d_y(e) \leq x_B$ (where the last step comes from the first observation). Since $||e|| \leq d_x(e) + d_y(e)$, thus $\sum_{e \in A} ||e|| \leq x_A + \sum_{e \in A} d_y(e) \leq x_A + x_B$.

Putting these together, the length of the path is bounded by $\sum_{e \in A} ||e|| + \sum_{e \in B} ||e|| \leq x_A + x_B + \sqrt{2} x_B \leq (1 + \sqrt{2})(x_A + x_B)$. Finally, by Lemma 3, $x_A + x_B = ||st||$, so this proves that the routing ratio is at most $(1 + \sqrt{2})$.

An example to show that this analysis is tight is given in the full version. $\square$

We conclude this section with two results on the behaviour of the routing algorithm on other triangulations. Proofs can be found in the full version.

Theorem 3. *In a triangulation with maximum angle $\alpha < 120°$ Algorithm 1 has a routing ratio of $(\sin \alpha + \sin \frac{\alpha}{2})/\sin \frac{3\alpha}{2}$ and this bound is tight.*

Theorem 4. *The routing ratio of Algorithm 1 on Delaunay triangulation is greater than 5.07.*

We believe that the routing ratio of Algorithm 1 on Delaunay triangulations is close to 5.07, but leave that as an open question. We remark that Algorithm 1 is different from the generalization of Chew's Routing Algorithm for Delaunay triangulations [10] (cf. the algorithm described in [7]).

4.3   Limits of Local Routing Algorithms on Gabriel Triangulations

In this section we prove some limits on local routing on Gabriel triangulations. Proofs are deferred to the full version.

A routing algorithm on a geometric graph G has a *competitive ratio* of c if the length of the path produced by the algorithm from any vertex s to any vertex t is at most c times the length of the shortest path from s to t in G, and c is the minimum such value. (Recall that the routing ratio compares the length of the path produced by the algorithm to the Euclidean distance between the endpoints. Thus the competitive ratio is less than or equal to the routing ratio).

A routing algorithm is *k-local* (for some integer constant $k > 0$) if it makes forwarding decisions based on: (1) the k-neighborhood in G of the current position of the message; and (2) limited information stored in the message header.

Theorem 5. *Any k-local routing algorithm on Gabriel triangulations has routing ratio at least 1.4966 and competitive ratio at least 1.2687.*

Although Gabriel triangulations are angle-monotone [12], Theorem 5 shows that no local routing algorithm can compute angle-monotone paths since that would give routing ratio $\sqrt{2}$. The following theorem tells us that even less constrained paths cannot be computed locally:

Theorem 6. *There is no k-local routing algorithm on Gabriel triangulations that always finds self-approaching paths.*

5 Conclusions

We conclude this paper with some open questions.

1. What is the minimum angle γ for which every point set has a plane geometric graph that is angle-monotone with width γ (and thus has spanning ratio $1/\cos\frac{\gamma}{2}$)? We proved $\gamma \leq 120°$, and it is known that $\gamma > 90°$.
2. Is there a local routing algorithm with bounded routing ratio for any angle-monotone graph? Any increasing-chord graph?
3. We bounded the routing ratio of our local routing algorithm on triangulations based on the maximum angle in the triangulation, but how does this relate to the property of being generalized angle-monotone? If a triangulation has bounded maximum angle, is it generalized angle-monotone? The only thing known is that maximum angle 90° implies angle-monotone with width 90° [12].
4. Is the standard Delaunay triangulation generalized angle-monotone? In particular, proving that the Delaunay triangulation is angle-monotone with width strictly less than 120° would provide a different proof that the Delaunay triangulation has spanning ratio less than 2 [21]. It is known that the Delaunay triangulation is not angle-monotone with width 90° (see Sect. 1).
5. How does our local routing algorithm behave on standard Delaunay triangulations? We proved a lower bound of 5.07 on the routing ratio. We believe the routing ratio is close to this value, but have no upper bound.

Acknowledgements. This work was begun at the CMO-BIRS Workshop on Searching and Routing in Discrete and Continuous Domains, October 11–16, 2015. We thank the other participants of the workshop for many good ideas and stimulating discussions. We thank an anonymous referee for helpful comments.

Funding acknowledgements: A.L. thanks NSERC (Natural Sciences and Engineering Council of Canada). S.V. thanks NSERC and the Ontario Ministry of Research and Innovation. N.B. thanks French National Research Agency (ANR) in the frame of the "Investments for the future" Programme IdEx Bordeaux - CPU (ANR-10-IDEX-03-02). I.K. was supported in part by the NWO under project no. 612.001.106, and by F.R.S.-FNRS.

References

1. Aichholzer, O., Aurenhammer, F., Icking, C., Klein, R., Langetepe, E., Rote, G.: Generalized self-approaching curves. Discrete Appl. Math. **109**(1–2), 3–24 (2001)
2. Alamdari, S., Chan, T.M., Grant, E., Lubiw, A., Pathak, V.: Self-approaching Graphs. In: Didimo, W., Patrignani, M. (eds.) GD 2012. LNCS, vol. 7704, pp. 260–271. Springer, Heidelberg (2013). doi:10.1007/978-3-642-36763-2_23

3. Angelini, P.: Monotone drawings of graphs with few directions. In: 6th International Conference on Information, Intelligence, Systems and Applications (IISA), pp. 1–6. IEEE (2015)
4. Angelini, P., Colasante, E., Battista, G.D., Frati, F., Patrignani, M.: Monotone drawings of graphs. J. Graph Algorithms Appl. **16**(1), 5–35 (2012)
5. Angelini, P., Frati, F., Grilli, L.: An algorithm to construct greedy drawings of triangulations. J. Graph Algorithms Appl. **14**(1), 19–51 (2010)
6. Bern, M., Eppstein, D., Gilbert, J.: Provably good mesh generation. In: Proceedings of 31st Symposium on Foundations of Computer Science (FOCS), pp. 231–241. IEEE (1990)
7. Bonichon, N., Bose, P., Carufel, J.-L., Perković, L., Renssen, A.: Upper and lower bounds for online routing on Delaunay triangulations. In: Bansal, N., Finocchi, I. (eds.) ESA 2015. LNCS, vol. 9294, pp. 203–214. Springer, Heidelberg (2015). doi:10.1007/978-3-662-48350-3_18
8. Bonichon, N., Gavoille, C., Hanusse, N., Ilcinkas, D.: Connections between theta-graphs, delaunay triangulations, and orthogonal surfaces. In: Thilikos, D.M. (ed.) WG 2010. LNCS, vol. 6410, pp. 266–278. Springer, Heidelberg (2010). doi:10.1007/978-3-642-16926-7_25
9. Bose, P., Fagerberg, R., van Renssen, A., Verdonschot, S.: Optimal local routing on Delaunay triangulations defined by empty equilateral triangles. SIAM J. Comput. **44**(6), 1626–1649 (2015)
10. Chew, L.P.: There is a planar graph almost as good as the complete graph. In: Proceedings of 2nd Annual Symposium Computational Geometry (SoCG), pp. 169–177 (1986)
11. Chew, L.P.: There are planar graphs almost as good as the complete graph. J. Comput. Syst. Sci. **39**(2), 205–219 (1989)
12. Dehkordi, H.R., Frati, F., Gudmundsson, J.: Increasing-chord graphs on point sets. J. Graph Algorithms Appl. **19**(2), 761–778 (2015)
13. Dumitrescu, A., Ghosh, A.: Lower bounds on the dilation of plane span-ners. In: Govindarajan, S., Maheshwari, A. (eds.) CALDAM 2016. LNCS, vol. 9602, pp. 139–151. Springer, Heidelberg (2016). doi:10.1007/978-3-319-29221-2_12. http://arxiv.org/pdf/1509.07181v3.pdf
14. He, X., He, D.: Monotone drawings of 3-connected plane graphs. In: Bansal, N., Finocchi, I. (eds.) ESA 2015. LNCS, vol. 9294, pp. 729–741. Springer, Heidelberg (2015). doi:10.1007/978-3-662-48350-3_61
15. Icking, C., Klein, R., Langetepe, E.: Self-approaching curves. Math. Proc. Camb. Philos. Soc. **125**, 441–453 (1995)
16. Leighton, T., Moitra, A.: Some results on greedy embeddings in metric spaces. Discrete Comput. Geom. **44**, 686–705 (2010)
17. Mulzer, W.: Minimum dilation triangulations for the regular n-gon. Master's thesis, Freie Universität Berlin (2004)
18. Narasimhan, G., Smid, M.: Geometric Spanner Networks. Cambridge University Press, Cambridge (2007)
19. Papadimitriou, C.H., Ratajczak, D.: On a conjecture related to geometric routing. Theor. Comput. Sci. **344**, 3–14 (2005)
20. Rote, G.: Curves with increasing chords. Math. Proc. Camb. Philos. Soc. **115**, 1–12 (1994)
21. Xia, G.: The stretch factor of the Delaunay triangulation is less than 1.998. SIAM J. Comput. **42**(4), 1620–1659 (2013)

Simultaneous Orthogonal Planarity

Patrizio Angelini[1], Steven Chaplick[2], Sabine Cornelsen[3],
Giordano Da Lozzo[4(✉)], Giuseppe Di Battista[4], Peter Eades[5],
Philipp Kindermann[6], Jan Kratochvíl[7], Fabian Lipp[2], and Ignaz Rutter[8]

[1] Universität Tübingen, Tübingen, Germany
`angelini@informatik.uni-tuebingen.de`
[2] Universität Würzburg, Würzburg, Germany
`{steven.chaplick,fabian.lipp}@uni-wuerzburg.de`
[3] Universität Konstanz, Konstanz, Germany
`sabine.cornelsen@uni-konstanz.de`
[4] Roma Tre University, Rome, Italy
`{dalozzo,gdb}@dia.uniroma3.it`
[5] The University of Sydney, Sydney, Australia
`peter@it.usyd.edu.au`
[6] FernUniversität in Hagen, Hagen, Germany
`philipp.kindermann@fernuni-hagen.de`
[7] Charles University, Prague, Czech Republic
`honza@kam.mff.cuni.cz`
[8] Karlsruhe Institute of Technology, Karlsruhe, Germany
`rutter@kit.edu`

Abstract. We introduce and study the ORTHOSEFE-k problem: Given k planar graphs each with maximum degree 4 and the same vertex set, do they admit an OrthoSEFE, that is, is there an assignment of the vertices to grid points and of the edges to paths on the grid such that the same edges in distinct graphs are assigned the same path and such that the assignment induces a planar orthogonal drawing of each of the k graphs? We show that the problem is NP-complete for $k \geq 3$ even if the shared graph is a Hamiltonian cycle and has sunflower intersection and for $k \geq 2$ even if the shared graph consists of a cycle and of isolated vertices. Whereas the problem is polynomial-time solvable for $k = 2$ when the union graph has maximum degree five and the shared graph is biconnected. Further, when the shared graph is biconnected and has sunflower intersection, we show that every positive instance has an OrthoSEFE with at most three bends per edge.

1 Introduction

The input of a simultaneous embedding problem consists of several graphs $G_1 = (V, E_1), \ldots, G_k = (V, E_k)$ on the same vertex set. For a fixed drawing style $\mathcal{S}$, the simultaneous embedding problem asks whether there exist drawings $\Gamma_1, \ldots, \Gamma_k$

This research was initiated at the Bertinoro Workshop on Graph Drawing 2016. Research was partially supported by DFG grant Ka812/17-1, by MIUR project AMANDA, prot. 2012C4E3KT_001, by DFG grant SCHU 2458/4-1, by the grant no. 14-14179S of the Czech Science Foundation GACR, and by DFG grant WA 654/21-1.

Y. Hu and M. Nöllenburg (Eds.): GD 2016, LNCS 9801, pp. 532–545, 2016.
DOI: 10.1007/978-3-319-50106-2_41

of $G_1, \ldots, G_k$, respectively, in drawing style $\mathcal{S}$ such that for any i and j the restrictions of Γ_i and Γ_j to $G_i \cap G_j = (V, E_i \cap E_j)$ coincide.

The problem has been most widely studied in the setting of topological planar drawings, where vertices are represented as points and edges are represented as pairwise interior-disjoint Jordan arcs between their endpoints. This problem is called SIMULTANEOUS EMBEDDING WITH FIXED EDGES or SEFE-k for short, where k is the number of input graphs. It is known that SEFE-k is NP-complete for $k \geq 3$, even in the restricted case of *sunflower instances* [25], where every pair of graphs shares the same set of edges, and even if such a set induces a star [3]. On the other hand, the complexity for $k = 2$ is still open. Recently, efficient algorithms for restricted instances have been presented, namely when (i) the shared graph $G_\cap = G_1 \cap G_2$ is biconnected [4,18] or a star-graph [4], (ii) $G_\cap$ is a collection of disjoint cycles [12], (iii) every connected component of $G_\cap$ is either subcubic or biconnected [10,25], (iv) G_1 and G_2 are biconnected and $G_\cap$ is connected [13], and (v) $G_\cap$ is connected and the input graphs have maximum degree 5 [13]; see the survey by Bläsius et al. [11] for an overview.

For planar straight-line drawings, the simultaneous embedding problem is called SIMULTANEOUS GEOMETRIC EMBEDDING and it is known to be NP-hard even for two graphs [17]. Besides simultaneous intersection representation for, e.g., interval graphs [13,19] and permutation and chordal graphs [20], it is only recently that the simultaneous embedding paradigm has been applied to other fundamental planarity-related drawing styles, namely simultaneous level planar drawings [2] and RAC drawings [5,7].

We continue this line of research by studying simultaneous embeddings in the planar *orthogonal* drawing style, where vertices are assigned to grid points and edges to paths on the grid connecting their endpoints [28]. In accordance with the existing naming scheme, we define ORTHOSEFE-k to be the problem of testing whether k input graphs $\langle G_1, \ldots, G_k \rangle$ admit a simultaneous planar orthogonal drawing. If such a drawing exists, we call it an OrthoSEFE of $\langle G_1, \ldots, G_k \rangle$. Note that it is a necessary condition that each G_i has maximum degree 4 in order to obtain planar orthogonal drawings. Hence, in the remainder of the paper we assume that all instances have this property. For instances with this property, at least when the shared graph is connected, the problem SEFE-2 can be solved efficiently [13]. However, there are instances of ORTHOSEFE-2 that admit a SEFE but not an OrthoSEFE; see Fig. 1(a).

Unless mentioned otherwise, all instances of ORTHOSEFE-k and SEFE-k we consider are sunflower. Notice that instances with $k = 2$ are always sunflower. Let $\langle G_1 = (V, E_1), G_2 = (V, E_2) \rangle$ be an instance of ORTHOSEFE-2. We define the *shared graph* (resp. the *union graph*) to be the graph $G_\cap = (V, E_1 \cap E_2)$ (resp. $G_\cup = (V, E_1 \cup E_2)$) with the same vertex set as G_1 and G_2, whose edge set is the intersection (resp. the union) of the ones of G_1 and G_2. Also, we call the edges in $E_1 \cap E_2$ the *shared edges* and we call the edges in $E_1 \setminus E_2$ and in $E_2 \setminus E_1$ the *exclusive edges*. The definitions of *shared graph*, *shared edges*, and *exclusive edges* naturally extend to sunflower instances for any value of k.

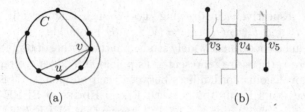

Fig. 1. (a) A negative instance of ORTHOSEFE-2. Shared edges are black, while exclusive edges are red and blue. The red edges require 270° angles on different sides of C. Thus, the blue edge (u, v) cannot be drawn. Note that the given drawing is a SEFE-2. (b) Examples of side assignments for the exclusive edges incident to degree-2 vertices of $G_\cap$: orthogonality constraints are satisfied at v_4 and v_5, while they are violated at v_3. (Color figure online)

One main issue is to decide how degree-2 vertices of the shared graph are represented. Note that, in planar topological drawings, degree-2 vertices do not require any decisions as there exists only a single cyclic order of their incident edges. In the case of orthogonal drawings there are, however, two choices for a degree-2 vertex: It can either be drawn *straight*, i.e., it is incident to two angles of 180°, or *bent*, i.e., it is incident to one angle of 90° and to one angle of 270°. If v is a degree-2 vertex of the shared graph with neighbors u and w, and two exclusive edges e, e', say of G_1, are incident to v and are embedded on the same side of the path uvw, then v must be bent, which in turn implies that also every exclusive edge of G_2 incident to v has to be embedded on the same side of uvw as e and e'. In this way, the two input graphs of ORTHOSEFE-2 interact via the degree-2 vertices. It is the difficulty of controlling this interaction that marks the main difference between SEFE-k and ORTHOSEFE-k. To study this interaction in isolation, we focus on instances of ORTHOSEFE-2 where the shared graph is a cycle for most of the paper. Note that such instances are trivial yes-instances of SEFE-k (provided the input graphs are all planar).

Contributions and Outline. In Sect. 2, we provide our notation and we show that the existence of an OrthoSEFE of an instance of ORTHOSEFE-k can be described as a combinatorial embedding problem. In Sect. 3, we show that ORTHOSEFE-3 is NP-complete even if the shared graph is a cycle, and that ORTHOSEFE-2 is NP-complete even if the shared graph consists of a cycle plus some isolated vertices. This contrasts the situation of SEFE-k where these cases are polynomially solvable [4,9,18,25]. In Sect. 4, we show that ORTHOSEFE-2 is efficiently solvable if the shared graph is a cycle and the union graph has maximum degree 5. Finally, in Sect. 5, we extend this result to the case where the shared graph is biconnected (and the union graph still has maximum degree 5). Moreover, we show that any positive instance of ORTHOSEFE-k whose shared graph is biconnected admits an OrthoSEFE with at most three bends per edge. We close with some concluding remarks and open questions in Sect. 6.

Complete proofs can be found in the full version of the paper [1].

2   Preliminaries

We will extensively make use of the NOT-ALL-EQUAL 3-SAT (NAE3SAT) problem [24, p.187]. An instance of NAE3SAT consists of a 3-CNF formula ϕ with variables $x_1, \ldots, x_n$ and clauses $c_1, \ldots, c_m$. The task is to find a NAE *truth assignment*, i.e., a truth assignment such that each clause contains both a true and a false literal. NAE3SAT is known to be NP-complete [26]. The *variable–clause graph* is the bipartite graph whose vertices are the variables and the clauses, and whose edges represent the membership of a variable in a clause. The problem PLANAR NAE3SAT is the restriction of NAE3SAT to instances whose variable–clause graph is planar. PLANAR NAE3SAT can be solved efficiently [22, 27].

Embedding Constraints. Let $\langle G_1, \ldots, G_k \rangle$ be an ORTHOSEFE-k instance. A *SEFE* is a collection of embeddings $\mathcal{E}_i$ for the G_i such that their restrictions on $G_\cap$ are the same. Note that in the literature, a SEFE is often defined as a collection of drawings rather than a collection of embeddings. However, the two definitions are equivalent [21]. For a SEFE to be realizable as an OrthoSEFE it needs to satisfy two additional conditions. First, let v be a vertex of degree 2 in $G_\cap$ with neighbors u and w. If in any embedding $\mathcal{E}_i$ there exist two exclusive edges incident to v that are embedded on the same side of the path uvw, then any exclusive edge incident to v in any of the $\mathcal{E}_j \neq \mathcal{E}_i$ must be embedded on the same side of the path uvw. Second, let v be a vertex of degree 3 in $G_\cap$. All exclusive edges incident to v must appear between the same two edges of $G_\cap$ around v. We call these the *orthogonality constraints*. See Fig. 1(b).

Theorem 1. *An instance $\langle G_1, \ldots, G_k \rangle$ of ORTHOSEFE-k has an OrthoSEFE if and only if it admits a SEFE satisfying the orthogonality constraints.*

For the case in which the shared graph is a cycle C, we give a simpler version of the constraints in Theorem 1, which will prove useful in the remainder of the paper. By the Jordan curve theorem, a planar drawing of cycle C divides the plane into a bounded and an unbounded region – the *inside* and the *outside* of C, which we call the *sides* of C. Now the problem is to assign the exclusive edges to either of the two sides of C so that the following two conditions are fulfilled.

Planarity Constraints. Two exclusive edges of the same graph must be drawn on different sides of C if their endvertices alternate along C.

Orthogonality Constraints. Let $v \in V$ be a vertex that is adjacent to two exclusive edges e_i and e_i' of the same graph G_i, $i \in \{1, \ldots, k\}$. If e_i and e_i' are on the same side of C, then all exclusive edges incident to v of all graphs $G_1, \ldots, G_k$ must be on the same side as e_i and e_i'.

Note that this is a reformulation of the general orthogonality constraints. Further, the orthogonality constraints also imply that if e_i and e_i' are on different sides of C, then for each graph G_j that contains two exclusive edges e_j and e_j' incident to v, with $j \in \{1, \ldots, k\}$, e_j and e_j' must be on different sides of C.

The next theorem follows from Theorem 1 and from the following two observations. First, for a sunflower instance $\langle G_1, \ldots, G_k \rangle$ whose shared graph is a cycle, any collection of embeddings is a SEFE [21]. Second, the planarity constraints are necessary and sufficient for the existence of an embedding of G_i [6].

Theorem 2. *An instance of* ORTHOSEFE-k *whose shared graph is a cycle C has an OrthoSEFE if and only if there exists an assignment of the exclusive edges to the two sides of C satisfying the planarity and orthogonality constraints.*

3   Hardness Results

We show that ORTHOSEFE-k is NP-complete for $k \geq 3$ for instances with sunflower intersection even if the shared graph is a cycle, and for $k = 2$ even if the shared graph consists of a cycle and isolated vertices.

Theorem 3. ORTHOSEFE-k *with $k \geq 3$ is NP-complete, even for instances with sunflower intersection in which (i) the shared graph is a cycle and (ii) $k-1$ of the input graphs are outerplanar and have maximum degree 3.*

Proof sketch. The membership in NP directly follows from Theorem 2. To prove the NP-hardness, we show a polynomial-time reduction from the NP-complete problem POSITIVE EXACTLY-THREE NAE3SAT [23], which is the variant of NAE3SAT in which each clause consists of exactly three unnegated literals.

Let $x_1, x_2, \ldots, x_n$ be the variables and let $c_1, c_2, \ldots, c_m$ be the clauses of a 3-CNF formula ϕ of POSITIVE EXACTLY-THREE NAE3SAT. We show how to construct an equivalent instance $\langle G_1, G_2, G_3 \rangle$ of ORTHOSEFE-3 such that G_1 and G_2 are outerplanar graphs of maximum degree 3. We refer to the exclusive edges in G_1, G_2, and G_3 as red, blue, and green, respectively; refer to Fig. 2.

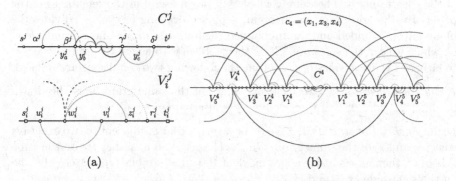

(a)   (b)

Fig. 2. (a) A clause gadget C_j (top) and a variable-clause gadget V_i^j (bottom); solid edges belong to the gadgets, dotted edges are optional, and dashed edges are transmission edges. (b) Illustration of instance $\langle G_1, G_2, G_3 \rangle$, focused on a clause c_4. Black edges belong to the shared graph $G_\cap$. The red, blue, and green edges are the exclusive edges of G_1, G_2, and G_3, respectively. (Color figure online)

For each clause c_j, $j = 1, \ldots, m$, we create a *clause gadget* C^j as in Fig. 2(a) (top). For each variable x_i, $i = 1, \ldots, n$, and each clause c_j, $j = 1, \ldots, m$, we create a *variable-clause gadget* V_i^j as in Fig. 2(a) (bottom). Observe that the (dotted) green edge $\{w_i^j, r_i^j\}$ in a variable-clause gadget is only part of V_i^j if x_i does not occur in c_j. Otherwise, there is a green edge $\{w_i^j, y_x^j\}$ connecting w_i^j to one of the three vertices y_a^j, y_b^j, or y_c^j (dashed stubs) in the clause gadget. Observe that these three *variable-clause edges* per clause can be realized in such a way that there exist no planarity constraints between pairs of them. In Fig. 2(b), the variable-clause gadgets V_1^4, V_3^4, V_4^4 are incident to variable-clause edges, while V_2^4 and V_5^4 contain edges $\{w_2^4, r_2^4\}$ and $\{w_5^4, r_5^4\}$, respectively.

The gadgets are ordered as indicated in Fig. 2(b). The variable-clause gadgets V_i^j, with $i = 1, \ldots, n$, always precede the clause gadget V^j, for any $j = 1, \ldots, m$. Further, if j is odd, then the gadgets $V_1^j, \ldots, V_n^j$ appear in this order, otherwise they appear in reversed order $V_n^j, \ldots, V_1^j$. Finally, V_i^j and V_i^{j+1}, for $i = 1, \ldots, n$ and $j = 1, \ldots, m-1$, are connected by an edge $\{w_i^j, w_i^{j+1}\}$, which is blue if j is odd and red if j is even. We call these edges *transmission edges*.

Assume $\langle G_1, G_2, G_3 \rangle$ admits an OrthoSEFE. Planarity constraints and orthogonality constraints guarantee three properties: (i) If the edge $\{u_i^j, v_i^j\}$ is inside C, then so is $\{u_i^{j+1}, v_i^{j+1}\}$, $i = 1, \ldots, n$, $j = 1, \ldots, m - 1$. This is due to the fact that, by the planarity constraints, the two green edges incident to w_i^j lie on the same side of C and hence, by the orthogonality constraints, the two transmission edges incident to w_i^j also lie on this side. We call $\{u_i^1, v_i^1\}$ the *truth edge* of variable x_i. (ii) Not all the three green edges $a = \{\alpha^j, \beta^j\}$, $b = \{\beta^j, \gamma^j\}$, and $c = \{\gamma^j, \delta^j\}$ lie on the same side of C. Namely, the two red edges of the clause gadget C^j must lie on opposite sides of C because of the interplay between the planarity and the orthogonality constraints in the subgraph of C^j induced by the vertices between β^j and γ^j. Hence, if edges a, b, and c lie on the same side of C, then the orthogonality constraints at either β^j or γ^j are not satisfied. (iii) For each clause $c_j = (x_a, x_b, x_c)$, edge $a = \{\alpha^j, \beta^j\}$ lies on the same side of C as the truth edge of x_a. This is due to the planarity constraints between each of these two edges and the variable-clause edge $\{w_a^j, y_a^j\}$. Analogously, edge b (edge c) lies on the same side as the truth edge of x_b (of x_c). Hence, setting $x_i = \mathsf{true}$ ($x_i = \mathsf{false}$) if the truth edge of x_i is inside C (outside C) yields a NAE3SAT truth assignment that satisfies ϕ.

The proof for the other direction is based on the fact that assigning the truth edges to either of the two sides of C according to the NAE3SAT assignment of ϕ also implies a unique side assignment for the remaining exclusive edges that satisfies all the orthogonality and the planarity constraints.

It is easy to see that G_1 and G_2 are outerplanar graphs with maximum degree 3, and that the reduction can be extended to any $k > 3$.   □

In the following, we describe how to modify the construction in Theorem 3 to show hardness of ORTHOSEFE-2. We keep only the edges of G_1 and G_3. Variable-clause gadgets and clause gadgets remain the same, as they are composed only of edges belonging to these two graphs. We replace each transmission edge in G_2 by a *transmission path* composed of alternating green and red edges, starting and ending with a red edge. This transformation allows these paths to traverse the transmission edges of G_1 and the variable-clause edges of G_3 without introducing crossings between edges of the same color. It is easy to see that the properties described in the proof of Theorem 3 on the assignments of the exclusive edges to the two sides of C also hold in the constructed instance, where transmission paths take the role of the transmission edges.

Theorem 4. ORTHOSEFE-2 *is NP-complete, even for instances* $\langle G_1, G_2 \rangle$ *in which the shared graph consists of a cycle and a set of isolated vertices.*

4 Shared Graph is a Cycle

In this section, we give a polynomial-time algorithm for instances of ORTHOSEFE-2 whose shared graph is a cycle and whose union graph has maximum degree 5 (Theorem 5). In order to obtain this result, we present an efficient algorithm for more restricted instances (Lemma 1) and give a series of transformations (Lemmas 2–3) to reduce any instance with the above properties to one that can be solved by the algorithm in Lemma 1.

Lemma 1. ORTHOSEFE-2 *is in P for instances* $\langle G_1, G_2 \rangle$ *such that the shared graph C is a cycle and G_1 is an outerplanar graph with maximum degree 3.*

Proof. The algorithm is based on a reduction to PLANAR NAE3SAT, which is in P [22,27]. First note that, since G_1 is outerplanar, there exist no two edges in E_1 alternating along C. Hence, there are no planarity constraints for G_1.

We now define an auxiliary graph H with vertex set $E_2 \setminus E_1$ and edges corresponding to pairs of edges alternating along C; see Fig. 3(a). W.l.o.g. we may assume that H is bipartite, since G_2 would not meet the planarity constraints otherwise [6]. Let $\mathcal{B}$ be the set of connected components of H, and for each component $B \in \mathcal{B}$, fix a partition B_1, B_2 of B into independent sets (possibly

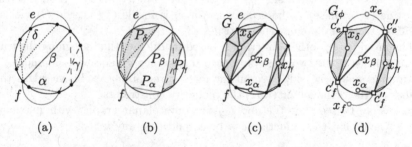

(a)   (b)   (c)   (d)

Fig. 3. (a) Instance $\langle G_1, G_2 \rangle$ satisfying the properties of Lemma 1, where the edges in E_2 belonging to the components α, β, γ, and δ of H have different line styles. (b) Polygons for the components of H. (c) Graph $\widetilde{G}$. (d) Variable–clause graph G_ϕ. (Color figure online)

$B_2 = \emptyset$ in case of a singleton B). Note that in any inside/outside assignment of the exclusive edges of G_2 that meets the planarity constraints, for every $B \in \mathcal{B}$, all edges of B_1 lie on one side of C and all edges of B_2 lie on the other side.

Draw the cycle C as a circle in the plane. For a component $B \in \mathcal{B}$, let P_B be the polygon inscribed into C whose corners are the endvertices in V of the edges in E_2 corresponding to the vertices of B; refer to Fig. 3(b). If B only contains one vertex (i.e., one edge of G_2), we consider the digon P_B as the straight-line segment connecting the vertices of this edge. If B has at least two vertices, we let P_B be open along its sides, i.e., it will contain the corners and all inner points (in Fig. 3(b) we depict this by making the sides of P_B slightly concave). One can easily show that, for any two components $B, D \in \mathcal{B}$, their polygons P_B, P_D may share only some of their corners, but no inner points. Hence, the graph $\widehat{G}$ obtained by placing a vertex x_B inside the polygon P_B, for $B \in \mathcal{B}$, making x_B adjacent to each corner of P_B and adding the edges E_1, is planar; see Fig. 3(c).

We construct a formula ϕ with variables x_B, $B \in \mathcal{B}$, such that ϕ is NAE-satisfiable if and only if $\langle G_1, G_2 \rangle$ admits an inside/outside assignment meeting all planarity and orthogonality constraints. The encoding of the truth assignment will be such that x_B is true when the edges of B_1 are inside C and the edges of B_2 are outside, and x_B is false if the reverse holds. Every assignment satisfying the planarity constraints for G_2 defines a truth-assignment in the above sense.

Let $e = (v, w)$ be an exclusive edge of E_1 and let e_v^1, e_v^2 (e_w^1, e_w^2) be the exclusive edges of E_2 incident to v (to w, respectively); we assume that all such four edges of E_2 exist, the other cases being simpler. Let $B(u, i)$ be the component containing the edge e_u^i, for $u \in \{v, w\}$ and $i \in \{1, 2\}$. Define the literal ℓ_u^i to be $x_{B(u,i)}$ if $e_u^i \in B_1(u, i)$ and $\neg x_{B(u,i)}$ if $e_u^i \in B_2(u, i)$. With our interpretation of the truth assignment, an edge e_u^i is inside C if and only if ℓ_u^i is true. Now, for the assignment to meet the orthogonality constraints, if $\ell_v^1 = \ell_v^2$, say both are true, then e must be assigned inside C as well, which would cause a problem if and only if $\ell_w^1 = \ell_w^2 = \text{false}$. Hence, the orthogonality constraints are described by NAE-satisfiability of the clauses $c_e = (\ell_v^1, \ell_v^2, \neg \ell_w^1, \neg \ell_w^2)$, for each $e \in E_1$. To reduce to NAE3SAT, we introduce a new variable x_e for each edge $e \in E_1 \setminus E_2$ and replace the clause c_e by two clauses $c_e' = (\ell_v^1, \ell_v^2, x_e)$ and $c_e'' = (\neg x_e, \neg \ell_w^1, \neg \ell_w^2)$. A planar drawing of the variable–clause graph G_ϕ of the resulting formula ϕ is obtained from the planar drawing $\widetilde{\Gamma}$ of $\widehat{G}$ (see Figs. 3(c) and 3(d)) by (i) placing each variable x_B, with $B \in \mathcal{B}$, on the point where vertex x_B lies in $\widetilde{\Gamma}$, (ii) placing each variable x_e, with $e \in E_1$, on any point of edge e in $\widetilde{\Gamma}$, (iii) placing clauses c_e' and c_e'', for each edge $e = (v, w) \in E_1$, on the points where vertices v and w lie in $\widetilde{\Gamma}$, respectively, and (iv) drawing the edges of G_ϕ as the corresponding edges in $\widetilde{\Gamma}$. This implies that G_ϕ is planar and hence we can test the NAE-satisfiability of ϕ in polynomial time [22,27]. □

The next two lemmas show that we can use Lemma 1 to test in polynomial time any instance of ORTHOSEFE-2 such that $G_\cap$ is a cycle and each vertex $v \in V$ has degree at most 3 in either G_1 or G_2.

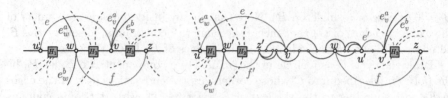

Fig. 4. Instances (left) $\langle G_1, G_2 \rangle$ and (right) $\langle G'_1, G'_2 \rangle$ for the proof of Lemma 2. Edges of $G_\cap$ ($G'_\cap$) are black. Exclusive edges of G_1 (G'_1) are red and those of G_2 (G'_2) are blue. (Color figure online)

Lemma 2. *Let $\langle G_1, G_2 \rangle$ be an instance of* ORTHOSEFE-2 *whose shared graph is a cycle and such that G_1 has maximum degree 3. It is possible to construct in polynomial time an equivalent instance $\langle G_1^*, G_2^* \rangle$ of* ORTHOSEFE-2 *whose shared graph is a cycle and such that G_1^* is outerplanar and has maximum degree 3.*

Proof sketch. We construct an equivalent instance $\langle G'_1, G'_2 \rangle$ of ORTHOSEFE-2 such that $G'_\cap$ is a cycle, G'_1 has maximum degree 3, and the number of pairs of edges in G'_1 that alternate along $G'_\cap$ is smaller than the number of pairs of edges in G_1 that alternate along $G_\cap$. Repeatedly applying this transformation yields an equivalent instance $\langle G_1^*, G_2^* \rangle$ satisfying the requirements of the lemma.

Consider two edges $e = (u, v)$ and $f = (w, z)$ of G_1 such that u, w, v, z appear in this order along cycle $G_\cap$ and such that the path $P_{u,z}$ in $G_\cap$ between u and z that contains v and w has minimal length. If G_1 is not outerplanar, then the edges e and f always exist. Figure 4 illustrates the construction of $\langle G'_1, G'_2 \rangle$.

By the choice of e and f, and by the fact that G_1 has maximum degree 3, there is no exclusive edge in G_1 with one endpoint in the set H_2 of vertices between w and v, and the other one not in H_2. Further, observe that in an OrthoSEFE of $\langle G'_1, G'_2 \rangle$ edges f and f' (edges e and e') must be on the same side. Further, e and f must be in different sides of $G'_\cap$. It can be concluded that $\langle G'_1, G'_2 \rangle$ has an OrthoSEFE if and only if $\langle G_1, G_2 \rangle$ has an OrthoSEFE.   □

The proof of the next lemma is based on the replacement illustrated in Fig. 5. Afterwards, we combine these results to present the main result of the section.

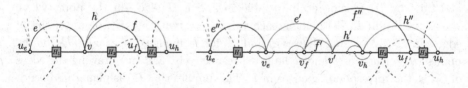

Fig. 5. Illustration of the transformation for the proof of Lemma 3 to reduce the number of vertices incident to two exclusive edges in G_1. Edges e', f' of G_2 and h' of G_1 (right) take the role of edges e, f of G_1 and h of G_2 (left), respectively. Thus, the orthogonality constraints at v' are equivalent to those at v. (Color figure online)

Lemma 3. *Let* $\langle G_1, G_2 \rangle$ *be an instance of* ORTHOSEFE-2 *whose shared graph is a cycle and whose union graph has maximum degree 5. It is possible to construct in polynomial time an equivalent instance* $\langle G_1^*, G_2^* \rangle$ *of* ORTHOSEFE-2 *whose shared graph is a cycle and such that graph* G_1^* *has maximum degree 3.*

Theorem 5. ORTHOSEFE-2 *can be solved in polynomial time for instances whose shared graph is a cycle and whose union graph has maximum degree 5.*

5   Shared Graph is Biconnected

We now study ORTHOSEFE-k for instances whose shared graph is biconnected. In Theorem 6, we give a polynomial-time Turing reduction from instances of ORTHOSEFE-2 whose shared graph is biconnected to instances whose shared graph is a cycle. In Theorem 7, we give an algorithm that, given a positive instance of ORTHOSEFE-k such that the shared graph is biconnected together with a SEFE satisfying the orthogonality constraints, constructs an OrthoSEFE with at most three bends per edge.

We start with the Turing reduction, i.e., we develop an algorithm that takes as input an instance $\langle G_1, G_2 \rangle$ of ORTHOSEFE-2 whose shared graph $G_\cap = G_1 \cap G_2$ is biconnected and produces a set of $O(n)$ instances $\langle G_1^1, G_2^1 \rangle, \ldots, \langle G_1^h, G_2^h \rangle$ of ORTHOSEFE-2 whose shared graphs are cycles. The output is such that $\langle G_1, G_2 \rangle$ is a positive instance if and only if all instances $\langle G_1^i, G_2^i \rangle$, $i = 1, \ldots, h$, are positive. The reduction is based on the SEFE testing algorithm for instances whose shared graph is biconnected by Bläsius et al. [9,10], which can be seen as a generalized and unrooted version of the one by Angelini et al. [4].

We first describe a preprocessing step. Afterwards, we give an outline of the approach of Bläsius et al. [10] and present the Turing reduction in two steps. We assume familiarity with SPQR-trees [15,16]; for formal definitions, see [1].

Lemma 4. *Let* $\langle G_1, G_2 \rangle$ *be an instance of* ORTHOSEFE-2 *whose shared graph is biconnected. It is possible to construct in polynomial time an equivalent instance* $\langle G_1^*, G_2^* \rangle$ *whose shared graph is biconnected and such that each endpoint of an exclusive edge has degree 2 in the shared graph.*

We continue with a brief outline of the algorithm by Bläsius et al. [10]. First, the algorithm computes the SPQR-tree $\mathcal{T}$ of the shared graph. To avoid special cases, $\mathcal{T}$ is augmented by adding S-nodes with only two virtual edges such that each P-node and each R-node is adjacent only to S-nodes and Q-nodes. Then, necessary conditions on the embeddings of P-nodes and R-nodes are fixed up to a flip following some necessary conditions. Afterwards, by traversing all S-nodes, a global 2SAT formula is produced whose satisfying assignments correspond to choices of the flips that result in a SEFE. We refine this approach and show that we can choose the flips independently for each S-node, which allows us to reduce each of them to a separate instance, whose shared graph is a cycle.

We now describe the algorithm of Bläsius et al. [10] in more detail. Consider a node μ of $\mathcal{T}$. A *part* of skel(μ) is either a vertex of skel(μ) or a virtual edge of skel(μ), which represents a subgraph of G. An exclusive edge e has an *attachment* in a part x of skel(μ) if x is a vertex that is an endpoint of e or if x is a virtual edge whose corresponding subgraph contains an endpoint of e. An exclusive edge e of G_1 or of G_2 is *important* for μ if its endpoints are in different parts of skel(μ). It is not hard to see that, to obtain a SEFE, the embedding of the skeleton skel(μ) of each node μ has to be chosen such that for each exclusive edge e the parts containing the attachments of e share a face. It can be shown that any embedding choice for P-nodes and R-nodes that satisfies these conditions can, after possibly flipping it, be used to obtain a SEFE [4, Theorem 1]. The proof does not modify the order of exclusive edges around degree-2 vertices of $G_\cap$, and therefore applies to ORTHOSEFE-2 as well.

Now, let μ be an S-node. Let ε be a virtual edge of skel(μ), G_ε be the subgraph represented by ε, and ν be the corresponding neighbor of μ in the SPQR-tree of G. An *attachment* of ν with respect to μ is an interior vertex of G_ε that is incident to an important edge e for μ. If ν has such an attachment, then it is a P- or R-node. It is a necessary condition on the embedding of G_ε that each attachment x with respect to μ must be incident to a face incident to the virtual edge twin(ε) of skel(ν) representing μ, and that their clockwise circular order together with the poles of ε is fixed up to reversal [10, Lemma 8].

For the purpose of avoiding crossings in skel(μ), we can thus replace each virtual edge ε that does not represent a Q-node by a cycle C_ε containing the attachments of ε with respect to μ and the poles of ε in the order O_ε. We keep only the important edges of μ. Altogether, this results in an instance $\langle G_1^\mu, G_2^\mu \rangle$ of SEFE modeling the requirements for skel(μ); see Figs. 6(a) and 6(b).

Lemma 5. *Let $\langle G_1, G_2 \rangle$ be an instance of ORTHOSEFE-2 whose shared graph is biconnected. Then $\langle G_1, G_2 \rangle$ admits an OrthoSEFE if and only if all instances $\langle G_1^\mu, G_2^\mu \rangle$ admit an OrthoSEFE.*

Next, we transform a given instance $\langle G_1^\mu, G_2^\mu \rangle$ of ORTHOSEFE-2 as above into an equivalent instance $\langle \overline{G_1^\mu}, \overline{G_2^\mu} \rangle$ whose shared graph is a cycle. Let C_{ε_i} be the cycles corresponding to the neighbor ν_i, $i = 1, \ldots, k$ of μ in $\langle G_1^\mu, G_2^\mu \rangle$. To

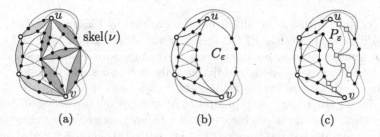

     (a)   (b)   (c)

Fig. 6. (a) Skeleton of an S-node μ in which the R-node ν corresponding to the virtual edge $\varepsilon = (u, v)$ is expanded to show its skeleton. (b) Replacing ε with cycle C_ε. (c) Replacing C_ε with path P_ε; vertices $a_1, a_2, x_1, \ldots, x_4, b_1, b_2$ are green boxes. (Color figure online)

obtain the instance $\langle \overline{G_1^\mu}, \overline{G_2^\mu} \rangle$, we replace each cycle C_{ε_i} with poles u and v by a path P_{ε_i} from u to v that first contains two special vertices a_1, a_2 followed by the clockwise path from u to v (excluding the endpoints), then four special vertices $x_1, \ldots, x_4$, then the counterclockwise path from u to v (excluding the endpoints), and finally two special vertices b_1, b_2 followed by v. In addition to the existing exclusive edges (note that we do not remove any vertices), we add to G_1 the exclusive edges (a_2, x_3), (x_1, x_3), (x_2, x_4), (x_2, b_1), and to G_2 the exclusive edges (a_1, x_3) and (x_2, b_2) to G_2; see Fig. 6(c).

The above reduction together with the next lemma implies the main result.

Lemma 6. $\langle G_1^\mu, G_2^\mu \rangle$ *admits an OrthoSEFE if and only if* $\langle \overline{G_1^\mu}, \overline{G_2^\mu} \rangle$ *does.*

Theorem 6. ORTHOSEFE-2 *when the shared graph is biconnected is polynomial-time Turing reducible to* ORTHOSEFE-2 *when the shared graph is a cycle. Also, the reduction does not increase the maximum degree of the union graph.*

Corollary 1. ORTHOSEFE-2 *can be solved in polynomial time for instances whose shared graph is biconnected and whose union graph has maximum degree* 5.

Observe that, from the previous results, it is not hard to also obtain a SEFE satisfying the orthogonality constraints, if it exists. We show how to construct an orthogonal geometric realizations of such a SEFE.

Theorem 7. *Let* $\langle G_1, \ldots, G_k \rangle$ *be a positive instance of* ORTHOSEFE-k *whose shared graph is biconnected. Then, there exists an OrthoSEFE* $\langle \Gamma_1, \Gamma_2, \ldots, \Gamma_k \rangle$ *of* $\langle G_1, \ldots, G_k \rangle$ *in which every edge has at most three bends.*

Proof sketch. We assume that a SEFE satisfying the orthogonality constraints is given. We adopt the method of Biedl and Kant [8]. We draw the vertices with increasing y-coordinates with respect to an s-t-ordering [14] $v_1, \ldots, v_n$ on the shared graph. We choose the face to the left of (v_1, v_n) as the outer face of the union graph. The edges will bend at most on y-coordinates near their incident vertices and are drawn vertically otherwise. Figure 7 indicates how the ports are assigned. We make sure that an edge may only leave a vertex to the bottom if it is incident to v_n or to a neighbor with a lower index. Thus, there are exactly three bends on $\{v_1, v_n\}$. Any other edge $\{v_i, v_j\}$, $1 \leq i < j \leq n$ has at most one bend around v_i and at most two bends around v_j. $\square$

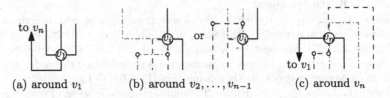

(a) around v_1 (b) around $v_2, \ldots, v_{n-1}$ (c) around v_n

Fig. 7. Constructing a drawing with at most three bends per edge. (Color figure online)

6   Conclusions and Future Work

In this work, we introduced and studied the problem ORTHOSEFE-k of realizing a SEFE in the orthogonal drawing style. While the problem is already NP-hard even for instances that can be efficiently tested for a SEFE, we presented a polynomial-time testing algorithm for instances consisting of two graphs whose shared graph is biconnected and whose union graph has maximum degree 5. We have also shown that any positive instance whose shared graph is biconnected can be realized with at most three bends per edge.

We conclude the paper by presenting a lemma that, together with Theorem 6, shows that it suffices to only focus on a restricted family of instances to solve the problem for all instances whose shared graph is biconnected.

Lemma 7. *Let $\langle G_1, G_2 \rangle$ be an instance of ORTHOSEFE-2 whose shared graph $G_\cap$ is a cycle. An equivalent instance $\langle G_1^*, G_2^* \rangle$ of ORTHOSEFE-2 such that (i) the shared graph $G_\cap^*$ is a cycle, (ii) graph G_1^* is outerplanar, and (iii) no two degree-4 vertices in G_1^* are adjacent, can be constructed in polynomial time.*

References

1. Angelini, P., Chaplick, S., Cornelsen, S., Da Lozzo, G., Di Battista, G., Eades, P., Kindermann, P., Kratochvíl, J., Lipp, F.: Simultaneous Orthogonal Planarity. ArXiv e-prints, abs/1608.08427 (2016)

2. Angelini, P., Da Lozzo, G., Di Battista, G., Frati, F., Patrignani, M., Rutter, I.: Beyond level planarity. In: Hu, Y., Nöllenburg, M. (eds.) GD 2016. LNCS, vol. 9801, pp. 482–495. Springer, Heidelberg (2016)

3. Angelini, P., Da Lozzo, G., Neuwirth, D.: Advancements on SEFE and partitioned book embedding problems. Theoret. Comput. Sci. **575**, 71–89 (2015)

4. Angelini, P., Di Battista, G., Frati, F., Patrignani, M., Rutter, I.: Testing the simultaneous embeddability of two graphs whose intersection is a biconnected or a connected graph. J. Discrete Algorithms **14**, 150–172 (2012)

5. Argyriou, E.N., Bekos, M.A., Kaufmann, M., Symvonis, A.: Geometric RAC simultaneous drawings of graphs. J. Graph Algorithms Appl. **17**(1), 11–34 (2013)

6. Auslander, L., Parter, S.V.: On embedding graphs in the sphere. J. Math. Mech. **10**(3), 517–523 (1961)

7. Bekos, M.A., van Dijk, T.C., Kindermann, P., Wolff, A.: Simultaneous drawing of planar graphs with right-angle crossings and few bends. J. Graph Algorithms Appl. **20**(1), 133–158 (2016)

8. Biedl, T., Kant, G.: A better heuristic for orthogonal graph drawings. Comput. Geom. **9**(3), 159–180 (1998)

9. Bläsius, T., Karrer, A., Rutter, I.: Simultaneous embedding: edge orderings, relative positions, cutvertices. In: Wismath, S., Wolff, A. (eds.) GD 2013. LNCS, vol. 8242, pp. 220–231. Springer, Heidelberg (2013). doi:10.1007/978-3-319-03841-4_20

10. Bläsius, T., Karrer, A., Rutter, I.: Simultaneous embedding: Edge orderings, relative positions, cutvertices. ArXiv e-prints, abs/1506.05715 (2015)

11. Bläsius, T., Kobourov, S.G., Rutter, I.: Simultaneous embedding of planar graphs. In: Tamassia, R., (ed.) Handbook of Graph Drawing and Visualization. CRC Press (2013)

12. Bläsius, T., Rutter, I.: Disconnectivity and relative positions in simultaneous embeddings. Comput. Geom. **48**(6), 459–478 (2015)
13. Bläsius, T., Rutter, I.: Simultaneous PQ-ordering with applications to constrained embedding problems. ACM Trans. Algorithms **12**(2), 16 (2016)
14. Brandes, U.: Eager *st*-ordering. In: Möhring, R., Raman, R. (eds.) ESA 2002. LNCS, vol. 2461, pp. 247–256. Springer, Heidelberg (2002). doi:10.1007/3-540-45749-6_25
15. Di Battista, G., Tamassia, R.: On-line maintenance of triconnected components with SPQR-trees. Algorithmica **15**(4), 302–318 (1996)
16. Di Battista, G., Tamassia, R.: On-line planarity testing. SIAM J. Comput. **25**(5), 956–997 (1996)
17. Estrella-Balderrama, A., Gassner, E., Jünger, M., Percan, M., Schaefer, M., Schulz, M.: Simultaneous geometric graph embeddings. In: Hong, S.-H., Nishizeki, T., Quan, W. (eds.) GD 2007. LNCS, vol. 4875, pp. 280–290. Springer, Heidelberg (2008). doi:10.1007/978-3-540-77537-9_28
18. Haeupler, B., Jampani, K.R., Lubiw, A.: Testing simultaneous planarity when the common graph is 2-connected. J. Graph Algorithms Appl. **17**(3), 147–171 (2013)
19. Jampani, K.R., Lubiw, A.: Simultaneous interval graphs. In: Cheong, O., Chwa, K.-Y., Park, K. (eds.) ISAAC 2010. LNCS, vol. 6506, pp. 206–217. Springer, Heidelberg (2010). doi:10.1007/978-3-642-17517-6_20
20. Jampani, K.R., Lubiw, A.: The simultaneous representation problem for chordal, comparability and permutation graphs. J. Graph Algorithms Appl. **16**(2), 283–315 (2012)
21. Jünger, M., Schulz, M.: Intersection graphs in simultaneous embedding with fixed edges. J. Graph Algorithms Appl. **13**(2), 205–218 (2009)
22. Moret, B.M.E.: Planar NAE3SAT is in P. ACM SIGACT News **19**(2), 51–54 (1988)
23. Moret, B.M.E.: Theory of Computation. Addison-Wesley-Longman, Reading (1998)
24. Papadimitriou, C.H.: Computational Complexity. Academic Internet Publ., London (2007)
25. Schaefer, M.: Toward a theory of planarity: Hanani-Tutte and planarity variants. J. Graph Algorithms Appl. **17**(4), 367–440 (2013)
26. Schaefer, T.J.: The complexity of satisfiability problems. In: Lipton, R.J., Burkhard, W.A., Savitch, W.J., Friedman, E.P., Aho, A.V. (eds.), STOC 1978, pp. 216–226. ACM (1978)
27. Shih, W., Wu, S., Kuo, Y.: Unifying maximum cut and minimum cut of a planar graph. IEEE Trans. Comput. **39**(5), 694–697 (1990)
28. Tamassia, R.: On embedding a graph in the grid with the minimum number of bends. SIAM J. Comput. **16**(3), 421–444 (1987)

Monotone Simultaneous Embeddings of Paths in d Dimensions

David Bremner[1], Olivier Devillers[2], Marc Glisse[3], Sylvain Lazard[2],
Giuseppe Liotta[4], Tamara Mchedlidze[5(✉)], Sue Whitesides[6],
and Stephen Wismath[7]

[1] University of New Brunswick, Fredericton, Canada
bremner@unb.ca
[2] Inria, CNRS, University of Lorraine, Nancy, France
{olivier.devillers,sylvain.lazard}@inria.fr
[3] Inria, Saclay, France
marc.glisse@inria.fr
[4] University of Perugia, Perugia, Italy
giuseppe.liotta@unipg.it
[5] Karlsruhe Institute of Technology, Karlsruhe, Germany
mched@iti.uka.de
[6] University of Victoria, Victoria, Canada
sue@uvic.ca
[7] University of Lethbridge, Lethbridge, Canada
wismath@uleth.ca

Abstract. We study the following problem: Given k paths that share
the same vertex set, is there a simultaneous geometric embedding of these
paths such that each individual drawing is monotone in some direction?
We prove that, for any dimension d, there is a set of $d + 1$ paths that
does *not* admit a monotone simultaneous geometric embedding.

1 Introduction

Monotone drawings and simultaneous embeddings are well studied topics in
graph drawing. Monotone drawings, introduced by Angelini et al. [2], are planar
drawings of connected graphs such that, for every pair of vertices, there is a
path between them that monotonically increases with respect to some direction.
Monotone drawings of planar graphs have been studied both in the fixed and
in the variable embedding settings and both with straight-line edges and with
bends allowed along edges; recent papers on these topics include [3,10,12,13].

The simultaneous (geometric) embedding problem was first described in a
paper by Braß et al. [7]. The input is a set of planar graphs that share the same
labeled vertex set (but the set of edges differs from one graph to another); the
output is a mapping of the vertex set to a point set such that each graph admits

Research supported in part by the MIUR project AMANDA "Algorithmics for
MAssive and Networked DAta", prot. 2012C4E3KT_001, and NSERC.

Y. Hu and M. Nöllenburg (Eds.): GD 2016, LNCS 9801, pp. 546–553, 2016.
DOI: 10.1007/978-3-319-50106-2_42

a crossing-free drawing with the given mapping. The simultaneous embedding problem has also been studied by restricting/relaxing some geometric requirements; for example, while every pair of planar graphs sharing the same labeled vertex set admits a simultaneous embedding where each edge has at most two bends (see, e.g., [9,11]), not even a tree and a path always admit a geometric simultaneous embedding (such that the edges are straight-line segments) [4]). See the book chapter on simultaneous embeddings by Bläsius et al. [6] for an extensive list of references on the problem and its variants.

In this paper, we combine the two topics of simultaneous embeddings and monotone drawings[1]. Namely, we are interested in computing geometric simultaneous embeddings of paths such that each path is monotone in some direction. Let $V = 1, 2, \ldots, n$ be a labeled set of vertices and let $\Pi = \{\pi_1, \pi_2, \ldots, \pi_k\}$ be a set of k distinct paths each having the same set V of vertices. We want to compute a labeled set of points $P = \{p_1, p_2, \ldots, p_n\}$ such that point p_i represents vertex i and for each path $\pi_i \in \Pi$ $(1 \leqslant i \leqslant k)$ there exists some direction for which the drawing of π_i is monotone.

It is already known that any two paths on the same vertex·set admit a monotone simultaneous geometric embedding in 2D, while there exist three paths on the same vertex set for which a simultaneous geometric embedding does not exist even if we drop the monotonicity requirement [7]. An example of three paths that do not have a monotone simultaneous geometric embedding in 2D can also be derived from a paper of Asinowski on suballowable sequences [5]. On the other hand, it is immediate to see that in 3D any number of paths sharing the same vertex set admits a simultaneous geometric embedding: Namely, by suitably placing the points in generic position (no 4 coplanar), the complete graph has a straight-line crossing-free drawing; however, the drawing of each path may not be monotone. This motivates the following question: Given a set of paths sharing the same vertex set, does the set admit a monotone simultaneous geometric embedding in d-dimensional space for $d \geqslant 3$?

Our main result is that for any dimension $d \geqslant 2$, there exists a set of $d + 1$ paths that does not admit a monotone simultaneous geometric embedding in d-dimensional space. Our proof exploits the relationship between monotone simultaneous geometric embeddings in d-dimensional space and their corresponding representation in the dual space. Our approach extends to d dimensions the primal-dual technique described in a recent paper by Aichholzer et al. [1] on simultaneous embeddings of upward planar digraphs in 2D.

2   Definitions

Let $\vec{v}$ be a vector in $\mathbb{R}^d$ and let G be a directed acyclic graph with vertex set V. An embedding Γ of the vertex set V in $\mathbb{R}^d$ is called $\vec{v}$-*monotone* for G if the vectors in $\mathbb{R}^d$ corresponding to oriented edges of G have a positive scalar product with $\vec{v}$. Let $\mathcal{V} = \{\vec{v}_1, \ldots, \vec{v}_k\}$ be a set of $k > 1$ vectors in $\mathbb{R}^d$ and let

---

[1] We reference the reader to [8] for the full version of this paper.

$\mathcal{G} = \{G_1, G_2, \ldots, G_k\}$ be a set of k distinct acyclic digraphs on the same vertex set V. A $\mathcal{V}$-*monotone simultaneous embedding* of $\mathcal{G}$ in $\mathbb{R}^d$ is an embedding Γ of V that is $\vec{v}_i$-monotone for G_i for any i. A *monotone simultaneous embedding* of $\mathcal{G}$ is a $\mathcal{V}$-monotone simultaneous embedding for some set $\mathcal{V}$ of vectors.

If a graph is a path on n (labeled) vertices, it can be trivially identified with a permutation of $[1, n]$. We look at the monotone simultaneous embedding problem in the dual space, by mapping points representing vertices to hyperplanes in $\mathbb{R}^d$. The dual formulation of monotone simultaneous embeddings is as follows (the equivalence of these formulations is shown in the next section). Let $\Pi = \{\pi_1, \pi_2, \ldots, \pi_k\}$ be a set of k permutations of $[1, n]$. A *parallel simultaneous embedding* of Π in $\mathbb{R}^d$ is a set of n hyperplanes $H_1, H_2, \ldots, H_n$ and k vertical lines $L_1, L_2, \ldots, L_k$ such that the set of n points $L_j \cap H_{\pi_j(1)}, \ldots, L_j \cap H_{\pi_j(n)}$ is linearly ordered from bottom to top along L_j, for all j.

3   The Dual Problem and Non-existence Results

The first two lemmas give duality results between monotone simultaneous embeddings and parallel simultaneous embeddings.

Lemma 1. *If a set of k permutations of $[1, n]$ admits a parallel simultaneous embedding in d dimensions, it also admits a monotone simultaneous embedding in d dimensions.*

Proof. Consider the following duality between points and hyperplanes, where we denote by $H^\star$ the dual of a non-vertical hyperplane H:

$$H : x_d = \left(\sum_{i=1}^{d-1} \alpha_i x_i \right) - \alpha_0, \qquad H^\star = (\alpha_1, \ldots, \alpha_{d-1}, \alpha_0).$$

This duality maps parallel hyperplanes to points that are vertically aligned (and vice-versa). Let $(H_i)_{1 \leqslant i \leqslant n}$, $(L_j)_{1 \leqslant j \leqslant k}$ be a parallel simultaneous embedding and refer to Fig. 1 By definition, line L_j crosses hyperplanes $H_1, \ldots, H_n$ in the order $H_{\pi_j(1)}, H_{\pi_j(2)}, \ldots, H_{\pi_j(n)}$. The intersection points $L_j \cap H_{\pi_j(1)}, L_j \cap$

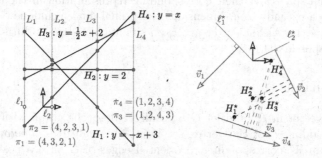

Fig. 1. Duality between monotone simultaneous embeddings and parallel simultaneous embeddings for $k = n = 4$ and $d = 2$.

$H_{\pi_j(2)}, \ldots, L_j \cap H_{\pi_j(n)}$ are collinear and therefore represent parallel hyperplanes in the dual plane. Consider the vector line $\vec{v}_j$ perpendicular to these hyperplanes and pointing downward. This line crosses them in the order $(L_j \cap H_{\pi_j(1)})^\star, (L_j \cap H_{\pi_j(2)})^\star, \ldots, (L_j \cap H_{\pi_j(n)})^\star$. Since point $H_i^\star$ lies in hyperplane $(L_j \cap H_i)^\star$, points $H_i^\star, 1 \leqslant i \leqslant n$, project on $\vec{v}_j$ in the order $H_{\pi_j(1)}^\star, H_{\pi_j(2)}^\star, \ldots, H_{\pi_j(n)}^\star$. Therefore $(H_i^\star)_{1 \leqslant i \leqslant n}$ is an embedding such that path π_j is $\vec{v}_j$-monotone, for all j. $\qquad\square$

Lemma 2. *If a set $(\pi_j)_{1 \leqslant j \leqslant k}$ of k permutations of $[1, n]$ admits a monotone simultaneous embedding in d dimensions, there is a set $(\pi'_j)_{1 \leqslant j \leqslant k}$ that admits a parallel simultaneous embedding in $\mathbb{R}^d$ where, for every j, π'_j is either equal to π_j or to its reverse.*

Proof. As in the proof of Lemma 1, we consider point-hyperplane duality. Let $(p_i)_{1 \leqslant i \leqslant n}$ be an embedding $\vec{v}_j$-monotone for π_j, and $(p_i^\star)_{1 \leqslant i \leqslant n}$ the corresponding set of dual hyperplanes. Let H_j be a hyperplane with normal vector $\vec{v}_j, 1 \leqslant j \leqslant n$. Define L_j to be the vertical line through point $H_j^\star$. By construction, the points $\left(L_j \cap p_{\pi_j(i)}^\star\right)_i$ appear in order on L_j for one of the two possible orientations of L_j. In particular, when $\vec{v}_j$ points downward, L_j lists the points $L_j \cap p_{\pi_j(i)}^\star$ from bottom to top and vice versa. $\qquad\square$

We now prove results of existence and non-existence of parallel simultaneous embeddings, starting with a very simple result of existence.

Proposition 1. *Any set of d permutations on n vertices admits a monotone simultaneous embedding and a parallel simultaneous embedding in d dimensions.*

Proof. Choose d points in general position in the hyperplane $x_d = 0$ and draw a vertical line through each of these points. For each vertical line, choose a permutation and place on the line n points numbered according to the permutation. Fit a hyperplane through all the points with the same number. By construction, this set of hyperplanes is a parallel simultaneous embedding. Going to the dual, by Lemma 1, gives a monotone simultaneous embedding. Alternatively, the monotone embedding can be seen directly by considering the rank in the i-th permutation as the i-th coordinate. $\qquad\square$

We now turn our attention to non-existence. For proving that there exists $k = d + 1$ permutations that do not admit a parallel simultaneous embedding in d dimensions, observe that we can consider any generic placement of the d first lines L_j since all such placements are equivalent through affine transformations. We then construct permutations for n big enough that cannot be realized with any placement of L_{d+1}. Similarly, constructing $k = d+1$ permutations that cannot be realized even up to inversion, yields the non-existence of a monotone simultaneous embedding in d dimensions by Lemma 2. We start with dimension 2, then move to dimension 3 and only then, generalize our results to arbitrary dimension. Observe that 2D results also follow from [5, Lemma 1 and Prop. 8], but we still present our proofs as a warm up for higher dimensions.

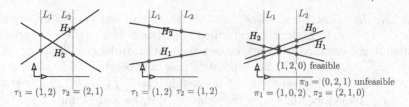

Fig. 2. Non-existence of two-dimensional parallel simultaneous embeddings.

Lemma 3. *There exists a set of 3 permutations on $\{0,1,2\}$ that does not admit a parallel simultaneous embedding in 2D.*

Proof. Let L_1 and L_2 be two vertical lines, H_1 and H_2 two other lines, and let $\tau_1 = (1,2)$ and $\tau_2 = (2,1)$ be two permutations of $\{1,2\}$. As in Fig. 2-left, if L_1 is left of L_2 and the intersections of H_1 and H_2 with L_j are ordered according to τ_i, we can deduce that $H_1 \cap H_2$ is between L_1 and L_2. It follows that a vertical line crossing H_1 below H_2 is to the left of that intersection point and thus to the left of L_2. Similarly, a vertical line crossing H_1 above H_2 is to the right of L_1. If we now consider $\tau_1 = \tau_2 = (1,2)$ we have that a vertical line crossing H_1 above H_2 is not between L_1 and L_2 (Fig. 2-center). Consider now $\pi_1 = (1,0,2)$, $\pi_2 = (2,1,0)$ and $\pi_3 = (0,2,1)$. Restricting the permutations to $\{1,2\}$ gives that L_3 must be right of L_1, restricting to $\{0,2\}$ gives that L_3 must be left of L_2, and restricting to $\{0,1\}$ gives that L_3 cannot be between L_1 and L_2 (Fig. 2-right). We deduce that no placement for L_3 can realize π_3. Notice that the reverse order $(1,2,0)$ can be realized and thus the dual of this construction is not a counterexample to simultaneous monotone embeddings. $\square$

Lemma 4. *There exists a set of 3 permutations on 6 vertices that does not admit a monotone simultaneous embedding in 2D.*

Proof. Let $\pi_1 = (f,b,d,e,a,c)$, $\pi_2 = (d,f,c,b,e,a)$, and $\pi_3 = (f,a,d,c,e,b)$. The sub-permutations of π_1, π_2 and π_3 on $\{a,b,c\}$ are (by matching (a,b,c) to $(0,1,2)$) the 3 permutations that do not admit a parallel simultaneous embedding in the proof of Lemma 3. The same is obtained by reversing only π_1 (resp. π_2, π_3) and considering sub-permutations on $\{a,c,d\}$ (resp. $\{d,b,e\}$, $\{b,f,d\}$). Other possibilities are symmetric and Lemma 2 yields the result. $\square$

Lemma 5. *There exists a set of 4 permutations on 5 vertices that does not admit a parallel simultaneous embedding in 3D.*

Proof. As in the proof of Lemma 1 we consider 3 points ℓ_1, ℓ_2, ℓ_3 in general position in the hyperplane $x_3 = 0$ and the 3 vertical lines L_1, L_2, L_3 going through these points. Let L be a vertical line (candidate position for L_4) and ℓ its intersection with $x_3 = 0$. We consider the 3 permutations $\tau_1 = (1,2,3)$, $\tau_2 = (2,3,1)$, $\tau_3 = (3,1,2)$ defining the vertical order of the intersections of L_1, L_2, L_3 with hyperplanes $(H_i)_{1 \leqslant i \leqslant 3}$. We denote by $h_{i,j}$ the projection of the line $H_i \cap H_j$, $1 \leqslant i \neq j \leqslant 3$, onto the plane $x_3 = 0$. Since the three planes H_i, $1 \leqslant i \leqslant 3$ meet

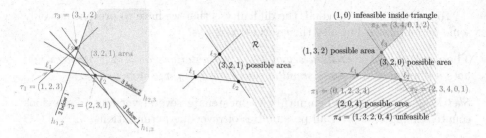

Fig. 3. Non-existence of three-dimensional parallel simultaneous embeddings for 5 vertices.

in one point, the lines $h_{1,2}$, $h_{2,3}$ and $h_{1,3}$ meet at the projection of that point onto the plane $x_3 = 0$.

Refer to Fig. 3. For L to cut H_2 below H_1, ℓ must be in the half-plane limited by $h_{1,2}$ and containing ℓ_2, and, similarly, for L to cut H_3 below H_2, ℓ must be in the half-plane limited by $h_{2,3}$ and containing ℓ_3. Thus, ℓ must be in a wedge with apex $h_{1,2} \cap h_{2,3}$ (Fig. 3-left). Since $h_{1,2}$ separates ℓ_2 from ℓ_1 and ℓ_3, and $h_{2,3}$ separates ℓ_3 from ℓ_1 and ℓ_2, the union of all wedges, for all possible positions of $h_{1,2}$ and $h_{2,3}$, is the union, $\mathcal{R}$, of triangle $\ell_1\ell_2\ell_3$ and the half-plane limited by $\ell_2\ell_3$ and not containing ℓ_1 (Fig. 3-center). To summarize, if $\tau_1 = (1,2,3)$, $\tau_2 = (2,3,1)$, $\tau_3 = (3,1,2)$, and $\tau_4 = (3,2,1)$ then ℓ_4 (the intersection point of L_4 with the hyperplane $x_3 = 0$) must lie in this region $\mathcal{R}$.

Next, we build the permutations π_1, π_2, π_3 and π_4 by repeating this example as follows: $\pi_1 = (0,1,2,3,4)$, $\pi_2 = (2,3,4,0,1)$, $\pi_3 = (3,4,0,1,2)$, and $\pi_4 = (1,3,2,0,4)$. The restriction of these permutations to $\{0,2,3\}$ yields that ℓ_4 must be in the triangle or in the half-plane limited by $\ell_2\ell_3$ and not containing ℓ_1. The restriction to $\{1,2,3\}$ yields that ℓ_4 must be in the triangle or in the half-plane limited by $\ell_1\ell_3$ and not containing ℓ_2. The restriction to $\{0,2,4\}$ yields that ℓ_4 must be in the triangle or in the half-plane limited by $\ell_1\ell_2$ and not containing ℓ_3. Finally, considering $\{0,1\}$ yields that ℓ_4 must be outside the triangle (Fig. 3-right). Thus there is no possibility for placing L_4. □

Lemma 6. *There exists a set of 4 permutations on 40 vertices that does not admit a monotone simultaneous embedding in 3D.*

Sketch of Proof. The idea is to concatenate several versions of the counterexample of the previous lemma to cover all possibilities of reversing permutations. Note that the number of 40 vertices is not tight. □

Lemma 7. *There exists a set of $d + 1$ permutations on $3 \cdot 2^d$ vertices that does not admit a parallel simultaneous embedding in d dimensions.*

Sketch of Proof. As in Lemma 5, the idea is to consider the simplex $(\ell_j)_{1 \leqslant j \leqslant d}$ and to construct the permutations for the L_i in order to prevent all possibilities for placing ℓ_{d+1}. □

To get a result in the dual, the difficulty is that we have to prevent not only some permutations but also their reverse versions.

Theorem 1. *There exists a set of $d + 1$ permutations on $3 \cdot 2^{2d}$ vertices that does not admit a monotone simultaneous embedding in d dimensions.*

Sketch of Proof. As for Lemma 6 we concatenate several versions of previous counter-example to cover all possibilities of reversing permutations.   □

Acknowledgements. This work was initiated during the 15th *INRIA–McGill–Victoria Workshop on Computational Geometry* at the Bellairs Research Institute. The authors wish to thank all the participants for creating a pleasant and stimulating atmosphere.

References

1. Aichholzer, O., Hackl, T., Lutteropp, S., Mchedlidze, T., Pilz, A., Vogtenhuber, B.: Monotone simultaneous embeddings of upward planar digraphs. J. Graph Algorithms Appl. **19**(1), 87–110 (2015)
2. Angelini, P., Colasante, E., Battista, G.D., Frati, F., Patrignani, M.: Monotone drawings of graphs. J. Graph Algorithms Appl. **16**(1), 5–35 (2012)
3. Angelini, P., Didimo, W., Kobourov, S.G., Mchedlidze, T., Roselli, V., Symvonis, A., Wismath, S.K.: Monotone drawings of graphs with fixed embedding. Algorithmica **71**(2), 233–257 (2015)
4. Angelini, P., Geyer, M., Kaufmann, M., Neuwirth, D.: On a tree and a path with no geometric simultaneous embedding. J. Graph Algorithms Appl. **16**(1), 37–83 (2012)
5. Asinowski, A.: Suballowable sequences and geometric permutations. Discret. Math. **308**(20), 4745–4762 (2008)
6. Blaäsius, T., Kobourov, S.G., Rutter, I.: Simultaneous embedding of planar graphs. In: Tamassia, R. (ed.) Handbook on Graph Drawing and Visualization. Chapman and Hall/CRC, Boca Raton (2013)
7. Braß, P., Cenek, E., Duncan, C.A., Efrat, A., Erten, C., Ismailescu, D.P., Kobourov, S.G., Lubiw, A., Mitchell, J.S.B.: On simultaneous planar graph embeddings. Comput. Geom. **36**(2), 117–130 (2007)
8. Bremner, D., Devillers, O., Glisse, M., Lazard, S., Liotta, G., Mchedlidze, T., Whitesides, S., Wismath, S.: Monotone simultaneous embeddings of paths in $\mathbb{R}^d$ (2016). http://arxiv.org/abs/1608.08791
9. Erten, C., Kobourov, S.G.: Simultaneous embedding of planar graphs with few bends. J. Graph Algorithms Appl. **9**(3), 347–364 (2005)
10. Felsner, S., Igamberdiev, A., Kindermann, P., Klemz, B., Mchedlidze, T., Scheucher, M.: Strongly monotone drawings of planar graphs. In: Fekete, S., Lubiw, A. (eds.) 32nd International Symposium on Computational Geometry (SoCG 2016). Leibniz International Proceedings in Informatics (LIPIcs), vol. 51, pp. 37:1–37:15. Dagstuhl, Germany (2016). Schloss Dagstuhl-Leibniz-Zentrum fuer Informatik
11. Giacomo, E.D., Didimo, W., Liotta, G., Meijer, H., Wismath, S.K.: Planar and quasi-planar simultaneous geometric embedding. Comput. J. **58**(11), 3126–3140 (2015)

12. Hossain, M.I., Rahman, M.S.: Straight-line monotone grid drawings of series-parallel graphs. Discret. Math. Algorithms Appl. **7**(2), 1550007-1–1550007-12 (2015)
13. Kindermann, P., Schulz, A., Spoerhase, J., Wolff, A.: On monotone drawings of trees. In: Duncan, C., Symvonis, A. (eds.) GD 2014. LNCS, vol. 8871, pp. 488–500. Springer, Heidelberg (2014). doi:10.1007/978-3-662-45803-7_41

Dynamic Graphs

Evaluation of Two Interaction Techniques for Visualization of Dynamic Graphs

Paolo Federico(✉) and Silvia Miksch

Institute of Software Technology and Interactive Systems, TU Wien,
Favoritenstrasse 9-11, 1040 Vienna, Austria
{federico,miksch}@ifs.tuwien.ac.at

Abstract. Several techniques for visualization of dynamic graphs are based on different spatial arrangements of a temporal sequence of node-link diagrams. Many studies in the literature have investigated the importance of maintaining the user's mental map across this temporal sequence, but usually each layout is considered as a static graph drawing and the effect of user interaction is disregarded. We conducted a task-based controlled experiment to assess the effectiveness of two basic interaction techniques: the adjustment of the layout stability and the highlighting of adjacent nodes and edges. We found that generally both interaction techniques increase accuracy, sometimes at the cost of longer completion times, and that the highlighting outclasses the stability adjustment for many tasks except the most complex ones.

Keywords: Network visualization · Dynamic graphs · Interaction · Evaluation · User-study · Time-oriented data

1 Introduction

Dynamic graphs can be used to model different complex real-word phenomena and, therefore, are collected and analysed in various disciplines. Visualization is an indispensable mean to make sense of this type of time-oriented data and gain valuable insights about the phenomena they represent. In recent years, the research about visualization of dynamic graphs has seen a rapid growth, with many novel approaches, techniques, and systems, as surveyed by recent reviews. Likewise, many evaluation studies have investigated those visualizations, to understand which are the key design factors, how they are perceived by users, and how they can support users in analysing data and solving their tasks. The evaluation of dynamic graph visualization has focused mainly on two aspects: comparing animation versus static views, and assessing the importance of the mental map preservation. These studies have often found conflicting results, or a high variability of results depending on different tasks. The use of interaction, in order to control the amount of mental map preservation, or to switch from animation to static views, has been proposed as a means to increase the applicability of a given visualization to diverse tasks. Nevertheless,

© Springer International Publishing AG 2016
Y. Hu and M. Nöllenburg (Eds.): GD 2016, LNCS 9801, pp. 557–571, 2016.
DOI: 10.1007/978-3-319-50106-2_43

few evaluation studies have focused on the role of interaction in dynamic graph visualization: usually static views are considered as noninteractive, while for animated views the most contemplated interaction is the playback control. To fill this gap, we focused on the mental map preservation and its interactive control, which we empirically evaluated in comparison with another common interaction such as highlighting. Thus, the contribution of this paper is a controlled task-based experiment to quantitatively evaluate two interaction techniques for dynamic graph visualization, namely the interactive control of the mental map and the interactive highlighting of adjacent nodes and links. In the following, we review the related work; list the hypotheses, the design, and the settings of our study; present the results and discuss their implications.

2   Related Work

Herman et al. [23] provide an early survey on graphs in information visualization, focusing on layout algorithms for both the general case and special subclasses (e.g., planar graphs, trees) as well as on techniques for interactive navigation, in particular focus+context and clustering. The state of the art report by von Landesberger et al. [41] offers an updated and extensive review of the field; it has a particular focus on issues and solutions for large scale graphs, but contains sections on dynamic graphs as well as interactions. Kerracher et al. [25] explore, and outline a structure of, the design space of dynamic graph visualization. Archambault et al. [3] also review the field, discussing in particular multivariate and temporal aspects of networks. A comprehensive survey on visualizing dynamic graphs is found in Beck et al. [10].

The Layout Stability. Many existing algorithms for drawing dynamic graphs ensure the layout stability in order to preserve the user's mental map of the graph [30]. This stability can be seen as an additional aesthetic criterion for dynamic graphs, prescribing that the placement of nodes should change as little as possible [16]. The utility of this dynamic aesthetics has been highly disputed in literature and several evaluations have been conducted, from both the algorithmic and the perceptual perspective. As for the algorithmic evaluation, Brandes and Mader [14] compare three approaches: *aggregation* (fixed nodes positions are obtained from the layout of an aggregate of all graphs in the sequence, achieving maximum stability), *anchoring* (nodes are attracted by reference positions), and *linking* (nodes are attracted by instances of themselves in adjacent time slices of the sequence). Their results suggest that the generally preferable approach is linking, that is also the most computationally demanding; a faster alternative is anchoring to an aggregate layout initialized with the previous one in the sequence. Many user studies have been performed to empirically assess the importance of mental map preservation for readability, memorability, or interpretability of dynamic graphs. Archambault and Purchase [6] review many of them. In an early study about readability of direct acyclic graphs (DAGs), Purchase et al. [34] found that the layout stability is beneficial for several tasks. Conversely, a similar study about readability of DAGs by Zaman et al. [46] found no significant effect of the layout stability. Purchase and Samra [35] tested several interpretation tasks for directed graphs and found that extremes (no stability or maximum stability) are better than a medium stability. Conversely, Saffrey and Purchase [37], by investigating readability and interpretability of directed graphs, found that the layout stability does not provide any advantage and can be even harmful for certain tasks. While all the evaluations mentioned so far were conducted on timeline based visualization, Ghani et al. [20] studied the effects of layout stability in readability of animated node-link diagrams, finding that a fixed layout (maximum stability) outperforms a force-directed layout with no stability. Analogously, by studying the memorability of animated node-link diagrams, Archambault and Purchase [5] found that maximum layout stability was the best condition.

User Interaction in Dynamic Graph Visualization. User interaction is, by definition, a crucial aspect of Information Visualization [15, page 6]. Various motivated calls have been issued to establish a "Science of Interaction" to complement Information Visualization [32]. Yi et al. [45] propose a taxonomy of interaction in Information Visualization based on the notion of user intent; Lam [27] introduces a theoretical framework to understand and possibly reduce the costs of interaction. Nevertheless, the importance of user interaction in dynamic graph visualization is generally underestimated in literature [10]; timeline approaches are generally considered as sequences of static (i.e., non interactive) drawings, while the most discussed interaction for animation approaches deals with animation control (e.g., play/pause, or time seeker). Wybrow et al. [44] review interaction techniques for multivariate graphs and propose a classification based on the Information Visualization Reference Model [15], distinguishing among view level, visual-representation level, and data level. Notable examples include a technique for selecting and manipulating subgraphs [29] or a network-aware navigation integrating "Link sliding" (guided panning when dragging along a visible link) and "Bring and Go" (bringing adjacent nodes nearby when pointing to a node), with the latter having the best performance [31]. Another example of evaluating interaction in dynamic graph visualization is provided by Rey and Diehl [36], who investigate the effects of two interaction techniques for animated visualization: interactive control of the animation speed and a tooltip showing details on demand. They found that the speed control does not provide a significant benefit, and the tooltip is outperformed by a visualization having labels always visible.

Given the high variability in the importance of the mental map (depending on tasks, user preferences, and possibly other factors), some scholars have proposed an interactive control of the layout stability, to let the user fine-tune it [8,19]. According to the taxonomy of interaction by Yi et al. [45], interactive control of stability can be understood as a *Reconfigure* interaction, corresponding to the user's intent: "Show me a different spatial arrangement". It falls into the class of user-controlled adjustments of the layout, which are very common visual-level interactions for graphs [44]. Bach et al. [8] evaluated this stability slider in the context of a specific technique (GraphDiaries) featuring staged animations of node-link diagrams, but they did not consider it as an independent factor in the study design. Smuc et al. [38] also evaluated a graph visualization featuring a stability slider, but without a specific focus on it.

Layout stability has been also described as a form of spatial highlighting, where position is used to identify different instances of the same node over time [7]. *Highlighting*, in the stricter sense, is a *brushing* interaction technique, originally developed for scatter plots [11], and then extended and applied to other visualization techniques. Brushing is a "change in the encoding of one or more items essentially immediately following, and in response to, an interaction with another item" [39, p. 235]. In particular, in the case of highlighting, the change may affect hue, brightness, or color. Brushing operates within a view or across multiple views; in the latter case, the interaction technique is better known as *linking and brushing* [24]. Highlighting makes some information stand out from other information; it effectively exploits pre-attentive processing [42], which is the human capability to process visual information prior to, or in the early stage of, focusing conscious attention. Linking and brushing techniques support two user's intents: *Select*, i.e., "Mark something as interesting" and *Connect*, i.e., "Show me related items", according to the interaction taxonomy by Yi et al. [45]. In the context of graph visualization, highlighting of adjacent nodes upon selection of a certain node (for example, by mouse hovering) is a common interaction technique, also known as *connectivity highlighting* [21]. An experiment by Ware and Bobrow has shown that interactive highlighting can efficiently support visual queries on graphs [43]. In the case of timeline visualization of dynamic graphs, the highlighting technique can be extended in order to fulfil the need of visually linking and synchronizing multiple instances of the same graph entities in different time slices [10], by considering adjacency not only across the graph structure, but also along the time dimension.

Evaluation of Other Aspects in Graph Visualization. Besides the importance of preserving the mental map in node-link diagrams, another issue which attracts the interest of scholars is the comparison between animation approaches and timeline approaches, the latter being usually based on small multiples in a juxtaposition arrangement. The controlled experiment by Farrugia and Quigley [17] found that static drawings outperform animation in terms of task completion time. Archambault et al. [4], in an analogous user study, found that small multiples are generally faster, but more error-prone for certain tasks; moreover, mental map preservation has little influence on both response time and error rate. Boyandin et al. [13] also conducted a comparative evaluation of animation versus small multiples in the context of flow maps. They found that with animation there were more findings of changes in adjacent time slices, where with small multiples there were more findings about longer time periods. Moreover, they suggest that switching from one view to the other might lead to an increase in the numbers of findings; we see this consideration in analogy with the mental map case, where the introduction of a stability slider might allow the user to adapt the layout to a particular task and possibly increase the overall visualization effectiveness.

Task Taxonomies. A profound understanding of analytical tasks is a necessary prerequisite to design novel visualization techniques as well as evaluate existing ones. Ahn et al. [1] propose a task taxonomy for dynamic graphs along three different axes: graph entities, temporal features, and properties (structural and domain-specific). According to Bach et al. [8], each task can be understood as a question containing references to two dimensions and requiring an answer in the third one. In this way, they distinguish between topological tasks, temporal tasks, and behavioural tasks. Archambault and Purchase [3] structure their taxonomy along two dimension, mostly aiming at assessing the importance of the mental map. They distinguish between local and global tasks, and between distinguishable and undistinguishable tasks (depending on whether graph entities need to be distinguished from each other or can be aggregated). A task taxonomy for multivariate networks can be found in [33].

Open Challenges. Summarizing our review of related work on visualization of dynamic graphs, we can observe that many techniques have been designed and evaluated, but the research community lacks final and well-established results about highly disputed issues, such as the importance of the layout stability, or the comparison between animation and timeline approaches. In both cases, it has been suggested that enabling users to interactively switch between different views might broaden the number of tasks that they can efficiently solve. Hence, there is a commonly recognised need of understanding the role of interaction in the context of dynamic graph visualization, but only few studies have specifically focused on the evaluation of interaction techniques.

3   Our Evaluation

Addressing the aforementioned need, we performed a user study to explore interaction in the context of dynamic graphs visualization. In particular, we considered a timeline visualization with the juxtaposition approach, where several node-link diagrams are arranged along a horizontal timeline (Fig. 1). We evaluated two interaction techniques. The first one is the interactive control of the layout stability, which is executed by the means of a slider control (thus, for the sake of brevity, we will refer to it as the *Slider*). The second interaction is the highlighting of adjacent nodes, adapted for dynamic graphs (in the following, *Highlighting*). In this section we detail the study design, the stimuli, the tasks, and the settings of our empirical experiment.

Study Design. We designed our user study as a quantitative controlled task-based evaluation, with two observed dependent variables: time and error. We considered two factors, i.e. independent variables: the presence of the *Slider* interaction (2 levels: off/on), and the presence of the *Highlighting* interaction (2 levels: off/on). In other words, we considered four different interfaces: no-interaction, only *Slider*, only *Highlighting*, and both interactions. We chose this design in order to compare the two interaction with each other and with the non-interactive baseline, and also to assess how the two interactions work together.

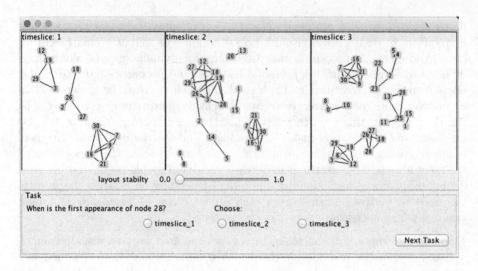

Fig. 1. The remote evaluation software displays stimuli, provides instructions, and measures time and error.

We tested 6 *task* types. The full factorial design led to a total amount of $N = Task \times Slider \times Highlighting = 6 \times 2 \times 2 = 24$ conditions. To mitigate the effects of personal skills and preferences, we chose a within-subject design; each subject tested 24 conditions, by solving a different task for each of the six task type and for each of the four interfaces. In order to lower the cognitive effort of switching between different interfaces, we grouped conditions by interface. To mitigate the effects of learning and fatigue, we used a Latin square arrangement of the interfaces and we randomized the order of the tasks within each interface. Moreover, we randomized the initial slider position.

Stimuli. We selected as baseline a spring-embedder layout as implemented in the Prefuse visualization toolkit [22] (Fig. 1). According to the *linking* approach [14], we added inter-time links to the graph in order to ensure layout stability. The *Slider* controls the amount of stability by interactively changing the relaxed lengths of inter-time springs. In the implementation of the *Highlighting* technique used for our experiment, when the user moves the mouse pointer over a node, a different combination of fill and stroke colors is used to highlight each different type of "adjacent" graph item, as shown in Fig. 2.

For the experiment we used real-world datasets: the dynamic graphs of social relationship between university freshmen collected by van Duijn, consisting of 38 nodes and 5 time points, and the one collected by van de Bunt, consisting of 49 nodes and 7 time points [40]. Through a threshold mechanism we derived two dynamic unweighted (i.e., binary) graphs from the original dynamic weighted graphs. Each task involved only subsets of 3 time slices.

Tasks. We selected six different types of tasks (Table 1). As a criterion for the selection of the tasks, we considered existing studies on the importance

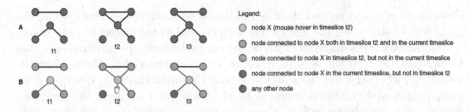

Fig. 2. A: a dynamic graph over three time slices; B: the same graph highlighted on a mouse-over event.

of layout stability and tried to elicit a set of similar tasks, in order to have comparable results. Furthermore, we considered the task taxonomy by Ahn et al. [1] in order to have a meaningful and representative set of tasks along its three axes. As for the graph entities, we included tasks referring to all the levels: the *entity* level (nodes, links), the *group* level (paths, components), and the *graph* level (the entire graph). As for the properties, we disregarded tasks referring to *domain properties* and only considered tasks referring to *structural properties*, which are specific aspects for graphs. As for the temporal features, we included tasks referring to *individual events* and *contraction & growth*, scoping out more complex tasks, which can be investigated in a follow-up study. In order to better describe the nature of our tasks and to enable a better interpretation of results, we also categorized our tasks according to other existing taxonomies [6,8], as shown in Table 1.

Table 1. Task types, examples, and classifications

Task	Description	by [1]	[6]	[8]
1.NO	Node occurence	Event/Node	Local/Distinguishable	When
	e.g., *When is the first appearance of node 27?*			
2.LO	Link occurence	Event/Link	Local/Distinguishable	When
	e.g., *When is the last appearance of link 6–9?*			
3.ND	Node degree	Event/Group	Local/Indistinguishable	When
	e.g., *When does the smallest degree (number of connections) of node 10 occur?*			
4.SP	Shortest path	Event/Group	Local/Distinguishable	When
	e.g., *When does the largest geodesic distance between node 7 and node 9 occur?*			
5.CC	Connected components	Growth/Graph	Global/Indistinguishable	What
	e.g., *Is the number of connected components increasing, decreasing, or stable?*			
6.AL	All links	Growth/Graph	Global/Indistinguishable	What
	e.g., *Is the total number of edges increasing, decreasing, or stable?*			

Subjects' Pool and Study Settings. We conducted the experiment remotely by using the Evalbench toolkit [2] (Fig. 1). In order to assess the technical setup, the estimated overall length of the evaluation session, and the understandability of textual descriptions of our tasks, we performed two pilot tests with direct observation of subjects, and then we implemented small adjustments before the main remote study. For the main study we recruited 64 volunteer subjects among undergraduate students at the fifth semester of a bachelor programme in Visual Computing. All the subjects had normal or corrected vision. Right after the recruiting, we instructed the subjects with a 15 minute briefing, describing the visualization and the interactions to be evaluated, and recalling the necessary concepts from graph theory (e.g., the notion of geodesic distance as shortest path, or the notion of connected components). The subjects were instructed to be fast and accurate in solving the tasks, without assigning any priority between speed and accuracy. The evaluation software included a training session for each of the four interfaces. During the training sessions, the software does not collect data; it shows the correct answer after completion of each task and allows repetitions until the subject feels confident of having understood the task types and the interface. The test, including the training sessions, had an average duration of 20 min.

Hypotheses. We designed our experiment to test three hypotheses: **(A)** the *Slider* reduces error rates at the cost of longer completion times, in comparison with the non-interactive interface; **(B)** the *Highlighting* reduces error rates at the cost of longer completion times, in comparison with the non-interactive interface; **(C)** the *Highlighting* outperforms the *Slider*.

We hypothesize that each interaction reduces error rates in comparison with the non-interactive interface, because both interactions comply with the *rule of self-evidence* and address the *adjacency task*. The *rule of self-evidence* for multiple views prescribes the use of "perceptual cues to make relationships among multiple views more apparent to the user" [9]. The *Highlighting* complies with this rule, by drawing attention to different instances of the same node across different time slices; the *Slider* also complies with this rule, by allowing the user to select the maximum stability and fix node positions across different time slices. The *adjacency task* (i.e., "Given a node, find its adjacent nodes") has been identified as the only graph-specific task [28]. The *Highlighting* obviously addresses this task, as well as the *Slider* does, by allowing the user to select the minimum stability and exploit the proximity *Gestalt* principle [12]. Conversely, we hypothesize that both interactions increase the task completion time in comparison with the non-interactive interface. We make this hypothesis in analogy with the existing comparative evaluations between animation and (static) timeline views [4,17], while we consider interactive timeline views as a middle way. More specifically, in terms of interaction costs [27], the *Highlighting* might increase the completion time because of the physical-motion cost of tracking elements with the mouse, while the *Slider* might imply view-change costs of reinterpreting the perception when the layout rearranges. For both techniques, there might be the decision costs of forming goals, such as deciding whether the available interaction is useful to solve the given task, and how. Moreover, the simple fact that the GUI provides an interactive option might lead users to explore its use, in order to form a solving strategy before solving a task, or to possibly increase the confidence about the solution afterwards. Furthermore, we hypothesize that the

Highlighting will have better performance than the *Slider*. We derive this hypothesis from the observation that the *Highlighting* is a common and relatively simple interaction, which at least partially exploits pre-attentive processing, while the *Slider* is based on a novel and complex concept. In other words, while the *Highlighting* directly addresses the issue of connecting entities along two dimensions (time and graph structure), the *Slider* implicitly introduces another dimension, since the stability lies in the parameter space of the layout algorithm.

4   Analysis

We preprocessed data collected from 64 subjects in order to assess whether they were eligible for analysis and we had to discard one subject whose logs were corrupted. The analysis was then performed on data from 63 subjects, consisting of 3024 samples in total. We checked the completion times for normality with the Shapiro-Wilk goodness-of-fit test but the check failed. We then applied a logarithmic transformation to the completion times and checked again the normality with a positive result. The verification of the Gaussian condition assured the applicability of parametric tests; we could perform the analysis of variance through an ANOVA with the subject as a random variable. When the ANOVA found a factor to have a statistically significant effect, we compared the two levels of that factor with a pairwise post-hoc Student's t test; when the ANOVA found the interaction between factors to be statistically significant, we performed an all-pairs Tukey's honestly significant difference (HSD) post-hoc test. The error can be understood as a dichotomous (i.e. binary) variable, since there are only two possible outcomes for each data sample (correct, not correct). Hence, we analysed the error by logistic regression as a generalized linear model (GLM) with a binomial distribution and a logit transformation as the link function, computing likelihood ratio statistics. When a factor was found to be a significant effect, we analysed the contrast between its levels in terms of pairwise comparisons between estimated marginal means.

5   Results and Discussion

Figure 3 shows time and error by *Highlighting* and *Slider*, grouped by *Task*; time is represented by box-plots with first, second (median), and third quartile, while error is represented by bars (mean) and error bars (standard error). Figure 4 shows statistically significant differences. In light of these results, we can verify our hypotheses.

Hypothesis A is partially confirmed. The *Slider* decreases the error rate for all tasks but the easiest one (1.NO), and it increases the completion time for tasks 3.ND and 4.SP only.

Hypothesis B is partially confirmed. The *Highlighting* decreases the error rate for all tasks but the most difficult one (6.AL); it increases completion times for

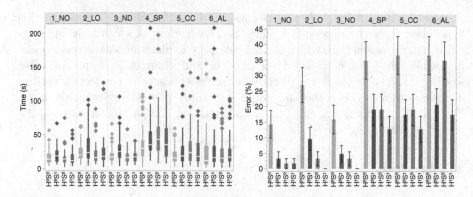

Fig. 3. Time (left-hand side, as box plots) and error (right-hand side, as bars representing means and error bars representing standard error) by *Highlighting* and *Slider*, grouped by *Task*. ■ H^0S^0 ■ H^0S^1 ■ H^1S^0 ■ H^1S^1

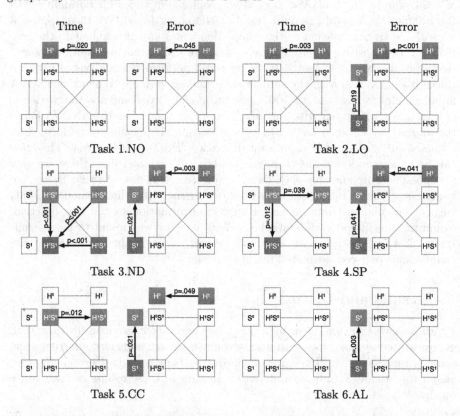

Fig. 4. Statistically significant differences for time and error, by *Task*. An arrow means that the source is faster or, respectively, more accurate than the destination, with the reported probability. Lines represent all-pairs comparisons between factor combinations (■ H^0S^0, ■ H^0S^1, ■ H^1S^0, and ■ H^1S^1), as well as pairwise comparisons by *Highlighting* (H^0–H^1, top) and *Slider* (S^0–S^1, left).

tasks 4.SP and 5.CC, but it reduces it for task 2.LO, and does not affect the remaining tasks.

Hypothesis C is partially confirmed. The *Highlighting* outperforms the *Slider* for tasks 1.NO, 2.LO, and 3.ND. For task 4.SP, the *Highlighting* and the *Slider* score equally: each of them decreases the error rate (by the same amount) and also increases the completion time if used alone, but when used together they do not increase the completion time, showing a desirable effect interaction. For task 5.CC, both factors reduce the error rate, but the *Highlighting* also increases the completion time when used alone. For task 6.AL, the only significant effect is that the *Slider* reduces the error rate.

Besides the verification of our hypothesis, which are mostly confirmed, our user study provides interesting insights about the differences between tasks. First of all, we observe that the differences in error rate and completion time among the tasks are significant, hence we can confirm that in general our tasks have different levels of difficulty. Secondly, we observe that the effectiveness of the tested interaction techniques varies with the levels of difficulty of the tasks. In a very brief but accurate summary we can say that, for easier tasks, the *Highlighting* decreases error rates and in some cases even decreases completion times; conversely, for more difficult tasks, it is the *Slider* that decreases error rates. Moreover, for tasks 3.ND, 4.SP and 5.CC, one technique increases completion times if used alone, but it does not if used in combination with the other one. Looking back at the classification of our tasks (Table 1), we can also identify the relevant aspects. We can observe that, for those tasks involving simpler temporal features of distinguishable single entities (1.NO and 2.LO), or indistinguishable groups (3.ND), the *Highlighting* is more effective. For those tasks that refer to more complex temporal features at the graph level, even if indistinguishable (5.CC and 6.AL), the *Slider* is more effective. Task 4.SP is about the changes of the geodesic distance between two nodes and requires the distinct identification of several nodes and links along the shortest path. In this case both techniques are equally accurate; if (and only if) they are used together, they do not even slow down the analysis. We can conjecture (by also considering our observations during the pilot experiments) that during the completion of a such complex task, the *Slider* can be used to switch back and forth between the minimum stability (to guess geodesic distances and shortest paths based on Euclidean distances) and the maximum stability (to identify instances of nodes and links across different time slices), while the *Highlighting* helps with tracking objects. As for task 6.AL, the *Slider* resulted to be effective; we hypothesize (by also considering our pilot observations) that subjects simply set the minimum stability and looked at the total graph area as an estimator of the density. We would have expected the *Highlighting* to be also effective, since the analysis of the degree a few central nodes might provide a good estimator of the graph density, given the power-law distribution of real-world networks. The results show that the test subjects did not exploit this expert strategy.

Design Implications. Both the *Slider* and the *Highlighting* are effective interaction techniques for dynamic graph visualization, and their use generally improves user performances. In those circumstances where it might be not possible to include them both (for example, if the color channel is employed to encode attributes of multivariate graphs, or if the GUI is already overloaded with many controls), our evaluation provides an indication to designers according to the user tasks to be supported. Our results suggest that the *Highlighting* is indicated for tasks involving temporal features of distinguishable single entities or indistinguishable groups, while the *Slider* is indicated for tasks involving complex temporal behaviours at the graph level. The joint use of both interactions is beneficial for the most complex task involving temporal behaviours of connectivity paths.

6   Conclusion

We have presented an evaluation of two interaction techniques for dynamic graph visualization, namely the interactive control of the layout stability and the interactive highlighting of adjacent nodes and links. The results mostly confirm our hypotheses: both interactions decrease the error rate, in some cases at the cost of a longer completion time. We observed significant differences between tasks, with the highlighting performing better for some tasks, and the stability control performing better for others. We acknowledge the limitations of our experiment, whose findings might not be directly generalizable to large-scale datasets. The highlighting interaction for dynamic graphs is much more complex then the standard connectivity highlighting for static graphs, and may require training to be understood and used effectively. The stability control might have a different effect when combined with 3D visualization and interaction techniques (e.g., the vertigo zoom [18]). However, our study provides preliminary clues for visualization designers who need to choose the most appropriate interaction technique for their users' tasks. Further studies are needed to obtain a comprehensive understanding of the role of interaction in visualization of dynamic graphs.

Acknowledgements. The authors wish to thank the anonymous study subjects for their participation, as well as Theresia Gschwandtner, Simone Kriglstein, and Margit Pohl for their feedback on the manuscript.

This work was partially supported by the Austrian Research Promotion Agency (FFG), project *Expand*, grant 835937.

References

1. Ahn, J.W., Plaisant, C., Shneiderman, B.: A task taxonomy for network evolution analysis. IEEE Trans. Visual. Comput. Graph. **20**(3), 365–376 (2014)
2. Aigner, W., Hoffmann, S., Rind, A.: EvalBench: a software library for visualization evaluation. Comput. Graph. Forum **32**(3pt1), 41–50 (2013)
3. Archambault, D., Abello, J., Kennedy, J., Kobourov, S., Ma, K.L., Miksch, S., Muelder, C., Telea, A.: Temporal multivariate networks. In: Kerren et al. [26], pp. 151–174

4. Archambault, D., Purchase, H., Pinaud, B.: Animation, small multiples, and the effect of mental map preservation in dynamic graphs. IEEE Trans. Visual. Comput. Graph. **17**(4), 539–552 (2011)

5. Archambault, D., Purchase, H.C.: The mental map and memorability in dynamic graphs. In: Proceedings of the Pacific Visualization Symposium, PacificVis 2012, pp. 89–96. IEEE, Washington, DC (2012)

6. Archambault, D., Purchase, H.C.: The map in the mental map: experimental results in dynamic graph drawing. Int. J. Hum.-Comput. Studies **71**(11), 1044–1055 (2013)

7. Archambault, D., Purchase, H.C.: On the application of experimental results in dynamic graph drawing. In: Proceedings of the 1st International Workshop on Graph Visualization in Practice, GViP 2014, vol. 1244, pp. 73–77. CEUR-WS (2014)

8. Bach, B., Pietriga, E., Fekete, J.D.: GraphDiaries: animated transitions and temporal navigation for dynamic networks. IEEE Trans. Visual. Comput. Graph. **20**(5), 740–754 (2014)

9. Baldonado, M.Q.W., Woodruff, A., Kuchinsky, A.: Guidelines for using multiple views in information visualization. In: Proceedings of the Working Conference on Advanced Visual Interfaces, AVI 2000, pp. 110–119. ACM, New York (2000)

10. Beck, F., Burch, M., Diehl, S., Weiskopf, D.: The state of the art in visualizing dynamic graphs. In: Borgo, R., Maciejewski, R., Viola, I. (eds.) EuroVis - STAR, pp. 83–103. The Eurographics Association (2014)

11. Becker, R.A., Cleveland, W.S.: Brushing scatterplots. Technometrics **29**(2), 127–142 (1987)

12. Bennett, C., Ryall, J., Spalteholz, L., Gooch, A.: The aesthetics of graph visualization. In: Cunningham, D.W., Meyer, G., Neumann, L. (eds.) Computational Aesthetics in Graphics, Visualization, and Imaging. The Eurographics Association (2007)

13. Boyandin, I., Bertini, E., Lalanne, D.: A qualitative study on the exploration of temporal changes in flow maps with animation and small-multiples. Comput. Graph. Forum **31**(3pt2), 1005–1014 (2012)

14. Brandes, U., Mader, M.: A quantitative comparison of stress-minimization approaches for offline dynamic graph drawing. In: Kreveld, M., Speckmann, B. (eds.) GD 2011. LNCS, vol. 7034, pp. 99–110. Springer, Heidelberg (2012). doi:10.1007/978-3-642-25878-7_11

15. Card, S.K., Mackinlay, J.D., Shneiderman, B. (eds.): Readings in Information Visualization: Using Vision to Think. Morgan Kaufmann, San Francisco (1999)

16. Coleman, M.K., Parker, D.S.: Aesthetics-based graph layout for human consumption. Softw.: Pract. Exp. **26**(12), 1415–1438 (1996)

17. Farrugia, M., Quigley, A.: Effective temporal graph layout: a comparative study of animation versus static display methods. Inf. Visual. **10**(1), 47–64 (2011)

18. Federico, P., Aigner, W., Miksch, S., Windhager, F., Smuc, M.: Vertigo zoom: combining relational and temporal perspectives on dynamic networks. In: Proceedings of the International Working Conference on Advanced Visual Interfaces, AVI 2012, pp. 437–440. ACM, New York (2012)

19. Federico, P., Aigner, W., Miksch, S., Windhager, F., Zenk, L.: A visual analytics approach to dynamic social networks. In: Proceedings of the International Conference on Knowledge Management and Knowledge Technologies, i-KNOW 2011, pp. 47:1–47:8. ACM, New York (2011)

20. Ghani, S., Elmqvist, N., Yi, J.S.: Perception of animated node-link diagrams for dynamic graphs. Comput. Graph. Forum **31**(3pt3), 1205–1214 (2012)

21. Heer, J., Boyd, D.: Vizster: visualizing online social networks. In: IEEE Symposium on Information Visualization, INFOVIS 2005, pp. 32–39, October 2005
22. Heer, J., Card, S.K., Landay, J.A.: Prefuse: a toolkit for interactive information visualization. In: Proceedings of the SIGCHI Conference on Human Factors in Computing Systems, CHI 2005, pp. 421–430. ACM, New York (2005)
23. Herman, I., Melançon, G., Marshall, M.S.: Graph visualization and navigation in information visualization: a survey. IEEE Trans. Visual. Comput. Graph. 6(1), 24–43 (2000)
24. Keim, D.A.: Information visualization and visual data mining. IEEE Trans. Visual. Comput. Graphics 8(1), 1–8 (2002)
25. Kerracher, N., Kennedy, J., Chalmers, K.: The design space of temporal graph visualisation. In: Elmqvist, N., Hlawitschka, M., Kennedy, J. (eds.) EuroVis - Short Papers, pp. 7–11. The Eurographics Association (2014)
26. Kerren, A., Purchase, H.C., Ward, M.O. (eds.): Multivariate Network Visualization. LNCS, vol. 8380. Springer, Heidelberg (2014)
27. Lam, H.: A framework of interaction costs in information visualization. IEEE Trans. Visual. Comput. Graph. 14(6), 1149–1156 (2008)
28. Lee, B., Plaisant, C., Parr, C.S., Fekete, J.D., Henry, N.: Task taxonomy for graph visualization. In: Proceedings of the AVI Workshop on Beyond Time and Errors: Novel Evaluation Methods for Information Visualization, BELIV 2006, pp. 1–5. ACM, New York (2006)
29. McGuffin, M., Jurisica, I.: Interaction techniques for selecting and manipulating subgraphs in network visualizations. IEEE Trans. Visual. Comput. Graph. 15(6), 937–944 (2009)
30. Misue, K., Eades, P., Lai, W., Sugiyama, K.: Layout adjustment and the mental map. J. Vis. Lang. Comput. 6(2), 183–210 (1995)
31. Moscovich, T., Chevalier, F., Henry, N., Pietriga, E., Fekete, J.D.: Topology-aware navigation in large networks. In: Proceedings of the SIGCHI Conference on Human Factors in Computing Systems, CHI 2009, pp. 2319–2328. ACM, New York (2009)
32. Pike, W.A., Stasko, J., Chang, R., O'Connell, T.A.: The science of interaction. Inf. Visual. 8(4), 263–274 (2009)
33. Pretorius, A., Purchase, H., Stasko, J.: Tasks for multivariate network analysis. In: Kerren et al. [26], pp. 77–95
34. Purchase, H.C., Hoggan, E., Görg, C.: How important is the "mental map"? – an empirical investigation of a dynamic graph layout algorithm. In: Kaufmann, M., Wagner, D. (eds.) GD 2006. LNCS, vol. 4372, pp. 184–195. Springer, Heidelberg (2007). doi:10.1007/978-3-540-70904-6_19
35. Purchase, H.C., Samra, A.: Extremes are better: investigating mental map preservation in dynamic graphs. In: Stapleton, G., Howse, J., Lee, J. (eds.) Diagrams 2008. LNCS (LNAI), vol. 5223, pp. 60–73. Springer, Heidelberg (2008). doi:10.1007/978-3-540-87730-1_9
36. Rey, G.D., Diehl, S.: Controlling presentation speed, labels, and tooltips in interactive animations. J. Media Psychol.: Theor. Methods Appl. 22(4), 160 (2010)
37. Saffrey, P., Purchase, H.: The mental map versus static aesthetic compromise in dynamic graphs: a user study. In: Proceedings of the Conference on Australasian User Interface, AUIC 2008, pp. 85–93. Australian Comp. Soc. (2008)
38. Smuc, M., Federico, P., Windhager, F., Aigner, W., Zenk, L., Miksch, S.: How do you connect moving dots? Insights from user studies on dynamic network visualizations. In: Huang, W. (ed.) Handbook of Human Centric Visualization, pp. 623–650. Springer, New York (2014)

39. Spence, R.: Information Visualization: Design for Interaction. Addison Wesley, Harlow, New York (2007)
40. Van De Bunt, G.G., Van Duijn, M.A.J., Snijders, T.A.B.: Friendship networks through time: an actor-oriented dynamic statistical network model. Comput. Math. Organ. Theory **5**(2), 167–192 (1999)
41. von Landesberger, T., Kuijper, A., Schreck, T., Kohlhammer, J., van Wijk, J., Fekete, J.D., Fellner, D.: Visual analysis of large graphs: state-of-the-art and future research challenges. Comput. Graph. Forum **30**(6), 1719–1749 (2011)
42. Ware, C.: Information Visualization: Perception for Design. Morgan Kaufman, San Francisco (2004)
43. Ware, C., Bobrow, R.: Supporting visual queries on medium-sized nodelink diagrams. Inf. Visual. **4**(1), 49–58 (2005)
44. Wybrow, M., Elmqvist, N., Fekete, J.D., von Landesberger, T., van Wijk, J., Zimmer, B.: Interaction in the visualization of multivariate networks. In: Kerren et al. [26], pp. 97–125
45. Yi, J.S., ah Kang, Y., Stasko, Y., Jacko, J.: Toward a deeper understanding of the role of interaction in information visualization. IEEE Trans. Visual. Comput. Graph. **13**(6), 1224–1231 (2007)
46. Zaman, L., Kalra, A., Stuerzlinger, W.: The effect of animation, dual view, difference layers, and relative re-layout in hierarchical diagram differencing. In: Proceedings of the Conference Graphics Interface, GI 2011, pp. 183–190. Canadian Human-Computer Comm. Soc. (2011)

Offline Drawing of Dynamic Trees: Algorithmics and Document Integration

Malte Skambath[1(✉)] and Till Tantau[2]

[1] Department of Computer Science, Kiel University, Kiel, Germany
malte.skambath@email.uni-kiel.de
[2] Institute of Theoretical Computer Science, Universität Zu Lübeck,
Lübeck, Germany
tantau@tcs.uni-luebeck.de

Abstract. While the algorithmic drawing of static trees is well-understood and well-supported by software tools, creating animations depicting how a tree changes over time is currently difficult: software support, if available at all, is not integrated into a document production workflow and algorithmic approaches only rarely take temporal information into consideration. During the production of a presentation or a paper, most users will visualize how, say, a search tree evolves over time by manually drawing a sequence of trees. We present an extension of the popular TEX typesetting system that allows users to specify dynamic trees inside their documents, together with a new algorithm for drawing them. Running TEX on the documents then results in documents in the SVG format with visually pleasing embedded animations. Our algorithm produces animations that satisfy a set of natural aesthetic criteria when possible. On the negative side, we show that one cannot always satisfy all criteria simultaneously and that minimizing their violations is NP-complete.

1 Introduction

Trees are undoubtedly among the most extensively studied graph structures in the field of graph drawing; algorithms for drawing trees date back to the origins of the field [26,40]. However, the extensive, ongoing research on how trees can be drawn efficiently, succinctly, and pleasingly focuses on either drawing a single, "static" tree or on interactive drawings of "dynamic" trees [11,12,27], which are trees that change over time. In contrast, the problem of drawing dynamic trees *non*interactively in an *offline* fashion has received less attention.

It is this problem that lies at the heart of our paper.

Consider how an author could explain, in a paper or in a presentation, how a tree-based data structure such as a search tree works. In order to explain the dynamic behavior, our author might wish to show how the data structure evolves for a sequence of update operations. A typical drawing of the evolving

---

Animations in this document will only be rendered in the SVG version [32], see Sect. 2.3 for a discussion of the reasons.

© Springer International Publishing AG 2016
Y. Hu and M. Nöllenburg (Eds.): GD 2016, LNCS 9801, pp. 572–586, 2016.
DOI: 10.1007/978-3-319-50106-2_44

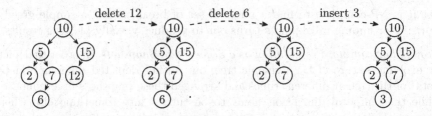

Fig. 1. A "manually" created drawing of a dynamic tree: each tree in the sequence has been drawn using the Reingold–Tilford [29] algorithm.

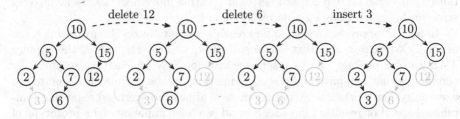

Fig. 2. The dynamic tree from Fig. 1, redrawn by drawing a "supergraph" (the union of all trees in the sequence) and then using the positions of nodes in this supergraph for the individual drawings.

sequence might look as in Fig. 1, which has been created "manually" by running the Reingold–Tilford algorithm [29] on each tree in the sequence independently. While the result is satisfactory, there are (at least) two shortcomings:

First Shortcoming: Flawed Layout. In the first step, the layout of the root's children changes (their horizontal distance decreases) even though there is no structural change at the root. While in the present graph the effect is small, one can construct examples where a single node removal causes a change in distances on all levels, obscuring where the actual structural change occurred. Since the whole sequence of trees (the whole "dynamic tree") is given by the author, the problem can be addressed by not running the Reingold–Tilford algorithm on each tree individually, but by running it on the "supergraph" resulting from uniting all trees in the sequence, resulting in the visualization in Fig. 2.

Unfortunately, this simple supergraph approach introduces new problems: First, the nodes "2" and "7" are unnecessarily far apart – the nodes "3" and "6" could use the same space since they are never both members of the same tree. Second, it is easy to construct sequences of trees whose union is not a tree itself.

We address these problems in Sect. 3, where we present a new algorithm for computing layouts of dynamic trees that addresses the above problems. For dynamic trees whose supergraphs are trees or at least acyclic, the algorithm finds an optimal layout (with respect to natural aesthetic criteria) of the dynamic tree in linear time. For cyclic supergraphs, which are also important in practice since they arise for instance from the rotations necessary to balance search trees in data structures such as AVL trees [1], we show that one has to break the cycles in order to layout the graph according to the criteria we develop. While we show that it is NP-complete to find a minimal set of break points, a simple greedy heuristic for finding breakpoints turns out to produce visually pleasing results.

Second Shortcoming: Presentation as a Sequence of Snapshots. In order to depict the evolving nature of her dynamic tree, our author depicted different "snapshots" of the tree at different times and arranged these snapshots in a sequence. While the temporal dimension *needs* to be turned into something else when our medium of communication is printed paper, for documents presented using appropriate electronic devices we can visualize dynamic trees using *animations.* Such an animation needs much less space on a page and, perhaps more importantly, our visual system is *much* better at spotting movement than at identifying structural changes between adjacent objects.

In Sect. 2 we present a system for creating animations on-the-fly during a run of the TEX program on a text document: First, we have augmented the popular TikZ graphic package [37] (a macro package for TEX for creating graphics) by commands that compute and embed animations in the output files. Due to the way the system works, these commands have almost no overhead regarding compilation speed or resulting file size. Second, we have implemented a prototype of our algorithm from Sect. 3 for drawing dynamic trees that uses these animation commands. In result, when an author specifies the above dynamic graph appropriately in a TEX document and then runs TEX on it to convert it, the resulting file will contain the normal text and graphics as well as an embedded animation of the dynamic tree. When the document is viewed on electronic devices with a modern browser, the animation runs right inside the document.

Related Work. Approaches to drawing *static* trees date back to the early 1970s, namely to the work of Knuth, Wetherell and Shannon, and Sweet [26,35,40]. A standard algorithm still in use today is due to Reingold and Tilford [29], see also [38]. They suggested that symmetric tree structures should be drawn symmetrically and provided an algorithm that supports this objective well and runs in linear time. Instead of visualizing trees as node-link diagrams, one can also use tree maps [25], three dimensional cone trees [30], or sunburst visualizations [33].

Approaches to drawing general *dynamic* graphs are more recent. The sequence-of-snapshot visualizations sketched before as well as animations are standard ways of visualizing them [19]. One can also generally treat time as another spacial dimension, which turns nodes into tubes through space [23]. There are many further techniques that are not restricted to node-link diagrams [8,9,22,28]; for an extensive overview of the whole state of the art including a taxonomy of different visualization techniques see Beck et al. [5], or [21] for a more tree-specific overview. Diehl, Görg and Kerren [14,15] introduced a general concept, called *foresighted layout,* for drawing dynamic graphs offline. They propose to collapse nodes in the supergraph that never exist at the same time and to then draw the supergraph. While this approach produces poor results for trees, the results are better for hierarchical graphs [20].

Approaches tailored specifically to drawing dynamic *trees* are currently almost always *online* approaches. The algorithms, which expect a sequence of update operation as input [12,27], are integrated into interactive software and create or adjust the layout for each change. An early algorithm designed for dynamic trees was developed by Moen [27]. Later Cohen et al. [11,12] presented algorithms for different families of graphs the includes trees.

Concerning the integration of tree drawing algorithms into text processing software, first implementations for the typesetting system TEX date back to Eppstein [18] and Brüggemann and Wood [6]. A more recent implementation of the Reingold–Tilford algorithm by the second author is now part of the graph drawing engine in TikZ [36].

Organisation of This Paper. This paper is structured into two parts: In the first part, Sect. 2, we present the system we have developed for generating animations of dynamic graphs that are embedded into documents. Our core argument is that the system's seamless integration into a widely used system such as TEX is crucial for its applicability in practice. In the second part, Sect. 3, partly as a case study, partly as a study of independent interest, we investigate how a dynamic tree can be drawn using animations. We derive aesthetic criteria that animations and even image sequences of dynamic trees *should* meet and present an algorithm that *does* meet them. Full proofs and pseudo-code can be found in the appendix of the full version, which also contains a gallery of dynamic trees drawn using our prototype.

2 Dynamic Trees in Documents

The problem for which we wish to develop a practical solution in the rest of this paper is the following: Visualize one or more dynamic trees inside a document created by an author from some manuscript. To make the terminology precise, by *dynamic graph* we refer to a sequence $(G_1, \ldots, G_n)$, where each $G_i = (V_i, E_i, \phi_i)$ is a directed, annotated graph with vertex set V_i, edge set E_i, and an annotation function $\phi_i \colon V_i \cup E_i \to A$ that assigns additional information to each node and edge from some set A of annotations like ordering or size information. A *dynamic tree* is a dynamic graph where each T_i is a tree with the edges pointing away from the root. A *manuscript* is a plain text written by an *author* that can be transformed by a program into an (output) *document,* a typically multi-page text document with embedded graphics or embedded animations. Note that the problem is an *offline* problem since the manuscript contains a full description of the dynamic graph and algorithms have full access to it. In rest of this section we explain how the practical obstacles arising from the problem are solved by the system we have developed, in Sect. 3 we investigate algorithmic questions.

In the introduction we saw an example of how a dynamic tree can be visualized using a series of "snapshots" shown in a row. While this way of depicting a dynamic tree is a sensible, traditional way of solving the problem (drawings on printed paper "cannot change over time"), documents are now commonly also read on electronic devices that are capable of displaying changing content and, in particular, animations. We claim that using an animation instead of a sequence of snapshots has two major advantages: First, sequences of snapshots need a lot of space on a page even for medium-sized examples. We did a cursory survey of standard textbooks on computer science and found that typically only three to four snapshots are shown and that the individual trees are often rather small. For an animation, the length of the sequence is only limited by the (presumed) attention span of the reader and not by page size. Second, our visual system is *much* better at spotting movement than at identifying structural changes between adjacent objects. When operations on trees such as adding or deleting a leaf or moving whole subtrees are visualized using movements, readers can identify and focus on these operations on a subconscious level.

Given the advantages offered by animations, it is surprisingly difficult to integrate animations into documents. Of course, there is a lot of specialized software for creating animations and graphics output formats like PDF or SVG allow the inclusion of movie files in documents. However, this requires authors to use – apart from their main text processor like TEX or Word – one or more programs for generating animations and they then have to somehow "link" the (often very large) outputs of these different programs together. The resulting workflows are typically so complicated that authors rarely employ them. Even when they are willing to use and integrate multiple tools into their workflow, authors face the problem that using different tools makes it next to impossible to keep a visually consistent appearance of the document [36]. Very few, if any, animation software will be able to render for instance TEX formulas inside to-be-animated nodes correctly and take the sizes of these formulas into account.

We have developed a system that addresses the above problems; more precisely, we have augmented an existing system that is in wide-spread use – TEX – by facilities for specifying dynamic trees, for computing layouts for them, and for generating animations that are embedded into the output files. Our extensions are build on top of Ti*k*Z's graph drawing engine [36], which has been part of standard TEX distributions since 2014.

Authors first specify the dynamic trees they wish to draw inside TEX manuscripts using a special syntax, which we describe in Sect. 2.1 (conceptually, this is similar to specifying for instance formulas inside the TEX manuscripts). Next, authors apply a graph drawing algorithm to the specified dynamic graph by adding an appropriate option to the specification and then running the TEX program as explained in Sect. 2.2. Lastly, in Sect. 2.3, we discuss which output formats are supported by our system, how the output can be viewed on electronic devices, and how a fallback for printed paper can be generated.

2.1 The Input: Specifying Dynamic Trees

In order to make dynamic trees accessible to graph drawing algorithms, we first have to specify them. For dynamic graphs and, in particular, for dynamic trees, there are basically two different methods available to us: We can specify each graph or tree in the dynamic graph sequence explicitly. Alternatively, we can specify a sequence of update operations that transform one graph into the next such as, for the dynamic trees of search trees, the sequence of insert and delete operations that give rise to the individual trees. Besides being easy and natural to use, the second method also provides algorithms with rich semantic information concerning the change from one graph to the next in the sequence.

Despite the fact that the second method is more natural in several contexts and more semantically rich, for our prototype we use the first method: Authors specify dynamic graphs by explicitly specifying the sequence of graphs that make up the dynamic graph. We have two reasons for this choice: First, specifying the sequence of graphs explicitly imposes the least restrictions on what kind of dynamic graphs can be drawn, in principle. In contrast, the set of update operations necessary to describe the changes occurring just for the standard data structures balanced search trees, heaps, and union–find trees is large and hard to standardize. For instance, should the root rotation occurring in AVL trees be considered a standard update operation or not? Second, it easy to convert a sequence of update operations into a sequence of graphs, while the reverse direction is harder and, sometimes, not possible. Our system can easily be extended to accept different sequences of update operations as input and convert them on-the-fly into a sequence of graphs that is then processed further.

There are different possible formats for specifying individual graphs and, in particular, trees of graph sequences, including GRAPHML, an XML-based markup language; the DOT format, used by GRAPHVIZ [17]; the GML format, used by the Open Graph Drawing Framework [10]; or the format of the \graph command of TikZ [37], which is similar to the DOT format. As argued in [36], it is not purely a matter of taste, which format is used; rather, good formats make it easy for humans to notate all information about a graph that is available to them. For instance, for static graphs the *order* in which vertices are specified is almost never random, but reflects information about them that the author had and that algorithms should take into account.

Since the algorithm and system we have implemented are build on top of the graph drawing engine of TikZ [36], we can use all of the different syntax flavors offered by this system, but authors will typically use the \graph command. Each graph in the sequence of graphs is surrounded by curly braces and, following the opening brace, we say [when=i] to indicate that we now specify the ith graph in the sequence. The graph is then specified by listing the edges, please see [36] and [37] for details on the syntax and its use in TikZ. The result is a specification of the dynamic graph such as the following for the example graph from Figs. 1 and 2:

```
\tikz \graph { {[when=1] 10->{ 5->{ 2,          7->6 },  15->12 } },
               {[when=2] 10->{ 5->{ 2,          7->6 },  15 } },
               {[when=3] 10->{ 5->{ 2,          7 },     15 } },
               {[when=4] 10->{ 5->{ 2->{ , 3 }, 7 },     15 } } };
```

2.2 Document Processing and Algorithm Invocation

Once a dynamic graph has been specified as part of a larger TEX document, we need to process it. This involves both running a dynamic graph drawing algorithm to determine the positions of the nodes and the routing of the edges as well as producing commands that create the desired animation.

The framework provided by the graph drawing engine [36] of TikZ is well-suited for the first task. All the author has to do is to load an appropriate graph drawing library and then use a special key with the \tikz command:

```
\tikz [animated binary tree layout]
  \graph { {[when=1] 10->{ ... } };
          {[when=2] 10->{ ... } },
          {[when=3] 10->{ ... } },
          {[when=4] 10->{ ... } } };
```

The key `animated binary tree layout` causes the graph drawing engine to process the dynamic graph. It will parse the dynamic graph, convert it to an object-oriented model, and pass it to an algorithm from the `evolving` library, which is written in the Lua programming language [24].[1] The framework also handles the later rendering of the nodes and edges and their correct scaling and embedding into the output document. Thus, the algorithm's implementation only needs to address the problem of computing node positions from an object-oriented model of the dynamic graph. The implementation need not (indeed, cannot) produce or process graphical output and primitives.

Once the algorithm has computed the positions for nodes and edges of the graphs in the sequence, actual animations need to be generated. For this, TikZ itself was extended by a new animation subsystem, which can be used independently of the graph drawing engine and allows users to specify and embed arbitrary animations in their documents. The animation subsystem adds *animation annotations* to the output file, which are statements like "move this graphics group by 1 cm to the right within 2 s" or "change the opacity of this node from opaque to transparent within 200 ms." More formally, they are XML elements in the Synchronized Multimedia Integration Language [7]. For the animation of dynamic graphs, the graph drawing engine can now map the computed positions of the nodes at different times to TikZ commands that add appropriate movement and opacity-change annotations to the output.

2.3 The Output: Scalable Vector Graphics

The annotation-based way of producing animations has two important consequences: Firstly, adding the annotations to the output does not have a noticeable effect on the speed of compilation (computing the necessary XML statements is quite easy) nor on the file size (annotations are small). However, secondly, the job of rendering the graph animations with, say, 30 frames per second does not

---

[1] When the algorithm is also implemented in the Lua language, it can be used directly by TEX without special configurations or runtime linking, but it can also be implemented in C or C++ at the cost of a more complicated deployment.

lie with TEX, but with the viewer application and we need both a format and viewer applications that support this.

Currently, there is only one graphics format that supports these annotation-based animations: The *Scalable Vector Graphics* (SVG) format [13], which is a general purpose graphics language that is in wide-spread use. All modern browsers support it, including the parsing and rendering of SVG animations. The DVISVGM program, which is part of standard TEX distributions, transforms arbitrary TEX documents into SVG files that, when viewed in a browser, are visually indistinguishable from PDF files produced by TEX – except, of course, for the animations of the dynamic graphs.

While we argued that animations are a superior way of visualizing dynamic graphs, there are situations where they are not feasible: First, documents *are* still often printed on paper. Second, the popular PDF format does not support annotation-based animations and, thus, is not able to display TikZ's animations. Third, it is sometimes desirable or necessary to display "stills" or "snapshots" of animations at interesting time steps alongside the animation. In these situations, authors can say `make snapshot of=`t to replace the animation by a static picture of what the animated graphic would look like at time t. Since the computation of the snapshot graphic is done by TEX and since no animation code is inserted into the output, this method works with arbitrary output formats, including PDF.

3 Algorithmic Aspects of Drawing Dynamic Trees

Given a dynamic tree $T = (T_1, \ldots, T_k)$ consisting of a sequence of trees $T_i = (V_i, E_i, \phi_i)$, we saw in the introduction that neither drawing each tree independently and then "morphing" the subsequent drawings to create an animation nor laying out just the supergraph $\mathrm{super}(T) = (\bigcup_i V_i, \bigcup_i E_i)$ and then animating just the opacity of the nodes and edges will lead to satisfactory drawings of dynamic trees. Our aim is to devise a new algorithm that addresses the shortcomings of these approaches and that meets a number of sensible *aesthetic criteria* that we formulate in Sect. 3.1. The algorithm, presented in Sect. 3.2, has been implemented as a prototype [31] and we have used it to create the animations of dynamic trees in the present paper. While the prototype implementation does not even run in linear time (as would be possible by Theorem 3.2), it only needs fractions of a second for the example graphs from this paper.

3.1 Aesthetic Criteria for Drawing Dynamic Trees

Already in 1979, Wetherell and Shannon [40] explicitly defined aesthetic criteria for the layout of trees. Two years later Reingold and Tilford [29] refined these *static criteria* towards more symmetric drawings in which isomorphic subtrees must have the same layout. While the criteria were originally formulated for binary trees only, one can allow any number of children when there is an ordering on the children of each node.

Criterion (Ranking). *The vertical position of a node equals its topological distance from the root.*

Criterion (Ordering). *The horizontal positions of a node's children respect their topological order in the tree.*

Criterion (Centering). *Nodes are horizontally centered between their leftmost and rightmost child if there are at least two children.*

Criterion (Symmetry). *All topologically order-isomorphic subtrees are drawn identically. Topologically mirrored subtrees are drawn horizontally mirrored.*

As numerous applications show, these rather sensible criteria lead to aesthetically pleasing drawings of static trees. We extend these criteria to the dynamic case. Ideally, we would like to keep all of the above criteria, but will see in a moment that this is not always possible.

Our first dynamic criterion forbids the unnecessary movement of nodes in drawings like the one shown on the right, which shows the same problem as the example in the introduction did: The horizontal offset between n and c changes from T_i to T_{i+1} even though there is no struc-

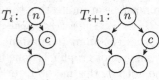

tural change at n. (Note that when a node disappears in the step from T_i to T_{i+2} and then reappears in T_{i+2}, the stability criterion does no require it to appear at the same position as before.)

Criterion (Stability). *The horizontal offset between a node n and a child c may not change between the layouts of trees T_i and T_{i+1} if c does not change its position in the ordering of the children of n.*

While the stability criterion forbids relative movements of connected nodes, it allows whole subtrees to move without changing their inner layout. This emphasizes the important parts of changes since multiple objects moving with the same speed are percieved as one connected group [4,39]. The criterion reduces movements and draws common structures identically, thereby reducing errors in understanding [2] and making it easier for viewers to correctly recognize the changes in the tree sequence [3].

While all of the above criteria are reasonable, unfortunately, there is no way of meeting all of them simultaneously, see the appendix for the proof:

Lemma 3.1. *No drawing of the dynamic tree $T = (T_1, T_2)$ from Fig. 3 meets all of the criteria Ranking, Ordering, Centering, Symmetry, and Stability.*

In view of the lemma, we will need to weaken one or more of our criteria, while still trying to meet them at least in "less problematic" cases than the dynamic tree from Fig. 3. Furthermore, even when the criteria *can* be met, this may not always be desirable.

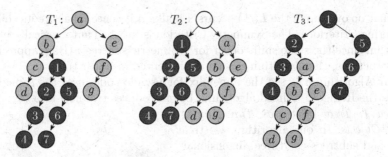

Fig. 3. A "problematic" dynamic tree. Already the dynamic tree $T = (T_1, T_2)$ cannot be drawn while meeting all of the criteria Ranking, Ordering, Centering, Symmetry, and Stability, as shown in Lemma 3.1. The whole dynamic tree $T = (T_1, T_2, T_3)$ cannot even be drawn when the Symmetry Criterion is replaced by the Weak Symmetry Criterion, see Lemma 3.3.

Consider the right example, which seems like a "reasonable" drawing of a dynamic tree. The Stability Criterion enforces the large distance between b and c already in T_1, but the Symmetry Criterion would now actually enforce the same distance between 2 and 3, which seems undesirable here. As a replacement of the Symmetry Criterion we propose a *Weak Symmetry Criterion* that our

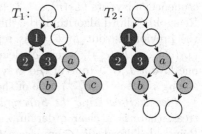

algorithm will be able to meet in many important cases, including the troublesome example from Lemma 3.1. Nevertheless, there are *still* dynamic trees that cannot be drawn in this way, see Lemma 3.3, which also turn out to be the algorithmically difficult cases.

Criterion (Weak Symmetry). *Let n and n' be nodes such that for all $i \in \{1, \ldots, n\}$ the subtrees rooted at n and at n' in T_i are order-isomorphic (or all mirrored). Then in all drawings of the T_i the subtrees rooted at n and at n' must all be drawn identically (or all mirrored).*

3.2 An Algorithm for Drawing Arbitrary Dynamic Trees

Our starting point for an algorithm that meets the aesthetic criteria just formulated is the classical Reingold–Tilford algorithm [29]. It will be useful to review this algorithm briefly, formulated in a "bottom-up" fashion: While there is a node that has not yet been processed, pick a node n whose children $c^1, \ldots, c^m$ have all already been processed (this is immediately the case for all leafs, where $m = 0$). For each child c^r a layout $L(c^r)$ will have been computed for the subtree $T(c^r)$ of T rooted at c^r. The algorithm now shifts the $L(c^r)$ vertically so that all c^r lie on the same horizontal line (Ranking Criterion), then shifts them horizontally so that the c^1 comes first, followed by c^2, and so on (Ordering Criterion), such that no overlap of the $L(c^r)$ occurs. Finally, n is centered above its children (Centering Criterion). The Symmetry Criterion is satisfied automatically by this algorithm since the same shifts occur for symmetric subtrees. Using appropriate data structures, the algorithm can be implemented in linear time.

Our Algorithm A.1, see the appendix for pseudo-code, uses the same basic idea as the Reingold–Tilford algorithm, but introduces two new ideas.

Idea 1: Treat Nodes as Three-Dimensional Objects. In our algorithm, we treat nodes and subtrees as "three dimensional" objects with time as the third dimension. Given a dynamic tree $T = (T_1, \ldots, T_k)$, the algorithm does not process the T_i one at a time (as online algorithms have to do), but instead considers for each node n of the supergraph super(T) the sequence

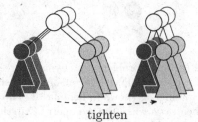

tighten

$(T_1(n), \ldots, T_k(n))$ of trees rooted at n in the different T_i and computes a whole sequence of layouts $(L_1(n), \ldots, L_k(n))$ for these trees: The core operation of the Reingold–Tilford algorithm, the shifting of a layout $L(c^r)$ until it almost touches the previous layout $L(c^{r-1})$, is replaced by a shifting of the whole sequence $(L_1(c_1^r), \ldots, L_k(c_k^r))$, where c_j^i denotes the ith child of n in T_j, until at least one layout $L_j(c_j^r)$ (one of the gray layouts in the example) almost touches its sibling's layout $L_j(c_j^{r-1})$ (one of the dark layouts).

Idea 2: Processing the Supergraph Using a Topological Ordering. For static trees, there is a clear order in which the nodes should be processed by the Reingold–Tilford algorithm: from the leafs upwards. For a dynamic tree, this order is no longer clear – just consider the example from Fig. 3: Should we first process node 1 or node a? Our algorithm address this ordering problem as follows: We compute the supergraph super(T) and then check whether it is acyclic. If so, it computes a topological ordering of super(T) and then processes the nodes in this order. Observe that this guarantees that whenever a node is processed, complete layouts for its children will already have been computed.

Theorem 3.2. *Let T be a dynamic tree whose supergraph is acyclic. Then Algorithm A.1 draws T in linear time such that all of the criteria Ranking, Ordering, Centering, Weak Symmetry, and Stability are met.*

Theorem 3.2 settles the problem of drawing dynamic trees with acyclic supergraphs nicely. In contrast, for a cyclic supergraph, things get *much* harder:

Lemma 3.3. *No drawing of $T = (T_1, T_2, T_3)$ from Fig. 3 meets all of the criteria Ranking, Ordering, Centering, Weak Symmetry, and Stability.*

The lemma tempts us to just "give up" on cyclic supergraphs. However, these arise naturally in prune-and-regraft operations and from rotations in search trees – which are operations that we would like to visualize. We could also just completely ignore the temporal criteria and return to drawing each tree individually in such cases – but we might be able to draw everything nicely except for a single "small" cycle "somewhere" in the supergraph.

We propose to deal with the cycle problem by "cutting" the cycles with as few "temporal cuts" as possible. These are defined as follows: Let $G = (G_1, \ldots, G_k)$ be a dynamic graph and let n be a node of the supergraph $\text{super}(G)$ and let $i \in \{1, \ldots, k-1\}$. The *temporal cut* of G at n and i is a new dynamic graph G' that is identical to G, except that for all $j \in \{i+1, \ldots, k\}$ in which G_j contains the node n, this node is replaced by the same new node n' (and all edges to or from n are replaced by edges to or from n').

Temporal cuts can be used to remove cycles from the supergraph of a dynamic graph, which allows us to then run our Algorithm A.1 on the resulting graph; indeed, simply "cutting everything at all times" turns every supergraph into a (clearly acyclic) collection of non-adjacent edges and isolated nodes. However, we wish to minimize the number of temporal cuts since, when we visualize G' using an animation, the different locations that may be assigned to n and n' will result in a *movement* of the node n to the new position of n'.

By the above discussion, we would like to find an algorithm that solves the following problem TEMPORAL-CUT-MINIMIZATION: Given a dynamic tree T, find a minimal number of temporal cuts, such that the resulting dynamic tree T' has an acyclic supergraph. Unfortunately, this problem turns out to be difficult:

Theorem 3.4. *The decision version of* TEMPORAL-CUT-MINIMIZATION *is NP-complete.*

In light of the above theorem, we have developed and implemented a simple greedy heuristic, Algorithm A.2, for finding temporal cuts that make the supergraph acyclic, which our prototype runs prior to invoking Algorithm A.1: Given a dynamic tree, the heuristic simply adds the trees T_i and their edges incrementally to the supergraph. However, whenever adding an edge $e = (v, w)$ of T_i to the supergraph creates a cycle, we cut w at $i - 1$.

4 Conclusion and Outlook

We have presented a system for offline drawings of dynamic trees using animations that are embedded in (text) documents. The system has been implemented [31] as an extension of the popular TEX system and will become part of future version of TikZ.[2] The generated animation are light-weight both in terms of file size and generation time, but require 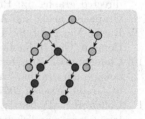 that the documents (or, at least, the graphic files) are stored in the SVG format. Our new algorithm is a natural extension of the Reingold–Tilford algorithm to the dynamic case, but while the original algorithm runs in linear time on all trees, we showed that the dynamic case leads to NP-complete problems. Fortunately, in practice, the hard subproblems can be solved satisfactorily using a greedy strategy – at least, that has been our finding

---

[2] Currently available in the development version at http://pgf.cvs.sourceforge.net.

for a limited number of examples such as the above animation; a perceptual study of animated drawings of dynamic graphs has not (yet) been conducted.

We see our algorithm as a first step towards a general set of algorithms for drawing dynamic graphs using animations, which we believe to have a great (and not yet fully realized) potential as parts of text documents. A next logical step would be a transferal of the Sugiyama method [16,34] to the dynamic offline case.

References

1. Adelson-Velsky, G.M., Landis, E.M.: An algorithm for the organization of information. Dokl. Akad. Nauk USSR **3**(2), 1259–1263 (1962)
2. Archambault, D., Purchase, H., Pinaud, B.: Animation, small multiples, and the effect of mental map preservation in dynamic graphs. IEEE Trans. Visual. Comput. Graph. **17**(4), 539–552 (2011)
3. Archambault, D., Purchase, H.C.: The mental map and memorability in dynamic graphs. In: Proceedings of Visualization Symposium (PacificVis) 2012, pp. 89–96. IEEE Press (2012)
4. Bartram, L., Ware, C.: Filtering and brushing with motion. Inf. Visual. **1**(1), 66–79 (2002)
5. Beck, F., Burch, M., Diehl, S., Weiskopf, D.: The state of the art in visualizing dynamic graphs. In: State of the Art Reports of the 16th Eurographics Conference on Visualization, EuroVis 2014, pp. 83–103. Eurographics Association (2014)
6. Brüggemann-Klein, A., Wood, D.: Drawing trees nicely with TEX. Electron. Publishing **2**(2), 101–115 (1989)
7. Bulterman, D., Jansen, J., Cesar, P., Mullender, S., Hyche, E., DeMeglio, M., Quint, J., Kawamura, H., Weck, D., García Pañeda, X., Melendi, D., Cruz-Lara, S., Hanclik, M., Zucker, D.F., Michel, T.: Synchronized multimedia integration language (SMIL 3.0), W3C Recommendation 01 December 2008. Technical report REC-SMIL3-20081201, The World Wide Web Consortium (W3C) (2008). http://www.w3.org/TR/2008/REC-SMIL3-20081201
8. Burch, M., Beck, F., Weiskopf, D.: Radial edge splatting for visualizing dynamic directed graphs. In: Proceedings of the International Conference on Computer Graphics Theory and Applications, IVAPP 2012, pp. 603–612. SciTe Press (2012)
9. Burch, M., Diehl, S.: TimeRadarTrees: visualizing dynamic compound digraphs. Comput. Graph. Forum **27**(3), 823–830 (2008)
10. Chimani, M., Gutwenger, C., Jünger, M., Klein, K., Mutzel, P., Schulz, M.: The open graph drawing framework. In: Poster at the 15th International Symposium on Graph Drawing 2007 (GD 2007) (2007)
11. Cohen, R.F., Di Battista, G., Tamassia, R., Tollis, I.G.: Dynamic graph drawings: trees, series-parallel digraphs, and planar st-digraphs. SIAM J. Comput. **24**(5), 970–1001 (1995)
12. Cohen, R.F., Di Battista, G., Tamassia, R., Tollis, I.G., Bertolazzi, P.: A framework for dynamic graph drawing. In: Proceedings of the 8th Annual Symposium on Computational Geometry, SCG 1992, pp. 261–270. ACM Press (1992)
13. Dahlström, E., Dengler, P., Grasso, A., Lilley, C., McCormack, C., Schepers, D., Watt, J.: Scalable vector graphics (SVG) 1.1 (2nd edn.), W3C Recommendation 16 August 2011. Technical report REC-SVG11-20110816, The World Wide Web Consortium (W3C) (2011). http://www.w3.org/TR/2011/REC-SVG11-20110816

14. Diehl, S., Görg, C., Kerren, A.: Foresighted graphlayout. Technical report A/02/2000, FB Informatik, University Saarbrücken, Saarbrücken, Germany (2000)
15. Diehl, S., Görg, C., Kerren, A.: Preserving the mental map using foresighted layout. In: Proceedings of the 3rd Joint Eurographics-IEEE TCVG Conference on Visualization, vol. 1, pp. 175–184. The Eurographics Association (2001)
16. Eades, P., Sugiyama, K.: How to draw a directed graph. J. Inf. Process. **13**(4), 424–436 (1990)
17. Ellson, J., Gansner, E.R., Koutsofios, E., North, S.C., Woodhull, G.: Graphviz and dynagraph - static and dynamic graph drawing tools. In: Junger, M., Mutzel, P. (eds.) Graph Drawing Software. Mathematics and Visualization, pp. 127–148. Springer, Heidelberg (2004)
18. Eppstein, D.: Trees in TEX. TUGboat **6**(1), 31–35 (1985)
19. Federico, P., Aigner, W., Miksch, S., Windhager, F., Zenk, L.: A visual analytics approach to dynamic social networks. In: Proceedings of the 11th International Conference on Knowledge Management and Knowledge Technologies, i-KNOW 2011, pp. 47:1–47:8. ACM Press (2011)
20. Görg, C., Birke, P., Pohl, M., Diehl, S.: Dynamic graph drawing of sequences of orthogonal and hierarchical graphs. In: Pach, J. (ed.) GD 2004. LNCS, vol. 3383, pp. 228–238. Springer, Heidelberg (2005). doi:10.1007/978-3-540-31843-9_24
21. Graham, M., Kennedy, J.: A survey of multiple tree visualisation. Inf. Visual. **9**(4), 235–252 (2010)
22. Greilich, M., Burch, M., Diehl, S.: Visualizing the evolution of compound digraphs with TimeArcTrees. Comput. Graph. Forum **28**(3), 975–982 (2009)
23. Groh, G., Hanstein, H., Wörndl, W.: Interactively visualizing dynamic social networks with DySoN. In: Proceedings of the Workshop on Visual Interfaces to the Social and the Semantic Web, VISSW 2009 (2009)
24. Ierusalimschy, R.: Programming in Lua, 2nd edn. Lua.org, San Francisco (2006)
25. Johnson, B., Shneiderman, B.: Tree-maps: a space-filling approach to the visualization of hierarchical information structures. In: Proceedings of the 2nd IEEE Conference on Visualization 1991, VIS 1991, pp. 284–291. IEEE Press (1991)
26. Knuth, D.E.: Optimum binary search trees. Acta Informatica **1**(1), 14–25 (1971)
27. Moen, S.: Drawing dynamic trees. IEEE Softw. **7**(4), 21–28 (1990)
28. Reda, K., Tantipathananandh, C., Johnson, A., Leigh, J., Berger-Wolf, T.: Visualizing the evolution of community structures in dynamic social networks. Comput. Graph. Forum **30**(3), 1061–1070 (2011)
29. Reingold, E.M., Tilford, J.S.: Tidier drawings of trees. IEEE Trans. Softw. Eng. **7**(2), 223–228 (1981)
30. Robertson, G.G., Mackinlay, J.D., Card, S.K.: Cone trees: Animated 3D visualizations of hierarchical information. In: Proceedings of the SIGCHI Conference on Human Factors in Computing Systems, CHI 1991, pp. 189–194. ACM Press (1991)
31. Skambath, M.: Algorithmic drawing of evolving trees. Master's thesis, Institute of Theoretical Computer Science, Universität zu Lübeck, Germany (2016)
32. Skambath, M., Tantau, T.: Offline drawing of dynamic trees: algorithmics and document integration. SVG Version of this document. http://www.informatik.uni-kiel.de/~msk/pub/2016-dynamic-trees/main.html
33. Stasko, J., Zhang, E.: Focus+context display and navigation techniques for enhancing radial, space-filling hierarchy visualizations. In: Proceedings of the IEEE Symposium on Information Vizualization 2000, INFOVIS 2000, pp. 57–65. IEEE Press (2000)

34. Sugiyama, K., Tagawa, S., Toda, M.: Effective representations of hierarchical structures. Technical report 8, International Institute for Advanced Study of Social Information Science, Fujitsu (1979)

35. Sweet, R.E.: Empirical estimates of program entropy. Report Stan-CS-78-698, Department of Computer Science, Stanford University, Stanford, CA, USA (1978)

36. Tantau, T.: Graph drawing in TikZ. J. Graph Algorithms Appl. **17**(4), 495–513 (2013)

37. Tantau, T.: The TikZ and pgf packages, manual for version 3.0.0 (2015). http://sourceforge.net/projects/pgf/

38. Walker II, J.Q.: A node-positioning algorithm for general trees. Softw.: Pract. Exp. **20**(7), 685–705 (1990)

39. Ware, C., Bobrow, R.: Motion to support rapid interactive queries on node-link diagrams. ACM Trans. Appl. Percept. **1**(1), 3–18 (2004)

40. Wetherell, C., Shannon, A.: Tidy drawings of trees. IEEE Trans. Softw. Eng. **5**(5), 514–520 (1979)

Contest Report

Graph Drawing Contest Report

Philipp Kindermann[1], Maarten Löffler[2(✉)], Lev Nachmanson[3],
and Ignaz Rutter[4]

[1] FernUniversität in Hagen, Hagen, Germany
philipp.kindermann@fernuni-hagen.de
[2] Utrecht University, Utrecht, The Netherlands
m.loffler@uu.nl
[3] Microsoft, Redmond, USA
levnach@microsoft.com
[4] Karlsruhe Institute of Technology, Karlsruhe, Germany
rutter@kit.edu

Abstract. This report describes the 23rd Annual Graph Drawing Contest, held in conjunction with the 24th International Symposium on Graph Drawing (GD'16) in Athens, Greece. The purpose of the contest is to monitor and challenge the current state of graph-drawing technology.

1   Introduction

This year, the Graph Drawing Contest was divided into two parts: the *creative topics* and the *live challenge*.

The creative topics had two graphs: the first one was a graph about country relations in the Panama papers, and the second one was a family tree of figures in Greek mythology. The data sets for the creative topics were published months in advance, and contestants could solve and submit their results before the conference started. The submitted drawings were evaluated according to aesthetic appearance, domain-specific requirements, and how well the data was visually represented.

The live challenge took place during the conference in a format similar to a typical programming contest. Teams were presented with a collection of challenge graphs and had one hour to submit their highest scoring drawings. This year's topic was to minimize the number of crossings in book layouts with a fixed number of pages.

Overall, we received 15 submissions: 6 submissions for the creative topics and 9 submissions for the live challenge.

2   Creative Topics

The two creative topics for this year were a graph about the Panama papers, and a Greek mythology family tree. The goal was to visualize each graph with complete artistic freedom, and with the aim of communicating the data in the graph as well as possible.

We received 2 submissions for the first topic, and 4 for the second. For each topic, we selected up to three contenders for the prize, which were printed on large poster boards and presented at the Graph Drawing Symposium. Finally,

© Springer International Publishing AG 2016
Y. Hu and M. Nöllenburg (Eds.): GD 2016, LNCS 9801, pp. 589–595, 2016.
DOI: 10.1007/978-3-319-50106-2_45

out of those contenders, we selected the winning submission. We will now review the top three submissions for each topic (for a complete list of submissions, refer to http://www.graphdrawing.de/contest2016/results.html).

2.1 Panama Papers

The *International Consortium of Investigative Journalists* (ICIJ)[1] is a global network of more than 190 investigative journalists in more than 65 countries who collaborate on in-depth investigative stories. Recently, the ICIJ released an *Offshore Leaks Database*[2] of almost 320,000 offshore companies and trusts from the Panama papers and the Offshore Leaks investigations.

For the first creative topic, we processed the database to create a weighted directed graph that shows the relationships between countries. A directed edge from country A to country B with weight w means that there are w Offshore Entities in country B that are linked to a company in country A.

The resulting layout of the graph should contain the names of the countries and should give a good overview on their correlation.

Runner-Up: Evmorfia Argyriou, Anne Eberle, and Martin Siebenhaller (yWorks). The committee likes the combination of clustering with radial layouts and organic edges, and a circular layout for the clusters that are connected with bundled edges. The representation of edge weights and (weighted) in-degrees via edge thickness and node sizes help a lot to grasp the underlying data.

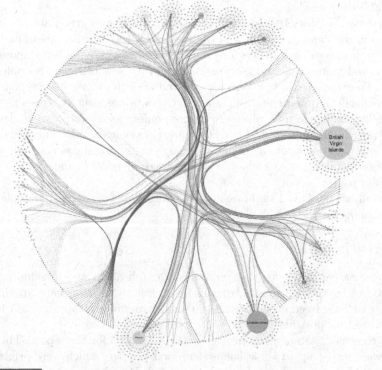

---

[1] https://www.icij.org/.
[2] https://offshoreleaks.icij.org/.

Winner: Fabian Klute (TU Wien). The committee likes the approach of this submission that derivates severely from standard approaches. The drawing is split into two parts. On the left, a highly connected subgraph consisting of eleven nodes is represented with different drawing styles depending on whether an edge exists in both directions or only in one. On the right, nodes without incoming edges are placed in treemaps that also represent vertices and are connected by various edge styles. The problem of label sizes was solved by using three-letter country codes and a different color for each country.

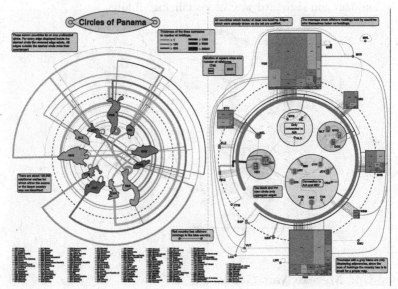

2.2   Greek Mythology Family Tree

The following data comes from the *Greek Mythic Genealogy Project*[3].

> 66 Greek myth contains a large amount of genealogical information. Various characters are related to each other in ways that are difficult for the non-specialist to keep track of, if for no other reason than that there are such a large number of gods, heroes, and other characters who appear in the various myths, epics, lyrics, legends, comedies, and other material. The Greek Mythic Genealogy Project is a fragmentary attempt to keep track of some of these relationships. 99

For the second creative topic, participants were asked to draw a family "tree" of popular characters in Greek Mythology. We created a subgraph of the large database by extracting only the most popular (by the number of Google results) names and their parents. This reduced the number of nodes to 118.

---

[3] http://patrickbrianmooney.nfshost.com/~patrick/greek-myth/greek-genealogy.html.

Runner-Up: Thom Castermans, Tim Ophelders, and Willem Sonke (TU Eindhoven). The first runner-up used a very interesting strategy to lay out siblings in order to reduce the complexity of the drawing: instead of drawing one edge from a parent to each of its children, a single (bundled) path is used that connects the childrin with the parent; in some cases (e.g., if there are too many children), more than one bundled path was used. The edges are colored by with the same color as the parent, which makes it easy to find the relationship between two nodes. The committee especially likes the metro map style of the drawing and the non-standard way of visualizing siblings.

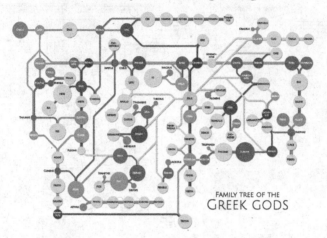

FAMILY TREE OF THE
GREEK GODS

Runner-Up: Mereke van Garderen (University Konstanz). The second runner-up drew the graph completely manually without using existing algorithms. The vertices are nicely layed out and both keep the number of crossings small and keep immediate family members close together. The edges are drawn with a mix of straight lines (with few slopes) and splines. In order to still show the hierarchical layout of the graph, the nodes and edges are colored by their generation with respect to the goddess Chaos. The committee finds the visual appearance, the colors, and the edge styles very appealing.

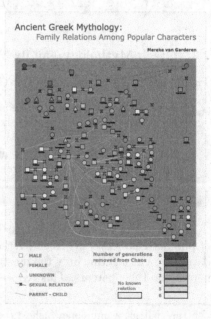

Ancient Greek Mythology:
Family Relations Among Popular Characters

Mereke van Garderen

Winner: Jonathan Klawitter and Tamara Mchedlidze (Karlsruhe Institute of Technology). The committee was impressed by the aesthetic appeal of this submission. The layout is very nicely done and clearly shows the hierarchy in the family tree. The nodes are carefully placed on several circular arcs and the edges are drawn as curves; in order to represent families, the father–children and the mother–children edges form bundles. The committee especially liked the coloring of the vertices that represent different types of entities in the Greek myth, such as titans, sea gods, or muses. Similar types are grouped closely together. For the twelve Olympian gods (and some other important figures), the authors also added their symbols inside the nodes. Below the drawing of the whole graph, there is a second drawing that only shows the partners and children of Zeus.

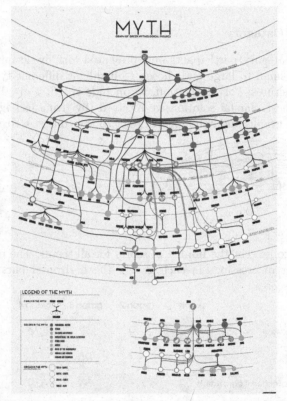

3   Live Challenge

The live challenge took place during the conference and lasted exactly one hour. During this hour, local participants of the conference could take part in the manual category (in which they could attempt to solve the graphs using a supplied tool), or in the automatic category (in which they could use their own software to solve the graphs). At the same time, remote participants could also take part in the automatic category.

The challenge focused on minimizing the number of crossings in a book embedding with k pages. The input graphs are arbitrary undirected graphs and a maximum number of pages that may be used.

A book with k pages consists of k half-spaces, the pages, that share a single line, the spine of the book. A k-page book embedding of a graph is an embedding of a graph into a book with k pages such that all the vertices lie at distinct positions of the spine and every edge is drawn in one of the pages such that only its endpoints touch the spine.

Note that edges may only cross if they are assigned to the same page. We are looking for drawings that minimize the number of crossings. The results are judged solely with respect to the number of crossings; other aesthetic criteria are not taken into account. This allows an objective way to evaluate a drawing.

3.1 Manual Category

In the manual category, participants were presented with five graphs. These were arranged from small to large and chosen to highlight different types of graphs and graph structures. For illustration, we include the first graph in its initial state and the best manual solution we received (by team JetLagged). For the complete set of graphs and submissions, refer to the contest website.

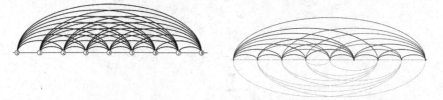

We are happy to present the full list of scores for all teams. The numbers listed are the number of crossings in each graph; the horizontal bars visualize the corresponding scores.

	graph 1	graph 2	graph 3	graph 4	graph 5
Bookembedder	17	9	47	1074	21
studs	21	97	95		
Senior Application Coordinators	19	7	59	659	4
noname	14	0	69	285	7
∫	15	6	39	310	16
yWorks	16	12	34	614	14
Jetlagged	13	0	33	487	47

The winning team is team noname, consisting of Michael Bekos, Thanasis Lianeas, and Chrysanthi Raftopoulou!

3.2   Automatic Category

In the automatic category, participants had to solve the same five graphs as in the manual category, and in addition another five—much larger—graphs. Again, the graphs were constructed to have different structure.

Once more, for illustration, we include the best solution (by team Ruhrpott) of the first large graph as it looks in the tool. The graphs themselves can be found on the contest website.

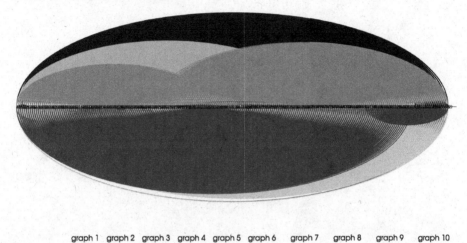

	graph 1	graph 2	graph 3	graph 4	graph 5	graph 6	graph 7	graph 8	graph 9	graph 10
Ruhrpott	8	0	6	71	1	2538783	2630	7009	234238	100343
Johan de Ruiter	8	0	3	56	0	2683608	3	10219	191716	122311

The winning team is team Johan de Ruiter, consisting of Johan de Ruiter!

> ❝ I used a Simulated Annealing approach with a collection of K 2D Fenwick trees as the underlying data structure in which the edges of the graph were stored according to their assigned pages in the book embedding. This allowed for logarithmic time counting of the number of crossings per edge, edge insertion and edge removal. Swapping two vertices, and moving a vertex into an empty spot, were realized by removing and reinserting all the incident edges. ❞
>
> *Johan de Ruiter*

Acknowledgments. The contest committee would like to thank the generous sponsors of the symposium, Dennis van der Wals for programming most of the tool for the manual category, and all the contestants for their participation. Further details including all submitted drawings and challenge graphs can be found at the contest website: http://www.graphdrawing.de/contest2016/results.html.

Ph.D. School Report

Ph.D. School on Visualization Software

Antonios Symvonis[✉]

School of Applied Mathematical and Physical Sciences,
National Technical University of Athens, Athens, Greece
symvonis@math.ntua.gr

Following the recent tradition of previous Symposia, the Organizing Committee of the 24th *International Symposium on Graph Drawing and Network Visualization* hosted a *Ph.D. School on Visualization Software*. The Ph.D. School was held at the National Technical University of Athens campus, September 22–23, 2016, and was attended by 16 participants.

The Ph.D. School on Visualization Software consisted of two one-day tutorials on Tom Sawyer's *Perspectives* graph and data visualization system and yWorks' *yFiles for Java* diagramming software. *Tom Sawyer Software* and *yWorks* are two of the longstanding sponsors of the Graph Drawing Symposia.

On September 22, 2016, Brendan Madden, Chief Executive Officer and Founder of Tom Sawyer Software, gave an in-depth introduction to Tom Sawyer Software's research and practical work on graph drawing software. He demonstrated the broad capabilities of Tom Sawyer Perspectives graph and data visualization system and discussed how their software is now used in numerous industries including the public sector, in finance and fraud detection, IT, energy, telecommunications and networking, and in aerospace and manufacturing.

On September 23, 2016, Christian Brunnermeier and Jasper Möller, Software Developers at yWorks, gave an overview over the yFiles product family, its main features and the underlying architecture. Some of the most prominent layout algorithms were presented, including their customization options to optimize the layout result and typical challenges when a theoretical algorithm meets real-life requirements. In a tutored practical session the participants worked in groups to fill a given demo application with life, building upon the presented topics e.g. finding a good layout for a given data set, presenting the business data in an appealing way and reducing visual complexity to make the data more approachable.

We thank the National Technical University of Athens and the School of Applied Mathematical and Physical Sciences for their support. We also thank the speakers, Brendan Madden, Christian Brunnermeier and Jasper Möller, as well as all participants who turned the Ph.D. School on Visualization Software into a successful event.

© Springer International Publishing AG 2016
Y. Hu and M. Nöllenburg (Eds.): GD 2016, LNCS 9801, p. 599, 2016.
DOI: 10.1007/978-3-319-50106-2

Poster Abstracts

A Simple Quasi-planar Drawing of K_{10}

Franz J. Brandenburg[✉]

University of Passau, 94030 Passau, Germany
brandenb@informatik.uni-passau.de

Abstract. We show that the complete graph on ten vertices K_{10} is a simple quasi-planar graph, which answers a question of Ackerman and Tardos [E. Ackerman and G. Tardos, On the maximum number of edges in quasi-planar graphs, J. Comb. Theory, Series A 114 (2007) 563–571] and shows that the bound $6.5n - 20$ on the maximum number of edges of simple quasi-planar graphs is tight for $n = 10$.

A graph is k-*quasi-planar* if it can be drawn in the plane so that there are no k pairwise crossing edges. A k-quasi-planar graph is *simple* if each pair of edges meets at most once, either at a common endpoint or at a crossing point, and is *geometric* if each edge is drawn as a straight line segment. 3-quasi-planar graphs are called *quasi-planar*. Quasi-planar graphs were introduced by Agarwal et al. [2] and have intensively been studied since then, with a focus on Turán-type problems. Do k-quasi-planar graphs have at most a linear number of edges? Which is the maximum complete quasi-planar graph?

Ackerman and Tardos [1] proved that simple quasi-planar graphs have at most $6.5n - 20$ edges and this bound is tight up to a constant. There are no simple quasi-planar graphs with exactly $6.5n - 20$ edges. In consequence, K_{11} is not a simple quasi-planar graph. It is known that K_9 is a quasi-planar geometric graph whereas K_{10} is not a quasi-planar geometric graph [3]. Here, we draw K_{10} so that each pair of edges meets at most once and three edges do not pairwise cross (Fig. 1), and thereby solve the open problem on K_{10}.

We partition K_{10} into two K_5 which are each drawn as an encircled pentagram (with solid black and dotted blue edges). Let $U = \{u_1, \ldots, u_5\}$ ($V = \{v_1, \ldots, v_5\}$) be the vertices of the outer (inner) K_5 in clockwise order. The remaining $K_{5,5}$ is drawn so that the edges $\{u_i, v_{i-1}\}, \{u_i, v_i\}$ and $\{u_i, v_{i+1}\}$ in circular order are drawn straight-line. Finally, (pink) edges $\{u_i, v_{i+2}\}$ go around u_{i+1} in the outer face, and (red) edges $\{u_i, v_{i-2}\}$ traverse the inner circle. The drawing is simple quasi-planar, since no three edges cross pairwise. For an inspection, first observe that the (dashed black) edges $\{u_i, v_i\}$ and the (black) edges on the circles of U and V are crossed at most once. Figure 2(a) shows K_{10} after their removal. Thereafter, the edges $\{v_i, u_{i+1}\}$ (thick and green) are crossed by two edges which do not cross pairwise, and similarly for $\{v_i, u_{i-1}\}$ (cyan). Finally, in the remainder, each red edge crosses a pink, two red and two dotted blue edges, which do not cross mutually, and similarly for the pink edges. Alternatively, consider the edge intersection graph which is triangle-free.

Supported by the Deutsche Forschungsgemeinschaft (DFG), grant Br835/18-2.

Y. Hu and M. Nöllenburg (Eds.): GD 2016, LNCS 9801, pp. 603–604, 2016.
DOI: 10.1007/978-3-319-50106-2

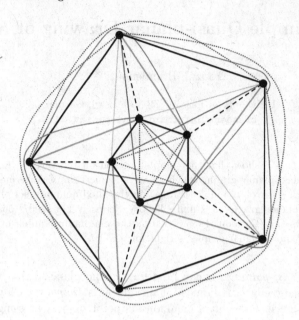

Fig. 1. A quasi-planar drawing of K_{10}

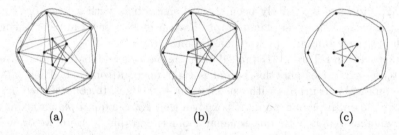

(a)   (b)   (c)

Fig. 2. Inspection of K_{10} after the removal of (a) black edges, (b) green and cyan edges, and (c) red and pink edges

References

1. Ackerman, E., Tardos, G.: On the maximum number of edges in quasi-planar graphs. J. Comb. Theory, Ser. A **114**(3), 563–571 (2007)
2. Agarwal, P.K., Aronov, B., Pach, J., Pollack, R., Sharir, M.: Quasi-planar graphs have a linear number of edges. Combinatorica **17**(1), 1–9 (1997)
3. Aichholzer, O., Krasser, H.: The point set order type data base: a collection of applications and results. In: Proceedings of the 13th CCCG, Waterloo, Ontario, Canada, 2001, pp. 17–20 (2001)

A Labeling Problem for Symbol Maps
of Archaeological Sites

Mereke van Garderen[✉], Barbara Pampel, and Ulrik Brandes

University of Konstanz, Konstanz, Germany
{mereke.van.garderen,barbara.pampel,ulrik.brandes}@uni-konstanz.de

Introduction

Given a set of n rectangles embedded in the Euclidian plane, we consider the problem of modifying the layout to avoid intersections of the rectangles. The objective is to minimize the total displacement under the additional constraint that the orthogonal order of the rectangles must be preserved. We call this problem MINIMUM-DISPLACEMENT OVERLAP REMOVAL (MDOR). We define the *total displacement* in the new layout as the sum of the Euclidian distances between the initial position (x, y) and the final position (x', y') of the centers of all rectangles. A layout adjustment is *orthogonal-order preserving* if the order of the rectangles with respect to the x- and the y-axis does not change. More formally, the order is preserved if and only if for any pair of rectangles r_i and r_j it holds that $x_i \leq x_j \Rightarrow x'_i \leq x'_j$ and that $y_i \leq y_j \Rightarrow y'_i \leq y'_j$.

Motivation

Our interest in this problem is motivated by the application of displaying metadata of archaeological sites. The most popular way of representing data of this kind is to use a *symbol map*, where each site is represented by a symbol that conveys (a selection of) the metadata about the site, and these symbols are placed on the map at the site's geographical coordinates. Overlap needs to be removed so all symbols are visible, but the symbols need to stay close to the corresponding sites, and because cardinal relations between sites are often important the orthogonal order should be maintained. Many GIS packages commonly used in archaeology do offer automated map production, but when it comes to the arrangement and scaling of objects they generally perform poorly [1]. Figure 1 shows examples as published in [2] (left) and [4] (right).

Contribution

MDOR is closely related to MINIMUM-AREA LAYOUT ADJUSTMENT (MALA), which is known to be NP-hard [3]. The difference is that the objective in MALA is to minimize the area of the drawing, rather than the total node displacement. We show by reduction from MONOTONE ONE-IN-THREE SAT that MDOR is NP-hard.

Theorem 1. MINIMUM-DISPLACEMENT OVERLAP REMOVAL *is NP-hard, even for equal-size squares at integer coordinates.*

© Springer International Publishing AG 2016
Y. Hu and M. Nöllenburg (Eds.): GD 2016, LNCS 9801, pp. 605–607, 2016.
DOI: 10.1007/978-3-319-50106-2

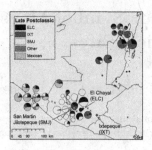 

Fig. 1. Archaeological symbol maps produced by commonly used GIS software.

Because MDOR is NP-hard, we turn to heuristic approaches to find a feasible solution to our problem. The objectives of existing overlap removal algorithms are not ideal for our application, therefore we present a new heuristic for solving the problem. The core of this algorithm is a loop that contains three steps:

1. Compute all pairs of overlapping nodes using a sweep line
2. Remove the overlap for each pair with local minimum displacement
3. Repair the orthogonal order using a variation of MERGESORT

These three steps are repeated until there are no more overlapping pairs (Fig. 2).

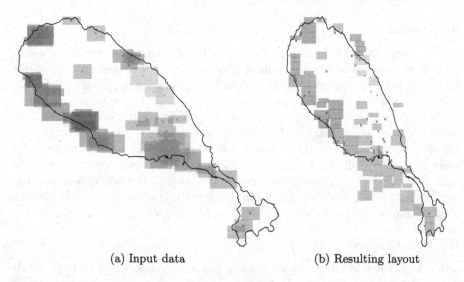

(a) Input data   (b) Resulting layout

Fig. 2. Input (a) and result (b) of the heuristic for a dataset of 70 cultural heritage sites on St. Kitts. In (b), the island map is distorted to fit the new bounding box.

Acknowledgments. This research received funding from the European Union's 7th Framework Programme for research, technological development and demonstration under grant agreement n° 1133 (Project CARIB) and ERC grant agreement n° 319209

(Project NEXUS1492). The project CARIB is financially supported by the HERA Joint Research Programme (www.heranet.info) which is co-funded by AHRC, AKA, BMBF via PT-DLR, DASTI, ETAG, FCT, FNR, FNRS, FWF, FWO, HAZU, IRC, LMT, MHEST, NWO, NCN, RANNS, RCN, VR and The European Community FP7 2007–2013, under the Socio-economic Sciences and Humanities programme.

References

1. Conolly, J., Lake, M.: Maps and Digital Cartography. In: Geographical information systems in archaeology. Cambridge University Press, pp. 263–279 (2006)
2. Golitko, M., Meierhoff, J., Feinman, G.M., Williams, P.R.: Complexities of collapse: the evidence of Maya obsidian as revealed by social network graphical analysis. Antiquity 86, 507–523 (2012)
3. Kunihiko, H., Michiko, I., Toshimitsu, M., Hideo, F.: A layout adjustment problem for disjoint rectangles preserving orthogonal order. In: Whitesides, S.H. (ed.) GD'98. LNCS, vol. 1547, pp. 183–197. Springer, Heidelberg (1998)
4. Morin, J.: Near-infrared spectrometry of stone celts in precontact British Columbia. Canada. Am. Antiq. 80(3), 530–547 (2015)

Capturing Lombardi Flow
in Orthogonal Drawings by Minimizing
the Number of Segments

Md. Jawaherul Alam$^{(\boxtimes)}$, Michael Dillencourt, and Michael T. Goodrich

Department of Computer Science, University of California, Irvine, CA, USA
{alamm1,dillenco,goodrich}@uci.edu

1 Introduction

An aesthetic property prevalent in Lombardi's art work is that he tends to place many vertices on consecutive stretches of linear or circular segments that go across the whole drawings. This creates a metaphor of a "visual flow" across a drawing. Inspired by this property, we study the following problems for orthogonal drawings of planar graphs (see Fig. 1)[1]:

1. A *minimum segment orthogonal drawing*, or *MSO-drawing*, of a planar graph G is an orthogonal drawing of G with the minimum number of segments.
2. A *minimum segment cover orthogonal drawing*, or *MSCO-drawing*, of G is one with the smallest set of segments covering all vertices of G.

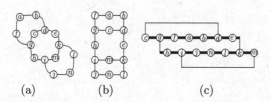

(a)   (b)   (c)

Fig. 1. (a) A planar graph G, (b) an MSO-drawing, (c) an MSCO-drawing of G.

There is a lot of prior work on minimizing the number of segments in straight-line drawings and on minimizing the number of circular arcs in planar drawings. However to the best of our knowledge, this problem has never been studied before in the context of orthogonal drawings.

---

[1] This article reports on work supported by DARPA under agreement no. AFRL FA8750-15-2-0092. The views expressed are those of the authors and do not reflect the official policy or position of the Department of Defense or the U.S. Government. This work was also supported in part by the NSF under grants 1228639 and 1526631. We thank Timothy Johnson and Michael Bekos for several helpful discussions.

Y. Hu and M. Nöllenburg (Eds.): GD 2016, LNCS 9801, pp. 608–610, 2016.
DOI: 10.1007/978-3-319-50106-2

A recent empirical study [4] concluded that orthogonal layouts generated by traditional algorithms focusing primarily on bend minimization lack aesthetic qualities compared to manual drawings. The study suggests that, like in the works by Lombardi, humans prefer drawings with linear "flow" that connect chains of adjacent vertices. Our specific interest here is to study such "Lombardi flow" for orthogonal graph drawings.

We present the following results.

- We give a polynomial-time algorithm for the MSO-drawing of an embedded plane graph, using the network-flow algorithms [2, 3, 5] for minimizing bends.
- We show that finding MSCO-drawing is NP-hard even for degree-3 graphs.
- For trees and series-parallel graphs with maximum degree 3, we provide polynomial-time algorithms for upward orthogonal drawings with the minimum number of segments covering the vertices.

2   Detailed List of Main Results

In this section, we give a detailed list of our main results; please see the full version of the paper [1] for proofs. Recall that the *2-factor* of a graph G is obtained by repeatively identifying each degree-2 vertex to one of its neighbors.

Lemma 1. *Let G be a plane graph with maximum degree 4 and let G' be the 2-factor of G. Then G' has an orthogonal drawing with b bends if and only if G has an orthogonal drawing with $b + k/2$ segments, where k is the number of odd-degree vertices in G.*

We find an MSO-drawing of G by computing a minimum-bend drawing of the 2-factor, using the $O(n^{1.5})$-time algorithm by Cornelsen and Karrenbauer [2].

Theorem 1. *For an embedded n-vertex planar graph G, with maximum degree 4, an MSO-drawing of G can be computed in $O(n^{1.5})$ time.*

Let Γ be an orthogonal drawing for a planar graph G with maximum degree 4. A set of segments S in Γ is said to *cover* G, if each vertex in G is on some segments in S. The *segment-cover number* of Γ is the minimum cardinality of a set of segments covering G. Given a planar graph G, with maximum degree 4, a *minimum segment cover orthogonal drawing* or *MSCO*-drawing of G is an orthogonal drawing with the minimum segment cover number.

Theorem 2. *Finding a minimum segment cover orthogonal drawing for a planar graph G is NP-hard, even if G is a planar graph with maximum degree 3.*

Theorem 3. *Let G be a series-parallel graph with maximum degree 3 with the SPQ-tree T and let $\#(P^*)$ be the number of P-nodes in T with at least two S-nodes as children. If the root of T is a P-node with three S-nodes as children, then $\#(P^*) + 2$ segments are necessary and sufficient to cover all vertices of G in any upward orthogonal drawing of G; otherwise $\#(P^*) + 1$ segments are necessary and sufficient.*

Algorithm for Rooted Trees. Let T be a rooted tree. Take two copies T_1, T_2 of T. Identify the two copies of each leaf, to obtain a series-parallel graph with the maximum degree 3. An upward orthogonal drawing of this graph by the above algorithm also gives an upward drawing of T with optimal segment-cover.

References

1. Alam, M.J., Dillencourt, M., Goodrich, M.T.: Capturing Lombardi flow in orthogonal drawings by minimizing the number of segments. CoRR abs/1608.03943 (2016)
2. Cornelsen, S., Karrenbauer, A.: Accelerated bend minimization. J. Graph Algorithms Appl. **16**(3), 635–650 (2012)
3. Garg, A., Tamassia, R.: A new minimum cost flow algorithm with applications to graph drawing. In: North, S.C. (ed.) GD 1996. LNCS, vol. 1190. Springer, Heidelberg (1997)
4. Kieffer, S., Dwyer, T., Marriott, K., Wybrow, M.: HOLA: human-like orthogonal network layout. IEEE Trans. Vis. Comp. Graph. **22**(1), 349–358 (2016)
5. Tamassia, R.: On embedding a graph in the grid with the minimum number of bends. SIAM J. Comput. **16**(3), 421–444 (1987)

Sibling-First Recursive Graph Drawing for Visualizing Java Bytecode

Md. Jawaherul Alam$^{(\boxtimes)}$, Michael T. Goodrich, and Timothy Johnson

Department of Computer Science, University of California, Irvine, CA, USA
{alamm1,goodrich,tujohnso}@uci.edu

1 Introduction

We describe a tool, the JVM abstracting abstract machine (Jaam) Visualizer, or "J-Viz" for short, which is intended for use by security analysts to perform such searches through the exploration of graphs derived from Java bytecode. The workflow for our tool involves taking a given program, specified in Java bytecode, and constructing a control-flow graph of the possible execution paths for this software, using a framework known as *control flow analysis* (CFA) [6]. Our tool then provides a human analyst with an interactive view of this graph, including heuristics for aiding the identification of the suspicious parts.

One of the main components of our J-Viz tool involves visualizing control-flow graphs in a canonical way based on a novel vertex numbering scheme that we call the *sibling-first recursive* numbering. This numbering scheme is essentially a hybrid between the well-known breadth-first and depth-first numbering schemes, but differs from both in a way that appears to be more useful for visualizing control-flow graphs. In particular, this tends to highlight areas in software where code is repeated and it also allows us this tends to highlight areas in software where code is repeated and allows us to provide visual highlights of code that is contained in deeply nested loops. We had the following goals in mind:

- We want users to recognize patterns in source code from our visualization; similar code sections should produce similar subgraphs, drawn similar way.

- We want to use a hierarchical visualization, in which users can collapse or expand sections of the graph to different levels of detail. But we also want to preserve a consistent mental model of the graph. Thus, drawings should not drastically shift the vertex positions when sections are collapsed or expanded.

- No matter what sequence of actions the user performs, drawings should be consistent. That is, the same view of a graph, in which the same set of nodes are collapsed and expanded, should always be drawn in the same way.

- Our system should rank sections of the graph by how likely they are to produce vulnerabilities, and display this information visually to the user.

We believe that J-Viz makes substantial progress in achieving these goals, and we provide several case studies in this paper that support this conclusion. Please see the full version of the paper [1] for more details.

© Springer International Publishing AG 2016
Y. Hu and M. Nöllenburg (Eds.): GD 2016, LNCS 9801, pp. 611–612, 2016.
DOI: 10.1007/978-3-319-50106-2

Related Work and Main Contributions. Visualization tools have also previously been applied to source code. Doxygen [5], a tool for automatically generating documentation, can produce various kinds of diagrams for visualizing code, including call graphs. It is generally configured to use the *dot* [4] tool from GraphViz to draw these graphs hierarchically. Similarly, Visual Studio can visualize call graphs to aid programmers in debugging applications [3]. In contrast with these systems, our J-Viz tool provides four main features that these tools do not provide. First, J-Viz shows a greater level of detail, since it analyzes code at the level of individual instructions rather than methods. Second, J-Viz allows the user to interact with a graph and produce multiple views of the same Java bytecode. Third, the layout algorithm used in J-Viz is designed to draw similar code fragments in the same (canonical) way, so as to highlight portions of repeated code. Fourth, J-Viz guides the user to potential security vulnerabilities, by highlighting nodes that are believed to be risky based on algorithmic complexity (or other factors), whereas these other systems were not focused on software security. Another tool, Jinsight [2], can be used to profile a Java program to provide various views of resource usage, such as highlighting which instances of a class have taken the most time or used the most memory. This tool does not provide a full graph of the program's possible execution paths, however, which we believe to be essential for detecting security vulnerabilities.

Acknowledgements. This reports on work supported by DARPA under agreement AFRL FA8750-15-2-0092. The views expressed are those of the authors and do not reflect the official policy or position of the Department of Defense or the U.S. Government. This work was also supported NSF grants 1228639 and 1526631. We also thank David Eppstein, Matthew Might, William Byrd, Michael Adams, and Guannan Wei for helpful discussions.

References

1. Alam, M.J., Goodrich, M.T., Johnson, T.: J-Viz: sibling-first recursive graph drawing for visualizing java bytecode. CoRR abs/1608.08970 (2016)
2. Pauw, W., Jensen, E., Mitchell, N., Sevitsky, G., Vlissides, J., Yang, J.: Visualizing the execution of Java programs. In: Diehl, S. (ed.) Software Visualization. LNCS, vol. 2269, pp. 151–162. Springer, Heidelberg (2002). doi:10.1007/3-540-45875-1_12
3. DeLine, R., Venolia, G., Rowan, K.: Software development with code maps. Commun. ACM **53**(8), 48–54 (2010)
4. Gansner, E.R., Koutsofios, E., North, S.C., Vo, G.P.: A technique for drawing directed graphs. IEEE Trans. Softw. Eng. **19**(3), 214–230 (1993)
5. van Heesch, D.: Doxygen: Source code documentation generator tool (2008). http://www.doxygen.org. Accessed 8 June 2016
6. Shivers, O.G.: Control-flow analysis of higher-order languages. Ph.D. thesis, Carnegie Mellon University (1991)

Tree-Based Genealogical Graph Layout

Radek Mařík[✉]

Faculty of Electrical Engineering, Czech Technical University,
Technicka 2, Prague, Czech Republic
Radek.Marik@fel.cvut.cz

Abstract. While a visual unconstrained tree structure planar layout design is easy to implement, a visualization of a tree with constraints on node ranks and their ordering within ranks leads to a difficult combinatorial problem. A genealogical graph can be taken as an example of such a case. We propose a new method of tree-driven graph node layout.

1 Introduction

Graph visualization can help to form an overview of relational patterns and detect data structure much faster than data in a tabular form [10, 18]. Working with genealogical graphs is no exception in this sense. Tree based drawing methods of genealogical graphs have been among the standard techniques for centuries [10, 24, 25]. Ancestor trees, descendant trees and Hourglass charts belong to a set of traditional tools implemented by a majority of freeware, shareware, or commercial tools [1–3, 14]. These tools provide a clear description of a situation when the user needs to investigate direct ancestors and/or descendants of a given person (often referred to as the main person). There are other more space-efficient representations such as fan charts or H-charts [4, 15, 26, 28]. However, such tree-based representations miss a broader context of relationships and do not allow the quick assessment of several interlinked families together. The situation with family members grouping changes significantly if the assumptions of one main person and direct ancestors/descendants are dropped. In a number of cases it is highly beneficial if the entire network of families, or at least a significant part, can be displayed in one layout. Then we face issues with challenges linked with edge crossing and preferences on node clustering [22, 23, 27]. Therefore, the standard techniques for planar graph layouts [5, 12, 13, 16, 20, 21] are not suitable in all cases. It is possible to group children or their parents (but not both). We support an approach that results in siblings of one family being clustered tightly while partnerships/parents might be mixed. Unfortunately, even state of art directed hierarchical drawing methods, such as those implemented in Graphviz [9], result in layouts with mixed generations and groups mixing several families. Recently, it was demonstrated using two simple propagated node order constraints that node layouts of such graphs can be improved significantly [17]. In this paper, we propose an even better technique using undirected tree structure properties.

Sponsored by the project for GAČR, No. 16-072105: Complex network methods applied to ancient Egypt data in the Old Kingdom (2700–2180 BC).

Y. Hu and M. Nöllenburg (Eds.): GD 2016, LNCS 9801, pp. 613–616, 2016.
DOI: 10.1007/978-3-319-50106-2

2   The Proposed Method

A genealogical graph is an acyclic bipartite directed graph $G(V_P, V_M, E)$ with two sorts of nodes, people V_P and marriages/partnerships V_M. The edges E are directed from parent nodes to marriage nodes and from marriage nodes to children nodes. Without loss of generality we can assume that the index of the generation layer (ranks) of parents is lower than the index of their marriage node, and further that the index of the marriage node is lower than the index of children nodes. The node rank can be computed easily using two DFS scans of the graph [17]. As cases when two individuals share two and more distinct subtrees are very rare in reality (just 4 cases in our database of 2100 individuals), we can transform the graph into an undirected tree by removing a few edges. We select suitable edges using blocks (biconnected components [7]) in linear time [11, 19]. These edges are drawn but not used by the node ordering algorithm. In the following algorithm we assume that the undirected processed graph is a tree.

The problem of a layout design might then be reduced to a determination of the order of people belonging to one generation. We propose that children belonging to a single family are ordered by their birth dates. Subtrees of the child descendants, including descendant marriage nodes, hold this order. In the opposite direction, i.e. from a marriage node to its spouses, the order of spouses can be determined according to birthdates of spouses. Starting from a node with the lowest rank we assign both time-stamps to each node using the DFS scanning [6]. Using the post order we can determine the minimum node rank of any subtree

Fig. 1. A visualization of the author's private family tree with 2117 people and 742 marriages created from the rank and node order constraints proposed in this contribution only. An ideal layout would result in edges creating "waves" only.

inside its time-stamp interval $(O(N))$. Again starting from the node with the lowest rank we assign nodes of time-stamp interval subtrees into rank arrays (initially empty for each rank). First, subtrees with a minimum rank higher than the rank of the current node are processed, then the remaining subtrees are processed according to their increasing size to minimize edge crossing.

A generated layout with uniformly placed nodes and family subtrees emphasized by colors is shown in Fig. 1. Experiments with families with up to 50 family members of the Egyptian database of 3057 people from 4th, 5th, and 6th dynasty created by Egyptologists [8] did not exhibit any layout deficiencies.

The experiments indicate the results provided by the state of the art tools are quite far from the optimum layout, at least for special sorts of graphs such as genealogical ones.

References

1. Gramps. Genealogical research software (2016). https://gramps-project.org/. Accessed 5 June 2016
2. Myheritage (2016). https://www.myheritage.cz. Accessed 5 June 2016
3. yed graph editor (2016). http://www.yworks.com/products/yed. Accessed 5 June 2016
4. Ball, R., Cook, D.: A family-centric genealogy visualization paradigm. In: 14th Annual Family History Technology Workshop. Provo, Utah (2014)
5. Booth, K.S., Lueker, G.S.: Testing for the consecutive ones property, interval graphs and graph planarity using PQ-tree algorithms. J. Comp. Syst. Sci. **13**(3), 335–379 (1976)
6. Cormen, T.H., Leiserson, C.E., Rivest, R.L., Stein, C.: Introduction to Algorithms, 3rd edn. The MIT Press (2009)
7. Diestel, R.: Graph Theory. Springer (2005)
8. Dulíková, V.: The Reign of King Nyuserre and its impact on the development of the Egyptian state. A multiplier effect period during the old kingdom. Ph.D. thesis, Charles University in Prague, Faculty of Arts, Czech Institute of Egyptology (2016)
9. Gansner, E.R., Koutsofios, E., North, S.C., Vo, K.P.: A technique for drawing directed graphs. IEEE Trans. Softw. Eng. **19**(3), 214–230 (1993)
10. Gibson, H., Faith, J., Vickers, P.: A survey of two-dimensional graph layout techniques for information visualisation. Inf. Vis. **12**(3–4), 324–357 (2013). http://ivi.sagepub.com/content/12/3-4/324.abstract
11. Hopcroft, J., Tarjan, R.: Algorithm 447: efficient algorithms for graph manipulation. Commun. ACM **16**(6), 372–378 (1973). http://doi.acm.org/10.1145/362248.362272
12. Hopcroft, J., Tarjan, R.: Efficient planarity testing. J. ACM **21**(4), 549–568 (1974). http://doi.acm.org/10.1145/321850.321852
13. Hsu, W.L., McConnell, R.: Handbook of Data Structures and Applications, chap. PQ Trees, PC Trees, and Planar Graphs, pp. 32-1–32-27. CRC Press (2004)
14. Keller, K., Reddy, P., Sachdeva, S.: Family tree visualization. Course project report (2010). http://vis.berkeley.edu/courses/cs294-10-sp10/wiki/images/f/f2/Family_Tree_Visualization_-_Final_Paper.pdf. Accessed 5 June 2016
15. Kieffer, S., Dwyer, T., Marriott, K., Wybrow, M.: Hola: Human-like orthogonal network layout. IEEE Trans. Vis. Comput. Graph. **22**(1), 349–358 (2016)
16. Lempel, A., Even, S., Cederbaum, I.: An algorithm for planarity testing of graphs. In: Rosenstiehl, P., Gordon, B. (eds.) Theory of Graphs. pp. 215–232. New York (1967)
17. Marik, R.: On large genealogical graph layouts (accepted for publication). In: WASACNA 2016: Workshop on Algorithmic and Structural Aspects of Complex Networks and Applications, September 17th, Tatranské Matliare, Slovakia, September 2016
18. McGrath, C., Blythe, J., Krackhardt, D.: Seeing groups in graph layouts. Connections **19**(2), 22–29 (1996)
19. Paton, K.: An algorithm for the blocks and cutnodes of a graph. Commun. ACM **14**(7), 468–475 (1971). http://doi.acm.org/10.1145/362619.362628
20. Reingold, E.M., Tilford, J.S.: Tidier drawings of trees. IEEE Trans. Softw. Eng. **SE–7**(2), 223–228 (1981)
21. Shih, W.K., Hsu, W.L.: A new planarity test. Theor. Comput. Sci. **223**(1–2), 179–191 (1999)

22. Sugiyama, K., Misue, K.: Visualization of structural information: automatic drawing of compound digraphs. IEEE Trans. Syst. Man Cybern. **21**(4), 876–892 (1991)
23. Sugiyama, K., Tagawa, S., Toda, M.: Methods for visual understanding of hierarchical system structures. IEEE Trans. Syst. Man Cybern. **11**(2), 109–125 (1981)
24. Tutte, W.T.: Convex representations of graphs. Proc. Lond. Math. Soc. Third Ser. (10), 304–320 (1960)
25. Tutte, W.T.: How to draw a graph. Proc. Lond. Math. Soc. Third Ser. (13), 743–768 (1960)
26. Tuttle, C., Nonato, L.G., Silva, C.: Pedvis: a structured, space-efficient technique for pedigree visualization. IEEE Trans. Vis. Comput. Graph. **16**(6), 1063–1072 (2010)
27. Warfield, J.N.: Crossing theory and hierarchy mapping. IEEE Trans. Syst. Man Cybern. **7**(7), 505–523 (1977)
28. Yoghourdjian, V., Dwyer, T., Gange, G., Kieffer, S., Klein, K., Marriott, K.: High-quality ultra-compact grid layout of grouped networks. IEEE Trans. Vis. Comput. Graph. **22**(1), 339–348 (2016)

An Experimental Study on the Ply Number of Straight-Line Drawings

F. De Luca[1], E. Di Giacomo[1(✉)], W. Didimo[1], S.G. Kobourov[2], and G. Liotta[1]

[1] Università degli Studi di Perugia, Perugia, Italy
felice.deluca@studenti.unipg.it,
{emilio.digiacomo,walter.didimo,giuseppe.liotta}@unipg.it
[2] University of Arizona, Tucson, USA
kobourov@cs.arizona.edu

Introduction. Let Γ be a straight-line drawing of a graph. For each $v \in \Gamma$, let C_v be the open disk centered at v whose radius is half the length of the longest edge incident to v. Denote by S_q the set of disks sharing a point $q \in \mathbb{R}^2$. The *ply number* of Γ is defined as $pn(\Gamma) = max_q|S_q|$. Based on the observation that real-world road networks have low ply number [5], this parameter has been recently proposed as a new quality measure of a graph layout [3] (see, e.g., Fig. 1). While theo-

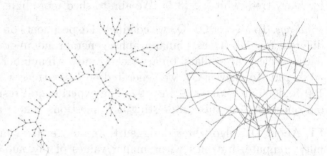

Fig. 1. Two drawings of the same graph with ply number 3 (left) and 12 (right).

retical results about computing drawings with low ply number have already appeared [1, 3], this work aims at experimentally assessing whether some of the most popular graph drawing methods actually compute drawings with low ply number. In addition, we want to experimentally study the theoretical gaps between upper and lower bounds of the ply number that have been established for some graph families [1, 3].

Experimental Questions. We address the following experimental questions:

Q1. *How good are the layouts computed by different drawing algorithms in terms of ply number?* We compare six force-directed algorithms (FM3 [9], FR [7], GEM [6], KK [10], SM [8], and LL [12]) and, for planar graphs, we also consider a canonical order based algorithm [2].

Q2. *How close is the ply number of drawings produced by existing algorithms to the ply number of the input graph (i.e., to the optimum value)?* For some graph families (e.g., paths, cycles, binary trees, caterpillars), the worst-case optimum

Research supported in part by the MIUR project AMANDA "Algorithmics for MAssive and Networked DAta", prot. 2012C4E3KT_001.

Y. Hu and M. Nöllenburg (Eds.): GD 2016, LNCS 9801, pp. 617–619, 2016.
DOI: 10.1007/978-3-319-50106-2

value of ply number is known. For these families, we evaluate the gap between the ply number of drawings computed by exiting algorithms and the optimum.

Q3. *Does ply number correlate with some other commonly used quality metrics?* We focus on three popular quality metrics that apparently have some theoretical connection with ply number: stress, edge-length uniformity, and number of crossings. For example, in a drawing with ply number one all edges have the same length, while the ply number can be sometimes reduced at the expenses of edge crossings [3]. Stress is an "energy" function of a graph layout, commonly adopted to establish how much the geometric distances between pairs of vertices reflect their theoretic distances in the graph; it implicitly assumes that all edges have about the same length (see, e.g., [4, 11]).

Q4. *Can we establish empirical upper bounds on the ply number of k-ary trees?* Every binary tree has ply number at most two [3] while the ply number of 10-ary trees is not bounded by a constant [1]. Hence, it is natural to ask what happens for k-ary trees whit $k \in [3, 9]$. We aim to shed more light on this problem.

Findings. To answer Q1–Q3 we conducted experiments on 288 graph instances of different families (trees, planar graphs, general random graphs, scale-free graphs, paths, cycles, caterpillars, binary trees), and with number of nodes ranging from 50 to 200. To answer Q4, we generated 66 k-ary trees with nodes ranging from 100 to 4000 and with $k \in \{3, 6, 9\}$. The experimental results (partially) answer our questions and raise interesting new questions. These are the main findings:

F1. About Q1, algorithms designed to minimize stress and edge-length uniformity compute drawings with smaller values of ply number. In particular, we observed good performances for SM and KK. This finding is in favor of the intuition that low ply number is related to stress and edge uniformity optimization (see also F3). We also observed good performance of FM3. We suspect that this is a consequence of its coarsening technique, which indirectly tends to evenly distribute the nodes in the plane, thus producing drawings with good edge length uniformity, independently of the original placement of the nodes.

F2. Concerning Q2, the best performing algorithms in terms of ply number very often generate drawings whose ply number is close to the worst-case optimum for several graph families, such as paths, cycles, caterpillars, and binary trees.

F3. The experiments executed for answering Q3 show a clear correlation between ply, stress, and edge-length uniformity. For planar graphs and low density graphs, the correlation between ply and crossings is also observed, while ply number is definitely non-correlated with the number of edge crossings on denser graphs and, in particular, on scale-free graphs. We remark that the correlation between ply number and stress does not imply that low ply number equals low stress.

F4. About Q4, we computed, for each instance, the ply number of a single drawing produced with the SM algorithm (the best performing algorithm for ply number, according to F1). We could not observe any asymptotic trend of the ply number towards a constant upper bound for k-ary trees, where $k \in [3, 9]$. This indicates that the ply number for such graphs is likely unbounded, which should be formally confirmed by a theoretical proof.

References

1. Angelini, P., Bekos, M., Bruckdorfer, T., Hančl, J., Kaufmann, M., Kobourov, S., Kratochvíl, J., Symvonis, A., Valtr, P.: Low ply drawings of trees. In: Hu, Y., Nöllenburg, M., (eds.) GD 2016. LNCS, vol. 9801, pp. 236–248. Springer International Publishing, Switzerland (2016)
2. Chrobak, M., Kant, G.: Convex grid drawings of 3-connected planar graphs. Int. J. Comput. Geom. Appl. **07**(03), 211–223 (1997)
3. Di Giacomo, E., Didimo, W., Hong, S.-H., Kaufmann, M., Kobourov, S.G., Liotta, G., Misue, K., Symvonis, A., Yen, H.-C.: Low ply graph drawing. In: IISA 2015 - 6th International Conference on Information, Intelligence, Systems and Applications, pp. 1–6. IEEE (2015)
4. Dwyer, T., Lee, B., Fisher, D., Quinn, K.I., Isenberg, P., Robertson, G.G., North, C.: A comparison of user-generated and automatic graph layouts. IEEE Trans. Vis. Comput. Graph. **15**(6), 961–968 (2009)
5. Eppstein, D., Goodrich, M.T.: Studying (non-planar) road networks through an algorithmic lens. In: GIS 2008, 16th ACM SIGSPATIAL International Conference on Advances in Geographic Information Systems, pp. 1–10. ACM (2008)
6. Frick, A., Ludwig, A., Mehldau, H.: A fast adaptive layout algorithm for undirected graphs (extended abstract and system demonstration). In: Tamassia, R., Tollis, I.G. (eds.) GD 1994. LNCS, vol. 894, pp. 388–403. Springer, Heidelberg (1995). doi:10.1007/3-540-58950-3_393
7. Fruchterman, T.M.J., Reingold, E.M.: Graph drawing by force-directed placement. Softw. Pract. Experience **21**(11), 1129–1164 (1991)
8. Gansner, E.R., Koren, Y., North, S.: Graph drawing by stress majorization. In: Pach, J. (ed.) GD 2004. LNCS, vol. 3383, pp. 239–250. Springer, Heidelberg (2004). doi:10.1007/978-3-540-31843-9_25
9. Hachul, S., Jünger, M.: Drawing large graphs with a potential-field-based multilevel algorithm. In: Pach, J. (ed.) GD 2004. LNCS, vol. 3383, pp. 285–295. Springer, Heidelberg (2004). doi:10.1007/978-3-540-31843-9_29
10. Kamada, T., Kawai, S.: An algorithm for drawing general undirected graphs. Inf. Process. Lett. **31**(1), 7–15 (1989)
11. Kobourov, S.G., Pupyrev, S., Saket, B.: Are crossings important for drawing large graphs? In: Duncan, C., Symvonis, A. (eds.) GD 2014. LNCS, vol. 8871, pp. 234–245. Springer, Heidelberg (2014). doi:10.1007/978-3-662-45803-7_20
12. Noack, A.: Energy models for graph clustering. J. Graph Algorithms Appl. **11**(2), 453–480 (2007)

Edge Bundling for Dataflow Diagrams

Ulf Rüegg[(⊠)], Christoph Daniel Schulze, Carsten Sprung, Nis Wechselberg,
and Reinhard von Hanxleden

Department of Computer Science, Kiel University, Kiel, Germany
{uru,cds,csp,nbw,rvh}@informatik.uni-kiel.de

Edge bundling is a well-known technique to reduce visual clutter in node-link diagrams by having different links share the same path through the diagram [1–4]. *Dataflow diagrams* consist of functional blocks (*nodes*) that transfer data through channels (*links* or *edges*, usually routed orthogonally) that connect the blocks through dedicated connection points (*ports*). A natural concept of dataflow diagrams which is similar to edge bundling is the usage of *hyperedges* which can connect more than two nodes. Here we show how edge bundles can sensibly be incorporated into dataflow diagrams and how they compare to and can coexist with hyperedges. We briefly discuss methods to compute edge bundles as part of the *layer-based approach* to layout [7].

Edge Bundles vs. Hyperedges. Hyperedges are part of the diagram's structure and distribute the same data between the connected ports. Edge bundles on the other hand are a means of presentation and are formed by combining edges suitably. They abstract from port connections and instead emphasize which nodes are

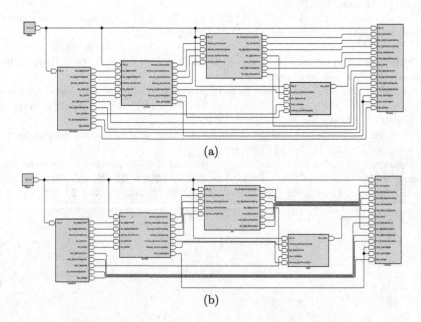

(a)

(b)

Fig. 1. (a) A dataflow diagram with a hyperedge between the top-left node and every other node. (b) The same diagram but with snuggling edge bundles.

© Springer International Publishing AG 2016
Y. Hu and M. Nöllenburg (Eds.): GD 2016, LNCS 9801, pp. 620–622, 2016.
DOI: 10.1007/978-3-319-50106-2

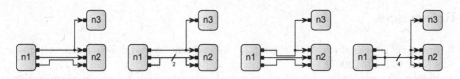

Fig. 2. Coexistence of edge bundles and hyperedges: four possible drawing styles.

connected, which is meant to reduce visual clutter possibly losing the more detailed port connectivity information. In the context of dataflow diagrams, we want to restrict this loss by putting constraints on which edges can be bundled: two edges can only be bundled if they connect the same nodes. An edge bundle then illustrates that *some* data is exchanged between these nodes, with the exact data not being further specified (Fig. 1).

Visual Representation. Hyperedges share as much of their path as possible and junction points are often emphasized using markers, e.g. little circles. As illustrated in Fig. 2, we propose four possibilities when drawing dataflow diagrams with edge bundles. We first distinguish between keeping hyperedges and edge bundles separate or combining them. Within each of these cases, we further distinguish whether to slightly separate the edges in each edge bundle (thus drawing them in a "snuggling" fashion), or to combine them. Separating hyperedges and edge bundles while drawing the bundles in a snuggling fashion retains the original connectivity information; the other drawing styles do not necessarily do so.

Methods. Given a finished drawing, we pursue two use cases: (a) Edges are bundled without moving the nodes, which preserves the mental map of a user and allows regular and bundled edge routing in the same diagram. The user can interactively switch between the routing styles or "un-bundle" a single bundle to see the explicit connections. Nevertheless, care has to be taken not to produce unfortunate edge overlaps in the latter case. (b) Node positions are allowed to be altered to produce smaller drawings by leveraging the space freed by combining edges.

The initial drawing can be computed using an existing layer-based method supporting ports and orthogonal edges [6]. An orthogonal edge consists of vertical and horizontal segments, the former of which are always placed in between *layers* and ordered to reduce edge crossings. For a bundle of edges a common route can directly be derived from the horizontal and vertical segments of the individual edges. A weighted shortest-path on an auxiliary graph determines the best suiting horizontal segments which induce the required height of the vertical segments. The order of (bundled) vertical segments between layers should be recomputed since the crossing number may change (we count a crossing with a bundle only once, even for snuggling bundles). Alternatively, a *constraint graph* can be formed from nodes and vertical segments and one-dimensional compaction techniques can be used to obtain a more compact drawing [5]. Appropriate implementations of the suggested methods are fast enough for interactive applications.

References

1. Eppstein, D., Goodrich, M.T., Meng, J.Y.: Confluent layered drawings. Algorithmica **47**(4), 439–452 (2007)
2. Holten, D., van Wijk, J.J.: Force-directed edge bundling for graph visualization. Comput. Graph. Forum **28**(3), 983–990 (2009)
3. Onoue, Y., Kukimoto, N., Sakamoto, N., Koyamada, K.: Minimizing the number of edges via edge concentration in dense layered graphs. IEEE Trans. Vis. Comput. Graph. **22**(6), 1652–1661 (2016)
4. Pupyrev, S., Nachmanson, L., Kaufmann, M.: Improving layered graph layouts with edge bundling. In: Brandes, U., Cornelsen, S. (eds.) GD 2010. LNCS, vol. 6502, pp. 329–340. Springer, Heidelberg (2011). doi:10.1007/978-3-642-18469-7_30
5. Rüegg, U., Schulze, C.D., Grevismühl, D., Hanxleden, R.: Using one-dimensional compaction for smaller graph drawings. In: Jamnik, M., Uesaka, Y., Elzer Schwartz, S. (eds.) Diagrams 2016. LNCS (LNAI), vol. 9781, pp. 212–218. Springer, Heidelberg (2016). doi:10.1007/978-3-319-42333-3_16
6. Schulze, C.D., Spönemann, M., von Hanxleden, R.: Drawing layered graphs with port constraints. J. Vis. Lang. Comput. **25**(2), 89–106 (2014). Special Issue on Diagram Aesthetics and Layout
7. Sugiyama, K., Tagawa, S., Toda, M.: Methods for visual understanding of hierarchical system structures. IEEE Trans. Syst. Man Cybern. **11**(2), 109–125 (1981)

Homothetic Triangle Contact Representations

Hendrik Schrezenmaier[(✉)]

Institut für Mathematik, Technische Universität Berlin, Berlin, Germany
schrezen@math.tu-berlin.de

A *triangle contact system* $\mathcal{T}$ is a finite system of triangles in the plane such that two triangles intersect in at most one point. Moreover such an intersection point has to be a corner of exactly one of the two involved triangles. We define $G^*(\mathcal{T})$ as the graph that has a vertex for every triangle of $\mathcal{T}$ and for every triangle contact an edge between the involved triangles. For a given plane graph G and a triangle contact system $\mathcal{T}$ with $G^*(\mathcal{T}) = G$ we say that $\mathcal{T}$ is a *triangle contact representation* of G.

Our main contribution is a novel proof of the following theorem:

Theorem 1. *Let G be a 4-connected triangulation. Then there is a triangle contact representation of G by homothetic triangles.*

This theorem has already been proved in [3]. There they make use of the Convex Packing Theorem by Schramm [5] which states that for a given triangulation G and for each vertex $v \in V(G)$ a given non-trivial convex set $\mathcal{P}_v$ in the plane there exists a contact representation of G where each vertex $v \in V(G)$ is represented by a homothetic copy of $\mathcal{P}_v$.

Our approach, however, makes use of the combinatorial structure of triangle contact representations in terms of Schnyder Woods. This approach has been mentioned in [1, 2] and studied in [4]. Felsner and Francis [2] even explicitly ask for a proof of Theorem 1 by this approach.

The crucial point is that we consider a larger class of triangle contact representations than the class of contact representations by homothetic triangles. A *right triangle contact representation* is a triangle contact representation by right triangles with a horizontal edge at the bottom and a vertical edge at the right hand side. The *aspect ratio* of such a triangle is the quotient of the lengths of its vertical and its horizontal edge. And the *aspect ratio vector* of a right triangle contact representation is the vector of the aspect ratios of its triangles (we assume the vertices of G have a fixed numbering $1, \ldots, n$). See Fig. 1 for an example of a right triangle contact representation. Instead of directly proving Theorem 1 we prove the following generalization:

Theorem 2. *Let G be a 4-connected triangulation and $\tilde{r} \in \mathbb{R}_{>0}^n$. Then there is a right triangle contact representation of G with aspect ratio vector $\tilde{r}$.*

Since the case $\tilde{r} = (1, \ldots, 1)$ of Theorem 2 is equivalent to Theorem 1, it is indeed a generalization.

We will now give a sketch of the proof of Theorem 2. Each right triangle contact representation $\mathcal{T}$ of G induces a Schnyder Wood S of G. If r is the aspect

© Springer International Publishing AG 2016
Y. Hu and M. Nöllenburg (Eds.): GD 2016, LNCS 9801, pp. 623–624, 2016.
DOI: 10.1007/978-3-319-50106-2

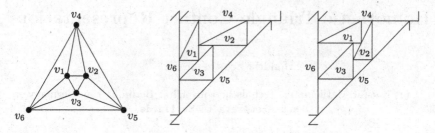

Fig. 1. Two right triangle contact representations of the same graph with aspect ratio vectors $(2, \frac{1}{2}, 1, 1, 1, 1)$ and $(\frac{1}{2}, 2, 1, 1, 1, 1)$

ratio vector of $\mathcal{T}$ we then say that S *realizes* r. On the other hand, if we are given a Schnyder Wood S of G and an aspect ratio vector $r \in \mathbb{R}_{>0}^n$, we can derive the edge lengths of the triangles of a right triangle contact representation with aspect ratio vector r that induces the Schnyder Wood S via a system of linear equations. This system is uniquely solvable, but the solution can contain negative variables. Thus our goal is to find a Schnyder Wood S such that the solution of the system corresponding to S and $\tilde{r}$ only contains nonnegative variables.

We start with an arbitrary pair of a Schnyder Wood S_1 and an aspect ratio vector $\hat{r}$ such that S_1 realizes $\hat{r}$, and examine the line segment $\{r_t := (1-t)\hat{r} + t\tilde{r} : 0 \le t \le 1\}$ of aspect ratio vectors. By studying the system of linear equations, we see that S_1 realizes a subsegment $\{r_t : 0 \le t \le t_1\}$ and that there is a Schnyder Wood S_2 realizing a subsegment $\{r_t : t_1 \le t \le t_2\}$, and so on (in fact, we sometimes have to perturb the starting point $\hat{r}$). Then we show that a line segment can only be divided into a bounded number of subsegments in this manner. Thus $\tilde{r}$ has to be reached at some point and that completes the proof.

This new proof might be useful in addressing two open questions: (1) Can homothetic triangle contact representations be computed efficiently? (2) Are homothetic triangle contact representations unique?

References

1. Felsner, S.: Triangle contact representations. In: Midsummer Combinatorial Workshop. Citeseer, Praha (2009)
2. Felsner, S., Francis, M.C.: Contact representations of planar graphs with cubes. In: Proceedings of the Twenty-seventh Annual Symposium on Computational Geometry, SoCG 2011, Paris, pp. 315–320. ACM (2011)
3. Gonçalves, D., Lévêque, B., Pinlou, A.: Triangle contact representations and duality. In: Brandes, U., Cornelsen, S. (eds.) GD 2010. LNCS, vol. 6502, pp. 262–273. Springer, Heidelberg (2011). doi:10.1007/978-3-642-18469-7_24
4. Rucker, J.: Kontaktdarstellungen von planaren Graphen. Diplomarbeit, Technische Universität Berlin (2011). http://page.math.tu-berlin.de/~felsner/Diplomarbeiten/dipl-Rucker.pdf
5. Schramm, O.: Combinatorially prescribed packings and applications to conformal and quasiconformal maps. Modified version of Ph.D. thesis from 1990 (2007). http://arxiv.org/abs/0709.0710v1

Flexible Level-of-Detail Rendering
for Large Graphs

Jan Hildenbrand(✉), Arlind Nocaj, and Ulrik Brandes

Department of Computer and Information Science, University of Konstanz,
Konstanz, Germany
{jan.hildenbrand,arlind.nocaj,ulrik.brandes}@uni-konstanz.de

Motivation

The visualization of graphs using classical node-link diagrams works well up
to the point where the number of nodes exceeds the capacity of the display.
To overcome this limitation Zinsmaier et al. [5] proposed a rendering technique
which aggregates nodes based on their spatial distribution, thereby allowing for
visual exploration of large graphs. Since the rendering is done on the graphics
processing unit (GPU) this process is reasonably fast. However, the connection
between input graph and visual image is partially lost, which makes it harder,
for instance, to process weights and labels of the input graph.

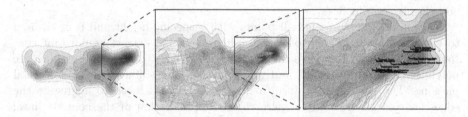

Fig. 1. Level-of-Detail rendering of the US air-traffic dataset.

We reproduce their approach with the goal of establishing a flexible structure
to improve the connection between input data and visualization. Additionally,
we control the layout features in a more direct way. For example, contour lines
are explicitly drawn in order to remove fuzziness of the density visualization.
Though the proposed CPU-based approach cannot render at interactive rates, it
can be computed as a preprocessing step and then interactively explored given
some predefined resolution constraints.

We would like to thank the German Research Foundation (DFG) for financial support within project B02 of SFB/Transregio 161.

Y. Hu and M. Nöllenburg (Eds.): GD 2016, LNCS 9801, pp. 625–627, 2016.
DOI: 10.1007/978-3-319-50106-2

Approach

The visualization consists of two main parts: the construction of the terrain and the aggregation of the edges depending on the underlying terrain. Our terrain is a triangulation, where the triangle corners consist of the nodes of the graph. The node clutter is reduced by a density visualization where each node gets a height assigned and the resulting terrain is visualized.

The heights of the nodes are computed by a kernel density estimation (KDE), which approximates unknown density distributions by overlaying kernel functions at different positions. The density of a particular point is the sum over all kernel functions evaluated at that point. We use the Improved Fast Gaussian Transform (IFGT) [4], which takes on average $\mathcal{O}(N)$ time for N sources and N evaluations.

We use a Delaunay triangulation to create a triangulated irregular network in $\mathcal{O}(N \log N)$ time [1]. Each node of the TIN has a height assigned by the KDE and we assume a linear interpolation between two nodes on the TIN. Large triangles can lead to a false depiction of the graph of the terrain. For instance, let us assume that two neighboring nodes in the Delaunay triangulation represent a hilltop with a valley between them. Without an additional point between the hilltops the edge between them represent a ridge. Therefore, Ruppert's Delaunay refinement algorithm [2] is used to insert points in the circumcenter of triangles which have a minimum angle of 15° or triangles which are particularly large. Additionally, a convex hull is created around the input to prevent confusing non-closing contour lines.

A contour line represents all points with a specific height and is often used to visualize the 3D terrain of topographic maps. We extract equidistant contour lines with van Kreveld's find-isolines algorithm [3]. The contour lines are polylines because of the TIN and get smoothed with splines to be more visual pleasing. The contour lines form a hierarchy, which is used to aggregate the edges: A contour tree (i.e. a hierarchical representation of the contour lines, where a parent has a child if the child is completely contained in the parent) is constructed and only edges between leaves of the contour tree are created.

In practice many of the contour trees are degenerated and consist of list-like substructures. Nevertheless, there could be nodes that are not represented by the edge visualization and therefore, edges of the non-represented nodes are moved to the nearest (in Euclidean distance) leaf of the contour tree. Additionally, the aggregated edges are scaled in width and opacity depending on the sum of the weights of the original edges.

In our implementation of this approach, graphs with up to 42 thousand nodes and 1.5 million edges (e.g., the net150 graph from the University of Florida sparse matrix collection[1]) can be handled in less than 10 s.

---

[1] http://www.cise.ufl.edu/research/sparse/matrices/Andrianov/net150.html.

References

1. Lee, D.T., Schachter, B.J.: Two algorithms for constructing a delaunay triangulation. Int. J. Comput. Inf. Sci. **9**(3), 219–242 (1980)
2. Ruppert, J.: A delaunay refinement algorithm for quality 2-dimensional mesh generation. J. Algorithms **18**(3), 548–585 (1995)
3. Van Kreveld, M.: Efficient methods for isoline extraction from a TIN. Int. J. Geogr. Inf. Syst. **10**(5), 523–540 (1996)
4. Yang, C., Duraiswami, R., Gumerov, N.A., Davis, L.: Improved fast gauss transform and efficient kernel density estimation. In: Proceedings of the Ninth IEEE International Conference on Computer Vision, vol. 1, pp. 664–671 (2003)
5. Zinsmaier, M., Brandes, U., Deussen, O., Strobelt, H.: Interactive level-of-detail rendering of large graphs. IEEE Trans. Vis. Comput. Graph. **18**(12), 2486–2495 (2012)

A Hexagon-Shaped Stable Kissing Unit Disk Tree

Man-Kwun Chiu[1,2], Maarten Löffler[3], Marcel Roeloffzen[1,2(✉)],
and Ryuhei Uehara[4]

[1] National Institute of Informatics (NII), Tokyo, Japan
[2] JST, ERATO, Kawarabayashi Large Graph Project, Tokyo, Japan
{chiumk,m}@nii.ac.jp
[3] Department of Information and Computing Sciences, Utrecht University,
Utrecht, The Netherlands
m.loffler@uu.nl
[4] School of Information Science, JAIST, Ishikawa, Japan
uehara@jaist.ac.jp

1 Introduction

A *disk contact graph* is a graph that can be represented by a set of interior-disjoint disks in the plane, where each disk represents a vertex and an edge between two disks exists if and only if the disks touch (or *kiss*). Many studies have been conducted to classify the types of graphs that can be represented as disk contact graphs as well as to design algorithms to find a set of disks that represent the graph (or to determine if this is even possible). A fundamental results in this area is Koebe's theorem, which states that every planar graph can be represented as a contact graph of disks [5].

The same question can be asked for unit disk graphs (or *coin graphs*). Breu and Kirkpatrick show that recognizing unit disk graphs is NP-hard [2]. Bowen et al. [1] study the problem of recognizing unit disk *trees*, and show that it is NP-hard to determine if a given tree can be represented as a unit-disk contact graph with a given embedding—that is, given the cyclic order of neighbors around each vertex. They claim the main obstacle in proving NP-hardness of unit disk tree recognition is creating a *stable* tree for which all embeddings are similar so the embedding can be used as a building block for an NP-hardness reduction.

A graph G is *ε-stable* if for any two embeddings of G as contact graphs of unit disks, there is a rigid transformation of one such that there is a matching between the resulting embeddings where the distance between matched disks is at most ε. Note that although each disk represents a tree-node, its matched node may not be the same node in the tree. Here, we show that arbitrarily large ε-stable trees exist.

2 The Construction

Since the embedding is free, our strategy will be to create a configuration that is stable because of its density: the circles will be so tightly packed, that there is simply no room to significantly change the embedding.

© Springer International Publishing AG 2016
Y. Hu and M. Nöllenburg (Eds.): GD 2016, LNCS 9801, pp. 628–630, 2016.
DOI: 10.1007/978-3-319-50106-2

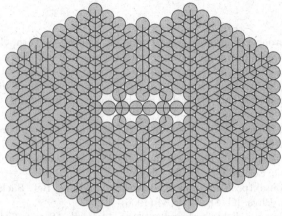

Tóth proved that a hexagonal lattice is the densest of all possible plane packings [6] (see also Chang and Wang [3]). This suggests that a hexagon-shaped graph, consisting of a central hub and six "feathers" growing out of it, would be very stable. However, we cannot use it, because its dual would not be a tree. We need enough room to move the disks slightly such as to make sure the dual graph is a tree.

To do this, we place *two* disks (the *hubs*) at distance d from each other, and center two hexagonal circle packings around them. If we choose d to be an integer which is slightly larger than an integer multiple of $\sqrt{3}$, the two packings will almost, but not quite, fit each other. Then we connect the hubs with a straight path to make sure they cannot drift further than d from each other. We then fill the space in the heart of the construction with as many disks as fit.

Concretely, we choose $d = 7$. Our construction (see above) consists of:

– two *hubs*: vertices of degree 5, connected each other by a *spine* of length 7;
– two times four *feathers*, connected to the hubs;
– two *arms*: paths of length k, connected to the 2nd and 5th spine-vertex;
– two *stubs*: paths of length 3, also connected to the 2nd and 5th spine-vertex.

Here, a *feather* consists of a *shaft* of length k: a path of mostly degree 4 vertices, with two *barbs* connected to each vertex of the shaft. A barb is a path of degree 2 vertices ending in a degree 1 vertex; the lengths of the barbs may vary. Due to space constraints, we sketch the main steps of the proof of stability.

– We show that it is possible to perturb the disks slightly so that their contact graph is indeed the correct tree.
– We show that the hubs must be placed at distance at least $4\sqrt{3}$ (otherwise, there is not enough space to fit all the disks).
– We show that in any valid embedding of G, all shaft vertices must be straight: they must have their barbs attached on opposite sides.
– We show that we cannot make enough room for any path to be compressed by "zigzagging" (because $7 - 4\sqrt{3}$ is smaller than $1 - \frac{1}{2}\sqrt{3}$).

We expect the ε-stable trees can be used to create a so-called logic engine [4] to show NP-hardness of unit disk tree recognition.

References

1. Bowen, C., Durocher, S., Löffler, M., Rounds, A., Schulz, A., Tóth, C.D.: Realization of simply connected polygonal linkages and recognition of unit disk contact trees. In: Di Giacomo, E., Lubiw, A. (eds.) GD 2015. LNCS, vol. 9411, pp. 447–459. Springer International Publishing, Cham (2015). doi:10.1007/978-3-319-27261-0_37
2. Breu, H., Kirkpatrick, D.G.: Unit disk graph recognition is np-hard. Comput. Geom. **9**, 3–24 (1998)
3. Chang, H.-C., Wang, L.-C.: A simple proof of thue's theorem on circle packing (2010). arXiv:1009.4322v1
4. Eades, P., Whitesides, S.: The logic engine and the realization problem for nearest. Theor. Comput. Sci. **169**(1), 23–37 (1996)
5. Koebe, P.: Kontaktprobleme der konformen abbildung. Ber. Sächs. Akad. Wiss. Leipzig, Math.-Phys. Kl. **88**, 141–164 (1936)
6. Tóth, L.F.: Uber die dichteste kugellagerung. Math. Z. **48**, 676–684 (1943)

Heuristic Picker for Book Drawings

Jonathan Klawitter and Tamara Mchedlidze[(✉)]

Institut für Theoretische Informatik, Karlsruhe Institute of Technology,
Karlsruhe, Germany
jo.klawitter@gmail.com, mched@iti.uka.de

Introduction. A *book* with k pages consists of a line (the *spine*) and k half-planes (the *pages*), each with the spine as boundary. In a *k-page book drawing* of a graph the vertices lie on the spine, and each edge is drawn as a circular arc in one of the k pages. The minimum number of edge crossings in a k-page book drawing of a graph is called its *k-page crossing number*, which, in general, is $\mathcal{NP}$-hard to determine [1]. Multiple heuristic approaches to compute a k-page drawing with a small number of crossings are available in the literature. On a very high level, they can be categorized as *simple* heuristics, those that consist of a single run, and *complex* ones, based on neural networks [6, 8, 13, 16], simulated annealing and evolutionary techniques [2, 5, 9, 14, 15]. Notice that a book drawing consists of two ingredients, an order of the vertices on the spine, and a distribution of the edges to the pages. Simple heuristics, given in literature, create vertex order and edge distribution independently. A complete book drawing is constructed by either applying a combination of a vertex order and an edge distribution heuristic, or by applying the mentioned above complex approaches that use simple heuristics as basis. As a result, the performance of the complex approaches depends on the performance of the applied simple heuristics. Up to our knowledge, every attempt to compare the performance of the existing simple heuristics is limited in some sense. These experiments are either limited to very few of them [15], or use very specific graph classes as benchmarks, or limit the experiments to one or two pages [6, 10, 11]. One of the goals of this work is to extend these experiments. Observe that complex heuristics, using advanced search patterns, almost always outperform the simple heuristics [6, 15]. But, since they use simple heuristics in their base, we believe that it is necessary to understand the relevant performance of the simple heuristics first. Thus, this work focuses on simple heuristics only. In particular the content of the poster is as follows: 1. We present several new heuristics and among them several *full drawing* heuristics, that create vertex order and edge distribution at the same time. 2. We present results of our extensive experimental study. The general target of the experiment was to provide an easy way to access the following information: given a graph class and the number of pages, which is the best combination of simple vertex order and edge distribution heuristics? Based on the experiments in the literature, our experimentation and intuition, we have chosen the most promising 7 heuristics from the literature, implemented them, as well as the new heuristics, and compared their performance based on the number of crossings they produce and the running time (complete experiment can be found in [12]).

Heuristics. We start with the vertex order heuristics. The heuristics randDFS [2] and smlDgrDFS [7] compute a vertex order based on a DFS traversal choosing

© Springer International Publishing AG 2016
Y. Hu and M. Nöllenburg (Eds.): GD 2016, LNCS 9801, pp. 631–633, 2016.
DOI: 10.1007/978-3-319-50106-2

the next vertex randomly and the one with the smallest degree, respectively. We introduce the heuristic `treeBFS`, which orders the vertices based on a crossing-free 1-page book drawing of a computed BFS spanning tree. The heuristic `conCro` [3] at each step selects the vertex with the most already placed neighbors and places it on one of the two ends of the current spine where it introduces the fewest new crossings. As an extension of this heuristic, we introduce `conGreedy` which considers not only the two ends of the spine but any position on it.

The edge distribution heuristics `eLen` [4], `ceilFloor` and `circ` [15] sort edges in some particular order and distribute them greedily to the page where they create fewest crossings. `eLen` and `ceilFloor` sort them by decreasing length in linear and circular spine, respectively. `circ` considers the order inspired by the construction of the book embeddings of complete graphs on their pagenumber. `slope` [11] considers a circular drawing and places the edges with similar slope to the same page. We introduce the heuristic `earDecomp`, which constructs the conflict graph of the edges in a circular drawing, and an ear decomposition of the conflict graph, and then alternates the vertices of each ear (edges of the original graph) between two or three pages.

Following the idea by He et al. [10], we extended the vertex order heuristics `randDFS`, `smlDgrDFS`, `conGreedy` to full heuristics `randDFS+`, `smlDgrDFS+`, `conGreedy+`, respectively, which distribute an edge to the best page greedily as soon as it gets closed, i.e. at the moment its second end-vertex appears on the spine. In contrast to `smlDgrDFS+` and `randDFS+`, `conGreedy+` decides for the position of a vertex based on the number of new crossings, and thus the order it computes is different from `conGreedy`. Thus, `conGreedy+` can also be used as an improved vertex order heuristic by discarding the edge distribution.

Experiment and Discussion. We tested the heuristics on graphs of different classes, size n and number of pages p. Among others our test suite includes random graphs of different densities, planar and 1-planar graphs, k-trees, cycle products and hypercubes. In each case we used 200 instances and measured the average number of crossings. A digest is given in the poster. The maximal used number of pages was determined either by the pagenumber of the graph, or when the best heuristic produced no more than 1 crossing on average.

From our experiments we concluded that the best heuristic combination depends not only on the density of the graphs, but remarkably also on the structural properties of the graphs. For example, the combination `conGreedy+-ceilFloor` performs best on planar and 1-planar graphs, while `conGreedy+` , as full drawing heuristic, performs best on random graphs with the same density.

In general, we observe that the extension of `conCro` to `conGreedy` as well as the full drawing heuristic `conGreedy+` often construct book drawings with fewer crossings, however, with the cost of higher running time, which was also clearly noticeable in the experiments.

Furthermore, we could observe that `conGreedy+-ceilFloor/eLen` achieved crossing-free book drawings of hypercubes Q_d when $p =$ pagenumber (tested up to $d = 10$).

References

1. Bannister, M.J., Eppstein, D.: Crossing minimization for 1-page and 2-page drawings of graphs with bounded treewidth. In: Duncan, C., Symvonis, A. (eds.) GD 2014. LNCS, vol. 8871, pp. 210–221. Springer, Heidelberg (2014). doi:10.1007/978-3-662-45803-7_18

2. Bansal, R., Srivastava, K., Varshney, K., Sharma, N.: An evolutionary algorithm for the 2-page crossing number problem. In: 2008 IEEE Congress on Evolutionary Computation (IEEE World Congress on Computational Intelligence). pp. 1095–1102, June 2008

3. Baur, M., Brandes, U.: Crossing reduction in circular layouts. In: Hromkovič, J., Nagl, M., Westfechtel, B. (eds.) WG 2004. LNCS, vol. 3353, pp. 332–343. Springer, Heidelberg (2004). doi:10.1007/978-3-540-30559-0_28

4. Cimikowski, R.: Algorithms for the fixed linear crossing number problem. Discrete Appl. Math. **122**(1), 93–115 (2002)

5. He, H., Newton, M., Sýkora, O.: Genetic algorithms for bipartite and outerplanar graph drawings are best! (2005)

6. He, H., Sălăgean, A., Mäkinen, E., Vrt'o, I.: Various heuristic algorithms to minimise the two-page crossing numbers of graphs. Open Comput. Sci. **5**(1) (2015)

7. He, H., Sýkora, O.: New circular drawing algorithms. In: Proceedings of the Workshop on Information Technologies - Applications and Theory (ITAT) (2004)

8. He, H., Sýkora, O., Mäkinen, E.: An improved neural network model for the twopage crossing number problem. IEEE Trans. Neural Netw. **17**(6), 1642–1646 (2006)

9. He, H., Sýkora, O., Mäkinen, E.: Genetic algorithms for the 2-page book drawing problem of graphs. J. Heuristics **13**(1), 77–93 (2007). doi:10.1007/s10732-006-9000-4

10. He, H., Sýkora, O., Salagean, A., Vrt'o, I.: Heuristic crossing minimisation algorithms for the two-page drawing problem (2006)

11. He, H., Sýkora, O., Vrt'o, I.: Crossing minimisation heuristics for 2-page drawings. Electron. Notes Discrete Math. **22**, 527–534 (2005). http://www.sciencedirect.com/science/article/pii/S1571065305052637. 7th International Colloquium on Graph Theory

12. Klawitter, J.: Algorithms for crossing minimisation in book drawings. Master's thesis, Karlsruhe Institute of Technology (2016). https://i11www.iti.uni-karlsruhe.de/en/members/tamara_mchedlidze/

13. López-Rodríguez, D., Mérida-Casermeiro, E., Ortíz-de-Lazcano-Lobato, J.M., Galán-Marín, G.: K-pages graph drawing with multivalued neural networks. In: Sá, J.M., Alexandre, L.A., Duch, W., Mandic, D. (eds.) ICANN 2007. LNCS, vol. 4669, pp. 816–825. Springer, Heidelberg (2007). doi:10.1007/978-3-540-74695-9_84

14. Poranen, T., Mäkinen, E., He, H.: A simulated annealing algorithm for the 2-page crossing number problem. In: Proceedings of International Network Optimization Conference (INOC) (2007)

15. Satsangi, D., Srivastava, K., Srivastava, G.: k-page crossing number minimization problem: An evaluation of heuristics and its solution using gesakp. Memetic Comput. **5**(4), 255–274 (2013). doi:10.1007/s12293-013-0115-5

16. Wang, J.: Hopfield neural network based on estimation of distribution for two-page crossing number problem. IEEE Trans. Circuits Syst. II Express Briefs **55**(8), 797–801 (2008)

1-Fan-Bundle-Planar Drawings

Patrizio Angelini[1], Michael Bekos[1]([⊠]), Michae Kaufmann[1],
Philipp Kindermann[2], and Thomas Schneck[1]

[1] Institut für Informatik, Universität Tübingen, Tübingen, Germany
{angelini,bekos,mk,schneck}@informatik.uni-tuebingen.de
[2] LG Theoretische Informatik, FernUniversität in Hagen, Hagen, Germany
philipp.kindermann@fernuni-hagen.de

Edge bundling [4, 7] is a powerful tool used in information visualization to avoid visual clutter. In fact, when the edge-density of the network is too high, the traditional techniques of graph layouts and flow maps become unusable. In this case, grouping together parts of edges that flow parallel to each other into a single bundle allows to reduce the clutter and improve readability.

We make a first attempt to combine edge bundling with previous theoretical considerations from the area of *nearly-planar* graphs, where in addition to a planar graph-structure some crossings may be allowed, if they are limited in locally-defined configurations. Examples include 1-*planarity* [6] and *fan-planarity* [5].

In a *fan-planar drawing* [1–3, 5] an edge is allowed to cross a set of edges if they belong to the same *fan*, that is, if they are all incident to the same vertex; refer to Fig. 1(a). The idea is that edges incident to the same vertex are somehow close to each other, and thus having an edge crossing all of them does not affect readability too much. In other words, edges of a fan can be grouped into a *bundle*, so that the crossings between an edge and all the edges of the fan become a single crossing between this edge and the corresponding bundle. However, the original definition of fan-planar drawings does not always allow for this type of bundling, as in the case of graph $K_{4,n-4}$, for large n (Fig. 1(a)).

We thus introduce 1-*fan-bundle-planar drawings*, in which edges of a fan can be bundled and crossings between bundles are allowed as long as each bundle is

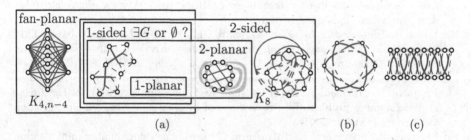

(a) (b) (c)

Fig. 1. (a) Relationships between classes; bundles are solid and non-bundled edge-parts dashed (uncrossed edges are black). (b) A 2-sided outer-1-fan-bundle-planar drawing of K_6. (c) A 1-sided 2-layer 1-fan-bundle-planar drawing.

Research partially supported by DFG grants Ka812/17-1 and SCHU 2458/4-1.

Y. Hu and M. Nöllenburg (Eds.): GD 2016, LNCS 9801, pp. 634–636, 2016.
DOI: 10.1007/978-3-319-50106-2

Table 1. Maximum number of edges of graphs in the considered classes

	2-layer	Outer	General
1-sided	$|E| \leq (5|V| - 7)/3$	$|E| \leq (8|V| - 13)/3$	$|E| \leq (13|V| - 26)/3$
2-sided	$|E| \leq 2|V| - 4$	$|E| \leq 4|V| - 7$	$|E| \leq 56|V| - 114$

crossed by at most one other bundle; see Fig. 1(a). This "1-planarity" restriction prevents an edge to cross edges of several fans, which is not allowed in a fan-planar drawing. However, to avoid that two bundle-parts of an edge are crossed by edges of different fans, we require that an edge is bundled with other edges only on one of its two endvertices. A drawing with this property is 1-*sided*, otherwise it is 2-*sided*. We remark that bundles are not allowed to branch into different sub-bundles, that is, each bundle has exactly two end-points: at one of them there is the vertex that is originating the fan, while at the other one all the edges in the fan are separated from each other to reach their other end-vertex.

We first discuss inclusion relationships with some relevant classes of nearly-planar graphs; Fig. 1(a) summarizes our findings. Relationships with 1-planar, 2-planar, and fan-planar graphs follow from definitions. The 2-sided 1-fan-bundle-planar graph K_8 is too dense to be fan-planar or 2-planar (they both have at most $5n - 10$ edges [5]). The fan-planar graph $K_{4,n-4}$ is not 2-sided 1-fan-bundle-planar (for $n \leq 18$ it is). Also, there exist 2-planar graphs that are not fan-planar [3] (and thus 1-sided 1-fan-bundle-planar). Finally, a tiling of pentagons with four edges in each face is 1-sided 1-fan-bundle-planar but not 1-planar. We are not aware of 1-sided 1-fan-bundle-planar graphs that are not 2-planar.

Then, we ask how dense these graphs may be. We study both the general case and two restricted cases that have been also studied for fan-planarity: In an *outer-1-fan-bundle-planar* drawing [1, 3], all vertices must be incident to the outer face (Fig. 1(b)), while in a *2-layer 1-fan-bundle-planar* drawing [2] vertices are placed on two parallel lines (Fig. 1(c)). Our results are summarized in Table 1. All the provided bounds are tight, except for the one for 2-sided 1-fan-bundle-planar graphs, for which we could only construct a graph with $7n - 18$ edges, by merging two maximally-dense outer-1-fan-bundle-planar graphs.

Finally, we prove that recognizing 1-fan-bundle-planar graphs is NP-complete, both in the 1-sided model and in the 2-sided model. On the other hand, we provide linear-time recognition algorithms for several classes of 2-layer and of outer-1-fan-bundle-planar graphs, in both models. Among the others, we remark a linear-time algorithm for recognizing maximal 1-sided 2-layer 1-fan-bundle-planar graphs.

References

1. Bekos, M.A., Cornelsen, S., Grilli, L., Hong, S.-H., Kaufmann, M.: On the recognition of fan-planar and maximal outer-fan-planar graphs. In: Duncan, C., Symvonis, A. (eds.) GD 2014. LNCS, vol. 8871, pp. 198–209. Springer, Heidelberg (2014). doi:10.1007/978-3-662-45803-7_17

2. Binucci, C., Chimani, M., Didimo, W., Gronemann, M., Klein, K., Kratochvíl, J., Montecchiani, F., Tollis, I.G.: 2-layer fan-planarity: from caterpillar to stegosaurus. In: Di Giacomo, E., Lubiw, A. (eds.) GD 2015. LNCS, vol. 9411, pp. 281–294. Springer, Heidelberg (2015). doi:10.1007/978-3-319-27261-0_24
3. Binucci, C., Di Giacomo, E., Didimo, W., Montecchiani, F., Patrignani, M., Symvonis, A., Tollis, I.G.: Fan-planarity: properties and complexity. TCS **589**, 76–86 (2015)
4. Holten, D.: Hierarchical edge bundles: visualization of adjacency relations in hierarchical data 12(5), 741–748 (2006)
5. Kaufmann, M., Ueckerdt, T.: The density of fan-planar graphs (2014). http://arxiv.org/abs/1403.6184
6. Ringel, G.: Ein Sechsfarbenproblem auf der Kugel. Abh. Math. Sem. Univ. Hamb. **29**, 107–117 (1965)
7. Telea, A., Ersoy, O.: Image-based edge bundles: simplified visualization of large graphs 29(3), 843–852 (2010)

Robust Genealogy Drawings

Fabian Klute[(✉)]

Algorithms and Complexity Group, TU Wien, Vienna, Austria
fklute@ac.tuwien.ac.at

Inspired by the GD2016 challenge[1] to draw a subset of the Greek gods ancestry graph we looked into the problem of drawing complex genealogies. Such graphs have still a hierarchical structure, but intermarriage and cross layer edges make it hard to use existing methods. We present a three step approach which is robust against these features. In the first phase we augment the graph, then an initial layout is calculated with Sugiyama's framework [5] and in the final phase we route the edges orthogonally and make them confluent wherever possible.

We can only give a short description of the genealogy graphs we are interested in. A proper characterization and study of genealogy graphs in general was done by McGuffin et al. [3]. Let $G = (V, E)$ be a directed acyclic graph. The central structure of any genealogy graph is a *family*. A set of nodes $F \subseteq V$ is called a family if and only if we can split F into a *parent* set P and a *children* set C such that all outgoing edges from nodes in C point to nodes in P and every node in C has an edge going to every node in P.

In contrast to normal genealogy graphs we allow the set of parents to be of arbitrary size, but nonempty. This means children can have more than two parents. Additionally we often find that parents are siblings. Layer-crossing edges are created by allowing parents from the same set to be in different layers. This especially enables parents and children to again have children of their own.

Before we turn to the algorithm lets describe what a good genealogy drawing is to us. The most central part is the relationship between family members. Consequently parents and children should be drawn close to each other. Further it is important to identify which nodes belong to which parent or children set. Here confluence can help by merging the edges after they leave the children or splitting them just before they enter the parents. Finally long edges need to be routed such that they only draw attention in their source and target layer.

Augmentation. As a first step the graph is augmented with a method taken from McGuffin et al. [3]. For every family we introduce one virtual *family node*. The node is then connected to its children and parents such that edges point from the children to the family node and from the family node to the parents.

The second type of virtual node is introduced to split edges crossing multiple layers. Since the hierarchy of the nodes is not altered anymore we can compute the layer in the Sugiyama framework at this point in time. Given an edge with target and source node on non-neighboring layers we then add one node for every layer it crosses, constructing a long path. Afterwards all edges run only between nodes in neighboring layers.

---

[1] http://graphdrawing.de/contest2016/topics.html.

© Springer International Publishing AG 2016
Y. Hu and M. Nöllenburg (Eds.): GD 2016, LNCS 9801, pp. 637–639, 2016.
DOI: 10.1007/978-3-319-50106-2

Using the augmentation with drawing methods described by McGuffin et al. [3] leads to problems with more complex graphs since the algorithms are not designed to work with edges spanning multiple layers or with a lot of intermarriage.

Sugiyama. We use the Sugiyama framework as implemented in the OGDF library [2] to produce a straight-line drawing of the augmented graph. Especially the in-layer order of the nodes is fixed after this step. The drawings generated by Sugiyama with the augmentation are already good in respect to the closeness of family members, but the straight-line edges make it very hard to track which parents and children belong to one family or which children are (half-)siblings.

Edge Routing. The edge routing consists of two steps. First we straighten long paths of nodes, then we route the edges orthogonally with two bends and calculate their confluent parts. For the path straightening we move nodes which are part of a long path, such that the nodes are on the same x-coordinate whenever possible. We do this from source towards target by moving the nodes in their layer without changing the node order. This method gets rid of long zig-zag paths and small bends introduced by Sugiyama.

The bends are calculated with a linear program, making it easy to extend this base method. Incidentally the result gives us the position of the confluent edge parts as well. Pupyrev et al. [4] present an edge-bundling approach which results in similar sets of confluent edges, but our algorithm is much simpler since we won't have to think about the x-position of the confluent parts. Bannister et al. [1] studied a more general concept of confluence in layered graph drawings without the need of prior graph augmentation. Again this approach is broader than necessary for our restricted case. Here we give just a rough sketch of the idea. On the poster more explanations on the created constraints can be found.

For every edge $e \in E$ one continuous variable $Y(e) \in [0.1, 0.9]$ is added. This variable encodes the y-coordinate of the bends, while the x-coordinates are already fixed by the incident nodes of the edge. Allowing only two bends restricts every edge to the same bounding box as in the case of a straight line, which prevents new crossings between far away edges in a layer.

The constraints we use are always between two edges $e, f \in E$. For every pair of edges there is maximum of one constraint. Either it is an equality constraint $Y(e) = Y(f)$ or it is an inequality of the form $Y(e) - Y(f) > c$ with c being a small constant appropriately chosen for the height of the layer. There are two main cases. In the first one edges share a common node. We then have to think about what this common node means for their relation and how we want this to be represented in the drawing. The second case contains edges with no common node. Here we have to sort them such that no unnecessary crossings are created. The condition under which a crossing between two edges is unavoidable is still the same as in the straight-line drawing.

In the future a formal study of the crossing number of the presented drawings would be interesting, as well as an improvement to the straightening step and the incorporation of more cross-layer criteria.

References

1. Bannister, M.J., Brown, D.A., Eppstein, D.: Confluent Orthogonal Drawings of Syntax Diagrams. Springer International Publishing, Cham (2015)
2. Chimani, M., Gutwenger, C., Jünger, M., Klau, G.W., Klein, K., Mutzel, P.: The open graph drawing framework (ogdf). In: Handbook of Graph Drawing and Visualization, pp. 543–569 (2011)
3. McGuffin, M.J., Balakrishnan, R.: Interactive visualization of genealogical graphs. In: 2005 IEEE Symposium on Information Visualization, INFOVIS 2005. pp. 16–23, October 2005
4. Pupyrev, S., Nachmanson, L., Kaufmann, M.: Improving Layered Graph Layouts with Edge Bundling, pp. 329–340. Springer, Heidelberg (2011)
5. Sugiyama, K., Tagawa, S., Toda, M.: Methods for visual understanding of hierarchical system structures. IEEE Trans. Syst. Man Cybern. $11(2)$, 109–125 (1981)

Author Index

Printed in the United States
By Bookmasters

Printed in the United States
By Bookmasters